BIOAEROSOLS HANDBOOK

Edited by
Christopher S. Cox
Christopher M. Wathes

LEWIS PUBLISHERS

Boca Raton   London   Tokyo

Library of Congress Cataloging-in-Publication Data

Bioaerosols handbook / edited by Christopher S. Cox, Christopher M. Wathes.
    p.   cm.
    Includes bibliographical references and index.
    ISBN 1-87371-615-9
    1. Air—Microbiology—Handbooks, manuals, etc.   2. Aerosols—Handbooks, manuals, etc.   3. Air—Pollution—Handbooks, manuals, etc.   I. Cox, C.S.   II. Wathes, C.M.
QR101.B55   1995
576'.190961—dc20   94-44749
                                                                                                                   CIP

This book contains information obtained from authentic and highly regarded sources. Reprinted material is quoted with permission, and sources are indicated. A wide variety of references are listed. Reasonable efforts have been made to publish reliable data and information, but the author and the publisher cannot assume responsibility for the validity of all materials or for the consequences of their use.

Neither this book nor any part may be reproduced or transmitted in any form or by any means, electronic or mechanical, including photocopying, microfilming, and recording, or by any information storage or retrieval system, without prior permission in writing from the publisher.

All rights reserved. Authorization to photocopy items for internal or personal use, or the personal or internal use of specific clients, may be granted by CRC Press, Inc., provided that $.50 per page photocopied is paid directly to Copyright Clearance Center, 27 Congress Street, Salem, MA 01970 USA. The fee code for users of the Transactional Reporting Service is ISBN 1-87371-615-9/95/$0.00+$.50. The fee is subject to change without notice. For organizations that have been granted a photocopy license by the CCC, a separate system of payment has been arranged.

CRC Press, Inc.'s consent does not extend to copying for general distribution, for promotion, for creating new works, or for resale. Specific permission must be obtained in writing from CRC Press for such copying.

Direct all inquiries to CRC Press, Inc., 2000 Corporate Blvd., N.W., Boca Raton, Florida 33431.

© 1995 by CRC Press, Inc.
Lewis Publishers is an imprint of CRC Press

No claim to original U.S. Government works
International Standard Book Number 1-87371-615-9
Library of Congress Card Number 94-44749
Printed in the United States of America 1 2 3 4 5 6 7 8 9 0
Printed on acid-free paper

Acknowledgments

The editors are extremely grateful to the authors and publishers for their patience, perseverance and enthusiasm for this Handbook, especially Jon Lewis and Sharon Ray of Lewis Publishers. Loraine Clark, assisted by Mavis Battey and Barbara Sydnor, Silsoe Research Institute, prepared all of the chapters up to proof stage as well as carrying out many of the administrative chores. These tasks were undertaken with great efficiency and good humor, for which we are especially thankful.

About the Editors

Christopher M. Wathes is an environmental scientist, originally trained in physics. After 10 years in the Department of Animal Husbandry at the University of Bristol, England, studying aerobiology in the context of animal houses, he moved to Silsoe Research Institute in 1990 to become Head of the Animal Science and Engineering Division. He has a long standing interest in the generation, dispersion and survival of bioaerosols in livestock buildings, including their effects on animals. Currently his two major interests are emissions of atmospheric pollutants within and from animal houses; and environmental influences on animal behavior and welfare.

Christopher S. Cox was educated at Bristol University, where in 1958 he obtained a B.SC. in special Honours Biochemistry and in 1961 his Ph.D. in Biophysical Chemistry. On completing his thesis, concerned with roles of water molecules in biological systems, he joined the Ministry of Defence, UK, initially as a Junior Research Fellow at the Microbiological Research Establishment (MRE), Porton Down, UK. On its closure in 1979, he transferred to the Chemical and Biological Defence establishment. His career has included also five years working in the U.S. as a Guest Scientist of the Department of Defense. Dr. Cox's published works include two books together with about 60 research papers all concerned with different aspects of bioaerosols and microbial survival following desiccation.

About the Authors

H.A. Burge, Harvard School of Public Health, Department of Environmental Health, 655 Huntington Avenue, Boston, MA 02115

R.P. Clark, Thermal Biology Research Unit, King's College London, Kensington, London W8 7AH, United Kingdom

C.S. Cox, Chemical and Biological Defence Establishment, Porton Down, Salisbury, Wiltshire SP4 0JQ, United Kingdom

B. Crook, Health and Safety Laboratory, Health and Safety Executive, Broad Lane, Sheffield, S3 7HQ, United Kingdom

A.J. Heber, Purdue University, Agricultural Engineering Department, 1146 Agricultural Engineering Building, West Lafayette, IN 47907-1146

A. Hensel, Institut für Bakteriologie und Tierhygiene der Veterinärmedizinischen Universität Wein, Linke Bahngasse II, A-1030 Wein, Austria

J.M. Hirst, FRS, The Cottage, Butcombe, Blagdon, Bristol, Avon BS18 6XQ, United Kingdom

J. Lacey, IARC - Rothamsted, Harpenden, Herts. AL5 2JQ, United Kingdom

J.M. Macher, Environmental Health Laboratory, California Department of Health Services, 2151 Berkeley Way, Berkeley, CA 94704-1011

T.M. Madelin, Bannerleigh House, Bannerleigh Road, Leigh Woods, Bristol BS8 3PF, United Kingdom

M.F. Madelin, Bannerleigh House, Bannerleigh Road, Leigh Woods, Bristol BS8 3PF, United Kingdom

J.P. Mitchell, 114 Sportsfield Court, Bryron, London, Ontario, Canada N6K 4K2

K.J. Morris, Biomedical Research, AEA Technology, 551 Harwell, Oxon. OX11 0RA, United Kingdom

K.W. Nicholson, AEA Technology, Culham, Abingdon, Oxfordshire OX14 3DB, United Kingdom

S.A. Olenchock, National Institute for Occupational Safety and Health, 1095 Willowdale Road, Morgantown, WV 26505-2845

A. Rantio-Lehtimäki, The Finnish Aerobiology Group, Department of Biology, University of Turku, FIN-20500 Turku, Finland

K.R. Spurny, Aerosol Chemist and Environmentalist, Eichenweg 6, 57392 Schamllenberg, Germany

A.J. Streifel, Department of Environmental Health and Safety, University of Minnesota, 410 Church Street Southeast, Minneapolis, MN 55455

J. Venette, Department of Plant Pathology, Walster Hall, North Dakota State University, Fargo, ND 58105

D. Vesley, School of Public Health, University of Minnesota, Minneapolis, MN 55455

C.M. Wathes, Animal Science and Engineering Division, Silsoe Research Institute, Wrest Park, Silsoe, Bedford MK45 4HS, United Kingdom

Table of Contents

Chapter 1
Editor's Introduction   3

Chapter 2
Bioaerosols: Introduction, Retrospect and Prospect, *J.M. Hirst*
    The Foundations 5
    The Tools of the Trade 7
    Handling the Catch 10
    Current Practice 11
    Overview 13
    Reference 14

Chapter 3
Physical Aspects of Bioaerosol Particles, *C.S. Cox*
    Introduction 15
    Brownian Motion 15
    Gravitational Field 16
    Electrical Forces 18
    Thermal Gradients 18
    Electromagnetic Radiation 19
    Turbulent Diffusion and Inertial Forces 19
        Inertial Impaction 22
    Particle Shape 23
    Relative Humidity and Hygroscopy 24
        Relative Humidity Measurement 24
    Conclusions 24
    References 25

Chapter 4
Physical Aspects of Bioaerosol Sampling and Deposition, *K.W. Nicholson*
    Introduction 27
    The Physics of Particle Sampling 27
    Practical Air Sampling of Atmospheric Bioaerosols 29
    Particle Size Analysis 30
    The Atmosphere-Surface Exchange of Particulate Materials
        in the Outdoor Environment 33
    Dry Deposition as a Removal Process 35

  Wet Deposition as a Removal Process 40
  The Resuspension of Material Into the Atmosphere 44
  Summary and Conclusions 46
  References 47

Chapter 5
Bioaerosol Particle Statistics, *A.J. Heber*
  Introduction 55
  Properties of Bioaerosol Size Distributions 55
    Shape 55
    Quantity 56
    Size 57
    Particle Grouping 57
    Averages 59
    Diameter of Average Quantity 59
    Moment Averages of Weighted Distributions 60
    Variance 60
  Size Distribution Functions 62
    Normal 62
    Lognormal 63
    Gaudin-Schuhman Power Law 65
    Rosin-Rammler 65
    Goodness of Fit 66
  Application 66
    Bioaerosol B1 66
    Bioaerosol B2 69
    Bioaerosol B3 69
    Open-Ended Classes 72
    Statistical Accuracy 72
  Summary 73
  Principal Symbols 73
  List of Subscripts 73
  References 74

Chapter 6
Stability of Airborne Microbes and Allergens, *C.S. Cox*
  Introduction 77
  Survival and Infectivity 77
    Most Probable Target Molecules 77
  Relative Humidity and Temperature 78
    Coliphages 78
    Viruses 79
    Bacteria 80
    Other Microbes 83
    Allergens 83
  Oxygen 84
  Open Air Factor and Other Pollutants 85
  Radiation 86
  Mathematical Models 86
    Dehydration and Temperature Model 86

Oxygen and Open Air Factor Models	89
Maillard Reactions	90
Membrane Phase Changes	93
Energetics	96
Repair	97
The Repair Problem	97
Summary	98
References	98

Chapter 7
Aerosol Generation for Instrument Calibration, *J.P. Mitchell*

Abstract	101
Introduction	101
Aerosol Generation Methods	102
Monodisperse Aerosols	103
Polymer Latex Particles	103
Vibrating Orifice and Related Aerosol Generators	105
Spinning Top/Disc Aerosol Generators	112
Condensation Aerosol Generators	116
Electrostatic Classifier	124
Other Sources of Monodisperse Particles	126
Polydisperse Aerosols	126
Compressed Air Nebulizers	126
Ultrasonic Nebulizers	127
Fluidized Bed Aerosol Generators	130
Other Dust Generators	131
Exploding Wire Aerosol Generators	135
Gas-Phase Reactions	136
Particle Standards	137
Instrument Calibration	141
Inertial Analyzers	142
Impactors and Cyclones	142
Inertial Spectrometers	150
Real-Time Aerodynamic Particle Sizers	153
Optical Particle Counters	157
Electrical Aerosol Analyzers	162
Condensation Nuclei Counters (CNCs)	166
Aerosol Mass Concentration Monitors	167
Conclusions	167
Acknowledgments	170
References	170

Chapter 8
Particle Size Analyzers: Practical Procedures and Laboratory Techniques, *J.P. Mitchell*

Abstract	177
Introduction	177
Sampling for Size Analyses	178
General Principles	178
Calm-Air Sampling	179

    Flowing Gas Streams 182
    Sample Line Losses 185
    Sampling from Ducts and Stacks 188
  Aerosol Particle Size Measurement Techniques 191
    Summary of Methods 191
    Sedimentation/Inertial Techniques 195
        Sedimentation Techniques (Elutriators) 195
        Cascade Impactors 197
        Cascade Cyclones 206
        Real-Time Analyzers 208
        Centrifugal Spectrometers 209
    Optical Techniques 213
        Optical Particle Counters 214
        Laser Diffractometers 221
        Laser (Phase)-Doppler Systems 222
        Intensity Deconvolution Technique 225
        GALAICIS Laser-Particle Interaction/Image Analyzer 227
    Electrical Mobility Techniques 230
        Operating Principle 230
        Electrical Aerosol Analyzer (EAA) 231
        Differential Mobility Analyzers (DMAs) 233
Microscopy 235
Summary 240
Acknowledgments 240
References 240

Chapter 9
Inertial Samplers: Biological Perspectives, *B. Crook*

Introduction 247
Gravitation/Settle Plates 247
Impactors, Sieve and Stacked Sieve Samplers 248
    Single-Stage Samplers 248
        Spore traps 248
        Whirling Arm or Rotorod Samplers 248
        Slit-To Agar Samplers 249
        Analysis 249
    Cascade Impactors 250
    Sieve Samplers 251
        Stacked Sieve 251
        Single-Stage Sieve Samplers 252
        Analysis 253
Impingers 255
  Size Fractionating Impingers 257
Centrifugal Samplers (Cyclones, Centrifuges) 259
    Cyclone Samplers 260
    Centrifugal Samplers 260
Practical Aspects 262
    High-Volume Sampling 262
    Sampling of Large Particles 263
    Other Practical Aspects 263

Summary 264
References 264

Chapter 10
Non-Inertial Samplers: Biological Perspectives, *B. Crook*

Introduction 269
Filtration 269
Filtration Media 269
    Fibrous Filters 270
    Membrane Filters 270
    Flat Filters 270
    Filter Holders 271
Analysis 273
    Microscopy 273
    Cultivation 275
    Airborne Allergens 276
Relative Performance of Filtration Sampling 276
Electrostatic Precipitation 277
    Collection Into Liquid 277
    Collection Onto Solid Surface 277
    Analysis 277
    Performance 278
Thermal Precipitation 279
Practical Aspects of Sampling 280
    High-Volume Sampling 280
    Sampling Large Particles 280
    Other Practical Aspects 280
Summary 280
References 281

Chapter 11
Modern Microscopic Methods of Bioaerosol Analysis, *K.J. Morris*

Introduction 285
Direct Visualization and Measurement of Spores and Pollens 286
    Sedimentation and Impaction Devices 286
    Filters 287
Measurement of Viable Microorganisms in the Air 287
    Sedimentation and Impaction Devices 288
    Gelatin Filters 289
Light Microscopy 289
    The Mono-Objective Bright Field Light Microscope 289
        Visualization of Cell Structures Using Traditional
        Histological Stains 291
        Counting Objects on a Slide 292
        Manually Measuring the Size of Objects on a Slide ... 294
        Counting Objects in a Liquid Suspension 295
    The Dark Field Illuminated Microscope 296
    The Phase Contrast Microscope 296
    The Polarizing Microscope 297
    The Interference Microscope 297

```
        The Fluorescence Light Microscope  . . . . . . . . . . . . . . . . . . . . . . . . . . . . 297
            Introduction. . . . . . . . . . . . . . . . . . . . . . . . . . . . . . . . . . . . . . . . . . . . 297
                Using the Fluorescence Microscope  . . . . . . . . . . . . . . . . . . . . . . . 298
        The Confocal Microscope . . . . . . . . . . . . . . . . . . . . . . . . . . . . . . . . . . . 301
        The Scanning Electron Microscope . . . . . . . . . . . . . . . . . . . . . . . . . . . . 301
        The Transmission Electron Microscope . . . . . . . . . . . . . . . . . . . . . . . . 304
    The Image Analyzer . . . . . . . . . . . . . . . . . . . . . . . . . . . . . . . . . . . . . . . . . . 304
        Introduction . . . . . . . . . . . . . . . . . . . . . . . . . . . . . . . . . . . . . . . . . . . . . . 304
        Video Cameras . . . . . . . . . . . . . . . . . . . . . . . . . . . . . . . . . . . . . . . . . . . 306
        Image Analysis  . . . . . . . . . . . . . . . . . . . . . . . . . . . . . . . . . . . . . . . . . . 306
        Colony Counting  . . . . . . . . . . . . . . . . . . . . . . . . . . . . . . . . . . . . . . . . 311
    Other Physical Methods for Detecting Microorganisms  . . . . . . . . . . . . . . . 312
    Summary . . . . . . . . . . . . . . . . . . . . . . . . . . . . . . . . . . . . . . . . . . . . . . . . . . 313
    References . . . . . . . . . . . . . . . . . . . . . . . . . . . . . . . . . . . . . . . . . . . . . . . . . 313

Chapter 12
**Chemical Analysis of Bioaerosols,** *K.R. Spurny*
    Introduction . . . . . . . . . . . . . . . . . . . . . . . . . . . . . . . . . . . . . . . . . . . . . . . . 317
    Collection and Specimen Preparation  . . . . . . . . . . . . . . . . . . . . . . . . . . . . 318
    Microscopical Detection and Identification  . . . . . . . . . . . . . . . . . . . . . . . 320
    Chromatographic Methods  . . . . . . . . . . . . . . . . . . . . . . . . . . . . . . . . . . . . 320
    Laser Spectroscopical Methods . . . . . . . . . . . . . . . . . . . . . . . . . . . . . . . . . 321
        Fluorescence and Luminescence Spectroscopy . . . . . . . . . . . . . . . . . . . 322
        Infrared and Raman Spectroscopy . . . . . . . . . . . . . . . . . . . . . . . . . . . . 323
        Mass Spectrometry . . . . . . . . . . . . . . . . . . . . . . . . . . . . . . . . . . . . . . . 323
    Flow Cytometry . . . . . . . . . . . . . . . . . . . . . . . . . . . . . . . . . . . . . . . . . . . . 325
    Electrochemical Methods and Sensors . . . . . . . . . . . . . . . . . . . . . . . . . . . 326
    Other Chemical Methods  . . . . . . . . . . . . . . . . . . . . . . . . . . . . . . . . . . . . . 327
    Future Developments . . . . . . . . . . . . . . . . . . . . . . . . . . . . . . . . . . . . . . . . 328
    Summary . . . . . . . . . . . . . . . . . . . . . . . . . . . . . . . . . . . . . . . . . . . . . . . . . . 330
    References . . . . . . . . . . . . . . . . . . . . . . . . . . . . . . . . . . . . . . . . . . . . . . . . . 331

Chapter 13
**Biological and Biochemical Analysis of Bacteria and Viruses,**
*Andreas Hensel and the late Klaus Petzoldt*
    Introduction . . . . . . . . . . . . . . . . . . . . . . . . . . . . . . . . . . . . . . . . . . . . . . . . 335
    Approach to a Microbiological Study  . . . . . . . . . . . . . . . . . . . . . . . . . . . . 336
        General Limitations  . . . . . . . . . . . . . . . . . . . . . . . . . . . . . . . . . . . . . . 336
        Choice of Methods . . . . . . . . . . . . . . . . . . . . . . . . . . . . . . . . . . . . . . . 336
        Microbiological Aspects . . . . . . . . . . . . . . . . . . . . . . . . . . . . . . . . . . . 338
        Effects of Collecting Devices on
        Microbiological Results  . . . . . . . . . . . . . . . . . . . . . . . . . . . . . . . . . . . 339
    Assaying Airborne Microorganisms and Their Components . . . . . . . . . . . . . 341
        Assaying Viability and Infectivity of
        Bacteria and Viruses . . . . . . . . . . . . . . . . . . . . . . . . . . . . . . . . . . . . . . 342
            Collecting Media . . . . . . . . . . . . . . . . . . . . . . . . . . . . . . . . . . . . . . 342
            Evaluation of Bacterial Viability . . . . . . . . . . . . . . . . . . . . . . . . . . 344
            Infectivity Assays  . . . . . . . . . . . . . . . . . . . . . . . . . . . . . . . . . . . . 344
            Cultivation in Tissue Cultures . . . . . . . . . . . . . . . . . . . . . . . . . . . . 344
            Cultivation in Embryonated Eggs . . . . . . . . . . . . . . . . . . . . . . . . . 346
```

    Animal Inoculation 346
   Non-Cultural Methods 347
    Direct Examination 347
    Staining 347
    Electron Microscopy 347
    Immune Electron Microscopy 350
    Antibody-Based Detection of Specific
    Microbial Antigens 350
    Immunofluorescence Tests 350
    Enzyme Immunoassays and Radioimmunoassays 350
    Immunoelectroblot Techniques 351
    Detection of Specific Microbial Nucleic Acids 351
    Detection of Specific Microbial Products 352
    Chromogenic and Fluorogenic Enzyme Substrate Tests 352
    Gas-Liquid Chromatography 352
    Luminescence 353
    Limulus Test for LPS Detection 353
  Summary 354
  References 354

Chapter 14
Biological Analysis of Fungi and Associated Molds, *T.M. Madelin and M.F. Madelin*
  Introduction 361
   General Characteristics of Fungi and Associated Molds 361
   Principles of Aeromycological Assays 363
  Total Assay of Aerosols of Fungi and Associated Molds 364
   Introduction 364
   Non-Volumetric Total Assays 364
   Volumetric Total Assays 365
   Total Assay of Rain- and Splash-Dispersed Air Spora 367
   Microscopic Identification of Air Spora Constituents 367
  Viability 368
   Introduction 368
   Culture Techniques 368
   Vital Staining 371
  Infectivity 372
   Plant Pathogens 372
   Animal Pathogens 373
    Airborne Mycoses 373
  Toxicity 375
  Allergenicity Assays 377
   Introduction 377
   Symptoms of Respiratory Allergy 377
   Sampling the Environment 378
   Identifying and Extracting the Antigens 378
   Testing for Allergy 379
   Species Implicated in Allergies 379
  Summary 380
  References 380

Chapter 15
Aerobiology of Pollen and Pollen Antigens, *Auli Rantio-Lehtimäki*

- Introduction ... 387
  - General Characteristics of Pollen Grains ... 387
  - Principles of Pollen Assays ... 387
- Total Assay of Pollen Aerosols ... 390
  - Introduction ... 390
  - Non-Volumetric Total Assays ... 390
  - Volumetric Total Assay ... 390
    - Intact Pollen Grains ... 390
    - Pollen Antigens ... 391
    - Pollen Grains and Pollen Antigens Indoors ... 392
- Pollen Information Networks ... 392
  - Local Networks ... 392
  - All-Continent Networks ... 393
- Viability ... 393
  - Introduction ... 393
  - Methods of Studying the Viability of Pollen Grains ... 394
- Pollen Allergies ... 395
  - Introduction ... 395
  - Allergenic Pollen ... 395
  - Allergic Reactions to Pollen Grains ... 396
  - Studies of Individual Pollen Allergens ... 397
    - Introduction ... 397
    - Methods of Studying Pollen Antigens ... 397
    - Individual Pollen Allergens ... 398
  - Allergens in Micron Size Fractions in the Air ... 398
  - Pharmacology ... 399
    - Detection of Pollen Allergy ... 399
    - Immunotherapy and Drugs ... 399
- Toxins ... 400
  - Toxic Pollen Grains ... 400
  - Pollen and Acid Rain ... 400
  - Changes in Pollen Chemistry Due to Atmospheric Pollutants ... 400
  - Pollen Grains in Toxin Tests ... 401
- Summary ... 401
- References ... 401

Chapter 16
Outdoor Air Sampling Techniques, *J. Lacey and J. Venette*

- Introduction ... 407
- Sources of Bioaerosols ... 408
- Ambient Air ... 410
  - Sedimentation Samplers ... 410
  - Impaction on Static Rods ... 412
  - Whirling Arm Impactors ... 412
  - Impaction in Suction Traps ... 415
  - Virtual Impactors ... 419
  - Impingers ... 420

      Cyclone and High-Volume Samplers 420
      Filtration Samplers 421
      Precipitation Sampling 422
      Biological Air Samplers 422
      Choice of Sampler 423
      Siting 424
      Collection Media 425
      Handling the Catch 427
  Under Plant Canopies and in Crops 430
  In and Over Cities 441
  Over Seas 444
  Upper Atmosphere 445
      Isokinetic Air Sampler 446
      Isokinetic Sampler for Light Aircraft 446
      Modified Isokinetic Impactor 448
      Andersen Drum Sampler 449
      Qualitative Collection of Airborne Microorganisms 449
      McGill G.E. Electrostatic Bacterial Air Sampler 449
      Filter Sampler 450
      Straight Tube Ram Jet Impactor 450
      Membrane Filtration 450
      Cyclone Sampler 450
      Sampling from Rockets 451
      Culture Media and Incubation Temperatures
      for Airborne Sampling 451
      Sampling Plans 452
  Large vs. Small Bioaerosol Particles: Choice of Samplers 452
  Summary 453
  Acknowledgments 454
  References 454

Chapter 17
Safety Cabinets, Fume Cupboards and Other Containment Systems,
R.P. Clark
  Introduction 473
  Sources of Bioaerosols 474
  Assessment of Hazard 474
  Containment Systems 475
      Microbiological Safety Cabinets 475
          Class I Microbiological Safety Cabinet 475
          Class II Microbiological Safety Cabinet 476
          Class III Microbiological Safety Cabinet 476
      Convertible Cabinets 476
      Safety Cabinets for Non-Microbiological Use 477
      Fume Cupboards (Hoods) 478
      Absorption Fume Cupboards 479
      Local Exhaust Systems (LEV) 479
      Rooms as Containment Facilities 479
  Containment Principles 480
  Operator Protection Factors 484

The Microbiological Method 486
The Potassium Iodide KI-Discus Method 487
Tests Based on Gas Tracers 488
Examples of the Use of Containment Measurements 490
Effect of Laboratory Ventilation Systems 493
Automated Analysis Equipment and Possible
Bioaerosol Hazards 497
Surface Contamination 500
Summary 500
References 503

Chapter 18
Problem Buildings, Laboratories and Hospitals, *J.M. Macher,*
A.J. Streifel and D. Vesley

Introduction 505
Sources and Consequences of Bioaerosol Exposures 506
    Problem Buildings 506
    Laboratories 507
    Hospitals 508
Deciding When Environmental Monitoring Is Appropriate 509
Investigating Building-Related Bioaerosol Problems and
Evaluating Exposures 511
    Visual Inspection 512
    Surface and Material Sampling 513
    Air Sampling 514
    Sampling Procedures 514
    Sampling Sites 515
    Sampling Frequency, Timing and Volume 516
    Detection Methods 517
Interpreting Findings 520
    Liquid and Material Samples 521
    Air Samples 521
        Problem Buildings 521
        Laboratories 522
        Hospitals 523
Modeling 523
Ventilation for Bioaerosol Control 524
Summary 525
Acknowledgments 525
References 525

Chapter 19
Industrial Workplaces, *B. Crook and S.A. Olenchock*

Introduction 531
Health Effects of Bioaerosols in the Workplace 531
    Microbial Infection 531
    Allergy to Airborne Microorganisms 532
    Allergy to Non-Microbial Airborne Materials 532
    Toxicosis 532
Sources of Bioaerosols 533

    Agriculture/Food Production 534
    Factory/Industrial 534
    Biotechnology 535
  Bioaerosol Samplers Used in Industrial Workplaces 535
    Airborne Microorganisms 535
    Non-Microbial Bioaerosols 535
    Endotoxins 537
  Practical Considerations 537
    Sampling Strategies 537
    Sampler Positioning 538
    When to Sample 538
    Handling the Catch 539
  Summary 539
  References 540

Chapter 20
Bioaerosols in Animal Houses, *C.M. Wathes*

  Introduction 547
  Why?—The Reasons for Bioaerosol Sampling in Animal Houses 547
  What?—The Nature of Bioaerosols in Animal Houses 548
    Airborne Pathogens in Animal Houses 549
    Typical Constituents of Bioaerosols in Animal Houses 552
  How?—General Principles of Bioaerosol
  Sampling in Animal Houses 555
    Physical Processes Involved in Sampling Bioaerosols 557
      First Step—Take a Representative Sample 557
      Second Step—Transport the Sample to the Collector 558
      Third Step—Condition the Sample 558
      Fourth Step—Collect the Bioaerosol With a Sampler 558
    Choice of Bioaerosol Samplers for Use in Animal Houses 563
    Optical Particle Counters 564
    Impactors 564
    Impingers 564
    Filtration Samplers 565
    Calibration for Particle Size and Flow
    Rate and Other Good Practices 566
    Supplementary Information to Be Reported 566
  Where?—Location of Bioaerosol Samplers in Animal Houses 567
    Location and Number of Sampling Sites 567
  When?—Frequency of Bioaerosol Sampling in Animal Houses 570
    Dynamic Behavior of Bioaerosols 570
    Duration and Frequency of Sampling 572
  Who?—Students of Bioaerosols in Animal Houses 572
  Conclusions 573
  References 573

Chapter 21
Bioaerosols in the Residential Environment, *Harriet A. Burge*

  Relationships Between Human Diseases and
  Bioaerosols Indoors 579

>   Infections 579
>   Hypersensitivity 580
>   Toxicoses 581
> Reservoirs, Amplifiers and Disseminators of Bioaerosols 582
>   Outdoor Air 582
>   Human Sources 583
>   House Dust 584
>   Other Substrates for Fungi 585
>   Water Reservoirs and Other Sites for Bacteria 586
>   Pets and Pests 586
> Sampling 587
>   Collecting a Valid Sample 587
>   Sample Analysis 588
> Data Analysis and Interpretation 589
> Prevention and Remediation 590
>   Prevention of Intrusion 590
>   Control of Potential Reservoirs 590
>   Remediation 591
> Summary 591
> References 591

Index 599

BIOAEROSOLS HANDBOOK

CHAPTER 1

Editors' Introduction

Many processes generate bioaerosols of diverse forms ranging from submicron allergens to much larger fungi, pollens, droplet nuclei, and dust rafts. Humans and animals are disseminators, e.g., during sneezing, while acting too as reservoirs and amplifiers. Indoor bioaerosol particles comprise respiratory pathogens, contaminated skin squames, dust mite fragments/faeces, fungal spores, hyphae and products, etc. Residential environments may present more serious risk through infection and allergy than those of lower bioaerosol concentration as occur outdoors. Reduced house ventilation rates may benefit energy conservation but concomitantly result in higher bioaerosol concentrations and risks of associated diseases. Offices, schools, hospitals and industrial workplaces, similarly are contaminated, while practices conducted therein contribute to bioaerosol burdens. Industrial workplaces provide further sources owing to contamination by, and the biological activity of, materials being handled, such as microbial and food allergens/toxins, industrial scale fermentation microbes and products, e.g., insulin. In animal houses, bioaerosols are produced from animals themselves, their foodstuffs, bedding and faeces. Within all buildings, particle diffusion and air currents ensure bioaerosols become distributed throughout and reach the most inaccessible places. But, during this process their biochemical properties (e.g., viability, infectivity, allergenicity) can be modified so that allergenic/immunological properties change and viability declines, while physical parameters such as size and shape alter.

Outdoors many processes cause bioaerosol liberation including air turbulence, spray irrigation, sewage treatment plants, breaking of waves, bursting of bubbles, and crop spraying. The purposeful release of pollen and spores provides a further example. The use and deliberate release of biological pesticides/genetically engineered microorganisms also represent potential hazards both during their application as well as manufacture in industrial fermenters. Our ability to successfully monitor and control any associated leaks seems essential while environmental impacts of these processes through colonization by released microbes, competition with indigenous species and genetic exchange through transformation also may present a hazard. Of concern too, is contamination via bioaerosols of food and pharmaceutical products especially where vaccines are concerned.

Bioaerosol hazards to man primarily arise from exposure to high concentrations or to unfamiliar forms, and comprise respiratory distress, microbial infection, allergenic reaction, respiratory sensitization, and toxicological reaction. Changing patterns of work and leisure have raised risks, while outdoors air pollution (particulate and gaseous) levels generally are increasing. Respiratory hazards, or our awareness of them, show an upsurge, one example being asthma in the UK. The costs to society of bioaerosol hazards are great, e.g., in the USA each year currently there are 250 million episodes of respiratory infection, i.e., 75 million physician visits/year, 150 million days lost from work with medical care costs of ca. $10 billion, plus loss of income of ca. $10 billion. There are too airborne plant and animal diseases, both causing substantial economic losses, e.g., respiratory diseases in pigs, exacerbated by poor air hygiene in animal houses with associated reduced feed efficiency, depressed growth rate and increased veterinary treatment, at an annual estimated cost in the UK of £20 million.

Exposure limits may be set for environments in which bioaerosols are found. In the UK, the Public Health Act (1936) and Control of Pollution Act (1990) are relevant, while in the USA, the Food and Drug Administration (FDA) sets stringent limits on levels of bioaerosols in pharmaceutical industries for example, especially when concerned with vaccine production. But, how are such limits to be monitored? Assessment requires the meaningful collection of representative samples, and their characterization. However, cursory examination of the aerosol literature often leads to confusing and conflicting perspectives on sampler choice, methodology and analytical procedures. In addition, what is the sampling efficiency and very importantly, how are samplers calibrated?

Even when these matters have been settled there are further questions concerning handling the catch to determine biological and physical characteristics, all of which can change in time and space. Microbial bioaerosols present additional special difficulties because of potential conflicts between their efficient sampling as particles and as viable entities. Furthermore, biological effects can be modified/exacerbated by simultaneous inhalation of other particles and/or pollutants, e.g., sulphur and nitrogen oxides.

Clearly, no single sampler/sampling protocol is likely to be adequate for all bioaerosols in their diverse environments. Two alternative strategies have been proposed: (1) a reference sampler (or protocol) can be specified with the advantage of consistency though there will be limitations to any one sampler; (2) performance standards may be set for a particular sampling objective against which alternative samplers (or protocols) can be tested. The latter encourages development of better, generic samplers/sampling procedures though gradual improvements may take time. Ultimately establishing performance standards for bioaerosol samplers/sampling is essential. The UK Department of Trade initiative in this area, as well as the UK Calibration Forum for Aerosol Analysis (NCFAA), are noteworthy.

The main objectives of the "Bioaerosols Handbook" are to provide up-to-date detailed descriptions, comparisons and calibration methods for bioaerosol samplers with appropriate sampling methodologies and analytical procedures. Physical and biological properties are considered from both practical and theoretical viewpoints. The Handbook represents a compilation of relevant, up-to-date knowledge and expertise of leading bioaerosol and aerosol scientists from six different countries in Europe and North America. The authors and editors are aware that there are other texts dealing with aerosols and measurement techniques. However, often these are theoretical and may fail to include essential practicalities of sampling, calibration or sample assay methodologies. This Handbook attempts to deal with the subject of bioaerosols on a broad yet in-depth basis, and provide guidance based on firm physical and biological principles plus practical experience of meaningful catching, assessing and monitoring bioaerosols encountered in many environments.

After two introductory chapters, the Handbook is divided into four parts. Chapters 3 through 6 examine the principles of bioaerosol sampling while emphasizing the essential foundations upon which good practice is built. Comprehensive descriptions of modern bioaerosol samplers, including direct reading instruments, are given in Chapters 7 to 10: many of the described calibration techniques are common to other aerosols samplers. Bioaerosols may be analyzed chemically, physically, and biologically, and current techniques are described in Chapters 11 to 15. Finally, Chapters 16 to 21 proffer the varied experiences of current practitioners of the 'art' of bioaerosol sampling in the workplace, home, specialized settings, e.g., laboratories and hospitals, and outdoors.

CHAPTER 2

Bioaerosols: Introduction, Retrospect and Prospect

J.M. Hirst[1]

THE FOUNDATIONS

'Bioaerosol' and the comparable term 'aerobiology' are compound words that describe studies relying on the interplay of disciplines primarily physical, chemical and biological. Neither the order of the syllables nor the history of the studies gives any guidance as to the precedence of the living or atmospheric component and attempts to identify which 'was the chicken and which the egg' would be fruitless; they must always be interactive and interdependent!

However, studies of how particles enter and travel in air are so characterized by diversity and diffusion that contributors and readers of the Handbook will need to recognize the same boundaries. The specialist will have no difficulty in distinguishing the scientific meaning of the word 'aerosol' from that thrust into common parlance by advertisements that lead the public to believe an 'aerosol' is the can bought at the chemist or supermarket and used to generate an aerosol of fly-spray, deodorant or the like. Scientists will support the definitions provided by the editors:

> An **aerosol** consists of material finely divided and suspended in air or other gaseous environment, with compositions as varied as matter itself.

> A **bioaerosol** is an aerosol comprising particles of biological origin or activity which may affect living things through infectivity, allergenicity, toxicity, pharmacological or other processes. Particle sizes may range from aerodynamic diameters of ca. 0.5 to 100 µm.

These definitions show how an aerosol can qualify to be a bioaerosol, yet they permit the great flexibility required to accommodate enormous biological diversity and to allow the subject to develop. Thus, life is not essential but the particles must have biological origin or activity. Also, sizing particles by *aerodynamic diameter* (see page 180) specifies airborne properties but allows the inclusion of great diversity in the dimensions and shape of biological propagules or fragments of organisms or their products formed mechanically.

---

[1] Editors' note.
This introduction to the Bioaerosols Handbook was written by Professor Hirst after he had been given sight of the text in draft. Through this unorthodox practice, he has gained the unique advantages of constructive comment based on hindsight and thereby has overcome the potential limitations of inadequate integration that can hamper multi-author volumes.

Inflexible definitions could also inhibit inclusion of desirable additions, for example, novel 'product allergens' comparable to those in the faeces of the house dust mite, or greater need for simultaneous study of biological and abiotic components that seem synergistic to biological activity.

Among the earliest recorded effects of aerosols and bioaerosols must have been types of occupational pneumoconioses that were mentioned in ancient Greek literature. Understanding grew slowly until the late 17th century when microscopic recognition became a reality. Thereafter study of airborne transmission accelerated, but erratically, for instance when important pathogens were shown to be transmitted by water or vectors rather than by dispersal in air. During the late 19th and 20th centuries the rate of development has become exponential, fostered by scientific progress, increased industrialization, and population growth (differentially causing either wealth or poverty), not to mention the consequences of international trade and conflicts. Later chapters show how all aspects of human activity are affected and exemplify the roles of viruses, bacteria, fungi, algae, plant microspores and many products with biological activity. The roles of many in causing the various plagues that afflict plants, animals and man are justifiably stressed in the Handbook as are the increasing concerns with the contamination of food, pharmaceutical products and effects related to environmental pollution. Currently these are the effects that most attract both interest and support. The hazard list (as exemplified in Chapter 1) is of course prodigious and very costly but judgement will not be balanced unless we also consider what may be receiving less attention than it should; has anything been neglected? In time, science will suffer if it forgets that bioaerosols are probably as beneficial as they are hazardous. Most of the benefits operate so effectively yet so surreptitiously that they are easy to overlook and although the costs of the hazards have to be met and so can usefully be estimated the same would not be possible or useful for the benefits. Nevertheless, life on earth could not survive for long without the re-mineralization of organic matter by the essential organisms of decay, many of which arrive by air. The mosses, ferns, and grasses (and many other plants and trees) could neither reproduce nor spread properly without airborne propagules. Surely not only 'green ecologists' should be concerned that there is so little attention (other than in polar regions) to the colonization of barren or environmentally altered substrates or in spreading genetic diversity (existing or engineered) through existing populations.

The real work of the Handbook begins in Chapters 3–6, each of which helps set the rules for studying bioaerosols and together they provide firm foundations, based on the wealth of knowledge and methods of aerosol science for understanding how particles behave in air and the 'ideals' of sampling. Some of these chapters may not be bed-time reading for biologists but they must persevere with the study because it is worth repeating that only thus can they build a better understanding of the proven behavior of particles in air. These chapters provide vital clues (if not always prescriptions) as to how physical principles may be applied to generate or to capture airborne particles. They not only explain the great diversity of forces that may operate but also how important it is to understand which may operate, when and on what. For example, even Brownian movement is shown to be important on sub-micron particles within the strict confines of the alveoli of the lungs. More familiar will be roles of gravitational settling, eddy diffusion, inertial impaction, filtration and other processes, here described separately but usually variously interacting in experimental conditions and during sampling both within buildings and in outdoor weather.

Gradually, but sometimes a bit erratically, principle gives way to practice with the introduction of less idealistic conditions. It is probably helpful and certainly realistic that this transfer cannot be seamless. Complex interactions between physical factors affecting sampling and deposition often make the path difficult to follow as complications of weather, topography and vegetation are introduced as modifiers of both liberation and deposition, and how samplers should be designed to try to match these circumstances. Consideration of how to attempt to bring some ordered measurement to the complexity of composite bioaerosols illustrates how difficult this process is, yet how necessary where total aerosol content may be important. The introduction of the concept of 'dispersal unit' raises the problem of when and what is the effective unit to estimate. When is one particle sufficient, or may a clump be required for effect or infection? How much are clumps fragmented on capture? When and in what circumstances is number most important? When should it be replaced by mass, surface area or other parameters from among those now often so much easier to measure electronically? Fortunately, perhaps, biological requirements can often be more precisely defined by selective trapping methods or catch treatment allowing cultural or visual identification. These chapters foreshadow some of the myriad complications inevitable with living things, for example, the difficulty of measuring effects of dispersal and the stresses of capture on viability, which form essential knowledge to many investigations but are of no consequence to others.

It is inescapable that the Bioaerosols Handbook, having stressed the importance of aerosol science, must gradually confront the no less difficult problems posed by the biotic components. Biologists must realize that the innumerable further complexities their target organisms will present in bioaerosol studies can only, as a last resort, justify any departure from the principles of good sampling. However, I fear that later chapters will confirm that in reality such 'last resorts' are all too common. It would also be wise for biologists to recall how often history has shown that methods first used in biology and medicine were later replaced by more capable and accurate instruments borrowed or developed from work by physicists, chemists or engineers engaged in industry, military studies, aerodynamics or meteorology. This is a point in the Handbook and in research where both biologists and physicists need, yet again, to remind themselves that their disciplines must interact and be interdependent. Success depends on both groups of discipline generating a common language and understanding of mutual aims and problems. Even when this is achieved the battle may not be won without joint attention to the problems and design of experimental sampling and of manufacture. Practitioners studying bioaerosols are often remarkably dependent on anecdotal episodes and on which instruments are available. Nevertheless, experiments should be designed to permit statistical analysis whenever possible. Also some teams need to forge close cooperative links with manufacturers jointly to develop and market convenient, reliable, robust, but accurate samplers that are well matched to biological techniques yet suited to use in laboratories, factories, forests, crops, poultry houses or patients' homes.

THE TOOLS OF THE TRADE

The diversity of particles, of aims and techniques involved in sampling bioaerosols ensures that there never will be one perfect, all-purpose sampler. That allows no excuse to abandon attempts to get as close as possible. The initial chapters indicated the enormous range of particle sizes and showed how differently their behavior is affected by the many forces active and the prevailing environment.

Chapters 7 to 10 examine how that information has been or could be used for sampling and measurement. Freely exposed surface samplers are often much more complex in action than their simple structure suggests. For those who seek it there is information to guide when and for what purposes they may reliably be used. Suction samplers are as difficult to calibrate; the first hurdle is to collect a known volume of the aerosol containing a representative sample of all airborne particles. By sampling uniform spherical monodispersed particles isokinetically in laminar air flow this perfection can be approached and there are many specialized circumstances for which it can and must remain the aim. However, in practice assorted shapes, sizes, densities, turbulent flow and gustiness create many problems, especially when suspected interactions between aerosol components seem increasingly to demand the simultaneous measurement of all. Wide ranges of size, shape and density make this difficult enough but when, in bioaerosols, they defy recognition unless grown, need to be measured over months or years, or are released intermittently in brief episodes in response to life-cycles, variable weather or treatment, then the full difficulties of measurement become evident.

Nevertheless, bioaerosol studies would have little validity without some standards. Perfection may be unattainable but at least more attempts must be made to estimate a theoretical sensitivity and to calibrate efficiency or, perhaps more realistically, estimate the deficiencies of sample collection. Once the project objectives are defined, accumulated experience must be used to help define the samplers best fitted to adopt for the particle spectra and environmental circumstances likely to be encountered. For testing to be believed, accurate and reproducible it must begin by comparing candidate samplers against near ideals using the defined conditions and standard particle spectra that are the tools of aerosol science. Unfortunately this implies access to facilities for generating standard aerosols and sampling them in specialized wind tunnels and settling chambers. Such facilities are rare and often prohibitively costly for the funding standards available to many small manufacturers and most biological projects. Desirable, indeed indispensable, though they may be, such tests should be but the first steps in a program that progressively introduces the difficulties and assesses errors of sampling real organisms in real environments.

The chapters on instrument calibration procedures (7 and 8) are essential to those who can approach perfection but no less important to those who must strive to know how far the methods they have to use will inevitably fall short. In very few instances is such information now available; at present it is difficult to do better than quote the advice in Chapter 8:

> "Whichever strategy is chosen, it is essential that due regard be given to the problem of obtaining representative samples and avoiding biases between the inlet and measurement/collection area. Viability may also be an important issue when sampling the various types of bioaerosol. The reader is encouraged to investigate both of these aspects before embarking on any quantitative sampling exercise."

The Bioaerosol Handbook will achieve much if it imprints this message on users. The need is urgent for the same degree of knowledge, skill and experience as applied to laboratory techniques to be applied to calibration of bioaerosol samplers that are to be used in 'less-than-ideal' conditions. Jolyon Mitchell provides evidence for aerobiologists to apply many of the rules but also has to make the honest if regrettable admission and exclusion that:

> "Measurements using the various types of directional and omni-directional samplers in the open environment are a separate subject in themselves."

Two succeeding chapters (9 and 10) list and comment on the inertial and non-inertial samplers used in studying bioaerosols. Both chapters give valuable, up-to-date guidance about accepted practice. Although well referenced, the number of instruments requiring mention greatly limits the detailed comment about each. This is a pity because correct choice of sampler depends so much on the environment, the particles sought, their concentration, trap exposure and sampling time. It is no fault of the author that, regrettably but quite usually, there is much less information available about the representativeness of the sample initially collected than of the effectiveness with which particles are retained after collection. Consequently real performance is often in doubt, a reminder of the urgency of studying Jolyon Mitchell's 'separate subject.'

Experience and available information indicates that efficiency is by no means constant, small differences can sometimes have large effects. It usually is essential to consider the environment and purpose of an investigation before deciding which trap to use and where to locate it. For example, if studying a wheat crop close to harvest, the best compromise for a prolonged, general survey of its exposure to airborne fungal spores might be made with some suction sampler but very different spectra of spore size would be found depending on height within or above the crop. However, deposition on the ear and its stalk might be measured most accurately by vertical sticky cylinders where the efficiency of capture would be related to wind speeds experienced, whereas in calm air near the soil, deposition (even of small spores) on basal leaves might best be represented by horizontal gravity slides, which would be misleading elsewhere.

There can be no doubting the correctness of the dictates of good laboratory volumetric sampling; some, such as short intakes, are consistently applicable but important ones, such as isokinetic sampling, seldom seem practical outdoors. The design of most suction samplers specifies some constant intake rates and often requires some defined orientation of the sampling inlet. These and earlier chapters give many clues about the principles conferring merits or failings of the samplers listed but too seldom provide estimates of the magnitude of their probable errors. The content of the Handbook indicates that bioaerosols are increasingly studied in enclosed environments, housing, factories and hospitals, with ever more interest in smaller particles, fortunately both factors which should incur smaller collection errors. Nevertheless, much important work with bioaerosols remains out-of-doors exposed to weather or in variable enclosures dictated by the environments and activities of the organisms under study.

It may sound heretic but having established so many of the ideals, there now seems a strong need for aerosol scientists to relinquish perfection and turn some attention to realistic imperfections. Could not some be enabled to help biologists further by defining the magnitude of the errors of samplers used or developing better ones? We need to know better the magnitude of collection errors incurred by sampling widely different particle sizes (and shapes), using a range of constant intake rates through variously shaped intakes oriented differently to air movements that vary rapidly both in speed and direction. Accurate definition may seldom be possible, but results and judgements would be much more valuable if experimenters knew what trust to place in them and could admit imperfections rather than ignore them, as is too often the case.

The value of bioaerosol sampling methods finally depends on whether they are practical, give reproducible results and effectively serve the sampling objectives. At present many investigators are forced (or worse, content) to use devices supported by few facts or advertised merely on the basis of "relative performance tests" in which a new device is compared only with some earlier but equally ill-defined instrument. Many honest investigators are aware they use inefficient devices that tell lies; at best they can hope that these are reproducible and that they have learned how to interpret their results so as to usefully assist the sciences they serve.

HANDLING THE CATCH

For particular groups of organisms (e.g., pollens and some fungi) that are difficult to grow, light microscopic recognition by experts may long offer unrivaled accuracy and specificity. However, anyone who has spent many hours (even months) scanning catches under the light microscope must fervently hope for the development of less tedious but equally effective assessments. The obstacles are formidable involving great diversity in composition, amount (size, shape, number), origin, targets, viability/infectivity, and means of detection and expressing effects. Even if there could be a single, universal catching method there could be no one means of assessing the presence and expressing the effects of bioaerosol components. There has certainly been progress with many techniques, efforts that must continue. However, the diversity of particles in bioaerosols, the minuteness of genetic differences that may affect specific functional capability, their constant susceptibility to change during and after dispersal and the selectivity required by project objectives often make successful automation capable of answering many biological problems seem still distant.

The five chapters (11 to 15) dealing with assay techniques contain much evidence of progress, much hope for future development and justifiable cautions that are expressed gently because authors quote correctly but seem loath to be prophets. There is some repeat reporting, inevitable in a multi-author volume, but here with some merit by offering the reader comment from different viewpoints. Some of the advances quoted, for example, in image analysis, programmed scanning, fluorescence marking and immunology can be adopted with little difficulty and great benefit. The practicality of others will no doubt depend on the selectivity required by project objectives and by technological improvements in methods for studying the various biological components or the activity of their products. When estimation of viability or infectivity is important, many methods are suggested but there are many limitations and an obvious need for less lethal methods of capture and retention.

In many respects the viruses and bacteria must be the most difficult groups to study. They are minute, relatively featureless and need high magnification or special methods for recognition. Because they replicate only in living (or host) cells, viruses are particularly difficult to assess. Liberation and dispersal of both bacteria and fungi is complicated by important rain-actuated processes that cannot be ignored by epidemiologists but which operate independently and contribute variably to bioaerosols. Toxicity is another important result of many fungal infections of both plants and animals. The basis of allergenicity has been studied most intensively in pollens, questions posed by current research are intriguing and suggest a need for comparable investigation among fungi and bacteria. For example, if there is some real association between anemophily and allergenicity, then is this coincidental or does it indicate some functional relationship and does the same apply to other airborne particles? Most pollens are of sizes expected to be deposited in the upper parts of the respiratory system but there is now much evidence that they may cause asthma rather than rhinitis which suggests deposition further into the respiratory system than their size would suggest. How does this happen? Is it due, as often suggested, to pollen fragments? But pollen grains seem strong, resistant structures difficult to fragment (except perhaps if eaten). Could pollen allergens exist in particles formed externally on pollen grains from which they may be detached to follow some independent route to respiratory deposition? Could the symptoms be explained by transport of the allergens within body tissues? Has it been proven not to be due to the inhalation of smaller unrelated propagules that may bear similar allergens?

Throughout this volume contributors and readers are confronted with a diversity bewildering in quantity and defying generalization. Certainly the authors' duty toward completeness leaves little room for expressing preferences, even if their experience and supporting evidence could allow this. How then are new experimenters to be helped select which methods best fit their objectives? These may, for example, range from interest in the whole aerosol entering a respiratory system to attempting to assess the gene-to-gene challenge to a newly-bred crop cultivar of some newly pathogenic variant. I admit to not knowing a ready answer which seems largely a task for the future. But I am confident that those who study the Handbook carefully will be warned off many blind alleys and find signs to many worthwhile trails.

CURRENT PRACTICE

Those who have read thus far will probably have realized from my comments and examples that although I have very nearly 50 years experience in aerobiology, my active bioaerosols research was concentrated in the first 25 years, centered on the larger-spored plant pathogenic fungi, on distant dispersal and latterly on samplers.[1] This further explanation is pertinent to my comments on the final group of chapters (16 to 21)

At first it may seem disproportionate that only Chapter 16 deals with 'the great outdoors.' However, this probably truly reflects the fact that contemporary interest and support are predominantly devoted to studying problems of enclosed spaces, special risks or processes. This results from the recognition of special risks, some technological advances, and greatly stimulated public concerns with health and environment (often powerfully bolstered by legislative imperatives). Defining how to study and meet these needs together with intense media interest has demanded attention. Although medical studies began much of bioaerosol research its intensity has recently developed greatly in homes and hospitals and added extra concerns such as new allergens, pollutants, toxins and a greater interest in opportunistic attack of patients experiencing immunological suppression. Increased processing of foods, detergents and materials has created extra needs for worker protection in laboratories as did farmers' lung disease and other hazards in animal housing.

While not arguing with this allocation of chapters, my experience and personal interest gives me reason to stress the heavy duty and responsibility placed upon the authors of Chapter 16. Bioaerosols varying enormously in kind and quantity occupy much of the global troposphere. Content is sparse in polar regions and may seldom reach the stratosphere but elsewhere is much affected by faunal and floral sources, responding to climate and hour by hour to weather in particle production, liberation, dispersal and deposition. Outdoor bioaerosols are the largest in quantity, variety and importance to the global environment. Furthermore, knowledge of interchanges between enclosed spaces and open air is essential to both. Popular myth has it that the contaminant spore of *Penicillium notatum* that led Sir Alexander Fleming toward the discovery of penicillin entered St. Mary's Hospital through his open window (they are usually more common indoors than outdoors). There can be less question about the pollen grains and basidiospores I know to have been the subject of fruitless treatment or research after deposition, initially unrecognized, on slides in histology laboratories. By contrast, ventilation and air conditioning from glasshouses, animal houses or offices have proved powerful routes for transfer of pathogens of plants, animals and man. The study of pollens in palynology offers excellent examples of multiple important actions, in cross-pollinating plants, in causing allergic rhinitis and being one of the few groups of airborne propagules resistant enough to be traced to their ultimate fate and adding evidence of great value to the historical and fossil records. Studies of outdoor bioaerosols have been the main route for accumulating the specialist botanical, mycological and agricultural knowledge essential to recognizing propagules which they have later helped to identify as causes of hitherto unrecognized problems, infections, toxins or allergens and then assist research towards control or avoidance. The history of research into farmers' lung disease well illustrates that interdependence within biological and medical disciplines is just as necessary as with the physicists!

Chapters 17 to 21 all share a primary concern with variously enclosed spaces although each is justifiably separated because of specialized conditions and purposes. The group share many features. Many of the closed environments (e.g., animal houses and some workplaces) are certainly characterized by very dense bioaerosol concentrations; by contrast contaminant detection in sterile rooms must be designed to detect very scarce particles. Much of the work is young and still defining its standards, testing methods and developing philosophies for action. Although the variety of problems, circumstances and bioaerosol components is so great that many studies need special methods or procedures, most investigators select samplers from among those already developed and on the market, perhaps too often this is done on the basis of reputation and convenience rather than detailed inquiry or test of suitability. For example, after 40 years of use, it may be excusable that few recall that the volumetric spore traps of the Hirst, Burkard, Lanzoni, Tilak sequence retain a standard orifice designed to catch the larger-spored plant pathogenic fungi (10–50μm) and, to avoid cluttering the microscopic image with dust, **intended not** to retain many particles <5μm dia. Studies indoors often involve small particles and slow air movements, both factors that decrease errors in collection efficiency. However, significant proportions of many aerosols sampled comprise larger fragments, clumps, or transport on relatively large rafts. On the basis of information in Chapter 8, air movements resulting from heating, ventilation and activities do seem strong and variable enough in velocity and direction to cause considerable sampling error with some particle sizes, intake volumes and orifice sizes and orientations. Some contributors do cautiously assume errors but there seem to be insufficient facts on most of the samplers used to justify the general, ready assumption that such errors are negligible. The 'comforting' idea that wearing some small sampler ensures accurate measurement of personal exposure may be most in need of confirmation and instrument calibration. Large errors may also be incurred through 'overloading' samplers or, to avoid this, by too brief exposures. Much of the sub-sampling of the catch and the subsequent manipulations to identify and assess special components must inevitably introduce considerable errors. However, estimating these rarely seems to get the attention that it deserves, it is by no means easy but accurate identification and precise and replicated assay (wherever possible) are of prime importance, particularly when used to formulate or satisfy legal regulations.

The safety of air in containment systems is subject to many different configurations. The standards required are defined by the purpose—some must maintain the highest standards, others may be required only for operator protection. Fortunately, as Chapter 17 shows, this variety has been well systematized and the categories defined and linked to permissible functions. Often the design of the containment ensures fail-safe protection, but sampling is a necessary routine check of correct operation. A lot can be achieved by standard challenge procedures. Much more difficult is to produce reliable continuous testing to indicate isolated episodes of contamination, added to which such collections would be especially subject to loss of viability of the catch during the long exposures necessary or large intakes required to give a high sensitivity of detection. The subsequent chapters exhibit great diversity in intensity, source and type of challenge, very different abilities to support costly precautions or controls. Again the diversity is great, ranging from the operating theater through food factory and home to the henhouse. Each chapter competently describes the current situation for its special circumstances, the nature of the threats and refers to their social or economic consequences. There seems to be much need for better information on the effectiveness of the samplers that have to be used to study various threats, and particularly is this necessary in relation to the perturbations of the environment by ventilation, feeding, movement and the daily rhythms of activity. Naturally attention is first given where the challenges seem greatest or perhaps least understood. Thus sick building syndromes perhaps now merit more research than the routines of hospital hygiene. There have been notable successes such as the discovery of ventilation systems as reservoirs for Legionella infection, for house dust mites (fragments and feces) being increasingly generated in centrally heated homes, foodstuffs and bedding as well as communicable diseases having been identified as hazardous to housed livestock.

OVERVIEW

The Handbook is too late to be laid as a foundation stone for bioaerosol research. Most of the pioneers are long dead; the work is too active and promising to merit a tombstone. Yet, without doubt the *Bioaerosol Handbook* provides very honest and significant milestones that deserve both congratulation and careful study. The prodigious quantity of information fulfills the definition of a 'handbook' to be both guide and manual. Were it to claim to be the 'last word,' it would be both less honest and less challenging. Careful inquiry within the well-referenced chapters and good index will allow scholars to identify where the science needs strengthening and how best to do this. Perhaps this introductory chapter may conclude with a suggestion or two?

Such is the diversity and complexity of the tasks, propagules and methods involved that nobody believes that there will ever be one perfect, all-purpose sampler or route in bioaerosol research. The physics of aerosol science has taught us much but has more to offer biologists, particularly by studying conditions that are less-than-ideal. We need devices that get as close to truth as possible but we also need to know the magnitude of errors incurred with particles of widely different size and shape. Isokinetic and into-wind sampling seem impossible counsels of perfection in gusty conditions outdoors or in the often chaotically disturbed flows within buildings. More effort is needed to establish the best attainable with omni-directional sampling at constant intake rates and defining the errors that have to be accepted with different particles and conditions. After that, solving problems depends increasingly on biologists, although they will still need help with designing gentler samplers capable of sterile handling and collecting the widest possible range of particle sizes as bulk samples (without overload). New and existing biological laboratory techniques must be used to identify and assay catches but it would be helpful if trap catches could be sub-sampled to provide for simultaneous sterile culture, microscopy or other tests.

Bioaerosol sampling is confronted with bewildering complexity of components, tasks, methods, extent and urgency. There will be strong temptations to attempt to calibrate speedy new techniques alongside laborious biology. However, this seems bound to be restricted to repeated tests in familiar circumstances. How can there be effective generalization between annual samples of pollen in Tauber traps in forests or the estimation of virus particles by electron microscopy of minute samples from brief catches in enclosures? Usually enlightenment must patiently await the publication and confirmation of investigations. The need for the most accurate possible assessment of the results of each observation is stressed if the wait is to be worthwhile. Some authors have estimated a detection threshold for their collection methods and bravely attempt to relate this to effect or impact.

The task is very difficult because detection must depend on the error introduced by every manipulation of each sample and then by how much repetition and statistical design can underwrite the conclusions. Cost, facilities, survival of the catch or the need for sampling unique episodes will often limit how much of what is desirable is practical. Nevertheless, there is need for more effort to convert 'single experiences' into replicated experiments whenever possible. A great advance would be made if every investigator strove to benefit from the advice of Christopher Wathes to begin their work by honestly posing and satisfactorily answering questions framed according to the names of Rudyard Kipling's "honest serving men," namely "... what and why and when and how and where and who."

Nobody is born to study bioaerosols; it is a practice so varied that recruits arrive from very different backgrounds by devious routes of interest or necessity. Such haphazard origins bring a source of strength. Similarly the Handbook authors were chosen for diverse special skills which their chapters reveal. Nevertheless, scanning chapter reference lists reveals a "Great Rift" separating subjects, sources language and thought processes that will form a serious barrier unless bridged. It seems that few physicists, chemists or engineers have been able or had opportunity to escape their disciplines enough to think 'biologically' about the environmental circumstances and components of bioaerosols. Equally there are few in medicine or biology who have the skills or facilities effectively to study and utilize the foundations of aerosol science. There have been notable exceptions—some individuals and particularly the mixed teams studying potential military threats (although, sadly if understandably, their ability to share their knowledge was for long much restricted). Even so, the contributions of these teams both to theory and practice have been and remain among the most important. Unfortunately, the teams assembled for the later emphasis on aerosols associated with atomic and nuclear problems had almost no need for a biological component.

Does this not suggest a great need and opportunity for some broad institution, perhaps a university embracing all the sciences involved, to develop a 'center of excellence' for research and advice in the bioaerosol sciences? Such an initiative must have the specialist facilities matched by grant finance available to support collaborative teams able to call upon physical, chemical, biochemical, electronic and engineering skills. The teams should also have available members with knowledge and experience in meteorology, medicine, pharmacology, agriculture, botany, zoology and the needs and contributions of these sciences.

If in addition to providing an effective guide and manual the *Bioaerosol Handbook* could stress the need for and help establish such a 'center of excellence' able to bridge the Great Rift with interdependent scientists talking a common language, then it would have set bioaerosol research on course into the 21st Century.

REFERENCE

1. Hirst, J.M. "Aerobiology at Rothamsted" Grana, 33:66–70 (1994).

CHAPTER 3

Physical Aspects of Bioaerosol Particles

C.S. Cox

INTRODUCTION

Finely divided matter when suspended in air or other gaseous environments generally is referred to as an aerosol, as opposed to a sol when suspended in solvents, and may have a composition as varied as matter itself. In contrast, the term 'bioaerosol' is more restrictive and taken to mean "an aerosol of biological origin which exerts a biological action in animals and plants by virtue of its viability, infectivity, allergenicity, toxicity, pharmacological or other biological properties, with an aerodynamic diameter in the range 0.5 to 100 µm" (see Chapter 1). The restriction in terms of upper and lower particle size limits in general is set by the circumstances of bioaerosol generation.

In one important aspect all aerosol particles have a commonality, namely, their aerodynamic behavior, and so bioaerosol particles are subject to the same physical laws as other aerosol particles. In addition, though, they are subject to laws concerned with their special biological properties. In this chapter bioaerosols are considered in terms of general physical attributes, whereas biological features are subjects for later chapters.

BROWNIAN MOTION

Bioaerosol particles are bombarded constantly by molecules of the surrounding medium, which move at random. Any imparted motion is also random and the net result is that smaller bioaerosol particles perturbate about a point that drifts. This form of motion is termed 'Brownian motion' and its intensity increases with temperature and with decreasing particle size.

It may be expressed by the Einstein equation, which, in simplified form, is

$$\bar{X} = 5 \times 10^{-6} \sqrt{(t/r)} \qquad (3.1)$$

where $\bar{X}$ = root mean square particle displacement,
  t = time, s
  r = particle radius, cm.

So, for particles greater than about 1 µm diameter, diffusion due to Brownian motion is less than gravitational settling.

An informative example, provided by Dimmick,[1] is that because the mean diameter of the human alveolus is 15×10^{-3} cm, then for a particle to travel the alveolus radius of 7 ×

10^{-3} cm in a 2-sec holding time, its displacement velocity would need to be greater than 3.5×10^{-3} cm/sec. Particles smaller than ca. 0.1 μm and greater than ca. 1 μm diameter consequently should be retained more efficiently than those within that size range.

Such is generally observed in practice.[2]

Another effect is that, owing to diffusion, very small particles tend to move away from regions of high concentration, but for the size range given above, gravitational settling is of more importance than Brownian diffusion.

GRAVITATIONAL FIELD

A particle in a parcel of still air falls owing to the gravitational field at a velocity dependent on its mass. As the rate of fall increases so does the drag or viscous frictional force acting on the particle. When the two forces are equal the particle attains its final or terminal velocity.

Stokes law relates terminal velocity to particle size, mass, etc., and can be derived by equating acceleration due to gravity with viscous drag force. For a spherical particle the terminal velocity, v (cm/sec), is

$$v = \frac{\rho d^2 gC}{18 \eta} \qquad (3.2)$$

where ρ = particle density, g/cm^3
 d = particle diameter, cm
 g = gravitational acceleration, cm/s^2
 η = air viscosity, g/cm s^{-1}
 C = Cunningham slip correction.

For ambient conditions Equation 3.2 becomes,

$$v = 3.2 \times 10^5 \rho d^2 \qquad (3.3)$$

so, a sphere of diameter 10 μm of unit density should fall at about 0.32 cm/sec. Such a prediction is accurate to within 1% for particles of this size but for smaller ones a discrepancy arises because as size decreases and approaches the mean free path of air molecules (ca. 6.5×10^{-6} cm at ambient conditions), particles slip between them. A simple form of the Cunningham slip correction factor is

$$C = 1 + \frac{2A\lambda}{d} \qquad (3.4)$$

where λ = mean free path of air molecules,
 A = constant of ca. unity.

Dimmick[1] suggests a useful approximation for ambient conditions, that is, to add 0.08 μm to the actual diameter before computing terminal velocity, or subtract this value from diameters calculated from such velocities. But for particles less than ca. 0.5 μm the terminal velocity is small and can be considered effectively zero.

Many bioaerosol particles have a density in the range 0.9 to 1.3 g/cm^3, and a value of 1.1 g/cm^3 is quite commonly used for computational purposes when the value has to be estimated.

A consequence of gravitational settling is that on storage in vessels particularly, and to some extent during downwind travel, bioaerosols tend to deposit on lateral surfaces. Their concentration therefore decreases with time and such losses are referred to as 'physical losses' or 'physical decay,' as opposed to 'biological decay,' which reflects loss of biological activity per se, e.g., viability, infectivity, allergenicity.

Because bioaerosol concentration can decline through both physical and biological decay processes, it is usually expedient to measure both effects independently. A common practice is to employ tracers to determine physical decay, preferably in the same experiment as biological decay. This is because physical losses can vary between experiments owing to other forces besides gravity, e.g., electrostatic, thermal gradients, etc., that can be difficult to maintain constant. Tracers have been discussed.[2]

A simplified approach is to model physical decay by a first-order decay process,

$$N_t = N_0 \exp(-k\,t) \tag{3.5}$$

where N_0 = number of particles at time t = 0,
       N_t = number of particles at time t,
       k = first-order decay rate constant.

And

$$\ln \frac{N_t}{N_o} = -kt$$

whence a plot of the logarithm of the airborne fraction as a function of time will be (approximately) a straight line of negative slope (-k). Values of k are functions of sedimentation behavior and chamber dimensions.

A convenient concept is that of aerosol half-life $t_{1/2}$ which is the time required for the bioaerosol concentration to become halved,

$$\ln(0.5) = k\, t_{1/2}$$

whence

$$t_{1/2} = \frac{0.69}{k} \tag{3.6}$$

This value of half-life is approximately given by[1]

$$k = \frac{v}{H} \tag{3.7}$$

where v = particle terminal velocity,
       H = height of a chamber with vertical walls, or an effective height otherwise.

A nomograph[2] is available that relates particle density and half-life to size for aerosols undergoing stirred settling in a chamber having an effective height of 100 cm. Polydispersity of particle size may be allowed for by calculating the decay curve for each size class of the distribution normalized to its frequency, then a composite curve derived by addition of separate decay curves. Even though the method is relatively simple it gives results close to those observed experimentally and clearly demonstrates how polydisperse bioaerosols change particle size distribution with time owing to preferential loss of larger particles.

Over the years various attempts have been made to prolong aerosol lifetimes to beyond that achievable in a stirred settling chamber. Most have been based on a rising column of air to counter particle fall but failed because isothermal and laminar flow could not be maintained. However, an elegant and simple solution[3] was to rotate slowly a settling chamber about a horizontal axis. One way of explaining how it prolongs half-life is that particles achieve a spiral path much longer than the drum diameter. Under ideal conditions a half-life of 12,000 min has been achieved for micron-sized particles.[4]

An alternative, that is especially useful for larger particles, is to attach them to extremely fine spider escape threads.[2]

ELECTRICAL FORCES

Effects owing to electrical forces tend to receive insufficient attention, or are even ignored, in bioaerosol sampling.[5] Airborne particles on generation are invariably charged unless purposely 'neutralized,' and dry-dissemination usually generates much higher charges than does wet-dissemination. The consequences of highly charged particles include rapid aerosol mass depletion owing to enhanced surface deposition and aggregation, as well as sampling artifacts.

Aerosol discharge can be enhanced by incorporating radioactive moieties or by passing aerosols through 'charge neutralizers' containing radioactive sources. While the resulting aerosols usually have overall zero charge, individual particles are still charged with (approximately) the Boltzmann's distribution in which there are equal numbers of particles carrying charges of +1,+2,+3,+4, etc., and -1,-2,-3,-4, etc. In certain cases, though, this distribution has not been observed.[6]

The proximity of charged surfaces by induction of opposite charges in aerosol particles similarly can reduce aerosol mass and affect sampling. Usage of insulating materials in the construction of aerosol holding vessels is best avoided, therefore, while sampling devices should be conducting. Consequently, filters (e.g., Millipore) and their 'plastic holders' are best avoided unless first made conducting, e.g., by gold film deposition.[5]

Additional aspects of electrostatic charges have been considered,[2,7] while mathematical models of effects of electrostatic charges on the processes of sampling and of filtration have been derived.[8]

THERMAL GRADIENTS

Thermal gradients can be responsible for aerosol movement as particles travel from, e.g., a warmer to a colder region. When a particle is warmer on one side than the other, there is a resultant force that causes particle motion. Transparent particles usually move toward the thermal source because they act as a lens thereby focusing energy on the distal side. On the other hand, opaque particles usually move away from the thermal source and 'down' the thermal gradient.

Thermophoretic velocity depends on the material properties of aerosol particles as well as the ratio of particle size to the mean free path of air molecules. Except in thermal precipitators where temperature gradients of at least 100°C operate over a few mm or less, thermophoresis for particles larger than ca. 3 µm is slight thereby imposing an upper size limit of ca. 5 µm in sampling with such devices. On the other hand, thermophoresis provides a gentle force that helps to ensure collection of samples undamaged by thermal precipitation.

ELECTROMAGNETIC RADIATION

Aerosol particles interact with electromagnetic radiation primarily through reflection, refraction, absorption and scattering. Capture of photons by absorption can lead to re-emission at other wavelengths, e.g., fluorescence, and/or to increase in particle temperature. Heat then is lost by convection and conduction. Photophoresis occurs with non-uniform particle heating as it causes one side of the particle to be hotter. Transparent particles tend to move toward the radiation source as they act as lenses thereby focusing the radiation on their distal side. Opaque particles in contrast tend to move away from the source as the nearest side is hottest and more air molecules strike that part of the aerosol particle.

Scattering of electromagnetic radiation can be a complex process with maximum scattered intensity at 0° and 180° to the incident radiation. Elastic Raleigh scattering occurs when particles are much smaller than the wavelength and scattering intensity changes smoothly with scattering angle. (The term 'elastic' refers to when incident and scattered wavelengths are the same.) Elastic Mie scattering occurs when particle and wavelength are comparable in size, and scattered intensity changes critically with scattering angle. Even though it has been modeled only for a few simple shapes, Mie scattering forms the basis of many particle sizing instruments (see, e.g., Chapter 8).

Inelastic scattering also can occur, and the wavelength difference between incident and scattered radiation (as in Raman scattering) is characteristic of the scattering material (see, e.g., Chapter 12). Also important is particle refractive index, both real and imaginary components. The real component is equal to the ratio of the speed of radiation in air to that in the substance, whereas the imaginary component is related to the degree of absorption of radiation by that substance. Both components change independently with wavelength in a manner characteristic of the substance, while for hygroscopic materials the values depend on relative humidity as well. Consequently, derived particle sizes determined with techniques relying in some way on scattered intensity are usually not unique values. Hence, such devices need to be calibrated with the test materials themselves *under* the conditions of test (see also Chapters 7 and 8).

TURBULENT DIFFUSION AND INERTIAL FORCES

When air moves rapidly, instead of flowing smoothly (i.e., laminar flow) the flow is unstable or turbulent, and the transition between these two regimes is governed by the Reynolds number, a unitless quantity (see below). In practice, turbulence causes a random motion to be superimposed on the mean air flow. Particle aerodynamic behavior and transport properties therefore can be very different in the two regimes. This is important during transport through the atmosphere and during transit through pipes, bioaerosol samplers, sizing instruments, etc. A useful exercise is to calculate the airflow Reynolds

20   BIOAEROSOLS HANDBOOK

number in each application so that potential problems owing to turbulence can be anticipated to some extent.

In laminar flow when air carries particles, the two are usually thought of as moving together. When neither changes direction and particles are less than about 10 μm diameter this is approximately so. But, on a change of direction of air flow, the heavier bioaerosol particles having higher inertia may be unable to follow the lower inertia air molecules. Hence, bioaerosol particles may deposit on the walls of curved pipes. The conditions under which this is likely can be estimated as follows.

For a 1 μm diameter particle of unit density in a pipe 0.6 cm diameter and curve of radius 5 cm, at a flow of 10 L/min, the linear air velocity, V, is given by,

$$V = \frac{F}{\pi \left(\frac{D}{2}\right)^2} \quad \text{(cm/sec)} \quad (3.8)$$

where F = flow rate, cm^3/s
       D = pipe diameter, cm
hence, V = $10^4/60 \times 1/\pi \times 1/0.3^2$
       V ≈ 600 cm/sec

and

$$v = \frac{\rho d^2 A}{18\eta} \quad (3.9)$$

where v = particle velocity, cm/sec
       ρ = particle density, g/cm^3
       d = particle diameter, cm
       A = acceleration other than gravitational, cm/sec^2
       η = viscosity of air, g/cm s^{-1}
hence, v = $1 \times [1 \times 10^{-4}]^2 \, A/[18 \times 18 \times 10^{-4}]$

and

$$A = \frac{V^2}{R} \quad (3.10)$$

where R = radius of pipe curvature, cm
hence, v = 0.22 cm/sec

For a curve of length L, the time spent in the curve is

$$t = \frac{L}{V}$$

and supposing the bend is a quarter of the circumference of circle radius R, then,

$$t = \frac{0.25 \times 2 \times \pi}{V} = 0.0135$$

Therefore, during the 0.013 spent in the pipe, the particle would continue in its original direction for that 0.013 and travel a distance of $0.013 \times v = 0.0029$ cm.

Hence, the particle moves 0.0029 cm nearer to the inside wall of the curved pipe, a distance insufficient to result in collision. But, for a 10 μm particle, the distance traveled becomes 0.29 cm and the probability of collision, p, is equal to this distance divided by the pipe diameter, viz.

$$p \approx 0.5$$

Consequently, only about half the number of 10 μm particles entering such a curved pipe may be expected to exit provided the air flow around the particle is not turbulent (see above).

The particle Reynolds number Re_p (unitless) is given by

$$Re_p = \frac{\rho_a v r}{\eta} \qquad (3.11)$$

where ρ_a = air density, g/cm^3
  v = particle velocity, cm/sec
  r = particle radius, cm
  η = viscosity of air, g/cm sec^{-1}

The ratio (η/ρ_a) is 0.15 and when v/r approaches the same value the air flow around the particle will be found turbulent, i.e., when the value of Re_p approaches unity. In the example above, for a 10 μm particle $Re_p = 0.1$ and the flow around the particle will have minimum turbulence, whereas for a 20 μm particle $Re_p > 1$ and it will experience turbulent flow.

Whether the air itself is turbulent can be deduced analogously, and the Reynolds number pertaining to that flow is given by

$$Re = \frac{\rho_a VD}{\eta} \qquad (3.12)$$

where D = pipe diameter, cm.

When Re is 2 to 3×10^3 or greater, turbulent air flow is likely and in the above example is 2.4×10^3, so the flow in the pipe would be likely to be turbulent. The effect would be to increase the probability of larger particles negotiating the bend whereas for smaller particles this probability would be slightly less. Hence, a polydisperse aerosol may change size distribution when flowing along a curved pipe.

One may see that in practice straight pipes are to be preferred for carrying bioaerosols and that gentle curves are better than sharp ones. This same principle applies also in devices that rely on inertial forces, e.g., impactors, impingers, cyclones and stacked sieve samplers. For any test system, it is a useful exercise to carry out a mass balance so that the fate of all bioaerosol particles can be sensibly accounted for.

Inertial Impaction

Inertial impaction will be described in terms of the way it operates in impactors. These devices basically consist of a jet (or series of jets) that may be tapered or not, and are rectangular or circular in cross-section. Below the jet(s) is an impaction plate (Figure 3.1), and particle-laden air is sucked through the jet such that as it approaches that plate it changes direction. Entrained particles, owing to their momentum and inertia, cross the streamlines and move toward the plate surface. The distance moved (D) will be the particle velocity (v) in the jet multiplied by time, viz.

$$D = v \times t$$

where $t = L / V$

where L = length of curved trajectory
V = air velocity.

Hence $v = DV/L$,

but from above,

$$v = \frac{\rho d^2 V^2}{18\eta R}$$

so,

$$\frac{DV}{L} = \frac{\rho d^2 V^2}{18\eta R}$$

or,

$$D_m = \frac{d^2 V}{18\eta} \cdot \frac{L}{R}$$

with L/R a dimensionless quantity equal to 1.56 (or the ratio of 1/4 circumference divided by radius).

The quantity D_m is known as the Sinclair stopping distance and is the vertical distance (Figure 3.1) a particle travels after a directional change. It defines the probability of a given particle colliding with the impaction plate. In the example the larger particle has a stopping distance greater than the diameter of the exit tube, so the particle impacts, unlike a smaller particle for which the stopping distance is smaller than the exit diameter. There will be a particle size for which the collision probability is 0.5, and equal to the probability of passage. This particle size is the characteristic diameter for that particular jet under the operating conditions pertaining. One consequence is that collection efficiency of an impactor depends both on particle size and air flow.

Collection efficiency also is a function of the strength of adhesion between the particle and surface compared to dislodging forces (e.g., particle bounce and blow-off). Common practice is for the impaction plate to carry an adhesive. Even so, problems of trapping can remain and the virtual impactor is one way of overcoming such difficulties. In this device, instead of impacting onto a solid surface particles do so into quiescent air from which they are subsequently and more gently collected.

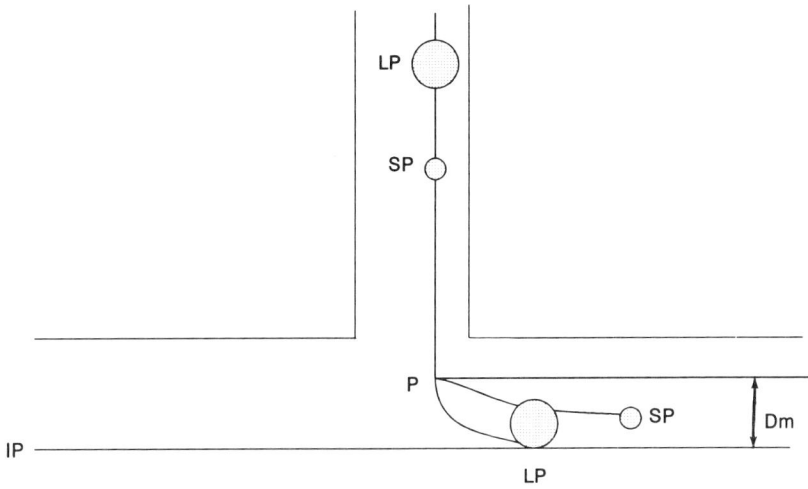

Figure 3.1. Idealized impaction in a simple jet. LP: large particle, SP: small particle, P: point of motion change, IP: impaction plate, Dm: stopping distance.

Another aspect is operating an impactor jet in the low-pressure regime. Under these conditions the pressure of the air as it exits the jet is far below atmospheric (e.g., a fine jet coupled to a high-capacity vacuum pump). At least two consequences follow: (i) the air molecules are widely separated so fine particles slip more easily between them than otherwise, and (ii) owing to the rapid expansion of the air, moisture condensation onto entrained particles can occur. Both these situations affect the collision probability of the particle with the impaction plate and can be applied to extend downward the range of particle sizes that can be sampled by impactors.

PARTICLE SHAPE

Bioaerosol particles have a wide diversity of shape ranging from spheres to spheroids to needle-like to irregular. One way of allowing for different shapes is through the concept of aerodynamic diameter, i.e., the size of a unit density sphere that behaves aerodynamically the same as the given non-spherical particle. Hence, in terms of sedimentation under gravity, two particles will fall at the same velocity even though they have different physical dimensions and density.

For sizing devices relying on particle aerodynamic behavior (e.g., impactors) this is a helpful concept especially for those applications where this is the property governing the particle behavior of interest, e.g., behavior in the respiratory tract. But, for light scattering based particle counters and sizers, etc., the concept is less helpful because the measured size depends on scattered light intensity and therefore orientation of a non-spherical particle at the time of measurement. Yet, common practice is to calibrate such instruments only with spherical particles (see Chapter 7).

There is increasing general interest in effects due to particle shape together with the availability of standard shaped particles (see also Chapter 7). For instance, these are essential for the proper evaluation of devices that try to measure both size and shape of particles.[9] One important area is how values of particle size as determined by various instruments are affected by shape, and an ensuing introduction of possible sizing errors. Another is how to formally describe particle shape and its effect on biological properties (see "Introduction") of bioaerosol particles.

One of the best ways to determine shape is by microscopy coupled to automated image analysis when formal descriptions of shape become possible.[2] But such systems can be expensive and labor intensive, hence the light scattering approaches mentioned above.

RELATIVE HUMIDITY AND HYGROSCOPY

Biological materials such as carbohydrates, proteins, nucleic acids and phospho-lipid membranes are hygroscopic, i.e., they attract water molecules. Likewise their assemblages (e.g., bacteria, viruses) are hygroscopic as are materials with which they are naturally associated (e.g., mucus). One result is that the amounts of water in these entities when airborne depend upon relative humidity (RH) and temperature. Their water sorption isotherms are S-shaped with multilayers of water molecules above about 60% RH and monolayers down to about 20% RH. They demonstrate hysteresis, i.e., there is a displacement of absorption and of desorption isotherms, and this reflects a difference in the energetics of the two processes.

Some materials undergo phase changes following dehydration, e.g., DNA, proteins, phospho-lipids, and on rehydration do not necessarily regain their original structures, e.g., protein denaturation and membrane re-organization. One reflection of this non-reversibility is viability loss (see Chapter 6).

Another consequence is that sizes of particles of such materials depend on RH, as can their shape. Of particular significance is the gross enlargement of fine hygroscopic particles on entering the respiratory tract where they are exposed to warm temperature and a water-saturated atmosphere. Whence, there is a corresponding changed deposition velocity and landing site compared to non-hygroscopic counterparts.

A further effect is that bioaerosol particle refractive index and density can be RH dependent. Then, devices utilizing such properties correspondingly have an RH- and temperature-dependent performance.

Relative Humidity Measurement

The wet and dry bulb method, or psychrometry, is probably the most common for determining relative humidity. The principle is that water, when exposed to an airstream, will cool until the vapor pressure at its surface equals that of the surrounding air. Comparing its equilibrium temperature with that of the ambient airstream permits derivation of relative humidity. In practice, provided that care is taken, the method can be both accurate and convenient. Alternatively, the dew point method provides an absolute method, but care is again required if meaningful values are to be obtained.

CONCLUSIONS

Bioaerosol behavior depends on both physical and biological attributes. Physical parameters mainly affect the where, how and in what quantities particles reach a particular location or landing site. They include diffusional, gravitational, thermal and electrostatic field effects as well as those due to relative humidity and temperature. Inertial forces and fluid dynamics are more concerned with the landing process whereas interactions with electromagnetic radiation are of interest mainly in terms of particle sizing, observation and analysis. Most if not all are functions of temperature and relative humidity.

REFERENCES

1. Dimmick, R.L. "Mechanics of aerosols." In *An Introduction to Experimental Aerobiology*, eds. R.L. Dimmick and Anne. B. Akers, (New York: Wiley Interscience, 1969), pp. 3–21.
2. Cox, C.S. *The Aerobiological Pathway of Microorganisms* (Chichester: John Wiley & Sons, 1987).
3. Goldberg, L.J., Watkins, H.M.S., Boerke, E.E. and Chatigny, M.A. "Use of a rotating drum for the study of aerosols over extended time intervals," *Amer. J. Hyg.* **68**:85–93 (1958).
4. Dimmick, R.L. and Wang, L. In *An Introduction to Experimental Aerobiology*, eds. R.L. Dimmick and Anne B. Akers, (New York: Wiley Interscience, 1969), pp. 164–176.
5. Cox, C.S. "Principles of aerosol sampling and samplers." In *Aerosol Sampling in Animal Houses*, eds. C.M. Wathes and J.M. Randall. Proceedings of a Workshop, University of Bristol, Department of Animal Husbandry, 26–28 July 1988, pp. 3–13. Sponsored by the Commission of the European Communities. EUR 11877 en.
6. Wake, D., Thorpe, A. and Brown, R.C. "Measurements of the electric charge on laboratory generated aerosols," In *Proceedings of the 3rd Annual Conference of the Aerosol Society* (West Bromwich, 20–22 March 1989), pp. 71–75.
7. Vincent, J.H. *Aerosol Sampling, Science and Practice* (Chichester: John Wiley and Sons, 1989).
8. Lui, B.Y.H., Pui, D.Y.H., Rubow, K.L. and Szymanski, W.W. "Electrostatic effects in aerosol sampling and filtration," *Ann. Occup. Hyg.* **29**:251–269 (1985).
9. Clark, J.M., Reid, K., Kaye, P.A. and Eyles, N.A. "Measurement of the asymmetry of aerosol particles by light scattering." In *Proceedings of the 4th Annual Conference of the Aerosol Society* (University of Surrey, 9–11 April 1990), pp. 203–208.

CHAPTER 4

Physical Aspects of Bioaerosol Sampling and Deposition

K.W. Nicholson

INTRODUCTION

The physical aspects of bioaerosol deposition and sampling are dominated by effects that relate to particle size. These incorporate the effects of inertia and sedimentation, as well as Brownian motion and sometimes electrical and phoretic effects (see also Chapter 3). Ideally, particulate materials should be sampled quantitatively, irrespective of their size, although this is difficult to perform in practice. In some cases, a knowledge of particle size is required, so that downwind air concentrations and deposition can be predicted, and this is, likewise, difficult to achieve. Meanwhile, there remain uncertainties relating to atmosphere-surface exchange rates and the interception of particles by various surface elements. The sections in this chapter aim to elucidate some of the uncertainties relating to particle sampling and transport and provide a description of the physical mechanisms involved. The reader is referred also to previous reviews in this subject area on air sampling,[1,2] particle deposition[3,4,5] and particle resuspension[6,7,8] and related material in Chapter 8.

THE PHYSICS OF PARTICLE SAMPLING

The effects of inertia usually dominate the efficiency of particle samplers, either through inlet losses or by undercollection or overcollection of particles at the inlet (see also Chapter 3). Inertial effects become increasingly important with increasing particle size and exist in any system in which there is a disturbance of the air flow. Consequently, quantitative particle sampling in the outdoor environment, in which there is a constantly changing wind flow, with regard to both wind speed and wind direction, is an especially difficult problem. Ideally, wind flow should not be disturbed and the velocity of any sampled air entering an orifice or nozzle should be equal to the local wind velocity. Such sampling is termed isokinetic (as opposed to anisokinetic) and this condition is illustrated in Figure 4.1. In areas of well-defined flow regimes (e.g., duct or stack sampling), isokinetic sampling can be achieved and the use of sharp-edged nozzles is preferred, because the body of sampler does not disturb the wind flow. In the outdoor environment, despite attempts by various workers to align a particle sampler into the wind automatically and to match the sampling velocity to the local wind speed,[9] results have been disappointing[10] and anisokinetic correction factors usually have to be applied. Such correction factors are described in many standard texts (e.g., Reist[11]) and are summarized here.

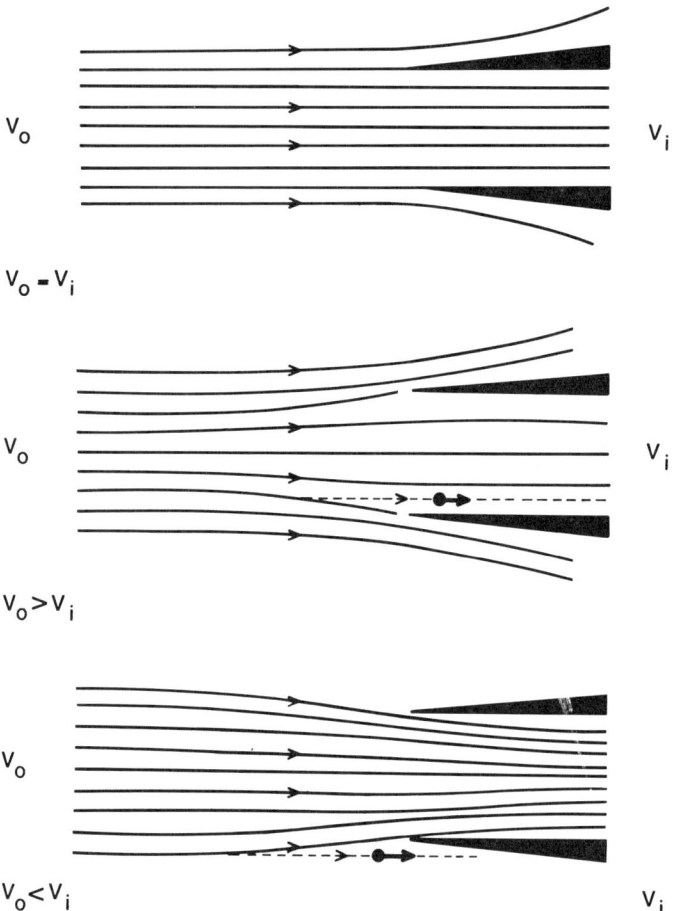

Figure 4.1. Conditions for isokinetic sampling.

A review by Stevens[12] brings some of the information together on air sampling, although much of this relates to sampling using sharp-edged nozzles. Usually, sampling efficiency is related to outside and inlet air velocities (V_o and V_i), the Stokes Number (St) and the radius of the sampling nozzle (R). The Stokes Number is defined as:

$$\text{St} = \frac{\tau V_o}{R} \quad (4.1)$$

τ, the relaxation time, can be calculated from

$$\tau = \frac{d^2 \rho_p C_c}{18\mu} \quad (4.2)$$

In Equation 4.2, d is the particle diameter, ρ_p is the particle density, μ is the viscosity of air ($\approx 1.8 \times 10^{-5}$ kg m^{-1}s^{-1} at stp) and C_c is the Cunningham correction factor. For d larger than 2λ, where λ is the mean free path of the gas molecules (≈ 0.07 μm at stp), Reist[11] approximated C_c as

$$C_c = \left[1 + \frac{2\lambda}{d}(1.257)\right] \quad (4.3)$$

Consequently, it can be seen from Equation 4.3 that in most practical cases of air sampling where inertial effects are important, C_c can be considered to be near unity.

Finally, errors due to anisokinetic operation can be estimated, according to Reist,[11] from

$$\frac{C}{C_o} = 1 - \alpha + \frac{\alpha V_o}{V_i} \quad (4.4)$$

where C and C_o are the measured and real concentrations

and

$$\alpha = \frac{2St}{1 + 2St}. \quad (4.5)$$

PRACTICAL AIR SAMPLING OF ATMOSPHERIC BIOAEROSOLS

Many practical air samplers have designs based on robustness and resistance to malicious damage, rather than on any aerodynamic considerations. In many cases, this is a necessity, since their operation may be at a remote or insecure site (see also Chapter 8). Even these designs of air sampler, however, generally are quite efficient at sampling small (less than around 10 μm diameter) particles, which are subject to limited inertial effects. A potential problem with the operation of these samplers is that their sampling characteristics, especially for the collection of large particles, usually are unknown. Furthermore, the collection efficiencies of these samplers are difficult to assess, due to the wide range and variability of meteorological conditions affecting their operation. Large particles, having the greatest inertia, have consequently the potential to act as sources of contamination, and a range of air samplers that incorporate a size-selective inlet has been designed. The upper diameter cut-off for particle entry into the sampler typically has been chosen as 10 μm, although there is no clear rationale for this decision. An assessment of performance for such size selective air samplers is relatively straightforward and can be assessed in terms of the characteristics of the inlet according to various meteorological conditions.[13] The measurement of particles smaller than around 10 μm diameter only, however, might not reflect the atmospheric concentrations of material that are available for inhalation and it is the larger particles that are most likely to be dominant in deposition processes.

Because sampling efficiency is related to the Stokes Number, an expedient for overcoming the difficulties of atmospheric sampling of large particles is to use a sampler with large inlet dimensions. Various techniques have been developed that involve the collection of material inside a wind tunnel, which in effect, acts as a large diameter sampler. Particle collection within the wind tunnel can be well controlled and defined, while the entry of even large particles into the mouth of the tunnel is efficient. The collection techniques within the wind tunnel have included the isokinetic operation of an aspirated sampler[14,15] or the use of an impaction surface such as a cylinder[16] or a ribbon.[17] The determination of the collection efficiencies and target areas of such impactors has been studied by May and Clifford[18] and can be used to determine atmospheric concentration. Nevertheless, such samplers exist only as research instruments and have not, so far, been used in routine sampling or monitoring.

Sampling from still air (see also Chapter 8) can provide good results if the sampling velocity, V_i, is many times higher than the sentimentation velocity of the largest particles present, but not so high that inertial effects are important. Davies[19] identified two criteria:

$$\frac{V_s}{V_i} \leq 0.04 \tag{4.6}$$

where V_s is the sedimentation velocity of the sampled aerosols, and

$$St \leq 0.032. \tag{4.7}$$

Several other workers have suggested that the second criterion of Davies is too restrictive.[2,20,21] Ogden[21] reviewed the work of Yoshida et al.,[22] ter Kuile[23] and Agarwal and Liu,[24] noting the restrictive nature of Davies' criteria in relation to the results of these authors. While a number of commercial instruments fail to meet Davies' second criterion by a large margin, Figure 4.2 illustrates a range of some experimental results that demonstrate that this might not be of such major significance.

PARTICLE SIZE ANALYSIS

Typically, devices used for size fractionation studies consist of an array of impaction surfaces (see also Chapter 8). The most widely known are cascade impactors with several impaction stages, placed in series, and each stage having a lower cut-off diameter than the previous one. Each stage consists of either a single or multiple slot[25] or nozzle,[26,27] which causes particulate material to impact on a collection surface. The cut-off is determined by the velocity of air through the nozzle and the distance of the nozzle from the collection surface.

Several types of material, including glass and various metals, have been used as collection surfaces in cascade impactors. Problems of particle bounce on the initial impaction surfaces have been widely noted and this results in a distortion of the measured size range to the lower end.[28] In an attempt to overcome the effects of particle bounce, surface coatings are found to be valuable[29] and there are various greases that are commercially available. Filters placed over the impaction stages can reduce particle bounce,[30,31] although their inclusion can affect the interception characteristics of the impaction stage. May[32] has summarized a number of factors relating to impactor efficiency including gravitational effects, the occurrence of haloes and edge effects in slot impactors.

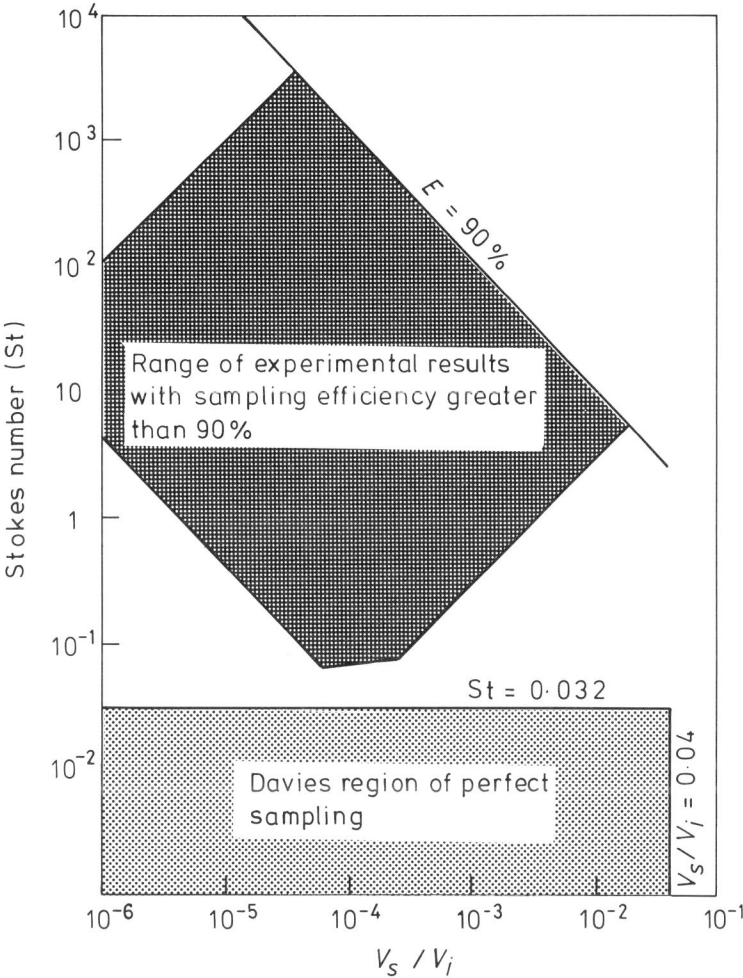

Figure 4.2. Criteria for sampling from still air (after Garland and Nicholson, 1991).

A potential problem in the use of cascade impactors is that the impaction stages can become overloaded with continual running. This may be important if lengthy sampling periods are necessary in order to measure low atmospheric concentrations or if sampling is necessary in locations of high particulate concentrations. Particle bounce from the overloaded stages may become a problem and, in order to prevent this, frequent replacement of the impaction stages is necessary. Hounam and Sherwood[33] produced a multistage virtual impactor (called a centripeter) with a high loading capacity. The principle of the centripeter is that of a traditional cascade impactor, where the impaction plate is replaced by a cup and the projected particles are captured in an enclosed void. The ability of each stage to sustain high loadings has been commented on by Yeh et al.,[34] although Biswas[35] has noted the complexity of flow in a device in which multiple virtual impactors are incorporated. Hounam and Sherwood, themselves, commented on significant wall losses within the device.

Cyclones have been used in dichotomous sampling arrangements,[36] and Liu and Rubow[37] have designed a five-stage cascade cyclone for size fractionation studies. Such cyclone devices are suitable for collecting large masses of material without overloading.

An inertial spectrometer has been designed that involves the injection of particles into a winnowing stream of air which then flows around a 90° bend.[38,39] Beyond the turn, adjacent to the outside line of the bend, is a porous plate supporting a filter medium. Sampled air is aspirated through the filter and the largest particles with the greatest inertia are sampled closest to the bend.

All of the described size fractionation devices have been found to be suitable for collecting particles up to only around 10 to 15 μm diameter.[40–42] For the size fractionation of larger particles, several devices have been proposed, although their performances in the outdoor environment have yet to be fully assessed.

Vawda et al.[43] used the principle of the wind tunnel sampler described above to collect large particles. They collected size fractionated samples, by operating a cascade impactor within a wind tunnel. Other workers[16,17] have used different impaction surfaces in various designs of wind tunnel, to get some indication of the size distribution of large particles.

Burton and Lundgren[44] described the **W**ide **R**ange **A**erosol **C**lassifier (WRAC sampler), which is designed with the intention of satisfying the criteria of Agarwal and Liu.[24] The entrance to the instrument is a 2 m long, 60 cm diameter vertical duct with a 1.5 m diameter rain shield. Size selection is achieved by four impactors operating in parallel and an additional sampler which draws air directly through a filter. The flow down the duct is controlled at 2.3 m s^{-1} and this effectively acts as a wind tunnel in which the samplers are enclosed. The large dimensions of the sampler suggest that the efficiency for large particle collection should be high, although the extreme size of the WRAC sampler means that calibration tests in a wind tunnel are impractical and are very difficult in the field.

An alternative to the sampling of particles by aspiration through an inlet, is by their collection on a rotating surface that operates in the free atmosphere. According to May et al.,[10] such a rotating device, termed the Rotorod, was first described by Perkins in 1957 (see also Chapter 9). This device consists of a square section rod which was bent in a 'U' shape and attached to a shaft of a small 12 V electric motor. When operating, the arms sweep through the air at 10 m s^{-1} and all particles larger than approximately 10 μm are impacted onto the leading surfaces of the rod, which should be coated with an adhesive. The Rotorod has been employed as a secondary standard in field experiments including other types of samplers by May et al.[10] and Vrins et al.[45] Larger rotating rod samplers have been utilized by Jaenicke and Junge[46] and Noll.[47,48] Jaenicke's device consisted of an arm with glass plates, 1 cm and 4 cm wide, rotating around a horizontal axis at 375 r.p.m. With a 30 cm radius, the glass plate passed through the air at 11.7 m s^{-1}. The axis of rotation was pointed into the wind by a vane and the plates were covered with a high viscosity silicone oil. The 50% (lower) cut-off diameters for the two plates were calculated to be 7 μm and 14 μm, respectively. Noll's sampler consisted of four sizes of impaction surface, assembled with four different radii on a rotating arm. The complete array of 16 collectors covered a range of cut-off diameters between 6 and 108 μm. Noll and Fang[49] developed a simplified device with four sizes of surface positioned on four rods of equal lengths, giving size selection over a range 7–60 μm.

Dust deposition gauges have been used to measure the deposition of large particles with significant sedimentation velocities. Usually, the gauges are employed to assess local pollution and consist of an up-facing collection disc and a drainage collecting bottle.[50] The gauges usually are not protected from precipitation and, consequently, measure material deposited in rain as well as dustfall. It is important to note, however, that such deposition gauges are suitable only for the collection of large particles and it is difficult to determine their collection characteristics, so that atmospheric concentration and deposition to surrounding surfaces are impossible to quantify from measurements made using them. Dust deposition gauges incorporating one or more vertical slots[51,52] have been utilized to assess the horizontal flux (i.e., product of wind speed and air concentration) of material from a source. Ralph and Hall[53] have measured the collection characteristics of this type of gauge noting a 25% collection efficiency for 87 μm diameter particles and an increasing collection efficiency with increasing particle diameter. While deposition gauges, designed to measure either horizontal or vertical flux, provide a cheap and simple measurement technique, they are only efficient at collecting large particles (at least several tens of micrometers in diameter). Many environmental aerosols are too small to be effectively collected by such gauges. Although collection of depositing material by a gauge might give some indication of the deposition flux to surrounding surfaces, it is unlikely that the results would, at best, be more than semi-quantitative.

Generally, the quantitative collection of large particles requires a specialized technique. There are few commercial instruments specifically designed for this purpose and their use, anyhow, is restricted to research. The greatest problem in the operation of many routine air samplers might be due to their unknown collection efficiencies for large particles and the inclusion of a size selective inlet may be a suitable, although less than fully satisfactory solution to this problem. The application of correction factors, for anisokinetic operation in the outdoor environment, is fraught with difficulties because of the complexities of wind flow, including turbulence.

THE ATMOSPHERE-SURFACE EXCHANGE OF PARTICULATE MATERIALS IN THE OUTDOOR ENVIRONMENT

All atmospheric aerosols undergo exchange processes with outdoor surfaces. Such processes include removal or deposition from the atmosphere and resuspension or re-entrainment from underlying surfaces. Removal can occur via wet or dry deposition processes. Dry deposition is the direct interaction of a material with a surface and this process occurs continuously. Wet deposition comprises the removal of material from the atmosphere in any falling hydrometeors, such as raindrops, snowflakes or hailstones. The interception at the surface of fog and cloud droplets, which are not sufficiently large to fall readily under gravity, is a process that is not categorized easily as either wet or dry. Since these droplets are not collected efficiently by precipitation collectors, the deposition of atmospheric materials associated with them sometimes has been termed 'occult deposition.' The processes of atmosphere-surface exchange are illustrated in Figure 4.3.

The relative importance of wet and dry deposition reflects the influence of meteorological conditions, the proximity of the receptor region in relation to the source and the nature of the atmospheric material. Dry deposition is non-episodic and will occur during intervals of precipitation, as well as during dry conditions. The occurrence of certain meteorological conditions (e.g., high wind speeds) might be expected to result in higher than average dry deposition rates and the nature of the surface onto which dry deposition is occurring also is an influencing factor. Wet deposition, on the other hand, is episodic and generally increases with increasing distance from a source as materials become mixed within clouds. The wet deposition of atmospheric materials therefore can be described as resulting from either in-cloud or below-cloud scavenging. The processes of wet and dry deposition are considered in the following sections.

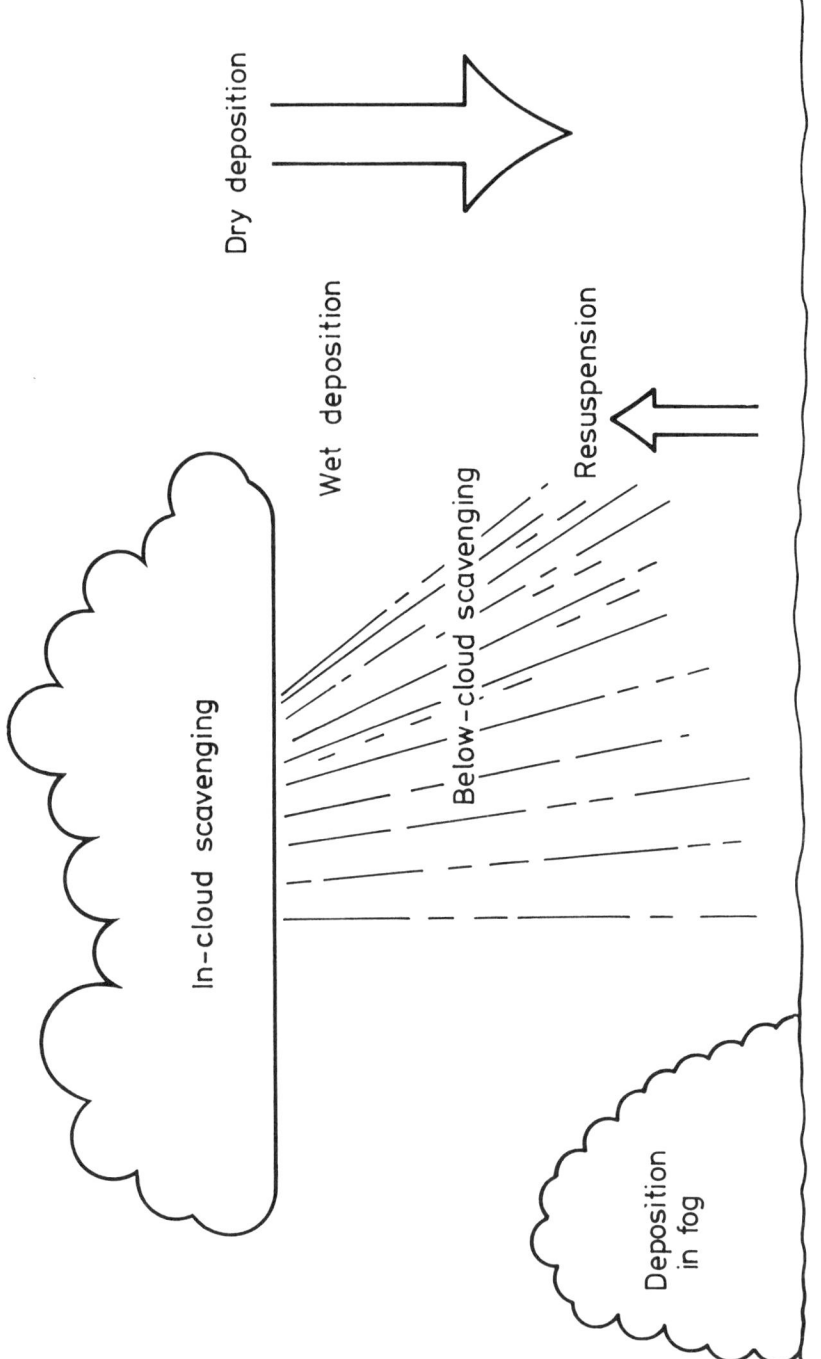

Figure 4.3. Atmosphere-surface-exchange processes.

DRY DEPOSITION AS A REMOVAL PROCESS

Dry deposition usually is expressed as a deposition velocity,[54] V_g, defined as:

$$V_g = \frac{\text{downward flux}}{\text{atmospheric concentration}} \qquad (4.8)$$

The dry deposition velocity is a measure of the rate of atmospheric removal processes and is considered to be independent of atmospheric concentration for particles (although not necessarily for gases). Deposition velocity is useful for the prediction of deposition flux in the constant flux layer, which exists above any surface, the depth of which depends on the undisturbed upwind fetch and might typically extend to a few meters above the top of the surface elements, given an upwind fetch of some hundreds of meters. A concentration gradient exists in the constant flux layer and, hence, deposition velocity is a function of height. The reference height to which reported deposition velocities are quoted is somewhat arbitrary, but typically might be 1 m above the underlying surface.

To consider the atmospheric regimes that affect dry deposition, it is useful to describe the types of boundary layer that can be built up over a surface. The planetary boundary layer can extend up to 1 or 2 km, although it may only extend up to 100 m or less at night. In this layer, flow is influenced directly by large-scale surface characteristics, thermal buoyancy is significant and mixing is important. The lower part of this layer is termed the turbulent surface layer and may extend up to about 100 m during the day and, like the planetary boundary layer, subsides diurnally, to perhaps as low as several meters at night. Vertical mixing in the turbulent surface layer is controlled by the surface elements and convection. The lowest atmospheric layer, which underlies the turbulent surface layer, is described as the laminar sub-layer. This, usually at most of the order of a millimeter or so in depth, is relatively slow moving and transport to the surface is impeded by limited vertical mixing. In reality, the laminar sub-layer periodically is penetrated by turbulent eddies and it is these that may be dominant in controlling the deposition of, especially, small particles. Nevertheless, the description of the atmosphere in terms of turbulent and laminar sub-layers is a useful way of describing deposition.

In the turbulent surface layer, eddy diffusion controls transport and sedimentational effects often can be considered to be small. Like molecular diffusion, turbulent diffusion is driven by a concentration gradient, and the level of turbulence, dependent on wind speed and surface roughness, determines the efficiency of transport. It is important to note that while certain conditions might favor turbulent diffusion, transport through the laminer sub-layer can dominate particle deposition and this will be reflected in the concentration gradient in the turbulent surface layer.

The deposition flux of particulate materials in the turbulent surface layer can be equated to a diffusion coefficient K_p, where

$$\text{flux} = -K_p \, dc/dz + V_s c \qquad (4.9)$$

dc/dz is the vertical concentration gradient and $V_s c$ (the product of atmospheric concentration and sedimentation velocity, see Equation 4.21) is included to allow for the effects of sedimentation. The negative sign is included on the right side of Equation (4.9) because the deposition flux is in the opposite direction to the concentration gradient. There are a number of analogies that have been drawn between heat, momentum and particle or gas diffusion. In thermally neutral atmospheric conditions (i.e., when thermal buoyancy or damping effects are small), it is assumed generally that the eddy diffusivities of momentum (K_m), heat (K_h) and gases or particles (K_p) are identical. Here:

$$K_m = K_h = K_p = ku_*(z-d) \tag{4.10}$$

where k is the von Karman constant (≈ 0.41), z is height above ground level, d is the zero plane displacement (usually taken to be around 0.7 of the individual surface element height, h) and u_* is the friction velocity. u_* is a measurement of the vertical gradient of horizontal wind speed and is approximately equal to the tangential velocity of turbulent eddies within the atmosphere. The variation of wind speed, u, with height, z, above a surface can be given in thermally neutral conditions as:

$$u(z) = \frac{u_*}{k} \ln[(z-d)/z_0] \tag{4.11}$$

where z_0 is the roughness length and is of order 0.1 h. In practice, u_* usually is found from the gradient of a plot of u against ln(z–d), where z_0 is determined from the intercept. In such a plot, d is taken either to be equal to 0.7 h or is determined from an iterative approach.

Outside thermally neutral atmospheric conditions, the logarithmic wind profile must be corrected for the effects of convection or damping.[55] A dimensionless shear of momentum (ϕ_m) is introduced, which is defined as:

$$\phi_m = \frac{k(z-d)}{u_*} \frac{du}{dz} \tag{4.12}$$

A similar term (ϕ_h) also is introduced into the equations for heat and gas transfer:

$$\phi_h = \frac{-k(z-d)\rho u_* C_p}{H} \frac{dT}{dz} \tag{4.13}$$

where ρ = density of air, C_p = specific heat capacity of air at constant pressure, H = sensible heat flux and dT/dz is the vertical temperature gradient. The equality of K_h and K_p (but not K_m) generally is assumed, so that Equation (4.10) then can be rewritten:

$$K_h = K_p = ku_*(z-d)/\phi_h \tag{4.14}$$

ϕ_m and ϕ_h usually are determined from empirical relationships with the Richardson Number (Ri), where:

$$Ri = \frac{g\frac{d\theta}{dz}}{T\left(\frac{du}{dz}\right)^2} \qquad (4.15)$$

g is the acceleration due to gravity, T is the absolute temperature and $d\theta/dz$ is the potential temperature gradient. Alternatively, ϕ_m and ϕ_h can be related to the Monin Obukhov length, L, where:

$$L = -\frac{u_*^3}{(kg\ H/\rho C_p T)} \qquad (4.16)$$

(see, for example, the review by Pasquill and Smith[56]). Typically adopted empirical relationships are those of Dyer and Hicks[57] for unstable atmospheres, and of Webb[58] for stable atmospheres:

$$\text{(stable)}\ \phi_m = \phi_h = (1 - 5.2\,Ri)^{-1} = (1 + 5.2\ z/L) \qquad (4.17)$$

$$\text{(unstable)}\ \phi_m^2 = \phi_h = (1 - 16\,Ri)^{-0.5} = (1 - 16\ z/L)^{-0.5} \qquad (4.18)$$

Garland[55] combined Equations 4.9, 4.12 and 4.14 and equated the deposition velocity as:

$$V_g = k^2 \left\{ \frac{dc/d\ln(z-d).du/d\ln(z-d)}{\phi_h \phi_m c} \right\} \qquad (4.19)$$

The above relationships between deposition flux and concentration gradient have led to a series of experiments in which deposition flux is determined from measurements of the vertical gradients of wind speed and concentration.[59-61] In practice, the deployment of such techniques for measuring particle deposition is, at least, extremely difficult.[3] Other measurements of particle deposition have been based on measurements of the covariance of instantaneous measurements of vertical wind speed (w) and atmospheric concentration (c).[62-64] Here:

$$\text{deposition flux} = \overline{w'c'} \qquad (4.20)$$

where the prime denotes values above or below the mean value. The method of measurement, termed the eddy correlation method, requires rapid response instrumentation, however. It is important to add that both measurement techniques are based on turbulent diffusion and if the particles have a significant sedimentation velocity, this will not be reflected in the measured deposition velocity.

As previously noted, particle transport through the laminar sub-layer can be dominant in controlling deposition and this is strongly dependent on particle diameter. For small particles (≤ 0.1 μm diameter), Brownian diffusion is the dominant transport mechanism, and this increases with decreasing particle diameter. For large particles (> 5 μm diameter), sedimentation is an important transport mechanism, and this increases with increasing particle diameter. The sedimentation velocity (V_s) can be calculated from:

$$V_s = \tau g \qquad (4.21)$$

where τ is the relaxation time as given in Equation 4.2. Equation 4.21 can be derived by equating the force on the particle due to gravity to the drag force calculated from Stokes' Law and holds true for particle Reynold's Numbers (Re_p) less than unity, where:

$$Re_p = \frac{d \, V_s \, \rho_p}{\mu} \qquad (4.22)$$

In practice, this means that sedimentation velocities of particles up to around 80 μm diameter can be calculated this way. As in the case of air sampling (see "The Physics of Particle Sampling"), the Cunningham correction factor should be included in the calculation of the sedimentation velocity of small particles, although this is a minor consideration in the practical applications of bioaerosol deposition. For values of Re_p greater than unity, V_s can be calculated[2] from Equation 4.21, where τ is given by:

$$\tau = \left(\frac{d^2 \, \rho_p}{18 \mu}\right) \left(\frac{24}{C_D \, Re_p}\right) \qquad (4.23)$$

where

$$C_D = \frac{24}{Re_p} \left(1 + \frac{3}{16} Re_p\right) \qquad (4.24)$$

Equation 4.24 is valid for $1 < Re_p < 5$, and for $Re_p > 5$ the calculation of V_s is more complicated. $Re_p < 5$, however, represents the normal range of atmospheric aerosols.

The deposition velocities of aerosols usually are considered to be greater than their sedimentation velocities. While there are considerations of particle bounce-off and blow-off from surfaces, most theoretical and semi-empirical considerations have V_g always greater than V_s (e.g., Sehmel[7]).

The declining importance of Brownian motion with increasing particle size and the increasing importance of sedimentational effects, result in a 'V' shaped curve of V_g against particle diameter, with the minimum for particles in the range 0.1 to 1.0 μm diameter (see Figure 4.4). In this intermediate range inertial effects can dominate and individual particles may impact on a surface. This may arise because the particles acquire sufficient momentum from the free-stream turbulent eddies that they are able to cross the laminar sub-layer of the underlying surface. This process is termed turbulent inertial impaction. Alter-

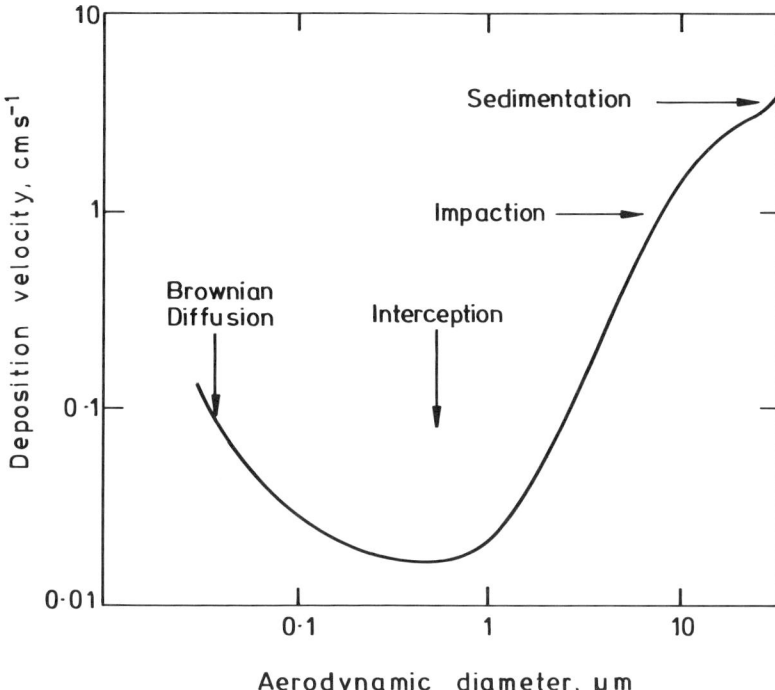

Figure 4.4. Schematic representation of deposition velocity as a function of particle size (note: values of deposition velocity are strongly dependent on environmental conditions).

natively, as is quite often the case when considering natural surfaces, part of a surface element may protrude into the free air stream and particles might impact it due to their momentum gained from movement with the wind. Particles also might be intercepted by surface elements protruding into the free air stream. Such interception occurs when a particle passes so closely to a surface element that it touches it, although it does not have sufficient inertia to cross the streamlines around the surface element. The processes of impaction and interception are illustrated in Figure 4.5.

It is for the intermediate particle diameters, not strongly influenced by sedimentation or Brownian diffusion, that the greatest uncertainties in V_g lie.[3] Both theoretical and semi-empirical predictive models have predicted values of V_g that have been near 0.01 cm s^{-1} for particles in this size range.[65–67] These values have to some extent been supported by laboratory experiments,[68–70] while field measurements of dry deposition have given anything from slightly higher values[60,63,71] to deposition velocities up to 1 cm s^{-1}.[61,72,73] The differences between the results of the laboratory and field measurements have been attributed to the different scales of atmospheric turbulence. The greater differences between the theoretical predictions and field measurements, however, probably reflect the over-simplistic approaches to modeling. Since passage through the laminar sub-layer is likely to present the greatest resistance to particle deposition in the intermediate size range, the calculation of its characteristics is crucial in determining the predicted deposition velocity. While inertial impaction onto rigid cylinders or plates can be calculated in terms of Stokes Number, the extrapolation of such calculations to outdoor conditions, where surface elements are complex in structure and may themselves not be smooth, is likely to be speculative. The fact that surface elements might move in the wind provides another complicating factor.

A resistance analogy often has been quoted for gases[55,74] to model their dry deposition, and can also be useful for describing particle deposition. In this approach, the total resistance to transport through the turbulent surface layer (r) is considered to consist of three components: r_a, the resistance to transport through the turbulent surface layer, r_b, the resistance to transport through the laminar sub-layer and r_c, the resistance to surface uptake. r is given as the reciprocal of V_g.

$$1/V_g = r = r_a + r_b + r_c \quad (4.25)$$

For particles, the total resistance to transport (r_p) includes the effects of sedimentation, so that:

$$1/r_p = 1/(r_a + r_b + r_c) + 1/r_g \quad (4.26)$$

where

$$1/r_g = V_s \quad (4.27)$$

Generally, r_a is assumed to be equal for heat and for particles, so that in neutral atmospheric conditions,

$$r_a = u/u_*^2 \quad (4.28)$$

Outside thermally neutral conditions, an empirical correction term (ψ_c) is included. Arranging Equation 4.28 with Equation 4.13 gives:

$$r_a = \frac{1}{ku_*} \left(\ln \frac{z-d}{z_0} - \psi_c \right) \quad (4.29)$$

r_b can be related to the sub-layer Stanton Number (B)

$$r_b = \frac{1}{Bu_*} \quad (4.30)$$

where B is determined empirically and the final resistance to transport (r_c) is assumed to be equal to zero for particles. There remains the possibility of particle bounce,[75] which means that r_c is not necessarily equal to zero, although the presence of surface moisture, leaf hairs or surface stickiness would help reduce this effect.[76]

WET DEPOSITION AS A REMOVAL PROCESS

Removal of material from the atmosphere in precipitation differs from dry deposition in that precipitation is episodic and occurs over a limited fraction of time. Nevertheless,

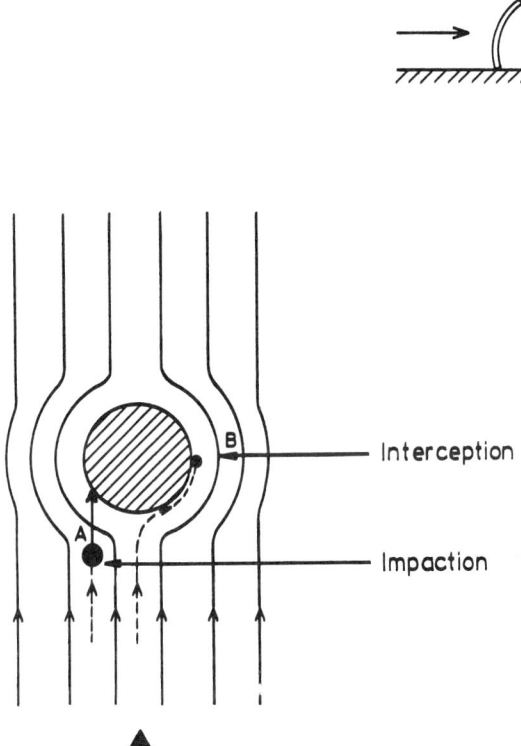

Figure 4.5. Interception and impaction of particles. Impaction is represented by particle A, which crosses the streamlines due to its inertia. The lighter particle, B, follows the streamlines but comes sufficiently close to the surface to make contact.)

the potential importance of wet deposition as a removal process was demonstrated clearly in the aftermath of the Chernobyl nuclear reactor accident, when levels of deposition in the UK were closely related to the amounts of precipitation falling during the passage of the main plume.[77,78]

Wet deposition can be described as resulting from either in-cloud or below-cloud scavenging processes. The importance of each of these processes depends on the release height and the level of the cloud base, but generally in-cloud scavenging tends to dominate with increasing distance from a source.

Measurements of wet deposition often have been described using a scavenging ratio, where:

$$\text{scavenging ratio} = \frac{\text{concentration of material in rain}}{\text{concentration of material in air}} \quad (4.31)$$

The concentrations in Equation 4.31 usually are measured in terms of mass per unit mass, and typical ratios are of the order of several hundred to a thousand or more for atmospheric trace species.[79] It is important to note that some authors express concentrations in terms of mass per unit volume and these result in scavenging ratios that are higher by a factor of approximately 840.

The scavenging ratio is a useful tool for predicting wet deposition levels around a source or receptor region. However, it gives little insight into the factors that affect wet deposition, not least since atmospheric concentration is measured near ground level and this may bear no relation to air concentrations in clouds or at altitude.

It is helpful to consider the processes that result in wet deposition and the formation of precipitation. On most occasions in temperate latitudes, precipitation usually occurs via the Bergeron process. This process describes the formation and growth of ice crystals at the expense of super-cooled water droplets and, hence, precipitation begins as snow. The occurrence of rain or sleet depends on whether the precipitation has melted totally or partially on collection at ground level. Thus, in clouds, ice crystals and super-cooled water droplets coexist with water vapor exchanging between each phase. It is important to note, however, that the Bergeron process does not describe all forms of precipitation formation and there are clear examples in storm systems, especially in tropical latitudes, in which the entire cloud is below freezing level. In these cases, rain is formed by the agglomeration of droplets.

If we consider the formation of in-cloud elements (i.e., ice crystals and water droplets), each is formed by an individual nucleus. There are two categories: the ice nuclei which form ice crystals, and cloud condensation nuclei, which form water droplets. Ice nuclei are relatively scarce, tend to be insoluble and might include particulate materials derived from a soil source. Cloud condensation nuclei include hygroscopic materials and often contain or adsorb pollutants. Although larger particles are the more efficient cloud condensation nuclei, most are smaller than 1 μm diameter.

Concentrations of material in-cloud are much greater in the droplet phase than in the ice phase. Consequently, any droplets that are intercepted by falling ice crystals might contribute considerably to wet deposition. This has been noted when rimed snowflakes (i.e., snowflakes that have frozen droplets associated with them) have been found to contain much greater amounts of pollutants than unrimed ones.[80,81]

In addition to the nucleation processes that occur in-cloud, particulate material can be incorporated into pre-existing droplets. Diffusion and phoretic effects dominate the efficiency of this process, although the relative importance of these processes is poorly understood. In-cloud scavenging experiments are notoriously difficult to undertake and involve the injection of material into clouds via either rockets[82] or aircraft.[83,84]

Below-cloud scavenging occurs due to the collection of material by falling hydrometeors. The processes that result in the collection of atmospheric aerosols can include the effects of Brownian motion, interception and impaction, and there are, consequently, analogies between the mechanisms of below-cloud scavenging and dry deposition (with the exception that sedimentation does not play a significant role in the capture of particles by falling precipitation).

The scavenging coefficient (Λ) often has been used to describe below-cloud scavenging,[85] where

$$\frac{d\chi}{dt} = -\Lambda \chi \qquad (4.32)$$

in which χ is the concentration of material available for scavenging. It can be easily shown that

$$\chi_t = \chi_0 \exp(-\Lambda t) \qquad (4.33)$$

where χ_t and χ_0 are the atmospheric concentrations of material at a time equal to 0 and t. For the simple case of uniformly sized rain droplets, Λ can be related to the removal efficiency of a monodisperse aerosol[86] as

$$\Lambda = N V_T E \pi R^2, \qquad (4.34)$$

in which N is the number of raindrops per unit volume, with a terminal velocity equal to V_t and radius R, which have a collection efficiency (i.e., fraction of material collected in the volume swept out by the falling hydrometer) equal to E. For a range of raindrop sizes, Λ can be given[87] as:

$$\Lambda(a, R_s) = \int_0^\infty \pi R^2 V_t(R) E(a,R) N(R) \, dR \qquad (4.35)$$

in which a is particle radius, N(R) dR is the number of raindrops per unit volume with radii R to R + dR and R_s refers to a particular raindrop spectrum.

Measurements of below-cloud scavenging have been based to a large extent on laboratory studies,[88] while field measurements have included measurements of the depletion of adventitious atmospheric species[89,90] and the measurement of released materials in collected precipitation.[91,92]

Precipitation intensity is an important parameter when evaluating below-cloud scavenging because both the number and size of raindrops depend on this. May[92] calculated scavenging coefficients for various precipitation intensities using a raindrop spectrum determined by Best.[93] Slinn[94] recommended that the result of Mason,[95] regarding the mean size of drops, could be used to calculate Λ in fairly steady rain, thus eliminating the need for a detailed raindrop size distribution. Here

$$R_m = 3.5 \times 10^{-4} \left(\frac{J}{1 \text{ mm h}^{-1}} \right)^{0.25} \qquad (4.36)$$

where J is the rainfall intensity (mm h^{-1}) and R_m is the mass mean raindrop radius (m). From Equation (4.34), it can be shown that

$$\Lambda \approx \frac{0.75 \, J}{R} \qquad (4.37)$$

since

$$J = \frac{4}{3} \pi R^3 N V_t \qquad (4.38)$$

Slinn,[94] using R_m, approximated

$$\Lambda = \frac{c \, J E(a, R_m)}{R_m} \qquad (4.39)$$

where c is a constant that may have a value around 0.5 for frontal rain.[96]

For typical precipitation intensities of 1 to 2 mm h^{-1}, it may be anticipated that values of Λ, for particles of several micrometers diameter, will be in the range 10^{-4} to 10^{-3} s^{-1}. However, there are few supportive data and the apparent observation, in some cases,[97] of collection efficiencies greater than 100% have led to increased uncertainty. There are even greater uncertainties in appropriate values of Λ to be used for <1 μm diameter particles.

Some predictive models for below-cloud scavenging adopt values of Λ/J. These are often assumed to be a constant [typically 10^{-4} s^{-1} (mm h^{-1})$^{-1}$], although there are few supportive experimental data for such an assumption.

As previously noted, the deposition of fog or cloud droplets is a potentially important process.[98,99] The formation of the droplets will include some of the nucleation processes that result in wet deposition. However, the droplets are sufficiently small to be unaffected by gravity and the deposition of them can be described as being similar to the dry deposition of large particles. The importance of such deposition, in relation to dry and wet deposition, is strongly dependent on location. However, the potential contribution of deposition in fog and cloud droplets has been noted widely, not least because the concentration of trace species has been found to be high in the droplets.[100]

THE RESUSPENSION OF MATERIAL INTO THE ATMOSPHERE

The terms resuspension and re-entrainment are used generally to describe the processes by which any deposited material might become airborne. Strictly, only aerially deposited material can become resuspended, although the term often is used also to describe the suspension of material from spills, etc.

Early interest in entrainment was due to the effects of erosion and soil transport.[101,102] However, with the advent of nuclear technology, the resuspension of radionuclides has received wide attention.[103,104] Other interests in resuspension include the fate of industrial spills,[105] the transport of chlorinated hydrocarbon pesticides,[106] the spreading of crop disease[107] and the transmission of human disease.[108]

The interests in resuspension are two-fold. First, an inhalation hazard might occur and, second, there can be a resulting spread of contamination due to the deposition of resuspended material. It is important also to note that an inhalation hazard occurs primarily due to small particles, although it is large particles that are dominant in the spread of contamination. Furthermore, resuspended material often is associated with surface dust or soil particles and the particle size range of resuspended material may bear no resemblance to the particle size range of the depositing species.[109]

Because the inhalation of resuspended material has been of concern, the resuspension factor (RF) has been used to describe resuspension.

$$RF (m^{-1}) = \frac{\text{atmospheric concentration}}{\text{ground contamination}} \tag{4.40}$$

There have been difficulties in ascribing the height of the concentration measurement and the depth of surface material that should be sampled to determine ground contamination. In addition, there is an underlying assumption, in the use of the resuspension factor, that airborne material originates from the immediately underlying soil and that downwind transport need not be considered. This is unlikely to be the case, although the resuspension factor still can be used in inhalation dose assessments for areas of fairly uniform contamination.

An alternative to the resuspension factor is the resuspension rate, R_r, where

$$R_r = \frac{\text{resuspension flux}}{\text{ground contamination}} \tag{4.41}$$

The use of a resuspension rate, rather than a resuspension factor, has been supported because this can be incorporated into a diffusion model to enable the prediction of downwind air concentrations.[110] However, in practice it is difficult to measure. The use of it suffers the same disadvantages as that of the resuspension factor, with regard to determining ground contamination.

The availability of material for resuspension declines with time lapsed after deposition. This decline has been expressed in terms of a negative exponential function[111] and as an inverse relationship.[112] Reeks et al.[113] have considered the theoretical aspects of resuspension, where surface particles are resident in potential energy wells and are periodically resuspended by the penetration of turbulent eddies through the laminar sub-layer. They concluded that resuspension was inversely proportional to time and there has been further supporting evidence of this relationship in describing resuspension from grass and concrete surfaces in wind tunnel experiments.[114] It is important to note that the decline in resuspension with time is not due to a depletion in the amount of deposited material, but is due to material becoming less erodible.[115] Consequently, the time after the initial deposition episode is of fundamental importance in the interpretation of resuspension data, and variations of this might go some way toward explaining the extremely large range of resuspension results.[8]

Resuspension rate has been found by several authors to increase in proportion to the wind speed raised to a power,[116] although the exact values of this power have varied, being typically in the range 1 to 6, with the possibility of the power being time dependent.[117] Nevertheless, such relationships are strong evidence of the episodic nature of wind-generated resuspension.

A possible important factor in the resuspension process is the occurrence of saltating particles[118] (i.e., particles that leave the surface, but that have such high sedimentation velocities that they do not remain suspended). Saltating material has been found to be of great importance in erosion processes and there has been a threshold velocity found for erosion,[119] which may relate to the presence of such saltating material. Resuspension, on the other hand, has been found to occur in the absence of saltating material[117] and it is not clear, therefore, whether a threshold velocity is appropriate for resuspension.

It has been found that, generally, large particles resuspend more easily than small ones.[114] This is due to the greater areas on which lift forces can act. As particles increase in size so that sedimentation effects become important, then there is a minimum wind speed which is necessary for the material to remain in suspension. Gillette et al.[120] equated this threshold to the ratio of the sedimentation velocity of the particle to the friction velocity (a measure of the vertical wind velocity). They found that the threshold for suspension lay in the range $0.12 < V_s/u_* < 0.68$, above which suspension would not occur.

There is a wide range of mechanical actions that result in resuspension due to momentum being directly imparted onto surface material. These include vehicular activity,[121,122] pedestrian activity[123] and agricultural work.[124] A major limitation of the results of such studies has been due to the difficulty in quantifying the mechanical processes. Consequently, these experiments have further increased the range of reported resuspension results.[7]

While there has been some recent work carried out on resuspension processes, the wide range of reported measurements and the shortage of experimental work related to quantifying resuspension mechanisms, means that there are considerable uncertainties associated with resuspension. The reader is referred to recent reviews on resuspension,[6,8] which give a full summary of the range of reported resuspension results, along with some comments on the relevant influential factors.

SUMMARY AND CONCLUSIONS

There are various commercially available samplers suitable for the quantitative collection of atmospheric aerosols smaller than around 10 μm diameter. There are also many devices available that are suitable for the size fractionation of such particles. However, the quantitative collection of larger particles is extremely difficult in the outdoor environment and the available devices have been used as research tools only. Some of these devices might be suitable for the size fractionation of large particles, although more field validation data are required to endorse their routine use outdoors.

The processes that affect the dry deposition of environmental aerosols have been identified, although there remain significant uncertainties in quantifying their importance, especially for particles in the diameter range 0.1 to several micrometers. This range, however, represents an important size fraction of environmental aerosols. There are further uncertainties related to the dry deposition of material to rough surfaces, such as forests. This constitutes an unacceptable lack of knowledge.

Scavenging ratios have been widely used to model wet deposition to receptor regions over fairly lengthy time scales. These rely on the assumption that near-surface air concentrations can be related to wet deposition rates. Although this does not take into consideration that at altitude and in particular, in-cloud, atmospheric concentration of a material may differ from that at ground level, the use of scavenging ratio is probably justified over lengthy periods when environmental fluctuations will average out. A scavenging ratio on a short time scale is, however, totally inappropriate. A below-cloud scavenging coefficient has been used to calculate wet deposition in the proximity of sources, although there are considerable uncertainties over probable values, especially for particles in the size range 0.1 to several micrometers. Further work is required to elucidate these uncertainties.

The resuspension of material can be of concern because of the possibility of an inhalation risk and because there may be a spread of contamination. There remain major uncertainties in the adoption of appropriate resuspension parameters and this constitutes an area where there is a clear need for further work. The effects of mechanical disturbances, including vehicular, agricultural and other activities are unknown and must be assessed. Resuspension is of particular concern when considering the effects of persistent contaminants.

REFERENCES

1. Garland J.A. and Nicholson K.W. (1991) A review of methods for sampling large airborne particles and associated radioactivity. *Journal of Aerosol Science* 22, 479–499.
2. Vincent J.H. (1989) *Aerosol Sampling: Science and Practice.* Chichester, John Wiley.
3. Nicholson K.W. (1988) The dry deposition of small particles: a review of experimental measurements. *Atmospheric Environment* 22, 2653–2666.
4. Sehmel G.A. (1980) Particle and gas dry deposition: a review. *Atmospheric Environment* 14, 983–1011.
5. McMahon T.A. and Denison P.J. (1979) Empirical atmospheric deposition parameters—a survey. *Atmospheric Environment* 13, 571–585.
6. Nicholson K.W. (1988) A review of particle resuspension. *Atmospheric Environment* 22, 2639–2651.
7. Sehmel G.A. (1984) Deposition and resuspension. In *Atmospheric Science and Power Production* (edited by D. Randerson), pp 533–583. National Technical Information Service, U.S. Department of Commerce, Springfield, VA.
8. Sehmel G.A. (1980) Particle resuspension: a review. *Environmental International* 4, 107–127.
9. May K.R. (1960) A size selective total-aerosol sampler, the tilting pre-impinger. *Annals of Occupational Hygiene* 2, 93–106.
10. May K.R., Pomeroy N.P. and Hibbs S. (1976) Sampling techniques for large wind-borne particles. *Journal of Aerosol Science* 7, 53–62.
11. Reist P.C. (1984) *Introduction to Aerosol Science.* Macmillan, London.
12. Stevens D.C. (1986) Review of aspiration coefficients of thin-walled sampling nozzles. *Journal of Aerosol Science* 17, 729–743.
13. Hall D.J., Upton S.L., Marshland G.W. and Waters R.A. (1988) Wind tunnel measurements of the collection efficiency of two PM 10 samplers: the Sierra Andersen Model 321A Hi-Volume Sampler and the EPA Dichotomous Sampler. Warren Spring Laboratory LR 669, Department of Trade and Industry, Stevenage.
14. May K.R. (1961) Fog droplet sampling using a modified impactor technique. *Quarterly Journal of the Royal Meteorological Society* 87, 535–548.
15. Hofschreuder P., Vrins E. and van Boxel J. (1983) Sampling efficiency of aerosol samplers for large wind-borne particles—a preliminary report. *Journal of Aerosol Science* 14, 65–68.
16. McKay W.A., Walker M.I. and Cloke J. (1990) Transfer of radionuclides from sea to land to air. DoE Report No: DOE/RW/90.094, U.K. Dept. of the Environment, Marsham Street, London.
17. Regtuit H., Vrins E. and Hofschreuder P. (1988) The tunnel impactor: a multiple inertial impactor for coarse particles. *Journal of Aerosol Science* 19, 983–986.
18. May K.R. and Clifford R. (1967) The impaction of aerosol particles on cylinders, spheres, ribbons and discs. *Annals of Occupational Hygiene* 10, 83–95.
19. Davies C.N. (1968) The entry of aerosols into sampling tubes and heads. *British Journal of Applied Physics* (Series 2) 1, 921–930.

20. Gibson H. and Ogden T.L. (1977) Some entry efficiencies for sharp-edged samplers in calm air. *Journal of Aerosol Science* 8, 361–365.
21. Ogden T.L. (1983) Inhalable, inspirable and total dust. In *Aerosols in the Mining and Industrial Work Environments* (edited by V.A. Marple and B.Y.H. Liu), pp 185–204, Ann Arbor Science, MI.
22. Yoshida H., Uragami M., Masuda H. and Iinoya K. (1978) Particle sampling efficiency in still air. Kagaku Kogaku Ronburshu 4, 123–8 (HSE Translation No. 8586).
23. ter Kuile W.M. (1979) Comparable dust sampling at the work place. Report F1699, Instituut voor Milieuhygiene en Gezondheitstechniek, Delft.
24. Agarwal J.K. and Liu B.Y.H. (1980) A criterion for accurate aerosol sampling in calm air. *American Industrial Hygiene Association Journal* 41, 191–197.
25. May K.R. (1975) An 'ultimate' cascade impactor for aerosol assessment. *Journal of Aerosol Science* 6, 413–419.
26. Andersen A.A. (1966) A sampler for respiratory health hazard assessment. *American Industrial Hygiene Association Journal* 27, 160–165.
27. Berner A. and Reischl G. (1988) The observations of the wet atmospheric aerosols by AERAS cascade impactors. In *Lecture Notes in Physics: Atmospheric Aerosols and Nucleation* (edited by P.E. Wagner and G. Vali), pp 126–129, Springer-Verlag, Berlin.
28. Dzubay T.G., Hines L.E. and Stevens R.K. (1976) Particle bounce errors in cascade impactors. *Atmospheric Environment* 10, 229–234.
29. Schumann T., Gysi H. and Kaelin S. (1988) Coating of impaction surfaces of cascade impactors: necessary for sampling ambient aerosols in rival and suburban areas? *Journal of Aerosol Science* 19, 933–996.
30. Rao A.K. and Whitby K.T. (1978) Non-ideal collection characteristics of inertial impactors—I. Single stage impactors and solid particles. *Journal of Aerosol Science* 9, 77–86.
31. Rao A.K. and Whitby K.T. (1978) Non-ideal collection characteristics of inertial impactors—II. Cascade impactors. *Journal of Aerosol Science* 9, 87–100.
32. May K.R. (1975) Aerosol impaction jets. *Journal of Aerosol Science* 6, 403–411.
33. Hounam R.G. and Sherwood R.J. (1965) The cascade centripeter: a device for determining the concentration and size distribution of aerosols. *American Industrial Hygiene Association Journal* 26, 122–131.
34. Yeh H.C., Chen B.T. and Cheng Y.S. (1986) Experimental and theoretical study of a Lovelace virtual impactor. In *Aerosols: Formation and Reactivity*, 2nd International Aerosol Conference, Berlin, 1986, pp 562–565, Pergamon Press, Oxford.
35. Biswas P. (1986) The particle trap impactor. In *Aerosols: Formation and Reactivity*, 2nd International Aerosol Conference, Berlin, 1986, pp 542–545, Pergamon Press, Oxford.
36. Wedding J.B., Welgand M.A. and Carney T.C. (1982) A 10 μm cutpoint inlet for the dichotomous sampler. *Environmental Science and Technology* 16, 602–606.
37. Liu B.Y.H. and Rubow K.L. (1984) A new axial flow cascade cyclone for size classification of airborne particulate matter. In *Aerosols: Science, Technology and Industrial Applications of Airborne Particles* (edited by B.Y.H. Liu, D.Y.H. Pui and H.J. Fissan), pp 115–118, Elsevier, Amsterdam.
38. Prodi V., Melandri C., Tarroni G., De Zaiacomo T., Formignani M. and Hochrainer D. (1979) An inertial spectrometer for aerosol particles. *Journal of Aerosol Science* 10, 411–419.
39. Belosi F. and Prodi V. (1987) Particle deposition within the inertial spectrometer. *Journal of Aerosol Science* 18, 37–42.
40. McFarland R.A., Wedding J.B. and Cermak J.E. (1977) Wind tunnel evaluation of a modified Andersen Impactor and an all weather sampler inlet. *Atmospheric Environment* 11, 535–539.

41. Vaughan N.P. (1989) The Andersen impactor: calibration wall losses and numerical simulation. *Journal of Aerosol Science* 20, 67–90.
42. Prodi V. and Belosi F. (1988) Particle separation in a shallow channel inertial spectrometer. *Journal of Aerosol Science* 19, 979–982.
43. Vawda Y., Colbeck I., Harrison R.M. and Nicholson K.W. (1989) The effects of particle size on deposition rates. *Journal of Aerosol Science* 20, 1155–1158.
44. Burton R.M. and Lundgren D.A. (1987) Wide range aerosol classifier: a size selective sampler for large particles. *Aerosol Science and Technology* 6, 289–301.
45. Vrins E., Hofschreuder P., ter Kuile W.M., van Nieuwland R., Oeseburg F. and Alderliesten P.T. Evaluation of the sampling efficiency of aerosol samplers for large windborne particles in the open air. Part I—Experimental set-up and quality assurance. *Journal of Aerosol Science* (in press).
46. Jaenicke R. and Junge C. (1967) Studien zur oberen Grenzgrösse des natürlichen Aerosols. *Beitr. Phys. Atmos.* 40, 130.
47. Noll K.E. (1970) A rotary impactor for sampling giant particles in the atmosphere. *Atmospheric Environment* 4, 9–19.
48. Noll K.E. and Pilat M.J. (1971) Size distribution of atmospheric giant particles. *Atmospheric Environment* 5, 527–540.
49. Noll K.E. and Fang K. (1986) A rotary impactor for size selective sampling of atmospheric coarse particles. The 79th Annual Meeting of the Pollution Control Association, Minneapolis, June 1986.
50. Hall D.J. and Waters R.A. (1986) An improved readily available dustfall gauge. *Atmospheric Environment* 20, 219–222.
51. Lucas D.H. and Moore D.J. (1964) The measurement in the field of pollution by dust. *Int. J. Air Wat. Pollut.* 8, 441–453.
52. Lucas D.H. and Snowsill W.L. (1967) Some developments in dust pollution measurement. *Atmospheric Environment* 1, 619–636.
53. Ralph M.D. and Hall D.J. (1989) Performance of the BS directional dust gauge. In *Proceedings of the Aerosol Society 3rd Annual Conference,* West Bromwich, 20–22 March 1989, pp 115–120.
54. Chamberlain A.C. and Chadwick R.C. (1953) Deposition of airborne radioiodine vapor. *Nucleonics* 11, 22–25.
55. Garland J.A. (1977) The dry deposition of sulphur dioxide to land and water surfaces. *Proc. R. Soc. London* A354, 245–268.
56. Pasquill F. and Smith F.B. (1983) Atmospheric Diffusion. Chichester, John Wiley.
57. Dyer A.J. and Hicks B.B. (1970) Flux-gradient relationships in the constant flux layer. *Quarterly Journal of the Royal Meteorological Society* 96, 715–721.
58. Webb E.K. (1970) Profile relationships: the log-linear range and extension to strong stability. *Quarterly Journal of the Royal Meteorological Society* 96, 67–90.
59. Nicholson K.W. and Davies T.D. (1987) Field measurements of the dry deposition of particulate sulphate. *Atmospheric Environment* 21, 1561–1571.
60. Garland J.A. and Cox L.C. (1982) Deposition of small particles to grass. *Atmospheric Environment* 16, 2699–2702.
61. Everett R.G., Hicks B.B., Berg W.W. and Winchester J.W. (1979) An analysis of particulate sulphur and lead gradient data collected at Argonne National Laboratory. *Atmospheric Environment* 13, 931–934.
62. Weseley M.L., Cook D.R. and Hart R.L. (1985) Measurements and parameterization of particulate sulphur dry deposition over grass. *J. Geophys. Res.* 90, 2131–2143.
63. Sievering H. (1983) Eddy flux and profile measurements of small particle dry deposition velocity at Boulder Atmospheric Observatory (BAO). In *Precipitation Scavenging, Dry Deposition and Resuspension* (edited by H.R. Pruppacher, R.G. Semonin and W.G.N. Slinn), Vol. 2, pp 963–978, Elsevier, Amsterdam.

64. Neumann H.H. and den Hartog G. (1985) Eddy correlation measurements of atmospheric fluxes of ozone, sulphur and particles during the Champaign intercomparison study. *Journal Geophysical Research* 90, 2097–2110.
65. Slinn W.G.N. (1982) Predictions for particle deposition to vegetative canopies. *Atmospheric Environment* 16, 1785–1794.
66. Davidson C.I. and Friedlander S.K. (1978) A filtration model for aerosol dry deposition: application to trace metal deposition from the atmosphere. *Journal of Geophysical Research* 83, 2343–2352.
67. Slinn S.A. and Slinn W.G.N. (1980) Predictions for particle deposition on natural waters. *Atmospheric Environment* 14, 1013–1016.
68. Chamberlain A.C. (1967) Deposition of lycopodium spores and other small particles to rough surfaces. Proc. Royal Society (London), A296, 45–70.
69. Sehmel G.A. (1973) Particle eddy diffusivities and deposition velocities for isothermal flow and smooth surfaces. *Aerosol Science* 4, 125–138.
70. Clough W.S. (1975) The deposition of particles on moss and grass surfaces. *Atmospheric Environment* 9, 1113–1119.
71. Allen A.G., Harrison R.M. and Nicholson K.W. (1991) Dry deposition of aerosol to a short grass surface. *Atmospheric Environment* 25A, 2671–2676.
72. Hicks B.B., Wesley M.L., Durham J.L. and Brown M.A. (1982) Some direct measurements of atmospheric sulphur fluxes over a pine plantation. *Atmospheric Environment* 16, 2899–2903.
73. Sievering H. (1982) Profile measurements of particle dry deposition velocity at an airland interface. *Atmospheric Environment* 16, 301–306.
74. Nicholson K.W. and Davies T.D. (1988) The dry deposition of sulphur dioxide at a rural site. *Atmospheric Environment* 22, 2885–2889.
75. Wu Y-L., Davidson C.I. and Russell A.G. (1992) A stochastic model for particle deposition and bounce-off. *Aerosol Science and Technology* 17, 231–244.
76. Hosker R.P. and Lindberg S.E. (1982) Review: atmospheric deposition and plant assimilation of gases and particles. *Atmospheric Environment* 16, 889–910.
77. Cambray R.S., Cawse P.A., Garland J.A., Gibson J.A.B., Johnson P., Lewis G.N.J., Newton D., Salmon L. and Wade B.O. (1987) Observations on radioactivity from the Chernobyl accident. *Nuclear Energy* 26, 77–101.
78. Clark M.J. and Smith F.B. (1988) Wet and dry deposition of Chernobyl releases. *Nature* 332, 245–249.
79. Cawse P.A. (1974) A survey of atmospheric trace elements in the U.K. (1972–1973). AERE-R7669, HMSO, London.
80. Davidson C.I., Honrath R.E., Kadane J.B., Tsay R.S., Magewsk P.A., Lyons W.B. and Heidam N.Z. (1987) The scavenging of atmospheric sulphate by Arctic snow. *Atmospheric Environment* 21, 871–882.
81. Murakami M., Chikashi H. and Choji M. (1981) Observation of aerosol scavenging by falling snow crystals at two sites of different heights. *J. Met. Soc. Japan* 59, 763–771.
82. Burtsev I.I., Burtseva L.V. and Malakhov S.G. (1966) Washout characteristics of a ^{32}P aerosol injected into a cloud. In *Atmospheric Scavenging of Radioisotopes*, Proc. of a Conf. in Palanga, USSR, June 7–9, 1966 (edited by B. Styra, Ch. A. Garbalganskas and V. Yu. Luganas), pp 242–250, translated from Issledovanie Protsessor Samoochischeniga Atmosfery of Radioaktivngkh Izotopov by Lederman D. Available as TT-69-55099 from National Technical Information Service, Springfield, VA.
83. Dingle A.N., Gatz D.F. and Winchester J.W. (1969) A pilot experiment using indium as a tracer in a convective storm. *Journal of Applied Meteorology* 8, 236–240.
84. Warburton J.A. (1973) The distribution of silver in precipitation from two seeded Alberta hailstorms. *Journal of Applied Meteorology* 12, 677–682.

85. Nicholson K.W., Branson J.R. and Giess P. (1991) Field measurements of the below-cloud scavenging of particulate material. *Atmospheric Environment* 25A, 771–777.
86. Stensland G.S. (1977) Modelling rainwater contaminant concentrations due to hygroscopic particle washout. In *Precipitation Scavenging* (1974). Proc. of a Symp. at Champaign, Illinois, 14–18 October 1974, pp 794–812, CONF-741003, Nat. Tech. Inf. Serv., U.S. Dept. of Commerce, Springfield, VA.
87. Dana M.T. and Hales J.M. (1977) Washout coefficients for polydisperse aerosols. In *Precipitation Scavenging* (1974), Proc. of a Symp. at Champaign, Illinois, 14–18 October 1974, pp 247–257, CONF-741003, National Technical Information Service, U.S. Dept. of Commerce, Springfield, VA.
88. Barlow A.K. and Latham J. (1983) A laboratory study of the scavenging of sub-micron aerosol by charged raindrops. *Quarterly Journal of the Royal Meteorological Society* 109, 763–770.
89. Davenport H.M. and Peters L.K. (1978) Field studies of atmospheric particulate concentration changes during precipitation. *Atmospheric Environment* 12, 997–1008.
90. Graedel T.E. and Franey J.P. (1977) Field measurements of submicron aerosol washout by rain. In *Precipitation Scavenging* (1974), Proc. of a Symp. at Champaign, Illinois, 14–18 October 1974, pp 503–523, CONF-741003, National Technical Information Service, U.S. Dept. of Commerce, Springfield, VA.
91. Slinn W.G.N., Katen P.C., Wolf M.A., Loveland W.D., Radke L.F., Miller E.L. Ghannam L.J., Reynolds B.W. and Vickers D. (1979) Wet and dry deposition and resuspension of AFCT/TFCT fuel processing radionuclides. SR0980-10, Oregon State University, Corvallis. Available from National Technical Information Service, U.S. Dept. of Commerce, Springfield, VA.
92. May F.G. (1958) The washout of lycopodium spores by rain. *Quarterly Journal of the Royal Meteorological Society* 84, 451–458.
93. Best A.C. (1950) The size distribution of raindrops. *Quarterly Journal of the Royal Meteorological Society* 76, 16–36.
94. Slinn W.G.N. (1977) Precipitation scavenging: some problems, approximate solutions and suggestions for future research. In *Precipitation Scavenging* (1974), Proc. of Symp. at Champaign, Illinois, 14–18 October 1974, pp 1–60, CONF-741003, National Technical Information Service, U.S. Dept. of Commerce, Springfield, VA.
95. Mason B.J. (1971) The Physics of Clouds, Clarendon Press, Oxford.
96. Slinn W.G.N. (1984) Precipitation scavenging. In *Atmospheric Science and Power Production* (edited by D. Randerson), pp 466–532, National Technical Information Service, U.S. Dept. of Commerce, Springfield, VA.
97. Engelmann R.J. (1965) Rain scavenging of zinc sulphide particles. Journal of *Atmospheric Science* 22, 719–727.
98. Lovett G.M. and Reiners W.A. (1983) Cloud water: an important vector of atmospheric deposition. In *Precipitation Scavenging, Dry Deposition and Resuspension* (edited by H.R. Pruppacher, R.G. Semonin and W.G.N. Slinn), Vol. 1, pp 171–179, Elsevier, Amsterdam.
99. Dollard G.J. and Unsworth, M.H. (1983) Field measurements of turbulent fluxes of wind driven fog drops to a grass surface. *Atmospheric Environment* 17, 775–780.
100. Mrose, H. (1966) Measurements of pH and chemical analyses of rain-snow-and fog-water. *Tellus* 18, 266–270.
101. Bagnold R.A. (1941) *The Physics of Blown Sand and Desert Dunes*. Chapman and Hall, London.
102. Chepil W.S. (1943) Relation of wind erosion to the water stable and dry clod structure of soil. *Soil Science* 55, 275–287.
103. Romney E.M. and Wallace A. (1977) Plutonium contamination of vegetation in dusty field environments. In *Transuranics in Natural Environments* (edited by M.G. White

and P. B. Dunaway), pp 287–302, Proc. of Symp. at Gatlinburg, October 1976, NVO-178. Nevada Applied Ecology Group, USERDA, Las Vegas, Nevada.

104. Shinn J.H., Homan D.N. and Gay D.D. (1983) Plutonium aerosol fluxes and pulmonary exposure rates during resuspension from bare soils from a chemical separation facility. In *Precipitation Scavenging, Dry Deposition and Resuspension* (edited by H.R. Pruppacher, R.G. Semonin and W.G.N. Slinn), Vol. 2, pp 1131–1143, Elsevier, Amsterdam.

105. Langer G. (1983) Activity, size and flux of resuspended particles from Rocky Flats soil. In *Precipitation Scavenging, Dry Deposition and Resuspension* (edited by H.R. Pruppacher, R.G. Semonin and W.G.N. Slinn), Vol. 2., pp 1161–1174, Elsevier, Amsterdam.

106. Orgill M.M., Peterson M.R. and Sehmel G.A. (1976) Some initial measurements of DDT resuspension and translocation from Pacific Northwest forests. Proc. of the Atmosphere-Surface Exchange of Particulate and Gaseous Pollutants, Richland, WA, 4–6 September 1974, pp 813–834, Energy Res. and Dev. Admin. Symp. Ser., CONF-740921. National Technical Information Service, Springfield, VA.

107. Aylor D.E. (1976) Resuspension of particles from plant surfaces by wind. Proc. of the Atmosphere-Surface Exchange of Particulate and Gaseous Pollutants, Richland, WA, 4–6 September 1974, pp 791–812, Energy Res. and Rev. Admin. Symp. Ser., CONF-740921, National Technical Information Service, Springfield, VA.

108. Hereim A.T. and Ritchie B. (1976) Resuspended bacteria from desert soil. Proc. of the Atmosphere-Surface Exchange of Particulate and Gaseous Pollutants, Richland, WA, 4–6 September 1974, pp 835–845, Energy Res. and Dev. Admin. Symp. Ser., CONF-740921, National Technical Information Service, Springfield, VA.

109. Sehmel G.A. (1976) Airborne ^{238}Pu and ^{239}Pu associated with larger than 'respirable' resuspended particles at Rocky Flats during July 1973. BNWL-2119, Pacific Northwest Laboratory, Richland, Washington.

110. Horst T.W. (1976) The estimation of air concentrations due to the suspension of surface contamination by the wind. BNWL-2047, Pacific Northwest Laboratory, Richland, WA.

111. Linsley G.S. (1978) Resuspension of the transuranium elements—a review of existing data. NRPB-R75, HMSO, London.

112. Garland J.A. (1979) Resuspension of particulate matter from grass and soil. AERE-R9452, HMSO, London.

113. Reeks M.W., Reed J. and Hall D. (1985) The long-term resuspension of small particles by a turbulent flow—Part III. Resuspension from rough surfaces. TPRD/B/0640/N85, CEGB Berkeley Nuclear Laboratories, Gloucestershire, U.K.

114. Nicholson K.W. (1993) Wind tunnel experiments on the resuspension of particulate material. *Atmospheric Environment* 27A, 181–188.

115. Anspaugh L.R., Shinn J.H., Phelps P.L. and Kennedy N.C. (1975). Resuspension and redistribution of plutonium in soils. *Health Physics* 29, 571–582.

116. Sehmel G.A. (1983) Resuspension rates from aged inert tracer sources. In *Precipitation Scavenging, Dry Deposition and Resuspension* (edited by H.R. Pruppacher, R.G. Semonin and W.G.N. Slinn), Vol. 2, pp 1073–1086, Elsevier, Amsterdam.

117. Garland J.A. (1983) Some recent studies of the resuspension of deposited material from soil and grass. In *Precipitation Scavenging, Dry Deposition and Resuspension* (edited by H.R. Pruppacher, R.G. Semonin and W.G.N. Slinn), Vol. 2, pp 1087–1097, Elsevier, Amsterdam.

118. Gillette D.A. (1976) Production of fine dust by wind erosion of soil: effect of wind and soil texture. Proceedings of the Atmosphere—Surface Exchange of Particulate and Gaseous Pollutants, Richland, WA, 4–6 September 1974, pp 591–609, Energy Res. and

119. Dev. Admin, Symp. Ser., CONF-740921, National Technical Inforamtion Service, Springfield, VA.
119. Gillette D.A. (1983) Threshold velocities for wind erosion on natural terrestrial arid surfaces (a summary). In *Precipitation Scavenging, Dry Deposition and Resuspension* (edited by H.R. Pruppacher, R.G. Semonin and W.G.N. Slinn), Vol. 2, pp 1047–1058, Elsevier, Amsterdam.
120. Gillette D.A., Blifford I.H. and Fryrear D.W. (1974) The influence of wind velocity on the size distributions of aerosols generated by the wind erosion of soils. *J. Geophys. Res.* 79, 4068–4075.
121. Nicholson K.W. and Branson J.R. (1990) Factors affecting resuspension by road traffic. *Science of the Total Environment* 93, 349–358.
122. Sehmel G.A. (1973) Particle resuspension from an asphalt road caused by car and truck traffic. *Atmospheric Environment* 7, 291–301.
123. Sehmel G.A. (1986) Resuspension research with tracers at PNL. PNL-5750—Part 3, pp 30–34, Pacific Northwest Laboratory, Richland, WA.
124. Milham R.C., Schubert J.F., Watts J.R., Boni A.L. and Corey J.C. (1976) Measured plutonium resuspension and resulting dose from agricultural operations on an old field at Savannah River Plant in the South-Eastern United States of America. In Transuranium Nuclides in the Environment, pp 409–421, International Atomic Energy Agency, Vienna.

CHAPTER 5

Bioaerosol Particle Statistics

A.J. Heber

INTRODUCTION

Bioaerosols generally are dispersed over a wide range of size characteristics and, hence, their measurement and representation usually demand a working knowledge of the fundamentals of particle statistics. Recommendations for avoiding certain pitfalls need to be followed to prevent data misrepresentation.

Techniques for measuring and analyzing bioaerosol data have evolved over the last several decades primarily from advances in modern instrumentation and the development of powerful computers and software (Chapter 8). Simplifying analytical shortcuts, such as the Hatch-Choate equations,[1] are no longer necessary; and will be deemphasized in favor of required mathematical equations.

The variables that describe particle "amounts" will be reviewed; and a discussion of "measures of dispersion," such as equivalent diameters, which represent ensembles of airborne particles will follow. Next, size distribution functions, moment and weighted averages (count, area, volume), and statistical accuracy will be presented.

Data from three real samples of bioaerosols are presented to demonstrate statistical techniques and illustrate their properties, limitations, and utility. These samples, referred to as B1, B2, and B3, were sized with a Coulter counter, a scanning electron microscope, and a cascade impactor, respectively. Statistical analyses of the sample data demonstrating the use of presented equations give readers an opportunity to construct and evaluate their own software for particle analysis.

PROPERTIES OF BIOAEROSOL SIZE DISTRIBUTIONS

Shape

Solid particles exhibit a variety of shapes, such as spheres, regular polyhedrons, straight and curved fibers, flakes, and irregular particles.[2] Spherical equivalent diameters based on aerodynamic drag forces often are used for solid particles to characterize their size and will be discussed later in this chapter.

The shape of bioaerosol particles is a fundamental property and is important in assessing health hazards. For example, angular particles or fibers (aspect ratio > 3) may be more detrimental to health than rounded (spherical) ones, even if the compound itself is inert. However, particle shape varies significantly because of natural variance and agglomeration. Possible relationships between size and shape have been analyzed by numerical and statistical techniques. The shape frequency distribution of a particle has been developed using Fourier analyses of its signature waveform. The signature waveform is generated by plotting the vector magnitudes from the centroid of the projected area of the particle to its perimeter as a function of angle.[3] The first five harmonics contributing to the waveform describe the basic shape of the profile, whereas the higher harmonics describe the texture of the profile.[4] The limitation of Fourier analysis of geometric signature waveforms is the indeterminacy of the vector with extremely rugged profiles.[4]

The fractal dimension of a surface boundary is useful in characterizing a particle's surface structure, especially for particles with profiles too rugged for Fourier analysis of geometric signature waveforms.[4] The fractal dimension is defined as 1.0 plus the absolute value of the slope of the dataline in a Richardson plot. The calculation of a particle's fractal dimension begins with an estimate of its perimeter by the structured walk technique. This technique involves "striding" around the profile with a compass with a step size or resolution and yields a polygon. The Richardson plot is the profile perimeter estimate against the measurement resolution on a log-log scale and results in a straight line if the profile boundary can be described with a fractal dimension.[4] The resolution limits of the Richardson plot should be stated clearly.

The fractal dimension of a particle's projected area profile (two-dimensional) ranges from 1 to 2 and is related directly to the convolution, ruggedness, or space-filling ability of the profile.[4] The fractal dimension of a perfectly smooth-textured boundary is 1.0. Fractal dimensions of 1.15 and 1.3 represent mild and severe convolution, respectively. The fractal dimensions of three-dimensional irregular surfaces enclosing a volume range from 2 to 3.

Quantity

The quantity of particles is expressed in terms of the property of interest or the property measured. These properties may be number, length, aspect ratio, surface area, volume or mass; and biochemical, biological, or toxicological activity.[5] The fact that the "quantity" of a particle can be one of several properties, or combinations thereof, makes particle statistics unique in comparison to number statistics. Particle quantity or amount in general will be denoted by the letter 'a' for a single particle and by 'A' for the total of more than one particle.

It can be shown (Equation 5.1) that, in general, "quantity" properties are proportional to particle diameter or side length, d, raised to an integer power, r, referring to number, length, area, and volume (or mass) when r=0,1,2, and 3, respectively. The quantity a_0, proportional to d^0, is dimensionless unity or merely a count of the particle, whereas the quantities a_1 and a_3 of a spherical particle are proportional to d^1 and d^3, respectively. Quantities a_0 and a_3, number and volume, are most commonly used in analyzing bioaerosols, though surface area is important for biochemical and biological activity. The length, surface area, and volume shape factors, α_1, α_2, and α_3, are 1, π, and $\pi/6$ for spherical particles and 1, 6, and 1 for cubical particles (where d=side length), respectively, and are determined experimentally for irregular shapes. The total quantity for a group of particles is given by Equation 5.2.

$$a_r = \alpha_r d^r \tag{5.1}$$

$$A_r = \sum a_r = \sum_{i=1}^{k} N_i a_{ri} = \alpha_r \int_0^\infty fr(d)dd^r \tag{5.2}$$

where dd^r = differential of d^r
 $f_r(d)$ = frequency function, fraction per cm
 i = class or group number
 k = total number of size classes
 N = number of particles

It is assumed that the full amount of the bioaerosol is known for meaningful statistical representations that follow in this chapter. Practical difficulties in bioaerosol sampling sometimes make it impossible to collect all sizes, especially for number distributions obtained with optical particle counters.[5]

Size

The measurable geometric size parameters are its diameter or length, surface area, and volume and these generally are used to classify bioaerosols. The quantity A usually is distributed according to the magnitude of the chosen geometric size parameter. However, diameter, surface area, or volume of aerosol particles of irregular shape cannot be measured directly. Thus, equivalent diameters (Table 5.1), the diameter of a spherical particle of the same material that generates the same measurement effect, e.g., settling speed, are utilized for nonspherical particles. The variation between different equivalent diameters increases as the particles become more nonspherical.[6] The symbol d for particle diameter will have a subscript, e, which denotes an equivalent diameter of one of the types described in Table 5.1. However, for convenience, the subscript will be omitted in most instances and an equivalent diameter is assumed in all references to particle diameters.

The quantity A_r is measured directly (r depending on the measurement method and indicating the weight) and is distributed according to equivalent diameters based on calibration with spheres of known size. The equivalent diameter used will depend on the measurement instrument. For example, a microscope can be used to measure particle number (A_0), length (A_1), or projected area (A_2), and each amount is size classified according to the projected area equivalent diameter, d_a (Table 5.1), which is proportional to d^2.

Particle Grouping

The total or average value of the desired particle quantity could be determined with Equation 5.2, if individual particle quantities were available. However, the range of the measured geometric size parameter, usually an equivalent diameter (Table 5.1), is commonly divided into several size classes or groups, and the particle quantity A_{ri} of particles in each class i is tallied or measured in sufficient quantities to be representative and statistically meaningful.

Particle mass and the number of bacterial colony-forming particles are two particle quantities that can be determined by means of a cascade impactor using regular filters and agar-treated impaction plates, respectively. However, the aerodynamic equivalent diameter, d_d (Table 5.1), is used to group each particle quantity. The aerodynamic equivalent diameter also is used to group particle numbers measured with an aerodynamic particle sizer.

Particle statistics in this chapter are based on grouped data only. The class mark or midpoint diameter d_i (arithmetic, geometric or harmonic) of each class represents its size. Assuming that the diameters of the upper and lower limits of each class are denoted by d_u

Table 5.1. Definitions of Equivalent Diameters

Symbol[a]	Equivalent Diameter	Definition
d_S	Surface	Diameter of sphere with same surface area
d_V	Volume	Diameter of sphere with same volume
d_d	Aerodynamic	Diameter of unit-density sphere with same resistance to motion
d_a	Projected area	Diameter of sphere with same projected area as viewed through a microscope
d_f	Free-falling	Diameter of sphere with same density and free-falling speed in air
d_{St}	Stokes	Free-falling diameter in laminar flow (Re < 0.2)
d_A	Sieve	Width of minimum square aperture through which particle will pass
d_{vs}	Specific surface	Diameter of sphere with same ratio of surface area to volume
d_{sca}	Optical	Diameter of sphere reflecting, scattering or absorbing light at the same intensity
d_{el}	Electrical	Diameter of sphere causing same change in electrical resistance (Coulter counter)

[a]The subscripts of each symbol substitute for e in d_e.
Source: Allen, 1968.[9]

and d_l, respectively, the arithmetic midpoint diameter, d_{im}, the geometric midpoint diameter, d_{ig}, and the harmonic midpoint diameter, d_{ih}, are defined as:

$$d_{im} = \frac{d_l + d_u}{2} \tag{5.3}$$

$$d_{ig} = \sqrt{d_l d_u} = e^{\left(\frac{\ln d_l + \ln d_u}{2}\right)} = 10^{\left(\frac{\log d_l + \log d_u}{2}\right)} \tag{5.4}$$

$$d_{ih} = \left[\frac{1}{2}\left(\frac{1}{d_u} + \frac{1}{d_l}\right)\right]^{-1} \tag{5.5}$$

The number of classes is preferably 10 or more, otherwise information tends to be lost, and false or distorted distributions are computed, especially for samples with large numbers of particles. Little additional information is gained with more than 20 classes.[7] Techniques for optimizing the number and widths of class intervals of histograms include the maximum entropy, phi, arithmetic, log arithmetic, Z-scores, and log Z-score methods.[8]

Averages

Average sizes of bioaerosols simply and concisely represent a group of particles by indicating their central tendency.[9] The amount is the quantity of measured A_r, and the weight is the size parameter, r, for which the mean is desired. The average size greatly depends on the weighting factor such as number (r=0), length (r=1), surface area (r=2), volume (r=3), or others. The arithmetic mean diameter, d_{mr}, weighted by property r is the sum of the products of the particle quantity, A_{ri}, and the arithmetic midpoint diameter, d_{im} (Table 5.1), in each class divided by the total particle quantity A_r:

$$\bar{d}_{mr} = \frac{\sum_{i=1}^{k} A_{ri} d_i}{A_r} \qquad (5.6)$$

The geometric mean diameter (Equation 5.7) is utilized with the lognormal distribution to indicate the particle size with the greatest frequency of particle amount. Whereas $\bar{d}_{mr}$ (Equation 5.6) is associated with discontinuous distributions, $\bar{d}_{gr}$ (Equation 5.7) is associated with continuous distributions. The harmonic mean diameter, $\bar{d}_{hr}$ (Equation 5.8), is important where the arithmetic and geometric mean diameters are not satisfactory in representing the size with the greatest contribution to particle quantity, but is rarely seen in the literature.

$$\bar{d}_{gr} = e^{\left(\frac{\sum_{i=1}^{k} A_{ri} \ln(d_{ig})}{A_r}\right)} \qquad (5.7)$$

$$\bar{d}_{hr} = \left(\frac{\sum_{i=1}^{k} A_{ri}\left(\frac{1}{d_{ih}}\right)}{A_r}\right)^{-1} \qquad (5.8)$$

The median diameter is defined as the diameter for which 50% of the total quantity of particles are smaller and 50% are larger and, thus, corresponds to a cumulative fraction of 0.50. The four most commonly used median diameters are the number and mass medians of actual and aerodynamic equivalent particle diameters.[5,10] The mode is the most frequent size or that with the greatest frequency of particle quantity A.

Diameter of Average Quantity

The average property $\bar{a}_r$ of a particle population, from Equations 5.1 and 5.2, is

$$\bar{a}_r = \frac{A_r}{N} = \alpha_r \overline{d^r} \qquad (5.9)$$

The arithmetic mean particle quantity $\overline{d^r}$ for different weights, r, is

$$\overline{d^r} = \frac{A_r}{N\alpha_r} = \frac{\sum_{i=1}^{k} A_{ri}}{N\alpha_r} = \frac{\sum_{i=1}^{k} N_i a_{ri}}{N\alpha_r} \qquad (5.10)$$

The diameter of average quantity[1] r is

$$d_{\overline{r}} = \left[\frac{\sum_{i=1}^{k} A_{ri}}{N\alpha_r}\right]^{\frac{1}{r}} = \left[\frac{\sum_{i=1}^{k} N_i d_i^r}{N}\right]^{\frac{1}{r}} \qquad (5.11)$$

Moment Averages of Weighted Distributions

The moment average of weighted distributions is

$$d_{\overline{rp}} = \left[\frac{\sum_{i=1}^{k} A_{ri} d_i^p}{A_r}\right]^{1/p} \qquad (5.12)$$

where p = moment.

When r=3, the average is weighted by volume, or by mass, if the density can be assumed constant with particle size. When r=2, the average is weighted by surface area. These averages can be calculated for any polydisperse bioaerosol from its frequency distribution, regardless of how the particle sizes are distributed.

Several moment averages of weighted distributions can be defined for the same set of particles, one of which probably best describes the process of interest (Table 5.2). Bioaerosols may need to be described by more than one, e.g., the mass moment mean diameter, d_{MP}, is related to respiratory deposition[11] and the surface volume mean diameter, d_{SV}, to physiological response from gases absorbed on particles.[10]

Variance

Along with the mean particle size, the variance of size about the mean is an equally important representation of a bioaerosol, if the size distribution is closed-ended,[7] meaning that virtually all of the particles are within the measured size range. The standard description of the dispersity of any aerosol is the standard deviation (Equation 5.13), which is defined as the square root of the variance, σ^2. The coefficient of variation is the standard deviation divided by the mean (Equation 5.14). The geometric standard deviation (Equation 5.15) is often reported, especially when the bioaerosol is distributed lognormally. The geometric standard deviation is more convenient to calculate when the bioaerosol is lognormal, but can be computed regardless of the type of distribution.[5]

Table 5.2. Weighted Mean Diameters and Moment Ratios of Various Aerosol Distribution Characteristics

Amount, Weight	Equation	Moment Ratio	B2 Data
number, length	$d_{NL} = \dfrac{\sum L_i}{\sum N_i}$	$\dfrac{\sum d\, N_i}{\sum N_i}$	6.79 μm
number, surface area	$d_{NS} = \left[\dfrac{\sum S_i}{\sum N_i}\right]^{\frac{1}{2}}$	$\left[\dfrac{\sum d^2 N_i}{\sum N_i}\right]^{\frac{1}{2}}$	9.51 μm
number, volume	$d_{NV} = \left[\dfrac{\sum V_i}{\sum N_i}\right]^{\frac{1}{3}}$	$\left[\dfrac{\sum d^3 N_i}{\sum N_i}\right]^{\frac{1}{3}}$	12.12 μm
length, surface area	$d_{LS} = \dfrac{\sum S_i}{\sum L_i}$	$\dfrac{\sum d^2 N_i}{\sum d\, N_i}$	13.33 μm
length, volume	$d_{LV} = \left[\dfrac{\sum V_i}{\sum L_i}\right]^{\frac{1}{2}}$	$\left[\dfrac{\sum d^3 N_i}{\sum d\, N_i}\right]^{\frac{1}{2}}$	16.19 μm
surface area, volume	$d_{SV} = \dfrac{\sum V_i}{\sum S_i}$	$\dfrac{\sum d^3 N_i}{\sum d^2 N_i}$	19.66 μm
volume, moment	$d_{VP} = \dfrac{\sum P_i}{\sum V_i}$	$\dfrac{\sum d^4 N_i}{\sum d^3 N_i}$	24.56 μm
mass, moment	$d_{MP} = \dfrac{\sum d\, W_i}{\sum W_i}$	$\dfrac{\sum d^4 N_i}{\sum d^3 N_i}$	24.56 μm

Source: Allen, 1968.[9]

$$\sigma_r = \left[\dfrac{\sum_{i-1}^{k} A_{ri}(d_{im} - \bar{d}_{mr})^2}{A_r}\right]^{\frac{1}{2}} \tag{5.13}$$

$$\gamma_r = \dfrac{\sigma_r}{\bar{d}_{mr}} \tag{5.14}$$

$$\sigma_{gr} = \exp\left[\dfrac{\sum_{i-1}^{k} A_{ri} \ln^2(d_{ig}/\bar{d}_{gr})}{A_r}\right]^{\frac{1}{2}} \tag{5.15}$$

where d_{im} = arithmetic midpoint diameter, cm

$\bar{d}_m$ = arithmetic mean, cm
$\bar{d}_g$ = geometric mean, cm
r = weight of the distribution
σ = standard deviation, cm
σ_g = geometric standard deviation
γ = coefficient of variation.

For number distributions, A_r in the denominator of Equation 5.15 is sometimes defined as N-1, especially for small numbers of particles. The subscript r appears with σ, γ, and σ_g because different values will result from the same set of particles depending on the weight of the distribution.

Another measure of dispersity is the inequality of weighted averages. For a polydisperse aerosol, the weighted averages, $\bar{d}_r$, differ with the diameters of larger r versus smaller r, and the differences become larger with greater dispersity.

SIZE DISTRIBUTION FUNCTIONS

One advantage of size distribution functions is that the dispersion of a bioaerosol can be adequately described and characterized by two parameters with a simple equation, thus facilitating comparisons. Four commonly used functions are the normal, lognormal, Gaudin-Shuhman power law, and Rosin-Rammler distributions.[12]

Normal

The Gaussian or normal distribution describes systems produced by random interactions of many causes of similar magnitude. The distribution of particle amount about the average measurements of particle size is Gaussian in the case of some pollens and spores and also some artificially generated monodisperse aerosols. The frequency function of the Gaussian distribution is

$$f_r(d) = \left[\frac{1}{\sigma_r\sqrt{2\pi}}\right] \exp\left[\frac{-(d-\bar{d}_{mr})^2}{2\sigma_r^2}\right], \quad -\infty < d < \infty \quad (5.16)$$

where $f_r(d)$ = frequency function.

The frequency function is used satisfactorily only if $\gamma_r \geq 1/3$, otherwise the function predicts negative diameters. If $\gamma_r > 1/3$, the mean will be sufficiently greater than zero such that the probability of negative diameters is negligible.

A discontinuous size frequency curve or histogram is obtained when the average value of the relative frequency in each size class is plotted. A frequency polygon[13] is obtained if the midpoints of the histogram are connected (Figure 5.1). The cumulative distribution (Equation 5.17) predicts the proportion of the particle amount that will be below a given diameter.

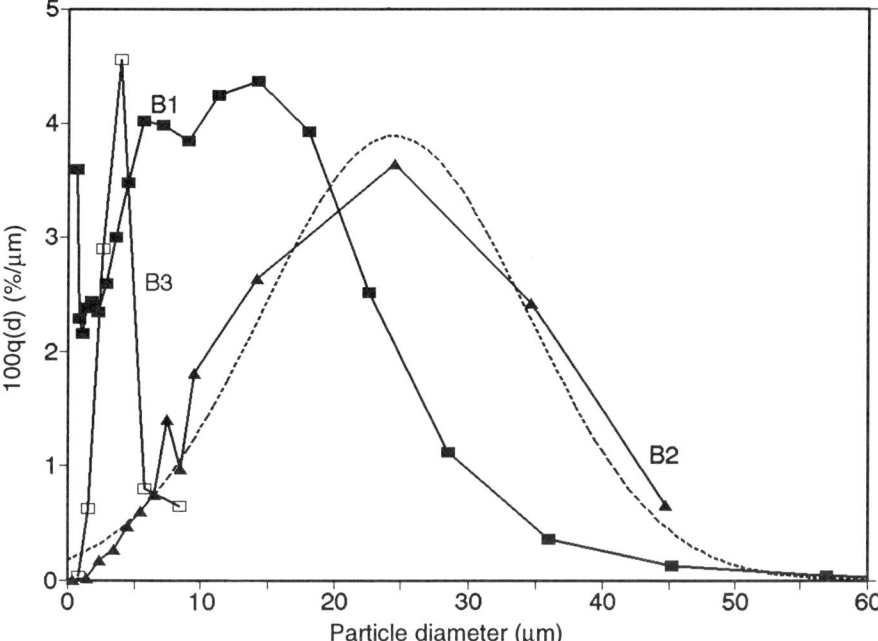

Figure 5.1. Frequency polygons of bioaerosols B1 (■), B2 (▲) and B3 (□) on normal axes (Equation 5.18). The q(d) for B3 was divided by 7 for this graph. The normal distribution function (Equation 5.17) using the sample mean and standard deviation is shown with dashed line for B2.

$$F_r(d) = \int_0^d f_r(d)dd \qquad (5.17)$$

where $F_r(d)$ is the cumulative distribution function.

Skew and kurtosis indicate the degree of distortion and flatness of the frequency distribution, respectively. A frequency distribution whose mode is greater than the mean has a positive skew. A negatively skewed distribution has a mode that is smaller than the mean. Curves with larger tails to the right or left have positive or negative skews, respectively. Skew and kurtosis are 0.0 and 3.0 for perfectly normal data. In Gaussian distributions, the mode equals the arithmetic mean diameter (Equation 5.6) and the median diameter. An advantage of the cumulative curve over the frequency polygon is that the class interval is eliminated and, thus, unequal intervals do not influence its shape. Also, the median value[7] is conveniently located with the 50% cumulative percentage line. Although the median value can be found where the 50% cumulative percentage line crosses a linear regression line, linear interpolation between the nearest data points on either side of the 50% cumulative percentage line will be more accurate. The frequency function (Equation 5.16) is obtained by differentiating F(d) with respect to d. A straight line results when Gaussian data are plotted on a normal probability graph (Figure 5.2).

Lognormal

Lognormal distributions are generated when given values of a particular particle property require relatively rare favorable combinations of several interacting causes.[14] The

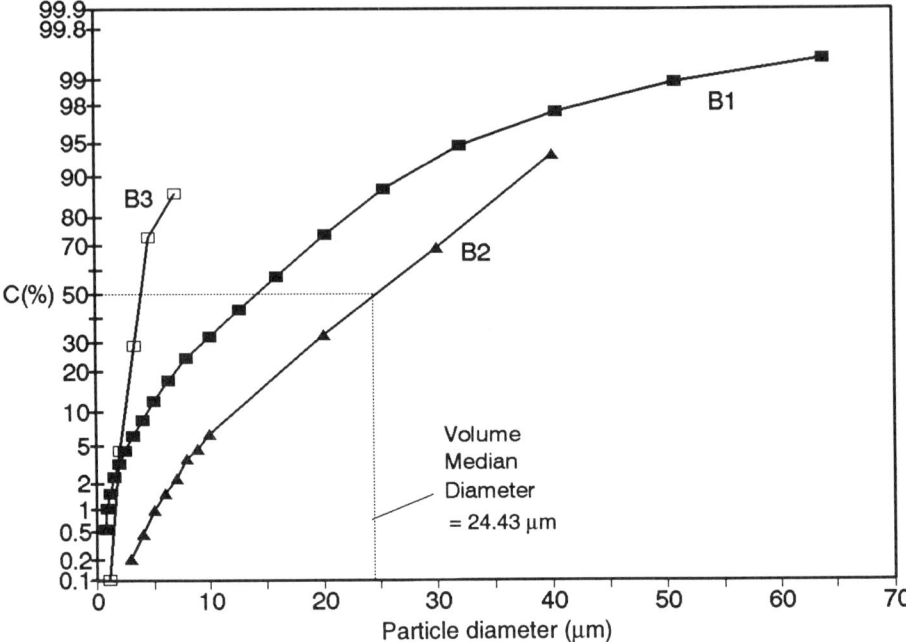

Figure 5.2. Normal probability plots of bioaerosols B1 (■), B2 (▲) and B3 (□). The volume mean diameter is shown for B2.

lognormal distribution fits single-source bioaerosols very well, especially when the ratio of maximum to minimum diameter is greater than ten[1] and, thus, is often more suitable for bioaerosol particles. A bell-shaped curve forms when the frequency density is plotted against the log of the diameter (Equation 5.18). Multimodal distributions showing more than one maximum result when two or more bioaerosols with widely separated means are mixed. A mixture of two lognormal distributions will not be lognormal.[1]

$$f_r(d) = \left[\frac{1}{d \ln \sigma_{gr} \sqrt{2\pi}}\right] \exp\left[\frac{-(\ln d - \ln \bar{d}_{gr})^2}{2 (\ln \sigma_{gr})^2}\right], \quad 0 \leq d < \infty \quad (5.18)$$

A straight line on the log-probability graph is convenient for analysis of the distribution. Though they have little physical significance in interpreting experimental results,[10] the geometric median diameter and geometric standard deviation describe completely the lognormal distribution on that graph. The geometric median diameter is readily determined from the 50% point of the graph (Figure 5.4). The geometric standard deviation is calculated by dividing the diameter taken from the 84% probability point by the geometric median diameter, or dividing the geometric median diameter by the diameter taken from the 16% probability point. The mode, median, and mean are always unequal in a lognormal distribution, whereas they are equal in a normal distribution. However, the geometric mean diameter (Equation 5.7) is equal to the geometric median diameter in a lognormal distribution.

A linear regression of the probit[15] versus log diameter is sometimes used to determine the equation of the cumulative probability line. However, using the linear regression to calculate the median diameter is not recommended. Classes near the ends of the distribution, containing relatively few particles and affected by size detection limitations of the measurement instrument, have equal influence on the straightness of the line and will lead to errors. Probit weighting coefficients that give more importance to the central classes than the end classes are recommended.[15] Instrument limitations that tend to cause an "S" shaped cumulative curve at one or both ends should be avoided in fitting a straight line.

Gaudin-Schuhman Power Law

The Gaudin-Schuhman power-law distribution[12] (Equation 5.19) is very useful for representing measured particle distributions and making calculations with them. The exponent, t, of the ratio of the diameter over a diameter, d_z, is a measure of the dispersity of the aerosol, whereas d_z locates the curve.

$$C(d) = \left[\frac{d}{d_z}\right]^t, \quad 0 \leq d \leq d_z \tag{5.19}$$

where C(d) = percentage of particle quantity with diameters less than d.
The linearized form of the power law function is:

$$\log C(d) = t \log\left[\frac{d}{d_z}\right] = t(\log d - \log d_z) \tag{5.20}$$

The power law distribution produces a straight line on log-log paper for C less than about 80%.[12]

Rosin-Rammler

The Rosin-Rammler distribution (Equation 5.21), a form of the Weibull distribution that was first used for pulverized coal,[12] provides good fit where the measured size distribution is more skewed than the lognormal or normal distribution.

$$R(d) = \exp\left[-(d)^n\right] \tag{5.21}$$

where R(d) = percent of total particle quantity with diameters greater than d.

The linear form of the Rosin-Rammler equation is:

$$y = \log \log\left[\frac{100}{R}\right] = n \log d \tag{5.22}$$

where y = linearized Rosin-Rammler function.

Goodness of Fit

The chi-square goodness-of-fit test can be employed to determine whether a particular frequency or number distribution is a good representation of the data. At least ten size classes are preferred when performing a chi-square analysis for goodness of fit to normal distributions.[16] Bias in the chi-square analysis can be reduced adequately by assuring that no $f(d)_i$ values are less than 1.0. Bias can be avoided by combining classes in the tails of the distribution or by using one of several other tests of normality that are not biased by $f(d)_i < 1.0$, such as the log-likelihood method.[16]

Visual examination of the normal or lognormal probability plots of cumulative number frequency also can be utilized to check normality or lognormality.[16] Graphical analysis should not be used exclusive of algebraic methods to test goodness of fit,[13] but rather as a first approximation and start of the analysis.[17] Proper display of particle size distributions also can provide insights into the formation of the aerosol,[18] and the following types of size-limited lognormal distributions have been identified:

1. Unlimited
2. Upper size limit
3. Lower size limit
4. Both upper and lower size limits
5. Bimodal
6. Upper discontinuous limit
7. Lower discontinuous limit
8. Both upper and lower discontinuous limits

Deviations from lognormality give the greatest insight into the particle-forming process and the nature of the resulting particle size distribution.[18] Histograms or frequency polygons are preferred to normal or log probability plots because multimodal distributions can be "hidden" as normal or log probability plots. However, cumulative distributions overcome the problems of unequal grouping that may occur with frequency distributions.

APPLICATION

Bioaerosol B1

Bioaerosol B1 was collected from inside a swine finishing house onto membrane filters.[19] The collected particles were washed off the filters for subsequent measurement with a Coulter counter utilizing the electrical resistance technique. Because the Coulter counter was calibrated to latex spheres, the electrical resistance equivalent diameter, d_{el}, proportional to particle volume, was used as the measure of dispersity. The particles were grouped into 21 geometrically progressive size classes of d_{el} ranging from $d_l = 0.63$ μm to 64 μm. In geometric progressions, the ratio of d_u to d_l is the same for each class (Table 5.3) and will produce equal width size classes on a log scale, which is advantageous for plotting a lognormal distribution. An arithmetic progression (same $d_u - d_l$ over all size classes) is advantageous for narrowly distributed bioaerosols, but geometric progression is more common because most aerosols are closer to a lognormal distribution. The midpoint diameter, d_{ig}, was calculated with Equation 5.4 (Table 5.3). The largest class in B1 was open-ended with $d_l = 64$ μm, for which a $d_u = 80.6$ μm was assumed. Such open-ended classification usually cannot be avoided because of the limitation of the measurement technique.[7] Estimation of an endpoint was not critical in the case of B1, because the particle amount in this end class was relatively small.[5]

Table 5.3. Data B1 from Coulter Counter Analysis of Swine House Dust

d_l μm	d_{ig} Eqn. 5.4 μm	N_i	V_i Eqn. 5.2 %	V_i %/μm	C %	x	R %	y Eqn. 5.22
0.63	0.71	62,907	0.58	3.59			100.0	
0.79	0.90	26,196	0.48	2.28	0.6	-2.54	99.4	-2.60
1.00	1.13	15,187	0.56	2.15	1.1	-2.32	98.9	-2.34
1.26	1.43	10,647	0.79	2.38	1.6	-2.16	98.4	-2.15
1.59	1.80	6,767	1.00	2.44	2.4	-1.99	97.6	-1.98
2.00	2.26	4,139	1.22	2.34	3.4	-1.84	96.6	-1.82
2.52	2.85	2,869	1.69	2.59	4.6	-1.70	95.4	-1.69
3.17	3.59	2,119	2.49	3.00	6.3	-1.54	93.7	-1.55
4.00	4.52	1,536	3.62	3.48	8.8	-1.36	91.2	-1.40
5.04	5.70	1,116	5.26	4.02	12.4	-1.16	87.6	-1.24
6.35	7.18	697	6.57	3.98	17.7	-0.93	82.3	-1.07
8.00	9.04	424	8.00	3.85	24.2	-0.70	75.8	-0.92
10.08	11.39	294	11.11	4.24	32.2	-0.46	67.8	-0.77
12.70	14.35	191	14.41	4.37	43.4	-0.17	56.6	-0.61
16.00	18.08	108	16.30	3.92	57.8	0.19	42.2	-0.43
20.16	22.78	44	13.21	2.52	74.1	0.64	25.9	-0.23
25.40	28.70	12	7.34	1.11	87.3	1.14	12.7	-0.05
32.00	36.16	2	3.01	0.36	94.6	1.61	5.4	0.10
40.32	45.56	1	1.31	0.13	97.6	1.98	2.4	0.21
50.80	57.40	0	0.54	0.04	98.9	2.30	1.1	0.30
64.00	72.30	0	0.52	0.03	99.5	2.56	0.5	0.36
80.60					100.0			
Total		135,257	100.00					

Source: Heber et al., 1991.[19]

In each class, V_i was calculated from d_{ig} and N_i and expressed as a percentage of the total volume. Since only about 0.5% of the volume was accounted for in the largest and smallest classes, most of the total volume-weighted distribution was apparently measured. The entire number-weighted distribution was not measured, because the greatest frequency occurred in the smallest size class and, thus, the small-size "tail" of the distribution was not defined. Assuming constant density over the range of particle size, the mass- and volume-weighted distributions are the same. The cumulative percentages undersize (C) and oversize (R) were calculated for the volume-weighted distribution in the sixth and eighth columns of Table 5.3, respectively. A cumulative percentage curve of the incomplete number-weighted distribution leads to gross errors of mean and variance and therefore was avoided. It is important to specify which distribution is being reported and also to state clearly which particle property is being utilized.[20] In this example, it is particle volume as measured by the electrical resistance method.

Bioaerosol B1 had four maxima in its volume frequency polygon (Figures 5.1 and 5.3), indicating a complex multimodal distribution. Definite modes occurred in the size classes of d_l=12.7, 5.04, and 1.59 μm. Another mode equal to or less than 0.63 μm was apparently present in the bioaerosol. The existence of more than one mode was due to mixtures of bioaerosols with widely separated mode diameters. The authors[19] hypothesized that B1 was a mixture of squamous skin and epithelial cells, tiny microorganisms, and starch and grain particles.

The lognormal distribution was a better fit than the normal distribution based on visual observation of the relative linearity of the normal and lognormal cumulative probability curves (Figures 5.2 and 5.4). The probit function, which is available with major statistical

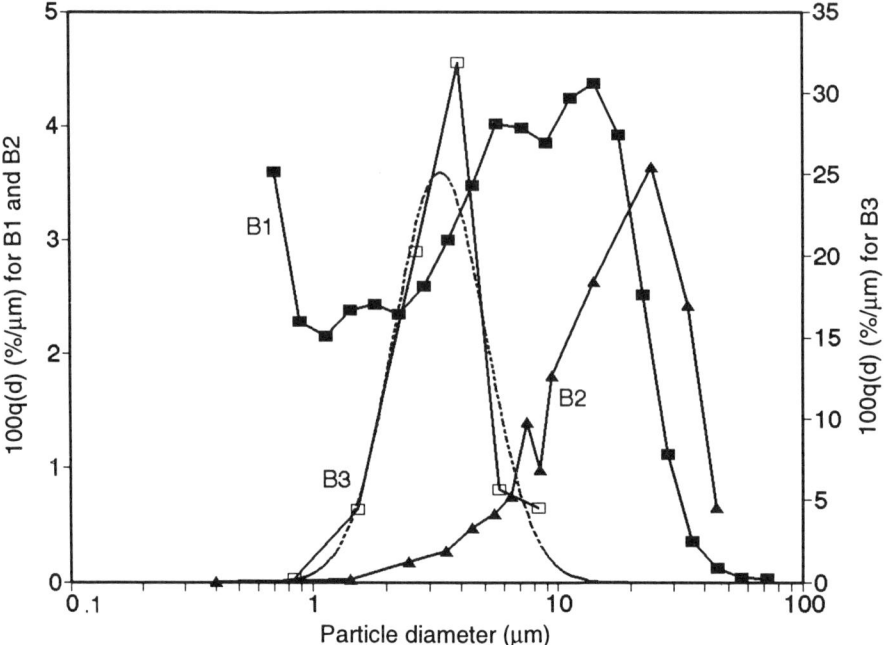

Figure 5.3. Frequency polygons of bioaerosols B1 (■), B2 (▲) and B3 (□) on lognormal axes. The lognormal distribution function (Equation 5.19) using the sample geometric mean and standard deviation is shown with dashed line for B3.

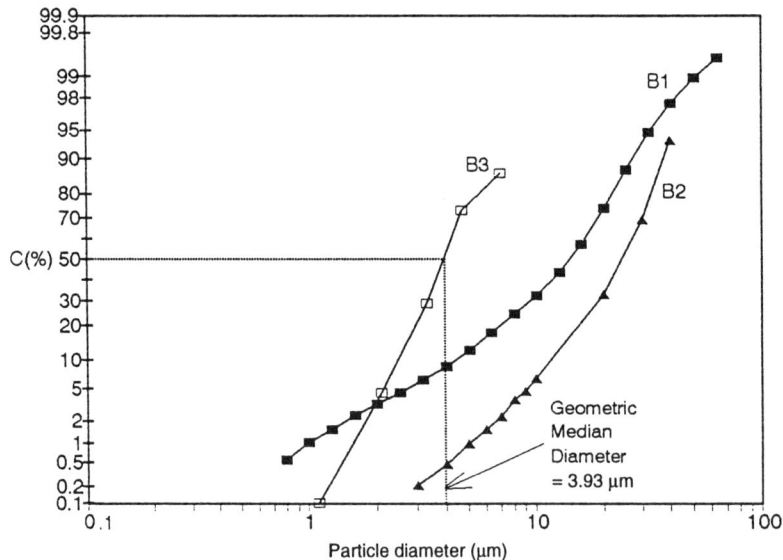

Figure 5.4. Log-probability plots of bioaerosols B1 (■), B2 (▲) and B3 (□). The geometric median diameter is shown for B3.

packages and linearizes the cumulative percentage oversize values is 0.00 at 50%, 1.00 at 84.1%, 2.05 at 98%, etc., and is symmetrical around C=50%, e.g., -2.05 at 2%. Because distribution functions are merely "best fit" approximations to the actual distribution,[9] the usefulness of the other less commonly used distribution functions was investigated.

The Rosin-Rammler distribution gave the very best fit to the data because the value of y=log(log 1/R) plotted against log d produced a nearly straight line on the Rosin-Rammler graph (Figure 5.5). The exponent n was 1.57 based on a linear regression of y vs. log d over all size classes. However, it is not recommended that classes be given equal weight in such a regression analysis to determine n. Rather, the best straight line should be drawn through the central classes of the distribution, because they contain most of the particles. Otherwise, the end classes, which contain large errors because of their smaller contributions, overinfluence the slope of the line.

If a bioaerosol is distributed according to the Gaudin-Schuhman power law function, a log-log plot of C vs (d/d_z) will result in a straight line except at the upper end (C>80%), where the curve will always become asymptotically horizontal. Ignoring those points where C≤80%, a linear regression of B1 resulted in t=1.42 and a correlation coefficient of 0.996 (Figure 5.6).

Bioaerosol B2

A scanning electron microscope was used to photograph dust (Bioaerosol B2) collected on membrane filters from a swine finishing house.[21] The projected area of each particle was compared to a circle of equal area, thus determining the projected area equivalent diameter d_a, which is useful when the area measurement must be converted into a volume.[12] Based on d_a, the particles were grouped in arithmetically progressive size classes, ten with 0.1 μm intervals from 0 to 10 μm and four with 10 μm intervals from 10 to 50 μm (Table 5.4). The end classes had definite boundaries and both number- and volume-weighted distributions appeared relatively complete. The mode of the number distribution was in the class with d_i=2 μm, whereas the mode of the volume distribution was in the class with d_i=20 μm (Figure 5.1).

The arithmetic volume frequency polygon for Bioaerosol B2 is roughly a bell-shaped curve (Figure 5.1) indicating a Gaussian population, whereas the lognormal distribution function shows a skew to the left (Figure 5.3). A second mode appearing between d_a=7 μm and d_a=8 μm indicates a bimodal distribution or a combination of two sources of particles (Figure 5.1). The second source is relatively minor (based on volume), so the combined distribution still may be Gaussian.

The volume moment mean diameter (d_{VP}) is 24.43 μm, whereas the volume weighted geometric mean diameter (d_{3g}) is 22.1 μm, and the standard deviation (σ_{3g}) is 1.62. A straight line occurs for a major part of the cumulative percentage distribution or normal probability graph (Figure 5.2), whereas the curve is extremely nonlinear on the log probability graph (Figure 5.4). The normal probability graphs will usually be curved asymptotically to the end limits and one is always advised to draw the best straight line through the central points of the distribution to characterize the distribution and to determine the constants for the function. The B2 data appear as a straight line on the Rosin-Rammler probability graph (Figure 5.5), but as a curved line on the Gaudin-Schuhman log-log plot (Figure 5.6). The normal distribution best fits the B2 data, followed by the Rosin-Rammler and lognormal distributions.

Bioaerosol B3

Bioaerosol B3 from dairy plant air was sampled with a six-stage Anderson sampler,[22] and colony forming units (CFU) were counted but expressed as a percentage of mass on

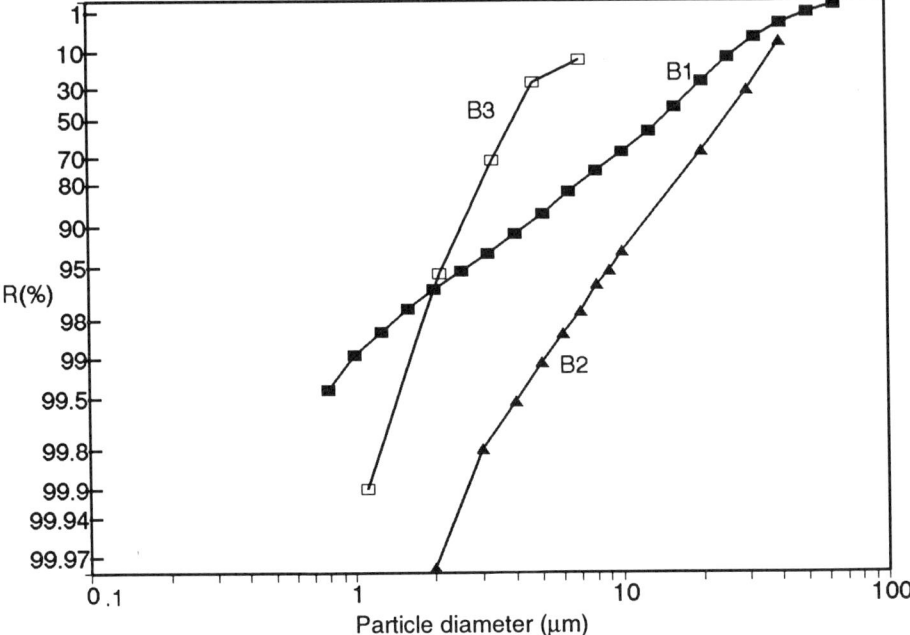

Figure 5.5. Rosin-Rammler distributions of bioaerosols B1 (■), B2 (▲) and B3 (□).

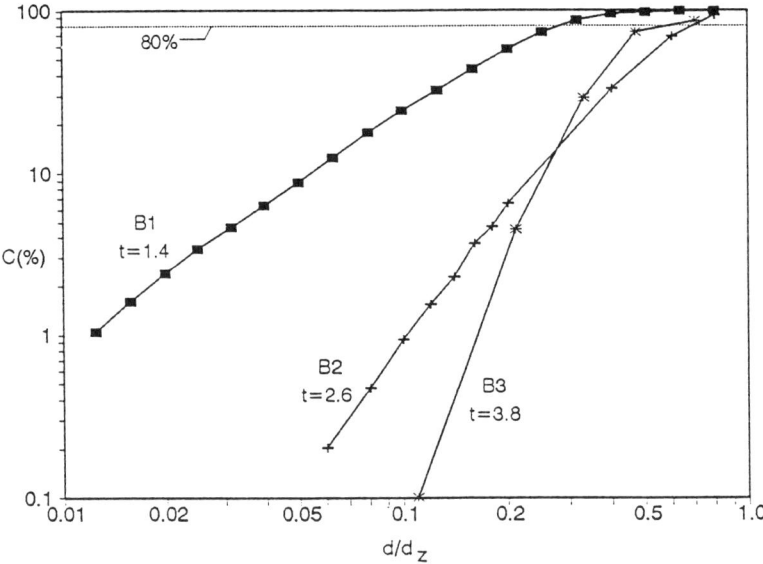

Figure 5.6. Gaudin-Schuhman distributions of bioaerosols B1 (■), B2 (▲) and B3 (□).

each stage. Based on the measurement technique, the aerodynamic diameter is the measure of dispersity and the number of CFUs is the amount A. The CFUs on each stage as a percent of the total are given in Table 5.5. The midpoint diameter was calculated as the geometric mean of the d_{50} values for successive stages of the Anderson sampler.[23] The distribution parameters calculated in Table 5.5 are graphed according to each distribution function (Figures 5.1 to 5.6). The best fit for B3 was the lognormal distribution (Figure 5.4).

Table 5.4. Data B2 from SEM Analysis of Swine House Dust

d_i μm	d_{ig} Eqn. 5.4 μm	N_i	V_i Eqn. 5.2 %	V_i %/μm	C %	χ	R %	y Eqn. 5.22
0	0.40	32	0.00	0.00			100.0	-6.19
1	1.41	120	0.02	0.02	0.00	-4.67	100.0	-3.97
2	2.45	166	0.18	0.18	0.02	-3.48	100.0	-3.06
3	3.46	89	0.27	0.27	0.20	-2.87	99.8	-3.06
4	4.47	73	0.48	0.48	0.47	-2.60	99.5	-2.69
5	5.48	50	0.60	0.60	0.95	-2.35	99.1	-2.38
6	6.48	38	0.75	0.75	1.55	-2.16	98.5	-2.17
7	7.48	46	1.40	1.40	2.30	-2.00	97.7	-2.00
8	8.49	22	0.98	0.98	3.70	-1.79	96.3	-1.79
9	9.49	29	1.80	1.80	4.68	-1.68	95.3	-1.68
10	14.14	128	26.37	2.64	6.49	-1.52	93.5	-1.54
20	24.49	34	36.40	3.64	32.9	-0.44	67.1	-0.76
30	34.64	8	24.22	2.42	69.3	0.50	30.7	-0.29
40	44.72	1	6.52	0.65	93.5	1.51	6.5	0.07
50					100			
Total		836	100.00					

Source: Nilsson, 1982.[21]

Table 5.5. Data B3: Bacterial Colony-Forming Units in Dairy Plant Bioaerosol

d_i μm	d_{ig} Eqn. 5.4 μm	Mass		Colony Forming Units (CFU)					
		M_i Eqn. 5.2 %	M_i %/μm	B_i Eqn. 5.2 %	B_i %/μm	C %	χ	R %	y Eqn. 5.22
0.65	0.85	1	2.2	0.1	0.2			100.0	
1.10	1.52	9	9.0	3.1	4.4	0.1	-3.09	99.9	-3.36
2.10	2.63	22	18.3	16.9	20.3	4.5	-1.70	95.5	-1.70
3.30	3.94	37	26.4	31.1	31.9	28.8	-0.56	71.2	-0.83
4.70	5.74	12	5.2	9.0	5.6	73.4	0.62	26.6	-0.24
7.00	8.37	19	6.3	9.5	4.6	86.4	1.10	13.6	-0.06
10.00									
Total		100		69.7					

Source: Kang and Frank, 1989.[22]

Because the data fit a lognormal distribution, the Hatch-Choate equations[1] enable one to easily calculate the average of the distribution by other combinations of weights and moments as a function of only σ_g and d_g. However, such simplified procedures are not necessary, because the equations for various combinations of weights and moments can be calculated with computers. Additionally, mean diameters can be calculated regardless of whether the data are lognormally distributed or not.

Table 5.6. Basic Statistics of Sample Bioaerosol Data

Data Set	Geometric Mean, μm	Geometric Standard Deviation	Skew	Kurtosis
B1 (Volume)	12.17	2.18	-0.469	1.61
B2 (Volume)	22.10	1.62	-0.106	0.23
B3 (CFU)	3.93	1.54	0.002	14.37

Table 5.7. Number of Particles to Be Counted to Achieve Any Given Accuracy

Expected Accuracy %	Weight % of Particles in Any Size Range				
	2	5	10	15	20
2.0	3	8	25	56	100
1.0	6	25	100	225	400
0.5	16	100	400	900	1,600
0.2	100	625	2,500	5,600	10,000
0.1	400	2,500	10,000	22,500	40,000

Source: Silverman et al., 1971.[25]

Open-Ended Classes

Bioaerosol samples B1 and B3 illustrate the problem of open-endedness, which has frustrated many particle analysts. The largest class of B1 has $d_l=64$ μm but no value for d_u, because this class contained all particles larger than 64 μm. The assignment of a midpoint diameter for the largest class is not straightforward. The particle analysts[19] assigned $d_u=80$ μm in keeping with the geometric progression of the smaller classes and calculated d_{ig} with Equation 5.4. The microscopist[21] who collected B2 established absolute limits for the end classes, $d_l=0$ for the smallest class and $d_u=50$ μm for the largest class.

Statistical Accuracy

The number of particles sampled determines the accuracy of a weighted distribution (surface area, volume, etc.) calculated from the number distribution.[24] The number of particles should be at least 1000 and preferably 10000. All conversions of count frequency distributions to mass or surface area distributions must be scrutinized for their validity. A statistically significant number of size measurements should be made in the upper size range, which constitutes a significant fraction of the weight in the calculated mass-weighted distribution.[25] Guidelines on the number of particles to be counted for any given accuracy are provided in Table 5.7. Equations have also been developed to calculate the number of particles required in an experiment.[26]

SUMMARY

Statistical representation of bioaerosol particles involves the use of size distributions. Particle shape, quantity, and size must be defined and represented for a set of particles. Bioaerosol size distributions, like those of other particles sets, involve particle grouping and the determination of numbers representing the size average and variance of the particles. The normal, lognormal, Rosin-Rammler, and Gaudin-Schuhman power-law distribution functions commonly are applied to represent bioaerosol size. Three example data sets were used to demonstrate these functions.

PRINCIPAL SYMBOLS

a	Particle amount
d	Particle diameter, cm
f(d)	Frequency function, fraction per cm
k	Total number of size classes
m	Mass of one particle, g
n	Number of particles
p	Moment
r	"Weight" of particle amount, which is proportional to d raised to power r (r=0, 1, 2 and 3 for count, length, surface area and volume or mass, respectively)
s	Surface area of one particle, cm^2
t	Exponent for the Gaudin-Schuhman power-law function
v	Volume of a particle, cm^3
y	log log (100/R) in linearized Rosin-Rammler function
A	Total particle amount
B	Biological activity, number of colony forming units
C(d)	Percent of particle quantity with diameters less than d
F(d)	Cumulative distribution function
L	Total particle length, cm
M	Total particle mass, g
N	Total number of particles
P	Moment of mass or volume, cm^4
R(d)	Percent of particle quantity with diameters greater than d
S	Total surface area of particles, cm^2
V	Total volume of particles, cm^3
W	Total particle weight, dynes
α	Shape factor
β	Amount of microbiological organisms
γ	Coefficient of variation
ρ	Particle density, g/cm^3
σ	Standard deviation
χ	Probit from normal probability function

LIST OF SUBSCRIPTS

e	equivalent, substituted by subscripts for definitions in Table 5.1
g	geometric
h	harmonic

i   class or group number
l   lower limit of size channel
m   arithmetic
r   particle weighting
$\bar{r}$ average property
u   upper limit of size channel
z   maximum d in Gaudin-Schuhman power-law function

REFERENCES

1. Hinds, W.C. *Aerosol Technology* (New York, NY: John Wiley and Sons, 1982), pp. 69–100.
2. Luerkens, D.W. *Theory and Applications of Morphological Analysis: Fine Particles and Surfaces* (Boca Raton, FL: CRC Press, Inc., 1991), pp. 1–15.
3. Ehrlich, R., P.J. Brown, and J.M. Yarus. "The Origin of Shape Frequency Distributions and the Relationship Between Size and Shape," *J. Sed. Petrol.* 50:475 (1980).
4. Kaye, B.H. "Fractal Dimension and Signature Waveform Characterization of Fine Particle Shape," *Amer. Lab.* April:55–63 (1986).
5. Knutson, E.O., and P.J. Lioy. "Measurement and Presentation of Aerosol Size Distributions" in *Air Sampling Instruments,* P.J. Lioy and M.J.Y. Lioy, Eds. (Cincinnati, OH: Amer. Conf. of Gov. Ind. Hyg., 1983), pp. G2-G12.
6. Harwood, C.F. "Problems in Particle Sizing: The Effect of Particle Shape," in *Particle Size Analysis,* J.D. Stockham and E.G. Fochtman, Eds. (Ann Arbor, MI: Ann Arbor Science Publishers, Inc., 1977), pp. 111–115.
7. Cadle, R.D. *The Measurement of Airborne Particles* (New York, NY: John Wiley and Sons, 1975), pp. 12–44.
8. Full, W.E., R. Ehrlich, and S. Kennedy. "Optimal Definition of Class Intervals of Histograms or Frequency Plots," in *Particle Characterization in Technology, Volume II: Morphological Analysis,* J.K. Beddow, Ed. (Chelsea, MI: Lewis Publishers, Inc., 1984), pp. 134–146.
9. Allen, T. *Particle Size Measurement* (London: Chapman and Hall, 1968), pp. 16–33.
10. Stockham, J.D. "What is Particle Size: The Relationship among Statistical Diameters," in *Particle Size Analysis,* J.D. Stockham and E.G. Fochtman, Eds. (Ann Arbor, MI: Ann Arbor Science Publishers, Inc., 1977), pp. 1–21.
11. Hamilton, R.J. "The Physics of Particle Size Analysis," in *Proceedings of Conference on the Physics of Particle Size Analysis* (London: British J. Appl. Phys., The Institute of Physics, 1954), pp. 90–95.
12. Rumpf, H. *Particle Technology,* F.A. Bull, Transl. (New York, NY: Chapman and Hall, 1975), pp. 8–29.
13. Kottler, F. "The Goodness of Fit and the Distribution of Particle Sizes, Part I," *J. Franklin Inst.* 251:499–514 (1951).
14. Kaye, B.H. "Fractal Description of Fineparticle Systems," in *Particle Characterization in Technology, Volume I: Applications and Microanalysis,* J.K. Beddow, Ed. (Chelsea, MI: Lewis Publishers, Inc., 1984), pp. 81–100.
15. Wardlaw, A.C. *Practical Statistics for Experimental Biologists* (London: John Wiley and Sons, Ltd., 1985), pp. 107–110.
16. Zar, J.H. *Biostatistical Analysis* (Englewood Cliffs, NJ: Prentice-Hall, 1984), pp. 88–96.
17. Kottler, F. "The Goodness of Fit and the Distribution of Particle Sizes, Part II," *J. Franklin Inst.* 251:617–641 (1951).
18. Ropp, R.C. "Display of Particle Size Distributions," *Amer. Lab.* July:76–83 (1985).

19. Heber, A.J., J.R. Dawson, V.A. Battams, and R.A.C. Nicol. "Effect of Surface Roughness and Angle on Indoor Dust Deposition," in *Proceedings of the 5th Annual Conference of the Aerosol Society*, Loughborough, U.K. (1991), pp. 41–46.
20. Cox, C.S. *The Aerobiological Pathway of Microorganisms* (New York, NY: John Wiley and Sons, 1987), pp. 88–107.
21. Nilsson, C. "Dust Investigations in Pig Houses," Swedish University of Agricultural Sciences, Report 25 (1982), pp. 1–9.
22. Kang, Y.J., and J.F. Frank. "Evaluation of Air Samplers for Recovery of Biological Aerosols," *J. Food Prod.* 52(9):655–659 (1989).
23. Hesketh, H.E. *Fine Particles in Gaseous Media* (Chelsea, MI: Lewis Publishers, Inc., 1986), pp. 1–22.
24. Willeke, K., and P.A. Baron. "Sampling and Interpretation Errors in Aerosol Monitoring," *Amer. Ind. Hyg. Assoc. J.* 51(3):160–168 (1990).
25. Silverman, L., C.E. Billings, and M.W. First. *Particle Size Analysis in Industrial Hygiene* (New York, NY: Academic Press, 1971), pp. 246–259.
26. Masuda, H., and K. Iinoya. "Theoretical Study of the Scatter of Experimental Data due to Particle-Size Distribution," *J. Chem. Eng. Japan* 4(1):60–66 (1971).

CHAPTER 6

Stability of Airborne Microbes and Allergens

C.S. Cox

INTRODUCTION

Bioaerosols can travel considerable distances owing to atmospheric dispersion; e.g., there are marked deposits of pollens at the Earth's poles and plants can become infected by *Puccinia graminis f.sp. tritici* when strong airflows carry stem rust disease from the Mississippi Valley where it is endemic to the central and northern regions of Canada. Even so, the ability of bioaerosols to initiate disease depends on other parameters, particularly the ability of microbes to survive and remain infective in susceptible hosts. But for many microbes and some allergens the airborne environment is hostile owing to desiccation, exposure to radiation, oxygen and pollutants.

Consequently, an important aspect of understanding how bioaerosols can be infective and/or allergenic, is untangling the numerous physical, chemical and biochemical factors known to influence these properties. This chapter tries to provide explanations for associated phenomena with emphasis on more major developments.

SURVIVAL AND INFECTIVITY

Following take-off, aerial transport and landing, the ability of infectious microbes to initiate and spread disease depends on how well they survive (ability to replicate) and maintain infectivity (ability to cause infection). Ability to survive is a pre-requisite for infectivity, whereas infectivity *per se* involving additional attributes can be lost more rapidly (Chapter 13).

Microbes and allergens following bioaerosol generation from liquid suspension (e.g., saliva) undergo desiccation, whereas those generated as dusts or powders partially rehydrate. A changed water content therefore occurs for all bioaerosol particles and represents the most fundamental potential stress. Other potential stresses including exposure to radiation, oxygen and pollutants depend on circumstances.

Most Probable Target Molecules

As explained later, various "building blocks" of microbes and allergens are not equally stable thermodynamically, with membranes being least stable and nucleic acids being considerably more so. Consequently, most probable target molecules may be related to energies associated with various stress factors. For example, desiccation *per se* involves less energy than does irradiation with UV, and therefore most probably will be more damaging for least stable molecules rather than for most stable molecules. In contrast, energetic UV radiation may damage both these moieties.

On this basis a summary of the most probable target molecules may be related to the various stress factors (Table 6.1), and *ipso facto* effects of relative humidity and temperature will be described first.

RELATIVE HUMIDITY AND TEMPERATURE

Many biological materials are hygroscopic and demonstrate hysteresis in water sorption isotherms. Therefore, most microbes (containing nucleic acids, proteins, and carbohydrates) like hygroscopic allergens (e.g., proteins) have water sorption isotherms similar to that shown in Figure 6.1 for *Serratia marcescens*.[1]

Following bioaerosol generation, rates of water transfer (dehydration and rehydration) depend on the prevailing relative humidity, but owing to hysteresis, equilibrium water contents depend on direction of water flow (i.e. into or out of the particle). Storage at a given RH therefore results, in practice, in higher equilibrium water contents (g water/g biomaterial) for dehydration than for rehydration.

Coliphages

T3 coliphage bioaerosols when wet-disseminated demonstrate best survival when stored above ca. 75% RH, whereas ability to survive declines progressively at lower values (Figure 6.2). However, on exposure to warm air saturated with water vapor before sampling, very much higher viabilities are observed.[2-4]

On exposure at less than 75% RH, the extent of desiccation is sufficient to cause bonds holding phage head and tail together to become weak and/or brittle. So, on collection of bioaerosols of T3 coliphage with an impinger (AGI-30 in this case), the high shear forces break the head-tail coliphage complex thereby separating head and tail. Since coliphage relies on its tail for injecting head nucleic acid into the host bacterium, those T3 without tails are unable to replicate to form plaques in bioassays. Employing vapor phase rehydration (rehumidification) before collection replaces water molecules lost on dehydration following bioaerosol generation, and the original strength and/or flexibility of head-tail complexes is restored sufficiently to withstand the shear stress imposed during subsequent collection by AGI-30 impinger.[3] Such phenomena occur also for lipid-free coliphage T7 and for *Francisella pestis* phage.[3,4]

The observed weakness of phage head-tail bonding may be related to their morphological mismatch.[5] Phages such as T7 have a connector protein (gp 8, MW 59,000) residing between head and tail. The attachment point of the latter to the former (capsid) occurs at one of the pointed vertices of an icosahedron of 5-fold rotational symmetry, whereas the connector protein has 12-fold rotational symmetry compatible with the 6-fold rotational symmetry of the tail. The presence of connector protein on isolated tails of T7 coliphage prepared by osmotic shock has been demonstrated.

This example is important for several reasons. For instance, it demonstrates how sampling, rather than aerosol generation or storage, can be responsible for observed losses of viability and infectivity, and therefore can be a source of serious experimental artifact. Another facet is the relative fragility of surface structures that is enhanced following desiccation. Consequently, a given biomaterial as a result of becoming airborne and landing may demonstrate different allergenicity from the native starting material.

A different aspect of coliphage fragility is the provision of an experimental basis for characterizing bioaerosol samplers in terms of stresses applied during operation (Chapters

Table 6.1. **Summary of Most Probable Target Molecules**

Stress	Most Probable Target
RH and temperature	Membranes phospholipids, proteins
Oxygen	Phospholipids, proteins
Ozone	Phospholipids, proteins
Open Air Factor (Ozone+olefins)	Phospholipids, proteins, nucleic acids
γ-rays, X-rays, UV	Phospholipids, proteins, nucleic acids

9 and 13). Even though it had been demonstrated in the 1950s that in operation various samplers killed vegetative microbes to different degrees, a reliable standardized test of this factor is lacking. It has been proposed[6] that such sampler efficacy possibly could be quantified on the basis of airborne desiccated T7 survival following sampling with and without rehumidification. The two derived survival values together with their ratio could provide some comparisons of samplers as well as indicating how they may perform in previously untested applications.

Viruses

Work during the late 1960s and early 1970s[7] demonstrated that viruses without structural lipids, such as mengovirus 37A, polio virus, foot-and-mouth disease (FMD) virus and encephalomyocarditis (EMC) virus are unstable as bioaerosols when wet-disseminated and stored in atmospheres below about 70% RH because of their denatured surface structures. The approach compared infectivity of whole virus with that of isolated infectious RNA[8] (Figure 6.3). For EMC virus there was evidence of a concomitant loss of hemagglutination activity and of affinity for hemagglutination inhibiting antibodies while viral RNA activity remained unimpaired.

Whether surface damage was caused by the sampling process as for the phages described above is uncertain, but for EMC virus experiments a relatively gentle subsonic sampler (bottom stage of a May 3-stage all-glass impinger) was employed. Also, rehumidification did not enhance survival of EMC virus.

In contrast to viruses without structural lipids and showing best survival at above ca. 70% RH, several viruses with structural lipids are least stable above ca. 70% RH,[7] e.g., Langat and Semliki Forest viruses, vesicular stomatitis virus, vaccinia and Venezuelan equine encephalomyelitis virus and influenza virus. Even so, at least for Semliki Forest virus bioaerosols, loss of infectivity still arises through damage to viral coat rather than nucleic acid. But, rehumidification seems to be without effect or even harmful.

DNA-containing pigeon pox virus and Simian virus 40 as bioaerosols at ambient temperatures maintain their infectivity at very high levels, independently of RH.[7] In view of the cause of loss of infectivity of RNA viruses, the effect would seem due to differences in viral coats rather than nucleic acid type.

The ability of airborne viruses to remain infective can be greatly affected by composition of suspending fluid prior to dehydration. Degree of solute hydrophobicity, and nature and concentration of salts, can be important, commensurate with viral coat being the primary target for desiccation damage.[7]

As discussed in Reference 9 and described in more detail later in this chapter, phospholipid-protein complexes usually denature most readily at mid to high RH, whereas proteins denature most readily at low RH. Consequently, the above observations fit well with the

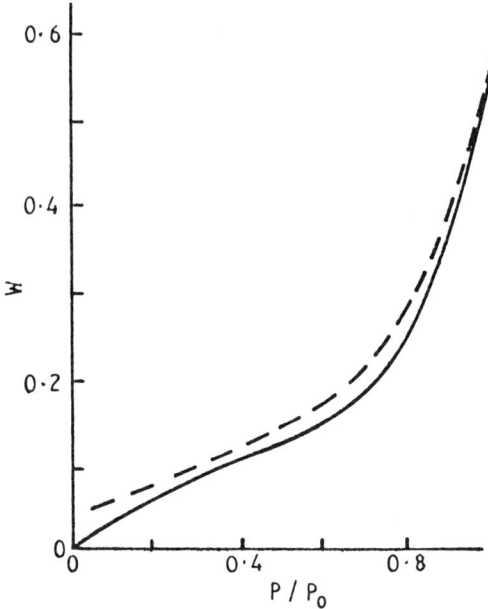

Figure 6.1. Water sorption isotherm for *Serratia marcescens*. W = g of water per g of bacteria; P/P_0 = water activity = %RH/100; - - - -, desorption; ———, absorption.

general pattern for stability at ambient temperatures of dehydrated phospholipid-protein complexes and of proteins. However, there is a particular need for experimental data on effects of temperature on virus infectivity, especially viruses having structural lipids. This is because loss of biological activity of phospholipid-protein complexes, e.g., membranes, depends on temperature as well as RH. It seems possible that at temperatures in the range of about -5 to +10°C virus, survival may be greatly enhanced compared to higher temperatures, and little affected by RH. If so, current data with regard to the airborne transmission of viruses may be misleading.

Usually there is a higher incidence of influenza (a lipid virus) during winter compared to summer both in northern and southern hemispheres. This observation, like that of higher incidence of polio (a lipid-free virus) during summer, may become more understandable in terms of prevailing temperatures being as, or more, important than RH in airborne transmission of viral diseases.

Bacteria

Being more complex biochemically, structurally, and in their organization, bacteria may be anticipated to have more involved mechanisms of inactivation than do phages and viruses. Consequently, conflicts of opinion and many apparent contradictions of results arising during earlier stages of aerobiological research perhaps were unavoidable.

Until publication of Hess's classic paper[10] in 1965 there was little real understanding of the aerosol survival of bacteria. We now know part of the difficulty arose because, unlike phages and viruses (so far tested), some (but not all) Gram-negative bacterial species on desiccation are inactivated by oxygen. Exacerbating this problem was collecting undamaged samples of bioaerosols, and roles in this of hypertonic collecting fluids. Perhaps (now) not surprisingly, the situation led to publication of many apparently conflicting results.

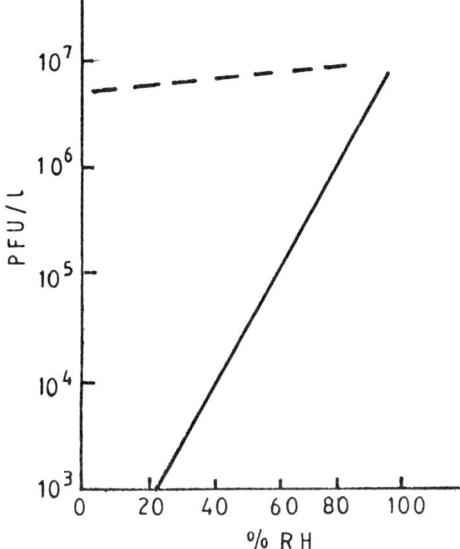

Figure 6.2. Aerosol survival of coliphage T3 in air collected by impinger (AGI-30). ———, direct sampling; - - - - - rehumidified before sampling.

Hess's contribution was to work with a single species, *Serratia marcescens*, and establish beyond reasonable doubt that its bioaerosol viability depended on RH, oxygen concentration and time. A 'benchmark' thereby was established and during ensuing years it became accepted that some Gram-negative bacterial species are inactivated by oxygen, e.g., *Escherichia coli*, *Klebsiella pneumoniae*, whereas others such as *Francisella tularensis* are little affected.

In turn, once it was established that oxygen can be toxic for some species, it was evident that bacterial bioaerosols stored in air may suffer simultaneous action of at least two entirely separate death mechanisms, e.g.,

(a) loss of viability induced by desiccation,
(b) loss of viability induced by oxygen toxicity.

Such arguments then led to many survival studies of microbial bioaerosols in highly purified inert atmospheres of nitrogen, argon and helium. Thence, effects of desiccation *per se* could be established.

In the late 1960s systematic studies[7] were begun of effects of RH and time on the survival of bioaerosols stored in inert atmospheres for several *E. coli* strains, for *F. tularensis* and for *K. pneumoniae*. A somewhat surprising pattern then emerged. Previously common practice had been to measure bioaerosol survival as a function of time at only three RH values corresponding to high (80%), mid (50%) and low (30%) RH, as general experience had indicated that relationships between survival and RH were smooth functions. In contrast, these studies in inert atmospheres were made over the RH range 20–100% in small RH increments, sometimes as little as 2%—an experimentally difficult task. Then the new and somewhat surprising pattern emerged that there are critical narrow RH bands where loss of viability of *E. coli* strains is much greater than that for adjacent regions (Figure 6.4). Also, in inert atmospheres *E. coli* strains were most stable at low RH rather than high RH as observed previously from experiments conducted in air. In the past,

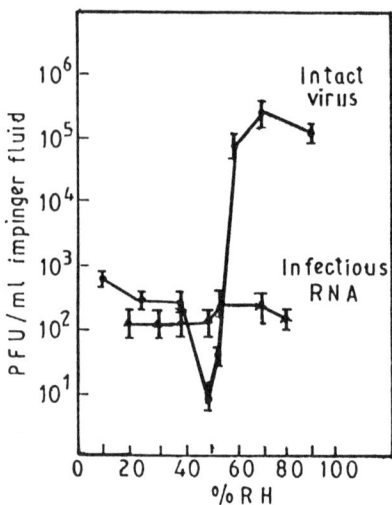

Figure 6.3. Aerosol survival in air of mengovirus.

poor survival of some bacterial strains at low RH had been attributed mistakenly to desiccation rather than toxic action of oxygen.

Another effect was that these bacteria could suffer a collection stress related to the nature of suspension fluid prior to bioaerosol wet-dissemination. Microbes have been sprayed from, or freeze-dried as, suspensions in solutions of all main classes of water-soluble or water-compatible materials including sugars, proteins, vitamins, dyes, feces, saliva, etc. Reasons are numerous, but one is that systematic variation of suspending fluid composition may lead to understanding behavior of microbes following dehydration.

Survival of coli bioaerosols when the fluid was triple-distilled water varies little with different collection methods. When sprayed from a raffinose solution collection into 1M sucrose enhances survival[7] as can rehumidification prior to bioaerosol collection by AGI-30 impinger.[7,9] The latter also is beneficial for *Klebsiella pneumoniae* (compare Figures 6.5 and 6.6), but usually, beneficial effects of rehumidification and 1M-sucrose collecting fluid are not additive because the two techniques probably represent two different ways of achieving a similar outcome.

Such behavior applies for other sugar-like compounds in spray fluids (and for *K. pneumoniae*) but it is not known how generally this occurs because few bacterial species have been studied in such detail. On the other hand, anhydrobiotic organisms, generally, synthesize large quantities of trehalose, etc.

Another facet of desiccation damage is survival of bacteria wet-disseminated compared to that dry-disseminated as freeze-dried powders. On the proposition that equilibrium water content controls survival, it could be argued that after allowing for hysteresis in water sorption isotherms, the same survival should be observed at a given RH irrespective of whether wet or dry dissemination was used. This is because at equilibrium both occasions should lead to nearly the same final water content.

Comparisons of coli survival data show that in terms of trends (i.e. maximum stability at low RH, marked instability at 80–85% RH) both wet and dry dissemination can lead to similar results, but in finer detail results differ. In the case of *F. tularensis* LVS, though, there are marked differences in bioaerosol survival depending on whether wet or dry dissemination is employed (Figure 6.7). Here the atmosphere is air, but as already pointed out oxygen toxicity is minimal for this bacterium. That the freeze-drying process *per se*

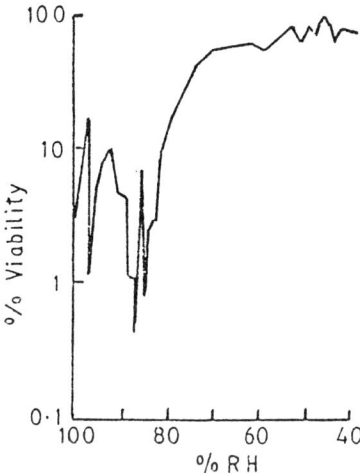

Figure 6.4. Aerosol survival of *Escherichia coli B* sprayed from suspension in distilled water in a nitrogen atmosphere.

is responsible, is excluded because when freeze-dried powders are reconstituted with water then sprayed, the observed patterns are as if freeze-drying had been omitted. Hysteresis in water sorption isotherm is an unlikely cause because of extent of differences, and for coli strains operates in the wrong 'direction.' Other possible causes are discussed later.

Other Microbes

Much less work seems to be concerned with causes of viability/infectivity loss of bioaerosols of psittacosis, chlamydia, rickettsia, pollens and fungi, etc. But, such features are studied in other environments.[11,12] Airborne mycoplasmas, bacterial L-forms and algae have received only slight attention, while spores of fungi and of actinomycetes, etc., are considered resistant to desiccation.

Mycoplasma pneumoniae and *M. gallisepticum* do not have cell walls and seem most unstable as bioaerosols in air in the range 40–60% RH, while the former also demonstrates a narrow range of instability at ca. 80% RH. Such behavior is similar to that for coli strains. Airborne streptococcal L-form (i.e. a Gram-positive bacterium lacking a cell wall) also is least stable at about 40% RH and most stable at ca. 20% RH. Algal species *Nannochloris atomus* and *Synechococcus sp.* (R3) are most stable at RH values above ca. 92% but below which no viable airborne cells of the former could be detected, whereas the latter was most unstable at ca. 80% RH.

These examples[7] show a somewhat analogous pattern to that for other microorganisms described above, but unfortunately there seems no information as to how survival patterns of these other microorganisms are affected by oxygen.

Allergens

Many materials, like bioaerosols, can cause a hypersensitive or allergic response in immune systems, and they include pollens, vegetable matter, house and paper dust mite

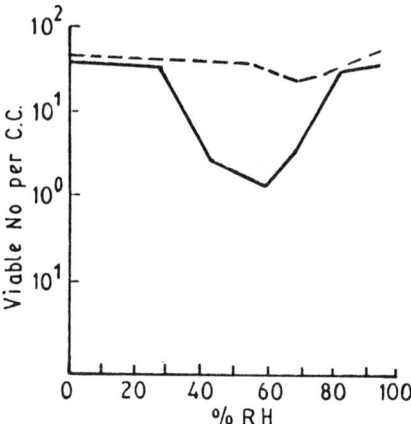

Figure 6.5. Aerosol survival of *Klebsiella pneumoniae* in a nitrogen atmosphere. ———, collection directly into phosphate buffer; - - - - -, rehumidified before sampling.

parts and excreta, insect parts and excreta, proteins and a wide variety of microorganisms, e.g., moldy hay actinomycetes, *Aspergillus niger, A. fumigatus, Penicillium,* etc. Immune systems respond whether such allergens are 'alive' or 'dead.' Consequently, 'the viability' of such materials is less important perhaps than the presence of particular molecular species. On the other hand, as indicated by viability studies described above, the precise nature of a molecular species can be modified by the airborne state, particularly through desiccation, and free radical reactions.

For example, a globular protein following desiccation can lose helical structure and thereby expose groupings previously hidden in molecular interstices, with concomitant generation of new antigenic sites.[13] More subtle perhaps are changes in surface structures of microbes on dehydration, as well as the ensuing enhanced fragility. In addition, osmotic lysis on collection can occur. Consequently, collected samples may have allergenic properties different to parent materials before aerosolization, as well as to materials as bioaerosols.

Furthermore, bioaerosol sampling may become more difficult when the concern is to preserve allergenicity as opposed to viability, because microorganisms to some extent can repair inflicted damage to their membranes, enzymes and nucleic acids-moieties perhaps more related to growth than to allergenicity.

OXYGEN

Oxygen toxicity for vegetative bacterial bioaerosols did not become widely accepted until after the work of Hess,[10] for the reasons given above, even though such toxicity was well known for hydrated anaerobes and had been established for freeze-dried materials in the 1960s. As far as is known there are no reports for oxygen toxicity for phages and viruses, whereas certain dehydrated pollen membranes can become inactivated,[14] as can the Ca-transporting activity of synthetic vesicles.[15]

Survival of airborne (and freeze-dried) *Serratia marcescens* 8UK, as for *E. coli B* and *Klebsiella pneumoniae,* depends on storage time, oxygen concentration and RH[7] (see also

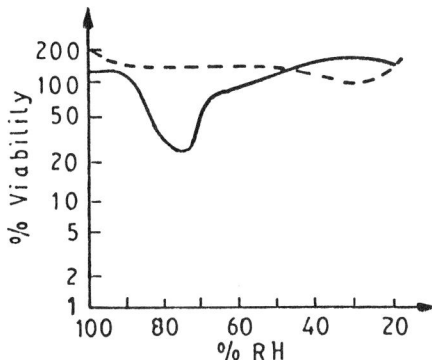

Figure 6.6. Aerosol survival of *Escherichia coli* commune in a nitrogen atmosphere sprayed from suspension in 0.13 M raffinose. ——— collection directly into phosphate buffer; -------- 1 M-sucrose + phosphate buffer.

Oxygen and Open Air Factor Models section). Oxygen toxicity is observed only at RHs below ca. 70%, and increases with oxygen concentration up to ca. 30% while higher concentrations produce no additional toxicity, i.e. the toxic effect 'saturates.' For dry disseminated *E. coli* B stored at low RH, oxygen likewise is toxic but when stored at high RH oxygen enhances survival. *Francisella tularensis* LVS similarly demonstrates slight oxygen protection at high RH when sprayed from suspension in spent culture fluid. In the latter case, spent culture medium when oxidized may have an enhanced stabilizing action as it may for dry-disseminated *E. coli* B at high RH. Effects of oxygen on other airborne microbes have been little studied.

OPEN AIR FACTOR AND OTHER POLLUTANTS

Pollutants such as SO_x and NO_x generally have less effect on bioaerosols than does ozone while toxicity of ozone is greatly enhanced following its reaction with atmospheric olefins present mainly as unburnt fuels. Ozone-olefin reaction products (probably at p.p.h.m. levels) are referred to as the 'open air factor' or OAF. The name arose because at the time of its discovery in the late 1960s it was ephemeral and of unknown nature. Now its high toxicity is considered due to ready condensation onto surfaces rather than especial chemical toxicity *per se*.[16] But, this feature makes it difficult to maintain OAF concentrations during transit in tubes and pipes or in vessels, unless replenished at an appropriate rate, hence the term 'open air factor.'

In practice, outdoor bioaerosol survival of many species of microorganism generally is much poorer than in inside air under otherwise similar conditions. Owing to OAF's oxidizing nature, susceptible allergen bioaerosols also may demonstrate analogous chemical modification. One effect is that airborne spread of diseases generally is more likely indoors than outdoors from this aspect alone, especially when ventilation is poor.

Sites of OAF action seem little studied except for *E. coli* and phage φx174, in which nucleic acids and coat proteins are seriously affected possibly through free radical reactions, perhaps $OH^·$.[7,9,16,17] In contrast, other microorganisms including spores, *Micrococcus radiodurans*, foot-and-mouth disease virus (FMDV) and swine vesicular disease virus are relatively resistant to OAF (and UV).[7]

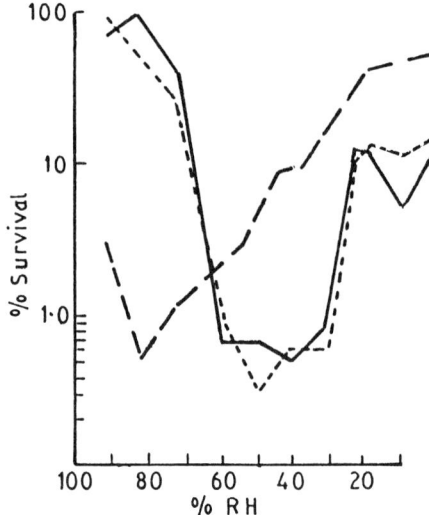

Figure 6.7. Aerosol survival of *Francisella tularensis* LVS in air disseminated from the ────── wet state; - - - - dry state; .... wet state, suspension prepared by reconstituting the freeze-dried powder with distilled water.

RADIATION

Radiation effects tend to follow those observed for entities in aqueous environments or when exposed on surfaces, but are exacerbated by dehydration and oxygen. More energetic radiation (gamma-rays, X-rays, UV) induces free-radical mediated reactions thereby damaging nucleic acids, proteins, sugars, lipids and membranes, including formation of thymine-thymine dimers, DNA strand breaks, protein-DNA and protein-protein cross-links, as well as fragmentation and polymerizing Maillard reactions. Highly energetic radiation causes breakage of covalent bonds directly.

One specific target is the cytoplasmic membrane because it contains the porphyrin components of the respiratory chain, absorbing strongly in the UV. Enhanced mutation frequency arises to an extent dependent on repair capability, while generally spores are more resistant than vegetative microorganisms—a feature attributed to their specialized spore coat and coloring.

Long wave radiation (infra-red, microwave), having much lower energies, is considered to be limited mainly to inducing temperature/dehydration effects.

MATHEMATICAL MODELS

Dehydration and Temperature Model

By the mid-1970s reliable data sets for effects of dehydration on bioaerosols of *E. coli*, *F. tularensis*, *K. pneumoniae*, *Flavobacterium*, Semliki Forest virus and Venezuelan equine encephalitis virus had been established. Mathematical models then were constructed[7] in an attempt to account for the experimental observations. Using a combination of probability theory and first order denaturation kinetics, a mathematical equation was derived that described dehydration effects.

Its basis is that microorganisms contain a key moiety, $B(nH_2O)$, the biological activity of which is essential for microbial cells to replicate, i.e. be viable. This key moiety when exposed to environments of lowered water activity (RH) forms a series of hydrates with some spontaneously denaturing through a first order process, e.g.,

$$-dx/dt = kx$$

where x is the concentration of the species that denatures.
Hence the model is

$$B(n)H_2O \rightleftharpoons B(n-x)H_2O + xH_2O \rightleftharpoons B(n-x-y)H_2O + yH_2O \quad (6.1)$$
$$\downarrow \qquad\qquad\qquad \downarrow$$
$$\text{denatured} \qquad\qquad \text{denatured}$$
$$\text{form} \qquad\qquad\quad \text{form}$$

and this situation can lead to the following relationship,

$$\ln V = K[B(n-x)H_2O]_0 (\exp(-kt) - 1) + \ln 100 \quad (6.2)$$

where $K[B(n-x)H_2O]_0$ = constant
k = first order denaturation constant
t = time
V = % viability at time t.

which was found to fit several hundred viability-time curves extremely well, two examples being given in Figures 6.8 and 6.9.

In 1987 a slightly different approach involving catastrophe theory (in place of probability theory) and denaturation kinetics yielded the following first order denaturation equation,[7]

$$V/100 = [K([B] - [B]_{min})]^{1/2} \quad (6.3)$$

which may be compared with equation 6.2 rewritten as,

$$V/100 = \exp[K([B] - [B]_0)] \quad (6.4)$$

Their similarity is quite marked at viable fractions close to unity but they differ fundamentally at fractions close to zero. The catastrophe approach indicates that viable fractions may be zero whereas the probability approach indicates a finite lower limit equal to $-K[B]_0$, and while this limit may closely approach zero it never can be exactly zero, thus potentially posing a difficulty.

The corresponding second order dimerization is

$$B + B \rightarrow 2B$$

with the dimer being the inactive form of the key moiety, whence,

$$V/100 = [(K[B]_0/(1 + K[B]_0 kt)) - K[B]_{min}]^{1/2} \quad (6.5)$$

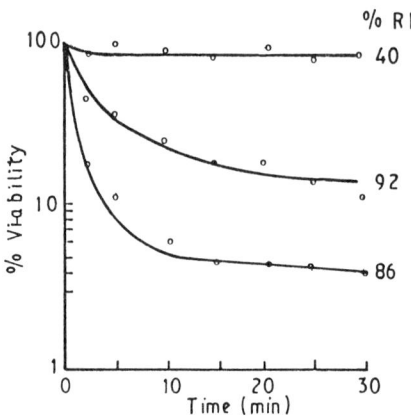

Figure 6.8. Aerosol survival of *Escherichia coli* Jepp in a nitrogen atmosphere. Points signify experimental data; lines calculated using probability theory.

Generally, equation 6.3 or its second order equivalent equation 6.5, fits desiccation damage data sets better than equation 6.2, but not always (Figure 6.10). Even so, all these analyses indicated that species B if it were to exist should do so on the outermost surface of coli cells.[7,9]

Availability of these equations then permitted analyses of temperature effects thereby providing an activation energy for desiccation damage. In turn, such values provide important indicators of possible underlying mechanisms, as well as providing clues as to possible natures of species B.

Ehrlich and co-workers[7] provide the few data sets appropriate for analyzing temperature effects on microbial bioaerosol viability, and in the example of Figure 6.11 lines are calculated using equation 6.5. Natural logarithms of derived values of the constants $K[B]_0$ and k (the second order denaturation constant) are plotted against reciprocal absolute temperature for *Flavobacterium sp.* in Figure 6.12 and for *F. tularensis* in Figure 6.13. Further details of this analysis are to be found in references (7) and (9).

One conclusion is that *Flavobacterium sp.* has optimum aerosol survival at about +8°C, compared to +11°C for *F. tularensis*. At higher temperatures the activation energy is positive, respectively, +28.9 and +43.1 kcal/mole, whereas below the optimum the activation energy is negative, respectively, -21.0 and -22.3 kcal/mole. (Corresponding values for VEE virus are +17.6 kcal/mole, 0 kcal/mole and +24°C.) These findings suggest that two different mechanisms cause desiccation damage, one occurring predominantly above optimum temperature and the other predominantly below it. Also, the optimum survival/infectivity temperature corresponds with gel ⇌ liquid-crystalline phospholipid phase transition temperature that in turn is dependent upon microbial fatty acid complement, as explained in more detail later in the chapter and in references (7) and (9).

Derived activation energies of +28.9, +43.1 and +17.6 kcal/mole are lower than those usually observed for heat denaturation of proteins, e.g., +100 kcal/mole. On the other hand, they compare with those for enzyme catalyzed reactions, and for non-enzymatic Maillard reactions (amino-carbonyl).

Other experimental findings seem consistent with Maillard reactions being involved in desiccation damage at temperatures greater than optimum. However, their full description is not possible here because they represent a subject in their own right with thousands of scientific papers devoted to them. They seem ubiquitous in nature and involved in myriad

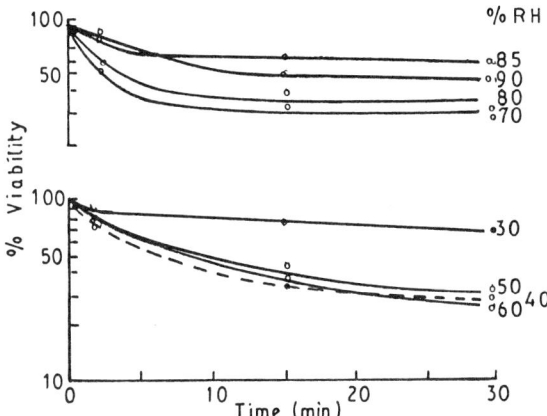

Figure 6.9. Aerosol survival of Semliki Forest virus in air; points are experimental data; lines calculated using probability theory.

reactions, many associated with life and disease processes, as discussed[7,9,13] and in the next main section ("Maillard Reactions").

Oxygen and Open Air Factor Models

Data sets for effects of RH, oxygen concentration and time on bioaerosol survival of *Serratia marcescens* 8UK are available.[7] A model[7,9,13] that fits their data extremely well is based on an analogous approach to that for desiccation damage. A key moiety A, essential for viability, forms a series of hydrates,

$$A(n)H_2O \rightleftharpoons A(n-x)H_2O + xH_2O \rightleftharpoons A(n-y)H_2O + yH_2O \rightleftharpoons A + iH_2O$$

A carrier X is postulated to combine reversibly with oxygen,

$$X + O_2 \rightleftharpoons XO_2$$

and XO_2 can react with each hydrate without forming a tertiary complex $A.XO_2$ to give oxidized species AO_2 and free carrier X. Integration of corresponding differential equations and use of probability theory leads to an equation that in simplified form is

$$\ln V = K[A]_0\{\exp(-k[X]_0[O_2]t/(K_x + [O_2])) - 1\} + \ln 100. \tag{6.6}$$

Figure 6.14 provides examples of results for *S. marcescens* and the degree to which the full equation fits experimental data. These analyses indicated that if species A were to exist, it should reside within the bacterial cell, unlike species B, which should reside on the outermost surface of coli cells.[7] The possible involvement of free radicals in oxygen toxicity has been discussed.[7,9,13]

The model for OAF activity is similar to that for oxygen except that the equilibrium reaction between the carrier and OAF is established more slowly.[7] Effects are to cause a lag before onset of viability loss, as often observed in practice. Because of the ephemeral

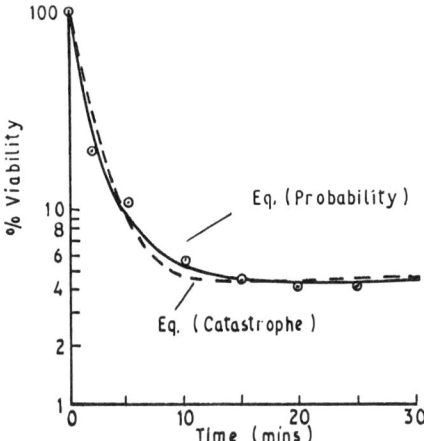

Figure 6.10. Comparison of viability-time curves for *Escherichia coli* in a nitrogen atmosphere. Points experimental data, lines calculated.

nature and relatively low concentrations (p.p.h.m.) of OAF in the atmosphere, there is no known analytical technique for directly measuring OAF concentrations. Instead, its highly variable concentration is estimated from standard viable decay curves obtained using the hardy strain *Escherichia coli* commune and measured at the same time and location as the test organism. These standard coli decay curves then may be converted for comparative purposes into arbitrary units (approximately equivalent to p.p.h.m.) by comparison with published decay curves calculated using the derived equation[7] for OAF action (Figure 6.15). This technique is described in additional detail.[7,17] It proved important in the 'identification' of the OAF as, predominantly, reaction products of ozone and olefins derived from unburnt hydrocarbon fuels.

MAILLARD REACTIONS

As described in more detail in reference (13), Louis-Carmille Maillard between 1912 and 1917 originally described these reactions as amino-carbonyl reactions of reducing sugars with amino-groups of proteins. Their condensation with elimination of a water molecule *reversibly* forms a Schiff's base subsequently undergoing an Amadori rearrangement followed by numerous other rearrangements, fragmentations and subsequent reactions to give a wide range of products. Nowadays, the term Maillard reaction has broadened to include many other carbonyl and nitrogen containing compounds such as sugar fragments and peroxides on the one hand and nucleic acid bases and N-containing vitamins on the other.

Maillard reactions are non-enzymatic and can be of low activation energy comparable to that of a strong H-bond, e.g., 6–10 kcal/mole, and consequently are ubiquitous. They are important both *in vitro* and *in vivo,* being involved in some stages of metabolism, e.g., binding of reducing sugars to lysine residues of enzymes thereby causing concomitant changes in enzyme folding, structure and activity. While their early stages are comparatively simple and understood, later stages are not, with reaction of one amino-acid with one reducing sugar leading to as many as 300 identified reaction products.

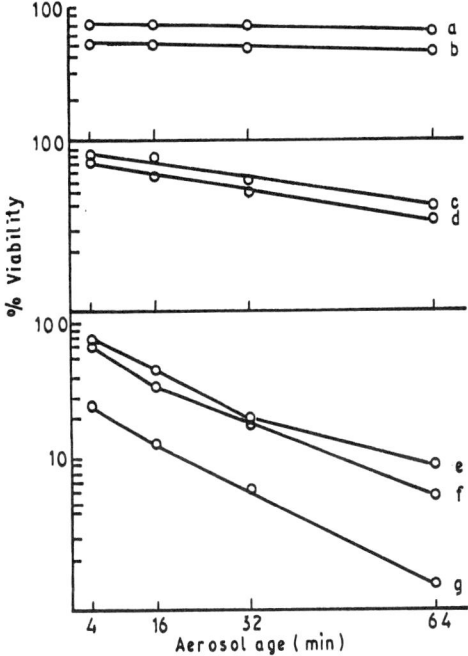

Figure 6.11. Effects of temperature on the aerosol survival of *Flavobacterium sp.* in air. Points are experimental data, lines are calculated using catastrophe theory applied to a second order denaturation process. a = -18; b = -40; c = 24; d = -2; e = 29; f = 38; g = 49°C.

Typically a Maillard reaction scheme is,

$$R_1R_2\text{-CHO} + H_2N\text{-}R_3 \rightleftharpoons R_1R_2C=N\text{-}R_3 + H_2O \rightarrow \text{products} \quad (6.7)$$
$$\text{free reactants} \qquad \text{Schiff's base}$$

with the Schiff's base undergoing *slow* Amadori rearrangement and subsequent reactions.

The initial forward reaction involving elimination of a water molecule is enhanced therefore by desiccation and usually is optimum at 60–80% RH. Slowing below 60% RH is by decreased diffusion of reactants and products, owing to increased viscosity on concentration.

Maillard reactions are sometimes referred to as 'browning reactions' owing to the color of many products; for example, this color develops when a creamy paste of *E. coli* dries in a vacuum desiccator. Many other properties of Maillard reactions seem shared with desiccation of microorganisms and allergens. For instance, the protecting action of raffinose is consistent with the conclusion that species B (see above), if it were to exist for coli bacteria, should do so on the surface of coli cells, since raffinose and other substances can protect under conditions when they are confined to outside the cell wall. Raffinose, like many sugars and related compounds (e.g., sorbitol), on dehydration forms an extremely viscous glass (or supercooled liquid) rather than crystallizing, and being a non-reducing sugar does not undergo Maillard reactions. Consequently, raffinose, like trehalose (found in high concentrations in desiccation-resistant higher life-forms[11]), is well suited for slowing

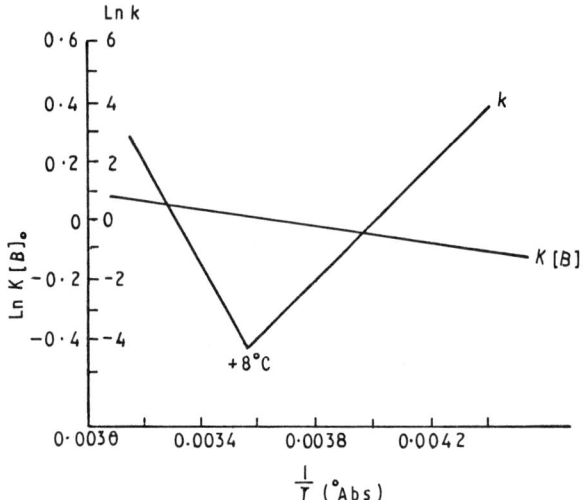

Figure 6.12. Values of the constants derived by fitting catastrophe theory equation to data of Figure 11 as a function of reciprocal absolute temperature.

through diffusion control Maillard reactions of surface components. In addition, such non-reducing sugars stabilize dehydrated phospholipid membranes (see later) as well as inhibit membrane fusion. The protection mechanism probably involves hydrogen bonding of sugar molecules with phospholipid polar head moieties. In this regard lower effectiveness of reducing sugars may be due to their high activity in Maillard reactions.

Effects of vapor phase rehydration also become reasonable as the initial step of Maillard reactions is reversed by addition of water molecules. Hence, coliphage desiccation could lead to a Schiff's base with concomitant loss of flexibility and/or strength of the head-tail complex. In turn, the head-tail complex being sensitive to shear on sampling suffers viability loss. Prior exposure to saturated water vapor could convert the Schiff's base back to free reactants (equation 6.7), thereby restoring the original flexibility and/or strength to the coliphage head-tail complex—and viability is preserved. Another effect of vapor phase rehydration is induced phospholipid phase changes, including some that may rely on the more ready diffusion of vapor water compared to liquid water.

A general conclusion concerning primary desiccation damage (in the absence of oxygen) of bioaerosols of lipid-containing microbes, e.g., *E. coli*, *S. marcescens*, *F. tularensis*, *M. gallisepticum*, *M. pneumoniae*, *Streptococcus* type A L-form and the viruses Langat, Semliki Forest, vesicular stomatitis, vaccinia, VEE, and influenza, is that lowest aero-stability occurs at mid to high RH. Such values correspond extremely well with those associated generally with maximum Maillard activity.[7,9,13] However, when desiccation is severe, DNA strand breaks, etc., also occur even for *B. subtilis* spores[17,18] and may involve other desiccation reaction mechanisms.[18,19]

Phosphoglycerides when isolated are waxy substances that oxidize readily in air and darken through peroxidation. Similar behavior is observed in oxygen-dependent free-radical mediated Maillard reactions of phospholipids. Furthermore, the kinetics of these reactions and of oxygen-induced viability loss of airborne Gram-negative bacteria *S. marcescens*, *E. coli*, are very similar. Both indicate involvement of a carrier species, its saturation at high oxygen tension, and reaction acceleration by dehydration. These aspects are discussed in greater detail in reference (13).

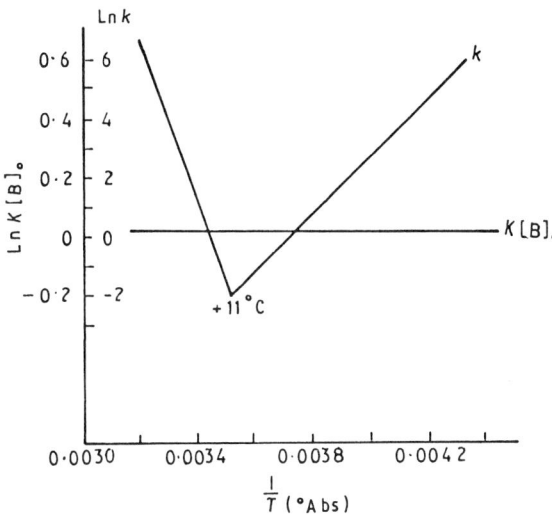

Figure 6.13. Similar plot as for Figure 6.12, but data for wet-disseminated *Francisella tularensis* LVS.

Toxic activity of OAF is considered to be mediated by a carrier[7] and possibly involves free radicals.[16,17] Consequently, one possible type of reaction scheme is

R + OAF• → ROAF•
ROAF• + C → COAF• + R

where OAF• may be a hydroxyl or a hydroperoxide radical, for example, and R a catalytic carrier, perhaps a metal ion as many Maillard reactions are catalyzed particularly by iron and copper ions.[13] Free radical reactions of this type often are characterized by a lag period while concentrations of products (COAF• in this case) build up. This possible explanation for observed lags in OAF action perhaps is more acceptable than a slow condensation of OAF onto bioaerosol particles, considering that OAF is highly condensable.[16]

Relatively high resistance of foot-and-mouth disease and of swine vesicular disease viruses may be related to being lipid-free, and an otherwise involvement of lipid-hydroperoxides known to be extremely reactive in Maillard reactions.

Ionizing radiation (e.g., gamma-rays, X-rays, UV), being energetic, gives rise to free radicals in irradiated entities and thereby may initiate and/or catalyze Maillard reactions because most involve free radical mechanisms.[13]

Generally, it seems virtually certain that Maillard reactions play important roles in the fates of bioaerosols particles, but not to the exclusion of other chemical reactions.[19]

MEMBRANE PHASE CHANGES

Membranes play essential parts in life processes yet are maintained by relatively weak non-covalent bonds, such as van der Waals forces, hydrophobic bonds, ionic forces, hydrogen bonds, etc. Gram-negative bacteria, such as *E. coli*, *K. pneumoniae*, *S. marcescens*, have a cytoplasmic membrane underlying their cell wall together with another membrane overlaying it. This outer membrane comprises a phospholipid bilayer periodically pierced by cylinders of helical proteins and other proteins, together with lipopolysacchar-

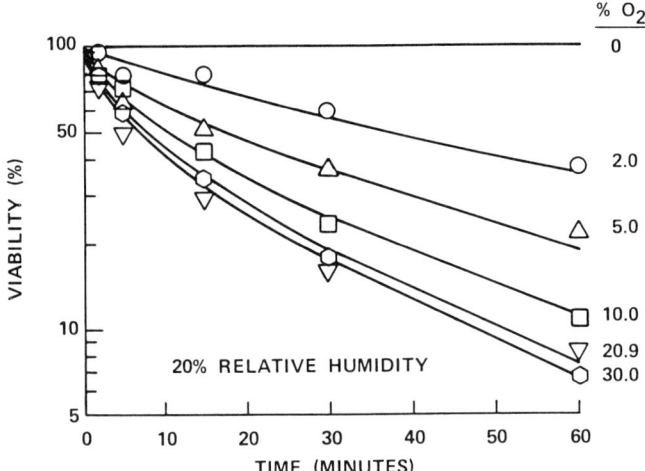

Figure 6.14. Effect of oxygen concentration (%) on the survival of bioaerosols of *Serratia marcescens* at an RH of 20%.

ides, and is attached to a peptidoglycan layer. Between -5°C and +10°C whole *E. coli* cells like isolated membranes demonstrate a single phase transition attributable to phospholipid crystallization; i.e., on cooling their membranes become more ordered. Temperatures at which such gel ⇌ liquid-crystalline phase transitions occur depend on chain length, degree of unsaturation and configuration of constituent fatty acids.[9,11]

In nature, microbes, animals and plants adapting to low temperatures incorporate increasing amounts of unsaturated fatty acids into their storage lipids and membranes.[21] In this manner fluidity and phase states of their membranes are conserved within certain limits and functionality of their membranes then is protected somewhat from direct effects of environmental change.

Phosphoglycerides characteristic of membranes exist as various hydrates owing to the polar nature of the 'head' of the molecule. One consequence is that the manner in which phospholipids pack together to form different structures depends on the particular hydrate present as well as temperature. Thus, phase transitions in phospholipids result from change in their water content and/or temperature. (Heat denaturation of biopolymers generally results from ensuing loss of water molecules; i.e., temperature rise is a dehydration stress.) As membranes are held together by relatively weak non-covalent bonding, relatively small perturbations induce marked phase transitions so that membrane structure inherently is unstable.

Even so, in aqueous environments hydrophobic hydrocarbon tails of phospholipid tails pack together to form a bilayer having polar groups on the outside. Depending on the number of water molecules associated with polar heads as well as those residing amongst hydrophobic tails, phospholipids pack differently. Hence, in crystals, stable phospholipid bilayers become separated by well-defined layers of water molecules of definite thickness, interchelated between two polar group (head) layers of adjacent bilayers so that repeat distances depend on levels of molecular hydration.

Protein molecules in membranes may be confined to aqueous environments or form integral parts of phospholipid bilayers. At least five distinct structures become possible that may interconvert with minimum energy expenditure. One result is membranes demonstrate

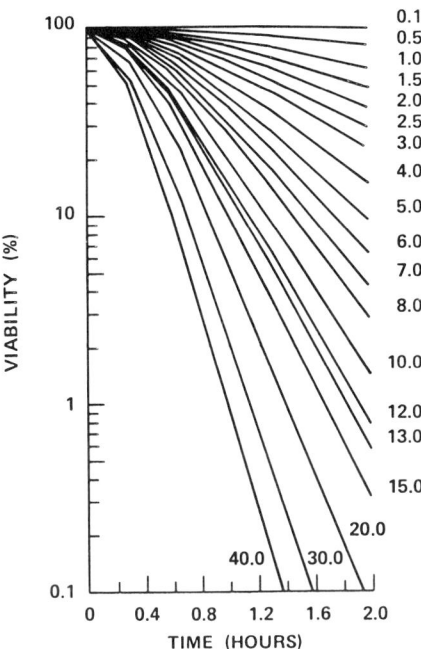

Figure 6.15. Calculated viable decay curves for bioaerosols of *Escherichia coli* commune exposed to different levels of the OAF. (Units approximately correspond with parts per hundred million).

many complex phase transitions, phase separations, phase aggregations and dimensional changes, both vertical and horizontal, readily induced by relatively small water content and/or temperature modifications.[9] Concomitantly there is changed biological function.

Because of complex membrane behavior, findings that there are comparatively narrow RH bands where bacteria such as *E. coli* or *S. marcescens* become particularly unstable are less surprising and more reasonable. However, at least in model systems such transitions are reversible and are of lower activation energy (<10 kcal/mole) than activation energies (ca. 30 kcal/mole) obtained for loss of viability of airborne bacteria and viruses induced by dehydration.

On the other hand, microbial outer membrane damage is well demonstrated by other findings such as dehydration-induced susceptibility to lysozyme and antibiotics, etc., and there is a positive correlation between levels of bacterial and viral survival and resistance to such agents.[7,9] Furthermore, beneficial effects of hypertonic sucrose collecting fluids may be related to activity in membrane fusion and reincorporation into membranes of membrane vesicles formed on dehydration. Also, there are model studies, e.g., Ca-transporting vesicles, that demonstrate dehydration destabilization and stabilization by trehalose,[15] plus other related work.[11,20] Hence, analogous damage may occur with cytoplasmic membranes, but this effect has been less well investigated.

How the smaller values for activation energies for desiccation-induced membrane damage may be reconciled with the observed larger values for viability loss provides the topic of the next section.

ENERGETICS

There is substantial evidence that Maillard reactions *and* membrane phase changes are involved in primary viability loss of dehydrated microorganisms at warmer temperatures and higher humidities. Yet, activation energies derived for desiccation-induced viability loss of Gram-negative bacteria are ca. +30 kcal/mole and considerably greater than for membrane phase changes, ca. +10 kcal/mole. Involved also is the concomitant loss of permeability barrier function and of energy production capability, as well as a change in death mechanism with temperature. This change occurs at below ca. +10°C for some species of Gram-negative bacteria, or +24°C for VEE virus. Whence, reactions become exothermic with an activation energy of ca. -20 kcal/mole for the former or 0 kcal/mole for the latter. Consequently, phospholipid crystallization seems very likely to be involved.

As discussed in greater detail[7,9,13] these observations may be explained along the following lines. On dehydration of, e.g., Gram-negative microorganisms such as *E. coli* held at temperatures above ca. +10°C, their membranes undergo phase changes. Concomitantly, some re-arrangement of membrane phospholipid and protein, etc., molecules occurs, including perhaps the trimer protein molecules (porins) of their outer membranes that form the water-filled pores required for transporting water soluble materials through hydrophobic membrane bilayers. In these changed orientations brought about by lipid phase changes, newly juxtaposed moieties, e.g., porins, then undergo cross-linking, etc., Maillard reactions, with an ensuing loss of permeability barrier function.

Carbohydrates, other proteins, and nucleic acids, etc., located in membranes and other regions of cells also may undergo Maillard reactions, accelerated *per se* by loss of water molecules and increased temperature.

At temperatures between ca. +10 and -50°C there are phospholipid gel⇌liquid-crystal phase transitions and this crystallization process being exothermic is accelerated by reduction in temperature. Together with this change, on cooling some protein molecules form subunits, e.g., chymotrypsin, cytochrome C, myoglobin, lactate dehydrogenase, etc. Such molecules demonstrate a two state equilibrium,

$$\text{N-state} \leftrightarrows \text{D-state}$$

with the D-state favored by subzero temperatures. Cryoinactivation of protein molecules accelerated by cooling generally seems to involve dissociation into subunits followed by relatively slow 'conformational drift' to conformationally different monomers. On warming, reaggregation gives reassociated forms different from original N-states. A tentative estimate of associated free energy changes[21] is in keeping with that given above (ca. -20 kcal/mole) for viability loss at lower temperatures.

On warming to ambient temperatures, membranes and associated protein molecules modified by drying at cool temperatures may be unable to regain original native configurations and biological functions, resulting in viability loss.

Hence, primary loss of viability of, e.g., Gram-negative bacteria may arise at warmer temperatures and higher humidities because of membrane phase changes coupled with concomitant, inactivating Maillard reactions, e.g., *association* of proteins and cross-linking of porin molecules. In contrast, at cooler temperatures, membrane phospholipid crystallization and protein molecular *dissociation* may give rise to biologically inactive forms on warming.

One consequence is that there seems likely to be an optimum temperature for airborne survival possibly corresponding with onset of the primary gel ⇌ liquid-crystalline phase transition,[9] thereby providing, for example, an outer membrane of minimum fluidity leading to some diffusion limitation of Maillard reactions of, e.g., protein molecules.

Oxygen, oxidizing pollutants and radiation exacerbate the situation probably owing to the generation of additional free radicals including hydroperoxide and hydroxyl. Such damage is additive to that imposed by dehydration *per se* (i.e. acts on additional targets).

Summing up events, it seems likely that for Gram-negative bacteria such as *E. coli* in the absence of oxygen, desiccation causes bacterial phospholipids to undergo phase changes, whereas in contrast, carbohydrates, proteins, nucleic acids, etc., undergo rate limiting, e.g., Maillard reactions. In oxidizing environments, additionally, phospholipids form hydroperoxides that are highly reactive in Maillard reactions[13] with, e.g., N-containing moieties that too may have become oxidized.

Analogous arguments seem likely for other microorganisms, including viruses, fungi and pollens, as well as for microbial and other allergens.

REPAIR

Whether damage inflicted to microbes by stress causes loss of biological function depends on the extent and whether it can be repaired[7] (Chapter 13). There are at least two types of repair process:

(1) physico-chemical, non-enzymatic,
(2) energy-dependent, enzyme-catalyzed reactions.

Physical repair includes vapor phase rehydration and, e.g., restoration of flexibility/strength of phage head-tail bonding, while another example is induced membrane phase transitions and conversions of non-native Schiff's bases back to native moieties through rehumidification and warming (annealing) of Gram-negative bacterial bioaerosols in 'artificial' and animal respiratory systems.[7,9,13,22]

Enzyme-catalyzed reactions include, for example, excision of thymine-thymine dimers, and repair of membranes and transport activity[11] and repair of DNA strand breaks, scission of desiccation-induced covalent DNA-protein cross-links.[18,19] But, owing to the fastidiousness of repair, microorganisms may appear non-viable under some sets of conditions but viable under others.[7,12,20]

The Repair Problem

Compounding the problem of repair is the feature that many such mechanisms rely on enzymes and thereby require energy. Yet, owing to their inherent low thermodynamic stability, membranes and associated energy generating assemblages provide primary targets for desiccation damage, with direct correlations observed between coli viability and oxygen utilization.[7]

Consequently, desiccation may lead only too easily to viability loss, while good reserves of energy -rich compounds (e.g., ATP, Ca dipicolinate, intracellular carbohydrate, etc.) together with membrane stabilizers (e.g., non-reducing trehalose) may be expected generally to enhance survival potential following drying and rehydration.[22,23] Such is found in some anhydrobiotic higher organisms,[11] while for airborne entities such effects can be masked by poor sampling protocols. Chatigny *et al.* offer advice[24] as to how these may be avoided, as do authors of chapters in this handbook.

SUMMARY

Lipid-containing microbes such as Gram-negative bacteria at warmer temperatures in the absence of oxygen tend to be least stable at mid to high RH. Their phospholipid membranes, having an inherent low thermodynamic stability, readily undergo phase changes induced by alterations in temperature and/or water content. Then, newly juxtaposed functional groups in, e.g., outer membrane porin molecules, can undergo endothermic Maillard cross-linking, etc., reactions to give biologically inactive forms with an ensuing parallel loss of energy production capability and viability. Other proteins, carbohydrates, nucleic acids at various locations also may undergo Maillard reactions that are accelerated *per se* by dehydration and are optimal at ca. 60–80% RH. At low Rh other desiccation reaction mechanisms may predominate (except for slow vapor phase rehydration).

In contrast, at cooler temperatures exothermic crystallization of lipid moieties together with protein subunit formation seems likely to lead to biological inactivation and viability loss in Gram-negative bacteria. In contrast, in VEE virus such a mechanism does not appear to occur, possibly reflecting a different structural organization.

Consequently, there can be an optimum airborne survival temperature possibly corresponding with the principal gel ⇌ liquid-crystal phase transition. One consequence is that temperature could be more important than RH for airborne disease spread.

Non-lipid microbes, e.g., viruses, tend to be least stable at low RH owing to denaturation of their coat proteins, etc., possibly also involving Maillard reactions.

Oxygen, ozone, the 'open air factor' and radiation exacerbate and add to desiccation damage partly because of the formation of free radicals, e.g., hydroperoxide, hydroxyl, highly reactive in Maillard reactions generally. Lipid-containing species, therefore, generally may be more susceptible than lipid-free counterparts.

Whether induced damage leads to viability/infectivity loss depends on extent as well as ability for repair. But, repair problems, because of their fastidiousness, etc., often are compounded by an energy requirement. Yet, the inherent thermodynamic instability of membranes and their associated energy generating assemblages, are primary targets for dehydration damage. Consequently, the strategy adopted by some anhydrobiotic organisms of having large reserves of energy-rich compounds (e.g., ATP, trehalose) and membrane stabilizers (e.g., trehalose) fits well with the above description of events leading to viability loss of microbes when as bioaerosols.

For microbial and non-microbial allergens, many parallels may be expected. However, ensuring their preservation on sampling from bioaerosols may be a more difficult task than for viability studies because microbes to varying degrees can repair inflicted damage.

REFERENCES

1. Bateman, J.B., C.L. Stevens, W.B. Mercer, and E.L. Carstensen. "Relative humidity and the killing of bacteria: the variation of cellular water content with external relative humidity or osmolality," *J. Gen. Microbiol.* (29):207–219 (1962).
2. Cox, C.S., and F. Baldwin. "A method for investigating the cause of death of airborne bacteria," *Nature (Lond.)* (202):1135 (1964).
3. Cox, C.S., J. Lee, and W.J. Harris. "Viability and electron microscope studies of phages T3 and T7 subjected to freeze-drying, freeze-thawing and aerosolization," *J. Gen. Microbiol.* (81):207–215 (1974).

4. Hatch, M.T., and J.C. Warren. "Enhanced recovery of airborne T3 coliphage and *Pasteurella pestis* bacteriophage by means of a pre-sampling humidification technique," *Appl. Microbiol.* (17):685–689 (1969).
5. Harris, J.R., and R.W. Horne, Eds. *Electron Microscopy of Proteins. Volume 5. Viral Structure.* (London, Academic Press Inc., 1986).
6. Cox, C.S. "Quantitative and qualitative analysis of airborne spora," *Grana* (30):407–408 (1991).
7. Cox, C.S. *The Aerobiological Pathway of Microorganisms.* (Chichester, John Wiley and Sons, 1987).
8. Akers, T.G., and M.T. Hatch. "Survival of a picornavirus and its infectious ribonucleic acid after aerosolization," *Appl. Microbiol.* (16):1811–1813 (1968).
9. Cox, C.S. "Airborne bacteria and viruses," *Science Prog.* (Oxon.) (73):469–500 (1989).
10. Hess, G.E. "Effects of oxygen on aerosolised *Serratia marcescens,*" *Appl. Microbiol.* (13): 781–787 (1965).
11. Leopold, A.C., Ed. *Membranes, Metabolism and Dry Organisms.* (Ithaca, Cornell University Press, 1986).
12. Andrew, M.H.E., and A.D. Russell, Eds. *The Revival of Injured Microbes.* (London, Academic Press, 1984).
13. Cox, C.S. *Roles of Maillard Reactions in Diseases.* (London, Her Majesty's Stationery Office, 1991).
14. Hoekstra, F.A. "Water content in relation to stress in pollen," in *Membranes, Metabolism and Dry Organisms,* A.C. Leopold, Ed. (Ithaca, Cornell University Press, 1986).
15. Crowe, J.H., and L.M. Crowe. "Stabilization of membranes in anhydrobiotic organisms" in *Membranes, Metabolism and Dry Organisms,* A.C. Leopold, Ed. (Ithaca, Cornell University Press, 1986).
16. Dark, F.A., and T. Nash. "Comparative toxicity of various ozonised olefins to bacteria suspended in air," *J. Hyg. Camb.* (68):245–252 (1970).
17. de Mik, G. "The Open Air Factor," Ph.D. Thesis, University of Utrecht, The Netherlands (1976).
18. Dose, K., A. Bieger-Dose, O. Kerz, and M. Gill. "DNA-strand breaks limit survival in extreme dryness," *Origins Life and Evolution of the Biosphere.* (21):177–187 (1991).
19. Dose, K., A. Bieger-Dose, M. Labusch, and M. Gill. "Survival in extreme dryness and DNA-single-strand breaks," *Adv. Space Res.* (12):221–229 (1992).
20. Mackey, B.M. "Lethal and sublethal effects of refrigeration, freezing and freeze-drying on microorganisms," in *The Revival of Injured Microorganisms,* M.H.E. Russell and A.D. Russell, Eds. (London, Academic Press, 1984).
21. Bowler, K., and B.J. Fuller, Eds. *Temperature and animal cells. Symposia of the Society of Experimental Biology.* Number XXXXI. (Cambridge, The Company of Biologists Ltd., 1987).
22. Cox, C.S. "Roles of water molecules in bacteria and viruses," *Origins of Life and Evolution of the Biosphere.* (23):29–36 (1993).
23. Israeli, E., Shaffer, B.T. and Lighthart, B. "Protection of freeze-dried *Escherichia coli* by trehalose upon exposure to environmental conditions," *Cryobiology* (30):519–523 (1993).
24. Chatigny, M.A., J.M. Macher, H.A. Burge, and W.R. Solomon. "Sampling airborne microorganisms and aeroallergens," in *Air Sampling Instruments,* S.V. Hering, Ed. (American Conference of Governmental Hygienists, Cincinnati, Ohio, 1989).

CHAPTER 7

Aerosol Generation for Instrument Calibration

J.P. Mitchell

ABSTRACT

Almost all aerosol analyzing instruments require frequent calibration, since small changes in experimental conditions can have a major influence on their performance. This chapter provides a review of the more important methods of generating calibration aerosols, emphasizing their strengths and limitations. The calibration of some commonly used aerosol analyzers is also described and advice is given about the choice of techniques.

> The naming of commercial equipment and supplies in this report is provided for the purposes of information and does not imply any criticism or endorsement by AEA Technology.

INTRODUCTION

The validity and value of particle size and concentration measurements made with aerosol analyzers depends on whether the instrument is being operated correctly. Although particle behavior in many aerosol samplers (e.g., impactors) can be described quite accurately from first principles, it is frequently necessary to check the operational characteristics of such instruments with well-defined particles in order to verify that they are working according to theory. This requirement is especially true where the measurement(s) being made are required to meet with quality assurance standards [e.g., BS 5750 (1987)] or to comply with procedures laid down by regulatory or accreditating bodies. While this chapter is concerned primarily with inert aerosols, the physical aspects of instrument calibration are applicable directly to bioaerosols.

Almost all properties of aerosols are strongly dependent on particle size, and the establishment of these relationships is the main task of many studies in which aerosol analyzers are used. Most environmental aerosols are reasonably polydisperse and it is necessary to assume that the properties of a polydisperse aerosol with a mean particle diameter (D_p) coincide with the corresponding properties of a monodisperse aerosol with the same particle diameter. Over the years there has been a tendency to use aerosols with the greatest monodispersity possible for most calibration studies. This report primarily is on aerosol generation for calibration purposes and is therefore biased heavily toward monodisperse aerosols. It is recognized, however, that calibration using a polydisperse aerosol that has been well characterized by another method (e.g., microscopy) can be valuable, especially if the instrument sensitivity is affected by a non–size-dependent property of the aerosol, such as refractive index. Some of the described generation methods can be applied also for viability studies, etc. (Chapter 6).

The particle size distribution of many aerosols approaches lognormal and the degree of dispersity conveniently can be characterized by the standard deviation (σ_g), given by the equation:

$$\ln(\sigma_g) = \left[\sum_i^N \frac{\left[\ln\frac{d_i}{d_g}\right]^2}{N} \right]^{1/2} \quad (7.1)$$

where N is the number of particles in the sample and d_g is the geometric mean diameter given by:

$$d_g = \left[\prod_i^N (d_i) \right]^{1/N} \quad (7.2)$$

For most practical purposes an aerosol is considered monodisperse if σ_g is less than 1.2, and many aerosol generators are capable of producing aerosols with σ_g values smaller than 1.1 within the optimum capabilities of their size range.

There are a number of excellent reviews on the subject of aerosol generation for calibration purposes.[1-5] A general review of the whole subject has been written by Dennis.[2] The paper by Fuchs and Sutugin[1] concentrates on condensation aerosol generators whereas the review by Liu[5] is concerned with the aerosol generators developed at the University of Minnesota.

Raabe has included much material concerned with the correct use of polymer latex particles.[3] The book edited by Willeke is concerned chiefly with the generation of aerosols for animal exposure facilities, but also contains a useful section on monodisperse aerosol generation methods.[4]

AEROSOL GENERATION METHODS

Most aerosol studies are carried out with particles ranging in size from about 0.01 to 100 μm diameter, and no single aerosol generation technique can produce particles spanning the entire range. Figure 7.1 shows the useful particle size ranges of various types of aerosol generator together with the operating ranges of commonly used analysis methods. As well as particle size range, factors such as aerosol concentration, liquid or solid particles, and solid particle morphology must all be considered in the choice of a suitable aerosol source. This section constitutes a brief summary of the main features of the more commonly used aerosol generators and is intended as a guide in the choice of calibration method.

A few general comments can be made before describing the individual techniques. Monodisperse particles that have D_p values larger than about 0.5 μm can be produced by controlled atomization of a liquid or by resuspension of pre-formed polymer latex particles. Condensation aerosol generators of the Sinclair-LaMer or MAGE type have a useful size range from 0.1 to about 5 μm diameter and methods based on classification by differences in electrical mobility are suitable only for producing sub-micron particles. Monodisperse

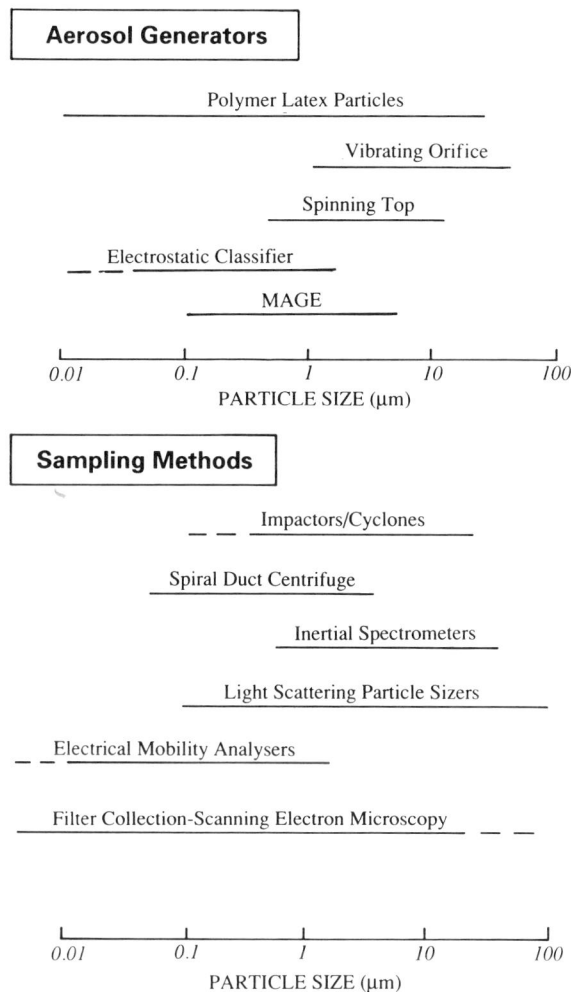

Figure 7.1. Size ranges of common calibration aerosol generators compared with various aerosol analyzers.

aerosols also commonly are produced by classification of a polydisperse aerosol on the basis of particle inertia or electrical mobility. Polydisperse aerosols normally are produced by direct dispersion from a dust generator, of which there are many types, as well as by liquid atomization.

MONODISPERSE AEROSOLS

Polymer Latex Particles

The polymer latex aerosol generator is probably the simplest technique to employ, since the particles are already formed and need only be suspended in air or other support gas. Monodisperse microspheres of polystyrene and related polymers ranging in size from 0.038 to 20 µm are currently available from several commercial sources including Seragen

Inc., Indianapolis, USA.; Duke Scientific (Europe), Hilversum, Netherlands (through Brookhaven Instruments Ltd, Stock Wood, Worcs. in the UK); Polysciences Inc., Warrington, PA, USA (through Park Scientific Ltd, Moulton Park, Northampton in the UK); Japan Synthetic Rubber Co. Ltd, Tokyo, Japan; and Dyno Industries a/s. Oslo, Norway. These particles can be supplied in certain sizes traceable to national standards of length (see section on "Particle Standards"). The particles usually are provided as an aqueous suspension containing between 0.1 and 10% w/v polymer together with a water-soluble inorganic emulsifying agent and an inorganic surfactant stabilizer (as much as 6% of the weight of polymer). After diluting the original suspension, the particles conveniently are suspended as an aerosol using a gas-driven nebulizer.[3] The amount of the dilution required to produce one polymer latex particle per water droplet depends chiefly on the size of the particles and to a smaller extent on the water droplet size distribution (Figure 7.2[3,6]). Thus to produce a singlet- to- multiplet ratio of 0.95, 2 μm diameter latex particles must be diluted from the original 10% w/v stock by a factor of 15, assuming the water droplet size distribution has a volume median diameter (VMD) of 5 μm and σ_g of 1.2. The dilution factor increases to 80 for the same VMD with a σ_g of 2.0. In comparison, 0.2 μm diameter latex particles must be diluted more than 15,000 times, assuming the same original water droplet size distribution. Although the formation of multiplet particles is undesirable for many calibrations, they can be used to provide additional information about the performance of many inertial devices, since the physical and aerodynamic size relationships between singlet and multiplet aggregates containing up to eight particles are well known (Table 7.1[7]).

Water of the highest purity always should be utilized for the dilution of suspensions since dissolved impurities will form separate sub-micron particles as well as coating the surfaces of the latex particles to make them larger.[8,9] This problem is difficult to overcome because the latex particles may agglomerate irreversibly if the stabilizer is removed. One possibility is to centrifuge the latex particles from the original suspension before re-suspending them in pure water. In practice, impurities do not cause a significant problem when working with particles larger than 0.5 μm diameter. Fuchs[10] has summarized the precautions to take when using polymer latex particles.

Many different designs of polymer latex aerosol generator exist and Figure 7.3 illustrates an example that has been constructed at the Aerosol Science Centre of AEA Technology. This equipment contains most of the features needed to ensure that a constant concentration of dry aerosol particles can be produced. The aqueous latex suspension is sprayed into the mixing chamber using a pneumatic nebulizer (see "Compressed Air Nebulizers" section) operating at an air pressure in the range 105 to 350 kPa (15 to 50 psig). The choice of nebulizer can be important, particularly for studies where constant aerosol concentration must be maintained for more than a few minutes, and the commonly used Retec X-70/N nebulizer is shown in Figure 7.3. If the liquid reservoir in the Retec is not agitated when in use, the particles sediment from suspension, resulting in a decrease in the generated aerosol concentration with time (e.g., a reduction of 10% in mass concentration in 15 minutes is typical for 2.8 μm diameter particles). The liquid is agitated continuously in the DeVilbiss type-40 nebulizer and this device therefore is likely to be more suitable for work in which aerosol concentration stability is important. There are many other types of nebulizer that are suitable for latex particle generation.

After leaving the mixing chamber, the polymer latex particles pass downward to the collection point through an electrostatic charge equilibrator (a line source containing radioactive krypton-85 gas), which reduces the high electrical charge imparted to the liquid droplets when they are atomized. The drying particles then pass through a short heated section (operated at about 100°C) to ensure that they are thoroughly dry. The vertical geometry of the aerosol generator is efficient at transporting particles larger than 5 μm diameter,

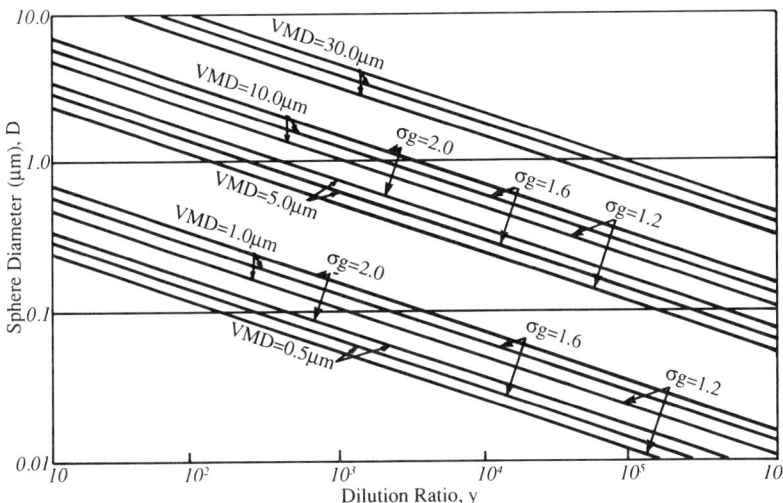

Figure 7.2. The dilution ratio required to generate a singlet ratio of 0.95 vs. sphere diameter from stock of 10% w/v, for various values of the volume median diameter and σ_g of the droplet distribution (O.G. Raabe. 1968[6] with permission Academic Press, Florida and AIHA).

which tend to deposit rapidly on horizontal surfaces. Other designs of polymer latex aerosol generator (Figure 7.4) with a horizontal particle-transport arrangement often include a diffusion drier filled with a suitable desiccant, such as silica-gel, to dry the particles.

Polymer latex particles have a well-defined refractive index (1.588 - 0i for various sizes of polystyrene, polyvinyltoluene and styrene-divinylbenzene particles[11]), and they have become recognized as the standard for the calibration of optical particle size analyzers (e.g., BS 3406 Part 7). These particles are virtually non-porous, with densities of 1.027 and 1.050 g cm^{-3} for polyvinyltoluene and polystyrene, respectively. Their aerodynamic size therefore can be determined readily from measurements of the physical size by microscopy (see next section).

Vibrating Orifice and Related Aerosol Generators

If a thin liquid stream is emitted under pressure from an orifice, this stream is by nature unstable and will disintegrate into discrete droplets by the action of external forces (gravity and surface tension). This phenomenon was first described in detail by Rayleigh.[12] The collapse of such an unstable stream into very uniform droplets is attained easily by the application of a periodic vibration of suitable amplitude and frequency. Several aerosol generators which work on this principle have been developed in recent years, including the Fulwyler droplet generator (Figure 7.5[13]) in which the oscillator is coupled to the orifice through the liquid reservoir, an aerosol generator reported by Hendricks and Babil[14] in which the droplets are produced by disruption of a liquid jet leaving a vibrating capillary tube, and the vibrating orifice aerosol generators (VOAG) described by Ström[15] and Berglund and Liu.[16] The last method has become the most popular technique of this type, and is available commercially [Thermosystems Inc. (TSI), St. Paul, MN, USA], is easy to use, and has become the dominant technique for monodisperse particle generation. Figure

Table 7.1. Relative Aerodynamic Diameters (D_{ae}rel) of Aggregates of n Uniform Spheres Having Different Configurations.

n	configuration	D_{ae}rel
1	○	1.00
2	∞	1.19
3	∞∞	1.28
3	▲	1.34
4	∞∞∞	1.38
4	∞⧗	1.42
5	∞∞∞∞	1.42
6	∞∞∞∞∞	1.45
4	⊗	1.47
7	∞∞∞∞∞∞	1.48
5	⧗∞∞	1.50
6	∞∞⧗	1.52
8	∞∞∞∞∞∞∞	1.52
8	∞∞⧗∞∞	1.56
5	⊗⊗	1.57
8	∞∞∞∞⧗	1.60
6	⊛	1.68

7.6 shows the TSI model 3450, which is the most recent version of the Berglund-Liu VOAG.

In the VOAG, highly monodisperse aerosols containing either solid or liquid particles can be prepared with σ_g values smaller than 1.05, by feeding either a solution of known concentration or a non-volatile liquid through the orifice plate, which is typically in the range 5 to 50 µm diameter. This orifice plate is mounted inside a piezoceramic crystal. When an a.c. signal is applied, the crystal vibrates in the vertical plane, disrupting the liquid jet as it emerges. The droplets formed are dispersed immediately in a small flow of air (typically 0.75 to 1.5 l min^{-1}) and the resulting calibration aerosol is produced after evaporating any solvent from the liquid aerosol in a flow of dilution air.

The mean diameter (D_p, µm) of the aerosol particles can be predicted directly from the operating parameters of the VOAG in accordance with the equation:

$$D_p = \left[\frac{10^{11} QC}{\pi F} \right]^{1/3} \quad (7.3)$$

where Q is the liquid feed rate (cm^3 min^{-1}), C is the fractional concentration of solute in the feed liquid, and F is the crystal vibration frequency (Hz). The absolute nature of this relationship makes the VOAG a primary calibration standard.

The optimum operating conditions of frequency and liquid flow rate using various orifice sizes were determined by Wedding and Stukel[17] and Wedding[18] to give the expression:

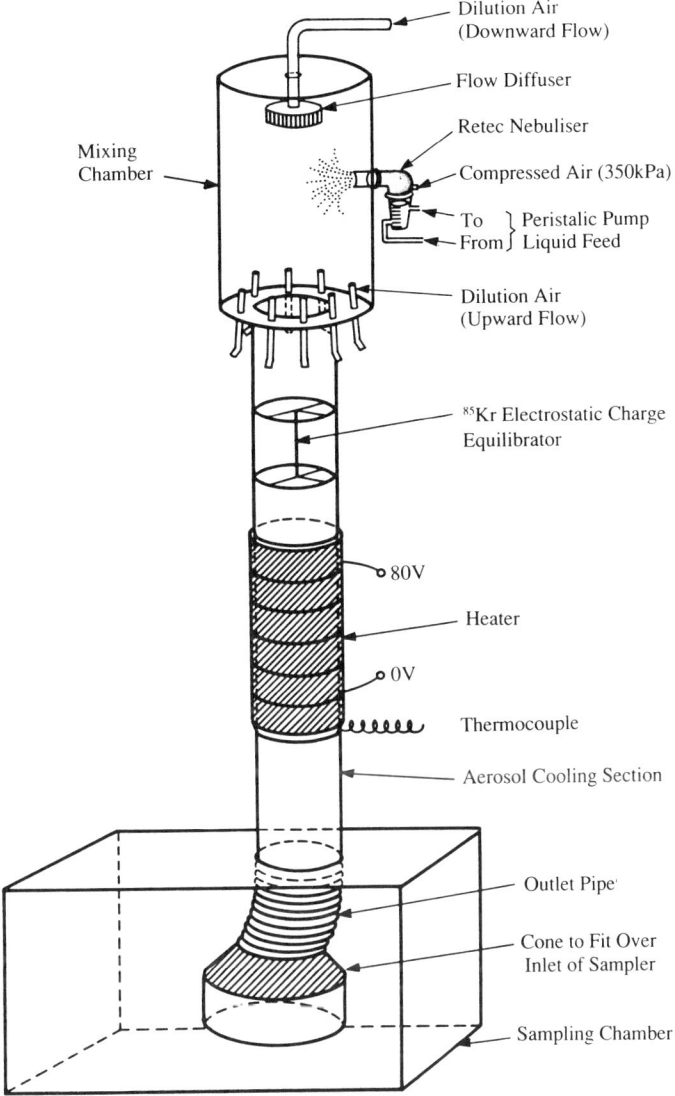

Figure 7.3. Polymer latex aerosol generator with downward dispersion of the aerosol (after O.G. Raabe, 1976[3]).

$$\frac{Q}{F_{max} A_j D_j} \leq \lambda_{min, max} \leq \frac{Q}{F_{min} A_j D_j} \tag{7.4}$$

that can be used to predict the limits for monodisperse particle generation, where A_j is the jet area (cm^2), D is the jet diameter (cm) and $\lambda_{min,max}$ are the minimum and maximum disturbance frequencies for monodisperse operation and apply to the left- and right-hand side of equation (4), respectively. In a series of tests, Wedding and Stukel were able to produce monodisperse di-octyl phthalate (DOP) droplets using a 21.5 µm orifice with a liquid feed rate of 0.191 cm^3 min^{-1} when F was in the range 82 to 103 kHz.

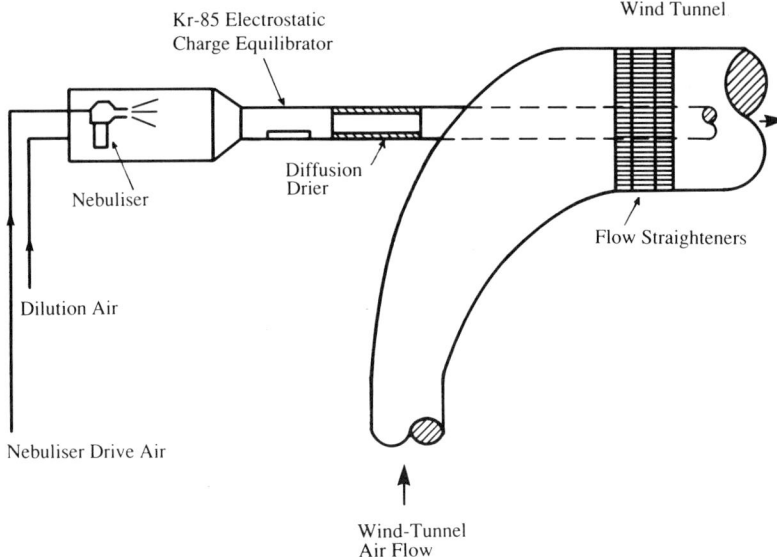

Figure 7.4. Polymer latex generator with horizontal dispersion of the aerosol.

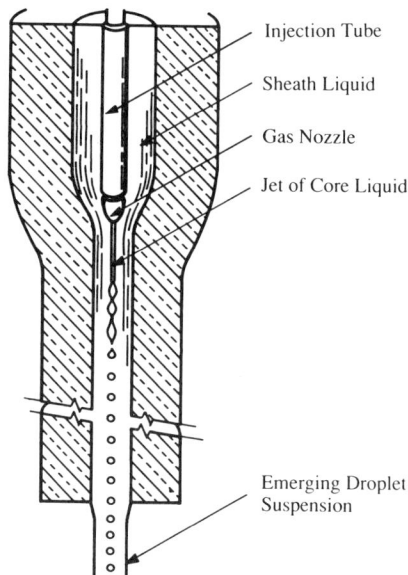

Figure 7.5. Fulwyler monodisperse droplet generator (M.J. Fulwyler, J.D. Perrings and L.S. Cram. 1973[13]).

Although the VOAG is easy to operate, several precautions need to be observed especially if it is used to make solid particles:

(a) the liquid pressure behind the orifice must be carefully controlled,
(b) the purest reagents must be used, particularly when producing small particles from dilute feed solutions,
(c) the droplets must be carefully dried in order to produce spheres,
(d) the resulting particles may be porous, particularly if the drying conditions are not well controlled, and this internal voidage will affect their aerodynamic properties.

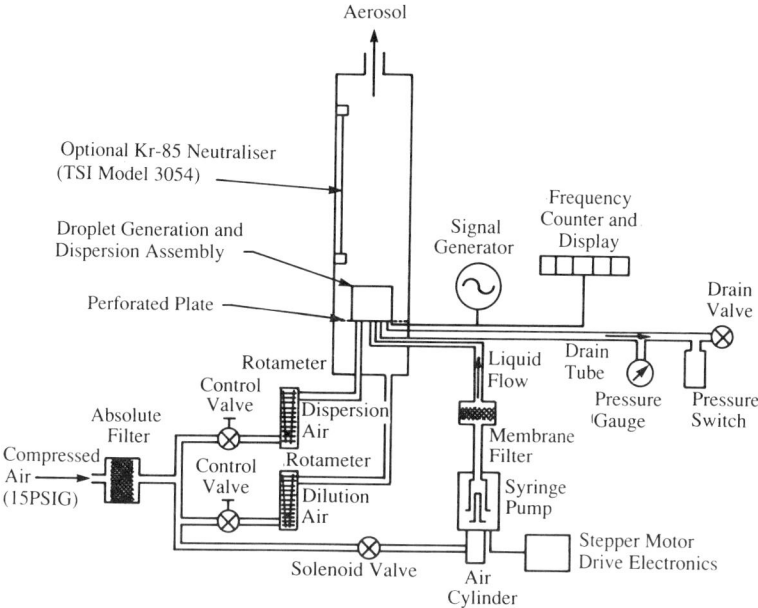

Figure 7.6. TSI model 3450 vibrating orifice aerosol generator (VOAG).

A modified version of the VOAG (Figure 7.7) has been developed at the Aerosol Science Centre of AEA Technology, paying particular attention to these points, and a significant improvement in reliability has been achieved.[19] A pneumatic liquid feed pump (Figure 7.8) has been constructed to enable the liquid pressure to be controlled to within 2 kPa in the range 100 to 200 kPa, equivalent to a feed rate stability of ±1% at 0.2 cm^3 min^{-1} using a 20 μm diameter orifice. Even better performance can be achieved using an isocratic pump of the type normally used with HPLC systems. The efficiency of particle production has been increased from <10% with the original equipment (the earlier TSI model 3050 VOAG), to more than 40% in the size range from 4 to about 15 μm diameter (Figure 7.9). This improvement has been achieved by inverting the vibrating orifice assembly and mounting it in a larger container with sloping or vertical surfaces to reduce deposition losses.

When this aerosol generator was used to prepare methylene blue particles from alcohol-water mixtures, as described by Berglund and Liu,[16] the resulting particles often were distorted in shape. Similar experiences have been encountered by other workers, and Leong[20] suggested that this effect might arise because the solvent evaporation rate was too rapid. Spherical particles were reliably formed when aqueous solutions were used and the dilution air flow rate was within the range 50 to 90 l min^{-1}. A small heater was found to be necessary at the later stage of drying, to ensure that the particles were fully dry by the time they reached the collection point. Solid particles as small as 0.8 μm diameter have been produced using pre-filtered, conductivity grade water and purified methylene blue.

The bulk density of the particles (ρ_p) can be obtained by comparing the aerodynamic (D_{ae}) and physical (D_p) diameters of the spherical particles whose dynamic shape factor (χ) is unity, since:

$$\rho_p = \left[\frac{D_{ae}}{D_p}\right]^2 \left[\frac{C(D_{ae})}{C(D_p)}\right] \rho_o \chi \qquad (7.5)$$

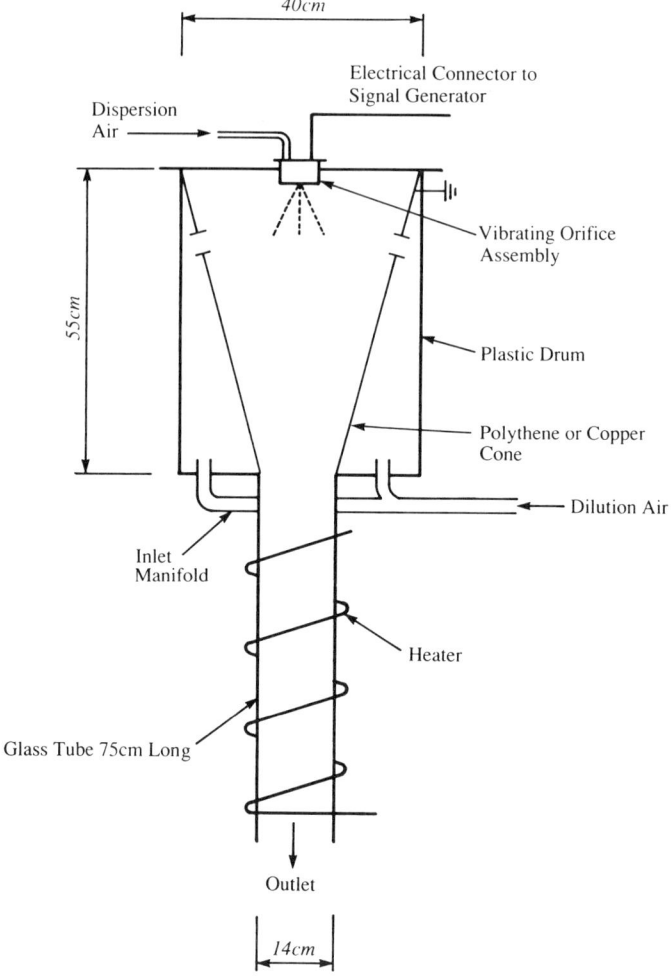

Figure 7.7. Inverted vibrating orifice aerosol generator (VOAG).

where ρ_o is unit density and where $C(D_{ae})$ and $C(D_p)$ are the Cunningham slip correction factors, which are close to unity for particles larger than 2 μm diameter. The aerodynamic sizes of the methylene blue particles were measured using an Aerodynamic Particle Sizer (model APS33B, Thermosystems Inc., St. Paul, MN) and compared with equivalent sizes obtained by microscopy. These measurements showed that the particles were slightly porous, as they had values close to 1.10 g cm^{-3}, compared with the bulk density of 1.26 g cm^{-3} quoted by Hinds.[21] A particle density of 1.10 g cm^{-3} also was obtained by O'Connor[22] for methylene blue particles prepared by a spinning top aerosol generator. Chen and Crow[23] have claimed that they can produce highly monodisperse ($\sigma_g < 1.02$), non-porous ammonium fluorescein microspheres with D_p values between 3 and 19 μm using an inverted VOAG, but working with lower dilution air flow rates (15 to 30 l min^{-1}).

The final shapes of the particles produced by the evaporation of solution droplets are determined by the presence of nuclei around which solid formation is initiated, the crystal-

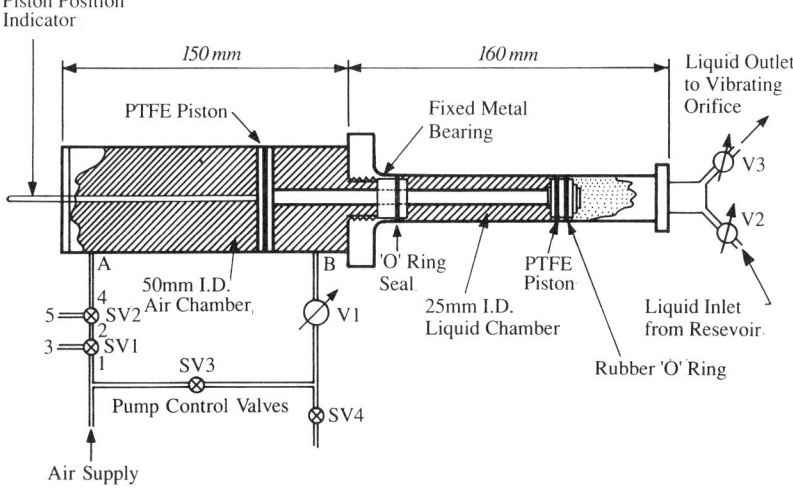

Operating Mode	Valve							Remarks
	SV1	SV2	SV3	SV4	V1	V2	V3	
Fast forward	1 - 2	2 - 4	C	O	O	C	O	Rapid liquid feed to VOAG
Slow Forward	1 - 2	2 - 4	C	O	A	C	O	For constant pressure
Reverse	1 - 2	4 - 5	O	C	O	O	C	To draw in liquid
Stop	1 - 2	2 - 4	O	O	A	O/C	O/C	
Over-pressure	1 - 3	2 - 4	C	O	A	C	O	Stops pump if pressure > 700 kPa

Notes A = Partially Open Needle Valve
C = Closed, O = Open
SV1 and SV2 are 3-way Solenoid Valves
SV3 and SV4 are 2-way Solenoid Valves
V1 is a high quality Needle Valve
V2 and V3 are Stop Valves

Figure 7.8. Pneumatic liquid feed pump for vibrating orifice aerosol generator (VOAG).

linity of the solid being formed, and the solvent evaporation rate. Leong[20] proposed that the optimum conditions for solid sphere formation should be a nuclei-free environment, a high solubility solvent and low drying rates. The optimum drying conditions vary for different particle types and solvents, and some exploratory studies are required almost always when using this technique. For instance, Vanderpool et al.[24] have extended the upper size limit to 70 μm diameter for ammonium fluorescein spheres by drying the aerosol slowly in humidified air. In contrast, when a VOAG was used to prepare sodium chloride aerosols[16] from aqueous solutions, the resulting dried particles were quasi-spherical aggregates, each composed of several smaller cubic crystals.

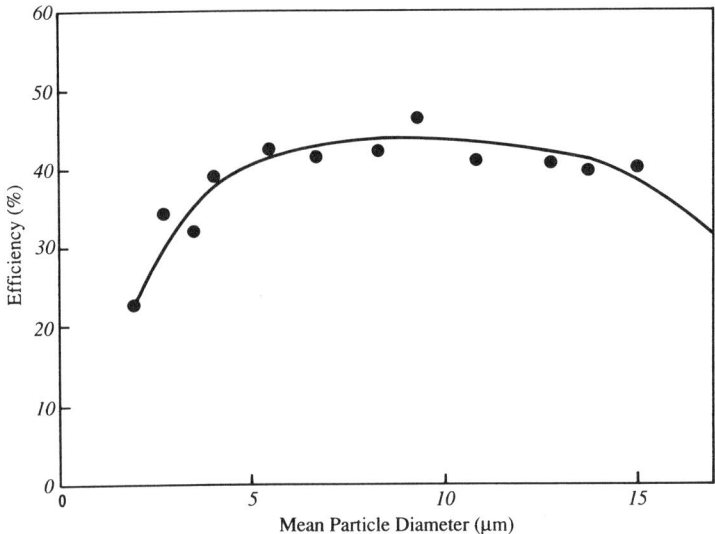

Figure 7.9. Particle production efficiencies with the inverted vibrating orifice aerosol generator (VOAG).

As mentioned earlier, the VOAG can be used also to generate monodisperse liquid droplets and many workers have fed non-volatile liquids such as DOP or di-2-ethyl hexyl sebacate (DEHS) dissolved in a suitable solvent through the vibrating orifice. Liquid aerosol containing a trace of fluorescent dye[25] is a popular method for calibrating cascade impactors, in which particle bounce can be a problem. The droplet diameters are unaffected by the small amount of dye incorporated in the feed solution and only small amounts of aerosol need be collected for the calibration, because the distribution of the droplets within the impactor can be obtained by highly sensitive fluorometric analysis.

Spinning Top/Disc Aerosol Generators

Monodisperse droplets can be generated when centrifugal forces disrupt a film of liquid that is allowed to spread from the centre of a horizontal disc rotating about a vertical axis. Several versions of this aerosol generator have been constructed and fall into two main classes: spinning disc generators in which the rotor is driven mechanically, and spinning top aerosol generators, where the rotor is driven by compressed air jets. A good example of the former type (Figure 7.10) was developed by Whitby et al.,[26] and the latter type (Figure 7.11) is represented by the May spinning top.[27,28] Both types of aerosol generator are alternatives to the VOAG, but they have the advantage that a higher output of particles is possible. In the case of the May spinning top aerosol generator (STAG), the liquid feed rate can be increased to about 1 cm^3 min^{-1} without affecting the monodispersity of the resulting aerosol significantly, although feed rates between 0.2 and 0.8 cm^3 min^{-1} are more usual.[29–31] A further advantage of the STAG is that it can be operated continuously for several hours as long as the rotor surface does not become heavily coated with deposits from the feed solution, which can affect the degree of wetting of the rotor by the liquid. In contrast, frequent cleaning of the orifice of the VOAG is necessary, since it is easily blocked, particularly when concentrated feed solutions are used.

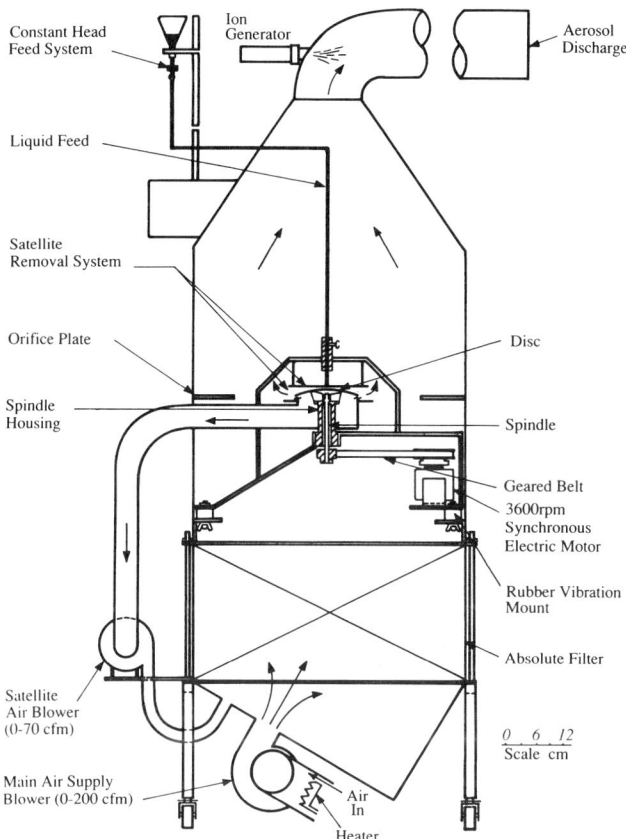

Figure 7.10. Model-III spinning disc generator for making homogeneous particles between about 3 and 30 μm. (K.T. Whitby, D.A. Lundgren and C.M. Peterson.)

Unlike the VOAG, which forms one size of droplets in normal operation, spinning top and disc aerosol generators produce smaller satellites as well as the monodisperse primary droplets. The satellites form as the thread of liquid drawn out behind each primary droplet collapses.[32] Several satellites are produced with every primary droplet and, because they are on average about a quarter of the size of the latter, they can be removed quite easily by inertial separation soon after they are formed. In the generators designed by Whitby et al.,[26] satellite removal is achieved by sucking the air close to the rotor through a separate exhaust. The primary droplets having greater inertia are able to escape this suction to be transported by a flow of carrier air to the collection point. A similar principle operates in the STAG, but the satellites are captured in the suction generated by the Bernouilli effect as the escaping spent rotor drive air passes through a small gap between the stator and cover dome. However, in practice, some additional suction is required,[33] particularly if the spent rotor drive air is vented some distance from the spinning top.

While the remainder of this section is concerned with the STAG, which is available commercially (Research Engineers Ltd, London), almost all the operating principles apply equally to spinning disc type generators. Aerosols that have the same dispersity, particle concentration and size ranges can be made with both systems. A final point is worth noting

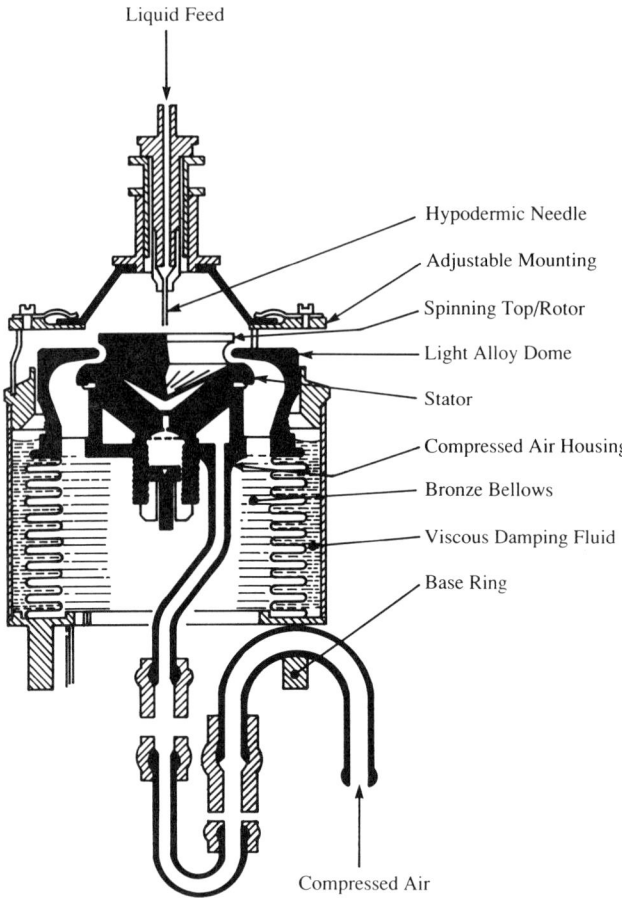

Figure 7.11. May spinning top homogeneous spray (K.R. May. 1966[28]).

with respect to the spinning disc generator: the rotor diameter can be made as large as convenient (within the limit of the mechanical strength of the rotor material), making it possible to increase the aerosol output without sacrificing monodispersity; in contrast, the STAG has a fixed rotor diameter of 2.54 cm, and size changes cannot be made without rebuilding the complete assembly.

The feed liquid is delivered to the rotor surface of the STAG via a hypodermic needle and careful alignment has been shown to be critical for correct operation of the generator.[32,34] The equation relating the mean diameter of the particles (D_p) to the operating conditions is:

$$D_p = \left[\frac{K}{2\pi F}\right] \left[\frac{T}{D\rho_l}\right]^{1/2} \left[\frac{C}{\rho_s}\right]^{1/3} \qquad (7.6)$$

where K is a numerical constant, F is the frequency of revolution of the rotor (Hz), T is the surface tension (dyn cm^{-1}), ρ_l is the density of the feed liquid (g cm^{-3}), D is the diameter of the rotor (cm), and ρ_s is the density of the solute or suspended solid in the feed liquid (g cm^{-3}). The sizes of particles produced cannot be predicted simply by applying Equation 7.6 unless the value of K is known for the system. K is a complex factor which depends on a number of ill-defined parameters such as the roughness of the rotor and type of material being atomized. Such effects cannot easily be quantified and vary considerably from one generator to another. Values of K obtained with a STAG at the Aerosol Science Centre of AEA Technology varied from 2.3 to 7.0 in tests where aqueous feed suspensions/solutions containing ferric oxide, methylene blue and sodium fluorescein were used to make particles in the range 1 to 11 μm diameter.[33]

One problem with the STAG is the severely reduced aerosol output that occurs when the primary droplet diameter is smaller than about 18 μm.[35] For many organic liquids that have low surface tension, the primary droplet diameter at normal rotational frequencies (ca. 1000 rev s^{-1}) is often smaller than this size. The stopping distance of the droplets (distance they are projected from the rotor before air resistance stops their outward motion) is therefore smaller than the radius of the stator support assembly. Cheah and Davies[35] used streak photography to show that most of these droplets are lost when they become entrained by the inflow of air created as the spent rotor drive air escapes. They modified their STAG so that the radius of the stator support assembly was slightly reduced, and at the same time the height of the cover above the stator could be adjusted to decrease the amount of suction without having to lower the cover far beneath the trajectories of the outward moving primary droplets (Figure 7.12). These modifications resulted in the production of primary droplets in the size range 5.2 to 18 μm diameter from DEHS solutions in hexan-l-ol, with yields several times greater than had been previously achieved. Yields exceeding 70% of the theoretical maximum were obtained when generating droplets between 8 and 18 μm diameter.

Two versions of the standard May STAG have been developed at the Aerosol Science Centre for slightly different calibration purposes. The first generator (Mk-I, Figure 7.13) is housed in a container in which the dilution air carries the drying aerosol droplets upwards to a heater. The needle height above the rotor can be controlled accurately by means of a spring-loaded mechanism which is linked to a micrometer gauge mounted on the outside of the container. The spent rotor-drive air is removed from the container under a slight vacuum to ensure that the satellites are always separated from the primary droplets. A few droplets, twice the diameter predicted by Equation 7.6, also are formed and these are eliminated by impacting them against the wall of the Perspex bell-jar. This aerosol generator has a low output efficiency, typically less than 15% of the theoretical maximum, but the monodispersity of the particles is excellent (σ_g values better than 1.08 are readily achieved, Figure 7.14), and is comparable with the monodispersity achieved using a VOAG.

A second version of the STAG (Mk-II, Figure 7.15) was developed to manufacture ceramic cerium oxide microspheres with D_p values in the range 1 to 7.5 μm. The aerosol output was optimized by minimizing the surface area on which the drying particles can impact or deposit, and yields in excess of 50% of the theoretical maximum have been achieved.[36] The mean droplet diameter in these studies was larger than 18 μm, making modifications to the spinning top unnecessary. The particles were calcined in a tube furnace maintained at 700°C in order to reduce their porosity so that their aerodynamic size could be determined accurately (more recent tests have shown that this porosity can be eliminated if the particles are heat treated at temperatures greater than 900°C). The particles were then collected in an impinger filled with propan-2-ol, and could be stored in acetone for at least two years without deterioration. It was observed that formation of carbon particles can occur by pyrolysis of the acetone, if the particles are sprayed into an inert atmosphere at high temperature. This complication can be avoided by resuspending them in water just before use. This design of STAG makes it possible to make milligram

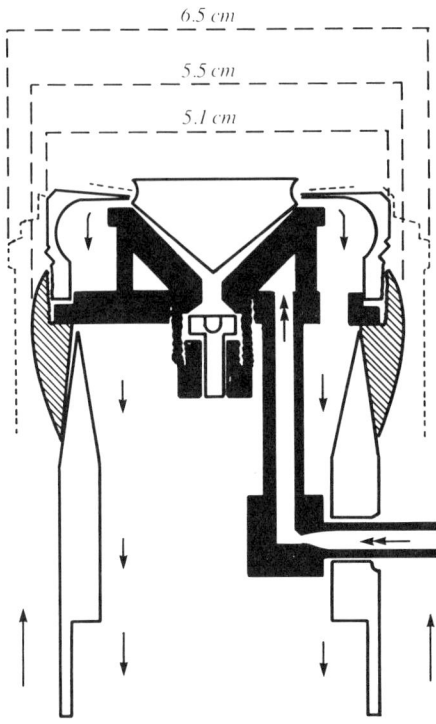

---- Outline of Standard (May) Generator

Figure 7.12. Modified May spinning top aerosol generator. (P.K.P. Cheah and C.N. Davies. 1984[35] with permission Pergamon Press Ltd, Oxford.)

quantities of particles in a single batch, but their σ_g values are often slightly larger than 1.08. Particles prepared with this aerosol generator are shown in Figure 7.16.

As well as the versions of the STAG described here, several other useful modifications to the basic design have been reported in the literature. For instance, the STAG developed by Bailey and Strong[37] includes a centripeter (virtual impactor) to concentrate the aerosol particles before use. Philipson[38] and Garland et al.[39] both suspended their spinning discs/tops in vertical columns to improve the particle transport. Factors such as particle size range, concentration, and the immediate or delayed use of the particles will influence the choice of STAG design. Further information on this technique is contained in the reviews by Fuchs and Sutugin[1] and Davies and Cheah.[40]

Condensation Aerosol Generators

The Sinclair-LaMer[41] and Rapaport-Weinstock[42] aerosol generators frequently are used to produce monodisperse particles by condensation of a vapor on foreign nuclei. The operation of both aerosol generators is described in detail by Fuchs and Sutugin[1] and Dennis,[2] and only the main features are discussed in this chapter. The Sinclair-LaMer generator constructed by Muir[43] is shown in Figure 7.17: a stream of oxygen-free nitrogen passes through a 2 l flask in which a high boiling-point liquid such as di-2-ethyl hexyl

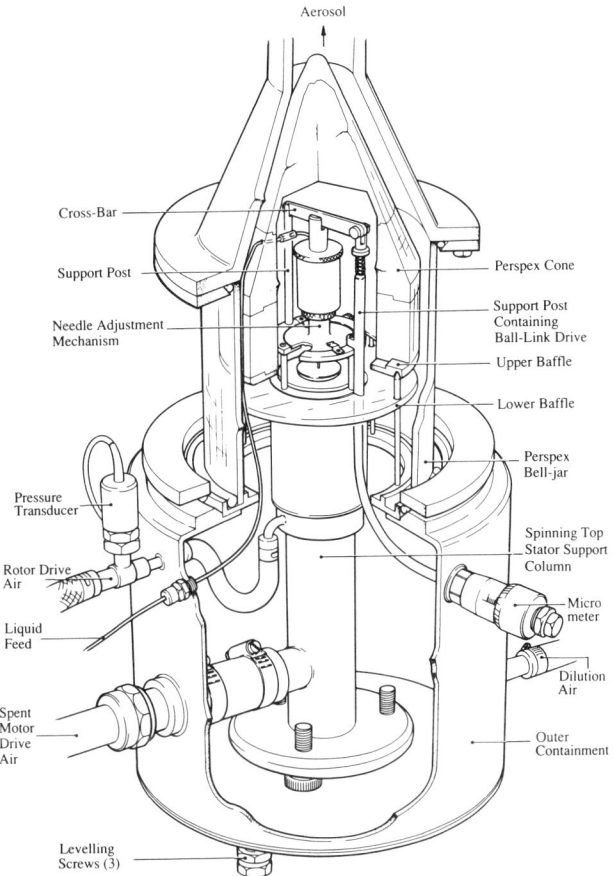

Figure 7.13. Mk-I Winfrith spinning top aerosol generator (STAG).

sebacate DEHS is heated and stirred continuously. The temperature is maintained with an accuracy of ± 0.5°C monitored by a contact thermometer. Nuclei are generated from several sources,[44] including heated wires dipped in NaCl, AgCl or KI, carbon electrode arc, and spark production between tungsten electrodes at high voltage. In contrast with the original design,[41] the carrier gas is not bubbled through the high-boiling point liquid, and the condensation chimney is directed downward to prevent droplets from running down the walls of the tube and producing instabilities in the heater. The vapor and nuclei enter the re-heater flask by a tangential jet to reduce rotation and mixing. The size and number of droplets produced by this generator depend on the concentration of nuclei, the boiler temperature, the rate of gas flow and the dimensions of the condensation chimney.[44]

The generation of aerosols containing droplets larger than 1 μm diameter requires both low nuclei and high vapor concentrations. Swift[44] used a Muir-type Sinclair-LaMer generator to show that D_p values from 0.3 to 1.0 μm could be obtained by increasing the boiler temperature from 110 to 150°C at constant nuclei concentration (8×10^5 nuclei cm^{-3}), the particle mass being proportional to the vapor pressure of DEHS. In other tests at constant nuclei concentration (3×10^5 nuclei cm^{-3}), the fraction of vapor condensing on the nuclei

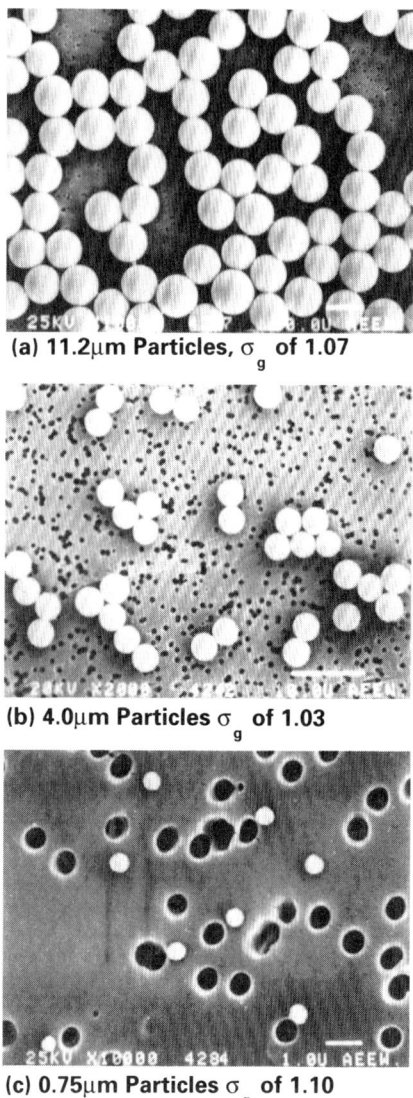

(a) 11.2μm Particles, σ_g of 1.07

(b) 4.0μm Particles σ_g of 1.03

(c) 0.75μm Particles σ_g of 1.10

Figure 7.14. Ferric oxide particles produced using the Mk-1 spinning top aerosol generator (STAG).

was greatest when the re-heater temperature was between 175 and 200°C. Thus D_p values increased sharply when the re-heater temperature was increased from 135 to 175°C and remained constant in the range 175 to 200°C before decreasing at higher temperatures. Polydisperse aerosols were produced when the re-heater temperature was close to the boiler temperature (130°C), but monodispersity improved at higher re-heater temperatures. Swift also showed that D_p decreased as the gas flow rate was increased; droplets generated with a volumetric flow rate of 100 cm^3 min^{-1} were three times larger than those obtained at 1000 cm^3 min^{-1}.

At very low nuclei concentrations ($< 10^4$ nuclei cm^{-3}), D_p is almost independent of the vapor concentration since nearly all condensation occurs on the walls of the chimney. Each

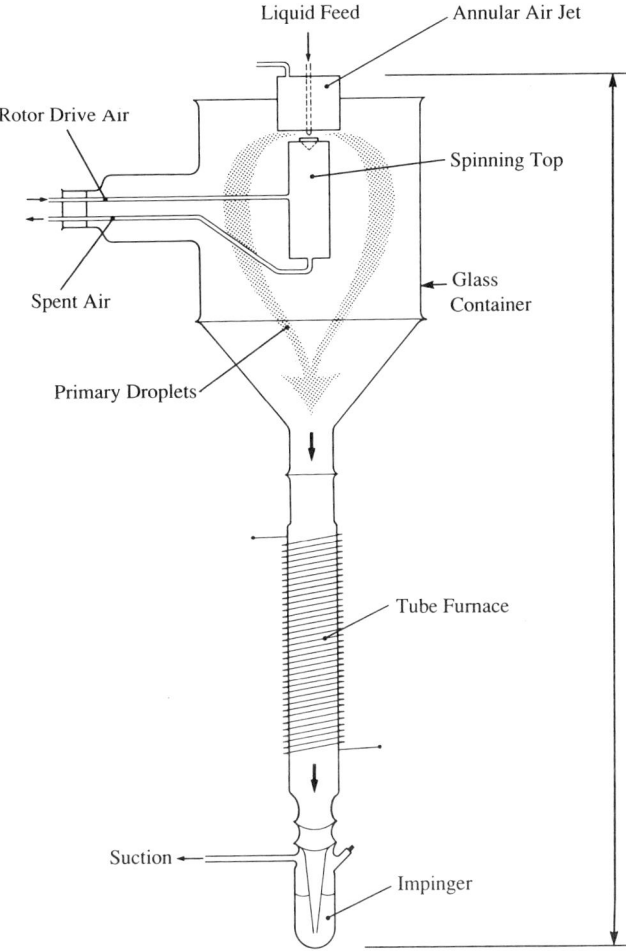

Figure 7.15. Mk-II Winfrith spinning top aerosol generator (STAG).

nucleus is exposed to the same vapor concentration and grows to a fixed size (0.8 to 1.0 μm diameter). At higher nuclei concentration (10^5 to 10^7 nuclei cm^{-3}), there is a weak dependence of D_p on concentration, whilst at even higher concentrations (> 10^7 nuclei cm^{-3}), almost all the vapor condenses on the nuclei, producing droplets with D_p values smaller than 0.4 μm. It is difficult to produce monodisperse aerosols with high nuclei concentrations because the coagulation rates of the nuclei and the droplets become significant. Nevertheless, the upper limit of aerosol concentration available with this type of aerosol generator can be more than 100 times greater than that achievable with either the VOAG or STAG.

The Rapaport-Weinstock aerosol generator (Figure 7.18) consists of three stages. A DeVilbiss nebulizer (see "Compressed Air Nebulizers" section) generates a polydisperse mist of the high boiling-point liquid (DOP or DEHS) in the first stage. The coarse droplets then are evaporated, and in the final stage they are re-condensed as a monodisperse aerosol. The condensation nuclei originate as non-volatile impurities in the liquid. This aerosol generator requires only a few minutes to attain thermal equilibrium in comparison with the Sinclair-LaMer type of generator which can take several hours to reach stability. Thermal

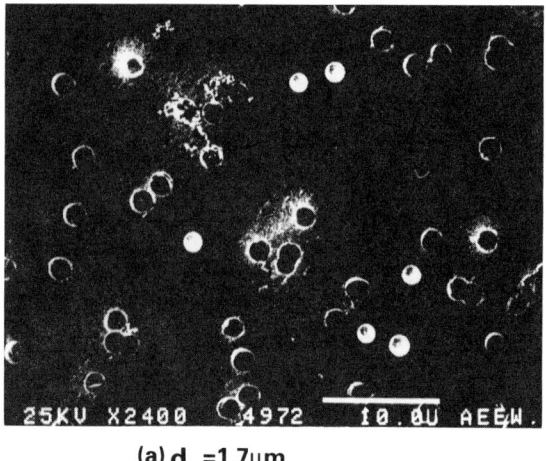

Figure 7.16. Cerium oxide particles after heat treatment at 750°C in a steam-argon atmosphere.

decomposition of the liquid, which can be a problem in the Sinclair-LaMer generator, is avoided because fresh liquid is continuously fed to the heated section. Another specific feature of the Rapaport-Weinstock generator is that the temperature of the vaporizer need not be rigorously kept constant; after evaporation of the droplets produced by the nebulizer, the vapor concentration ceases to be temperature dependent. Furthermore, since the nuclei and vapor concentrations are proportional to the concentration of the nebulized droplets, changes in the latter do not affect appreciably the droplet size of the generated aerosol. Aerosols with D_p values in the range 0.3 to 1.4 µm have been produced by changing the distance from the nozzle of the nebulizer to a special screen placed inside the nebulizer.[45] Limited changes in droplet size also can be achieved by introducing dilution air to the droplet stream entering the vaporizer.

The Monodisperse Aerosol Generator [(MAGE), Lavoro & Ambiente scrl, Bologna, Italy] is a condensation-type aerosol generator that has greater flexibility than the previous-

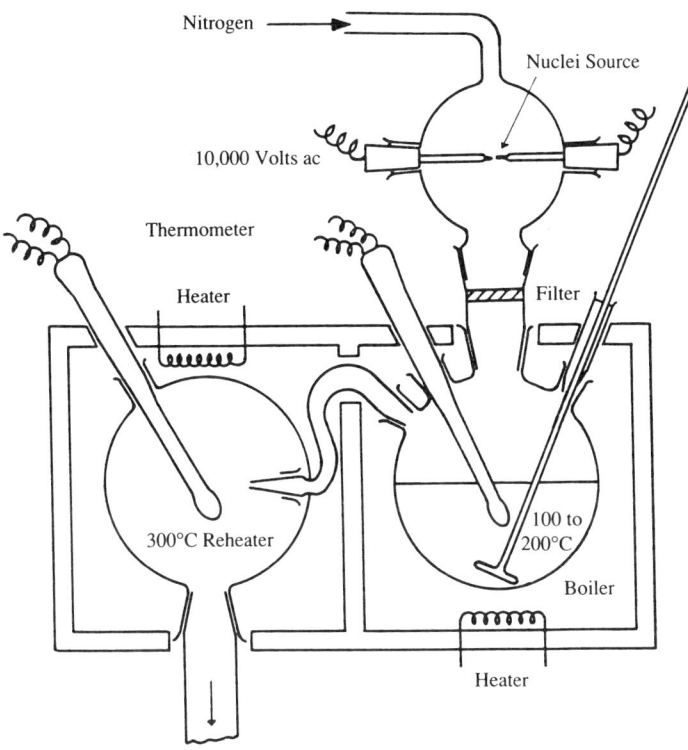

Figure 7.17. Sinclair-LaMer condensation aerosol generator. (D.C.F. Muir, 1965[43]).

ly mentioned systems. It is claimed to be capable of producing highly monodisperse particles ($\sigma_g < 1.1$) in the size range 0.2 to 8 μm geometric diameter.[46] The MAGE has many similarities with the Sinclair-LaMer and Rapaport-Weinstock generators; a stream of nuclei is exposed to the vapor of a low-volatile liquid at an elevated temperature, and the controlled heterogeneous condensation of the vapor into the nuclei results in the formation of the product aerosol. However, neither of the latter systems are readily suited to the rapid, continuous production of solid particles as well as liquid droplets. The MAGE was designed to meet this need by improving the temperature control of the vessel containing the high-boiling liquid (bubbler vessel). Most significantly, the size of the particles can be adjusted to a new value in a few seconds by altering the proportion of the gas-flow that bypasses the bubbler vessel.

The construction and operation of a prototype version of the MAGE has been described by Prodi,[46] and the gas flow path through the commercially available version of this aerosol generator is illustrated in Figure 7.19. A collison atomizer generates nuclei in a nitrogen atmosphere from a dilute solution of a substance which is insoluble in the low volatile liquid. Scheuch and Heyder[47] have used aqueous solutions containing high-purity sodium chloride in the atomizer. More recently, solutions containing a fluorescent tracer compound (sodium fluorescein) have produced nuclei that can be measured when coated by the low-volatile liquid.[48] Such aerosols can be detected at very low mass concentrations by fluorimetry, making it possible to minimize the time to calibrate aerosol analysis equipment.

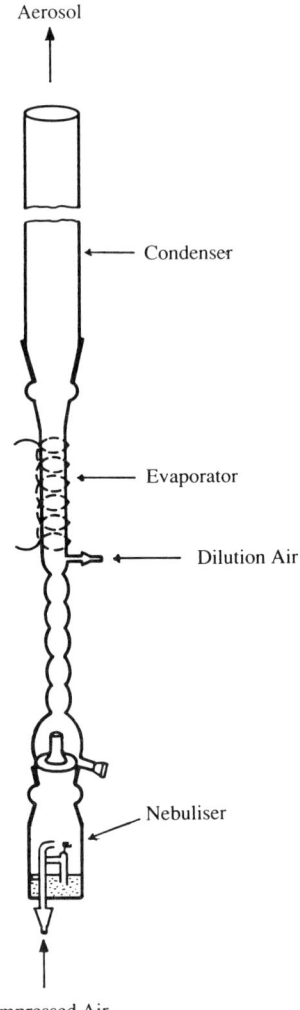

Figure 7.18. Rapaport-Weinstock aerosol generator (E. Rapaport and S. Weinstock, 1955[42]).

The droplets leaving the atomizer are dried by passing them through a vertical tube containing silica gel desiccant. The resulting nuclei stream then is split into two; part of the flow is passed directly at ambient temperature to the reheater section (by-pass flow), while the remaining flow is bubbled through the vessel containing the low-volatile liquid (typically DEHS) by partly closing a needle valve located at the entrance to the bypass line. The bubbler vessel is maintained at sufficient temperature to produce a moderate flow of vapor without significant thermal degradation. Temperature control typically is maintained within ± 1°C to ensure a constant supply of vapor. On leaving the bubbler vessel, the vapor flow is immediately diluted with the flow from the bypass line before entering the reheater section to ensure the complete vaporization of any remaining liquid. The reheater consists of a short vertically mounted resistance furnace containing a 10 mm bore glass tube through which the vapor-nuclei-gas mixture passes in the downward direction. Heterogeneous condensation occurs onto the nuclei at the outlet of the reheater where a well-

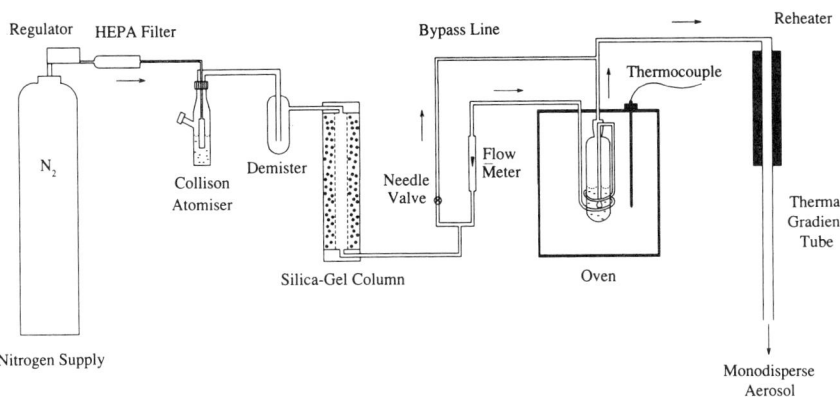

Figure 7.19. Prodi MAGE monodisperse condensation aerosol generator.

defined thermal-gradient exists from the reheater temperature to ambient temperature. The process is rapid once the vapor-nuclei-gas mixture is cooled to the point at which the vapor concentration exceeds the saturation limit. The process results in uniform-sized droplets that can be directed to the equipment requiring calibration. The outlet droplet concentration is typically close to 10^5 droplets cm^{-3} in a flow rate of 3.5 L min^{-1}.

The MAGE also can generate uniform crystals of materials, such as caffeine, that sublime in the bubbler vessel.[49] In these systems, vapor condensation takes place onto the nuclei directly from the vapor to the solid phase and the resulting particles have the crystal habit of the low-volatile substance. The process is analogous to the formation of hoar-frost, and these particles are a useful supply of particle shape standards (see "Particle Standards" section).

In a recent performance evaluation of the MAGE,[48] the following observations were made.

(1) Incomplete drying of the nuclei (NaCl) does not have any significant influence on aerosol monodispersity (DEHS) if the reheater temperature is sufficient to ensure complete evaporation.
(2) Premature condensation of the low-volatile vapor between the bubbler vessel and reheater must be avoided by trace-heating this part of the MAGE.
(3) The reheater temperature should be at least 50°C higher than the temperature in the bubbler vessel, and the establishment of optimum operating conditions for a new material is best achieved by experimental studies.
(4) The size of the aerosol droplets produced by the MAGE is influenced slightly by the amount of liquid in the bubbler vessel. Reproducible aerosols were produced when the depth of liquid was maintained within 0.5 cm of 5.0 cm.
(5) Adjustments to the ratio of bubbler vessel to bypass flow rates resulted in an almost immediate change to the size of the aerosol produced by the MAGE. Stable conditions (at the new size) were achieved within one minute of the alteration in the flow conditions.

Other condensation aerosol generators have been described by Nicolaon et al.[50] and Fuchs and Sutugin.[51]

Electrostatic Classifier

For particles carrying one elementary unit of electric charge, the electrical mobility (Z_p, cm^2 v^{-1} s^{-1}) is a unique function of the particle diameter (D_p, μm) given by:

$$Z_p = \left[\frac{3 \times 10^{15} \, C \, e}{\pi \, \mu \, D_p}\right] \qquad (7.7)$$

where e is the electronic charge constant (1.60×10^{-19} Coulombs), C is the Cunningham slip correction factor (dimensionless) and μ is the gas viscosity (dyn s cm^{-2}). The electrostatic classifier described by Liu and Pui[52] is a primary standard for producing calibration particles of known size, since the electrical mobility-particle size relationship can be determined precisely by Equation 7.7. An example of a commercially available electrostatic classifier (Thermosystems Inc., St. Paul, MN) is shown in Figure 7.20, and produces a monodisperse aerosol by precipitating all particles selectively except those within a narrow size (electrical mobility) range. A pneumatic nebulizer (see "Compressed Air Nebulizers" section) generates a polydisperse aerosol which then is dried and brought to a state of charge equilibrium with bi-polar ions by means of a krypton-85 radioactive source. The charges on the particles obey a Boltzmann distribution, and most particles are therefore electrically neutral or carry a single positive or negative charge. This aerosol is introduced in laminar flow to the classifier (sometimes referred to as a differential mobility analyzer, DMA), which is a cylindrical vessel containing concentric cylindrical electrodes. The inner electrode is maintained at a high voltage and the outer electrode is maintained at earth potential. Charged particles in the aerosol stream flow along the outer tube and are deflected by virtue of their electrical mobility. If a negative potential is applied to the centre electrode, positively charged particles move across the core of clean air before arriving at the rod. Other particles either are unaffected because they are neutral or are deflected to the outer electrode. In either case, they are swept to a filter in the air flow through the classifier. Only particles within a narrow band of electrical mobility have trajectories enabling them to arrive at the exit slot at the base of the centre electrode where they emerge from the classifier as a monodisperse aerosol.

This type of classifier (TSI model 3071) has a useful range from 0.01 to 1.0 μm diameter, and the monodisperse aerosols produced can have σ_g values in the range of 1.04 to 1.08 under the best operating conditions. However, the size distribution of the original polydisperse aerosol must be selected carefully so that the modal size is smaller than the desired size of monodisperse particles. If this precaution is not observed, the dispersity of the aerosol leaving the classifier will be increased by additional particles twice the desired size carrying twice the electrical charge (i.e., having the same electrical mobility). The problem of multiple-charged particles becomes more severe as the size increases,[53] as can be seen from Table 7.2, which shows the percentages of different sized particles carrying from 1 to 4 units of charge at equilibrium with bi-polar ions.

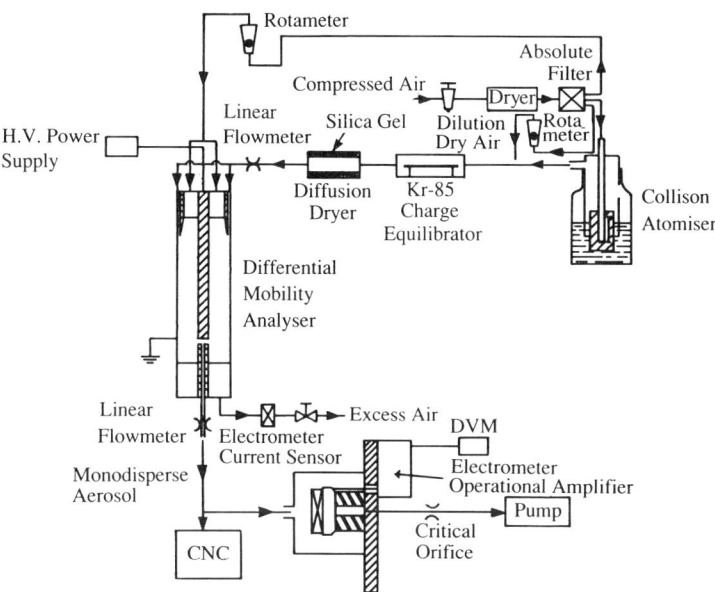

Apparatus for generating monodisperse aerosols by mobility classification of singly charged particles (particle size 0.02-0.1μm)

Figure 7.20. Electrostatic classifier (J.K. Agarwal and G.J. Sem. 1980[161] with permission Pergamon Press Ltd, Oxford).

Table 7.2. Distribution of charges on Aerosol Particles According to Boltzmann's Law

D_p	Percent of Particles Carrying n Elementary Charge Units								
(μm)	-4	-3	-2	-1	0	1	2	3	4
0.01	0	0	0	0.3	99.3	0.3	0	0	0
0.04	0	0	0.2	16.2	67.1	16.2	0.2	0	0
0.08	0	0.1	2.8	23.4	47.5	23.4	2.8	0.1	0
0.10	0	0.3	4.4	24.1	42.5	24.1	4.4	0.3	0
0.20	0.3	2.3	9.7	22.6	30.0	22.6	9.7	2.3	0.3
0.60	3.8	7.4	11.9	15.8	17.4	15.8	11.9	7.4	3.8
1.00	5.4	8.1	10.7	12.7	13.5	12.7	10.7	8.1	5.4

Source: Pui, D. Y. H. and Liu, B. Y. H., p 384 in *Aerosol Measurement,* ed. Lundgren, D.A., University of Florida Press, Gainesville, FL, 1979.[156] (With permission American Chemical Society and University of Florida Press.)

As well as providing a source of monodisperse particles, the classifier can be utilized to calibrate Condensation Nuclei Counters (see section on "Condensation Nuclei Counters"), since almost all the particles carry the same charge that can be measured in an electrometer. The electric current (amperes) is given by:

$$I = Q e N \tag{7.8}$$

where N is the particle number concentration (particles cm^{-3}) and Q is the volumetric flow rate into the electrometer (cm^{-3} s^{-1}).

Other Sources of Monodisperse Particles

Although not widely used today, several species of pollen are commercially available (Polysciences Inc., Warrington, PA, USA) and are highly monodisperse, with D_p values in the range 5 to 30 µm and typical σ_g values between 1.04 and 1.07.[2] Some pollens, such as clover, are spherical with rough surfaces, but most are non-spherical. A species of orchard grass (*Dactlyis glomerata*) has elliptical pollen grains. Certain types of fungi produce monodisperse spores; D_p values for Lycoperdon and Lycopodium spores are 2.09 and 15 µm, respectively, with σ_g values in the range 1.08 to 1.10. Dry dispersion methods (see "Other Dust Generators" section) allow re-suspension of these particles.

Bacteria, red blood cells and, more recently, inorganic powders prepared under carefully controlled conditions have become available as sources of monodisperse aerosol. Titanium dioxide particles (mostly in the anatase form) with a mean D_p of 0.6 µm and σ_g of 1.09 are available in aqueous suspension from Polysciences Inc. This material is formed by high-temperature hydrolysis of aerosol droplets of titanium tetrachloride.[54] Similar sized TiO_2 particles also have been made by controlled hydrolysis of dilute alcoholic solutions of titanium alkoxides.[55] Haggerty and Cannon[56] have reported a method of making sub-micron monodisperse ceramic powders (alumina) by laser-driven gas-phase reactions at extremely high temperatures.

POLYDISPERSE AEROSOLS

Compressed Air Nebulizers

The simplest way to generate a droplet aerosol is by means of a pneumatic nebulizer in which an aerosol of small droplet size is produced by impacting the larger droplets formed in the atomizer spray before they leave the device. The resulting aerosol is polydisperse (σ_g values are typically between 1.5 and 2.0), with a sharply defined cut-off at the upper end of the size distribution. Nebulizers produce aerosols with mass concentrations from 5 to 50 mg m^{-3}, and mass median diameters (MMD) in the range of 1 to 10 µm. The operating principle of most pneumatic nebulizers is illustrated by the DeVilbiss type-40 nebulizer (Figure 7.21). Compressed air at a pressure of 35 to 270 kPa (5 to 40 psig) enters the nebulizer at high velocity from a small-bore tube. The low pressure created by the Bernouilli effect causes liquid to be drawn from the reservoir through a second tube into the airstream. This liquid emerges as a thin filament which ruptures into droplets when accelerated in the airstream. The coarse spray is impacted on the wall of the aerosol outlet tube, where large droplets are deposited and drain back to the reservoir. In some nebulizers, for example, the Retec X-70/N, the impaction surface is a small plastic sphere which is mounted immediately in front of the liquid spray orifice. Most nebulizers produce a maximum number concentration in the range 10^6 to 10^7 droplets cm^{-3}. The performance data for several commercial nebulizers are given in Table 7.3, taken primarily from data provided by Raabe[3] for water droplet aerosols. Two of the nebulizers listed in this table

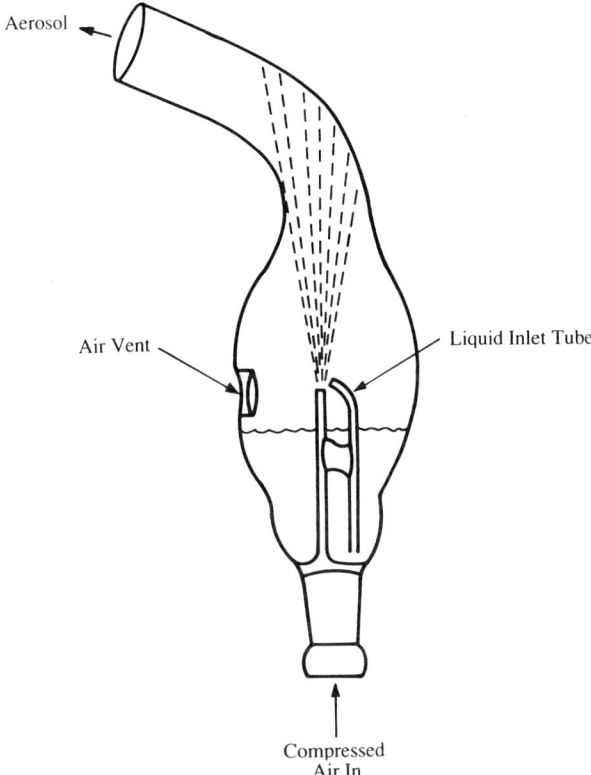

Figure 7.21. DeVilbiss type-40 pneumatic nebulizer.

operate slightly differently: the Laskin nebulizer is submerged in the liquid and the aerosol impinges on the liquid surface as it exits from the jet; compressed air escapes from the Babington nebulizer[57] at high velocity from a small slit or hole in a hollow sphere over which liquid flows to form a film that is shattered by the flow of compressed air to generate the aerosol.

Pneumatic nebulizers may be operated with pure liquids or with dilute solutions in a volatile solvent (e.g., NaCl in water). In the latter case, the resulting aerosol contains mainly sub-micron particles that can provide the starting material for producing a monodisperse aerosol by means of the electrostatic classifier (see "Electrostatic Classifier" section). The use of nebulizers to generate monodisperse polymer latex aerosols has already been described (see "Polymer Latex Particles" section).

Ultrasonic Nebulizers

The dispersing force in ultrasonic nebulizers is the mechanical energy produced by a piezoelectric crystal vibrating in an electric field induced by a high-frequency oscillator. These generators can produce a higher concentration of aerosol than achieved with pneumatic nebulizers, and the MMD of the particles is generally small (< 1 μm). The performance and operating conditions of two types of ultrasonic nebulizer are given in Table 7.4 and an experimental ultrasonic nebulizer designed by Raabe[3] is shown in Figure 7.22. The

Table 7.3. Characteristics of some compressed air nebulizers

Type	Operating Pressure (kPa)	Flow Rate (L min⁻¹)	Output Concn. (g m⁻³)	Droplet Size Distribution MMD (μm)	σ_g	Reservoir Volume (mL)
Babington (Solosphere)	350	15	5	4	—	200–400
Collison	140	7.1	7.7	2.0	2.0	20–1000
Three-jet Collision	210	9.4	5.9			
DeVilbiss type-40 (vent closed)	70	1	16	4.2	1.8	10
	140	16	14	3.2	1.8	
	210	20	12	2.8	1.9	
Laskin	70	48	3.8	0.7	2.1	500–5000
	140	84	4.8			
Lovelace	140	1.5	40	5.8	1.8	4
	210	1.6	31	4.7	1.9	
Retec X-70/N	140	5.4	53	5.7	1.8	10
	210	7.4	54	3.6	2.0	
Acorn Sidestream	140	6.0	70	2.8	1.5	10

Source: Data for water primarily from Raabe, O., p 57 in *Fine Particles*, ed. Liu, B.Y.H., Academic Press, NY, 1976.[3] (With permission Academic Press, Florida, and A.I.H.A.)

Table 7.4 Operating Characteristics of Some Ultrasonic Nebulizers for Water Aerosols

Generator	Operating Frequency (kHz)	Air Flow Rate (L min^{-1})	Aerosol Mass Concentration (mg m^{-3})	MMD	Size Parameters CMD	σ_g
DeVilbiss 800 Setting 2	1350	41.0	54	5.8	—	1.6
Setting 3	1350	41.0	95	6.9	—	1.5
Setting 4	1350	41.0	150	6.9	3.8	1.6
Mistogen 140 no reservoir	1400	24.7	35.6	6.5	3.7	1.5
with reservoir	1400	24.7	61.5	6.5	—	1.4
Mead	800			9.0	4.9	1.7

Source: Data primarily from Dennis, R., Handbook on Aerosols, U.S. Energy Research and Development Administration, 1976.[2]

130   BIOAEROSOLS HANDBOOK

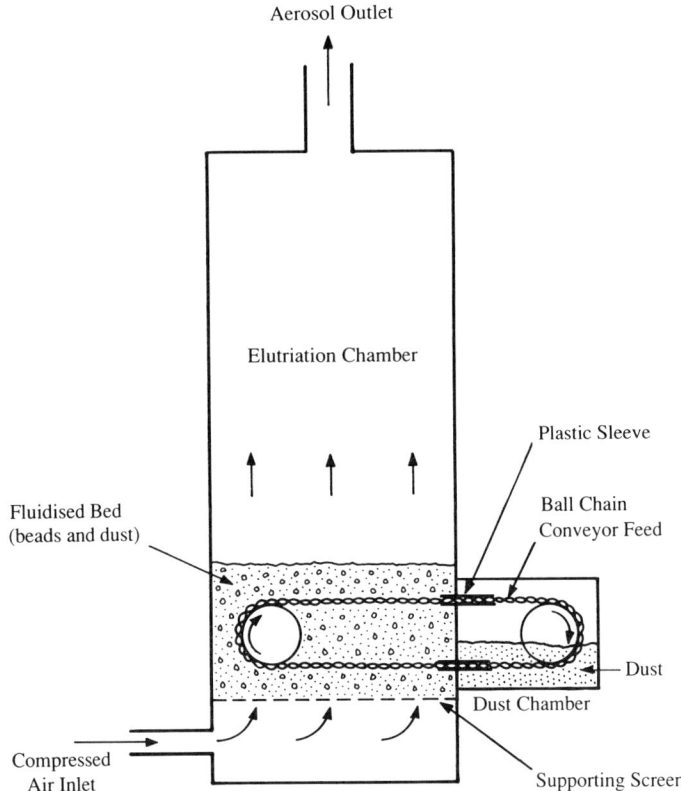

Figure 7.22.   Ultrasonic nebulizer (With permission: O.G. Raabe. U.S.A.E.C. Div. Tech. Info. Rep., 1970)

sizes of droplets formed in this generator depend on the frequency of the acoustic field and the degree of coupling between the piezoelectric crystal and the liquid being nebulized. Droplet coagulation may be a problem because of the high initial particle concentrations.

Fluidized Bed Aerosol Generators

The fluidized bed aerosol generator is one of the most convenient of several methods for the dry dispersion of dusts. Output concentrations in excess of 100 mg m^{-3} can be achieved with this type of generator. The technique has been reviewed by Guichard,[58] and an example of a generator based on the design of Marple et al.[59] is shown in Figure 7.23. A similar aerosol generator is available commercially (TSI model 3400 from Thermosystems Inc., St. Paul, MN).

The Marple aerosol generator has a 5.1 cm diameter fluidized bed chamber filled with 180 μm diameter bronze beads to a depth of 1.5 cm. The dust is eluted directly from the fluidized bed into a chamber containing a krypton-85 radioactive source to achieve electric charge equilibrium. The commercial systems use 100 μm diameter beads and the smaller model contains a cyclone at the outlet of the elutriator to remove any agglomerated particles. The smallest size of particles produced by this technique is about 0.5 μm diameter and particles up to about 50 μm diameter can be suspended by the Marple

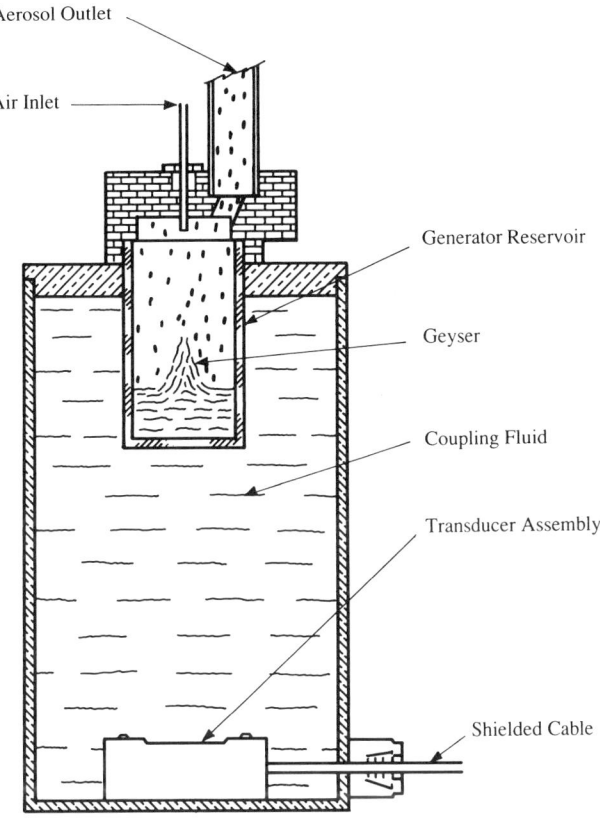

Figure 7.23. Fluidized-bed aerosol generator (V.A. Marple, B.Y.H. Liu and K.L. Rubow. 1978[59])

generator. The dust is metered into the fluidized bed by a conveyor feed, in which the space between each ball in the conveyor chain picks up a fixed volume of dust and transports it to the bed through a close-fitting tube. It takes several hours for the generator to achieve a constant output following the initial introduction of powder but the equipment provides a very stable source of aerosol once constant output has been attained.

Other Dust Generators

A wide variety of methods exist for dispersing dry powders, the mass concentrations varying from less than 1 to more than 100 mg m^{-3}. Many of these systems are described by Dennis[2] and therefore only the most important aspects of their use are presented here. A few of the more common types of dust generators are listed in Table 7.5.[60–66,69–70] Dust generators are not often used in calibration studies because they are not as controllable as other techniques.

The basic requirements for a dust generator are:

(a) a means of continuously metering the powder feed;
(b) a method of dispersing the powder to form the aerosol.

Table 7.5. Characteristics of selected dust feeders

Reference Source	Dust Feed Rate (mg min^{-1})	Comments
Battelle-Northwest [61]	20–80	Used with elutriator for asbestos fibers
Drew and Laskln [62]	50–200	Used with polyester fiber gas dust
Fuchs and Murashkevich [63]	1000–5000	Used with corundum and open-hearth dust
Hounam [64]	< 10	Used with various types of asbestos fibers
Knutson et al [65]	1–1000	Used to deliver two dusts simultaneously
SRI [66]	< 10	Used with carbon-black powders
Wright [60]	1–1000	Used with pulverized coal, silica and lead dust
Blackford and Rubow [69]	0.05–1.5	Used with polystyrene microspheres and other particles collected on filters or deposited on emery paper
Zahradnicek and Löffler [70]	700–7000	Used with wide variety of compact-shaped particles

The simplest self-metering systems are gravity feeds of loose powder into an airstream, usually assisted by vibration. They give an uneven delivery which causes fluctuations in the aerosol concentration. A more stable dust metering system is provided by a cylinder of compressed powder which is eroded or scraped away at a constant rate. This method is used in the Wright dust feed (Figure 7.24[60]), which is reliable for long-term studies because the bulk density of the powder can be controlled during packing.

Powder dispersibility depends on both the physical and chemical properties of the material, particle size distribution, shape and moisture content, and the presence of any electric forces. Incomplete dispersion results in an aerosol particle size distribution larger than that of the original powder. In general, hydrophobic materials such as talc are easier to disperse than hydrophilic substances. Dispersibility increases rapidly with particle size, since it is necessary to supply sufficient energy to a small volume of the bulk powder to separate the particles by overcoming the attractive forces between them. There is a size below which particles cannot be satisfactorily dispersed with a given generator. The most common method of dust dispersion is to feed the powder into a high-velocity airstream so that the shear force in the turbulent flow results in deagglomeration.

Electrostatic charge is a problem common to all dry-dispersion dust generators. Excessive charging results in extensive deposition on nearby surfaces, which reduces the aerosol concentration. Therefore, it is advisable to pass the aerosol as soon as possible after dispersion through an electrostatic charge equilibrator. The problem is exacerbated when dry powders are suspended in conditions of low humidity.

Often it is necessary to introduce a size-selective classifier at the outlet of the dust generator to remove agglomerates and particles larger than the chosen size. Cyclones, impactors, elutriators and sedimentation chambers have each been applied to modify the

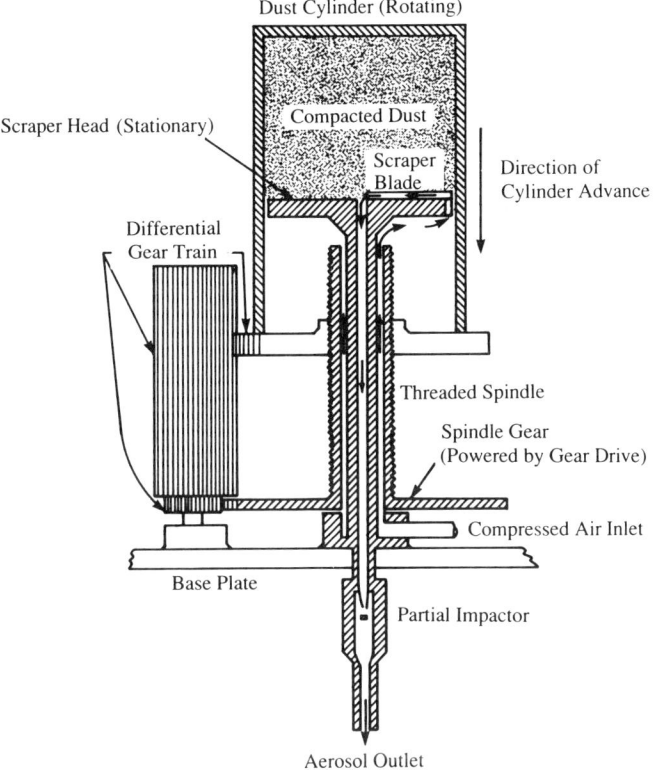

Figure 7.24. Wright dust feed mechanism (B.M. Wright, 1950,[60] with permission A.I.H.A.).

size distribution of powder aerosols. This is particularly important in the calibration of respirable dust monitoring devices, where the test aerosol should meet the size requirements, which define the health-related aerosol fractions as a function of aerodynamic diameter.[67,68]

Two other dry aerosol generators are worth mentioning because they are in fairly wide use. These are the TSI Small-Scale Powder Disperser [SSPD (model 3433)] and the Palas Dry Powder Disperser [DPD (model RBG-100), also marketed for a time by TSI Inc]. The SSPD (Figure 7.25[69]) employs the suction force generated when an air stream is expanded through a venturi to lift particles from a collection substrate such as a filter. The suspended aerosol is deagglomerated by the strong shear force encountered as the particles pass through the venturi section prior to sampling by the instrument being calibrated. The powder to be suspended is gently brushed over the surface of one of the three annular concentric rings of abrasive paper which, in turn, is glued to the upper surface of a rotating turntable (Figure 7.26a). The lower end of the capillary tube connected to the venturi is located just above the particle deposit, while the upper end of the tube floats in the venturi. When air is applied to the inlet, a region of low pressure is created in the venturi which draws gas flow through the capillary tube. The tube therefore acts like a small vacuum cleaner as the turntable rotates slowly beneath. The SSPD also may be operated with circular filters mounted on the turntable or with glass microscope slides cut to fit the holder (Figure 7.26b). The SSPD is most useful with dry powders and is only capable of resus-

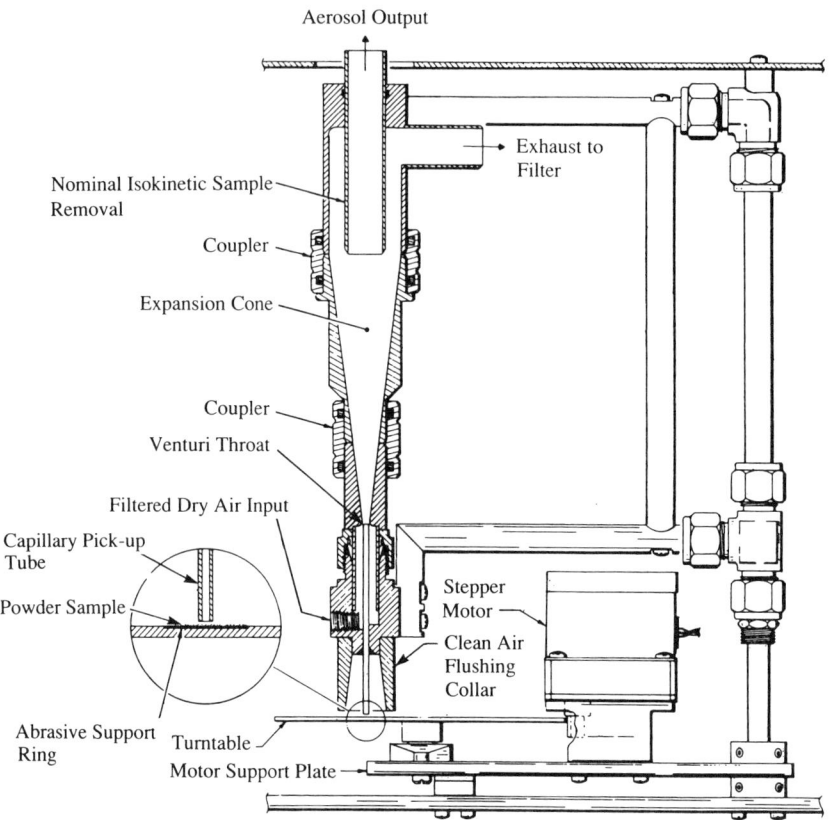

Figure 7.25. TSI small-scale powder disperser (SSPD).

pending small amounts of material (3 to 90 mg h^{-1}), resulting in aerosol mass concentrations between about 3 and 100 mg m^{-3}. It is highly effective at resuspending large-diameter polymer latex particles (> 5 μm diameter) that are not readily produced as an aerosol by atomization/nebulization (see "Polymer Latex Particles" section). Particles in the size range 1 to 50 μm diameter can be easily suspended by the SSPD; smaller particles tend to remain agglomerated after suspension, and larger particles are too heavy to suspend using the capillary aspiration technique.

The DPD (Figure 7.27[70]) operates by forcing a plug of powder into the path of a rotating wire-brush. The loosened powder is dispersed rapidly from the rear of the generator as an aerosol in a flow of clean air. The rate of removal of powder from the compacted mass is proportional to the product of the feed rate and the cross-sectional area of the compacted powder (piston diameter), and is independent of the properties of the powder as long as the material flows freely. Feed rates can be varied between about 40 and 400 g h^{-1}, making it possible to produce aerosol mass concentrations in the range 0.005 and 50 g m^{-3}. The DPD normally produces aerosols of available test dusts (see "Particle Standards" section) in reasonable quantities. The technique is very effective for wind-tunnel calibrations of aerosol analysis equipment where the aerosol produced is highly diluted by the large volumetric flow.

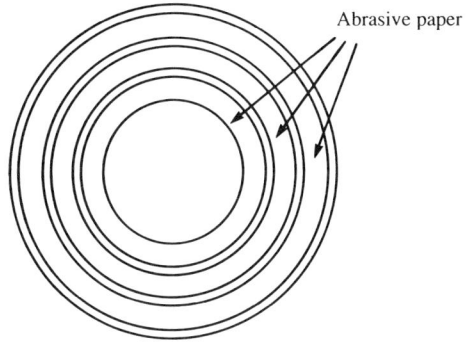

(a) Turntable containing concentric rings of abrasive paper

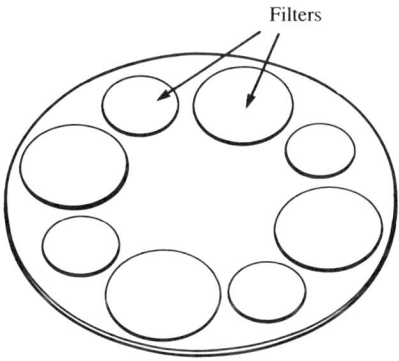

(b) Filter holder turntable

Figure 7.26.   Turntable types for use with TSI SSPD.

Exploding Wire Aerosol Generators

Ultrafine aerosol particles between 0.01 and 0.1 µm diameter can be generated by exploding a metal wire with a massive surge of thermal energy. Phalen[71] has reviewed the technique, and the exploding wire generator of Wegrzyn[72] demonstrates the main features of the method. This aerosol generator (Figure 7.28) was utilized to produce gold aerosols for the study of coagulation rates. The main features are:

(a) 2 cm long, 0.005 cm diameter gold wire, which is mounted between the electrodes,
(b) electrical energy (2000 V at 3 µF), which vaporizes the wire almost instantaneously,
(c) aerosol carrier gas, which is argon.

Aerosols produced by the exploding wire technique consist of chains of smooth, spherical primary particles, whose size distribution is near lognormal with σ_g values typically 2.0. The initial number concentration of primary particles can be extremely high ($>10^9$ particles cm^{-3}), thus promoting rapid agglomeration and the eventual formation of a highly polydisperse aerosol. The size range of the primary particles depends on the concentration of energy that can be passed through the wire. Increased energy results in

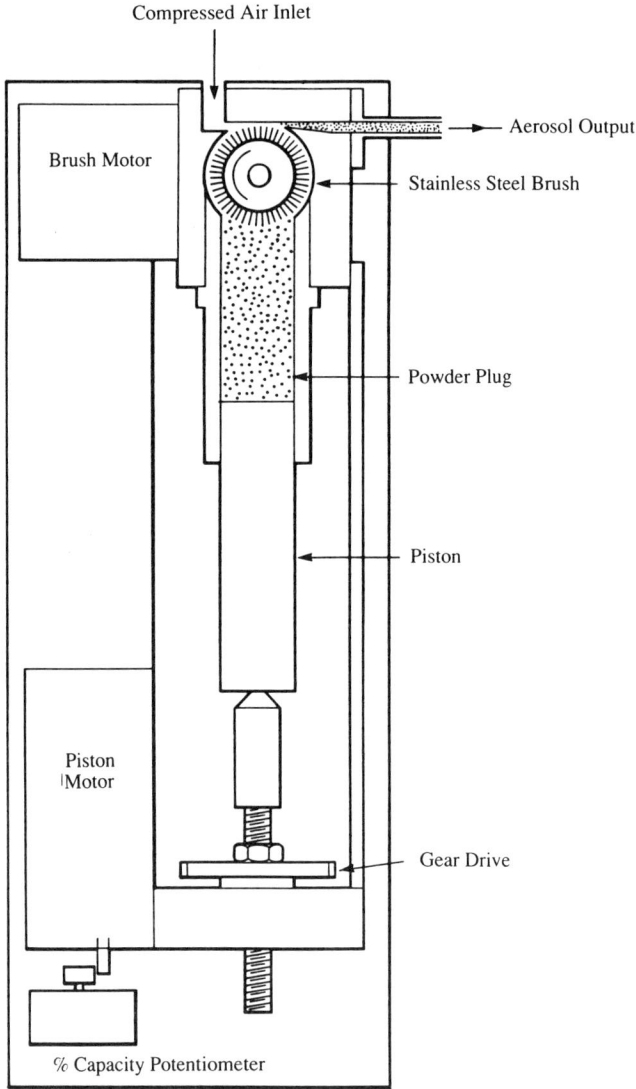

Figure 7.27. Dry powder disperser.

more extensive vaporization and the formation of smaller primary particles. The exploding wire technique also can be used to make various metal oxide aerosols from metal wires by introducing oxygen into the carrier gas.

Gas-Phase Reactions

A number of gases and vapors react chemically with each other to form a solid or liquid, and this process can be employed to generate test aerosols. A list of some of these reactions is given in Table 7.6 and further details are reported by Dennis.[2] One of the simplest methods is the reaction of ammonia gas with hydrogen chloride, which can be accomplished by placing side by side trays containing ammonium hydroxide and hydrochloric

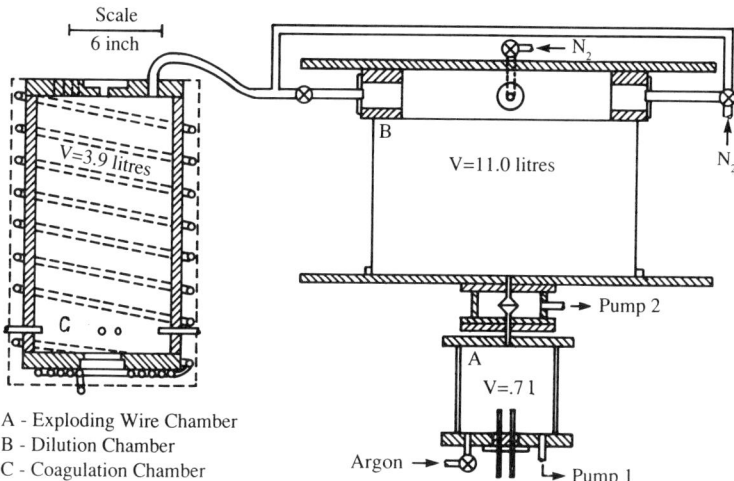

Figure 7.28. Exploding wire aerosol generator (J.E. Wegrzyn, 1976[72] with permission Academic Press, Florida).

acid. The resulting aerosol is polydisperse and consists of primary particles with D_p values in the range 0.1 to 10.0 μm. More careful control of the mixing of the gases results in aerosols that have more reproducible size and concentration characteristics. A simple ammonium chloride generator capable of producing particles with a count mean D_p of 0.7 μm and σ_g of 1.12 is described by Kotrappa and Bhanti.[73] A more sophisticated and controllable gas mixing arrangement was used by Dahlin et al.,[74] in which the ammonia and hydrogen chloride are diluted with nitrogen before mixing.

PARTICLE STANDARDS

Increasingly, it is becoming a requirement that aerosol analysis equipment is calibrated in a manner traceable to national length standards.[75] A variety of reference materials (RMs) have become available from certifying organizations, such as the Bureau of Community Reference (BCR) of the European Commission[76–78] and the U.S. National Institute of Standards and Technology [NIST (formerly NBS)[79–82]]. The current list of particulate Certified Reference Materials (CRMs or SRMs) covers the range of sizes that includes both aerosol and powder analysis equipment; the more important materials for aerosol analyzer calibration are listed in Table 7.7. They include monodisperse polystyrene latex microspheres,[77,80–82] that can be dispersed as described on pp. 103–105, as well as polydisperse particles (quartz powder[78] and glass beads) that usually are dispersed dry (see "Other Dust Generators" section). These materials generally are termed primary standards because of their direct link with national metrology standards.

There is an increasing demand to extend the number of RMs to include particles in the sub-micron size range, which have well-defined optical characteristics, to satisfy requirements for the calibration of optical particle measurement techniques (see "Optical Particle Counters" section). For example, NIST recently has announced the certification of a 0.1 μm diameter polystyrene latex microsphere-based SRM to meet this need.[82]

Table 7.6. Selected methods for generating aerosols from gas-phase reactions

Material	D_p (μm)	Chemical Reaction	Reactor Description	Comments
Ammonium chloride	0.1–1.0	HCl + NH3 → NH4Cl	Simple	Agglomerates rapidly
Carbon black	0.05–1.0	Incomplete combustion of hydrocarbon	Vapor passed over open flame	Forms chains of agglomerates
Silicon	0.1–1.0	Thermal decomposition of SiH_4	Laser heating	
Sulfuric acid	0.3–1.0	Vaporization of fuming H_2SO_4	Simple mixing	Humidity control necessary
Tungsten/ferric oxide	0.1–1.0	Decomposition of $W(CO)_6/Fe(CO)_5$	Heat to vapor to 140–160°C	Dense clouds of W/Fe_2O_3 particles
Various metals, e.g., Mg, Pb, In, Cpd	0.05–1.0	Thermal oxidation in furnace tube at moderate-high temperatures (> 200°C)	Condensation in thermal gradient	Agglomerates form rapidly
Group 1A metals, e.g., Li, Ma, K, Cs	0.5–5	Rapid oxidation at low temperatures (< 200°C)	Localized heat source	Metal oxide reacts rapidly with ambient water vapor and carbon dioxide

Table 7.7. Particle size standards in the aerosol range certified by governmental organizations

Code	Material	Source	Dispersity	RM Particle Size Range (μm)
SRM 1960	Polystyrene spheres	NIST	Monodisperse	10.0
SRM 1690	Polystyrene spheres	NIST	Monodisperse	0.9
SRM 1691	Polystyrene spheres	NIST	Monodisperse	0.3
SRM 1963	Polystyrene spheres	NIST	Monodisperse	0.10
CRM 167	Polystyrene spheres	BCR	Monodisperse	9.48
CRM 166	Polystyrene spheres	BCR	Monodisperse	4.82
CRM 165	Polystyrene spheres	BCR	Monodisperse	2.22
SRM 1003	Glass spheres	NIST	Polydisperse	3–5
CRM 067	Quartz powder	BCR	Polydisperse	2.4–32
CRM 070	Quartz powder	BCR	Polydisperse	1.2–20
CRM 066	Quartz powder	BCR	Polydisperse	0.35–3.5

NIST = U.S. National Institute for Standards and Technology, Gaithersburg, Maryland; BCR = EC Bureau of Community Reference, Brussels. Approximate sizes quoted for polydisperse reference materials.

Table 7.8. Planned BCR Certified reference materials (March 1994)

Size Range	Material Type (μm)
0.1–1.0	Opaque White - Oxide
0.1–1.0	Opaque Colored - Oxide
0.3–3.0	Opaque White - Oxide
0.3–3.0	Opaque Colored - Oxide
1–10	Transparent - Glass/Oxide
1–10	Opaque and Colored - Carbon
3–30	Transparent - Glass
3–30	Opaque and Colored - Carbon
10–100	Transparent - Glass
10–100	Opaque and Colored - Carbon
150–650	Transparent - Glass
150–650	Opaque and Colored - Carbon

Recently, there have been increasing requests for polydisperse RMs to calibrate the large number of different types of optical-based particle size analysis equipment on the

Table 7.9. Test dusts (secondary standards)

Designation	Material Description	Particle Size Range (μm)
BS 1701 (Coarse)	Quartz	0–150
BS 1701 (Fine)	Quartz	0–75
MIL 810	Quartz	0–125
DEFSTAN 0755	Quartz Sand	100–1000
MIL 810 - 5007 (Sand)	Quartz Sand	75–1000
SAE 726J (Coarse)	Quartz	0–125
SAE 726J (Fine)	Quartz	0–75
BS 4552 (MIRA Grade 1)	Fused Alumina	2.5–9
BS 4552 (MIRA Grade 2)	Fused Alumina	3–11
BS 4552 (MIRA Grade 3)	Fused Alumina	6.0–21
BS 4552 (MIRA Grade 4)	Fused Alumina	15.0–53
BS 4552 (MIRA Grade 5)	Fused Alumina	27.0–90
ISO 4020/1 E8	Fused Alumina	3.5 - 10.5
BS 2831 (No 3)	Fused Alumina	8–32
BS 2831 (No 2)	Fused Alumina	0–10

Source: Powder Products Ltd, P.O. Box 5, Spondon, Derby, UK.

market. Although the existing polydisperse RMs from the BCR span a very wide range (from 0.3 to 650 μm diameter), their suitability for calibrating optical particle analyzers is limited; they consist of irregular-shaped quartz particles with variable refractive index. The BCR therefore currently is coordinating a world-wide initiative to augment these standards with a range of spherical polydisperse particles of known refractive index (clear and opaque). These new standards also will find wide use in the calibration of aerodynamic as well as optical-based aerosol analyzers (see "Inertial Analyzers" section), because their density as well as refractive index will be established. Details of these materials are provided in Table 7.8. All RMs are certified by independent laboratories using sizing methods that can be traced to the international standards of length and mass.

The primary particle standards are augmented by a range of secondary standards such as Arizona Road Dust and the MIRA (Motor Industry Research Association) dusts. These test dusts have well-defined size distributions, but have not been tested as part of a formal certification program. Some example secondary standards are listed in Table 7.9. Test dusts generally are available in much larger quantities than RMs and therefore are used more widely in applications where significant amounts of aerosol must be generated (e.g., wind-tunnel testing of samplers).

Particle shape standards are in their infancy, but this topic is likely to become increasingly important as shape-related phenomena are investigated in aerosol analysis equipment.[83–84,86] Recent developments include the formation of compact micron-sized non-spherical particles by controlled crystal growth (Figure 7.29[85–87]), heterogeneous condensation from the vapor directly to the solid phase of materials such as caffeine (see "Condensation Aerosol Generators" section), and the formation of standard fibers by silicon micromachining.[88,89] None of these materials has yet been adopted for certification; they are therefore termed secondary standards.

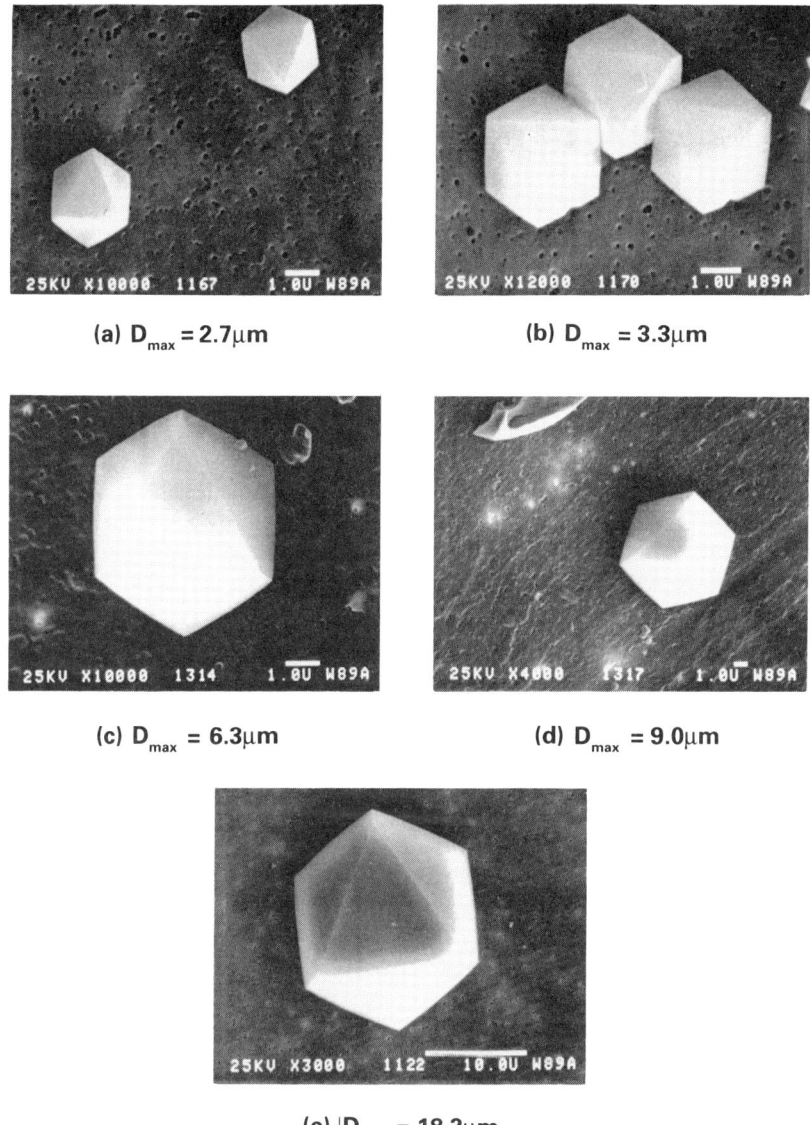

(a) $D_{max} = 2.7\mu m$

(b) $D_{max} = 3.3\mu m$

(c) $D_{max} = 6.3\mu m$

(d) $D_{max} = 9.0\mu m$

(e) $ID_{max} = 18.3\mu m$

Natrojarosite = Na Fe$_3$ (SO$_4$)$_2$(OH)$_6$

Figure 7.29. Natrojarosite particle shape standards.

INSTRUMENT CALIBRATION

The two parameters of greatest importance in the calibration of aerosol analyzers are particle concentration and size distribution. Aerosols can be analyzed in terms of particle number or mass/volume related parameters, and the calibration procedure must take this distinction into account. For instance, impactors and cyclones generally determine the mass concentration and size distribution of an unknown aerosol, since the deposition pattern is assessed more easily by gravimetric-based analytical methods than by microscopic examination of individual particles. However, optical particle counters monitor individual particles to derive number concentrations and size distributions. Number-size distributions can be converted easily to mass-size distributions and vice versa, but large errors can occur in such manipulations, particularly at the extremes of the size distributions. It is good practice, therefore, to choose a calibration method that is appropriate to the type of data expected from the aerosol analyzer. Mass-size calibrations require larger quantities of aerosol than number-size calibrations, unless a sensitive tracer technique is used, such as the addition of a fluorescent dye to the aerosol particles (see "Vibrating Orifice and Related Aerosol Generators" section). Calibrations based on mass analysis are more sensitive to fluctuations in the number of particles larger than the mean size; conversely, calibrations based on particle counting are more sensitive to variations in the number of particles smaller than the mean. Therefore, the removal of outlying particles significantly larger or smaller than the mean size may be a necessary requirement in the preparation of a calibration aerosol.

Aerosol analyzers need to be calibrated for four important reasons:

(a) to ensure that the instrument is functioning correctly (routine quality assurance),
(b) to compare performance with theoretical predictions,
(c) to compare performance with other analyzers of the same type,
(d) to compare measurements of a given aerosol with other instruments working on different principles.

Comparison (d) is often of major importance when undertaking several different measurements of an unknown aerosol, because the geometric diameters of many particles may be quite different from their aerodynamic diameters or the equivalent size parameters obtained by light scattering or electrical mobility measurements.

A few of the more important calibration techniques are described below. The operating principles and limitations of the various types of aerosol analyzers are beyond the scope of this chapter. The descriptions of the analyzers are not meant to be exhaustive, but provide a guide to the current method(s) of calibration. Review articles on the subject have been written by Fuchs (impactors only[90]), Fissan and Helsper,[91] and Liu.[5]

INERTIAL ANALYZERS

The description of the calibration of inertial analyzers is split into three sub-sections; the first section is concerned with impactors and cyclones, the second examines the various spectrometers that operate by inertial separation, and the third section is concerned with the special class of time-of-flight aerodynamic particle size analyzers. All inertial devices measure the aerodynamic rather than the physical diameter, and it is important to determine the aerodynamic properties of the calibration aerosol before use.

Impactors and Cyclones

Both impactors and cyclones fractionate the aerosol into discrete sizes, and the total aerosol concentration can be determined from the total mass of the material collected on all the separation stages. Almost all the operational requirements for impactor calibration apply equally to gas cyclones. Cyclones therefore only are mentioned specifically when it is necessary to point out some distinguishing feature of their performance compared with impactors.

Although single-stage impactors are employed mainly as pre-collectors for other aerodynamic analyzers, it is very common to use several impactors in series for size distribution analyses (a cascade impactor). Each stage must be calibrated, and this procedure can be time consuming for systems that contain as many as eight stages. For this reason, many workers have relied on the calibration curves provided by the manufacturer. This practice should be avoided, since minor changes to the geometry can have a marked influence on the calibration, especially for multi-orifice impactors.

Calibrations may be undertaken with monodisperse or polydisperse aerosols; the main advantage of the former is that the shape and size of the test particles are well defined, whilst the advantage of the latter is that the calibration of the whole impactor can be accomplished in a single experiment. Impactor calibrations have been reported extensively in the literature, and the list given here is rather selective. Good descriptions of calibrations using monodisperse aerosols may be found in References 92 to 98 (describes a cascade cyclone calibration[97]), and calibrations with polydisperse particles are given in References 99 to 101 (describes a cyclone calibration[101]).

The object of the calibration exercise is to characterize the size range within which the collection efficiency of the inertial separator varies from 0 to 100%. The curve of collection efficiency versus particle size often is referred to as the stage characteristic curve, and examples of such curves are given in Figure 7.30 for a cascade impactor and Figure 7.31 for three different gas cyclones. The main feature to note in both figures is the steep change in efficiency for a small change in particle size. An ideal separator would have single step from 0 to 100% efficiency and the size at which this occurred would define unambiguously the performance of the separator. However, in practice this behavior is never observed because of slight irregularities in the trajectories of the particles, particle bounce and re-entrainment.

When calibrating an impactor or cyclone with a polydisperse aerosol, size analysis of the particles on each stage of the sampler usually is made by optical or electron microscopy. Particles collected in definite, narrow intervals of size between d and d + δd are counted, and a similar analysis is made of the aerosol, which penetrates the stage and is collected on a similar flat surface or a membrane filter. The ratio (R) of the particle number within the size band from d to d + δd found in the deposit to that in the aerosol penetrating the stage is given by:

$$R = \left[\frac{\eta(d)}{1 - \eta(d)}\right] \quad (7.9)$$

where η is the collection efficiency for that size range. The procedure is repeated until the complete characteristic curve of the stage has been determined. This is relatively simple to carry out, especially if automated image analysis facilities are available to scan the deposits. However, significant systematic errors can arise because of the following factors:

(a) small particles may be missed (especially true for optical microscopy, where particles less than approximately 0.5 μm diameter are below the lower limit of resolution),
(b) aggregates may be erroneously counted as single particles,
(c) the sample may be unrepresentative (mainly applicable to slot impactors, which concentrate the deposit on the collection surfaces immediately beneath the slot-ends[101,102]).

If the particles are sized by microscopy, physical rather than aerodynamic diameters are measured and the relation between the two parameters may be difficult to determine because of irregular particle shapes and internal porosity. A more serious problem may arise

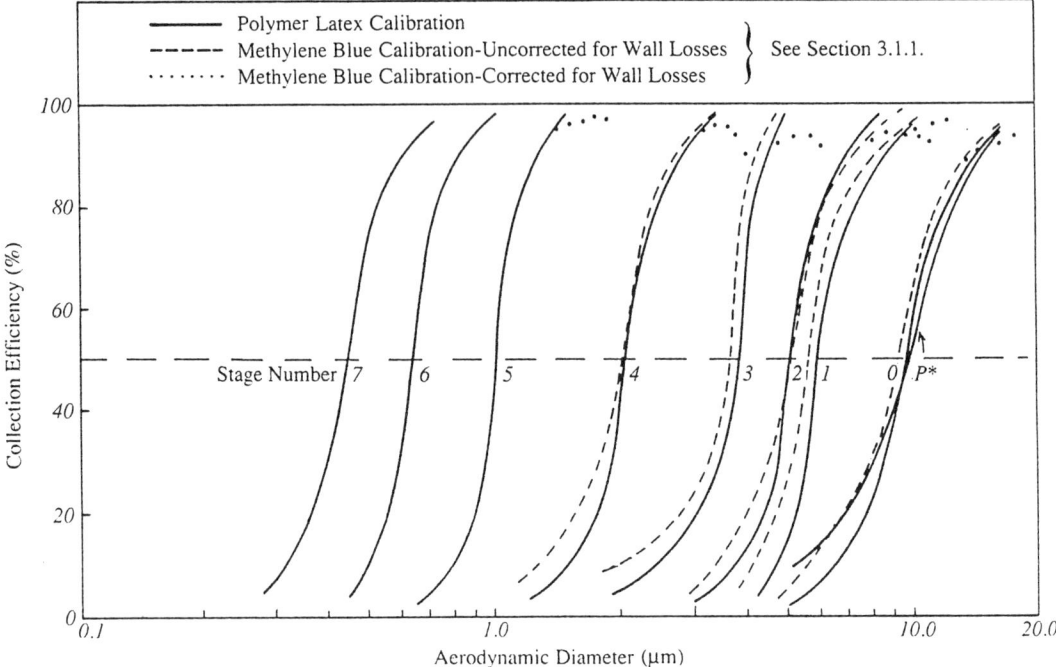

Figure 7.30. Cascade impactor calibration curves. Methylene Blue data only shown for the pre-separator (P). (Data from Mitchell et al.[106])

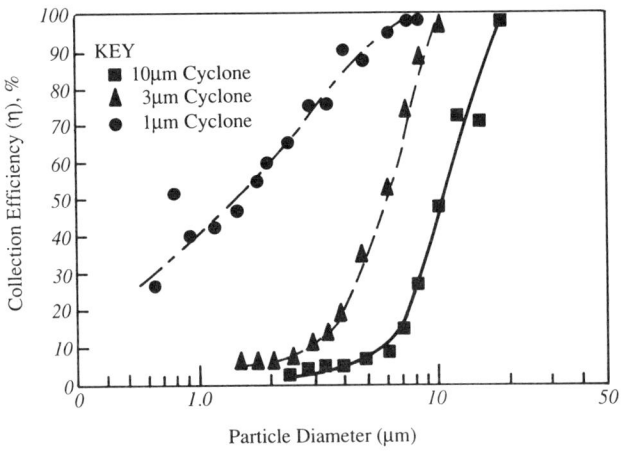

Figure 7.31. Calibration curves for three cyclones (J.C.F. Wang and M.A. Libkind, 1982[98] with permission Sandia National Laboratories).

if the particles are significantly elongated (e.g., fibrous), since the shear flow in the region of the impactor jet tends to orientate them so that their long axes are parallel with the direction of flow and perpendicular to the impaction plate.[103] Thus the measured aerodynamic size will be greater than the correct value would have been if they were orientated at random.

Optical particle analyzers (and the TSI Aerodynamic Particle Sizer) have been employed to count the particles during the calibration of impactors or cyclones. However, these monitoring devices may have appreciable cross-sensitivity; particles of equal size produce signals that spread across the neighboring channels of the multi-channel analyzer. The likelihood of cross-sensitivity increases with the narrowing of the size range corresponding to each channel, making it desirable to group several channels in an instrument that has more than 6 to 8 channels which span the size range of the impactor or cyclone. This procedure has the added advantage that statistical errors are reduced by increasing the count for each size band. On the other hand, reduced resolution can be a problem, particularly in the middle region of the efficiency curve, where the rate of change of efficiency with particle size is at its greatest.

In view of all the problems associated with the use of polydisperse calibration particles, many workers prefer to calibrate aerosol analyzers with monodisperse particles, despite the increased effort required. Data analysis and interpretation are much easier particularly for spherical, monodisperse particles, since their aerodynamic sizes usually are well defined and the mentioned orientation effects do not occur. Samplers have been calibrated with monodisperse solid and liquid particles; particle bounce is more likely to occur when solid particles are used,[92] whilst liquid droplets may shatter on impact with the collection plates. The recommended approach is to calibrate with solid particles if the sampler is for analyses of solid particles, and with liquid droplets when sprays and mists are being studied.

Two techniques have been developed to calibrate impactors and cyclones with monodisperse particles.[93,104-108]

(a) Well-characterized polymer latex particles are utilized, and the particle number concentration is measured upstream and downstream of the sampler. The procedure is repeated at different particle sizes until the collection efficiency curve has been derived. The size corresponding to a collection efficiency of 50% is an important parameter defining the performance of the sampler and is often referred to as the effective cut-off diameter (ECD). For a single-stage sampler, if N' and N are the upstream and downstream particle concentrations, respectively, the collection efficiency ($\eta(d_i)$) for particles of size d_i is given by:

$$\eta(d_i) = \left[1 - \frac{N}{N'}\right] \quad (7.10)$$

The experimental procedure for calibrating cascade samplers has been described in detail by Rao and Whitby,[104] and their arrangement is shown in Figure 7.32. The optical particle counter is located downstream of the impactor for all the measurements, and the airstream leaving the impactor is split so that any excess air is diverted away from the counter. Only the relevant stages of the impactor need to be assembled for calibration: the stage of interest and the stages that come before the stage (Figure 7.33). The upstream concentration (N') is obtained by making measurements without the collection plate immediately beneath the stage being calibrated, and the downstream concentration (N) is measured with this collection plate in position.

(b) The aerosol analyzer can be calibrated with reasonable ease using particles detected by gravimetric or chemical methods. Popular choices of solid calibration particles are methylene blue, ammonium fluorescein and nigrosine dye. Du Pont Pontamine Fast Turquoise 8GLP dye has been used for calibrating cyclones at moderate temperatures (177°C) that would result in the decomposition of other materials.[109] DOP and oleic acid often are chosen as liquid droplet calibrants. Both types of standard aerosol are generated usually by means of a dispersion-type aerosol generator (see "Vibrating Orifice and Related Aerosol Generators" sec-

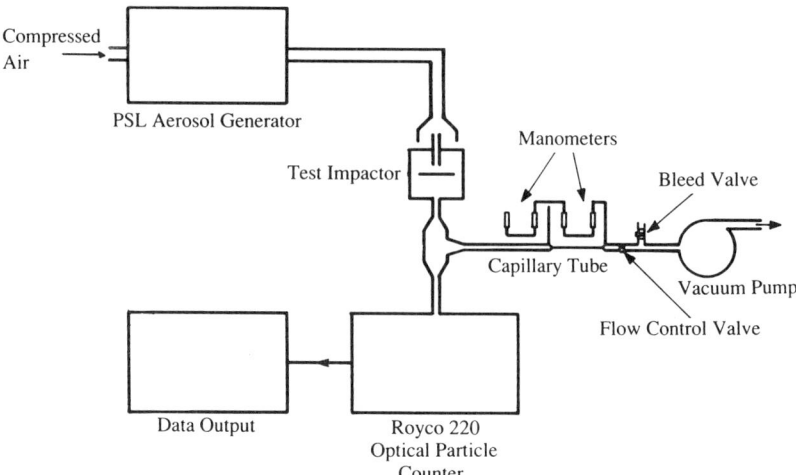

Figure 7.32. Impactor calibration with polymer latex particles (A.K. Rao and K.T. Whitby, 1978[104] with permission Pergamon Press Ltd, Oxford).

tion). The efficiency of stage i (η_i) in a cascade sampler containing x stages is given by:

$$\eta_i = \left[\frac{M_i}{\sum_i^x M_i} \cdot 100 \right] \quad (7.11)$$

where M_i is the mass of particles collected on stage i. The denominator on the right hand side of Equation 7.11 represents the sum of the mass loadings from stage i to the bottom stage (stage x).

Both methods have been employed at the Aerosol Science Centre of AEA Technology to calibrate an Andersen Mk-II eight-stage impactor. Good agreement was found with the manufacturer's calibration curves, which were obtained with oleic acid droplets (Figure 7.34), demonstrating the equivalence of these procedures.

If the impactor or cyclone is functioning correctly, the collection efficiency curve is a smooth monotonic function of particle size. However, the curve can be distorted by non-ideal collection behavior arising from:

(a) particle bounce, causing the curve to reach a plateau at an efficiency less than 100% and usually decreasing at larger sizes,
(b) particle filtration when glass fiber or paper filters are the collection surfaces, causing the curve to rise less steeply than for a non-porous collecting medium,
(c) surface irregularities, which cause the collection efficiency to increase at small particle sizes, producing a prolonged tail in the characteristic curve.

The effect of particle bounce is illustrated in Figure 7.35 and may cause a significant shift in the ECD for a given stage. Wall losses cannot be detected from the shape of this calibration curve and must be measured by mass balance studies of the aerosol deposited

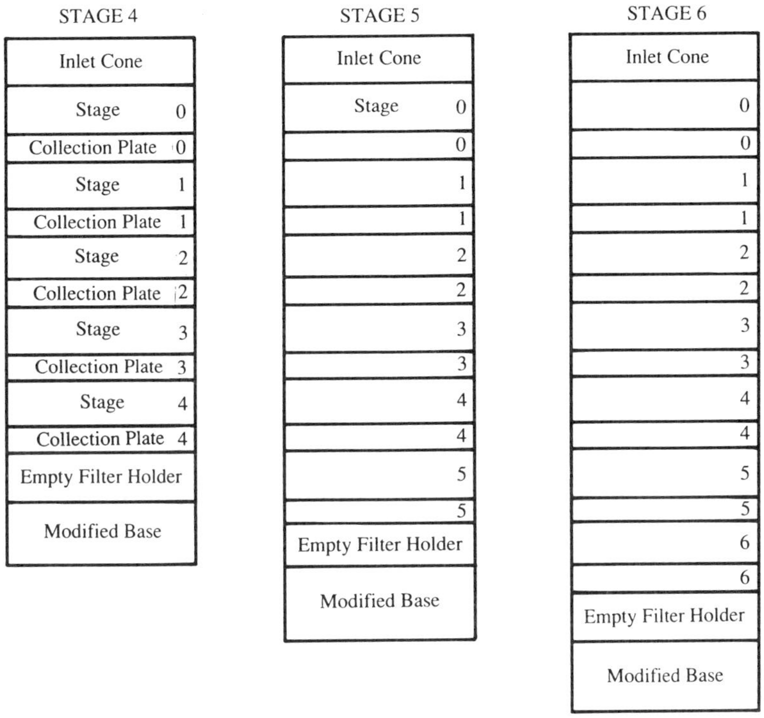

Figure 7.33. Assembly of cascade impactor for calibration with polymer latex particles.

throughout the sampler, including the internal walls.[95] For this reason it is useful to calibrate the sampler by the gravimetric/chemical method described above. Wall losses always should be expected in impactors when particles are being measured with aerodynamic diameters exceeding 5 µm.

Care is needed to ensure that electrostatic charges associated with the calibration particles are minimized, since particles produced by atomization methods frequently are highly charged. The process usually is carried out by passing the calibrant aerosol through a tube containing a Kr-85 radioactive source (for example, see Figure 7.6); the resulting particles emerge with a low charge level corresponding to the Boltzmann charge distribution (Table 7.2). In a recent calibration of a California Measurements QCM low-pressure cascade impactor with oleic acid droplets produced by a VOAG, charged 5 µm diameter particles deposited away from the sensitive region of the oscillating quartz crystal (Figure 7.36) to give erroneous results.[110] Similar particles after passage through a charge equilibrator deposited correctly (Figure 7.37) in the central region of the sensing crystal. The removal of excessive electrostatic changes also reduces parasitic deposition of calibration particles on internal walls of the impactor.

Virtual impactors do not have particle collection surfaces and both fractions of the aerosol remain airborne after inertial separation has occurred. The calibration of a virtual impactor has been described by Loo et al.,[111] O'Connor[95] and Chen et al.;[112] this last paper describes two approaches to calibration that are analogous to the methods for conventional impactors. Polystyrene latex particles were used, and particles in both the minor and major flows were counted alternatively with a TSI Aerodynamic Particle Sizer (Figure 7.38). The collection efficiency (η) was determined from the relationship:

148   BIOAEROSOLS HANDBOOK

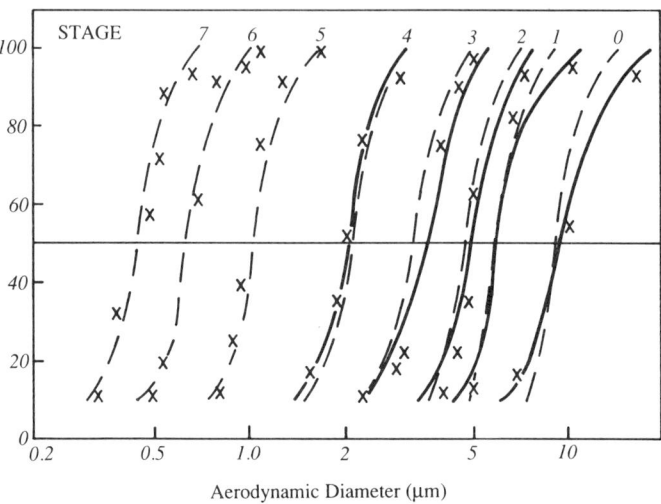

Figure 7.34. Andersen MK-II cascade impactor calibrations.

$$\eta = \frac{c_a}{c_a + \left[\dfrac{c_b\, Q_b}{Q_b - Q_c}\right]} \tag{7.12}$$

where c_a and c_b are the number concentrations measured in the minor and major outlet flows from the impactor, respectively, and Q_b and Q_c are the volumetric flow rates in the major and bleed-off airstreams, respectively. At low flow rates the bleed-off flow was replaced with dilution air to satisfy the requirements of the particle counter, and the collection efficiency of the virtual impactor was calculated according to:

$$\eta = \frac{c_a}{c_a + c_b} \tag{7.13}$$

Wall losses were measured in the second method used by Chen et al.,[112] and the impactor was calibrated by the liquid droplet fluorescent tracer technique (see "Vibrating Orifice and Related Aerosol Generators" section) using fluorescein-doped DOP. The major and minor flows from the impactor were drawn through filters to remove the DOP for subsequent analysis (Figure 7.39). The collection efficiency (η) was calculated for each particle size using the equation:

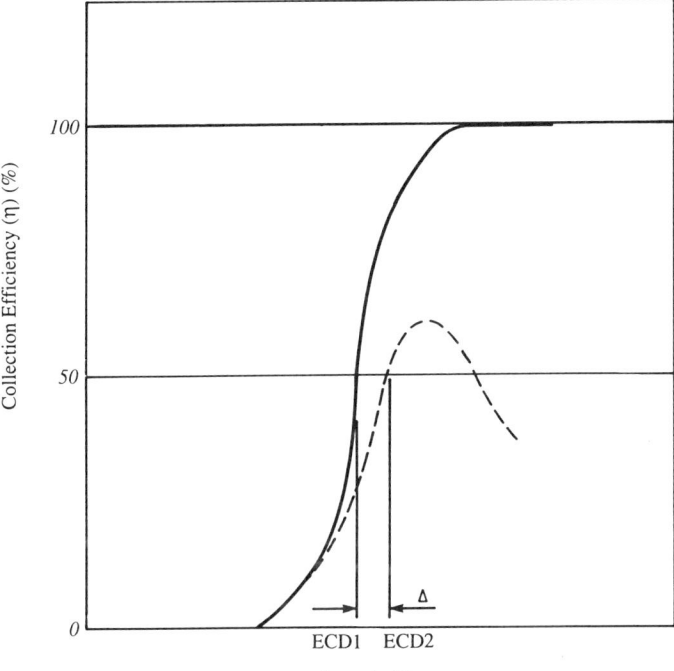

Figure 7.35. Effect of particle bounce on the shape of the collection efficiency curve of a single impactor stage.

$$\eta = \frac{m}{m + M} \qquad (7.14)$$

where m is the mass of fluorescein collected from the minor flow (large particles), and M is the mass of fluorescein deposited from the major flow (small particles). Wall losses were obtained by washing the inside surfaces of the impactor (Figure 7.40), and applying the expression:

$$WL_i = \frac{Mw_i}{M_w + M + m} \qquad (7.15)$$

where WL_i is the wall loss on the i th section of the impactor and Mw_i and Mw are the mass of fluorescein deposited on the i th section and the total mass collected on all internal surfaces, respectively.

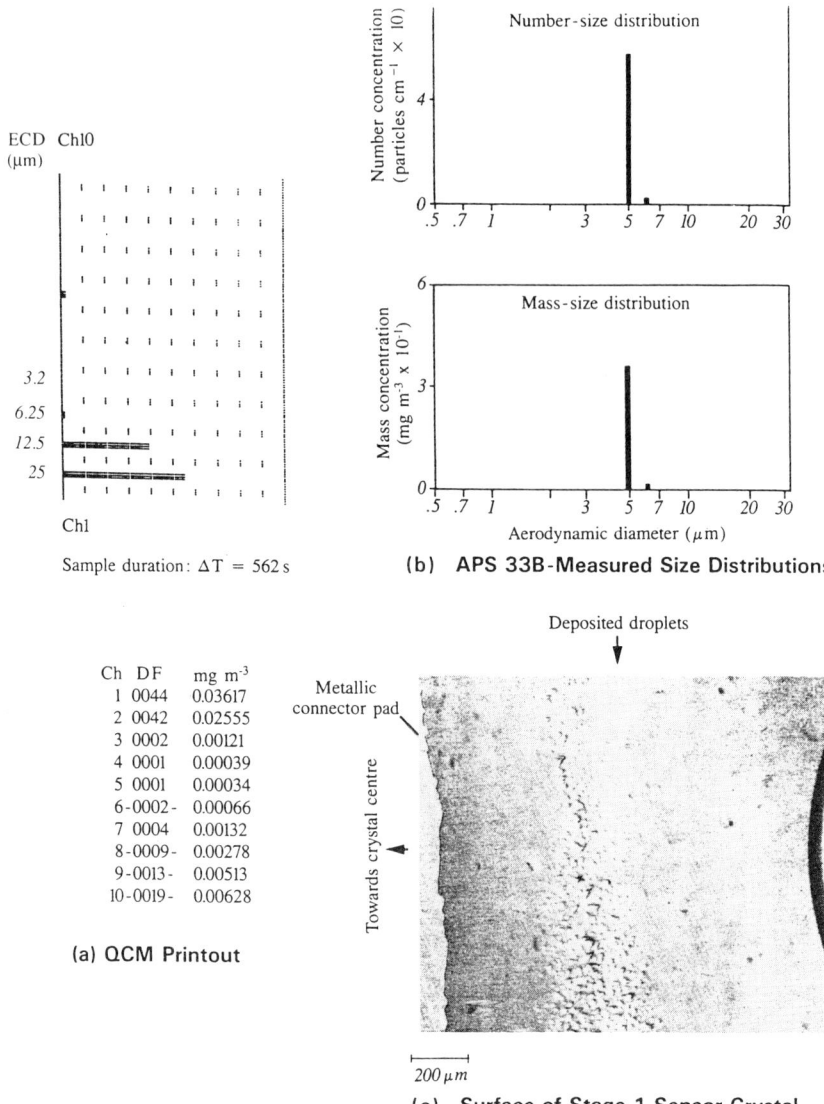

Figure 7.36. QCM impactor—collection of 5 μm aerodynamic diameter oleic acid droplets without electrostatic charge equilibration.

Inertial Spectrometers

All the particles that pass through an inertial spectrometer are collected on a single deposition surface (metal foil or membrane filter) after size separation has occurred. This process means that the entire deposit can be scanned by a microscope, and the aerodynamic size of the particles determined as a function of deposition distance from the inlet to the spectrometer. The aerosol collection medium after use can be sectioned also into smaller size fractions for higher resolution mass analysis.

Several instruments have been designed since 1955 that can be classified as inertial spectrometers. These include the Timbrell spectrometer,[113] various spiral duct centrifuges (Goetz,[114] Stöber,[7,115,116] Los Alamos[117] and Lovelace Foundation[118]) and the Inspec designed

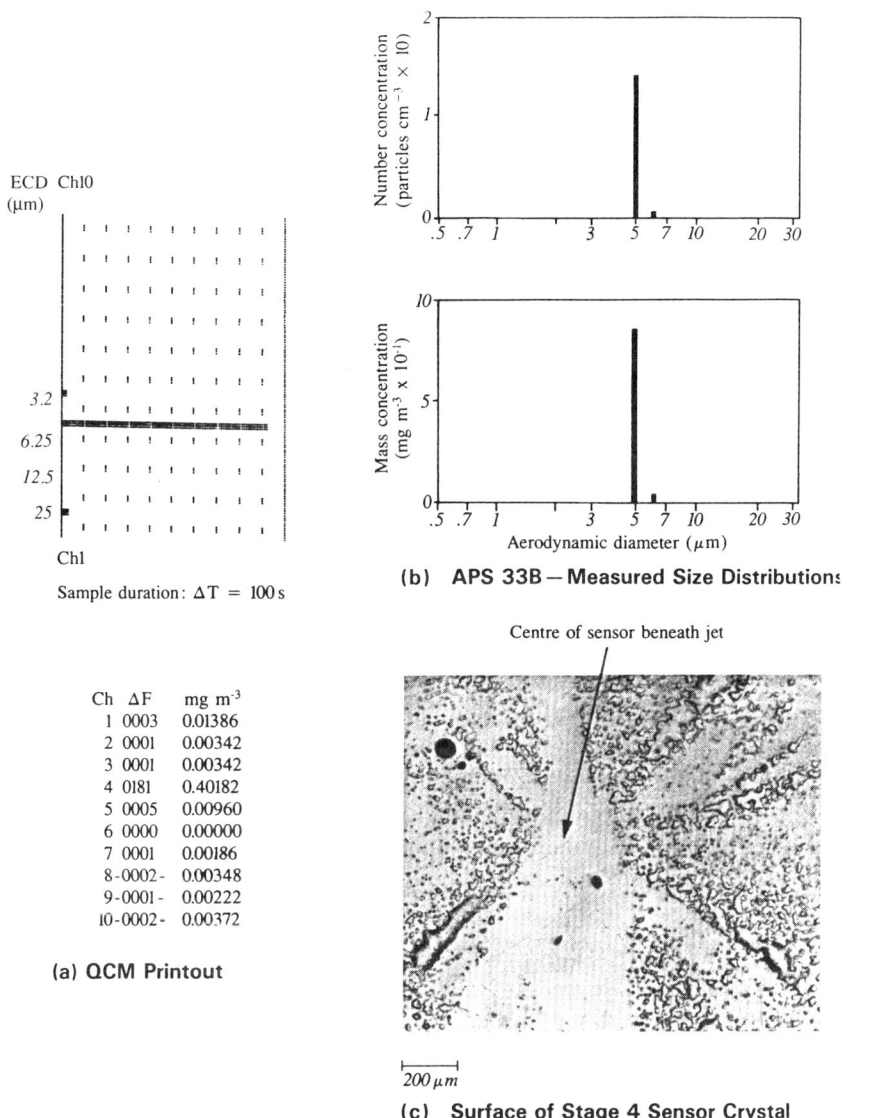

Figure 7.37. QCM impactor—collection of 5 μm aerodynamic diameter oleic acid droplets after electrostatic charge equilibration.

by Prodi et al.[119,120] Very careful control of the operating parameters is required with all of these instruments for high size resolution. These instruments are calibrated best with monodisperse polymer latex particles, since useful information can be obtained also from the deposition profiles of the multiplets (see "Polymer Latex Particles" section). Stöber and Flaschbart[116] calibrated their spiral duct centrifuge with latex microspheres, and were able to resolve aggregates consisting of twenty-three 1.8 μm diameter polystyrene particles from those containing twenty-two particles. Similar studies have been undertaken at the Aerosol Science Centre of AEA Technology to calibrate the Inspec with 2 μm diameter polyvinyltoluene microspheres,[121] showing that this instrument could resolve multiplet aggregates

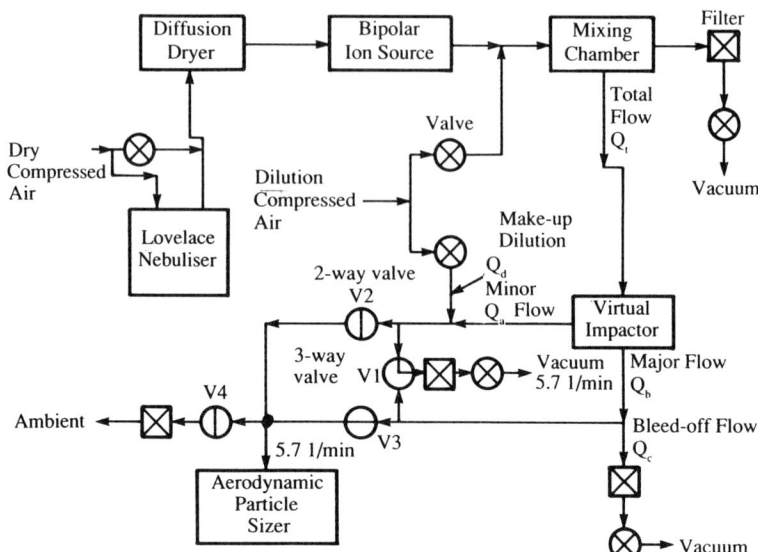

Figure 7.38. Calibration of a virtual impactor with polymer latex particles (B.T. Chen, H.C. Yeh and Y.S. Cheng, 1985[112] with permission Pergamon Press Ltd, Oxford).

containing as many as eight particles (Figure 7.41). A similar calibration has been undertaken also with a Timbrell spectrometer,[122] demonstrating that optimum size resolution occurs between 4 and 10 μm aerodynamic diameter (Figure 7.42).

The calibration curve for an inertial spectrometer relates the deposition distance along the collection medium to the aerodynamic sizes of the test particles. This curve is a function of the air flow rates through the spectrometer, and also may depend on other factors specific to the instrument such as the revolution rate of the rotor in a centrifuge. A Stöber spiral duct centrifuge has been calibrated at the Aerosol Science Centre (Figure 7.43[123]) using ten different sizes of latex particles, probably the minimum required to obtain adequate resolution. Figure 7.43 also indicates the effect of the rotor revolution rate on the calibration curve. A similar set of calibration curves is shown in Figure 7.44 for an Inspec, where the influence of total air flow rate through the spectrometer is also evident. Ideally, the central region of these curves requires more data points than the extremities, because of the greater sensitivity to particle size in this region.

Non-ideal collection behavior can occur in inertial spectrometers when they are used to sample highly concentrated aerosols.[123–125] The onset of this cloud-settling behavior depends on the fluid dynamics within the spectrometer and cannot be readily detected when using a monodisperse aerosol. Figure 7.45 shows the effect that cloud settling can have on the resolution of a Stöber spiral duct centrifuge. The upper photograph shows well-resolved sodium chloride particles of approximately 1 μm aerodynamic diameter, obtained at an aerosol concentration of 1 g m^{-3}. The lower photograph was taken at the same location along the deposition foil when the aerosol concentration exceeded 10 g m^{-3}, and shows no evidence of size resolution: the particles moved as a single body or cloud that deposited as unseparated particles on the collection foil. The test particles for these high concentration aerosol studies were produced by means of a salt-stick generator,[126] in which a magnesium oxide stick impregnated with sodium chloride was passed through an oxy-

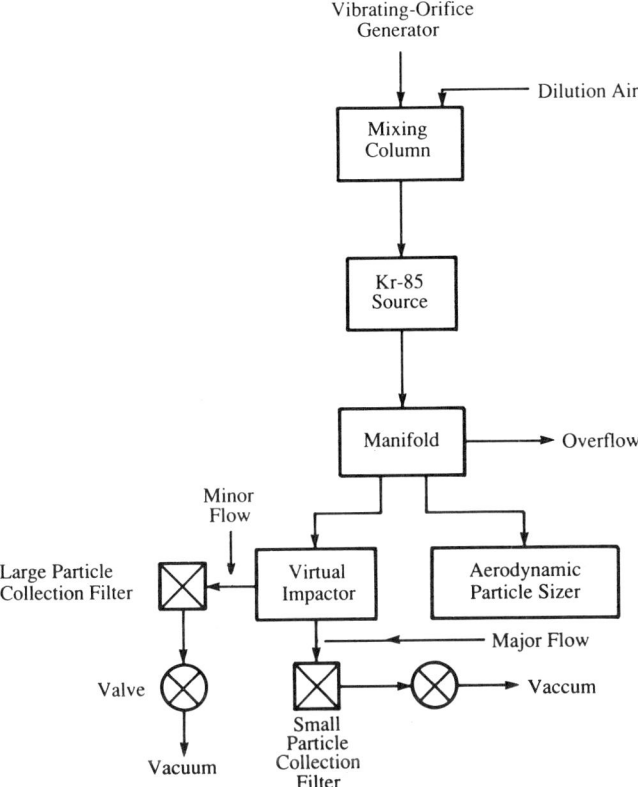

Figure 7.39. Calibration of a virtual impactor with monodisperse DOP droplets (B.T. Chen, H.C. Yeh and Y.S. Cheng. 1985[112] with permission Pergamon Press Ltd, Oxford).

propane flame. Sodium chloride evaporated in the flame and rapidly condensed to form a highly concentrated aerosol with a mass median aerodynamic diameter ranging from 1 to 2 μm and a σ_g of about 2.0.

Real-Time Aerodynamic Particle Sizers

Calibrations of the TSI Aerodynamic Particle Sizer (APS33B) and Amherst Aerosizer [Amherst Process Instruments Inc., Hadley, MA, USA (API)] differ from the previous two classes of inertial sampler, because they measure aerodynamic size without collecting or trapping the aerosol particles. In the APS33B the particles are separated on the basis of their inertia by accelerating them through a well-defined nozzle and timing their flight between the two portions of a split laser beam. This instrument detects the light scattered by the particles as they cross the laser beams, and the signal starts and stops a timer. Thus the time taken for a particle to travel a known distance at a fixed flow rate is measured, and can be related directly to the aerodynamic diameter. The APS33B has some of the characteristics of an optical particle counter, which enable full advantage to be taken of the excellent size resolution of the technique.

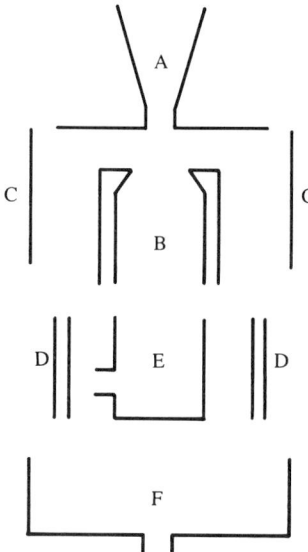

Figure 7.40. Parts of a virtual impactor that were washed separately to analyze for wall losses (B.T. Chen, H.C. Yeh and Y.S. Cheng. 1985[112] with permission Pergamon Press Ltd, Oxford).

The APS33B has been calibrated several times,[127-131] in which good size resolution has made the use of the highest quality monodisperse aerosols essential. A calibration curve obtained with polymer latex particles at the Aerosol Science Centre of AEA Technology is shown in Figure 7.46, and demonstrates the highly non-linear relationship between the channel number of the multi-channel analyzer and the aerodynamic diameter. As a consequence of this non-linearity, the resolving power at the lower end of the particle size range (≤ 1 µm) is reduced greatly in comparison with larger particles. Considerable care must be taken to ensure that the operating parameters (flow rates and pressures) are adjusted correctly before calibrating the instrument, because any small variations can have a significant effect on the calibration of this instrument.[130]

Some care is required also in the choice of calibration aerosol; Baron[128] and Griffiths et al.[130] have shown that the calibration curve is displaced when liquid droplets are calibrants rather than solid particles (Figure 7.47). It is generally accepted that liquid droplets can be distorted to form oblate spheroids in the high-velocity flow field within the region of the laser beams. The droplets appear smaller in aerodynamic size than if they were spherical, and the calibration curve can be displaced by as much as 20% for droplets of approximately 15 µm diameter. This effect decreases as the droplets are reduced in size and is probably insignificant for droplets smaller than 2 µm diameter. The magnitude of the distortion is affected by the surface tension of the liquid, and small differences can be expected in the calibration curves for different liquids.[128] The APS33B therefore should be calibrated with solid particles if it is to be used to study solid aerosols, and with the liquid of interest when studying droplets.

Both the APS33B and the Aerosizer which operates on a similar principle, have been shown to undersize solid non-spherical particle shape standards (see "Particle Standards"

Optimum Resolution of 2.02μm Diameter Particles (ref. 121)

$Q_a = 3.6 \text{ l h}^{-1}$ = Aerosol Flow Rate
$Q_w = 480 \text{ l h}^{-1}$ = Winnowing Air Flow Rate

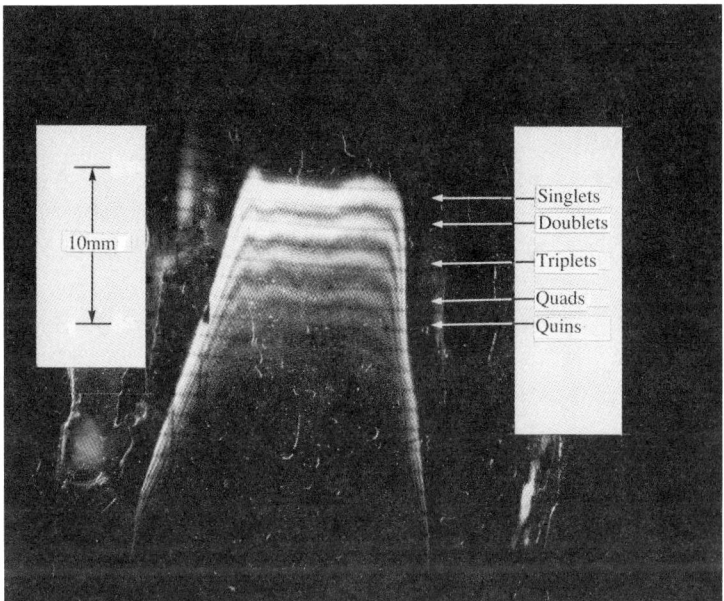

Figure 7.41. Resolution of polymer latex particles with the Prodi inertial spectrometer (Inspec).

section). The effect is substantial (Figure 7.48a and 7.48b); 15 μm aerodynamic diameter particles may be undersized by as much as 25% in the APS33B and 40% in the Aerosizer.[84,132] Recent theoretical studies indicate that the dynamic shape factor of the calibrant particles increases as they are accelerated; however, further work with a wide variety of different particle shapes is required to confirm this hypothesis.

Calculations by Wilson and Liu[133] indicate that, for particles of the same aerodynamic diameter, the particle with the higher density will be measured as larger by the APS33B. This effect was confirmed by Baron,[128] who measured an increase of about 7 to 9% for spherical particles with a density of 2 g cm^{-3} compared with spherical particles of equivalent size but density of 1 g cm^{-3}. Wang and John[134] have determined a density correction factor for the APS33B, and the manufacturer provides the option to allow for the effect of density in the operating software. This effect should be considered if the APS33B is to sample aerosols with densities appreciably different from that of the calibrant.

The APS33B also can count individual aerosol particles in a similar manner to optical particle counters. However, particle coincidence will occur if the aerosol number concentration exceeds a few hundred particles cm^{-3} (Table 7.10). Baron has suggested that coincidence may occur also between sub-micron particles not normally detected by the APS33B, and between any electronic noise and larger particles. There are indications that

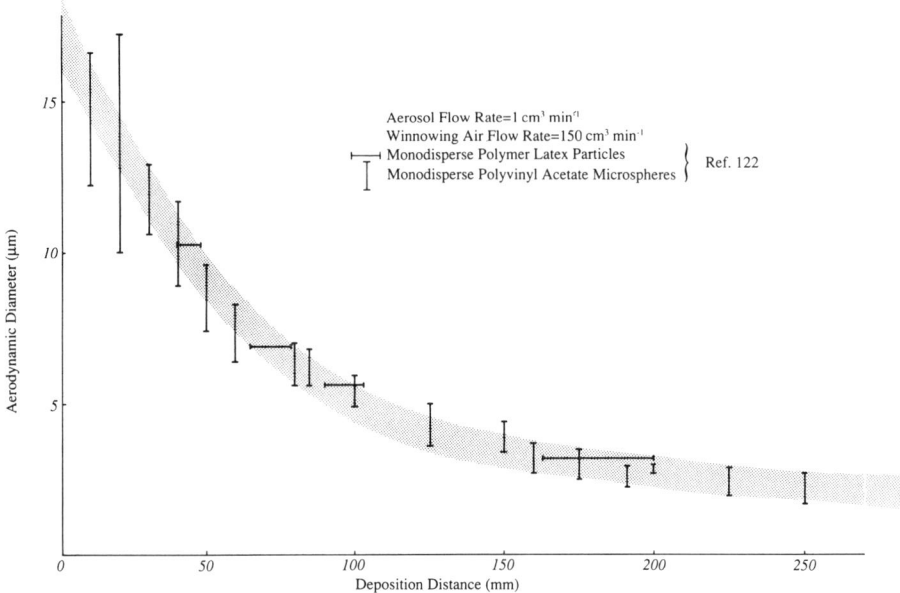

Figure 7.42. Calibration of a Timbrell spectrometer.

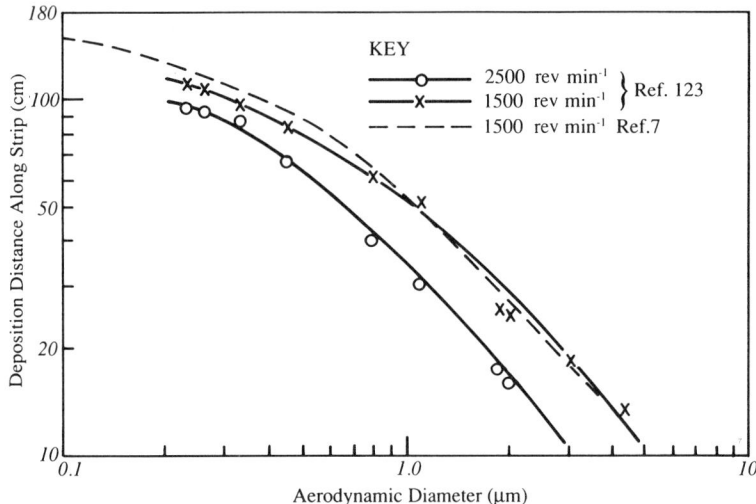

Polystyrene Latex Microspheres: Aerosol Inlet Flow Rate 0.5 l min^{-1}, Total Air Flow Rate from Centrifuge 10 l min^{-1}

Figure 7.43. Calibration curves for a Stöber spiral duct centrifuge.

particles smaller than 0.7 μm diameter may not be detected at 100% efficiency,[135] but to date there has been no rigorous study of this problem.

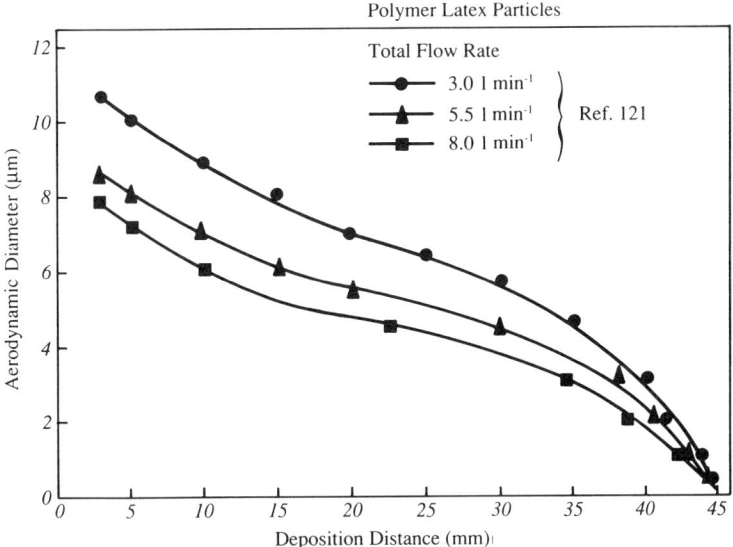

Figure 7.44. Calibration curves for a Prodi inertial spectrometer (Inspec).

OPTICAL PARTICLE COUNTERS

Optical particle counters (OPCs) are a class of aerosol analyzer that measure the quantity of light scattered by individual particles as they pass through a beam of intense monochromatic or white light. This distinguishes them from instruments that are based on Lorenz-Mie-Fraunhofer diffraction (Malvern-type particle sizers) and laser (phase) Doppler velocimetry, neither of which are considered here.

In a single-particle OPC, the beam of light is focused onto a "view-volume" through which the particles pass one at a time. The amount of scattered light is measured by a photosensitive detector such as a photomultiplier tube. The magnitude of the electrical signal is proportional to the particle size, and is fed into a multi-channel analyzer for data conversion. Several types of OPC are available commercially and their operating size limits vary from about 0.05 μm for open cavity laser-based instruments of the Knollenberg-type[136] to more than 50 μm for the Polytec.[137] The number of size channels varies from four to five in some instruments designed for clean-room monitoring, but can be as high as 128 for some counters.

The objective of calibrating an OPC is to determine the relationship between the intensity of scattered light and particle size. This relationship can be calculated by applying Mie theory,[138,139] which agrees quite well with experimental data for spherical particles of known refractive index. However, this theoretical approach often is unsuitable for more complex shaped particles and experimental calibrations are necessary for most applications.

Polymer latex particles (see "Polymer Latex Particles" section) are utilized by almost all workers as calibration standards for OPCs. Some calibration curves for a Knollenberg-type counter are shown in Figure 7.49, and illustrate the effect of refractive index on the instrument sensitivity.[140] Similar variations have been seen with other types of OPC,[141–145]

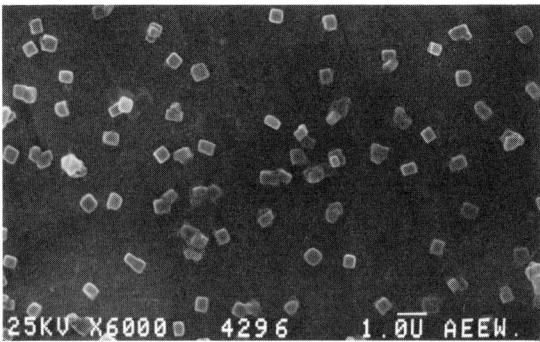

Sodium Chloride Aerosol After Size Separation in the Centrifuge:
Section of Collection Strip Showing Particles of 1µm Aerodynamic Diameter.

Cloud Settling Effect: Section of Collection Strip Showing
Polydisperse Aerosol

Figure 7.45. Cloud settling behavior in the Stöber spiral duct centrifuge.

and it is accepted generally that for the most accurate work the OPC should be calibrated with independently sized particles of the material to be studied. However, this procedure seldom is feasible since most aerosols of interest consist of ill-defined polydisperse particles of irregular shape. The practice has therefore arisen of calibrating the OPC with an inertial pre-collector of known ECD (see section on "Impactors and Cyclones"). The method has been described by Marple and Rubow[146] and Marple,[147] and also has been carried out successfully by Fissan and Helsper[143] and Büttner.[148] Single-stage impactors and cyclones have been used as the pre-collector; thus the OPC is calibrated with respect to the aerodynamic size characteristics of the pre-collector rather than to a size related to the light scattering cross-section of the calibrant. Aerodynamic size measurements are closely related to the dynamic behavior of particles throughout the size range of most OPCs, whereas the optical properties of aerosols have no significance in defining the more important aerosol transport properties. The calibration of an OPC with a pre-collector depends on generating a polydisperse test aerosol with a size distribution that matches the collection efficiency-particle size curve of the pre-collector. The particle size distribution

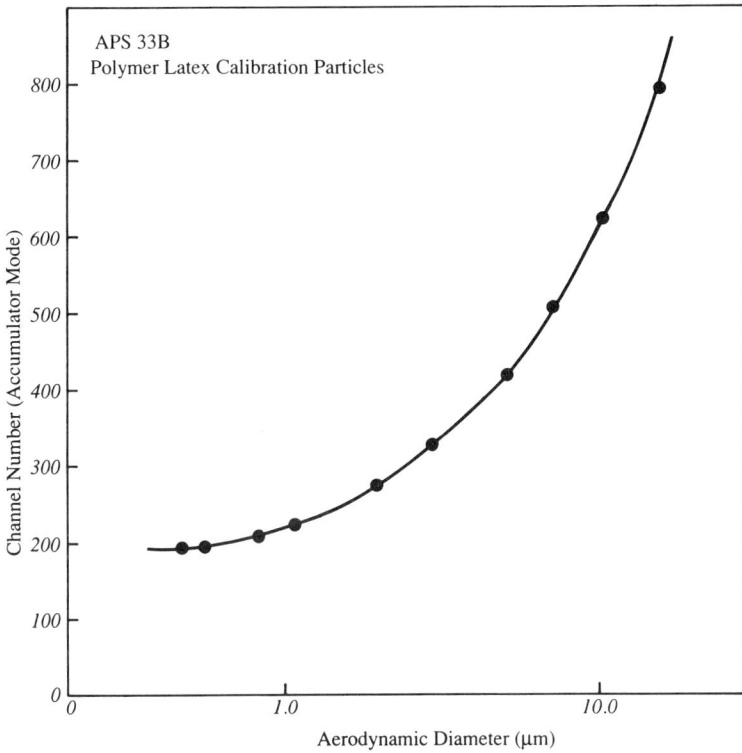

Figure 7.46. Calibration curve for TSI Aerodynamic Particle Sizer.

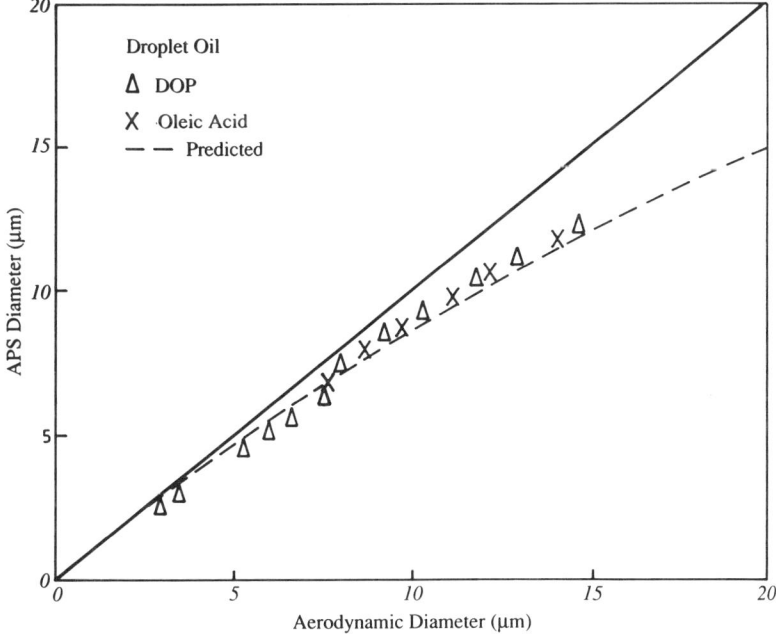

Figure 7.47. Effect of liquid droplet distortion on the calibration of TSI Aerodynamic Particle Sizer (P.A. Baron. 1986[128] with permission Elsevier Science Publishing Co., Inc.)

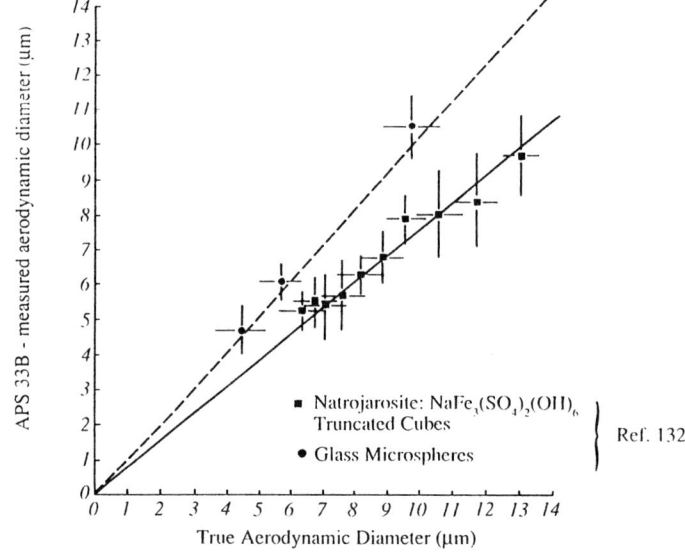

(a) TSI Aerodynamic Particle Sizer (APS 33B)

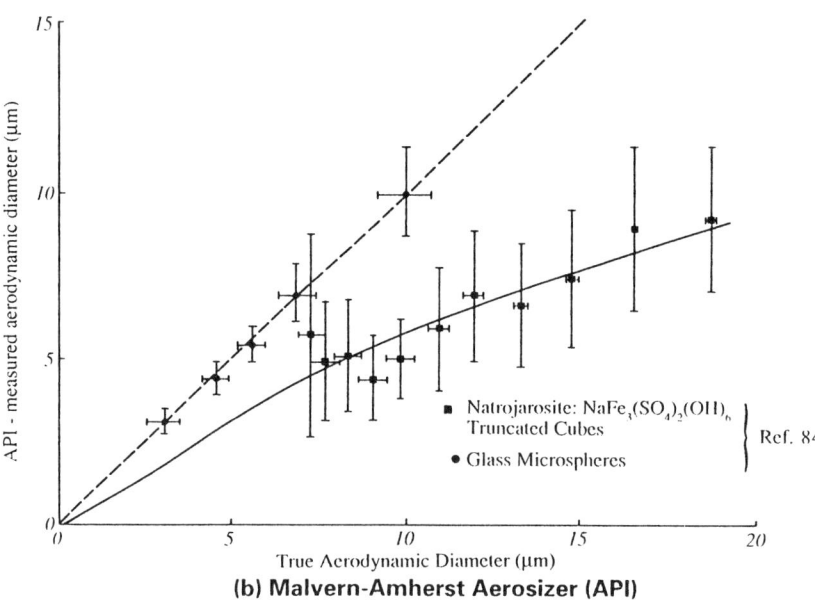

(b) Malvern-Amherst Aerosizer (API)

Figure 7.48. Influence of particle shape on APS33B and API responses.

is measured by the OPC without the pre-collector, and the process is repeated with the pre-collector placed immediately in front of the OPC. If the OPC has i channels, the number count in each channel from the first measurement can be represented by $n_1^1, n_2^1, n_3^1 ... n_i^1$, and for the second measurement by $n_1^2, n_2^2, n_3^2 ... n_i^2$. The attenuation of the particle concentration by the pre-collector (A) for the x th channel of the OPC is given by:

$$A_x = \left[1 - \frac{n_x^2}{n_x^1}\right] \quad (7.16)$$

Table 7.10. Number Concentrations (particles cm^{-3}) for 1, 5, and 10% Coincidence Error in the APS33B. Instruction Manual for the TSI Model APS33B Aerodynamic Particle Sizer, February 1988.

Timer	Particle Diameter (μm)	Maximum Concentration for Coincidence Error		
		1%	5%	10%
2 ns*	0.8	558	2850	5853
2 ns*	3.0	387	1973	4052
2 ns*	10.0	234	1193	2450
66.67 ns**	10.0	55	280	575
66.67 ns**	29.0	43	218	448

* small-particle timer
** large-particle timer

If the pre-collector and aerosol characteristics are chosen correctly, A varies from 0% in the lowest channels of the OPC and rises in a smooth curve to 100% in the higher channels. The channel number in which A is closest to 50% corresponds to the ECD of the pre-collector, providing a direct link between this channel number of the OPC and aerodynamic size. If the collection efficiency curve of the pre-collector is known, other calibration points can be determined by matching values of A with the corresponding collection efficiencies of the pre-collector. However, it is more usual to repeat the whole procedure several times with pre-collectors that have different ECD values, because ECDs can be more accurately defined than a complete efficiency curve. This procedure can be highly effective with stable aerosols as shown in Figure 7.50, which was obtained for several different types of particle using a Polytec HC-15 OPC.[148] The pulse amplitude in Figure 7.50 corresponds to channel number (i.e., low pulse amplitudes are assigned low channel numbers). The calibration of an OPC by this method is rapid and easy to perform, but the accuracy depends on the number concentration and size distribution of the test aerosol remaining constant for each set of measurements. This method has been applied successfully at the Aerosol Science Centre for the calibration of a Polytec HC-15 OPC with water droplets.[149] It was shown that this instrument sized water droplets as if they were twice the size of equivalent PSL microspheres, in close agreement with predictions from Mie theory (Figure 7.51).

Since OPCs detect individual particles, they are frequently used to measure number concentrations. These measurements are subject to errors caused by particle coincidence, but generally they are insignificant unless the aerosol concentration exceeds about 1000 particles cm^{-3}. The calibration of OPCs against an aerosol of known number concentration usually is carried out by comparing the total number concentration measured by the OPC with the number concentration obtained by counting the particles collected on selected areas of a membrane filter.[33,150] It is important to obtain a homogeneous aerosol deposit on the membrane filter, and this can be achieved easily by mounting the main filter on top of a second back-up filter. Although this procedure enables the efficiency of the OPC to be checked against an independent standard, it is time consuming unless an automated image analysis system scans the filter. Therefore, many workers prefer to obtain comparative performance data for an OPC against other similar instruments. The condensation nuclei counter provides an alternative to filter collection-microscopy,[151] especially if the OPC is sampling sub-micron particles.

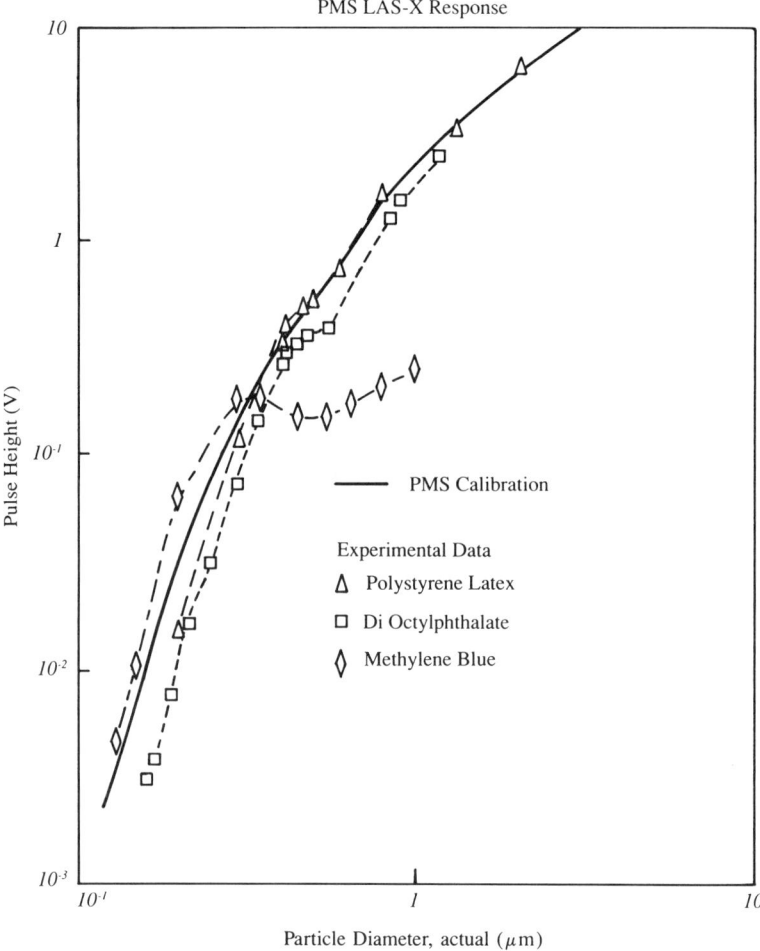

Comparison of experimental response of LAS-X counter and "calibration" curve used by manufacturer to establish the instrument response

Figure 7.49. Response of Knollenberg-type OPC to particles with different refractive indices (W.W. Szymanski and B.Y.H. Liu. 1986[140] with permission VCH Verlagsgesellschaft mbH, Weinheim).

ELECTRICAL AEROSOL ANALYZERS

Electrical aerosol analyzers classify aerosols by distinguishing between the electrical mobilities of singly charged particles in an applied electric field (see "Electrostatic Classifier" section). The most common electrical aerosol analyzers are derived from an instrument developed by Whitby and Clark[152] at the University of Minnesota. Two versions are available commercially (both from Thermosystems Inc.): the Electrical Aerosol Analyzer (EAA) and the Differential Mobility Particle Sizer (DMPS). The performance of the EAA was described first by Liu and Pui,[153] and that of the DMPS by Knutson and Whitby.[154] The EAA is a self-contained instrument with an optional computer for control and data analysis, whereas the DMPS consists of an electrostatic classifier and condensation nuclei counter together with electronics interface, interconnecting pipework and a computer for control and data interpretation.[155]

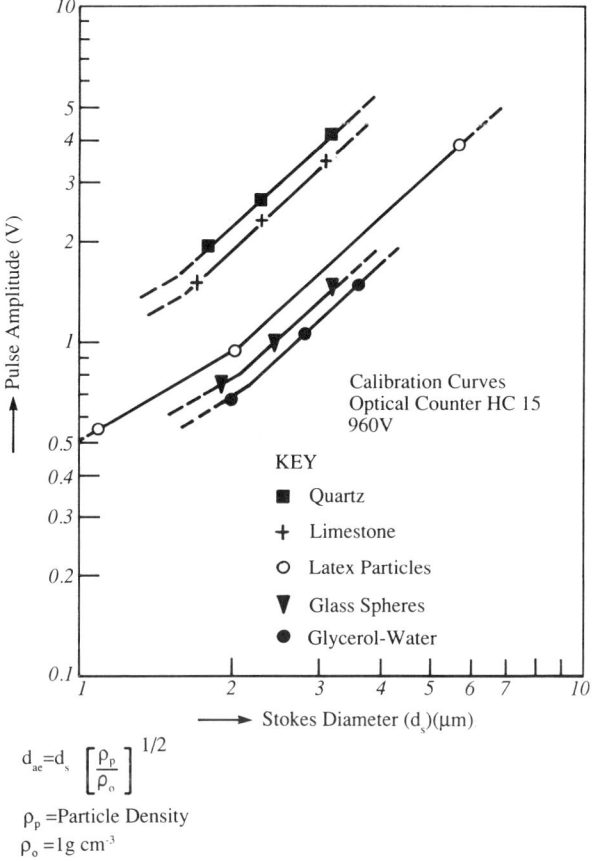

Figure 7.50. Calibration curves for Polytec OPC with various types of particle using the pre-collector method (H. Büttner. 1985[148] with permission VCH Verlagsgesellschaft mbH, Weinheim).

Both the EAA and the DMPS may be calibrated with either monodisperse or polydisperse aerosols. The calibration particles must be either solid or a low-volatile liquid, and their number concentration has to remain constant throughout the calibration period, which may be as long as 0.5 h when using dilute aerosols with the DMPS. Techniques for generating suitable test aerosols to calibrate the EAA (Figure 7.52) have been described by Pui and Liu,[156] and may also be used with the DMPS. The electrical classification procedure (see "Electrostatic Classifier" section) is probably the most popular method of generating calibrants, except for the problem of generating the smaller size particles from a nebulizer. However, such fine polydisperse test aerosols can be generated by other methods, such as a gas-phase reaction (see "Gas-Phase Reactions" section) or by atomization of a dilute solution of an aqueous salt [(e.g., NaCl), see "Compressed Air Nebulizers" section]. This method was chosen in a recent study to demonstrate the equivalence in measurements made by an EAA and DMPS (Figure 7.53[157]).

The object of the calibration procedure is to produce a response matrix of number concentrations (N_i) in channels i to j of the analyzer, for given mean particle diameters in the range D_p (l) to D_p (i). An example of such a matrix is given in Table 7.11, and should consist only of diagonal elements for an ideal system (i.e., $N_i = 1$ when i = j, and $N_i = O$ if i ≠ j). However, the data of Table 7.11 indicate that the response matrix of the EAA is

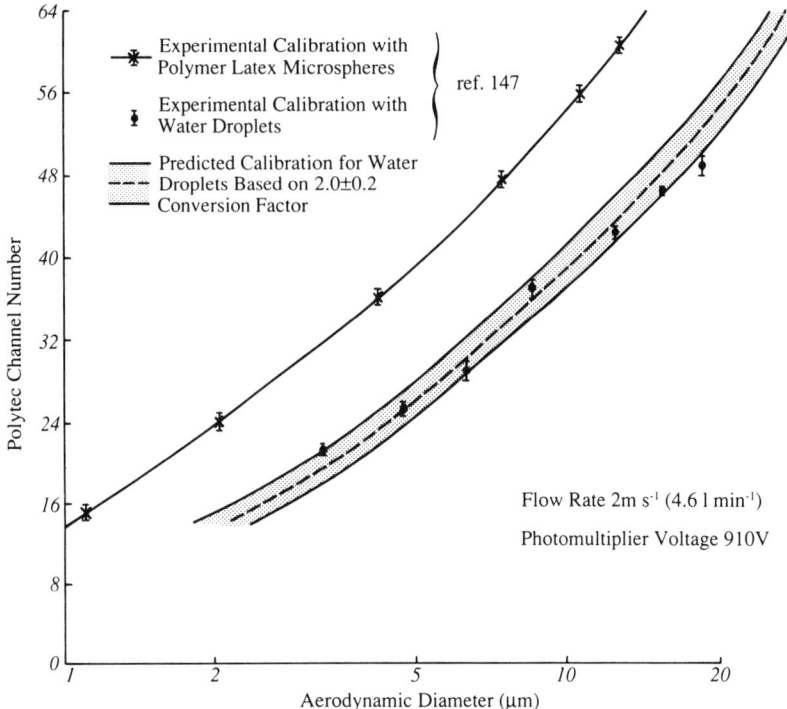

Figure 7.51. Calibration of a Polytec HC-15 optical particle counter with PSL particles and water droplets.

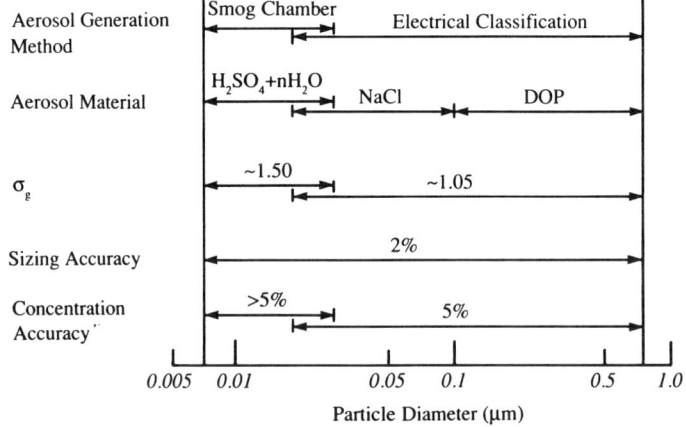

Figure 7.52. Aerosol generation techniques used to calibrate an electrostatic aerosol analyzer (D.Y.H. Pui and B.Y.H. Liu. 1979[156] with permission University of Florida Press)

not ideal, except at very small particle sizes. This spreading of the response into several channels causes the instrument to overestimate the σ_g values of test aerosols. The data in

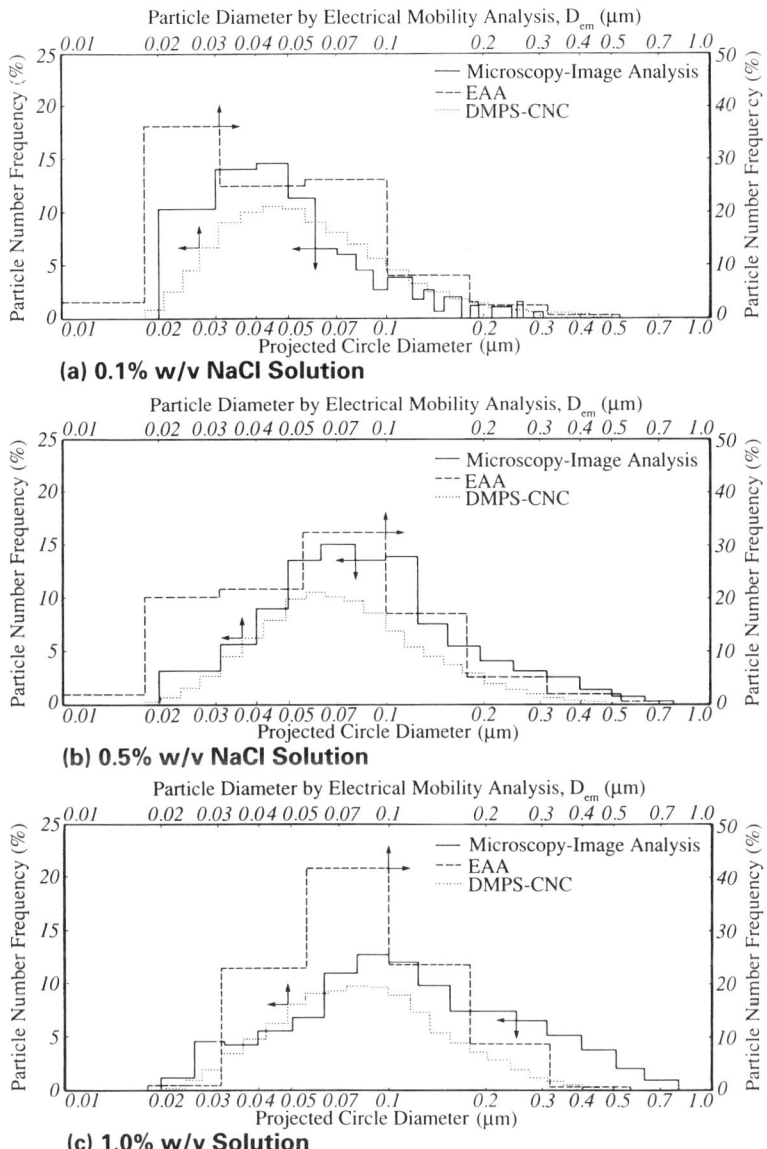

Figure 7.53. Intercomparison of a TSI EAA and DMPS with sodium chloride sub-micron particles generated by nebulization of a dilute aqueous solution of sodium chloride.[157]

Table 7.12 indicate this effect when the performance of an EAA was compared with other instruments in the analysis of a monodisperse aerosol.[158] Although the count mean diameter (CMD) was in the mid-range of the EAA, the DMPS would have been capable of greater size resolution[155] and estimates of σ_g would have been more likely to be closer to the correct values. Errors caused by multiply charged particles are reduced in the DMPS by means of a correctly greased single stage pre-impactor having an ECD substantially smaller than 1 μm.

A major criticism of the EAA calibration procedure is that the same electrical mobility technique is also used to produce the test aerosol. However, other methods for generating sub-micron calibrants cannot meet the stability and monodispersity requirements, especially when the CMD is smaller than 0.1 µm. Nevertheless, at some stage in the calibration procedure it is good practice to measure the size distribution of the test particles by an independent method, such as by filter collection-microscopy.

Electrical aerosol analyzers frequently are calibrated against each other, and this approach has been chosen by the manufacturer for maintenance purposes.[159] The response matrix of one electrical aerosol analyzer is compared with a standard instrument sampling the same polydisperse aerosol (tobacco smoke or metal fume). This type of comparison is easy to perform and useful in detecting the anomalous behavior of EAAs on a routine basis.

CONDENSATION NUCLEI COUNTERS (CNCs)

Several types of CNC have been developed to measure the number concentration of aerosols consisting of sub-micron particles. These include manually operated counters such as the Nolan-Pollack CNC[160] and the continuous-flow CNC developed by TSI Inc.[161] The Nolan-Pollack CNC is often used as a secondary standard for the calibration of other CNCs[162] because the decrease in light transmission across the expansion chamber is related directly to the number of particles present. The Nolan-Pollack CNC originally was calibrated against absolute Aitken-type CNCs, which operated by recording the aerosol concentration by photography.

The calibration of CNCs relies on the production of nuclei of known number concentration, which can be counted either by direct microscopic observation or from photographs of the droplets formed in the CNC. Alternatively, the droplets can be deposited on a coated glass slide and counted using a microscope.

The continuous flow CNC has been developed by Agarwal and Sem[161] and can be calibrated by counting the number of particles of known size and number concentration produced by an electrostatic classifier.[52] The number concentration measurements are made using a Faraday-cup electrometer, and applying Equation 7.8. The presence of multiply charged particles with the same electrical mobility as the calibration aerosol constitutes a source of error, and a correction factor has to be applied.

The continuous-flow CNC can be operated over a wide range of number concentration (from 1 to 10^7 particles cm^{-3}) and operates in two different modes: individual particles are counted when the concentration is below 10^3 particles cm^{-3}, whilst above this value the total light extinction caused by the droplets in the "view volume" (photometric mode) is measured. A correction factor is applied in the count mode to account for coincidences at concentrations approaching the upper limit. When calibrating the system in the count mode, monodisperse aerosol from the electrostatic classifier is diluted with clean excess air to reduce the particle concentration to the desired range. The amount of dilution air is reduced gradually until only the monodisperse aerosol stream is sampled, and calibration data have been obtained for aerosol number concentrations in excess of 5×10^5 particles cm^{-3}.[161]

It is important to replace the n-butanol working fluid regularly in the continuous-flow CNC: water vapor is absorbed from the air during use, diluting the n-butanol and reducing the size of the droplets formed in the condensation chamber. This error is present only when the CNC is operated in the photometric mode and is largest at the highest aerosol concentrations.

AEROSOL MASS CONCENTRATION MONITORS

A wide range of instruments has been developed to measure the mass concentration of aerosols in the working environment. Most measure the respirable fraction of these aerosols[67,68] and are operated in this mode to meet the requirements of government legislation for the working environment. Common samplers of this type include the 10 mm cyclone, MRE gravimetric dust sampler, Hexhlet, Simslin, TSI Piezobalance and GCA Miniram,[163-166] and they all require calibration against mass-based aerosol concentration standards.

The normal calibration procedure is to sample the test aerosol (which may be either monodisperse or polydisperse) in parallel with a pre-weighed filter. Several hours continuous sampling may be required for low aerosol mass concentrations (0.01 to 10 mg m^{-3}), and care is needed to ensure that the aerosol concentration remains stable. Sufficient mass of aerosol must be collected on the filter for accurate weighing, usually with a balance capable of microgram resolution. It is advisable to use membrane filters that are reasonably insensitive to variations in ambient humidity, and to condition these filters at a controlled humidity before each weighing, often two identical filters are placed in the same filter holder for these measurements; the upper filter collects the aerosol, whilst the lower filter acts as a blank that has been exposed to the same conditions. Care is needed to eliminate electrostatic charges on the filters, especially if a Cahn microbalance is used for the weighings. A pistol-type electrostatic charge remover (Zerostat) has been successful in this purpose at the Aerosol Science Centre.

If an aerosol mass monitor is being calibrated against a filter sampler or other instrument, it is very important to ensure that the aerosol in the test chamber is homogeneous. This is particularly important when large particles are generated, as these are more prone to losses in the test apparatus by impaction and deposition. Elaborate aerosol chambers have been constructed to try to minimize errors caused by non-representative sampling (Figure 7.54). The chamber shown in Figure 7.54 relies on the careful location of identical lengths of sampling tube to the instruments, and contains an electrostatic charge equilibrator to reduce the build-up of charge on the particles.

Most respirable aerosol samplers are calibrated with standard polydisperse test dusts, such as Arizona Road Dust (see section on "Particle Standards"). However, care must be taken with the choice of dust generator to ensure that the particles are fully dispersed (see "Ultrasonic Nebulizers" section). Monodisperse particles also may be used for more extensive calibration, if details are required on the size selectivity of these instruments.

CONCLUSIONS

A wide range of aerosol generation techniques is available for calibrating aerosol analysis equipment. The choice of a particular calibration method depends largely on the type of instrument being calibrated and the degree of accuracy required. Emphasis has been placed in this chapter on the appropriate calibration procedures for equipment that measures the most important physical parameters in assessing environmental consequences: the aerosol mass and number concentrations, and the corresponding particle size distributions.

Particle concentration standards are not too difficult to obtain provided that attention is paid to the degree of dispersion of the test aerosol. This is especially important with dry dispersion methods and with highly concentrated aerosols containing sub-micron particles, where significant particle agglomeration or coagulation may occur in the aerosol generator. Number concentrations normally are confirmed by counting the particles collected on a

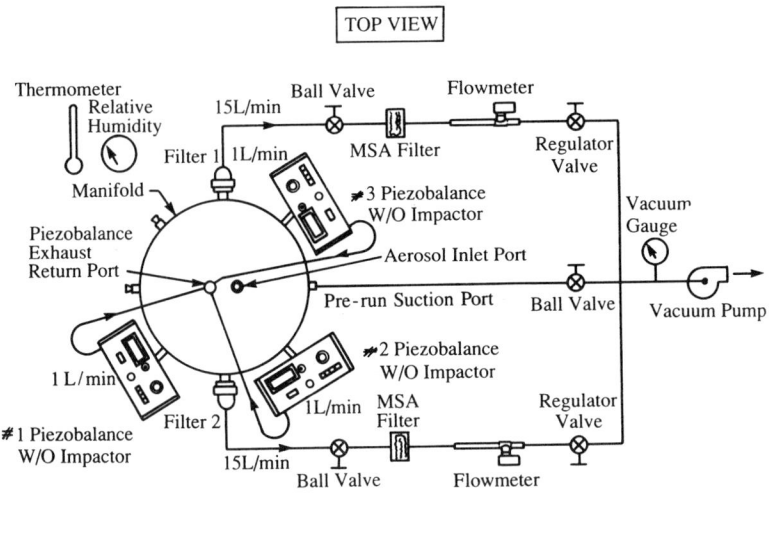

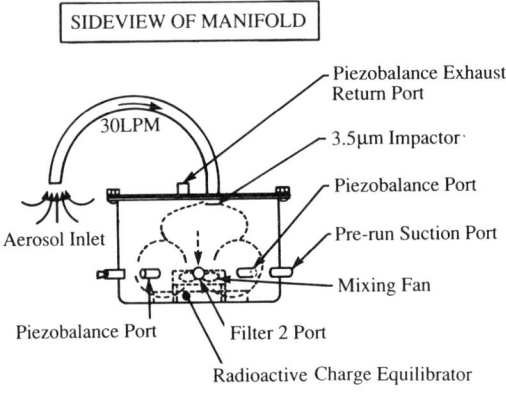

Figure 7.54. Technique for calibrating TSI Piezobalance aerosol mass sensors (G.J. Sem, K. Tsurubayashi and K. Homma. 1977[166] with permission AIHA).

filter or suitable flat surface, and mass concentrations conveniently can be checked by weighing the particles collected on a filter. Both procedures are well defined and are accurate, provided they are carried out carefully.

Calibrations using monodisperse particles are more frequently carried out than those with polydisperse aerosols despite the extra work involved, since the error introduced by the monodisperse technique almost always is smaller. This is particularly true with impactors and cyclones where the ECD values are determined from collection efficiency curves, which are usually sensitive functions of particle size. High-resolution aerosol size analyzers such as the APS33B, Aerosizer, most OPCs and electrical aerosol analyzers also are calibrated best with monodisperse particles in order to determine whether subtle changes have occurred in their calibration functions, which may indicate the onset of a degradation in performance. However, polydisperse calibration aerosols also have a place, particularly with the inertial spectrometers (which collect the entire aerosol for subsequent examination) and respirable mass concentration monitors that usually are assessed over the complete respirable size range.

Table 7.11. Calibration response matrix for an electrical aerosol analyzer

Channel	Channel Midpoint (nm)	Channel Boundary (nm)	Particle Diameter (nm)	4.2	7.5	13	24	42	75	130	240	420	750
			Sensitivity pA/(10^6 cm^{-3})	0.105	0.24	6.0	11.5	22.5	41.5	81.5	150	285	
			Collector-rod Voltage (V)										
1	4.2	3.2	19	0	0	0	0	0	0	0	0	0	0
2	7.5	5.6	59	1	1	0	0	0	0	0	0	0	0
3	13.3	10.0	186	0	0	0.08	0	0	0	0	0	0	0
4	23.7	17.8	588	0	0	0.92	0.49	0	0	0	0	0	0
5	42.2	31.6	1870	0	0	0	0.07	0.14	0.42	0	0	0	0
6	75	56.2	2600	0	0	0	0.44	0.65	0.49	0.22	0	0	0
7	133	100	4440	0	0	0	0	0.17	0.09	0.48	0.31	0.11	0
8	237	178	6600	0	0	0	0	0.02	0	0.20	0.36	0.28	0
9	422	316	8380	0	0	0	0	0.02	0	0.06	0.18	0.25	0
10	750	562	9600	0	0	0	0	0	0	0.04	0.15	0.36	0
		1000	10600										

Response Matrix of the EAA for nt = 1 × 10^7 (ions cm^{-3} sec^{-1}). With permission University of Florida Press.

Table 7.12 Size Comparison of Monodisperse Fused Aluminosilicate Particles

Instrument	Measured Particle Size	
	CMD (μm)	σ_g
Electrical Aerosol Analyzer (EAA)	0.107*	1.44
Electrostatic Classifier (EC)	0.104	—
Stöber Spiral Duct Centrifuge	0.108	1.10
Electron Microscope	0.105	1.08

* Data obtained using modified data reduction table.
Source: Yeh, H.C., Cheng, Y.S. and Kanapilly, G.M., Ch. 75 in "Aerosols in the Mining and Industrial Work Environment," ed. Liu, B.Y.H., Ann Arbor Science, Ann Arbor, Mich., 1983.

Other physical properties of the calibrant are often as significant as the degree of monodispersity. For instance, the refractive index is important when calibrating an OPC, and the choice of solid or liquid particles may influence the outcome of APS33B/Aerosizer calibrations. It can be highly advantageous to adopt the inertial pre-collector method for OPCs, thus avoiding the need to know the refractive indices of either the calibration aerosol or the particles used in any subsequent study. If the APS33B/Aerosizer is sampling liquid droplets, it may be necessary to refine the calibration technique so that the performance can be measured with a liquid that has a similar surface tension. Particle shape also may have a marked influence on the sensitivity of aerodynamic particle size analyzers (see "Real-Time Aerodynamic Particle Sizers" section).

Finally, the performance of ancillary equipment such as gas flowmeters, pressure gauges, etc., is also important, and care should be taken to achieve the optimum conditions for aerosol generation and sampling. It is recommended that every aerosol analyzer is calibrated on a regular basis as part of a checking and maintenance policy, since the performance can be influenced significantly by the previous treatment of the analyzer and small changes in the experimental conditions.

ACKNOWLEDGMENTS

The author wishes to thank his former colleagues at AEA Technology for their advice in the preparation of this chapter. The author is now with Trudell Medical, London, Ontario, Canada.

REFERENCES

The references are not exhaustive, and the reader is advised to refer to the review articles for fuller information about the various aerosol generation and instrument calibration techniques.

1. Fuchs, N.A. and Sutugin, A.G., Chapter 1 in *Aerosol Science* ed. Davies, C. N. Academic Press, NY, 1966.
2. Dennis, R., *Handbook on Aerosols,* Technical Information Centre, U.S. Energy Research and Development Administration, 1976.

3. Raabe, O., p 57 in *Fine Particles,* ed. Liu, B.Y.H., Academic Press, NY, 1976.
4. Willeke, K., Generation of Aerosols, Ann Arbor Science, Ann Arbor, MI, 1980.
5. Liu, B.Y.H., p 39 in *Fine Particles,* ed. Liu, B.Y.H., Academic Press, NY, 1976.
6. Raabe, O., *Am. Ind. Hyg. Assoc. J.,* 29, 439, 1968.
7. Stöber, W. and Flaschbart, H., *Environ. Sci. Technol.,* **3(12),** 1280, 1969.
8. Reist, P.C. and Burgess, W.A., *J. Colloid Interface Sci.,* **24,** 271, 1967.
9. Langer, G. and Lieberman, A., *J. Colloid Sci.,* **15,** 357, 1960.
10. Fuchs, N.A., *J. Aerosol Sci.,* **4,** 405, 1973.
11. Garvey, D.M. and Pinnick, R.G., *Aerosol Sci. Technol.,* **2,** 477, 1983.
12. Rayleigh, Lord, *Proc. London Math. Soc.,* **10,** 4, 1878.
13. Fulwyler, M.J., Perrings, J.D. and Cram, L.S., *Rev. Sci. Instruments,* **44(2),** 204, 1973.
14. Hendricks, C.O. and Babil, S., *J. Phys. E. (Scient. Instruments),* **5,** 905, 1972.
15. Ström, L., *Rev. Sci. Instruments,* **40,** 778, 1969.
16. Berglund, R.N. and Liu, B.Y.H., *Environ. Sci. Technol.,* **7,** 147, 1974.
17. Wedding, J.B. and Stukel, J.J., *Environ. Sci. Technol.,* **8**(5), 456, 1974.
18. Wedding, J.B.J., *Environ. Sci. and Technol.,* **9(7),** 673, 1975.
19. Mitchell, J.P., Snelling, K.W. and Stone, R.L., *J. Aerosol Sci.,* **18(3),** 231, 1987.
20. Leong, K.H., *J. Aerosol Sci.,* **12(5),** 417, 1981.
21. Hinds, W.C., *Aerosol Technology,* John Wiley and Sons, NY, 1982.
22. O'Connor, D.T., *Ann. Occ. Hyg.,* 16, 119, 1973.
23. Chen, B.T. and Crow, D.J., *J. Aerosol Sci.,* **17,** 963, 1986.
24. Vanderpool, R.W., Black Hall, A.P. and Rubow, K.L., *TSI Quarterly,* **X(I),** 3, 1984.
25. McFarland, A.R., Wedding, J.B. and Cermak, J.E., *Atmos. Environ.,* **11,** 535, 1977.
26. Whitby, K.T., Lundgren, D.A. and Peterson, C.M., *Int. J. Air Water Pollut.,* **9,** 263, 1965.
27. May, K.R., *J. Appl. Phys.,* **20,** 932, 1949.
28. May, K.R., *J. Scient. Instruments,* **43,** 841, 1966.
29. Stahlhofen, W., Gebhart, J., Heyder, J. and Stuck, B., *Staub Reinhalt. der Luft,* **39,** 73, 1979.
30. Maguire, B.A., Barker, D. and Wake, D., *Staub-Reinhalt. der Luft,* **33(3),** 93, 1973.
31. Hurford, M.J., *J. Aerosol Sci.,* **12(5),** 441, 1981.
32. Walton, W.H. and Prewett, W.C., *Proc. Phys. Soc. B,* **62,** 341, 1949.
33. Mitchell, J.P., *J. Aerosol Sci.,* **15(1),** 35, 1984.
34. Mitchell, J.P. and Stone, R.L., *J. Phys. E. (Scient. Instruments),* **15,** 565, 1982.
35. Cheah, P.K.P. and Davies, C.N., *J. Aerosol Sci.,* **15(6)** 741, 1984.
36. Jenkins, R.A., Mitchell, J.P. and Nichols, A.L., p 337 in *Particle Size Analysis,* University of Bradford, Sept 1985.
37. Bailey, M.R., and Strong, J.C., *J. Aerosol Sci.,* **11,** 557, 1980.
38. Philipson, K., *J. Aerosol Sci.,* **4,** 51, 1973.
39. Garland, J.A., Wells, A.C. and Higham, E.J., AEA Technology Report AERE-R 10491, 1982.
40. Davies, C.N. and Cheah, P.K.P., *J. Aerosol Sci.,* **15(6),** 719, 1984.
41. Sinclair, D. and Laner, V., *Chem. Rev,* **44,** 245, 1949.
42. Rapaport, E. and Weinstock, S., *Experimentia,* **11,** 363, 1955.
43. Muir, D.C.F., *Ann. Occup. Hyg.,* **8,** 233, 1965.
44. Swift, D.L., *Ann. Occup. Hyg.,* **10,** 337, 1967.
45. Lassen, L., *Angew. Phys.,* **12,** 157, 1960.
46. Prodi, V., p 169 in *Assessment of Airborne Particles,* ed., Mercer, T.T., Morrow, P.E. and Stöber, W., Charles C. Thomas Publ., Springfield, IL, 1972.
47. Scheuch, G. and Heyder, J., p 1057 in *Aerosols: Formation and Reactivity,* eds., Schikarski, W., Fissan, H.J. and Friedlander, S.K., Pergamon Press, Oxford, 1986.
48. Horton, D.K., Miller, R.D. and Mitchell, J.P., *J. Aerosol Sci.,* **22(3),** 347, 1991.

49. Vaughan, N.P., *J. Aerosol Sci.,* **21(3),** 453, 1989.
50. Nicolaon, G., Cooke, D., Kerker, M. and Matijevic, E., *J. Colloid Interface Sci.,* **34,** 534, 1970.
51. Fuchs, N. and Sutugin, A., *Kolloid Zh.,* **25,** 487, 1963.
52. Liu, B.Y.H. and Pui, D.Y.H., *J. Colloid Interface Sci.,* **47(1),** 155, 1974.
53. Fuchs, N.A., *Geofis. Pura. Appl.,* **56,** 185, 1963.
54. Matijevic, E., Budnik, M. and Meites, L., *J. Colloid Interface Sci.,* **61(2),** 302, 1977.
55. Barringer, E.A. and Bowen, H.K., *J. Am. Ceram. Soc.,* **65(12),** Cl99, 1982.
56. Haggerty, J.S. and Cannon, R.W., p 165 in *Laser Induced Chemical Processes,* ed. Steinfeld, J.I., Plenum Publ. Co., NY, 1981.
57. Babington, R.S., Slivka, W.R. and Yetman, A.A., U.S. Pat. 3,421,192, 1969.
58. Guichard, J.C., p 173 in *Fine Particles,* ed. Liu, B.Y.H., Academic Press, NY, 1976.
59. Marple, V.A., Liu, B.Y.H. and Rubow, K.L., *Am. Ind. Hyg Assoc. J.,* **39,** 26, 1978.
60. Wright, B.M., *J. Sci. Instruments,* **27,** 12, 1950.
61. Craig, D.K., Wehner, A.P. and Morrow, W.G., *Am. Ind. Hyg. Assoc. J.,* **33,** 283, 1972.
62. Drew, R.T. and Laskin, S., *Am. Ind. Hyg. Assoc. J.,* **32(5),** 327, 1971.
63. Fuchs, N.A. and Murashkevich, F.I., *Staub-Reinhalt. der Luft,* **30(11),** 1, 1970.
64. Hounam, R.F., *Ann. Occup. Hyg.,* **14,** 329, 1971.
65. Knutson, E.O., Pontinen, K.W. and Rees, L.W., *Am. Ind. Hyg. Assoc. J.,* **28,** 83, 1967.
66. Cadle, R.D. and Magill, P.L., *Ind. Eng. Chem.,* **43,** 1331, 1951.
67. Vincent, J.H., *Aerosol Sampling: Science and Practice,* John Wiley and Sons, Chichester, 1989.
68. International Standards Organisation (ISO), *Am. Ing. Hyg. Assoc. J.,* **42,** A64, 1981.
69. Blackford, D.B. and Rubow, K.C., Proc NOSA Symposium Solna, Sweden, November 1986.
70. Zahradnicek, A. and Löffler, F., *Staub Reinhalt. der Luft,* **11,** 36, 1976.
71. Phalen, R.F., *J. Aerosol Sci.,* **3,** 395, 1972.
72. Wegrzyn, J.E., p 254 in *Fine Particles,* ed Liu, B.Y.H., Academic Press, NY, 1976.
73. Kotrappa, P. and Bhanti, D.P., *Am. Ind. Hyg. Assoc. J.,* **28,** 171, 1967.
74. Dahlin, R.S., Su J., and Peters, L.K., *A.I.Ch.E. J.,* **27(3),** 404, 1981.
75. Scarlett, B., in *Particle Size Analysis 91 Royal Society of Chemistry,* John Wiley and Sons, Chichester, 1992.
76. Wilson, R., *Powder Technol.* **27,** 37, 1980.
77. Thom, R., Marchandise, H., and Colinet, E., European Commission Report EUR 9662 EN, Brussels, 1985.
78. Wilson, R., Leschonski, K., Alex, W., Allen, T., Koglin, B., and Scarlett, B., European Commission Report EUR 6825 EN, Brussels, 1980.
79. Dragoo, A.L., Robbins, C.R. and Hsu, S.M., U.S. National Institute of Science and Technology (NIST) Special Publication, Part 766, 1, 1989.
80. Mulholland, G., Hartman, A.W., Hembree, G.G., Marx, E., and lettieri, T.R., J. Res. Natl. Bur. Standards, **90(1),** 3, 1985.
81. Lettieri, T.R. and Hembree, G.G., *J. Colloid Interface Sci.,* **127(2),** 566, 1989.
82. Kinney, P.D., Pui, D.Y.H., Mulholland, G.W. and Bryner, N.P., *J. Res. Natl. Inst. Standards,* **96(2),** 2, 1991.
83. Cheng, Y.S., Chen, B.T. and Yeh, H.C., *J. Aerosol Sci.,* **21(5),** 701, 1990.
84. Marshall, I.A., Mitchell, J.P. and Griffiths, W.D., *J. Aerosol Sci.,* **22(1),** 73, 1991.
85. Marshall, I.A., Mitchell, J.P. and Griffiths, W.D., p 7 in *Aerosols: Their Generation, Behavior and Applications,* Proc. 4th Conf. of the (UK) Aerosol Society, Guildford, The Aerosol Society, 1990.
86. Marshall, I.A. and Mitchell, J.P., in *Particle Size Analysis 91,* Royal Society of Chemistry, John Wiley and Sons, Chichester, 1992.

87. Gowland, R.J. and Wilshire, B., in *Particle Size Analysis 91,* Royal Society of Chemistry, John Wiley and Sons, Chichester, 1992.
88. Hoover, M.D., Casalnvovo, S.A., Lipowicz, P.J., Hsu, C.Y., Hanson, R.W. and Hurd, A.J., *J. Aerosol Sci.,* **21(9)**, 569, 1990.
89. Kaye, P.H., Shepherd, J.N. and Clark, J.M., p 223 in *Aerosols: Their Generation, Behavior and Applications,* Proc. 5th Conf. of the (UK) Aerosol Society, Loughborough, The Aerosol Society, 1991.
90. Fuchs, N.A., Ch. 1 in *Fundamentals of Aerosol Science,* ed. Shaw, D.T., John Wiley and Sons, NY, 1978.
91. Fissan, H. and Helsper, C., VDI Berichte No. 429, 1982.
92. Franzen, H. and Fissan, H., *Staub-Reinhalt. der Luft,* **39**, 50, 1979.
93. May, K.R., *J. Aerosol Sci.,* **6**, 403, 1975.
94. Flesch, J.P., Norris, C.H. and Nugent, A.E., *Am. Ind. Hyg. Assoc. J.,* **28**, 507, 1967.
95. O'Connor, D.T. AHSB(RP)R108, 1971; O'Connor, D.T., *Ann. Occup. Hyg.,* **16**, 119, 1973.
96. Lundgren, D.A., *J. Air Pollut. Control Assoc.,* **17**, 225, 1967.
97. Smith, W.B., Wilson, R.R. and Harris, D.B., *Environ. Sci. Technol.,* **13(11)**, 1387, 1979.
98. Wang, J.C.F. and Libkind, M.A., Sandia National Laboratories Report SAND 82-8611, 1982.
99. Fairchild, C.I. and Wheat, L.D., *Am. Ind. Hyg. Assoc, J.,* **45(4)**, 205, 1984.
100. May, K.R., *J. Scient. Instruments,* **22**, 187, 1965.
101. Mercer, T.T. and Chow, H.Y., *J. Colloid Interface Sci.,* **27**, 75, 1968.
102. Chang, H., *Am. Ind. Hyg. Assoc. J.,* **35**, 538, 1974.
103. Fuchs, N.A., *Mechanics of Aerosols,* Pergamon Press, Oxford 1966.
104. Rao, A.K. and Whitby, K.T., *J. Aerosol Sci.,* **9**, 87, 1978.
105. Boulaud, D., Proc. 9th Conference of the European Association for Aerosol Research, Duisburg, 125, 1981.
106. Mitchell, J.P., Costa, P.A. and Waters, S., *J. Aerosol Sci.,* **19(2)**, 213, 1988.
107. Rao, A.K. and Whitby, K.T., *J. Aerosol Sci.,* **9**, 77, 1978.
108. Lee, K.W., Gieseke, J.A. and Piispanen, W., *Atmos. Environ.,* **19(6)**, 847, 1985.
109. Smith, W.B., Iozia, D.L. and Harris, D.B., in Proceedings of the 10th Conference of the European Association for Aerosol Research, *J. Aerosol Sci.,* **14(3)**, 402, 1983.
110. Horton, K.D. and Mitchell, J.P., *J. Aerosol Sci.,* **23(5)**, 505, 1992.
111. Loo, W.W., Jacklevic, J.M. and Goulding, F.S., p 311 in *Fine Particles,* ed. Liu, B.Y.H., Academic Press, NY, 1976.
112. Chen, B.T., Yeh, H.C. and Cheng, Y.S., *J. Aerosol Sci.,* **16(4)**, 343, 1985.
113. Timbrell, V., Ch. 15 in *Assessment of Airborne Particles,* eds., Mercer, T.T., Morrow, P.E. and Stöber, W., Charles C. Thomas Publ., Springfield, IL, 1972.
114. Goetz, A., Stevenson, H.J.R. and Preining, O., *J. Air Pollut. Control Assoc.,* **10**, 378, 1960.
115. Stöber, W. and Flaschbart, H. and Boose, C., *J. Colloid Interface Sci.,* **39(1)**, 109, 1972.
116. Stöber, W. and Flaschbart, H., *J. Aerosol Sci.,* **2**, 103, 1971.
117. Tillery, M.I., *Am. Ind. Hyg. Assoc. J.,* **35**, 62, 1974.
118. Kotrappa, P. and Light, M.E., *Rev. Sci. Instrum.,* **43**, 1106, 1972.
119. Prodi, V., Melandri, C., Tarroni, G., DeZaiacomo, T. and Formigniani, M., *J. Aerosol Sci.,* **10(6)**, 411, 1979.
120. Prodi, V., DeZaiacomo, T., Hochrainer, D. and Spurny, K., *J. Aerosol Sci.,* **13(1)**, 49, 1982.
121. Mitchell, J.P. and Nichols, A.L., *Aerosol Sci. Technol.,* **9**, 15, 1988.
122. Marshall, I.A., Mitchell, J.P. and Griffiths, W.D., *J. Aerosol Sci.,* **21(7)**, 969, 1990.

123. Smith, A.D., AEA Technology Report AEEW-M 2077, 1982; Mitchell, J.P., Nichols, A.L., Smith, A.D. and Snelling, K.W., *Filtration and Separation,* 345, 1984.
124. Martonen, T.B., PhD Thesis, University of Rochester, 1977.
125. Stöber, W., Martonen, T.B. and Osborne, S., Ch. 12 in *Recent Developments in Aerosol Science,* ed. Shaw, D.T., John Wiley and Sons, NY, 1978.
126. Edwards, J. and Kinnear, D.I., Proc. 13th AEC Air Cleaning Conference, CONF-740807, 1, 552, 1974.
127. Agarwal, J.K., Remiarz, R.J., Quant, F.R. and Sem, G.J., *J. Aerosol Sci.,* **13**, 222, 1982.
128. Baron, P.A., p 215 in Aerosols, ed Liu, B.Y.H., Pui, D.Y.H. and Fissan, H., Elsevier Science Publ., N.Y., 1984; Baron, P.A., *Aerosol Sci. Technol.,* **5(1)**, 55, 1986.
129. Chen, B.T., Cheng, Y.S. and Yeh, H.C., *Aerosol Sci. Technol.,* **4**, 189, 1985.
130. Griffiths, W.D., Iles, P.J. and Vaughan, N.P., *TSI Journal of Particle Instrumentation,* **1(1)**, 3, 1986.
131. Griffiths, W.D., Iles, P.J. and Vaughan, N.P., *J. Aerosol Sci.,* **17(6)**, 921, 1986.
132. Marshall, I.A. and Mitchell, J.P., AEA Technology Report, AEA RS 5167, 1991.
133. Wilson, J.L. and Liu, B.Y.H., *J. Aerosol Sci.,* **11(2)**, 139, 1980.
134. Wang, H-C. and John, W., *Aerosol Sci. Technol.,* **6(2)**, 191, 1987.
135. Blackford, D.B., Hanson, A.E., Pui, D.Y.H., Kinney, P. and Ananth, G.P., p. 311 in Proc 3rd Conf of the (UK) Aerosol Society, Bournemouth, the Aerosol Society, 1988.
136. Pinnick, R.G. and Auvermann, H.J., *J. Aerosol Sci.,* **11(2)**, 139, 1980.
137. Schegk, C., Umhauer, H. and Löffler, F., *Staub-Reinhalt. der Luft,* **44(6)**, 264, 1984.
138. Mie, G., *Ann. Physik,* **25**, 377, 1890.
139. Kerker, M., The Scattering of Light and Other Electromagnetic Radiation, Academic Press, NY, 1969.
140. Szymanski, W.W. and Liu, B.Y.H., *Part. Charact.,* **3**, 1, 1986.
141. Whitby, K.T. and Vomela, R.A., *Environ. Sci. Technol.,* **1**, 801, 1967.
142. Willeke, K. and Liu, B.Y.H., p 698 in *Fine Particles,* ed Liu, B.Y.H., Academic Press, NY, 1976.
143. Fissan, H. and Helsper, C., Ch. 58 in *Aerosols in the Mining and Industrial Work Environment,* Ann Arbor Science, Ann Arbor, MI, 1983.
144. Pinnick, R.G., Rosen, J.M. and Hofmann, D.J., *Appl. Optics,* **12(1)**, 37, 1973.
145. Liu, B.Y.H., Berglund, R.N. and Agarwal, J.K., *Atmos. Environ.,* **8**, 717, 1974.
146. Marple, V.A. and Rubow, K.L., *J. Aerosol Sci.,* **7**, 425, 1976.
147. Marple, V.A., p 207 in *Aerosol Measurement,* ed. Lundgren, D.A., University Press of Florida, Gainesville, FL, 1979.
148. Büttner, H., *Part. Charact.,* **2**, 20, 1985.
149. Mitchell, J.P., Nichols, A.L. and Van Santen, A., *Part. Part. Syst. Charact.,* **6**, 119, 1989.
150. Rimberg, D., p 321 in *Aerosol Measurement,* ed. Lundgren, D.A., University Press of Florida, Gainesville, FL, 1979.
151. Kruger, J. and Leuschner, A.H., *Atmos. Environ.,* **12**, 2011, 1978.
152. Whitby, K.T. and Clark, W.E., *Tellus,* **18**, 573, 1966.
153. Liu, B.Y.H. and Pui, D.Y.H., *J. Aerosol Sci.,* **6**, 249, 1975.
154. Knutson, E.O. and Whitby, K.T., *J. Aerosol Sci.,* **6**, 453, 1975.
155. Keady, P.B., Quant, F.R. and Sem, G.J., *TSI Quarterly,* **X(2)**, 3, 1983.
156. Pui, D.Y.H. and Liu, B.Y.H., p 384 in *Aerosol Measurement,* ed. Lundgren, D.A., University of Florida Press, Gainesville, FL, 1979.
157. Horton, K.D., Mitchell, J.P. and Nichols, A.L., *TSI J. of Particle Instrum.,* **3**, 1, 1989.
158. Yeh, H.C., Cheng, Y.S. and Kanapilly, G.M., Ch. 75 in *Aerosols in the Mining and Industrial Work Environment,* Ann Arbor Science, Ann Arbor, MI, 1983.
159. Sem, G.J., p 400 in *Aerosol Measurement,* ed. Lundgren, D.A., University of Florida Press, Gainesville, FL, 1979.

160. Nolan, P.J. and Pollack, L.W., Proc. Roy. Irish Acad., Sect. A, **51**, 9, 1946.
161. Agarwal, J.K. and Sem, G.J., *J. Aerosol Sci.,* **11**, 343, 1980.
162. Pollack, L.W. and Metnieks, A.L., *Geofis. Pura. Appl.,* **43**, 285, 1959.
163. Hinds, W.C., *Aerosol Technology,* John Wiley and Sons, NY, p. 225, 1982.
164. Simslin-II, Technical Specification, Rotheroe and Mitchell Ltd., London, 1985.
165. Lilienfeld, P. and Stern, R., U.S. Bureau of Mines Research Contract Report, H0308132, 1982.
166. Sem, G.J., Tsurubayashi, K. and Homma, K., *Am. Ind. Hyg. Assoc. J.,* **38**, 580, 1977.

CHAPTER **8**

Particle Size Analyzers: Practical Procedures and Laboratory Techniques

J.P. Mitchell

ABSTRACT

The development of an effective strategy to determine the size distribution of airborne particulates (aerosols) requires careful consideration of the sampling method, the choice of in-situ or extractive analysis as well as the selection of the most appropriate measurement technique(s). An inappropriate choice of sampler or analyzer can severely impair an otherwise useful assessment. This report contains guidance on all these aspects, based both on experience gained during the past few years and on an evaluation of the potential of the newer non-intrusive optical techniques that are likely to become more important.

> The descriptions of equipment and observations on operation/performance are given in good faith by the author, and do not imply any endorsement by AEA Technology.

INTRODUCTION

The measurement of bioaerosol emissions poses considerable difficulties for those faced with the task of assessing the environmental impact of such sources. Although significant strides currently are being made to develop non-invasive techniques, these analyzers are extremely expensive and the data require careful interpretation before they can be used. In-situ methods are almost always optical based and in most cases (laser diffractometers being the obvious exception) provide number-based size information as raw data. Volume/mass-based measurements usually are preferred, since such data can be compared directly with the limits specified by the regulatory authorities. In contrast, many of the more traditional methods of aerosol analysis, such as filter/impactor collection, provide mass-based measurements directly. However, with such extractive methods, sampling of the aerosol becomes as important as the measurement, and an incorrect sampling procedure can wreck an otherwise valid size distribution determination.

This review introduces some of the general principles of aerosol sampling from both static and flowing environments, and they are applicable both to the collection of bioaerosols and to size distribution measurement. The details of both calm-air and in-stack/duct measurement techniques are mentioned only briefly, since the object is not to concentrate on any one type of sampling environment. The bulk of the chapter is a description of a range of commonly encountered particle size analysis techniques, placing emphasis on those that can be deployed readily in the field, although laboratory-based methods are described where appropriate. Size-selective sampling for respirable dust and bioaerosols has been mentioned only briefly, as this topic is covered in Chapters 9 and 10 while bioaerosol particle statistics are considered in Chapter 5. Measurements using the various types of directional and omni-directional samplers in the open environment are a separate subject in themselves.

SAMPLING FOR SIZE ANALYSES

General Principles

An essential element in all bioaerosol research, but especially in any assessment of particle size, is the ability to collect a representative sample for analysis. Measurements made in a well-designed sampling environment should enable the user to obtain representative particle number- or mass-concentrations throughout the range of sizes present. The overall sampling efficiency (η) is usually defined in terms of the number concentration of particles of a given size at the sampler entrance (C_{N_0}) and the number concentration entering the sensitive region of the analyzer (C_N):

$$\eta = \frac{C_N}{C_{N_0}} \tag{8.1}$$

η therefore takes into account particle losses at the inlet, through the sampling line (if used), including any conditioning (dilution, temperature, humidity and pressure changes) as well as losses within the analyzer itself (Figure 8.1). The sampling process is not a trivial problem since η is particle-size dependent, and difficulties increase as the size distribution becomes broader. Unless careful attention is paid to the method used, large systematic biases can enter into the data. At best, such errors detract from the value of the results. At worst, the data can be highly misleading, since the presence of undetected particles may go unnoticed.

Sampling can be avoided by the use of so-called 'non-invasive' or 'in-situ' techniques. These methods are almost always optical, since light (or other electromagnetic radiation) can penetrate aerosols without disturbing the flow or particle motion. The development of on-line techniques is at the forefront of aerosol analyzer development at the present time, and a number of promising instruments are described later in this chapter. A non-invasive technique should be used ideally in preference to extractive sampling if the environment is suitable. However, these analyzers invariably are expensive, and a cost-benefit exercise is justified if it is believed that sampling errors with an extractive system can be kept small (usually true if the largest particles are smaller than about 15 µm).

For many purposes, non-invasive particle size analysis methods are unsuitable, and thought therefore must be given to sampling methodology. Airborne particulates may be sampled either from static conditions (calm-air sampling) or from a moving gas stream. In all cases, the disturbance of the flow of particles by the sampling device should be minimized. Furthermore, sampling lines should be kept as short as feasible with as few bends and restrictions to flow (control valves, etc.) as possible. The object is to avoid particle segregation either at the point of interception from the main flow or during transfer via the sampling line to the particle size analyzer. Although several forces can influence the motion of fine particles, in practice the most important are gravitational sedimentation

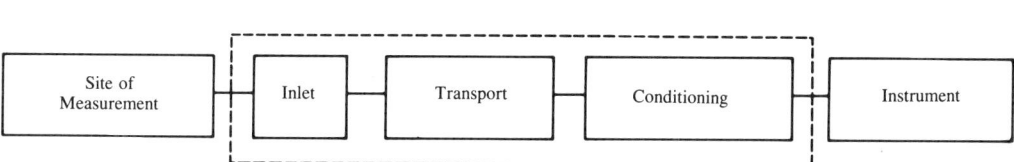

Figure 8.1. Principles of sampling.

and inertia (generally associated with particles larger than 0.5 μm) and Brownian diffusion (associated with smaller particles). The detailed physics of the operation of these processes is described in most aerosol science textbooks, such as those by Hinds[1] and Vincent.[2]

Calm-Air Sampling

The relationship between η and the variables governing sampling behavior (particle size, inlet geometry, gas/particle velocity in the vicinity of the entrance to the inlet) has been established for two conditions:

(a) sampling from a static aerosol (calm-air sampling),
(b) sampling from a flow of aerosol with a probe directed upstream.

Sampling from a flow of aerosol is discussed in the next section, "Flowing Gas Streams."

It is frequently necessary to be able to carry out particle size analysis of aerosols in calm environments (e.g., room sampling for occupational hygiene assessments). Under such circumstances, the only gas movement is that associated with the flow into the sampling device. The motion of micron-sized particles approaching the inlet therefore is influenced significantly by their settling velocity due to gravity as well as by their inertia. The current theories do not explicitly consider the influence of molecular motion on particles (Brownian diffusion) at this stage, assuming (reasonably) that diffusional behavior is not altered by the design of inlet. Hence, sub-micron particles are assumed to be collected perfectly by calm-air samplers; diffusion-based losses become important as the aerosol is transported from the inlet to the analyzer.

A detailed description of calm-air sampling theory is outside the scope of this chapter. However, a brief description of the most important features is given, together with expressions that can be used to assess the likelihood that a sampler will be size selective within the range of interest.

Davies[3] established a criterion for negligible sampling bias due to particle settling behavior, written in terms of the air velocity within the sampler probe (U, cm s⁻¹), the volumetric flow rate into the sampler (Q, cm³ s⁻¹) and the particle relaxation time (τ, s):

$$D_s \leq \frac{2}{5} \left[\frac{Q}{\pi \tau g} \right]^{1/2} \qquad (8.2)$$

where D_s is the probe diameter (cm) and g is the acceleration due to gravity (cm s^{-2}). τ is the time required for the particle to adjust its velocity to a change in the balance of forces governing its motion, and is related to the terminal settling velocity (V_{TS}, cm s^{-1}) by:

$$V_{TS} = \tau g \tag{8.3}$$

Settling velocities for particles of different sizes can be obtained from tables (e.g., Hinds[1]), and a correction is required for particle densities other than unity. Equation 8.2 can be re-written in terms of particle aerodynamic diameter (d_{ae}, μm):

$$D_s \leq 4.1 \left[\frac{Q^{1/2}}{d_{ae}} \right] \tag{8.4}$$

[Note that the term aerodynamic diameter is used to compare the motion of particles that have a wide range of densities and shapes. The aerodynamic diameter is the diameter of a spherical particle of unit density (c.g.s. system) that has the same settling velocity as the particle of interest. Aerodynamic diameter therefore incorporates the effect of shape and porosity on particle motion. The calculation of d_{ae} from the physical dimensions of a particle measured by microscopy is described by Hinds.[1]]

The criterion for negligible bias due to inertial effects also established by Davies[3] can be written similarly:

$$D_s \geq 10 \left[\frac{Q\tau}{4\pi} \right]^{1/3} \geq 0.062(Q)^{1/3} (d_{ae})^{2/3} \tag{8.5}$$

Equations 8.4 and 8.5 set lower and upper limits on the probe diameter that can be employed to sample a given particle size at a given flow rate. Figures 8.2 and 8.3 give the minimum and maximum allowable probe diameters for specified volumetric flow rates and particle aerodynamic size. Davies' criteria, especially that based on particle inertia, are considered by many to be overly restrictive; nevertheless, they define a range of conditions that ensures unbiased sampling. As a general guide, the suction velocity into the inlet probe should be just sufficient to ensure that particles enter. However, consideration of subsequent particle losses through sampling lines may necessitate that this velocity be increased. Agarwal and Liu[4] developed a less stringent criterion incorporating both particle settling and inertial behavior, based on a simulation of the flow field close to a sampling inlet. Their criterion:

$$D_s \geq 20 \tau^2 g \tag{8.6}$$

specifies the limit within which there will be less than a 10% error, and has the advantage of being independent of inlet velocity or volumetric flow rate. In practice, the Agarwal-Liu criterion predicts that there are few restrictions on the sampling of particles smaller than 10 μm aerodynamic diameter.

The maximum air velocity (U_0) for calm-air sampling is given by:

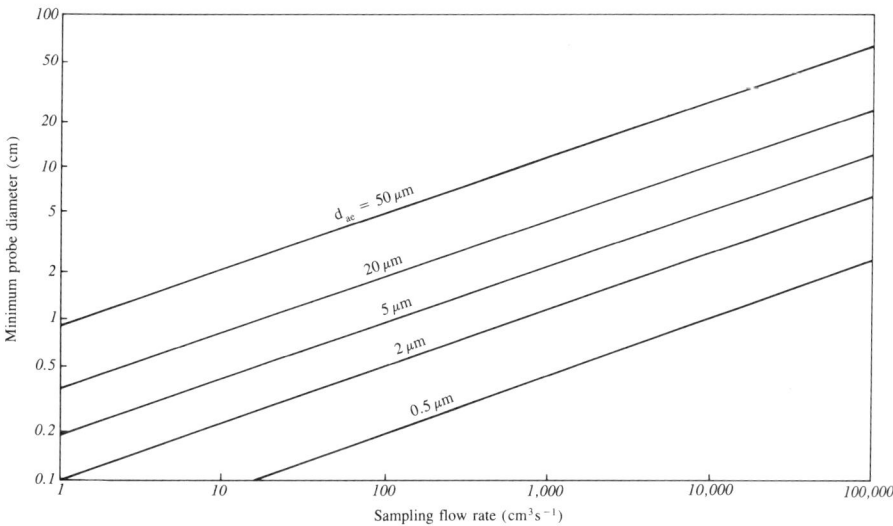

Figure 8.2. Minimum inlet diameter versus flow rate for unbiased calm air sampling (W.C. Hinds, 1982[1] with permission from John Wiley & Sons, Inc.).

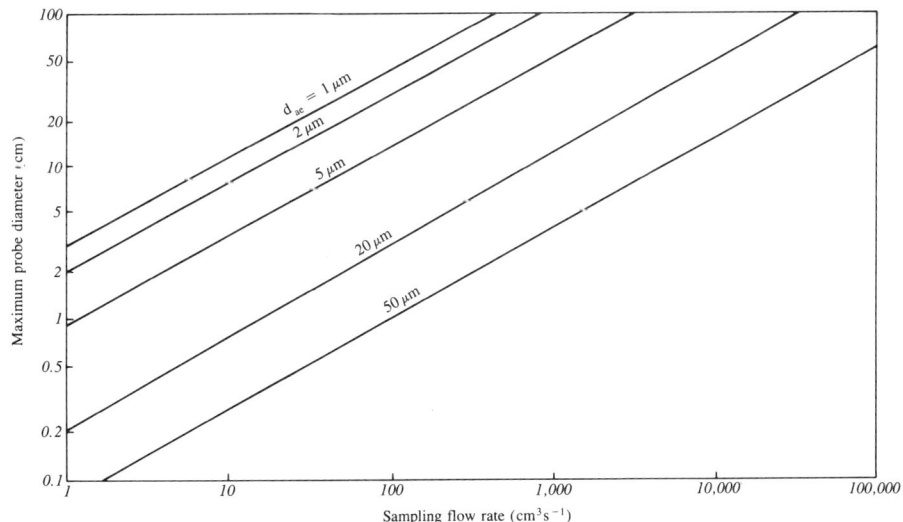

Figure 8.3. Maximum inlet diameter versus flow rate for unbiased calm air sampling (W.C. Hinds, 1982[1] with permission from John Wiley & Sons, Inc.).

Figure 8.4 indicates the maximum sampling velocities allowed for calm-air sampling at various values of Q. Isokinetic sampling methods (see next section) should be adopted if the air velocity is greater than the appropriate limit.

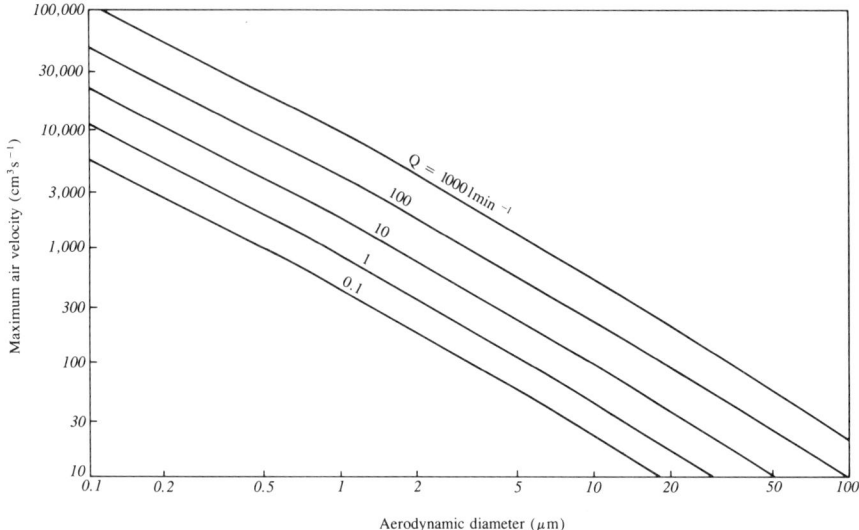

Figure 8.4. Maximum air velocity for use of calm-air sampling criteria (W.C. Hinds, 1982[1] with permission from John Wiley & Sons, Inc.).

Flowing Gas Streams

The second condition for exact values of η is isoaxial sampling from a moving flow under isokinetic conditions, using a thin-walled, sharp-edged nozzle. Various designs of nozzle exist (BS 3405 (1983)), but they are always circular in cross-section and the internal and external surfaces have an inclination not greater than 45° to the axis of the nozzle. This angle should ideally be 10° or smaller (BS 893 (1986)). The term 'isokinetic' describes the condition in which the free stream velocity (U_0) equals the velocity within the probe (U_s). The ideal condition is to be able to take the sample without disturbing the flow field (Figure 8.5a). A failure to sample isokinetically (anisokinetic sampling) will arise whenever U_0 is not equal to U_s.

If $U_s > U_0$ (Figure 8.5b), large particles are undersampled, since they cannot follow the streamlines because of the significant particle inertia and miss the probe entrance altogether. Likewise, if $U_s < U_0$ (Figure 8.5c), large particles are oversampled since they cross the streamlines and enter the probe. If isokinetic sampling is not achieved, it follows that there is no way to determine the true size distribution in the particle size analyzer, unless there is additional information known by another technique. Conditions for isokinetic sampling and for good probe design are provided in British Standards BS 893 (1986) and BS 5243 (1975), as well as in textbooks on sampling practice (e.g., Vincent[2]).

If anisokinetic sampling is unavoidable, the results of an empirical investigation by Belyaev and Levin[5] can be used to estimate the systematic errors by considering the ratio (U_0/U_s) and particle size, expressed in terms of the Stokes number (Stk):

$$\eta_{inl} = 1 + \left[\frac{U_0}{U_s} - 1\right] \left[\frac{2(U_0/U_s) + 0.62}{Stk^{-1} + 2(U_0/U_s) + 0.62}\right] \quad (8.8)$$

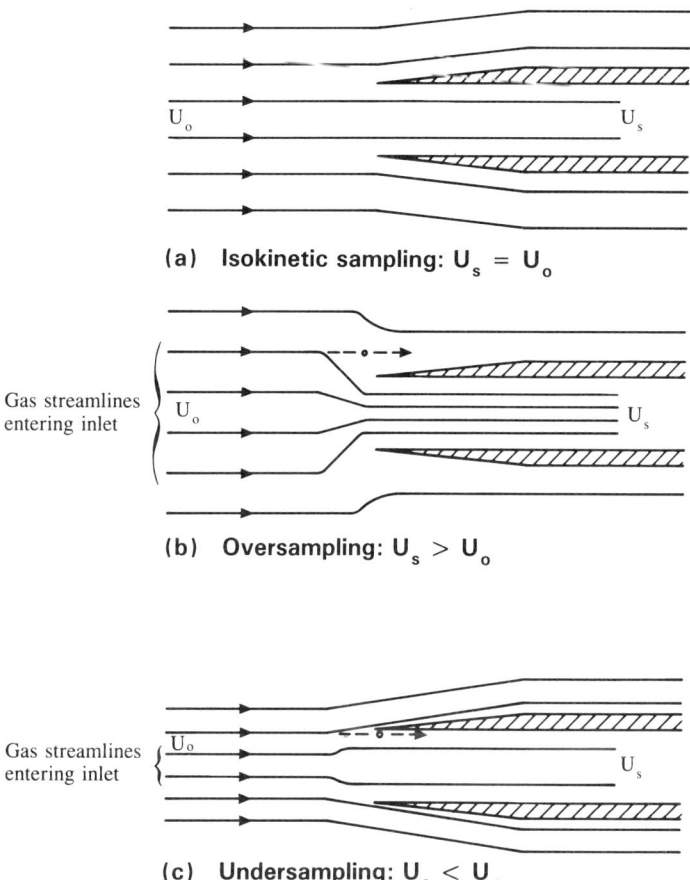

Figure 8.5. Iso-axial sampling from a flow containing aerosol.

where:

$$\text{Stk} = \frac{\tau U_0}{D_s/2} = \frac{d_{ae}^2 U_0 C_c}{9\mu D_s} \qquad (8.9)$$

μ is the gas viscosity (g cm^{-1} s^{-1}) and C_c is the Cunningham slip correction factor (Hinds[1]; C_c is close to unity for particles larger than about 1 μm aerodynamic diameter). Figure 8.6 can be employed to assess errors due to anisokinetic sampling and their effect on η_{inl}, the inlet sampling efficiency or aspiration coefficient. Alternatively, the correlation given in Table 8 of BS 5243 may be used to provide a quick estimate of errors. In general, significant over- or under-sampling can be tolerated if the particles are smaller than 10 μm aerodynamic diameter.

Sampling errors also will be introduced if the sampling probe is misaligned with the flow. Durham and Lundgren[6] provided an empirical correlation to correct for probe misalignment (θ from 0 to 90° in Figure 8.7) under isokinetic conditions:

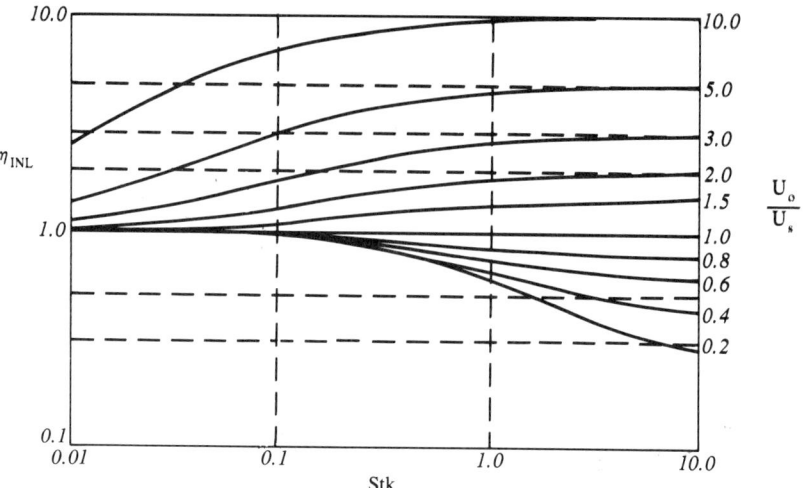

Figure 8.6. Prediction of anisokinetic sampling errors.

$$\eta_{inl} = 1 + (\cos\theta - 1)\left[1 - \frac{1}{1 + 0.55 Stk'\exp(0.25 Stk')}\right] \quad (8.10)$$

where:

$$Stk' = Stk\,\exp(0.022\theta) \quad (8.11)$$

for θ in degrees. Equation 8.10 applies for particles of moderate size ($0.01 < Stk < 6$). The motion of particles of less inertia is unaffected by the probe misalignment (they follow the streamlines into the probe). The expression reduces to:

$$\eta_{inl} = \cos\theta \quad (8.12)$$

for very large particles ($Stk > 6$), whose trajectories are not deflected by the change in direction of the streamlines entering the probe. In practice, care should be taken to avoid probe misalignment, since it is difficult to quantify θ, particularly if the sample is being extracted from a duct or stack.

So far, this description of sampling from a flow of aerosol has assumed laminar conditions. In many applications (e.g., open-air or large duct sampling), the free-stream flow is likely to be turbulent. As yet, there is no rigorous theoretical understanding of the influence of such turbulence on sampling behavior, although it is recognized that important processes such as diffusive transport of momentum in the flow are dependent upon turbulence intensity. Experiments performed at the Institute of Occupational Medicine with a disc-shaped blunt sampler operated over a wide range of turbulence intensity (up to 15%), indicated that free-stream turbulence exerted a negligible effect on the shape of the mean air flow pattern on the upstream side. However, the extent to which this finding can be applied to other inlets is not known.[2]

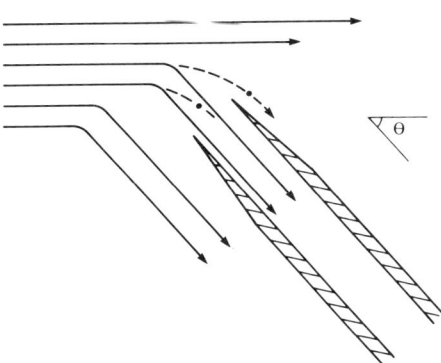

Figure 8.7. Misalignment of inlet to oncoming flow.

In duct/stack sampling with sharp-edged inlets, it is common practice to calculate an average value of U_0, either from measurements of velocity by traverses across the duct/stack or by simple derivation from the volumetric flow rate divided by the cross-sectional area of the air pathway. The equations relating to isokinetic sampling are then applied without correction for the effects of turbulence.

The influence of wind direction and probe orientation (pitch and yaw angles) must be allowed for in fugitive emission sampling (Vincent[2]). Open-air samplers therefore often are designed to operate only when the wind direction is closely aligned with the probe (directional samplers). Alternatively, several designs of in-field high volume (HI-VOL) samplers have purpose-built inlets that conform to a specific particle size-collection efficiency curve, such as the US PM_{10} standard for omni-directional size-selective sampling of particles smaller than 10 μm aerodynamic diameter.

Sample Line Losses

Sampling lines are almost always a source of particle losses, and therefore should be minimized in length (Chapter 3). However good the sampling technique, the particle size analysis will be marred if insufficient attention is paid to the particle transport arrangements. Table 8.1 is a summary of the processes that can affect the transport of aerosols through sampling lines; a full assessment of the problem involves determining if the flow is laminar or turbulent and assessing the inertial/diffusional/phoretic losses through each part of the sampling line. In practice, this rigorous approach cannot be justified, and assumptions therefore are made to simplify the assessment. In the simplest method, the tables contained in BS 5243 (1975) permit assessment of estimated particle losses in horizontal and vertical tubes, as well as 90° bends. These tables are based on simplified expressions and care must be taken to ensure that the regimes to which they apply (e.g., laminar/turbulent flow) describe the actual conditions within the sampling line. Laminar flow normally can be assumed if the flow Reynolds number (Re_f) is less than 2300. Re_f is a dimensionless quantity relating inertial to viscous forces in a flow, and is defined as:

Table 8.1. Sources of Particle Losses in Sampling Lines

Flow Regime	Loss Mechanism	Size Range Predominantly Affected
Laminar	Brownian Diffusion	Sub-micron
Laminar	Gravitational Settling	Super-micron
Turbulent	Brownian Diffusion	Sub-micron
Turbulent	Inertial Effects	Super-micron
Independent	Electrostatic Effects	Sub-micron
Independent	Thermo-Diffusio-Phoretic Effects	All sizes

$$Re_f = \frac{U_t D_t \rho_g}{\mu} \qquad (8.13)$$

where D_t is the internal diameter of the sample line ([cm], assuming circular cross-section), and U_t is the average flow velocity (cm s^{-1}). $D_t = D_s$ and $U_t = U_s$ if the internal diameter of the inlet probe and sampling line are constant.

Some of the more useful equations are given here; more details can be found in Hinds[1] or in the specific references quoted.

(a) *Sedimentation* — laminar flow only
— large particles ($d_{ae} > 1$ μm)

Thomas[7] has shown that:

$$L_{100} = \frac{8 D_t U_t}{6 V_{TS}} \qquad (8.14)$$

where L_{100} is the length of horizontal sample line (cm) for 100% particle deposition, The length for 50% deposition (L_{50}) is given by:

$$L_{50} = 0.354 \, L_{100} \qquad (8.15)$$

It is good practice to minimize gravitational deposition by the use of vertical sampling lines. If horizontal lines cannot be avoided, they should be kept as short as possible and of a size that maximizes the flow velocity without introducing turbulence (turbulent deposition is another important mechanism for sample line losses).

(b) *Brownian Diffusion*—small particles (< L μm volume equivalent diameter)

Gormley and Kennedy[8] demonstrated that:

$$\eta_{tube} = 0.819\exp(-3.657\phi) + 0.097\exp(-22.3\phi) + 0.032\exp(-57\phi) \quad (8.16)$$

where ϕ is the dimensionless group $\pi D_k L/Q$ in which L is the sample line length (cm) and Q is the volumetric flow rate (cm^3 s^{-1}), D_k is the diffusion constant (cm^2 s^{-1}) given by:

$$D_k = \frac{2.4 \times 10^{11}}{d_v}\left[1 + \frac{1.8 \times 10^5}{d_v}\right] \quad (8.17)$$

for air, and d_v is the particle volume equivalent diameter. Equation 8.16 is unsuitable for values of ϕ less than 0.01 (short sample lines or high volumetric flow rates), and a simpler expression:

$$\eta_{tube} = 1 - 2.56(\phi)^{2/3} \quad (8.18)$$

is adequate. It should be noted that diffusion losses are unaffected if the diameter of the sampling line is altered, assuming a constant sampling flow rate (Q). The extra distance particles must travel in a wider tube is exactly offset by the longer time for diffusion permitted by the wider tube.

(c) *Turbulent Deposition* — large particles
— $Re_f > 2300$

Particles carried in turbulent flow will be deposited by both diffusion across the laminar sub-layer, and those with sufficient inertia may penetrate through the boundary layer to reach the wall. The process is complex and only partly understood. Furthermore, particle resuspension may also occur if there is sufficient turbulent energy for vortices to penetrate the boundary layer and scavenge deposited particles from the surface. The equation given in BS 5243 is an empirical expression that does not take into account resuspension:

$$\eta_{tube} = \exp\left[\frac{-\pi K L D_t}{Q}\right] \quad (8.19)$$

from which:

$$L_{50} = \frac{0.693\, Q}{\pi\, D_t\, K} \quad (8.20)$$

D_t is the tube diameter (cm) and K is the deposition velocity (cm s^{-1}), which is a function of several variables and has been determined only for a few specific cases (e.g., Schwendiman et al.,[9] Stevens et al.[10]). The subject of turbulent deposition of particles in pipe bends recently has been addressed comprehensively by Pui et al.[11]

Sampling from Ducts and Stacks

The sampling of dusty gas streams is technically an aerosol-related problem, although it is encountered also in the analysis of powder process streams and in the determination of particle size in effluent gases. A detailed description of the various sampling techniques is beyond the scope of this chapter, and much information on the subject may be found in articles by Lewandowski[12] and Allen.[13] The standards published in the UK, USA and Germany may be taken as representing the state of the art, and are as follows:

BS 893: Method for the Measurement of the Concentration of Particulate Material in Ducts Carrying Gases (1978),

BS 1042: Measurement of Fluid Flow in Closed Conduits (1984); sections 2.1-2.3 are concerned with the use of Pitot tubes,

BS 3405 Part 1: Method for the Measurement of Particulate Emission Including Grit and Dust (Simplified Method) (1983),

BS 5243: General Principles for Sampling Airborne Radioactive Materials (1975),

BS 6069 Part 4: Characterization of Air Quality - Stationary Source Emissions (ISO 6069) (1992),

ASTM Standard D 2928 - 71: Sampling Stacks for Particulate Matter (1970),

ASME Power Test Code 27: Determining Dust Concentration in a Gas Stream (1957),

EPA Method V: Standards of Performance for New Stationary Sources (Federal Register, pp 41776, **43(166)**, 1977),

VDI 2066: Measurement of Particles in Flowing Gas Streams: Part 4 (Optical Transmission Method) and Part 6 (Photometer KTN) are both current (1989),

VDI 2463: Measurement of Particles in Air: Part 5 (Automated Filter Collection type FH 621), Part 6 (Automated Filter Collection by β-Attenuation type F703) and Part 9 (Use of Filter Device LIS/P) are current (1987),

These standard methods relate to extractive sampling, and provide detailed information about the design of inlet probes, sampling lines and conditioning (e.g., trace heating of the sampling train to prevent premature water vapor condensation). Although the exact sampling strategy chosen will depend on the nature of the particulate stream being sampled, there are certain guidelines that are recognized as essential for reliable results. These are summarized below.

1. Establish the location of the sampling point(s) as far as possible from any disturbances to the flow, such as pipe bends, valves or air moving devices. The optimum number of sampling locations depends on the geometry of the pipe (duct),

and the recommended locations for circular and rectangular cross-sections are given in most of the standards. Perhaps the most useful is that provided in the EPA regulations (EPA Method V, 1977), and illustrated in Figure 8.8a. Sampling locations for rectangular and circular pipes also are given in Figures 8.8b and 8.8c respectively.

Ideally, sampling should not be carried out at a distance of less than 7 duct diameters (or widths for rectangular cross-sections) from a flow disturbance. In practice this condition may be difficult to achieve, and it may be necessary to resort to flow straighteners to obtain a uniform flow distribution across the pipe.

2. Carry out at least one traverse across the width of the pipe to obtain the velocity profile at the sampling location(s). Velocity variations greater than ±10% should be avoided if particles larger than 5 μm aerodynamic diameter are being sampled, since anisokinetic errors will be significant (see section "Flowing Gas Streams"). Pitot tubes orientated so that they face the direction of flow are most commonly used to measure flow velocity. They operate on the principle that the force exerted by a flowing gas creates an impact pressure on an obstacle in its path. Pitot tubes can be used to sense the direction of flow, in contrast with non-directional devices such as hot-wire or hot-film anemometer probes. This is a particularly important requirement when sampling behind air moving devices such as fans, where a vortex can be set up with consequent large variations in velocity across the pipe width. BS 1042 (1984) describes the design of the commonly used ellipsoidal-nosed pitot tube, illustrated in Figure 8.9, together with an indication of the effect of misalignment on the accuracy of velocity measurements. The pitot tube measures both the impact, or velocity pressure (h, N m^{-2}) and the static pressure (P, N m^{-2}) in the duct. Equation 8.21:

$$U = k \left[\frac{T\,h}{S\,P} \right] \qquad (8.21)$$

is used to obtain the gas velocity (U, m s^{-1}), where S is the gas density ratio (specific gravity) compared with dry air at temperature T, and k is a constant with a value 23.96 that includes the pitot coefficient (close to unity for the BS 1042 design).

The ellipsoidal type of pitot tube may become blocked if the particle concentration is high, and the S-type (Stausscheibe) pitot tube (Figure 8.10) should be utilized instead. This type of tube requires calibration over the range of velocities of interest, and the EPA regulations (Method 2, 1977) state that calibration should be carried out against a standard design of pitot tube. The orientation of the S-type pitot tube to the flow is critical; permissible values of the rotation (θ_1) and tilt (θ_2) angles are 10° and 5°, respectively, and the flow must be parallel to the walls of the duct (EPA Method 2, 1977).

3. Assemble the sampling train. The minimum requirements are the isokinetic sampling probe, including inlet nozzle and probe, particle collection device (filter for simple total mass concentration, impactor/cyclone for particle size analysis) and a means for maintaining a predetermined volumetric flow rate and total flow through the device. A critical orifice flow-meter is the simplest kind of flow control device and in practice a needle valve is installed upstream of the air-tight pump, and a dry-gas meter is used to determine the volumetric flow (at ambient pressure and temperature) after the pump. Whichever method is used, corrections for temperature and pressure differences between the duct and ambient atmospher-

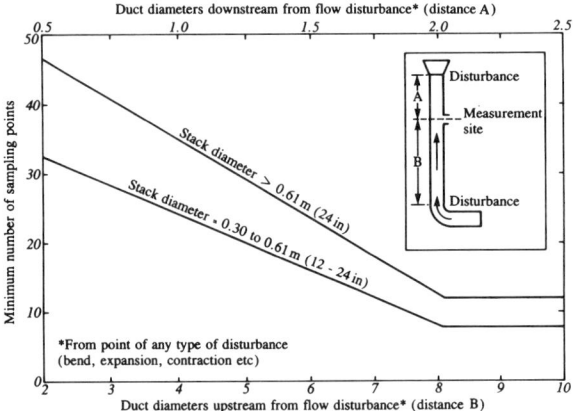

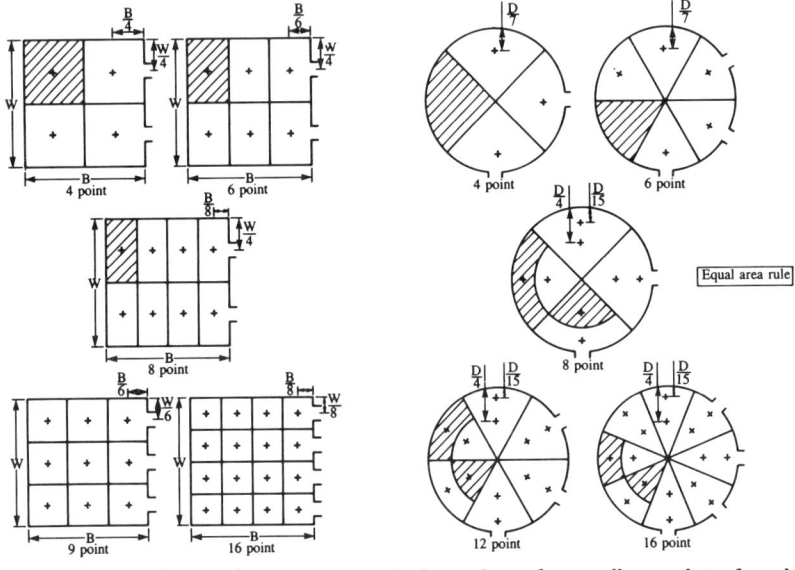

(a) **Minimum number of sample points for particulate traverses (from USEPA Method V Regulations. (1977))**

(b) **Location of sampling points for rectangular ducts**

(c) **Location of sampling points for circular pipes excluding centre point (D is internal diameter of pipe)**

Figure 8.8. Sampling locations in ducts and pipes of circular and rectangular cross-section (USEPA, 1977).

ic conditions are required to establish the gas velocity entering the probe (needed to establish isokinetic sampling conditions). If appreciable quantities of water vapor/steam or other condensable vapors are present, the sampling train must be modified to incorporate a condenser (e.g. ice-bath trap), which is installed behind the particle sampling device. Sampling lines upstream of the condenser therefore must be trace-heated to prevent premature condensation, and the volumetric flow rate of non-condensable gases (determined with the dry gas meter) must be cor-

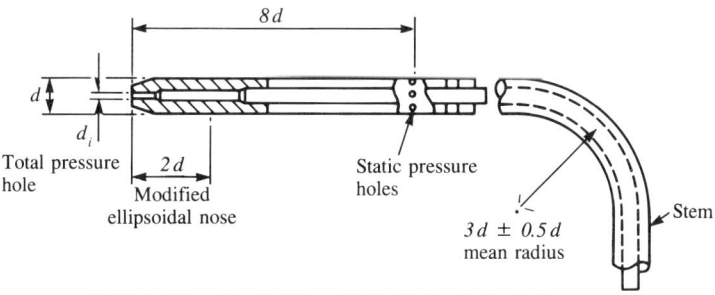
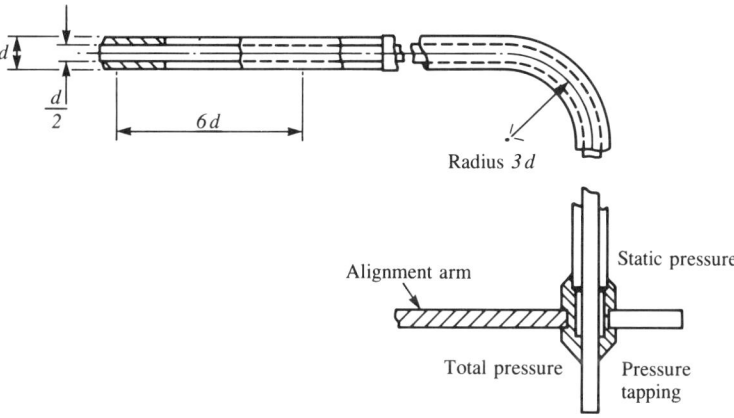

Figure 8.9. Pitot tube designed in accordance with BS 1042, British Standards Institute, Milton Keynes, UK (1984).

rected for the amount of condensate collected upstream. The EPA Method 5 sampling train is widely regarded as the most suitable method for particle size analysis, although the actual procedure relates to the use of a filter sampler for total particulate mass concentration, rather than a particle size analyzer. Note that the particle size analysis equipment may be installed in the heated box located immediately outside of the duct, as carried out with the Source Assessment Sampling System (Krill[14]). However because of the size-selective nature of sampling probes, it is recommended that the particulate be collected within the duct if possible. Various designs of in-stack impactor and miniature cyclone trains (see "Cascade Impactors" and "Cascade Cyclones", respectively) are available for this purpose.

AEROSOL PARTICLE SIZE MEASUREMENT TECHNIQUES

Summary of Methods

Particle sizing techniques for aerosols fall into five main categories:

(a) sedimentation/inertial analyzers,

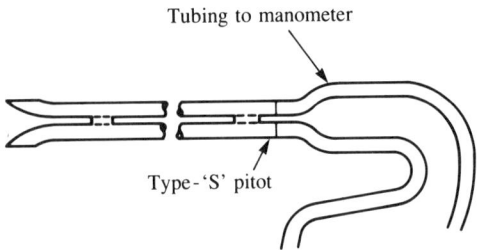

(a) General view

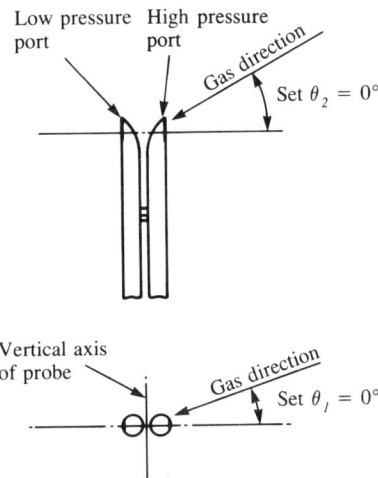

Figure 8.10. 'S'-type pitot tube.

- (b) optical techniques,
- (c) electrical mobility techniques,
- (d) diffusion methods,
- (e) microscopy-image analysis.

The approximate size ranges of each category of analyzer are provided in Table 8.2, together with comments concerning their limitations (Nichols and Mitchell[15]). Microscopy is the only technique capable of spanning the entire range of particle sizes associated with aerosols. Useful descriptions of the various techniques are to be found in Hinds,[1] Mercer[16] and Liu,[17] and the methods of calibration of almost all commonly used aerosol analyzers have been described in Chapter 7 (see also Mitchell[18]).

The various types of particle diameter mentioned in Table 8.2 are summarized in Table 8.3. Note that it is customary to use aerodynamic rather than Stokes diameter to describe the performance of sedimentation- and inertial-based aerosol analysis equipment.

Table 8.2. Techniques for Measurement of Particle Size Distribution

Principle	Technique	Approximate Size Range	Comments
Gravitational Sedimentation	Horizontal and Vertical Elutriators	1 to 100 μm Stokes diameter	Gentle flow conditions enable large agglomerates to be classified; flow rates are low and may result in severe inlet losses; iso-thermal sampling essential
Inertial Separation	Cascade Impactor	0.5 to 15 μm aerodynamic diameter (low-pressure designs extend the range down to 0.05 μm)	Rugged and easy to use; in-stack versions; operable at high temperatures; sample-line and wall losses, particle bounce and re-entrainment; easily overloaded
	Cascade Cyclone	0.5 to 15 μm aerodynamic diameter	Same as cascade impactor, but less prone to bounce and overloading
	Centrifuge	0.08 to 5 μm aerodynamic diameter	Spatial resolution for either mass- or number-size distributions; excellent resolution; drawbacks include fragility, sensitivity to temperature and pressure fluctuations. Inoperable at high concentrations; inlet losses
	Prodi Inspec	0.5 to 10 μm aerodynamic diameter	Same as centrifuge, but more robust
	TSI Inc. APS33B	0.5 to 30 μm aerodynamic diameter	Rapid analysis in real time; sensitivity below 0.7 μm diameter low; particle co-incidence effects
	API Aerosizer	0.5 to 200 μm aerodynamic diameter (powders > 30μm)	Good resolution; electrical mobility is the only high resolution method for 0.01 to 0.1 μm range
Light Scattering	Near-forward, right-angle, and wide-angle OPC	0.3 to 100 μm diameter	Semi-quantitative analysis in conjunction with condensation nuclei counter; cumbersome to use.
Microscopy	Optical SEM/TEM	2 to 1000 μm volume equivalent diameter (optical); 0.01 to 100 volume equivalent diameter	Careful sample preparation; unsuitable for sizing of complex agglomerates; special techniques for droplets

Table 8.3. Definitions of Particle Diameter

Symbol	Diameter	Definition	Comments
		Fundamental Diameters	
D_v	Volume equivalent diameter	Diameter of a sphere with same volume as the particle	Also known as spherical diameter
D_s	Surface equivalent diameter	Diameter of a sphere with same external area as particle	
D_d	Drag diameter	Diameter of a sphere with same resistance to motion as the particle in a gas of the same viscosity	
		Projected Diameters	
$D_A(D_g)$	Projected area diameter	Diameter of a circle with same area as the projected area of the particle	Also known as the geometric diameter (D_g); measured by microscopy-image analysis
D_p	Perimeter diameter	Diameter of a circle with same perimeter as the projected outline of the particle	
		Derived Diameters	
D_{st}	Stokes diameter	Diameter of a sphere with the same density and setting velocity as the particle in a gas of the same density and viscosity	
D_{ae}	Aerodynamic diameter	Diameter of a sphere with unit density (cgs system) and the same settling velocity as the particle in a gas of the same density and viscosity	
D_{em}	Electrical mobility diameter	Diameter of a sphere with the same mobility in an applied electric field as the particle in a gas of the same density and viscosity	Measured by electrical mobility analyzers ($D_{em} = D_{st}$) for singly charged particles
D_{diff}	Diffusion diameter	Diameter of a sphere with same velocity due to Brownian motion as the particle in a gas of the same density and viscosity	Measured by diffusion batteries

This situation arises because frequently the density and shape of aerosol particles are not known, and can be difficult to determine because of the small amount of material available for study. The concept of aerodynamic size ("Calm-Air Sampling" section) avoids the need to determine particle density, while at the same time providing a direct measure of the transport behavior of micron-sized particles, both in the open environment and with these classes of analyzer.

Sedimentation/Inertial Techniques

Sedimentation and inertial techniques form a single class of aerosol analyzers that determine particle aerodynamic diameter directly. Sedimentation devices are encountered in a limited number of specialized applications, such as occupational hygiene, where low aerosol flow rates are acceptable. However, general-purpose inertial aerosol classifiers are by far the widest used type of particle size analyzer. This category includes fractionating instruments, such as cascade impactors, multi-stage cyclones and centrifugal spectrometers. The latter techniques preserve the entire size-separated particulate on a single substrate, a feature shared with most types of sedimentometer.

Sedimentation Techniques (Elutriators)

The simplest sedimentation samplers consist of a chamber into which aerosols are introduced at a well-defined flow rate in laminar flow. The particles are size-separated under the influence of gravity, in accordance with Stokes Law. The terminal velocity (v_t) of a particle of volume equivalent diameter d_v is given by:

$$v_t = \frac{d_v^2 (\rho_p - \rho_0) \chi}{18 \mu} \tag{8.22}$$

where ρ_p and ρ_0 are the particle density and gas density (cgs system) respectively, μ is the gas viscosity, g is the acceleration due to gravity, and χ is the particle dynamic shape factor (unity for spheres). In practice $\rho_p >> \rho_0$ and the latter can be neglected. Equation 8.22 is only strictly valid if the particle motion is unaffected by wall interactions, terminal velocities have been reached and particle-particle interactions are negligible. The first two criteria are easy to satisfy with aerosol-based systems, but the third criterion limits the particle concentration below about 100 mg m^{-3}.

Elutriators occasionally are used to determine true aerodynamic diameter (d_{ae}) directly from a calibration of deposition distance versus particle size, using spherical particles of known density.[19] The Timbrell spectrometer[20] can achieve a high degree of size-grading of aerosols containing particles in the size range from about 2 to 20 μm aerodynamic diameter (Figure 8.11a). In this instrument, the particles are winnowed in a recirculating flow of clean air in a horizontal settling chamber; the size-separated particles are collected onto a series of microscope slides located along the bottom of the chamber. Although this instrument is easy to use, it has not been widely adopted because the incoming aerosol flow rate needs to be very small (typically 1 cm^3 min^{-1}) for good size resolution.[21] Typical calibration data[19] are shown in Figure 8.11b for polystyrene latex and polyvinyl acetate microspheres (both having densities close to unity). The optimum size resolution (close to 12%) was achieved in the range from 4 to 12 μm aerodynamic diameter, when the incoming aerosol flow rate was 1 cm^3 min^{-1} and the ratio of winnowing to aerosol flow rates was 150:1.

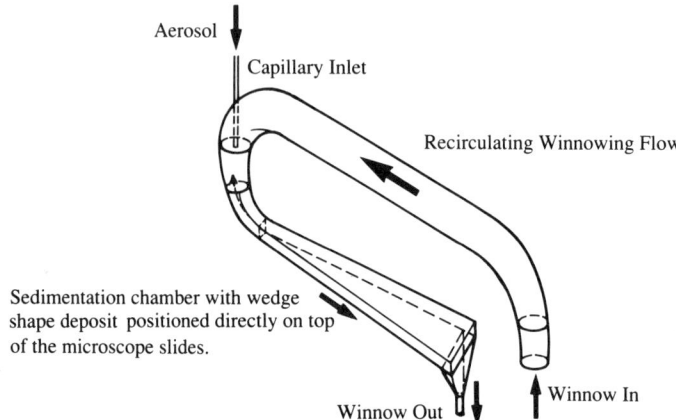

(a) Timbrell Aerosol Spectrometer [from ref. 20]

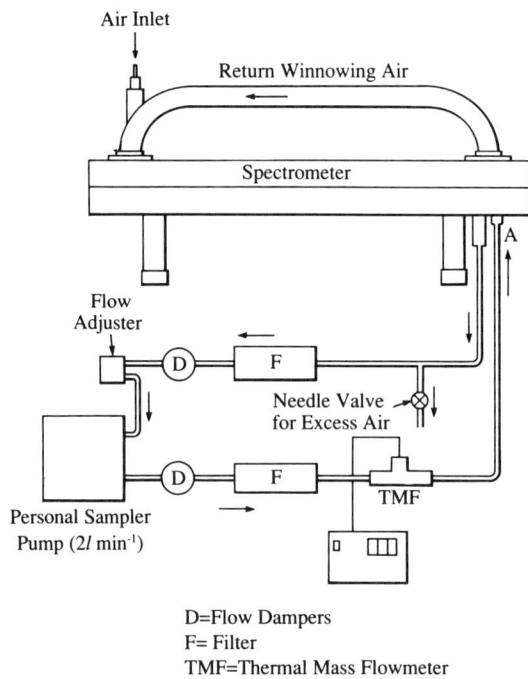

(b) Flow Circuit (from ref. 19)

Figure 8.11. Timbrell sedimentation aerosol spectrometer.

There are many individual designs of sedimentation devices, including the elutriator of McCormack and Hilliard,[22] which provided aerodynamic size distributions of sodium fire aerosols that were in good agreement with aerodynamic size distribution data from an Andersen Mark-II cascade impactor (see next section "Cascade Impactors"). The horizontal sedimentation battery of Boulaud et al.[23] is a development of the basic sedimentometer, and

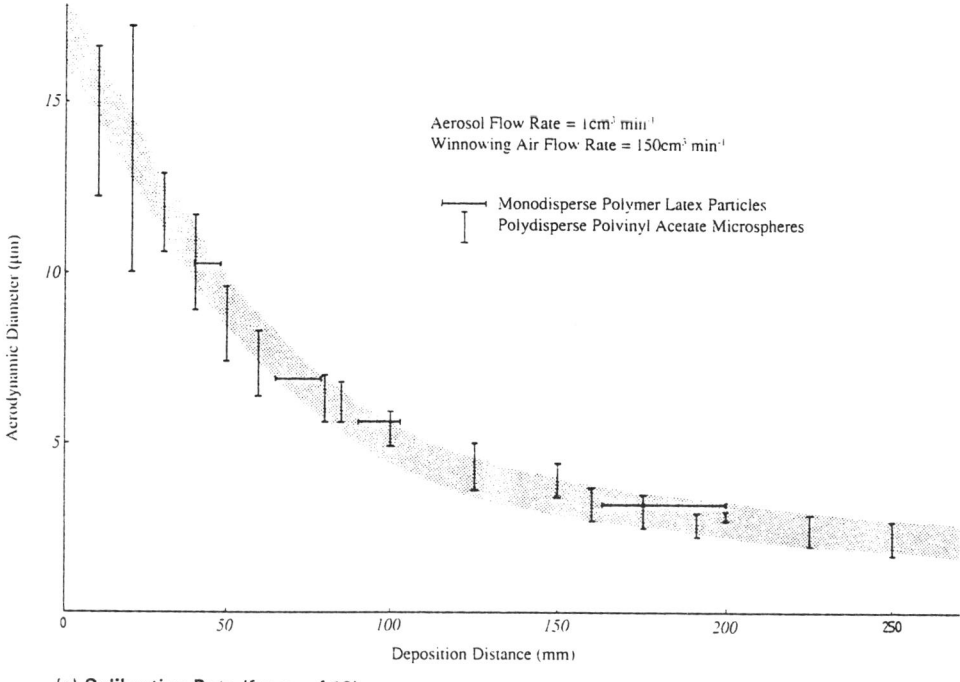

(c) Calibration Data (from ref.19)

Figure 8.11. Continued.

has been shown to provide reliable mass-aerodynamic size distribution data, but the size range available at a given aerosol flow is small (ca. 2 to 10 μm aerodynamic diameter) and the data analysis procedure is complicated. The incoming aerosol is introduced into the 10 rectangular channels via a 30 cm long cone whose purpose is to ensure that the aerosol is homogeneous as it enters the sedimentation battery (Figure 8.12a). The channels each contain moveable plates that enable the particles to be collected after travelling a given horizontal distance. The deposition distances 0–5, 5–10, 10–15, 15–20 and 20–30 cm were analyzed in the system described by Boulaud et al.,[23] which represented a battery with the following lengths 5, 10, 15, 20 and 30 cm (Figure 8.12b). The minimum aerodynamic diameters of the monodisperse ammonium fluorescein particles collected (100% efficiency) in the seven ranges were 9.1, 6.4, 5.2, 4.6 and 3.9 μm and these values defined the measurement range of the system. Their data analysis method involves taking into account the variation of collection efficiency with particle size for each stage of the sedimentation battery.

Cascade Impactors

Cascade impactors are the most commonly encountered aerosol sampling instrument for both stationary source and process stream evaluations. The design, calibration and use of commercially available devices have been recently reviewed by Lodge and Chan.[24] The simplest impactor consists of a series of stages, each comprising a jet plate that is located at a fixed distance from a horizontal collection surface (Figure 8.13). Aerosol passes through the jet plate (containing one or many orifices); the streamlines of the flow diverge on approach to the collection surface, whereas the inertia of the particles causes them to

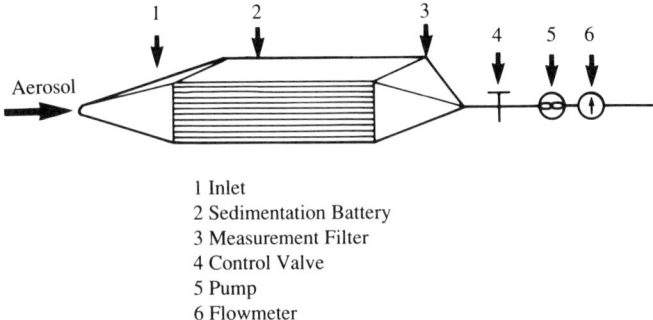

1 Inlet
2 Sedimentation Battery
3 Measurement Filter
4 Control Valve
5 Pump
6 Flowmeter

(a) Schematic Diagram of the Sedimentation Battery and Aerosol Flow Arrangement

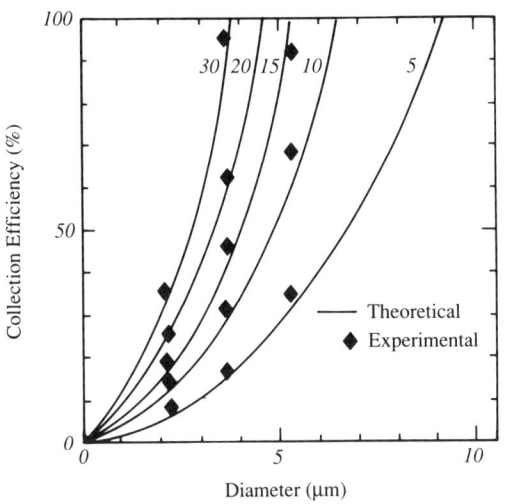

(b) Calibration Data for 5 different Battery lengths (from ref. 23)

Figure 8.12. Sedimentation battery of Boulaud et al. (b: with permission from Pergamon Press, Oxford. D. Boulaud et al. *J. Aerosol Sci.* 1983, **14**:421–424).

cross the streamlines. Particles of sufficient size (and inertia) describe trajectories that intersect the collection surface; smaller particles are able to remain gas-borne, passing to the next stage. The jet diameter(s) of the second plate are smaller, increasing the gas velocity so that smaller particles are collected. The process is repeated, collecting progressively smaller particles until the bottom of the impactor is reached.

The last stage may contain a filter to trap all particles that pass the final impaction stage.

In normal use, the fully assembled impactor is coupled to a vacuum pump operated at a pre-determined volumetric flow rate to meet both the sampling requirements, and achieve effective cut-off diameters (ECDs) for each stage (particle sizes corresponding to 50% collection efficiency) that span the range of sizes of interest. ECDs for some of the more common impactors are summarized in Table 8.4; these values are based on unit density

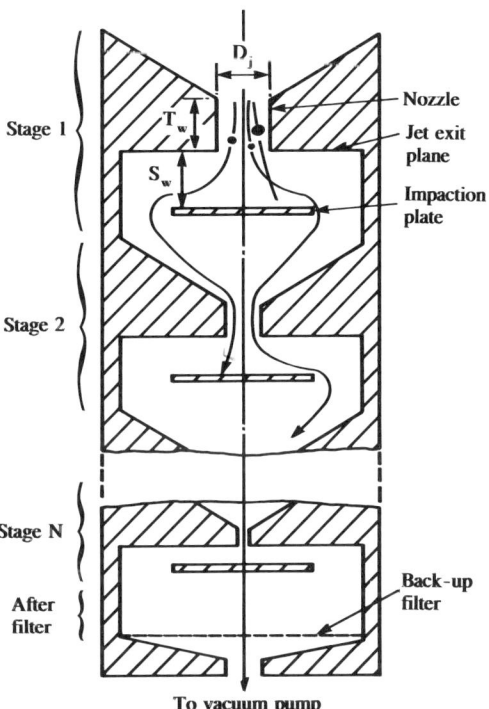

Figure 8.13. Principle of cascade impactor (with permission from W.C. Hinds. Data analysis. In *Cascade Impactor: Sampling and Data Analysis,* J.P. Lodge and T.L. Chan, Eds. Americam Industrial Hygiene Association, Akron, OH. 1986, 46).

spheres, and in most cases have been established by calibration with particles of well-defined aerodynamic size.

The inertial collection process may be defined in terms of the particle Stokes number (see "Flowing Gas Streams" section), scaled on the radius of the orifice ($D_j/2$) forming the jet plate (or average orifice radius if a multi-orifice design is being considered). By analogy with Equation 8.9:

$$\text{Stk} = \left[\frac{\rho_p d_v^2 U C_c}{9\chi\mu D_j}\right] = \left[\frac{d_{ae}^2 U C_c}{9\mu D_j}\right] \tag{8.23}$$

where χ is the dynamic shape factor of the particles (unity for spheres). Marple and Liu[25] and Rader and Marple[26] among others have determined the precise functional relationship between the stage collection (grade) efficiency curve and Stk, in expressions of the general form:

$$E_{\text{stage i}} = E_{\text{stage i}}(\text{Stk}, \text{Re}_f, S_w/D_j, T_w/D_j) \tag{8.24}$$

Table 8.4 ECDs for Commonly Encountered Cascade Impactors

Nominal Flow Rate (ft³/min)	Impactor Type	Approximate Cut-Off Size Corresponding to Nominal Flow Rate[a] Aerodynamic Diameter (μm)
	Multiple Orifices per Stage	
20 to 40	Sierra Hi-Vol, parallel slots, 1133 L min⁻¹ (40 ft³/min) Andersen Hi-Vol, round holes, 566 L min⁻¹ (20 ft³/min)	
0.5 to 1 (stack)	Metrology Research, Inc. 1502, round holes, 28.3 L min⁻¹ (1 ft³/min) University of Washington Mark III, round holes, 28.3 L min⁻¹ (1 ft³/min) Sierra 226, radial slots, 21.2 L min⁻¹ (0.75 ft³/min) Andersen Mark III, round holes, 14.2 L min⁻¹ (0.5 ft³/min) Sierra TAG, parallel slots, 14.2 L min⁻¹ (0.5 ft³/min)	
0.5 to 2	MSP Corp[e] Multi-Orifice Uniform Deposit Impactor (MOUDI) 30 L min⁻¹ (1.1 ft³/min) Andersen Ambient Mark II, b round holes, 28.3 L min⁻¹ (30 ft³/min) Sierra Ambient, radial slots, 14.2 L min⁻¹ (0.5 ft³/min) Berner[f] Low Pressure Impactor, 30 L min⁻¹ (1.1 ft³/min)	
	Single Orifice per Stage	
30	B. Gussman, Inc. (BGI) Hi-Vol, single slot, 850 L min⁻¹ (30 ft³/min)	
1 to 4	Sierra-Lundgren, slot, rotating drum, 113 L min⁻¹ (4 ft³/min)	
0.1 to 1	Casella Mark IIc (BGI), slots, 17.5 L min⁻¹ (0.62 ft³/min) Battelle DCI-6 (Delron, single round hole, 12.5 L min⁻¹ (0.4 ft³/min) Brink Model B (Monsanto), single round hole, 2.8 L min⁻¹ (0.1 ft³/min)	
0.003 to 0.1	Battelle DCI-5 (Delron), single round hole, 1.05 L min⁻¹ (0.037 ft³/min) Aries 04-001, single round hole, 0.65 L min⁻¹ (0.023 ft³/min) Aries 04-002, single round hole, 0.085 L min⁻¹ (0.003 ft³/min) California Measurements PC-2 Quartz Crystal Sensor,[d] single round hole, 0.24 L min⁻¹ (0.0085 ft³/min) - QCM impactor	

[a] The cut-off size is defined as the aerodynamic diameter corresponding to 50% collection efficiency; [b] The Andersen Ambient Mark II impactor can be used with a six-stage low-pressure extension: at 3 L min⁻¹ the size range of the complete impactor ranges from 0.08 to 35 μm aerodynamic diameter; [c] The Casella Mark II impactor is based on the original May four-stage impactor, but is no longer commercially available; [d] The California PC-2 impactor has two additional low-pressure stages with cut-off sizes of 0.05 and 0.1 μm aerodynamic diameter; [e] The MOUDI is available in UK from BIRAL -There is a choice of 8 stages from sizes shown, including stages with cut-off sizes of 0.056, 0.1 and 0.18 μm aerodynamic diameter; [f] The Berner LPI has additional stages with cut-off sizes of 0.06 and 0.125 μm aerodynamic diameter.

where S_w and T_w are the ratios of jet diameter to the clearance between jet and collection plates, and jet throat length to diameter, respectively. These data demonstrate that the typical impactor stage has a sharp, well-defined ECD; most particles smaller than this size will pass to the next stage. Stk is essentially independent of jet-to-plate spacing and flow Reynolds number over a wide range of conditions. Most commercially available impactors are designed to operate within this range, where Stk_{50} (Stokes number corresponding to 50% collection efficiency) is constant.

The effect on ECD of changing the operating parameters may be determined from:

$$ECD = d_v \left[(C_c'/C_c)(\rho_p'/\rho_p)(Q'/Q)(\mu'/\mu) \right]^{1/2} \qquad (8.25)$$

where primed and unprimed variables refer to standard (calibration) and actual measurement conditions respectively. The influence of Re_f on Stk_{50} should be considered if outside the range $10^2 < Re_f < 10^4$.

Cascade impactors can be grouped into four main types:

(a) ambient systems, designed to sample aerosol from a stationary source or slow-moving stream,
(b) in-stack systems to sample aerosols from flows in ducts and stacks,
(c) low-pressure and micro-orifice systems to extend the lower measurement limit from ca. 0.3 to <0.05 µm aerodynamic diameter,
(d) virtual systems, in which the collection plate is replaced by a large stagnant volume to eliminate wall losses and overloading problems.

Ambient impactors such as the Andersen Mark-II (Figure 8.14a), are fairly bulky,[27] and usually are designed to operate at a fixed volumetric flow rate [28.3 L min^{-1} (1 acfm)]. In-stack impactors are much more compact, since they are intended to be inserted into the duct to avoid sampling losses. For instance, the Andersen Mark-III 'in-stack' impactor (Figure 8.14b) can be inserted through a port as small as 73 mm (2.875 in). The reduction in size is achieved by using specially perforated collection substrates that permit the flow to pass to the next stage without having to pass around the periphery. Assembly of the collection stages can be awkward with in-stack impactors of this type. Some designs (e.g., University of Washington series of impactors) have been developed to overcome this difficulty (Pilat et al.[28,29]).

Low pressure impactors (Figure 8.14c) operate by passing the flow through a critical orifice, usually located toward the bottom of the stack of collection stages. The upper part of the analyzer therefore operates as a conventional impactor, whereas efficient sub-micron separation also can take place in the low pressure section because the slip correction factor is greatly increased, reducing the aerodynamic drag on the particles (Hering and Marple[30]). Several designs are in common use, such as those of (Pilat;[31,32] Hering et al.;[33,34] McFarland et al.;[35] and Berner[36]). One instrument (California Instruments QCM impactor) uses vibrating quartz crystal sensors to measure mass-size distributions in the range from 0.05 to 25 µm aerodynamic diameter in real-time. The performance of this device has been described by Fairchild and Wheat,[37] who commented on a susceptibility to diffusion-based losses of particles smaller than about 0.5 µm diameter. Recent work by Horton and Mitchell[38] also has shown that the behavior of electrostatically charged particles is unpredictable in this impactor. Low pressure impactors are useful to study low-volatile aerosols (metal fumes, etc.), but are not recommended if the particles have appreciable volatility. Micro-orifice impactors (Marple et al.[39]) operate at near-ambient atmospheric pressure, but

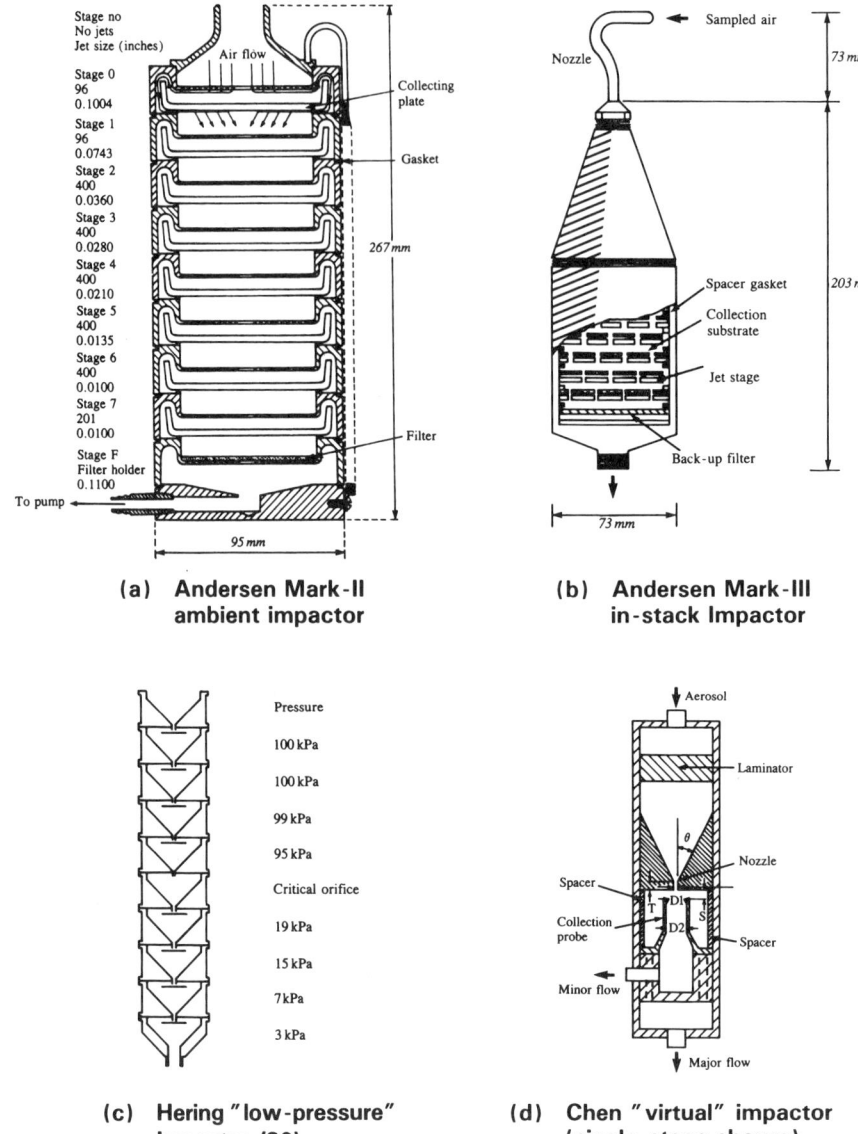

Figure 8.14. Various types of impactor (c: with permission from S.V. Hering and V.A. Marple. Low pressure and micro-orifice impactors. In "Cascade impactor: sampling and data analysis," J.P. Lodge and T.L. Chan, Eds. American Industrial Hygiene Association Journal, Akron, OH 1986, 110. d: with permission from Pergamon Press, Oxford. B.T. Chen and H.C. Yeh, *J. Aerosol Sci.* 1985, **16**, 343. f: with permission from Elsevier Science Publishing. V.S. Novick and J.L. Alvarez, *Aerosol Sci. Technol.* 1987, **6**, 63–70).

achieve a size range similar to that obtained in low-pressure impactors by reducing the size of the orifices in the jet plates from millimeters to values between 50 and 100 μm. As many as 2000 orifices per jet plate are required to operate at volumetric flow rates comparable with ambient or in-stack impactors. The use of these impactors to sample process

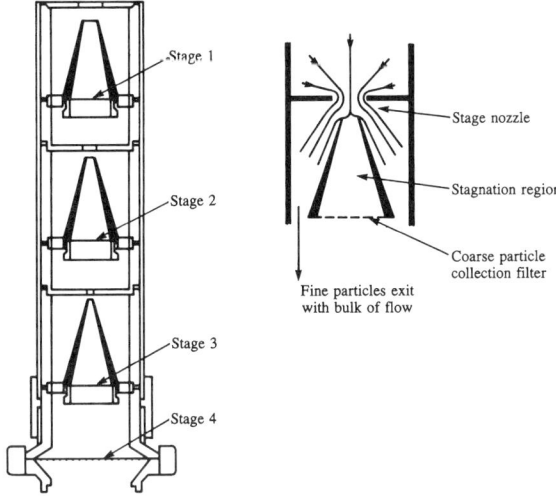

(e) UKAEA virtual impactor (40)

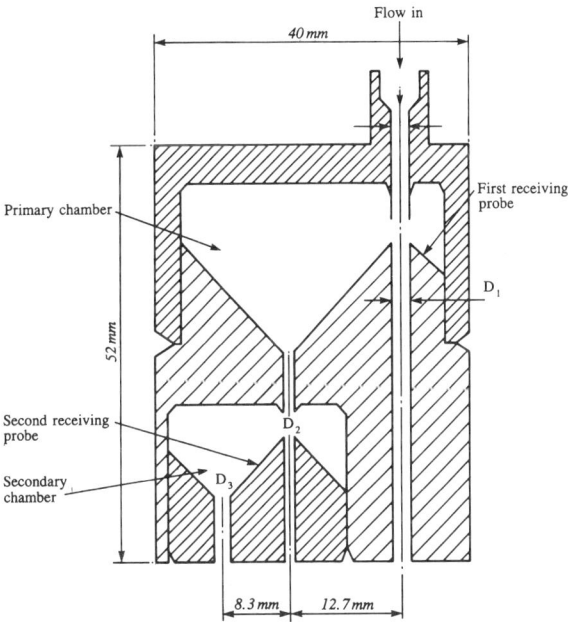

(f) Novick and Alvarez miniature virtual impactor (41)

Figure 8.14. (continued).

streams is in its infancy, but the technique is promising, since in contrast with low-pressure systems, multi-orifice impactors can be used effectively to size classify volatile particles.

Virtual impactors (Figure 8.14d) are used in situations where it is desirable to collect a larger mass of particles than would be possible with conventional designs. The UKAEA virtual impactor (Figure 8.14e) is a four-stage cascade design in which filters are used to collect the fractionated particles. This instrument was designed primarily to sample atmospheres containing minute amounts of particulate-borne radioactivity, and it fulfills this function well. However, it suffers from substantial losses due to deposition on internal surfaces, and the collection efficiency curves are not as steep as those of conventional impactor designs (O'Connor[40]). Novick and Alvarez[41] have described an extremely compact virtual impactor, in which two separation units are combined within a single body to provide three size fractions with minimal internal wall losses (Figure 8.14f). The aerosol flow through the system is controlled by a series of critical orifices located at the base of the receiving probes. This impactor is not yet available commercially, but has been operated successfully in hostile environments (high temperatures, high pressures, radioactive aerosols, etc.). Virtual impactors more usually are designed to split the incoming aerosol into two fractions (dichotomous sampling); the minor flow contains the larger particles that have penetrated the stagnant layer. However, as with all types of virtual impactor, the minor flow containing the coarse fraction is contaminated by some fine particles. The operation of dichotomous samplers has been described by McFarland et al.,[42] Masuda et al.,[43] and Willeke and Pavlik.[44]

The mass of particles collected on each impactor stage (or in the outlet flows of a virtual impactor) is normally measured by a standard microbiological chemical or radioanalytical technique. Alternatively, the particles may be collected on a pre-weighed substrate (usually glass-fiber or cellulose triacetate filters, or occasionally metal [aluminum] plates) and the mass determined gravimetrically. If the latter method is chosen, precautions must be taken to eliminate errors due to collection of moisture on the substrate/collected particles. All types of impactor therefore determine primarily particle mass- or volume-based size distributions, unless the deposits are so small that counting of individual particles on each collection plate is possible by microscopy (not recommended). The mass collected on each stage is plotted against the ECD relating to that stage in any of the four basic types of size distribution (absolute, differential, cumulative or normalised (Hinds[1]).

Several precautions should be taken when using impactors.

1. Many types of cascade impactor are easily overloaded with the sample; this is especially true with single orifice designs on which no more than 10 mg should normally collect on any stage. Multi-orifice designs can collect proportionately more material, although the deposits always should be visually inspected for signs that the particles have spread away from the collection locations immediately beneath the jets. The measurements should be rejected if overloading is suspected.
2. The high velocities attained within impactors makes them prone to particle bounce. This phenomenon is diagnosed if a significant fraction of the aerosol sample collects near to the bottom stages or on the back-up filter. Data should be discarded if particle bounce is suspected. Cheng and Yeh[45] have developed an empirical model that can be used to predict the onset of particle bounce. However, particle-related properties (coefficient of restitution, etc.) that influence the phenomenon are difficult to quantify and the normal practice is to select a suitable coating or collection substrate that is effective in trial measurements. There is a large body of literature on the subject of such coatings, and useful formulations are summarized in Table 8.5.[46-54] There is no ideal solution; coatings may interfere with the quantitative removal and microbiological or chemical analysis of the collected particles, and filter substrates can modify the stage collection efficiency curves and hence the impactor calibration. Any coating must be inert both at ambient temperature and at the temperature of the measurement. For instance, glass microfibre substrates should be avoided where SO_2/SO_3 is present. Furthermore, the material

Table 8.5. Surface Adhesives Used in Impactor Studies

Author	Adhesive
May (46)	Oil or vaseline
Davies et al. (47)	Glycerin jelly
Stern et al. (48)	Dow-Corning Hi-Vac silicone grease
McFarland and Zeller (49)	Dow-Corning silicone oil (DC-200 fluid; 2% solution in n-hexane)
Lundgren (50)	Viscous oils and sticky tapes Dow-Corning Hi-Vac silicone grease 3M Kel-F polymer wax
McFarland and Husar (51)	10^6 cs. Dow-Corning DC-200 silicone oil (1% solution in n-hexane)
Mitchell et al. (52)	Dow-Corning Hi-Vac silicone grease (10% solution in n-hexane)
Mercer and Chow (53)	Dow-Corning anti-foam A
Hogan (54)	silicone resin
Berner (36)	Vaseline dissolved in toluene

Source: Marple, V.A. and Liu, B.Y.H. Inertial impactors: theory, design and use. In *Fine Particles*. Ed. B.Y.H. Liu, Academic Press, NY, 1975, 432)

must have negligible volatility at the maximum temperature of use, especially if a gravimetric procedure is to be used to analyze the data. For this reason, grease coatings are not recommended for use at high temperatures.

3. The high velocities encountered in impactor jets also may cause particles to fragment, skewing the measured size distribution to smaller sizes. This effect is most likely to occur with loosely bound agglomerates (smoke, metal fumes, etc.), and an inertial spectrometer (see "Centrifugal Spectrometers" section) might therefore be used in preference to an impactor to investigate such aerosols.
4. If impactors are used to sample live micro-organisms, consideration should be given to the effect of impact at high velocities on viability. Again, inertial spectrometers may be preferred if viability is of importance.
5. Wall losses caused by inertial deposition are a significant source of error, particularly if particles larger than about 5 μm aerodynamic diameter are being sampled (Cushing et al.[55]). Not much can be done to reduce the magnitude of these losses, other than to select an alternative design that is less prone to this problem. Impactors with tortuous transport paths, or having dead volumes between stages are especially prone to large wall losses. Rader et al.[56] has described a procedure to correct impactor data for wall losses; this technique involves calibration with well-defined particles and is not recommended unless a high degree of precision (< ±5%) is required.
6. Diffusion-induced wall losses may be large for certain designs of low-pressure impactor (Fairchild and Wheat[37]). However, the problem is largely confined to particles smaller than 0.3 μm aerodynamic diameter, and is therefore not frequently encountered in bioaerosol sampling or in more general industrial or workplace emission measurements.
7. Particles with a significant electrostatic charge may not collect at the correct location in the impactor, particularly if the collection substrates are insulating materials (Horton and Mitchell[38]). Charge build-up can be reduced by passing the aerosol

through a charge equilibrator, such as a radioactive Kr-85 line source, but this approach is not practical for in-stack/duct sampling. It is better to collect the particles on metal substrates if electrostatic problems are suspected.

Finally, particle losses associated with in-stack/duct samplers should be reduced where possible by mounting the impactor within the pipe and using a sharp-edged nozzle. If an in-pipe measurement is impractical, the length of the sampling line should be minimized, avoiding bends and obstructions to the flow. Sampling should also be carried out in a vertical section of pipe if the particles have an appreciable settling velocity (particles > 10 μm aerodynamic diameter). A single point measurement at the center of the pipe will suffice if the velocity profile across the duct is uniform. A mixing baffle, such as a Stairmand disc, can be inserted upstream of the sampling location to homogenize the flow in cases where the duct is not vertical, or the velocity profile is not uniform. However, it should be noted that such devices may increase the loading on the fan or air-moving device supplying the pipe to an unacceptable degree. In addition to these sampling-related precautions, the impactor must be rigidly mounted to prevent vibration, and trace-heated if there is any possibility of condensation. A comprehensive guide to USEPA procedures for the operation of impactors for sampling process streams has been produced by Harris.[57]

Cascade Cyclones

Gas cyclones may be utilized for particle size analysis in the range from about 0.5 to 20 μm aerodynamic diameter, as an alternative to impactors. Although other designs of cyclone exist, the reverse-flow miniature cyclone (Figure 8.15a) is most likely to be encountered for sampling bioaerosols. The operating principles of industrial gas clean-up cyclones have been described comprehensively by Licht.[58] Miniature cyclones operate in a similar way, and only the more important aspects of their use and limitations will be addressed here.

Several cyclones may be operated in series to form a sampling system analogous to the cascade impactor (Figure 8.15b). Although bulkier in design, the cascade cyclone train has the distinct advantage that much larger masses of particulate (gram quantities) can be collected in each stage without the risk of carry-over into the next cyclone. Hence, cascade cyclones are used to sample highly concentrated dust streams (> 1 g m^{-3}), which would overload cascade impactors in seconds.

Unfortunately, there is no rigorous theory that links the collection efficiency of a reverse-flow cyclone to particle Stokes number, as is possible with impactors. Various correlations have been developed to predict ECD values from cyclone geometry and operating conditions. These correlations have the general form:

$$\text{ECD}_{\text{cyclone}} = k \left[\frac{D_c^3 \mu}{Q} \right]^{1/3} \tag{8.26}$$

where Q is the volumetric flow rate, μ is the gas viscosity, D_c is the diameter of the cyclone (Figure 8.15a). k is a constant of proportionality that depends on the cyclone geometry, and may be a function of at least seven dimensionless groups (Licht[58]). Beekmans[59] demonstrated that the grade efficiency curve was a function of both flow Reynolds and particle Stokes numbers:

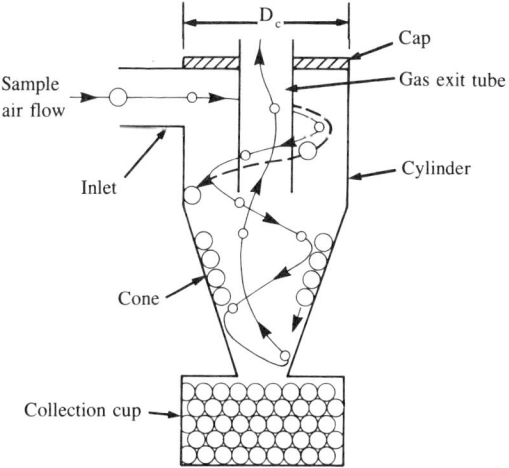

(a) Hypothetical flow through typical reverse flow cyclone

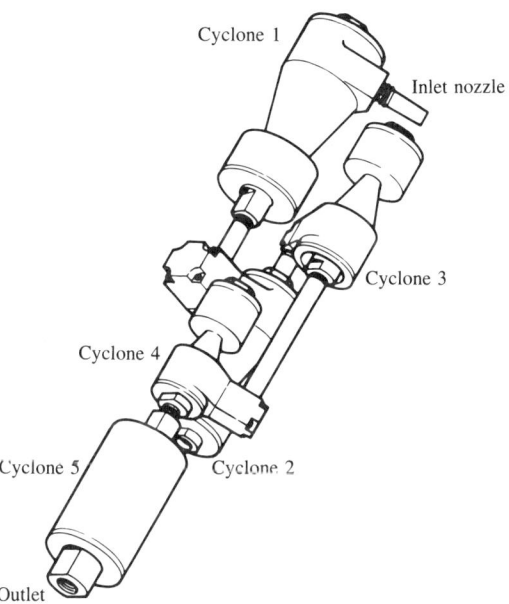

(b) Sierra-Andersen cyclone train

Figure 8.15. Reverse-flow gas cyclone: operating principle and example of a cascade cyclone train.

$$E_{cyclone} = E_{cyclone}(Re_f^a, Stk^b) \qquad (8.27)$$

where a and b are calibration coefficients. More recently, Lee et al.[60] found that the ECD values correlated best with Re_f alone for a train consisting of three of the five Sierra-Andersen miniature cyclones. Their data were obtained in both air and argon atmospheres at a range of flow rates, and agree with the calibration data for the same design of cyclone sampler reported by Smith et al.[61,62] Until further data are obtained that support the more general use of these correlations, calibration is recommended for all cyclone samplers, unless the manufacturer has already provided such data. This advice is especially important when cyclones are operated at high temperatures (unlikely for bioaerosols) and pressures, since the relationships between ECD and gas properties (density, viscosity, etc.) are not well understood far beyond ambient conditions.

The collection efficiency-particle size curves are less steep for cyclones than impactors. It is therefore more difficult to design a multi-cyclone system in which the collection efficiency curves for successive stages do not overlap. Thus, the size range from 0.3 to 15 μm aerodynamic diameter is resolvable into 5 parts with the Sierra-Andersen cyclone train and 8 parts with the Andersen in-stack impactor.

Real-Time Analyzers

Two new and powerful instruments have become available for the almost instantaneous measurement of aerosol particle size distributions based on aerodynamic diameter. The TSI APS33B was developed prior to the Amherst Process Instruments Aerosizer (Table 8.6), but both techniques are increasingly employed to monitor aerosol behavior. Their high initial cost is offset by savings made in analysis time if more conventional inertial particle size analyzers were used.

In the APS33B (Remiarz et al.[63]), the velocities of individual particles are measured after acceleration under sub-sonic conditions, as they pass across a twin-beam laser velocimeter (Figure 8.16a). The recorded transit times can be related directly to aerodynamic diameter through the equation of motion of the particles:

$$\frac{dU_p}{dt} = \frac{3\mu U_p}{D_{ae}^2} \left[6 + \left[\frac{\rho_g U_p D_{ae}}{\mu}\right]^{2/3} \left[\frac{\chi}{\rho_p}\right]^{1/3} \right] \qquad (8.28)$$

where dU_p/dt is the particle acceleration. The response of the instrument therefore is dependent upon particle density and shape as well as particle size. The density correction is well understood, and the manufacturers provide a correction algorithm in the software (in the case of the APS33B, this correction is based on the work of Wang and John[64]). The effect of particle shape on the behavior of the APS33B is the subject of current research (Marshall et al.[65]); present indications are that the instrument undersizes cube-shaped non-spherical particles by as much as 25%, depending on the value of the dynamic shape factor, (χ). The behavior of this analyzer with other well-defined particle shapes (e.g., oblate and prolate spheroids) and fibers is being investigated. Liquid droplets distort from sphericity as they pass through the acceleration nozzle, and significant undersizing has been reported both by Baron[66] and Griffiths et al.[67]

Although the quoted size range of the APS33B is from 0.5 to 30 μm aerodynamic diameter, the sensitivity decreases rapidly below 0.7 μm diameter, particularly with light absorbing particles (Blackford et al.[68]). Particles larger than about 10 μm aerodynamic diameter are also detected with reduced efficiency due primarily to inertial losses at the inlet. The APS33B may detect spurious large particles if used to measure aerosols with count mean size close to the lower detection threshold of the instrument, since individual particles may just scatter sufficient light to trigger the timer but fail to reset the system as they pass through the second laser beam. An additional larger particle arriving before the

Table 8.6. Operating Ranges of Selected Centrifugal and Real-Time Inertial Aerosol Analyzers

Analyzer	Nominal Volumetric Flow Rate (L min^{-1})	Nominal Size Range (μm)
Stöber spiral-duct centrifuge	0.5	0.08–5
Prodi inertial spectrometer	0.1–1	0.5–10
TSIPS33B	1.0 (5 total)	0.5–30
Amherst Process Instruments Aerosizer	1.0	0.5–30 (aerosols) 0.5 - 200 (powders)

time-out condition will then reset the timer, and the size recorded will depend on the elapsed time between the two separate particles. The problem can be alleviated but not entirely eliminated by reducing the sensitivity of the analyzer so that only larger (micron-sized) particles are detected.

As is the case with all single particle counters, coincidence will occur if the aerosol concentration is sufficiently high (> ca. 100 particles cm^{-3}). The manufacturer of the APS33B has provided well-characterized diluters to overcome the problem (Remiarz and Johnson[69]), enabling the user to reduce the incoming aerosol concentration by as much as a factor of 10,000.

The APS33B has been used both as an ambient air quality sampler and as a monitor of gaseous effluent streams. This instrument may also be employed to analyze small quantities (> 100 mg) of resuspended dry powder using a verturi-suction-based dispersion unit (TSI Small Scale Powder Disperser (Blackford and Rubow[70]). However, the sensor head is fragile, making it unsuitable for use in arduous environments.

The Aerosizer (Amherst Process Instruments) is based on work by Dahneke[71] and Dahneke and Cheng[72] to form particle beams when an aerosol stream is vented via a critical orifice into a partial vacuum. The operating principle is similar to that of the APS33B, but particle acceleration is much more rapid as the flow passes at near-sonic velocity through a critical orifice located immediately upstream of the laser velocimeter (Figure 8.16b). The instrument is claimed to have a much wider operating range (0.5 to 200 μm aerodynamic diameter) than the APS33B, but not all of this range is available for a given set of operating conditions. For instance, particles much larger than 15 μm aerodynamic diameter cannot be sampled efficiently without using the purpose-built particle delivery systems designed to handle dry powders. Little information is available about the performance of the Aerosizer: however, it is reasonable to expect that the operating limitations are quite similar to those of the APS33B.

Centrifugal Spectrometers

Aerosol spectrometers may be divided into two classes, centrifuges and fixed-geometry devices. Both types of instrument are used increasingly to size classify aerosol particles, although the centrifuges are suitable only for laboratory applications.

Aerosol centrifuges are not in the strictest sense inertial analyzers; rather they comprise a class of sedimentation analyzer in which the effect of gravity is replaced by centrifugal force. As a result, centrifuges are capable of good size resolution of sub-micron particles, making them the only class of sedimentometer likely to be encountered for routine aerosol

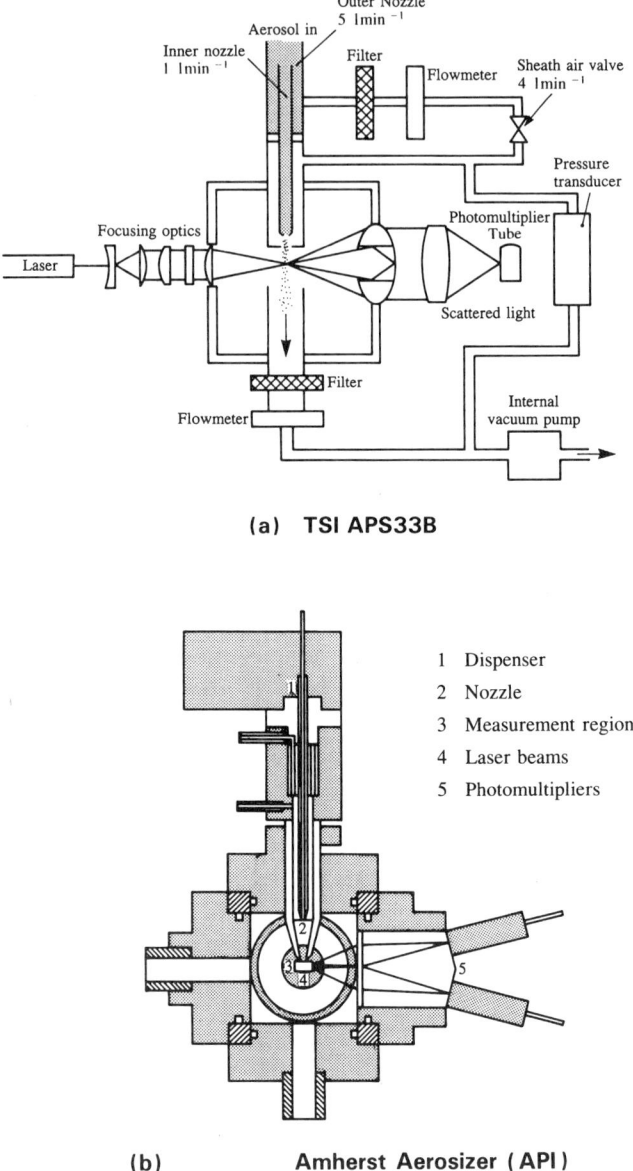

Figure 8.16. Real-time aerodynamic particle sizers (a: with permission from TSI Inc., St. Paul, MN. b: Amherst Process Instruments Inc., Hadley, MA).

analysis. The basic design consists of a circular- or spiral-shaped deposition channel that is rotated to exert a centrifugal force on the gas-borne particles as they travel through the channel in laminar flow. The earliest successful design was the 'conifuge' of Sawyer and Walton,[73] with an operating range from 0.5 to 30 μm aerodynamic diameter. More recent systems have extended the lower size limit by almost an order of magnitude at the expense of inlet designs that restrict the entry of particles larger than about 5 μm aerodynamic diameter. At terminal velocity:

$$\frac{dx}{dt} = \frac{(\rho_p - \rho_g) d_v^2 \omega^2 x}{18 \mu} \tag{8.29}$$

where x is the distance from the axis of rotation to the particle location, dx/dt is the outward (radial) velocity of the particle, ρ_p and ρ_o are the particle and gas densities respectively, μ is the gas viscosity and ω is the angular velocity of the rotor (= rpm x $2\pi/60$ radians per second). The collection surface (usually a metal foil or strip of membrane filter) is located along the outer wall of the channel. Equation 8.29 is complicated by the effect of the forward velocity imparted to the particles as they are swept into the channel in a flow of winnowing gas, and also by the channel geometry, which in most designs is formed in a spiral rather than a circle.

Aerosol centrifuges offer the highest degree of size resolution in the range from 0.2 to about 2 μm aerodynamic diameter (5% is typically achievable when the flow of aerosol entering the channel (Q_a) is less than 5% of the winnowing flow (Q_w)). The Stöber spiral-duct centrifuge (Stöber and Flachsbart[74,75]), which is one of the most widely encountered instruments of this type, has a channel length of 180 cm, and an operating size range from 0.08 to 5 μm aerodynamic diameter (Table 8.6). The design of the centrifuge is shown in Figure 8.17a together with calibration data obtained at two different values of rotor speed (Figure 8.17b). The size resolving power is shown in Figure 8.17c, which illustrates the degree of size separation of aggregate particles consisting of primary polystyrene microspheres 2.02 μm aerodynamic diameter. A modified, lower-cost version with a shorter channel length (45 cm) also has been developed by Kotrappa and Light.[76] In contrast with liquid-borne powder centrifuges, operation of these aerosol centrifuges is easy and the data reduction is relatively simple.

Aerosol mass-size distributions can be determined from analysis of the particles collected on a foil or filter substrate mounted along the outer wall of the settling channel. The substrate is sectioned into well-defined lengths and the aerosol deposits are either weighed or subjected to microbiological or chemical analysis. Sections of the substrate can be examined under the microscope, and if the deposit is not too densely packed, the particles can be counted and sized by microscopy. Thus aerosol spectrometers can be used to provide number- as well as mass-weighted size distribution data.

It is essential that the composition, pressure and temperature of the winnowing gas match those of the aerosol being sampled or aerodynamic size separation will not occur due to turbulent mixing when the two streams meet at the channel entrance (Martonen[77]). Pressure fluctuations in the sampled flow also must be avoided. Furthermore, particle-particle interactions will prevent size separation if the aerosol concentration exceeds about 1 g m^{-3} (Mitchell et al.[78]).

The Prodi inertial spectrometer (Inspec) represents one of a family of fixed-geometry centrifugal spectrometers in which inertial size separation takes place as the particles are transported around the curve of a 90° bend located immediately below the aerosol injection point (Prodi et al.[79]). The flow of winnowing gas (typically 6 L min^{-1}) is drawn into the compact sampling head through a rectangular channel, and the aerosol stream is introduced as a thin pencil close to the inner wall (Figure 8.18a). The size-separated particles travel through a small sedimentation chamber that magnifies the separation, and are collected on a substrate (usually a membrane filter) located at the base of the chamber on a porous support plate. The operating range of the Inspec is from 0.5 to 10 μm aerodynamic diameter (Table 8.6), depending on the value of Q_w (Figure 8.18b), and the smallest particles deposit the greatest distance along the length of the substrate (furthest from the 90° bend).

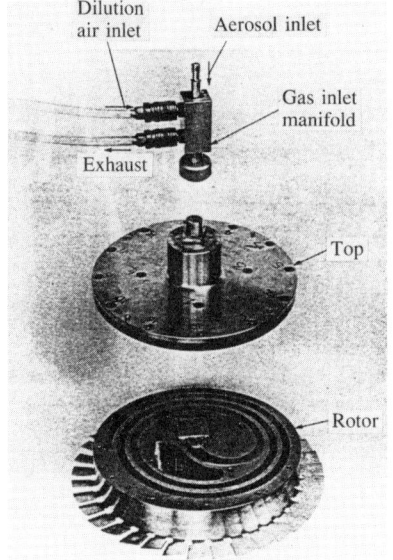

(a) Components of the Stöber spiral duct centrifuge

Atomiser pressure 14 p.s.i.
run time 5 h

Band		d_{ae} (μm)
1	Singlets	2.02
2	Doublets	2.40
3	Triplets	2.63
4	Quadruplets	2.79
5	Quintuplets	2.87

(c) Deposit of aggregates of polystyrene latex microspheres (2.02 μm diameter) on a stainless steel sampling foil: 1500 rev min^{-1}

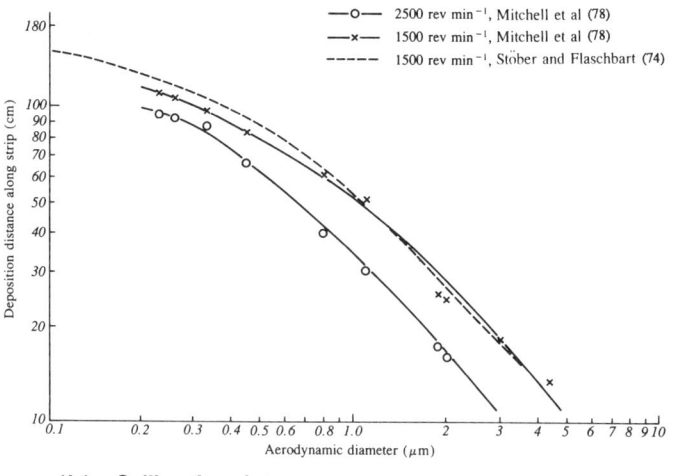

(b) Calibration data

Figure 8.17. Stöber spiral-duct aerosol centrifuge.

The size resolution approaches that of spiral duct centrifuges (Figure 8.18c), when Q_a is less than 2% of Q_w (Mitchell et al.[80]).

While it is important to match the winnowing and aerosol gas streams, the Inspec is much easier to operate and recently has been adapted for use at high temperatures (Prodi and Belosi[81,82]). The collection substrates can be removed and replaced within a few seconds, making it possible to carry out time-dependent analyses of aerosol behavior. Current developments include the provision of an automated loading and removal system, making this device an attractive alternative to cascade impactors/cyclones for in-stack particle size analyses.

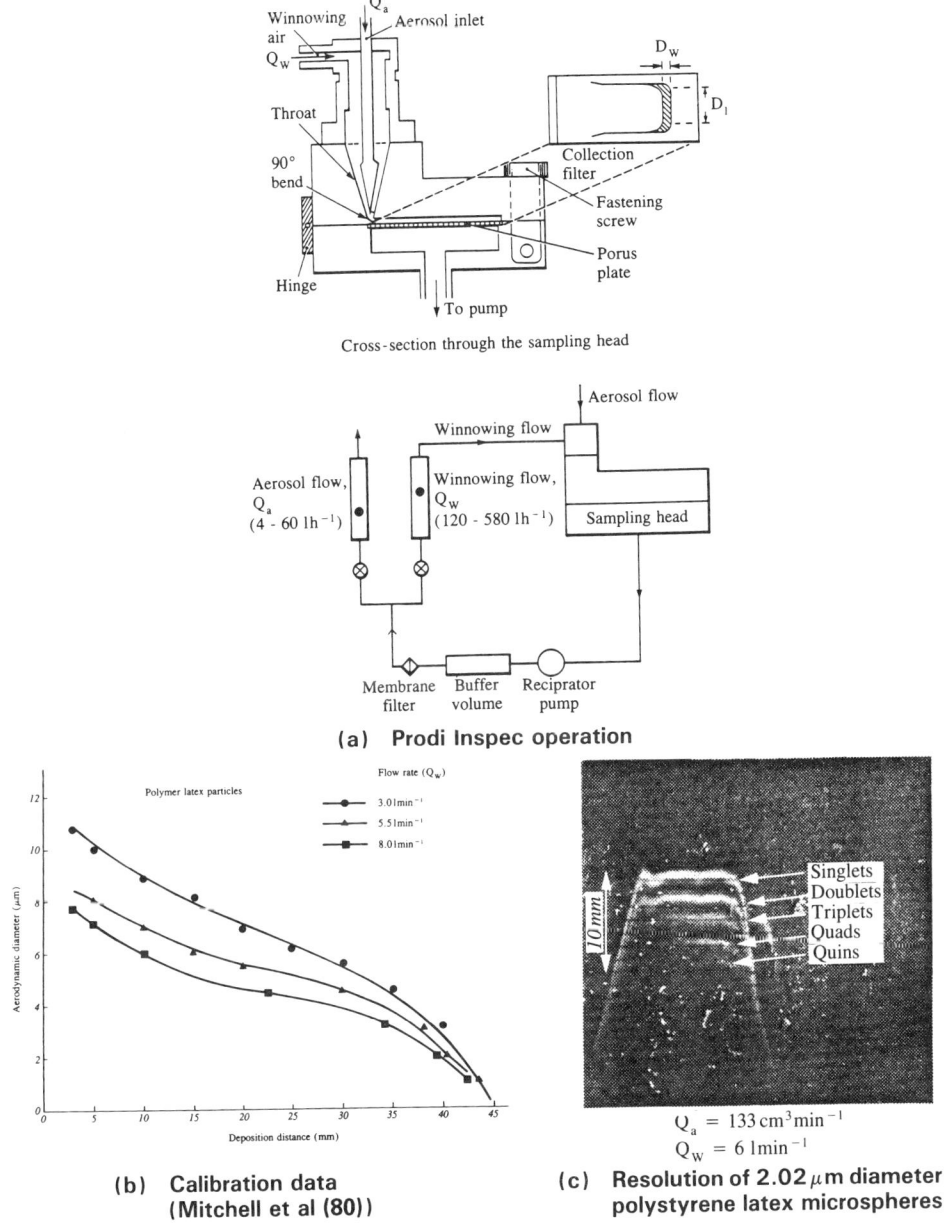

Figure 8.18. Prodi inertial spectrometer (Inspec).

Optical Techniques

A large group of aerosol analyzers operate on the basis of light interaction with particles. These instruments may be divided into three sub-classes:

(a) classical and active-cavity laser light scattering instruments (optical particle counters),
(b) laser(Fraunhofer/Mie) diffraction analyzers,
(c) laser(phase)-Doppler techniques,
(d) light intensity deconvolution analyzer,
(e) Galai CIS time-of-interaction analyzer.

The first four types of system are in wide use. The last system is beginning to become more common, but its potential remains to be fully exploited for aerosols. Single-particle light scattering is most useful to resolve particles in the range 0.1 to 100 μm diameter. Laser diffractometers are field scattering systems (measure the size distribution of a population of particles simultaneously), and have a wider operating range (0.1 to 1000 μm diameter). They can be operated in a non-invasive manner, making them very useful for many applications. Laser (phase)-Doppler systems are also non-invasive, but in general they are generally more expensive than diffractometers. However, such analyzers can provide additional information about particle velocity, and frequently are used to study complex processes, such as atomization, combustion and filtration, when fluid flow data are as important as the particle size measurements. A summary of the performance specifications is given in Table 8.7.

Optical Particle Counters

A detailed description of the large number of individual techniques is beyond the scope of this chapter. Much information on the various commercially available instruments has been compiled by Gebhart et al.[83] and Gebhart and Roth,[84] and a summary of their measured sensitivity data is presented in Table 8.8. The angles given under the heading 'optics' refer to the solid angle through which the scattered light is detected.

As defined in light scattering theory (Van de Hulst,[85] Kerker,[86] Bayvel and Jones[87]), the scattering function (i) depends on the scattering angle (θ), the polarization angle (ϕ), the wavelength of light (λ), the refractive index of the particle relative to that of the fluid (m) and particle diameter (d):

$$i = i(\theta, \phi, \lambda, m, d) \qquad (8.30)$$

d and λ are linked by the dimensionless size parameter α:

$$\alpha = \frac{\pi d}{\lambda} \qquad (8.31)$$

The flux of light (S) scattered by a particle per unit solid angle in direction θ is given by:

$$S(\theta, \phi, \lambda, m, d) = I_o \left[\frac{\lambda^2}{4\pi^2}\right] i(\theta, \phi, \lambda, m, d) \qquad (8.32)$$

where I_o is the incident light flux on the particle.

Light scattering by small particles is governed by Rayleigh's theory of dipole scattering (Bayvel and Jones[87]), for which the relationship:

Table 8.7. Performance of selected optical aerosol analyzers

Instrument	Collection Geometry	Size Range (μm)	Light Source	Max Concentration (particles cm^{-3})	Comments
Ensemble Measurement					
Malvern 2600	Near-forward spatial distribution of diffracted light	1.2 to 1876 μm in discrete ranges using one of six receiver lenses	He-Ne laser	Multiple diffraction limits use to ~50% obscuration	Non-invasive, measures volume (mass)-size distributions; does not measure individual particles
Malvern Master Sizer		sub-micron particles by reverse-Fourier optics			
Individual Particle Measurement					
Particle Measuring Systems (PMS) classical scattering aerosol spectrometer (CSAS)	Near forward scattering	0.3 to 20 μm, 4 ranges each containing 15 channels	He-Ne laser	1.5×10^5	Wide size range-good resolution a constant known flow rate is assumed
PMS-LASX	Wide-angle (35 to 120°)	(a) 0.09 to 3 μm (b) 0.10 to 7.5 μm 4 ranges with 15 channels 15 channels - whole range	He-Ne laser	1.7×10^4	Good resolution and counting efficiency for sub-micron particles; low flow rate; delicate
PMS-LAS 250X	Wide-angle (35 to 120°)	0.2 to 12 μm, 16 channels	He-Ne laser	3×10^3	Higher flow rate (2.8 1 min than LASX; good counting efficient submicron particles; delicate
HIAC-Royco 4102	Near-forward scattering	0.3 to 20 μm, 6 channels	White light	3×10^3	Flow rate similar to LAS250X insensitive around 1 μm; delicate
Climet CI-800	Wide-angle (15 to 150°)	0.3 to 10 μm, 6 channels	White light	$< 10^2$	High flow rate (28.3 1 min^1) resolution, delicate sensor

Instrument	Collection Geometry	Size Range (μm)	Light Source	Max Concentration (particles cm^{-3})	Comments
Polytec HC-15/70	Right-angle scattering	1.5 to 100 μm, 128 channels	White light	10^5 (HC-15) 10^3 (HC-70)	Moderate flow rate; rugged sensor
AEA Technology PD-LISATEK	Phase-Doppler Anemometry	1 μm - several mm	Argon-ion or He-Ne laser	10^5	High resolution; provides velocity as well as particle size; non-invasive
Aerometrics Inc., Phase-Doppler Particle Analyzer (P/DPA)	Phase-Doppler Anemometry	1 to 8000 μm He-Ne laser beams	He-Ne laser	10^6	High resolution; wide size range; non-invasive.
Dantec Particle Dynamics Analyzer (PDA)	Phase Doppler Anemometry	1 to 500 μm	He-Ne laser	10^6	See Aerometrics Analyzer
Galai Labs, Inc. CIS-I (Roth Scientific)	Rotating time-focused laser beam-particle size related to duration of scanning laser	0.5 to 1200 μm	He-Ne laser	not known, but assumed to be >10^3 on basis of size of measurement volume	High resolution; non-invasive independent of optical properties of particles and temperature changes; particle shape can be measured by internal microscope-camera-image analyzer
INSITEC PCSV series	Near-forward scattering intensity deconvolution	0.2 to 200 μm	He-Ne laser	10^7	High resolution; non-invasive independent of particle refractive index; can be used with sprays, aerosols and slurries.

Table 8.8. Efficiencies of selected optical particle counters**

Manufacturer	Model	Optics	Flow Rate (L min^{-1})	Counting Efficiency at Selected Particle Sizes (%)						
				0.16	0.22	0.32	0.47	0.72	0.95	2.02 μm
ROYCO	227	10–30°	2.8 0.28	—	—	12	66	—	62	39
KRATEL	Parto-scope 'A'	10–30°	0.28 2.8	—	—	16	92	—	105	—
CLIMET	CL 3060	15–150°	28	—	—	15	102	89	—	—
POLYTEC	HC-15	90°	0.5	—	—	—	9	77	89	106
PMS	LAS-X*	35–120°	0.28	37	104	110	107	106	—	90
PMS	LPC-555*	60–120°	28	—	36	~100	~100	~100	—	—
ROYCO	236*	35–120°	0.28	103	—	99	102	—	103	71
ROYCO	5100*	60–120°	28	—	23	65	75	93	—	—
TSI	3755*	15–88°	28	—	—	—	86	103	—	101
CLIMET	CL 6300*	45–135°	28	—	74	97	99	100	—	—
MET ONE	205*	45–125°	28	—	84	93	~100	~100	—	—

* Laser-based systems.
** Data from Gebhart et al. (ref. 83).

$$S(\theta, \lambda, m, d) = I_o \left[\frac{\lambda^2}{4\pi^2}\right] \left[\frac{m^2 - 1}{m^2 + 2}\right]^2 (1 + \cos^2\theta) \qquad (8.33)$$

holds for unpolarized light when $\alpha \ll 1$. Rayleigh scattering of visible light occurs with particles smaller than about 0.3 µm diameter. S is related to the amount of detected light by the geometry of the sensing arrangement, and is a function of d^6. Thus, very high light intensities are required to size particles much smaller than 0.3 µm diameter, and so-called 'active-cavity' laser spectrometers have been developed in which the aerosol stream passes within the lasing volume. The most sensitive of these instruments is capable of detecting particles as small as 0.05 µm diameter. These instruments (e.g., most of the PMS analyzers, see Table 8.7) are widely used as contamination monitors in 'clean' environments, and their aerosol sampling rate is generally restricted to less than a few hundred cm^3 min^{-1}. The aim of current research is to increase the sampling rate and at the same time preserve the detection efficiency for the smallest particles.

For particles larger than 0.3 µm, Mie theory must be used to determine the angular distribution of scattered light. In the limiting case where $\alpha \gg 1$, the scattered light can be considered to contain three components due to the effects of diffraction, reflection and refraction (if the particle is transparent). Mie theory predicts the relationship between the scattered light intensity at a distance R in the direction θ from the particle, and particle size (d) to be:

$$I(\theta) = \frac{I_o \lambda^2 (i_1 + i_2)}{8\pi^2 R^2} \qquad (8.34)$$

where i_1 and i_2 are the Mie intensity parameters for scattered light with perpendicular and parallel polarization, respectively. These terms are functions of θ, m and d, defined by an infinite series containing Legendre polynomials and Bessel functions. Mie theory provides an exact solution to quantify light scattering by large particles, with the following important consequences.

1. The quantity $I(\theta)$ is not a simple function of α and d. The size of the oscillations of $I(\theta)$ (expressed in terms of the Mie intensity parameter, $i_1 + i_2$) depend on θ, and can be damped to some extent by using a polychromatic rather than a monochromatic (laser) light source. Instruments that operate with near-forward scattering ($\theta < 30°$) where most scattered light is available have the highest sensitivity to sub-micron particles. However, those that use monochromatic light are generally insensitive to particle size in the region close to 1 µm diameter ($\alpha < 10$) where these oscillations are at their greatest (Cooke and Kerker[88]). The Polytec series of optical aerosol spectrometers (Umhauer[89]) were designed to remain capable of size resolution in this difficult region by illuminating the particles with white light and detecting the scattered light at right angles to the incident beam.
2. $I(\theta)$ depends on the particle refractive index (m). Cooke and Kerker[88] have compared the collected light intensity as a function of particle size for optical aerosol spectrometers of almost all configurations. Their observations confirm that many instruments have quite different responses to particles of the same nominal size depending on whether the particles have a low or high value of m. Light scattered in the near forward direction ($\theta < 20°$) is predominantly diffracted and can be shown to be a function of the projected area of the particle. However, at wider an-

gles, reflection and refraction predominate, and the relationship no longer corresponds in a simple way with a measurable particle dimension. The situation is further complicated by the behavior of absorbing particles. Such particles have a finite imaginary component to the refractive index; dark particles have larger imaginary components. Under extreme conditions where black, opaque particles (carbon dust) are being sampled, the response of the analyzer to particle size may decrease from the d^2 relationship predicted by Equation 8.34 to become almost independent of size. This behavior is particularly evident when micron-sized particles are sampled with wide-angle active-cavity laser spectrometers (Szymanski and Liu[90]). The 'diameter' measured by optical aerosol analyzers therefore cannot be unambiguously specified in terms of a well-defined size, and it is usual to calibrate these instruments with particles that have a 'standard' refractive index. BS3406 part 7 (1988) describes techniques using polystyrene latex spheres (m = 1.59 - 0i). Alternatively, the optical aerosol analyzer may be calibrated in terms of aerodynamic diameter using an inertial pre-separator of known effective cut-off size (see below).

3. The behavior of non-spherical particles is a highly complex issue, and at present there is no satisfactory description of light scattering in terms of shape-related parameter(s). Marple and Rubow,[91] Büttner[92] and Mitchell et al.[93] have described methods whereby optical aerosol analyzers can be calibrated in terms of d_{ae}, using an inertial preseparator of defined ECD to remove part of a polydisperse aerosol. This approach is recommended if an optical aerosol analyzer is being used to study irregular-shaped or inhomogeneous particles.

The optical sensing volume can be formed either by introducing the particles in a sheath of gas so that they are focussed aerodynamically within the incident light beam (Figure 8.19a), or by arranging the detector and light source so that the measurement volume is formed at their intersection without focussing the particles (Figure 8.19b). The former method, adopted by most manufacturers, has the advantage that all particles are fully illuminated, but often involves particle losses in the aerosol transport system. The aerodynamically un-focussed or 'optically defined' measurement approach (Figure 8.19c) largely avoids this problem, enabling near non-invasive measurements to be made under certain circumstances (low background light levels). However, instruments that utilize this method of detection (e.g., Polytec HC Series) suffer from border-zone error (Borho[94]), since particles that graze the measurement volume are undersized. Recent work (Mitchell et al.[95]) has indicated that border-zone error can dominate measurements of particles that are close to the upper size limit of the instrument. A new generation of optically-defined aerosol analyzers with an outer 'guard' sensing volume (Umhauer[96]) has appeared recently (Model PCS-2000, Palas GmbH., Karlsruhe, Germany).

As a class, optical particle counters are restricted by coincidence errors to the measurement of aerosols of moderate or low number concentration. Typical maximum concentrations are of the order of 10^3 particles cm^{-3}, although certain designs such as the Polytec HC-15 and Palas PCS-2000 have reduced the size of measurement volume to such an extent that particle concentrations two orders of magnitude higher can be detected with only 5% coincidence error (Umhauer[96]). Coincidence errors arise when two or more particles are in the sensitive volume at the same time, resulting in a spurious signal that leads to an underestimation of the total particle concentration and an overestimation of the average particle size. The magnitude of the error can be estimated readily from a knowledge of the capacity of the measurement volume (v_s) and the observed particle count (N_o) in a given time:

220   BIOAEROSOLS HANDBOOK

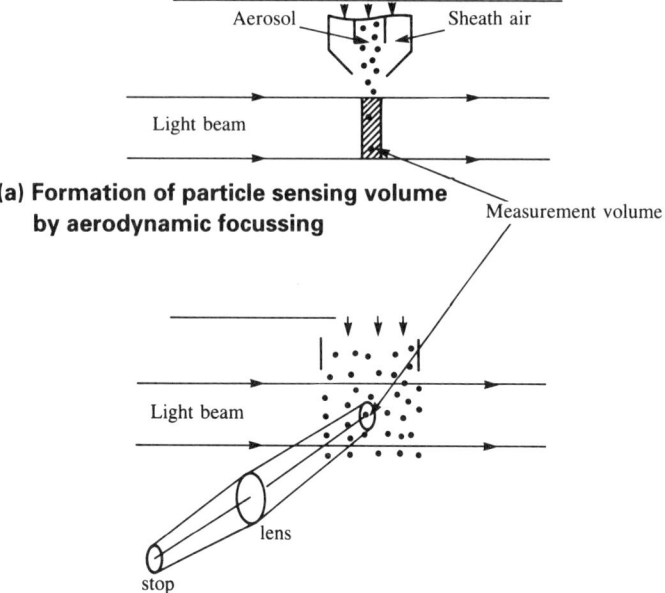

(a) **Formation of particle sensing volume by aerodynamic focussing**

(b) **Formation of particle sensing volume by optical means**

Location
1 Particle illuminated fully in full view of detector - sized correctly
2 Particle illuminated fully out of view of detector - not measured
3 Particle in full view of detector but unilluminated - not measured
4 Particle partly illuminated in full view of detector - undersized (border-zone error)
5 Particle unilluminated and out of view of detector - not measured

(c) **Origin of border zone error**

Figure 8.19.   Optical particle counters: detection methods.

$$\frac{N_t}{N_o} = \exp(-N_t v_s) \qquad (8.35)$$

where N_t is the true particle count. The value of v_s or the particle concentration associated with a defined (usually 5%) coincidence level is normally specified by the manufacturer of the optical particle counter. Aerosol dilution may be carried out if the particle concentration exceeds the capability of the analyzer, but precautions must be taken to minimize the risk of particle losses in the system. The diluter described by Remiarz and Johnson[69] for the APS33B, and shown schematically in Figure 8.20 is a good example of a well-characterized and efficient system for diluting aerosols consisting of particles smaller than 15 μm aerodynamic diameter.

Laser Diffractometers

For small scattering angles ($\theta < 20°$), the diffraction component of scattered light can be approximated by the Fraunhofer expression:

where J_1 is a first order Bessel function. I_0 is the light intensity at $\theta = 0$:

$$I_0 = \frac{E \, D_a}{\lambda^2} \qquad (8.37)$$

where $D_a = \pi d_v^2/4$, the cross-sectional area of the spherical particle, and E is the total light energy incident upon the particle (proportional to the projected area):

$$E = F \left[\frac{\pi \, d_v^2}{4} \right] \qquad (8.38)$$

A range of instruments based on the principle of laser diffraction has been developed from the original work of Swithenbank et al.[97] Early instruments made use of the Fraunhofer approximation to resolve the scattered light intensity data to a particle size distribution. However, most modern analyzers employ the full theory of angular light scattering developed by Lorenz-Mie, which provides more accurate results, especially where the relative refractive index (refractive index of the particles compared with that of the support medium) is small.

The basic arrangement of a laser diffractometer is shown in Figure 8.21a. An expanded laser beam provides a parallel beam of coherent, monochromatic light, and a Fourier transform lens is used to focus the diffraction pattern generated by the collection of particles entering the measurement zone onto a photodetector consisting of a series of concentric ring-diodes (Figure 8.21b). The undiffracted light thereby is reduced to a spot at the center of the detector plane, with the diffraction pattern forming circular rings. The multi-element detector determines the radial distribution of diffracted light intensity (Figure 8.21c), which can be related to the original size distribution by means of data inversion procedures. Early instruments produced data fittings to continuous functions, such as the Rosin-Rammler distribution, but further procedures have been developed that enable a more realistic 'model-independent' solution to be provided. However, these instruments generally are unsuitable for measurements of aerosols consisting of highly uniform sized particles.

Movement of the particles in the beam does not affect the diffraction pattern, since the light diffracted by a particle at a given angle will give the same radial displacement in the

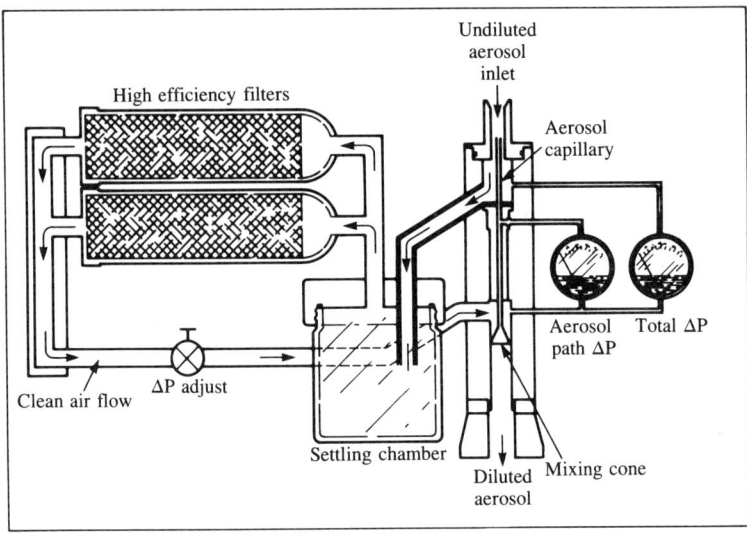

N.B. Capillary size selectable for 20 : 1 or 100 : 1 dilution

Figure 8.20. Commercially available aerosol diluter (with permission from TSI Inc., St. Paul, MN).

focal plane of the lens irrespective of the position of the particle in the beam. Laser diffractometers therefore determine the size distribution of a population of particles weighted by volume, rather than count individual particles entering a fixed measurement volume to provide a number-size distribution. They are not susceptible to coincidence problems, and can measure concentrated aerosols/sprays, etc. However, at very high concentrations, particle-particle scattering (obscuration) will lead to errors, and these instruments cannot be used reliably when this obscuration exceeds about 50%. Most commercially available instruments provide an indication of the obscuration.

There is a wide range of commercially available laser diffractometers, mainly developed for powder and spray analysis (Table 8.9, and also Kaye[98]). Not all of these instruments can be employed directly with aerosols (e.g., some CILAS systems) and advice should be sought from the manufacturers. An outstanding feature with many of these instruments is their ease-of-use combined with their ability in some cases to make nonintrusive measurements (e.g., Malvern 2600). The optical components can be mounted in the open for ambient measurements or outside of the chamber containing the aerosol, and the processes observed by means of optically clear windows.

Laser diffractometers have a wide operating size range. Two or more orders of magnitude is typical, although additional lens combinations may be required to span the entire range of a given instrument.

Laser (Phase)-Doppler Systems

Other non-invasive particle sizing instruments are based on the laser-Doppler effect. The scattered light is detected as individual particles traverse a series of interference fringes formed from intersecting laser beams (e.g., Yeoman et al.[99]). This technique was developed from laser-Doppler anemometry, and information about individual particle velocity as well as size can be obtained simultaneously.

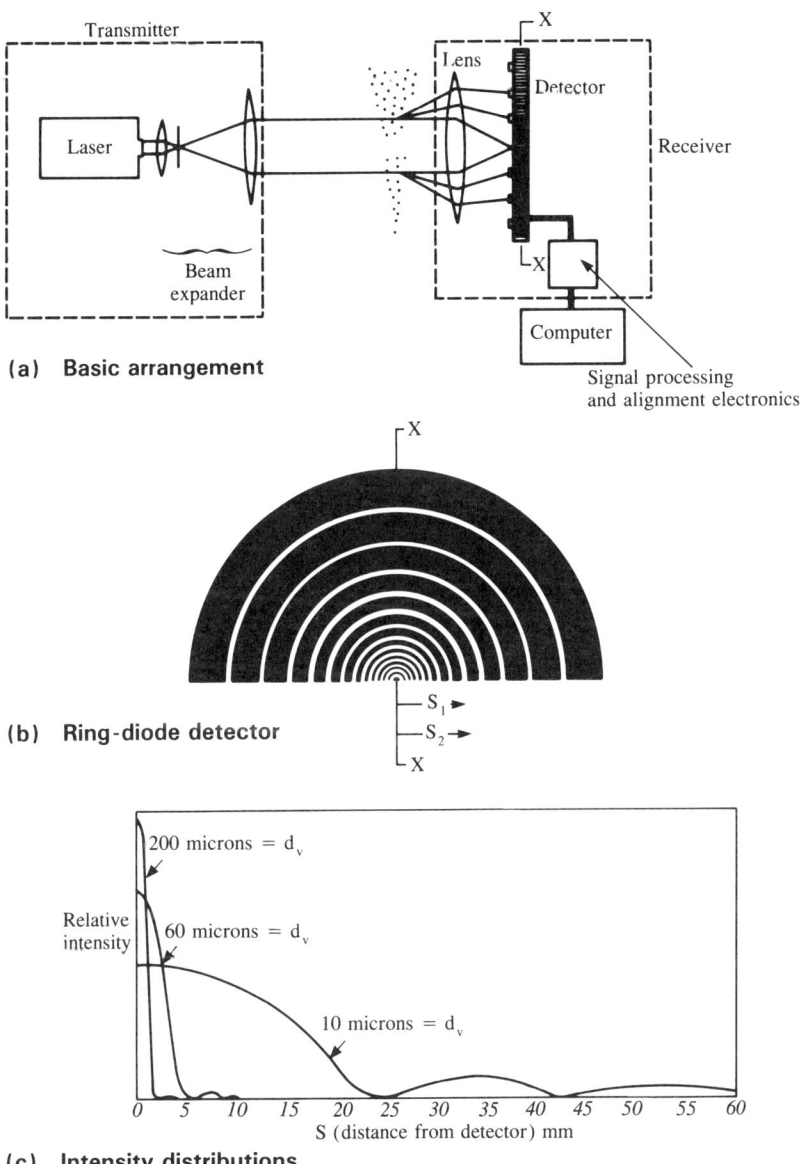

Figure 8.21. Laser diffractometer: principle of operation.

The principle of the technique is illustrated in Figure 8.22a. The two laser beams establish an interference pattern with fringe spacing λ_f

$$\lambda_f = \lambda/2 \sin(\gamma) \tag{8.39}$$

where γ is the half angle between the two laser beams. The general form of the scattered light signal from a particle crossing the fringe pattern with velocity u is:

Table 8.9. Operating Ranges of Selected Laser Diffractometers

Instrument	Range
Malvern 2600	1.2–1876 μm (6 ranges)
Malvern MasterSizer	0.1–600 μm (3 ranges) (dry/wet dispersions)
Leeds and Northrup Microtrac	0.1–700 μm (dry/wet dispersions)
Fritsch Analysette 22	0.2–1160 μm (dry/wet dispersions)
Sympatec Helos	0.1–2000 μm (dry/wet dispersions)

$$I_{sca} = A + B \cos(2\pi v_b t) \qquad (8.40)$$

where $v_b = u/\lambda_f$ and t is the transit time (Figure 8.22b). It is possible to measure either the time-averaged scattered intensity of the signal (A) or the signal visibility (v_i). The latter is defined (Farmer[100]) in terms of the amplitudes I_{max} and I_{min}:

$$v_i = \frac{I_{max} - I_{min}}{I_{max} + I_{min}} \qquad (8.41)$$

where v_i is uniquely related to the size of the particle producing the signal normalised to the fringe size, providing that the scattering angle and collection aperture are carefully chosen (Negus and Drain[101]). An example of a plot of v_i versus the non-dimensional parameter (particle circumference/wavelength of the light source) is shown in Figure 8.22c. The size ranges of laser-Doppler systems that use the visibility technique are therefore dependent on the form of this relationship. Simultaneous measurements of fringe visibility and the corresponding scattered light intensity for each particle (Figure 8.22d) passing through the center of the fringe pattern extend the dynamic size range of the instrument far beyond the range available using visibility alone (Yeoman et al.[99]). The various commercial instruments incorporate methods that ensure only particles passing through the central region of the fringe pattern are measured. Thus, a representative sample of particles are sized that meet rigid criteria for signal quality. The technique is independent of refractive index and has been used to study a wide variety of aerosols/sprays in the range 20 to 500 μm diameter. Particle concentrations approaching 10^6 particles cm^{-3} can be sampled, although care must be taken to minimize coincidence effects.

Despite the advantages outlined above, the visibility technique is limited in dynamic range, and cannot be extended to the study of particles much smaller than 20 μm. In recent years, an important development has enabled the basic laser-Doppler method to overcome this limitation, in instruments where the oscillatory signals from particle-beam interactions are recorded on detectors that are closely separated in space. The phase-difference between these signals at appropriate scattering angles has been found to be a linear function of particle size (Bachalo and Houser,[102,103] Saffman and Buchhave[104]). The principle of the phase-Doppler method is illustrated in Figure 8.23. At the time of writing, the Aerometrics P/DPA instruments (Table 8.7) have a much wider operating range (1 to 8000 μm diameter) than single-detector-based laser-Doppler systems, and the Dantec and Lisatek (AEA Technology) analyzers have similar capabilities to those of the Aerometrics systems. At present, this technique is very expensive, and can probably only be justified for specific cost-effective applications. Recent developments of the phase-Doppler technique reported

PARTICLE SIZE ANALYZERS 225

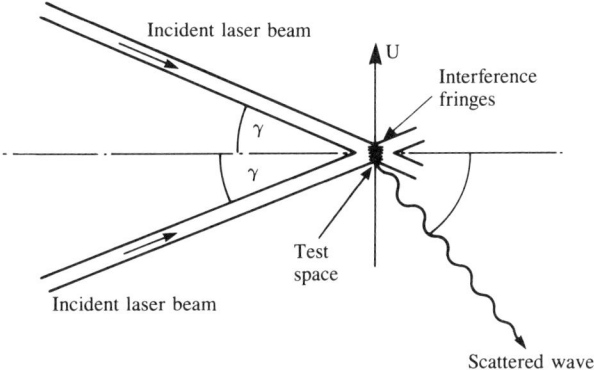

(a) Operating principle

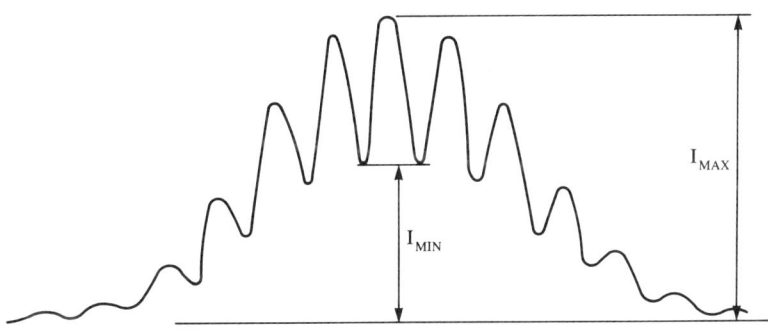

(b) Signal visibility profile

Figure 8.22. Laser-Doppler particle sizing: operating principles.

by Negus et al.[105] and Burton and Negus[106] are extending the capability of the phase-Doppler technique to particles smaller than 1 μm diameter.

Intensity Deconvolution Technique

In addition to the interferometric techniques described above, single-particle counting can be accomplished by measuring the absolute intensity of light scattered by particles traversing a focused laser beam. The Insitec PCSV family of instruments (Table 8.7) was developed from the work of Holve and Self,[107,108] and Holve,[109] and measures the absolute intensity of light scattered with a photomultiplier as particles pass through an optically-defined measurement volume (Figure 8.24). Since the sample volume has a non-uniform light intensity distribution, the intensity of the scattered light depends on both particle size and trajectory. The potential ambiguity in particle size is resolved using a deconvolution algorithm, whose solution has two requirements:

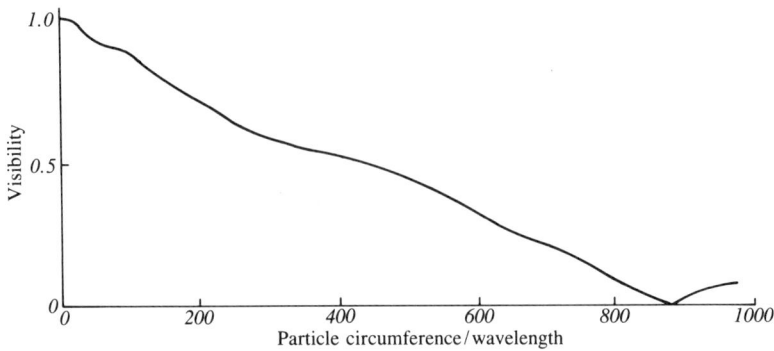

(c) Visibility function (depends on scattering angle and detector aperture)

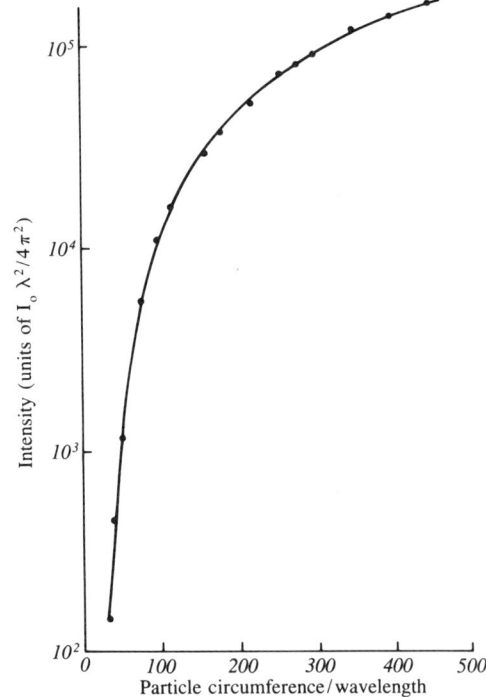

(d) Intensity function (spherical particles)

Figure 8.22. (continued).

(a) the absolute scattered light intensity of a large number of particles must be measured (typically requiring less than 1 minute),
(b) The sample volume intensity distribution must be known.

The deconvolution algorithm resolves the size-scattered light intensity ambiguity in a manner analogous to the way in which CAT-scan images are obtained in the medical field. Thus, the count-based spectrum of scattered light intensities is deconvoluted to yield the absolute particle number concentration as a function of size (hence the term 'intensity deconvolution technique'). The Insitec instruments are fully non-invasive light scattering

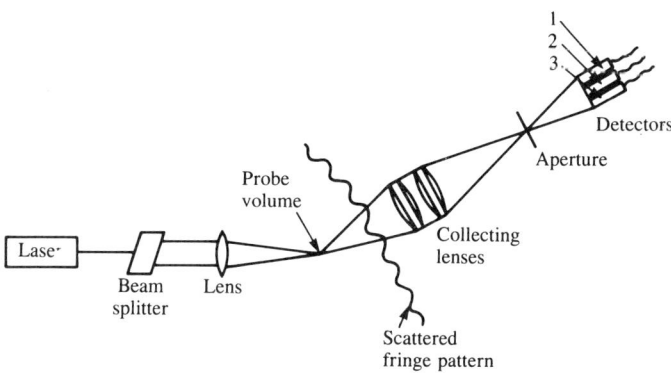

Figure 8.23. Phase-Doppler particle sizing technique.

techniques, claimed to be capable of determining particle concentrations up to 10^7 particles cm^{-3} (well in excess of conventional single particle counters); the claimed size range (0.2 to 200 µm) is also very wide, but not all of this range may be available without adjustments to the optical components. These instruments have been employed to study gas-borne particles, slurries and powders, and measurements have been made chiefly in arduous conditions, such as in fossil fuel combustion plant (Holve and Self[108]). Furthermore, the performance of this instrument has recently been compared favorably with in-stack impactor measurements of emissions from a pilot-scale combustion plant (Dahlen et al.[110]). The technique, like phase-Doppler analyzers, is probably too expensive for most bioaerosol applications where conditions are more likely to correspond to room ambient, and where simpler (if more invasive) techniques will suffice.

GALAI CIS Laser-Particle Interaction/Image Analyzer

The GALAI-CIS analyzer (Galai Labs) is unique in that it combines light blockage with image analysis in a single analyzer,[111] that can be used with almost any type of particle dispersion. The claimed operating size range for powders in liquid suspension is very wide (0.5 to 1200 µm), but not all of this is available in any one measurement, and the four ranges:

(a) 0.5 to 150 µm,
(b) 2 to 200 µm,
(c) 5 to 600 µm,
(d) 10 to 1200 µm,

are provided with the standard instrument (Table 8.7).

The laser-interaction part of the analyzer is based on the principle of time-of-transition, and has a dynamic range of 300:1. A focused low-power He-Ne laser beam of 1 µm width is scanned in a circle through the measurement zone that contains the particle suspension. The laser beam interacts with individual particles by light blockage, and a PIN photodiode detector located directly behind the measurement area senses the duration of each interaction (Figure 8.25a). The width of the pulse represents the time of interaction of the laser

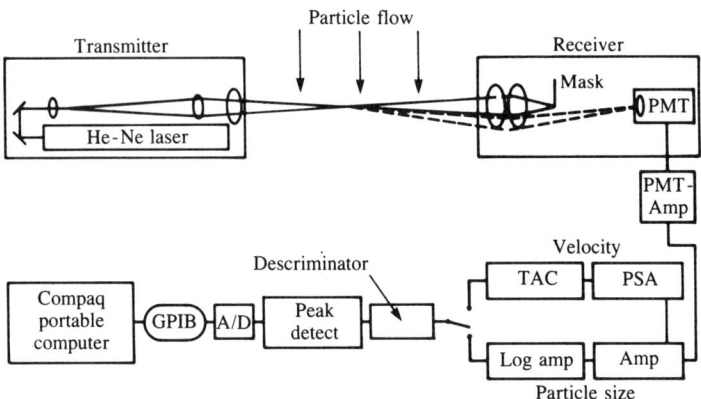

Figure 8.24. Intensity deconvolution particle sizing technique.

beam with the particle as the laser beam scans across the surface. This time of transition thus provides a direct parameter related to particle size that is independent of particle refractive index, the attenuation of the fluid supporting the particle and the output power of the laser. The size resolution is excellent (300 channels per range selected), and is limited by the sampling rate of the detecting electronics.

Whilst the time-size relationship is unique, the laser beam diameter varies along the length of the beam (Figure 8.25b), and ambiguity might arise from particles that interact with the laser beam out of the focal plane. However, such interactions produce longer pulse lengths for a given particle size, and the analyzer can therefore discriminate against these signals by pulse editing techniques. In addition, the laser beam may not interact with the particle along its diameter or longest dimension. This type of interaction produces a signal that is shorter in duration than expected for a given particle size. The problem is easily resolved with spherical particles, since the rise and decay times of the pulse are steepest for an interaction crossing the center of the particle and therefore measuring the true diameter (Figure 8.25b). Similar criteria also can apply with certain non-spherical, but regularly shaped particles (e.g., spheroids). However, there is an inherent ambiguity in the interpretation of signals from irregularly shaped particles, and it is for this reason that the GALAI instrument is equipped with an optical microscope coupled to a CCD TV camera-image analyzer. The microscope optics are arranged orthogonally with respect to the axis of the laser interaction system (Figure 8.25c). The manufacturer recommends that this facility be used to derive shape-related data for the particles[112] and a powerful suite of image analysis software is available to interpret the various aspects of particle shape and size. However, at first sight it is not easy to see how microscopy-image analysis data easily can be used to interpret size distributions determined by the laser interaction part of the equipment, particularly for complex-shaped particles.

The GALAI CIS-1 analyzer is designed for studying both aerosols and liquid-borne particulates. Aerosols can be sampled directly or after formation using a purpose-build nebulizer/atomizer[112,113] In addition, particles collected on membrane filters also can be examined by the image analysis mode.[114] At present there is little information concerning the performance of the GALAI analyzer, except for studies originating from the manufacturer/inventing laboratory. Recent studies using the time-of-transition mode[115] with the irregular-shaped certified reference materials (CRMs) from the European Bureau of Community Reference (Quartz CRMs 67, 68 and 69) indicate good agreement over a very

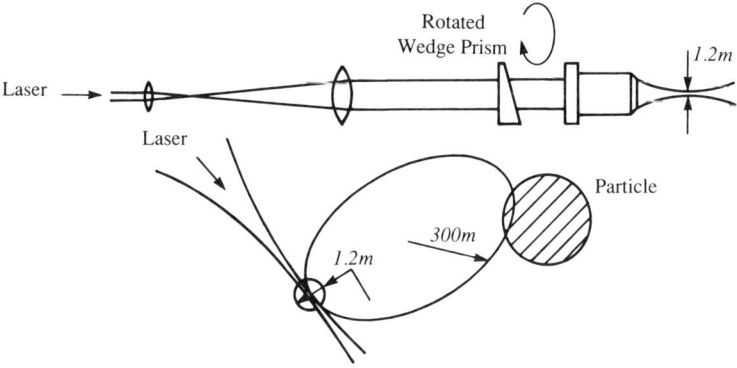

(a) Scanning laser beam for time-of-transition interaction with particles

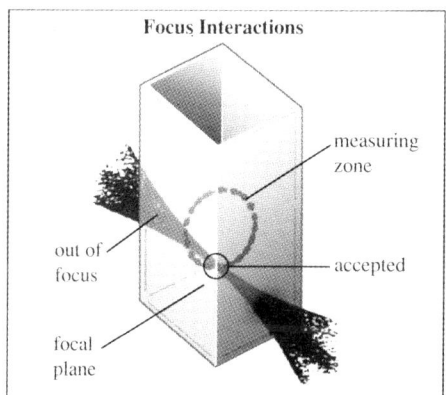

Out-of-focus or off-centre interactions generate rise times that are too long in proportion to the particle diameter. Using an algorithm based on the overall pulse signature with normalized rise-time criteria, each interaction is accepted or rejected. Only data from accepted signals are filed for analysis.

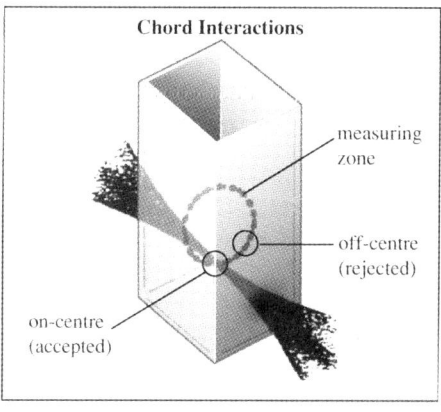

(b) Discrimination between focussed and off-centre particle-laser beam interactions

Figure 8.25. GALAI CIS-1 aerosol analyzer (with permission from Galai Production Ltd., Migdal Haemek, Israel).

wide size range (2 to 630 μm volume equivalent diameter) with the certification data (based on sedimentation). The CRMs were dispersed in liquid to enable the larger particles to be examined. In further work, the authors made the point that time-of-transition analysis pro-

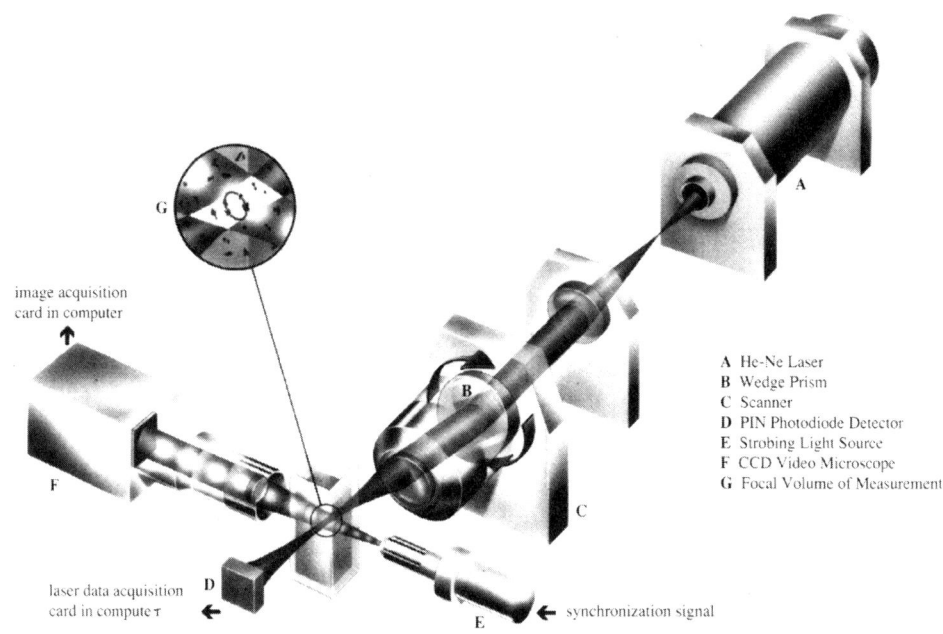

(c) Orthogonal arrangement of time-of-transition and image analysis optical axes

Figure 8.25. (continued).

duces data that are not dependent upon particle refractive index; thus, colored micron-sized monodisperse polystyrene latex microspheres were sized correctly by both time-of-transition and image analysis modes. It is notable, however, that the measured monodispersity of these polymer latex particles was not as good as might have been expected from typical specifications for these calibrants.

Electrical Mobility Techniques

Electrical mobility aerosol analyzers are based on the movement of gas-borne particles carrying a known electric charge toward an electrode of opposite charge. These techniques are applied widely in the laboratory, especially in particle formation studies, since they have the potential for very good size resolution in the range 0.01 to 1.0 µm diameter. However, their application in the field is largely restricted to the measurement of low concentrations of stable aerosols, such as may be found in environmental pollution monitoring. Electrical mobility analyzers are fairly expensive to purchase, and like many of the optical techniques just described, require skill in the operation and interpretation of the data.

Operating Principle

Electrical mobility (Z_p) is defined in terms of the velocity component (U_e) that a charged particle experiences under the influence of an external electric field of strength E_e:

$$Z_p = \frac{U_e}{E_e} \tag{8.42}$$

Z_p is expressed in terms of cm^2 V^{-1} s^{-1} if U_e and E_e are in units of cm s^{-1} and V cm^{-1}, respectively. When a charged particle attains terminal velocity in an electric field, the electric force is balanced by the drag force on the particle. Under Stokesian conditions (sub-micron particles), the expression:

$$Z_p = \frac{n_p\, e\, C_c}{3\, \pi\, \mu\, D_v\, \chi} \tag{8.43}$$

is obtained, where n is the number of charges on the particle, e is the elementary unit of charge (4.803 × 10^{10} statcoulombs), C_c is the Cunningham slip correction factor, μ is the gas viscosity, D_v is the volume equivalent diameter of the particle and χ is the dynamic shape factor (unity for spheres). Equation 8.43 provides a means of size-separating particles of known charge on the basis of their differing electrical mobilities.

Electrical mobility analyzers are used widely to size particles in the range 0.004 to 1 µm volume equivalent diameter. Particles much smaller than this lower limit are difficult to charge, whereas multiple charging becomes a problem with micron-sized and larger particles. Electrical mobility analyzers are the only high resolution techniques for particles smaller than about 0.1 µm volume equivalent diameter. Although most systems have been designed for research purposes, there are a few commercially available versions; the Electrical Aerosol Analyzer (EAA) and the Differential Mobility Particle Sizer (DMPS), both from TSI Inc., are the two most frequently encountered instruments in this class. The basic operating principle and limitations of each of these analyzers are given below; more information can be found in Liu and Pui[116] for the EAA, and Keady et al.[117] for the DMPS. The two techniques have been compared with measurements of sub-micron sodium chloride particles made by microscopy-image analysis; the superior size resolution of the DMPS was clearly evident, although this analyzer was shown to suffer greater internal losses upstream of the measurement section due to diffusional-based deposition (Horton et al.[118]).

The EAA has been used at low pressures (Yeh et al.[119]), from an aircraft (Sem[120]), and in conjunction with a cyclone preseparator to sample dry, dusty gases from stacks (Felix et al.[121]). Particulate contamination from so-called 'pure' solvents have been analyzed by the DMPS; droplets of the solvent were formed by atomization, and the residual size distribution was measured after evaporation had been completed (Blackford et al.[122]). The DMPS is less rugged than the EAA, particularly when used with a condensation nucleus counter, and therefore should only be regarded as a laboratory-based technique.

Electrical Aerosol Analyzer (EAA)

The EAA consists of three main components (Figure 8.26a): a unipolar diffusion charger, the mobility analyzer and a Faraday-cup electrometer detector. Gas-borne particles pass through the diffusion charger at a flow rate of 4 L min^{-1}, where they acquire a well-defined electrostatic charge that depends on the number of ions encountered during the time spent within the charger (ion product), as well as on particle size. The charged particles pass to the tubular mobility section, which consists of a central electrode surrounded by a core of clean (sheath) gas (Figure 8.26b). The aerosol flow is introduced so as to surround

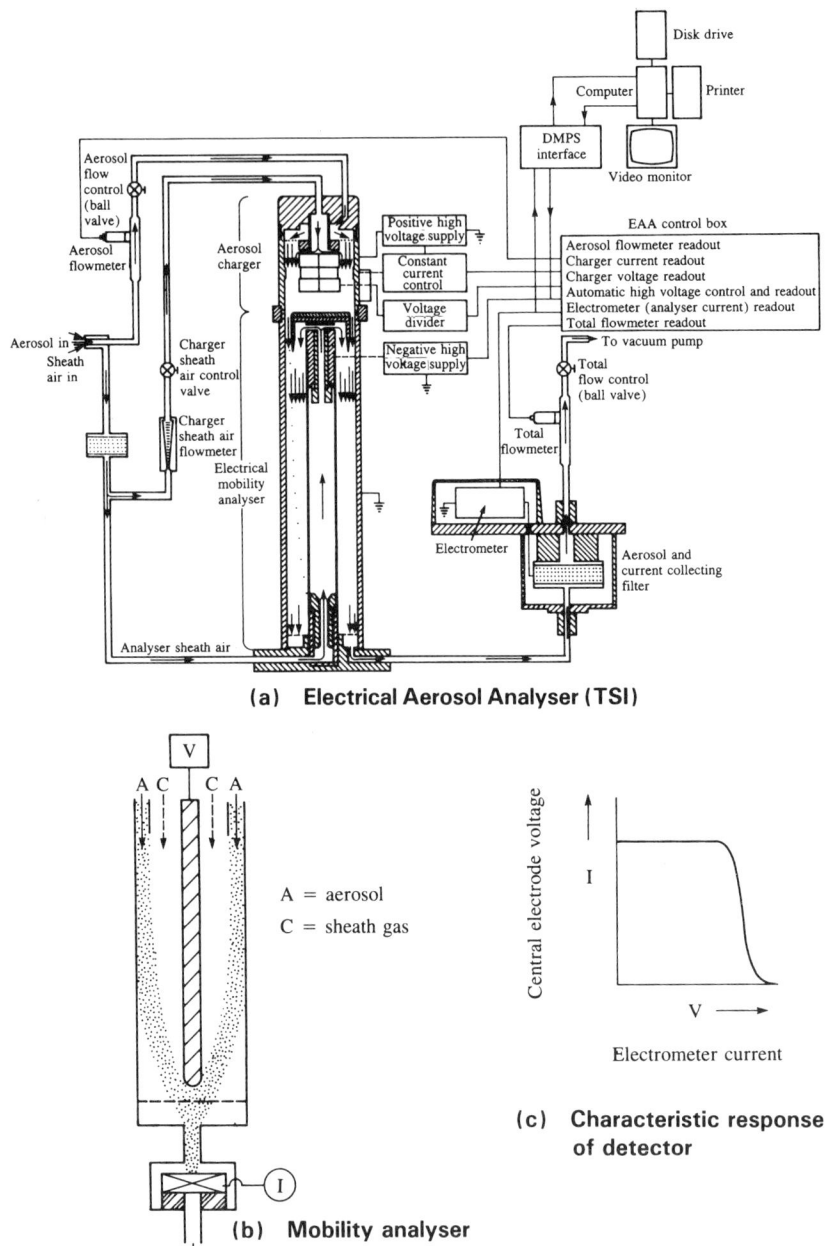

Figure 8.26. Electrical Aerosol Analyzer (EAA): Operating principle (with permission from TSI Inc., St. Paul, MN).

the sheath gas in laminar flow; any aerosol leaving the mobility analyzer is collected on the filter of the electrometer, and the electrostatic charge measured as the current drains to earth potential. Initially, the analyzer is operated with the central electrode at low voltage, so that all the aerosol collects on the electrometer filter. As this voltage is increased in a series of well-defined steps, progressively fewer particles penetrate through the mobility section, and the electrometer current falls to zero (Figure 8.26c). The measured decrease in electrometer current between two successive settings of electrode voltage can be related to the particle number concentration associated with a particular size band through a deconvolution technique (Kapadia[123]). The EAA has the capability to discriminate size bands in the range 0.013 to 0.75 μm diameter; this range can be extended to 0.004 μm, but the performance of the instrument below 0.013 μm is uncertain. Over the normal operating range, a size distribution may be obtained within 2 minutes if the aerosol source is stable. It should be noted that the density and viscosity of the sheath and aerosol gases should be the same for the size separation process to work correctly.

Differential Mobility Analyzers (DMAs)

DMAs have been developed to capture the narrow range of particles that have a common trajectory within an electrical mobility analyzer. The DMPS is the most commonly encountered instrument of this type, consisting of an Electrostatic Classifier (Liu and Pui[124]) as the mobility analyzer, which is coupled to either a continuous flow condensation nucleus counter—DMPS-CNC (Agarwal and Sem[125]), or a Faraday-cup electrometer—DMPS-ELE. The DMPS may be operated in either underpressure (Figure 8.27 a,d) or overpressure (Figure 8.27 b,c) modes; the former is used when the aerosol source is at or below ambient pressure, while the latter mode is adopted to sample from a pressurized source.

The aerosol flow rate entering the DMPS ranges from 0.1 to 1.0 L min^{-1}, although the instrument normally is operated at the flow rate of the CNC (0.3 L min^{-1}). Particles larger than 1 μm aerodynamic diameter are initially removed in a single-stage impactor, since they may carry more electrostatic charge than the data reduction procedure permits, resulting in the propagation of large errors throughout the measured size distribution (Keady et al.[117]). The aerosol then is passed through a bipolar charge equilibrator consisting of a Kr-85 radioactive source contained within the electrostatic classifier section. Emerging particles carry a Boltzmann distribution of charges (the overall charge is zero, but the aerosol contains well-defined proportions of particles carrying ±1, ±2, ±3 charges, etc.) (Table 8.10). The design of the mobility section (Figure 8.28a) is superficially similar to that of the EAA. However, the central electrode does not occupy the full length of the analyzer; a small gap exists near to the exit pipe (Figure 8.28b). The electrode voltage is initially set to a low positive potential; particles that have a narrow range of high electrical mobilities (smallest particles) enter the gap and are collected by the detector as a 'monodisperse' aerosol. As the electrode voltage is increased, the sizes of particles exiting the electrostatic classifier also increase, since the electrical mobility of the particles that enter the gap at the base of the electrode decreases. As the particle size increases above 0.05 μm volume equivalent diameter, the aerosol begins to consist of several monodisperse sub-fractions corresponding to the different negative charge levels allowed by the Boltzmann charge equilibrium (Table 8.10). Thus, the signal recorded by the detector during the measurement sequence corresponds to the actual number-size distribution, modified by the presence of a known proportion of multiply charged particles (Figure 8.28c). The analyzer software corrects for these multiply charged particles up to six charges per particle. The DMPS is capable of measuring as many as 32 size channels in the range 0.01 to 0.9 μm electrical mobility diameter; such a scan can take as much as 20 minutes to complete, so a facility is available to monitor alternate channels or one channel in four, with a concomitant loss of size resolution. This feature is useful if aerosol stability is poor.

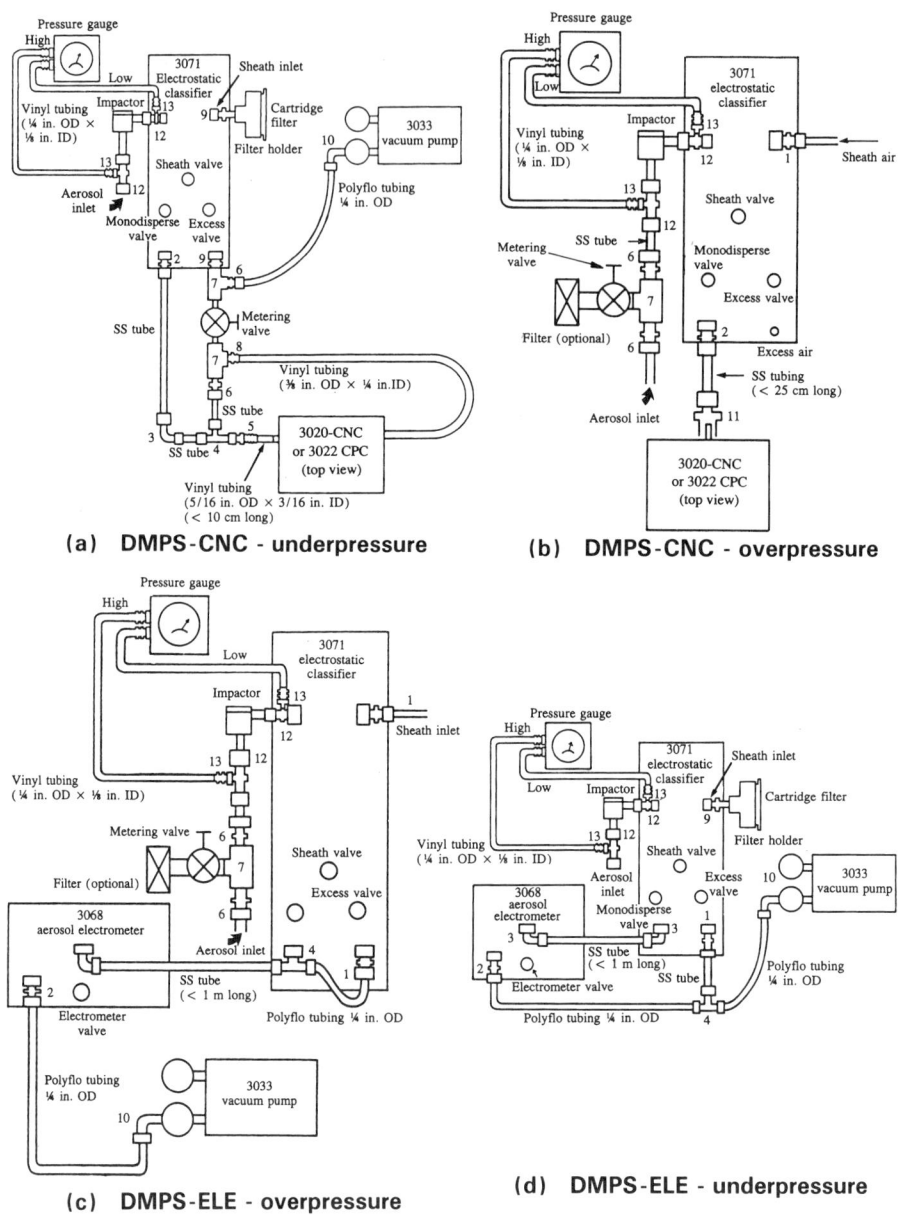

Figure 8.27. Differential Mobility Particle Sizer (DMPS): possible operating configurations (with permission from TSI Inc., St. Paul, MN).

Recently, TSI has introduced a modification to the operation of the DMPS so that complete size scans can be achieved in as little as 60 s at very high resolution (64 channels per decade of size). The Scanning Mobility Particle Sizer (SMPS) is based on the work of Wang and Flagan[126] in which the electric field strength in the electrical classifier section of

Table 8.10. Distribution of Electric Charges According to Boltzmann's Law

D_v (μm)	\multicolumn{9}{c}{Percentage of Particles Carrying n_p Elementary Charges}								
	$n_p = -4$	-3	-2	-1	0	1	2	3	4
0.01					0.3	99.3	0.3		
0.04			0.2	5.2	89.5	5.2	0.2		
0.08			2.8	23.4	47.5	23.4	2.8		
0.20		2.3	9.7	22.6	30.1	22.6	9.7	2.3	
0.60	3.8	7.4	11.9	15.6	17.4	15.6	11.9	7.4	3.8
1.00	5.4	8.1	10.7	12.7	13.5	12.7	10.7	8.1	5.4

the DMPS is varied monotonically, at the same time making particle number concentration measurements in rapid succession (as much as 10 times per second) using a condensation nuclei counter. The measurement cycle consists of repeated ramps of the central electrode voltage on an exponential scale, increasing from a defined minimum value to maximum field strength, then decreasing the field strength back to the minimum value. The particles entering the sample extraction slot of the classifier will have a monotonic variation in electrical mobility if the electric field strength is varied monotonically. Hence, after making allowances for the finite transit times of the particles within the classifier and from the extraction slot to the condensation nucleus counter, the entire size distribution of the incoming aerosol can be scanned both accurately and rapidly. The system is designed to be used with the most recently introduced range of condensation nucleus counters that contain the necessary interface electronics.

MICROSCOPY

Microscopy is a large and wide-ranging method of analysis, and an in-depth description of the available techniques is outside the scope of this chapter. Emphasis instead is placed on the preparation and study of aerosol particles. Although microscopic techniques are labor intensive, they should not be neglected, since they provide the only direct method of viewing particles with sufficient resolution to study shape and surface structure.

The basic guidelines for particle size analysis by optical microscopy are given in BS 3406 part 4 (1993) and US ASTM Standard D-2009-65 (reapproved 1979). The British Standard contains detailed guidelines on the use of automated image analyzers, by which some of the tedious operations associated with particle size analysis now can be accomplished rapidly with good precision and accuracy.[127] A full description of optical microscopy for particle characterization also is given in part 1 of the Particle Atlas (McCrone and Delly[128]), and basic measurement techniques are also described in *Microscopy Handbook 23* of the Royal Microscopical Society.[129]

Microscopy is still probably the most valuable tool for particle size analysis because this type of measurement relates directly to the physical dimensions of the particles. Furthermore, limited information about surface texture and shape also can be deduced from a careful study of good quality micrographs. Optical techniques are confined by the limit imposed on resolution due to the wavelength of the light source, i.e., for visible light particles larger than 3 μm projected area diameter. Smaller particles can be sized using electron microscopy, but such techniques operate in vacuo and may be inappropriate for volatile particles. At present, there is no satisfactory technique for handling sub-micron volatile

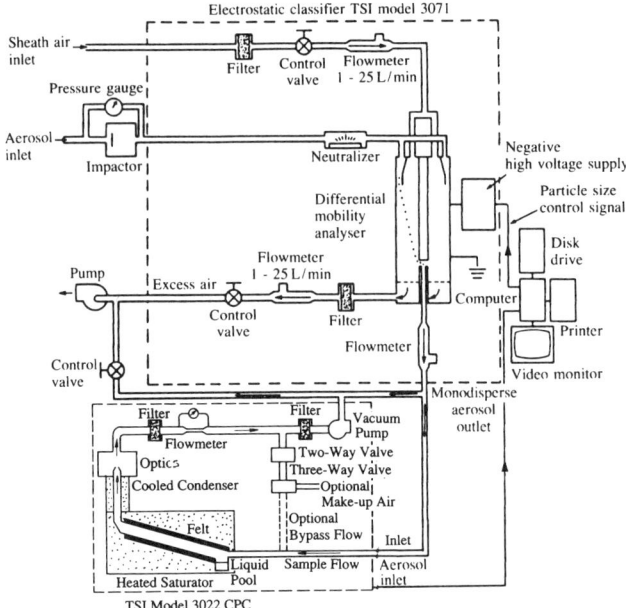

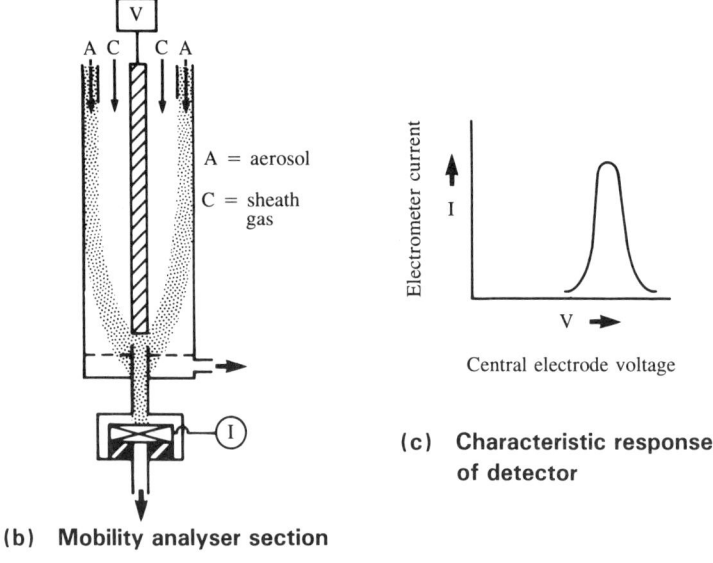

Figure 8.28. Differential Mobility Particle Sizer (DMPS): operating principle (with permission from TSI Inc., St. Paul, MN).

particles, although cryogenic stages can be fitted to electron microscopes in an attempt to minimize evaporation. If the particles are compatible with the conditions in an electron microscope, there are a growing number of electron-beam techniques available for probing chemical composition of the surface and within the particle (Spurny[130]). Combined physico-chemical analysis of particles is likely to become as important as particle sizing alone, since such a combination of information assists considerably in the understanding and control of chemical processes (Nichols and Bowsher[131]). Fuller descriptions of the wide range of microscopic techniques available to study aerosols can be found in the textbooks by Hinds[1] and Spurny.[130]

Sample collection is an important feature of control in obtaining good results (see "Sampling for Size Analyses" section). Microscopy is useful in the characterization of respirable dusts, micro-organisms, spores and particularly fibers; polarized-light microscopy is still the established method for both sizing and species identification for the various forms of asbestos. The collection of respirable dust from personal filter-based samplers is usually undertaken to obtain only mass concentration. However, the procedure can be adapted to include microscopic examination of the collected particles, provided that the filters are lightly loaded and individual particles can be identified. Care is needed to locate these samplers if meaningful results are to be obtained. As discussed in later chapters, care is required also in the selection of representative locations for aerosol collectors as environmental or workplace monitors, especially if a large volume is to be sampled such as a clean room or factory work area. The number of samples required for a given area will be based on the floor area, interruptions to the air flow, and the room volume. Samples must be large enough to be measurable and representative; for instance the volume of air sampled in a clean room might have to be 100 times greater than that required to monitor air quality in a factory environment. Filters may be mounted also as in-line samplers using special gas-tight holders to measure particle contaminants in gas lines. If the samples are to be taken from a duct or stack, the aerosol collector (usually a filter holder) should be mounted within the pipe (Figure 8.29). The problem is to prevent overloading the filter; precautions must be taken (reducing sample times, dilution, etc.) whenever concentrated aerosols are being sampled. When monitoring micron-sized particles, care needs to be taken to prevent overloading if the total particle concentration exceeds about 100 particles cm^{-3}. Transmission electron microscope grids can be mounted inside vessels and pipework as passive samplers, but care is required in the location of these collectors to avoid systematic biases due to sedimentation.

Since the particles are to be sized rather than merely counted or weighed for the determination of total number or mass concentration, it is important that the surface of the filter does not intrude into the image as viewed in the microscope. Thus, cellulose ester membrane filters are a good choice for optical microscopy (particles larger than 5 µm projected area diameter), since the opaque membrane can be cleared after the particles have been collected by exposure to the vapors of either acetone or a 50% v/v mixture of 1-4 dioxan and 1-2 dichloromethane. However, these filters are not a good choice for particle sizing by scanning electron microscope (SEM), as the fibrous surface of the membrane is sufficiently rough to merge with the outlines of the particulate debris (Figure 8.30a). A similar problem exists if glass microfibre filters are used. Image outline merging is especially acute if automated image analysis methods are being used to size the particles, since it may not prove possible to distinguish the particle contours from the background, even when using instruments with 256 grey levels of discrimination. Over the previous 10 years, Nuclepore porous polycarbonate membranes (Nuclepore Corp.) have become accepted widely as the standard particle collection surface for use in the SEM, since their microscopically smooth surfaces permit clear images of the collected particles (Figure 8.30b). It should be noted that these membranes are delicate and must be mounted on a support filter (usually a cellulose acetate filter) to prevent particles clustering around the pores. Recently, a new filter based on porous alumina (Anotec Separations) has become available (Figure 8.30c). Particles are not as easily distinguished, but the pressure drop across Anopore membranes is much lower and they are more efficient particle collectors

238   BIOAEROSOLS HANDBOOK

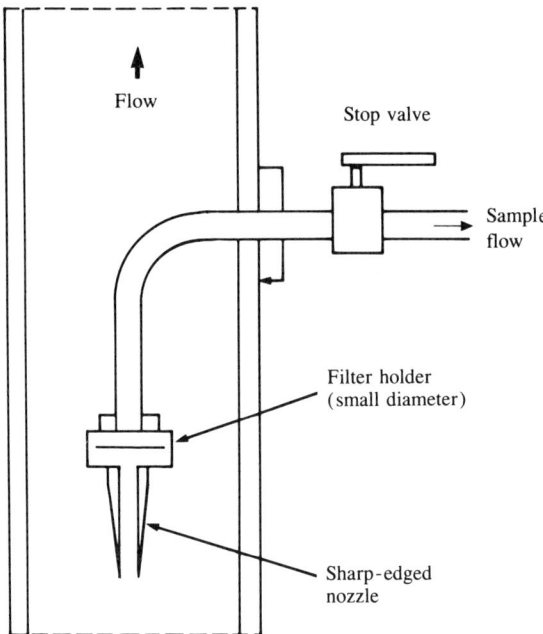

Figure 8.29. Aerosol sampling with filters for particle concentration/size measurement by microscopy.

(larger area filters, greater sampling flow rates). Anopore filters also can be used at high temperatures. All these filter membranes are non-conductive, and must be treated to prevent build-up of electrostatic charges in the SEM, with consequent loss of image quality. The usual procedure is to apply a thin film of gold to the surface of the filter membrane under vacuum in a sputter-coating device prior to examination. Carbon coating is suitable for elemental analysis but cannot be recommended for particle sizing, since image quality often is poor.

Whichever collection method is employed, the problem is to obtain representative numbers of particles throughout the actual size distribution present. Between 300 and 1000 particles may need to be sized to obtain representative data, depending on the required size resolution and the width of the size distribution. This may require the sizing of as many as 100 images, which can be tedious unless automated (Allen[13]). If an automated image analyzer is used, care is required to ensure that the particles are always in focus and that the discrimination techniques used to outline the particles (e.g., outline erosion and fill space within outline) do not introduce systematic biases. Particles located on the boundary of the image should be eliminated from the analyses; most image analyzers are equipped with software to perform this function automatically by introducing a narrow 'guard' area surrounding the main survey area. Detailed aspects of image analysis and particle shape and size have been reviewed in a book edited by Beddow.[132]

Recommended magnifications in the SEM are given in Table 8.11 for certain submicron particle sizes. These values are based on the expression:

$$N = 10^8 \left[\frac{l_m^2 \, A_f \, N_m}{m_m^2 \, L_m \, W_m \, n_m} \right] \qquad (8.44)$$

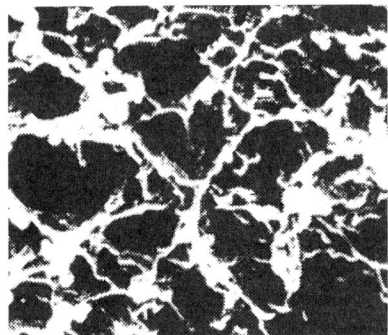

(a) Cellulose ester membrane (0.2 μm porosity)

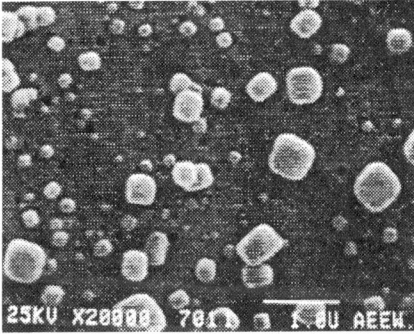

(b) Polycarbonate membrane (0.1 μm porosity)

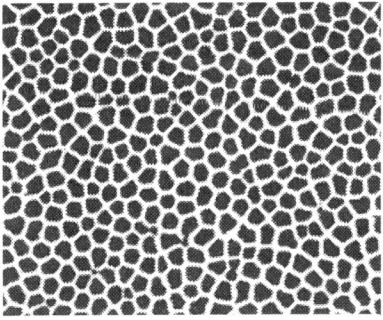

(c) Anopore membrane (0.2 μm porosity)

Figure 8.30. Surfaces of various filter collection media.

where N is the total number of particles forming the size distribution, N_m is the number of particles counted per micrograph, m_m is the calibrated length of the micron-scale marker on the micrograph (μm), l_m is the actual length of this marker (cm), L_m and W_m are the length and width respectively of the micrograph (cm), n_m is the number of micrographs examined and A_f is the collection area of the filter (cm^2). In all the examples given in Table 8.11, no particle image will appear smaller than 2 mm in the micrograph. It is good practice to locate the fields to be surveyed on the filter in a systematic manner.

Table 8.11. Recommended Magnifications for Specific Particle Sizes

Particle Size (μm)	Magnification	Size of Field (cm^2)	No. of Fields per cm^2
4.0	600	2.8×10^{-4}	3.6×10^3
1.0	2,500	1.6×10^{-5}	6.2×10^4
0.25	10,000	1.0×10^{-6}	9.9×10^5
0.15	15,000	4.5×10^{-7}	2.2×10^6
0.10	20,000	2.5×10^{-7}	3.9×10^6
0.07	35,000	8.3×10^{-8}	1.2×10^7

SUMMARY

The aim of this chapter has been to provide guidance on the selection of the most suitable technique(s) from a range of aerosol analysis equipment. When choosing a suitable analysis method, it is advisable to opt initially for simplicity and introduce more complex techniques at a later stage if the problem warrants such an approach. Emphasis has been placed on the measurement of particle size rather than particle concentration, although the latter is often implicit in these techniques. However, it should be noted that a simple pre-weighed filter (paper, glass microfibre or membrane) may be all that is required to collect aerosol samples in a known gas volume if a straightforward mass-concentration measurement is needed.

Whichever strategy is chosen, it is essential that due regard be given to the problem of obtaining representative samples and avoiding biases between the inlet and the measurement/collection area. Viability may also be an important issue when sampling the various types of bioaerosol. The reader is encouraged to investigate both of these aspects thoroughly before embarking on any quantitative aerosol sampling exercise.

ACKNOWLEDGMENTS

The author would like to acknowledge advice from several former colleagues of AEA Technology. He is now with Trudell Medical, London, Ontario, Canada.

REFERENCES

1. Hinds, W.C., *Aerosol Technology,* John Wiley and Sons, New York, 1982.
2. Vincent, J.H., *Aerosol Sampling and Practice,* John Wiley and Sons, Chichester, UK, 1989.
3. Davies, C.N., "The Entry of Aerosols into Sampling Tubes and Heads," *Brit. J. Appl. Phys. (J. Phys. D.),* **1**, 921–932, 1968.
4. Agarwal, J.K. and Liu, B.Y.H., "A Criterion for Accurate Aerosol Sampling in Calm Air," *Am. Ind. Hyg. Assoc. J.,* **41**, 191–197, 1980.
5. Belyaev, S.P. and Levin, L.M., "Techniques for the Collection of Representative Aerosol Samples," *J. Aerosol Sci.,* **5**, 325–338, 1974.
6. Durham, M.D. and Lundgren, D.H., "Evaluation of Aerosol Aspiration Efficiency as a Function of Stokes Number," *J. Aerosol Sci.,* **11**, 179–188, 1980.

7. Thomas, J.W., "Gravity Settling of Particles in a Horizontal Tube," *J. Air Pollut. Control Assoc.,* **8**, 32–34, 1958.
8. Gormley, P.G. and Kennedy, M., "Diffusion from a Stream Flowing Through a Cylindrical Tube," *Proc. Royal Irish Acad.,* **52(A)**, 136–169, 1949.
9. Schwendiman, L.C., Sehmel, G.A., and Postma, A.K., "Radioactive Aerosol Transport Systems," Proc. Int. Conf. on Radioactive Pollution of Gaseous Media, Saclay, France, II, 373; also authors as above (1963), "Radioactive Particle Retention in Aerosol Transport Systems," Hanford Laboratories, Richland, WA, Report HW-SA-3210, 1963.
10. Stevens, D.C., Bull, R.K., and Parker, R.C., "Penetration of Airborne Particles through Pipes Used for Sampling Radioactive Aerosols," AEA Technology, Report AERE-R 13131, 1988.
11. Pui, D.Y.H., Romay-Novas, F., and Liu, B.Y.H., "Experimental Study of Particle Deposition in Bends of Circular Cross Section," *Aerosol Sci. Technol.,* **7**, 301–315, 1987.
12. Lewandowski, G.A., "Stack Sampling," in *Air/Particulate Instrumentation and Analysis,* Cheremisinoff, P.N., Ed., Ann Arbor Science, Ann Arbor, MI, 1981.
13. Allen, T., *Particle Size Measurement,* Chapman and Hall, London, 4th ed., 1992.
14. Krill, W., "IERL-CRB Sampling Manual for Level 1 Environment Assessment," Aerotherm Report No. TM-77-160, Appendix E to Cooper, L., Measurement of High Temperature, High Pressure Processes, Annual Report, U.S. Environmental Protection Agency Report No. EPA-600-7-78-011, 1978.
15. Nichols, A.L. and Mitchell, J.P., "Measurements of the Physical Properties of Nuclear Aerosols," *Nuclear Technol.,* **81**, 205–232, 1988.
16. Mercer, T.T., "Aerosol Technology in Hazard Evaluation," Academic Press, New York, 1973.
17. Liu, B.Y.H., *Fine Particles,* Academic Press, New York, 1976.
18. Mitchell, J.P., "Aerosol Generation for Instrument Calibration," AEA Technology Report AEA-EE-0253, January 1992.
19. Marshall, I.A., Mitchell, J.P., and Griffiths, W.D., "The Calibration of a Timbrell Aerosol Spectrometer," *J. Aerosol Sci.,* **21(7)**, 969–975, 1990.
20. Timbrell, V., "An Aerosol Spectrometer and Its Applications," in *Assessment of Airborne Particles,* Mercer, T., Morrow, P.E. and Stober, W., Eds., Charles C. Thomas, Springfield, IL, 1972.
21. Griffiths, W.D., Patrick, S., and Rood, A.P., "An Aerodynamic Particle Size Analyzer Tested with Spheres, Compact Particles and Fibers Having a Common Settling Rate Under Gravity," *J. Aerosol Sci.,* **15**, 491–502, 1984.
22. McCormack, J.D. and Hilliard, R.K., "Aerosol Measurement Techniques and Accuracy in the CSTF," Proc. CSNI Specialists Meeting on Nuclear Aerosols in Reactor Safety, Gatlinburg, TN, NUREG/CR-1724, 249, 1980.
23. Boulaud, D., Chouard, J.C., Frambourt, C., and Madelaine, G., "Study of a Sedimentation Battery," *J. Aerosol Sci.,* **14(3)**, 421–424, 1983.
24. Lodge, J.P. and Chan, T.L., Eds., *Cascade Impactor: Sampling and Data Analysis,* American Industrial Hygiene Association, Akron, OH, 1986.
25. Marple, V.A. and Liu, B.Y.H., "Characteristics of Laminar Jet Impactors," *Environ. Sci. Technol.,* **8(7)**, 648–654, 1974.
26. Rader, D.J and Marple, V.A., "Effect of Ultra-Stokesian Drag and Particle Interception on Impaction Characteristics," *Aerosol Sci. Technol.,* **4**, 141–156, 1985.
27. Andersen, A., "A Sampler for Respiratory Health Assessment," *Am. Ind. Hyg. Assoc. J.,* **27**, 160–165, 1966.
28. Pilat, M.J., Ensor, D.S. and Bosch, J.C., "Source Test Cascade Impactor," *Atmos. Environ.,* **4**, 671–679, 1970.

29. Pilat, M.J., Ensor, D.S., and Bosch, J.C., "Cascade Impactor for Sizing Particulates in Emission Sources," *Am. Ind. Hyg. Assoc. J.,* **32(8)**, 508–511, 1971.
30. Hering, S.V. and Marple, V.A., "Low-Pressure and Micro-Orifice Impactors," 103–127 in Cascade Impactor: Sampling and Data Analysis, Lodge, J.P. and Chan, T.L., Eds., American Industrial Hygiene Association, Akron, OH, 1986.
31. Pilat, M.J., "Low Pressure Impactors for In-Stack Particle Sizing," Proc. Sem. on In-Stack Particle Sizing for Particulate Control Device Evaluation, USEPA Report EPA-6-00/2-77-060, 1977.
32. Pilat, M.J., "Sizing Sub-Micron Particles with a Cascade Impactor," Paper 4 in Proc. Advances in Particle Sampling and Measurement, Asheville, NC, USEPA Report EPA-600/7-79-065, 1979.
33. Hering, S.V., Flagan, R.C., and Friedlander, S.K., "Design and Evaluation of New Low-Pressure Impactor 1," *Environ. Sci. Technol.,* **12**, 667–673, 1978.
34. Hering, S.V., Friedlander, S.K., Collins, J.J., and Richards, L.W., "Design and Evaluation of New Low-Pressure Impactor 2," *Environ. Sci. Technol.,* **13**, 184–188, 1979.
35. McFarland, A.R., Nye, H.S., and Erickson, C.H., "Development of a Low Pressure Impactor," USEPA Report EPA-650/2-74-014, 1973.
36. Berner, A., "Practical Experience with a 20-Stage Impactor," *Staub Reinhalt. Luft,* **32**, 8.1, 1972.
37. Fairchild, C.I. and Wheat, L.D., "Calibration and Evaluation of a Real-Time Cascade Impactor," *Am. Ind. Hyg. Assoc. J.,* **45(4)**, 205–211, 1984.
38. Horton, K.D. and Mitchell, J.P., "The Calibration of Andersen Mark-II and California Measurements PC-2 Cascade Impactors," *J. Aerosol Sci.,* **23(5)**, 505-524, 1992.
39. Marple, V.A., Rubow, K.L., and Behm, S.M., "A Micro-Orifice Uniform Deposit Impactor (MOUDI): Description, Calibration and Use," *Aerosol Sci. Technol.,* **14(4)**, 434-446, 1991.
40. O'Connor, D.T., "Calibration of a Cascade Centripeter Dust Sampler," *Ann. Occup. Hyg.,* **16**, 119–125, 1971.
41. Novick, V.J. and Alvarez, J.L., "Design of a Multi-Stage Virtual Impactor," *Aerosol Sci. Technol.,* **6(1)**, 63–70, 1987.
42. McFarland, A.R., Ortiz, C.A., and Bertch, R.W., "Particle Collection Characteristics of a Single-Stage Dichotomous Sampler," *Environ. Sci. Technol.,* **12(6)**, 679–682, 1978.
43. Masuda, H., Hochrainer, D., and Stöber, W., "An Improved Virtual Impactor for Particle Classification and Generation of Test Aerosols with Narrow Size Distribution," *J. Aerosol Sci.,* **10(3)**, 275–287, 1979.
44. Willeke, K. and Pavlik, R.E., "Size Classification of Fine Particles by Opposing Jets," *Environ. Sci. Technol.,* **12(5)**, 563–566, 1978.
45. Cheng, Y-S, and Yeh, H-C., "Particle Bounce in Cascade Impactors," *Environ. Sci. Technol.,* **13(11)**, 1392–1396, 1979.
46. May, K.R., "The Cascade Impactor: An Instrument for Sampling Coarse Aerosols," *J. Sci. Instrum.,* **22**, 187–194, 1945.
47. Davies, C.N., Aylward, M., and Leacey, D., "Impingement of Dust from Air Jets," *Arch. Ind. Hyg. Occup. Med.,* **4**, 354–397, 1951.
48. Stern, S.C., Zeller, H.W., and Schekman, A.I., "Collection Efficiency of Jet Impactors at Reduced Pressures," *Ind. Eng. Chem. Fundam.* **1**, 273–277, 1962.
49. McFarland, A.R. and Zeller, H.W., "Study of a Large-Volume Impactor for High-Altitude Aerosol Collection," Report No. 2391, Contract AT(11-1), General Mills, Electronics Division., Minneapolis, 1963.
50. Lundgren, D.A., "An Aerosol Sampler for Determination of Particle Concentration of Size in Time," *J. Air Pollut. Control Assoc.,* **17**, 225–229, 1967.

51. McFarland, A.R. and Husar, R.B., "Development of a Multistage Inertial Impactor," Particle Technology Lab. Report No. 120, University of Minnesota, MN, 1967.
52. Mitchell, J.P., Costa, P.A., and Waters, S., "An Assessment of an Andersen Mark-II Cascade Impactor," *J. Aerosol Sci.*, **19(2)**, 213–221, 1988.
53. Mercer, T.T. and Chow, H.Y., "Impaction from Rectangular Jets," *J. Colloid Interface Sci.*, **27**, 75–83, 1968.
54. Hogan, A.W., "Evaluation of a Silicone Adhesive as an Aerosol Collector," *J. Appl. Meteorol.* **10**, 592–595, 1970.
55. Cushing, K.M., Lacey, G.E., McCain, J.D., and Smith, W.B., "Particulate Sizing Techniques for Control Device Evaluation: Cascade Impactor Calibrations," USEPA Report EPA-600/2-76-280, 1976.
56. Rader, D.J., Mondy, L.A., Brockmann, J.E., and Lucero, D.A., "Stage Response of an Andersen Cascade Impactor to Monodisperse Droplets," Sandia National Labs. Report SAND88-1085, 1989.
57. Harris, D.B., "Procedures for Cascade Impactor Calibration and Operation in Process Streams," USEPA Report EPA-600/2-77-004, 1977.
58. Licht, W., "Air Pollution Control Engineering," Marcel Dekker, New York, 1988.
59. Beekmans, J.M., "Analysis of the Cyclone as a Size-Selective Aerosol Sampler," 56–65 in *Aerosol Measurement,* Lundgren, D.A. et al., Eds. University Presses of Florida, Gainesville, FL, 1979.
60. Lee, K.W., Gieseke, J.A., and Piispanen, W.H., "Evaluation of Cyclone Performance in Different Gases," *Atmos. Environ.*, **19(6)**, 847–852, 1985.
61. Smith, W.B., Wilson, R.R., and Harris, D.B., "A Five-Stage Cyclone System for In-Situ Sampling," *Environ. Sci. Technol.*, **23**, 1307–1392, 1979.
62. Smith, W.B., Cushing, K.M., Wilson, R.R., and Harris, D.B., "Cyclone Samplers for Measuring the Concentration of Inhalable Particles in Process Streams," *J. Aerosol Sci.*, **13(3)**, 259–267, 1982.
63. Remiarz, R.J., Agarwal, J.K., Quant, F.R., and Sem, G.J., "Real-Time Aerodynamic Particle Size Analyzer," 879–895 in *Aerosols in the Mining and Industrial Work Environments,* Marple, V.A. and Liu, B.Y.H., Eds., Ann Arbor Science, Ann Arbor, MI, 1983.
64. Wang, H-C. and John, W., "Particle Density Correction for the Aerodynamic Particle Sizer," *Aerosol Sci. Technol.*, **6(2)**, 191–198, 1987.
65. Marshall, I.A., Mitchell, J.P., and Griffiths, W.D., "The Behavior of Non-Spherical Particles in a TSI Aerodynamic Particle Sizer," *J. Aerosol Sci.*, **22(1)**, 73–90, 1991.
66. Baron, P.A., "Calibration and Use of the Aerodynamic Particle Sizer," *Aerosol Sci. Technol.*, **5(1)**, 55–67, 1986.
67. Griffiths, W.D., Iles, P.J., and Vaughan, N.P., "The Behavior of Liquid Droplet Aerosols in an APS3300," *J. Aerosol Sci.*, **17(6)**, 921–930, 1986.
68. Blackford, D.B., Hanson, A.E., Pui, D.Y.H., Kinney, P., and Ananth, G.P., "Details of Recent Work Towards Improving the Performance of the TSI Aerodynamic Particle Sizer," 311–316 in Proc. 2nd Conf. Aerosol Society, Bournemouth, UK, 1988.
69. Remiarz, R.J. and Johnson, E.M., "A New Diluter for High Concentration Measurements with the Aerodynamic Particle Sizer," *TSI Quarterly,* **X(1)**, 7–12, 1984.
70. Blackford, D.B. and Rubow, K.L., "A Small-Scale Powder Disperser," pp. 13–27 in Proc. NOSA Aerosol Symposium, Solna, Sweden, 1986.
71. Dahneke, B.E., "Aerosol Beams," in "Recent Developments in Aerosol Science," Shaw, D.T., Ed., Wiley-Interscience, New York, 1978.
72. Dahneke, B.E. and Cheng, Y.S., "Properties of Continuum Source Particle Beams 1: Calculation Methods and Results," *J. Aerosol Sci.*, **10(3)**, 257–274, 1979.
73. Sawyer, K.F. and Walton, W.H., "The 'Conifuge': A Size-Separating Sampling Device for Airborne Particles," *J. Sci. Instrum.*, **27**, 272–276, 1950.

74. Stöber, W. and Flaschbart, H., "Size-Separating Precipitation of Aerosols in a Spinning Spiral Duct," *Environ. Sci. Technol.*, **3**, 1280–1296, 1969.
75. Stöber, W. and Flaschbart, H., "High Resolution Aerodynamic Size Spectrometry of Quasi-Monodisperse Latex Spheres with a Spiral Centrifuge," *J. Aerosol Sci.*, **2**, 103–116, 1971.
76. Kotrappa, P. and Light, M.E., "Design and Performance of the Lovelace Aerosol Particle Spectrometer," U.S. Atomic Energy Commission Report LF-44, 1971.
77. Martonen, T.B., "Theory and Verification of Formulae for Successful Operation of Aerosol Centrifuges," *Am. Ind. Hyg. Assoc. J.*, **43(3)**, 154–159, 1972.
78. Mitchell, J.P., Nichols, A.L., Smith, A.D., and Snelling, K.W., "Aerosol Particle Size Analysis Using the Stober Spiral Duct Centrifuge," Filtr. Sep., 345–348, Sept./Oct., 1984.
79. Prodi, V., DeZaiacomo, T., Melandri, C., Tarroni, G., Formigniani, M., Olivieri, P., Barilli, L., and Hochrainer, D., "An Inertial Spectrometer for Aerosol Particles," *J. Aerosol Sci.*, **10(4)**, 411–419, 1979.
80. Mitchell, J.P. and Nichols, A.L., "Experimental Assessment and Calibration of an Inertial Spectrometer," *Aerosol Sci. Technol.*, **9(1)**, 15–28, 1988.
81. Prodi, V. and Belosi, F., "A Time Resolved Particle Spectrometer for High-Temperature In-Situ Sampling," 571–574 in Proc. 2nd Int. Conf. Aerosols, Berlin, Pergamon Press, Oxford, UK, 1986.
82. Prodi, V. and Belosi, F., "A Particle Spectrometer for Elevated Temperatures," 575–578 in Proc. 2nd Int. Conf. Aerosols, Berlin, Pergamon Press, Oxford, UK, 1986.
83. Gebhart, J., Blankenberg, P., and Roth, C., "Counting Efficiency and Sizing Characteristics of Optical Particle Counters," 7–10 in Proc. 1st Int. Aerosol Conf., Minneapolis, Liu, B.Y.H., Pui, D.Y.H., and Fissan, H., Eds., Elsevier, New York, 1984.
84. Gebhart, J. and Roth, C., "Background Noise and Counting Efficiency of Single Particle Optical Counters," 607–611 in Proc. 2nd. Int. Conf., Aerosols Berlin, Schikarski, W., Fissan, H., and Friedlander, S.K., Eds., Pergamon Press, Oxford, UK, 1986.
85. Van De Hulst, H.C., "Light Scattering by Small Particles," Chapman and Hall, London, 1957.
86. Kerker, M., "The Scattering of Light and Other Electromagnetic Radiation," Academic Press, New York, 1969.
87. Bayvel, L.P. and Jones, A.R., "Electromagnetic Scattering and Its Applications," Applied Science Publishers, London, 1981.
88. Cooke, D.D. and Kerker, M., "Response Calculations for Light-Scattering Aerosol Particle Counters," *Appl. Optics*, **14(3)**, 734–739, 1975.
89. Umhauer, H., "Particle Size Measurements of Suspensions by Means of a Light Scattering Particle Counter," *Chem. Ing. Tech.*, **52(1)**, 55–58, 1980.
90. Szymanski, W.W. and Liu, B.Y.H., "On the Sizing Accuracy of Laser Optical Particle Counters," *Part. Charact. J.*, **3(1)**, 1–7, 1986.
91. Marple, V.A. and Rubow, K.L., "Aerodynamic Particle Size Calibration of Optical Particle Counters," *J. Aerosol Sci.*, **7**, 425–433, 1976.
92. Büttner, H., "Measurement of Particle Size Distributions in Gas Flows with an Optical Particle Counter," *Part. Charact. J.*, **2(1)**, 20–24, 1985.
93. Mitchell, J.P., Nichols, A.L. and Van Santen, A., "The Characterization of Water-Droplet Aerosols by Polytec Optical Particle Analyzers," *Part. Charact. J.*, **6(3)**, 119–123, 1989.
94. Borho, K., "A Light Scattering Measuring Instrument for High Dust Concentrations," *Staub Reinhalt. Luft,* **30(11)**, 479–483, 1970.
95. Mitchell, J.P., Ashcroft, J., Fromentin, A., Holmes, R., Marsault, P., McAughey, J.J., Patel, A., and Phillips, H., "Laboratory Intercomparison of Polytec Optical Aerosol Analyzers," pp. 643–646 in Proc. 3rd Int. Aerosol Conf., Kyoto, Japan, 1990.

96. Umhauer, H., "Particle Size Distribution Analysis by Scattered Light Measurements Using an Optically Defined Measurement Volume," *J. Aerosol Sci.,* **14(6)**, 765–770, 1983.
97. Swithenbank, J., Beer, J.M., Taylor, D.S., Abbot, D., and McCreath, C.G., "A Laser Diagnostic Technique for the Measurement of Droplet and Particle Size Distribution," *Prog. Astronaut. Aeronaut.,* **53**, 421–447, 1977.
98. Kaye, B.H., "Characterization of Fine Particle Systems by Utilising Diffraction Pattern Analysis," 626–644 in Proc. 2nd Europ. Symp. Particle Characterization, Nuremberg, Germany, 1979.
99. Yeoman, M.L., Azzopardi, B.J., White, H.J., Bates, C.J., and Roberts, P.J., "Optical Development and Application of a Two Colour LDA System for the Simultaneous Measurement of Particle Size and Particle Velocity," AEA Technology Report AERE-R 10468, 1982.
100. Farmer, W.M., "Measurement of Particle Size, Number Density and Velocity Using a Laser Interferometer," *Appl. Optics,* **11**, 2603–2612, 1972.
101. Negus, C.R. and Drain, L.E., "Mie Calculations of the Scattered Light from a Spherical Particle Traversing a Fringe System Produced by Two Intersecting Laser Beams," *J. Phys. D: Applied Physics,* **15**, 375–402, 1982.
102. Bachalo, W.D. and Houser, M.J., "Phase/Doppler Spray Analyzer for Simultaneous Measurements of Drop Size and Velocity Distributions," *Opt. Eng.,* **23(5)**, 583–590, 1984.
103. Bachalo, W.D. and Houser, M.J., "Spray Drop Size and Velocity Measurements Using the Phase-Doppler Particle Analyzer," 2, paper VC/2/1 in ICLASS-85, Proc. 3rd. Conf. Liquid Atomization and Spray Systems, London, UK, 1985.
104. Saffman, M. and Buchhave, P., "Simultaneous Measurement of Size, Concentration and Velocity of Spherical Particles by a Laser Doppler Method," Proc. 2nd Int. Symp. on Applications of Laser Anemometry to Fluid Mechanics, Lisbon, Portugal, 1984.
105. Negus, C.R., Bosley, R.B., Martin, S.R., and Wallace-Sims, G.R., "Use of the Phase-Difference Technique for Droplet Sizing in the Aerosol Range," 37–42 in Proc. 3rd Conf. of the Aerosol Society, West Bromwich, UK, March 1989.
106. Burton, G.R. and Negus, C.R., "Sizing of Small Particles Using a Miniaturised Phase-Difference Instrument," 211–216 in Proc. 4th Conf. of the Aerosol Society, Guildford, UK, 1990.
107. Holve, D.J. and Self, S.A., "An Optical Particle Sizing Counter for In-Situ Measurements: Parts I and II," *J. Appl. Optics,* **18(10)**, 1632–1652, 1979.
108. Holve, D.J. and Self, S.A., "Optical Measurements of Mean Particle Size in Coal-Fired MHD Flows," *Combustion and Flame,* **37**, 211–214, 1980.
109. Holve, D.J., "In-Situ Optical Particle Sizing technique," *J. Energy,* **4(4)**, 176–183, 1980.
110. Dahlen, R.S., Beittel, R., Morgan, B.A., and Holve, D.J., "Applications of an In-Situ Optical Counter for Combustion Measurements: Comparisons with Cascade Impactor Measurements," Paper presented at Proc. Meeting of the American Association for Aerosol Research (AAAR), Seattle, 1987.
111. Karasikov, N., Krauss, M., and Barazanii, G., "Measurement of Particles in the Range of 150–1200 μm using a Time-of-Transition Particle Size Analyzer," 51–62 in Particle Size Analysis 1988, Lloyd, P.J., Ed., John Wiley and Sons, Chichester, UK, 1988.
112. Aharonson, E.F., Karasikov, N., Roitberg, M., and Shamir, J., "GALAI CIS-1: A Novel Approach to Aerosol Particle Size Analysis," *J. Aerosol Sci.,* **17(3)**, 530–536, 1986.
113. Aharonson, E.F., "Shape Analysis of Airborne Aerosol Particles by the GALAI CIS-1," 398–401 in Proc. Ind. Inst. Aerosol Conf., Berlin, Schikarski, W., Fissan, H. and Friedlander, S.K., Eds., Pergamon Press, Oxford, UK, 1986.

114. Karasikov, N. and Krauss, M., "Automated Analyses of Particles Collected on Membrane Filters," paper presented at ANALYTICA, Munich, 1988 (available from GALAI Laboratories on request).
115. Karasikov, N. and Krauss, M., "Examining the Influence of Index of Refraction on Particle Size Measurements using a Time-of-Transition Optical Particle Sizer," *Filtr. Sep.*, 121–124, March/April 1989.
116. Liu, B.Y.H. and Pui, D.Y.H., "On the Performance of the Electrical Aerosol Analyzer," *J. Aerosol Sci.*, 6, 249–264, 1975.
117. Keady, P.B., Quant, F.R., and Sem, G.J., "Differential Mobility Particle Sizer: A New Instrument for High Resolution Aerosol Size Distribution Measurement Below 1 μm," *TSI Quarterly*, **IX(2)**, 3–11, 1983.
118. Horton, K.D., Mitchell, J.P., and Nichols, A.L., "Experimental Comparison of Electrical Mobility Aerosol Analyzers," *Tsi J. Particle Instrum.*, **4(1)**, 3–19, 1989.
119. Yeh, H-C., Cheng, Y-S., and Kanapilly, G.M., "Use of the Electrical Aerosol Analyzer at Reduced Pressure," 1117–1133 in "Aerosols in the Mining and Industrial Work Environments," Marple, V.A. and Liu, B.Y.H., Eds., Ann Arbor Science, Ann Arbor, MI, 1983.
120. Sem, G.J., "Electrical Aerosol Analyzer: Operation, Maintenance and Application," 400–430 in *Aerosol Measurement* Lundgren, D.A., et al., Eds., University Presses of Florida, Gainesville, FL, 1979.
121. Felix, L.G., Merritt, R.L., McCain, J.D., and Ragland, J.W., "Sampling and Dilution System Design for Measurement of Submicron Particle Size and Concentration in Stack Emission Aerosols," *TSI Quarterly*, **VII(4)**, 3–12, 1981.
122. Blackford, D.B., Belling, K.J., and Sem, G.J., "The Measurement of Non-Volatile Residue for UltraPure Solvents: A Rapid Method Employing Commercially Available Instrumentation," Proc. Inst. of Environmental Sciences, 360–365, 1987.
123. Kapadia, A., "Data Reduction Techniques for Aerosol Size Distribution Measuring Instruments," Ph.D. Thesis, University of Minnesota, MN, 1979.
124. Liu, B.Y.H. and Pui, D.Y.H., "A Sub-Micron Standard and the Primary, Absolute Calibration of the Condensation Nucleus Counter," *J. Colloid Interface Sci.*, **47(1)**, 155–171, 1974.
125. Agarwal, J.K. and Sem, G.J., "Continuous Flow, Single Particle-Counting Condensation Nucleus Counter," *J. Aerosol Sci.*, 11, 343–357, 1980.
126. Wang, S-C., and Flagan, R.C., "Scanning Electrical Mobility Spectrometer," *Aerosol Sci. Technol.*, **13(2)**, 230-240, 1990.
127. Germani, M.S. and Buseck, P.R., "Automated Scanning Electron Microscopy for Atmospheric Particle Analysis," *Anal. Chem.*, 63, 2232–2237, 1991.
128. McCrone, W.C. and Delly, J.G., "The Particle Atlas: Volume 1, Principles and Techniques," Ann Arbor Science Publishers, Ann Arbor, MI, 1973.
129. Bradbury, S., "Basic Measurement Techniques for Light Microscopy," in *Microscopy Handbook 23,* Royal Microscopical Society, Oxford University Press, Oxford, UK, 1991.
130. Spurny, K.R., "Physical and Chemical Characterization of Individual Airborne Particles," Ellis Horwood Press, Chichester, UK, 1986.
131. Nichols, A.L. and Bowsher, B.R., "Chemical Characterization of Nuclear Aerosols," *Nuclear Technol.*, 81, 233–245, 1988.
132. Beddow, J.K., *Particle Characterization in Technology: Vol. 2-Morphological Analysis,* CRC Press, Boca Raton, FL, 1984.

CHAPTER 9

Inertial Samplers: Biological Perspectives

B. Crook

INTRODUCTION

Many samplers are available for the collection of bioaerosols and for determining their size distributions (Chapter 8). These may be categorized according to the method of collection. Some are samplers that have been adapted from dust or particle samplers, while others have been designed specifically to collect bioaerosols such as airborne microorganisms. Development of methods to collect non-microbial bioaerosols, e.g., proteinaceous aerosols, has been limited, but it is generally accepted that methods used to collect microorganisms can be adapted successfully for this purpose.

This chapter is concerned with the collection of bioaerosols and will deal with gravitational samplers, i.e., devices that collect by settling (see Chapter 3) onto a collection surface, and inertial samplers, including impactors, impingers and centrifugation. Inertial samplers allow the collection of particles by size-selective sampling, that is, the collection of one particle size fraction, or by fractionation of the particles sampled into different sizes. Their calibration is described in Chapters 7 and 8, while the latter also contains some general principles of aerosol sampling.

GRAVITATION/SETTLE PLATES

This is the simplest form of collection of airborne biological particles. An adhesive substrate, such as a coated microscope slide or a petri dish containing agar medium, is exposed face upwards to the atmosphere to collect particles settling by gravity. This method does not require specialist equipment and because of its simplicity it has been used frequently, sometimes in preference to other aerobiological samplers[1] or has been recommended to give valuable additional information.[2,3] However, it is a passive, i.e., non-volumetric method, that does not give information on the volume of air from which the particles have been collected. In addition, it over-represents larger particles sampled from a larger volume of air during the exposure period because of their faster sedimentation rate. For example, the time taken for microorganisms associated with moldy hay to settle a distance of 1 cm by gravitation has been calculated to range from 10 s for large fungal spores such as *Humicola lanuginosa,* to 34 s for small fungal spores such as *Aspergillus fumigatus,* to 48 s for bacterial cells and 91 s for actinomycete spores such as *Faenia rectivirgula.*[4] Furthermore, collection in turbulent air is seriously affected by shadowing or turbulent deposition. Failure to recover airborne microorganisms of interest in a hospital environment, while a volumetric method was successful, led to a recommendation for the use of settle plates to be avoided.[5]

IMPACTORS, SIEVE AND STACKED SIEVE SAMPLERS

The principle of collection by impaction relies on the tendency of a particle to deviate from the airstream, because of inertia, when the air flow streamlines bend to bypass a solid or semi-solid surface. Particles thus leaving the airstream will impact on the surface (Chapter 3). Samplers of this type can be subdivided into sieve and stacked sieve samplers, slit samplers and cascade impactors (see also Chapter 8).

Single-Stage Samplers

Spore Traps

In this type of sampler air is drawn through a narrow slit and the airborne particles are impacted onto an adhesive surface. They include spore traps, first designed and described by Hirst[6] for the collection of spores outdoors to study the epidemiology of plant diseases. Airborne particles are impacted onto a glass microscope slide, coated with petroleum jelly, mounted on a motorized stage so that the slide is moved during sampling to give a time resolution of particles collected. Sampling at 10 L min^{-1} is achieved with a mains operated pump. A hand-held, battery-operated slit sampler has been described recently,[7] which collects airborne particles onto an adhesive coated microscope slide this time held in a static position in the sampler. The sampler can be operated at an air flow rate of 10 or 20 L min^{-1}.

For sampling over long periods of time, another version of the large spore trap collects particles onto a reel of adhesive tape. The tape is wound by a clockwork motor from one drum to another as sampling continues, to allow continuous sampling for up to a week. Suction is provided by a separate pump at rate of 10 L min^{-1}.

Whirling Arm or Rotorod Samplers

Rotorod samplers (see also Chapter 4) consist of a U-shaped brass rod, square in cross-section, with a pivot mounted centrally at the bottom of the U. This is connected to an electric motor so that the two vertical arms are rotated at a speed of approximately 3500 rpm (Figure 9.1). Spores are impacted onto a sellotape strip coated with adhesive, such as silicone grease,[8] and mounted on the rotating arms. After exposure, these strips can be removed and mounted on microscope slides for direct observation. This type of sampler is largely qualitative rather than quantitative, but an estimate of spore numbers m^{-3} air can be calculated by determining the volume of air sampled by the rotorod (two arms) as a function of rotation:

$$V = 2 (2 \pi r w l) s$$

where V = volume sampled cm^3 min^{-1}
    r = radius of arms (cm)
    w = width of trapping surface (cm)
    l = length of trapping surface (cm)
    s = speed of rotation (rpm)

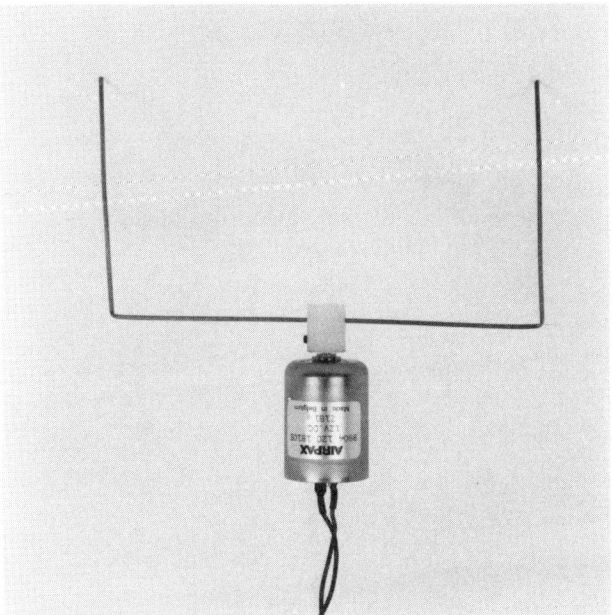

Figure 9.1. Rotorod sampler. A whirling arm impaction sampler with an adhesive collection substrate, e.g., double-sided tape, mounted on the leading edges of the arms. The sampling arms are turned by a battery driven motor.

This assumes a sampling efficiency of 100% for all particle sizes, which is unlikely to be achieved.

Slit-to-Agar Samplers

Air drawn through a narrow slit is accelerated and directed toward the surface of a petri dish containing agar media placed on a rotating turntable (Figure 9.2). Airborne microorganisms thus impact onto the agar surface and are separated spatially by the plate rotation. This sampler was one of the first described for use specifically to collect airborne microorganisms[9] and has been used extensively to collect microorganisms in a wide range of settings, including highly contaminated agricultural environments,[10] in sewage treatment plants,[11] in domestic environments,[12,13] sometimes with modification to allow automatic incubation.[14] It has also been used successfully to collect airborne virus particles[15,16].

Analysis

Petri dishes used in slit-to-agar samplers can be incubated to develop colonies from impacted microbial cells and viable numbers, or colony-forming units (cfu) per unit volume of air sampled can be calculated. Particles deposited onto glass slides in impactors can be observed directly under the microscope and portions of tapes can be mounted on glass slides for microscopic observation. The concentration of particles per unit volume of air can then be calculated as

Figure 9.2. Casella slit-to-agar single-stage impactor. An agar plate is mounted below the inlet on a turntable so that a time-resolved deposition of particles is achieved. Suction is provided by an external mains operated vacuum supply.

$$\frac{\text{No. spores counted}}{\text{Volume air sampled}} \times \frac{\text{Area of trapping surface}}{\text{Area counted}} \times \text{Sampling efficiency}$$

This type of sampler is useful for collecting larger airborne particles, such as pollen grains, large fungal spores and clumps of spores, and has been employed to sample airborne molds in domestic environments,[17] but tends to undersample smaller spores and bacterial cells. However, recent studies[18] have shown good agreement between yields of airborne fungal spores with a spore trap and with an Andersen viable impactor.

Cascade Impactors

Although the samplers described above collect without dividing the catch into different particle size fractions, the principle of collecting by impaction means that size discrimination is possible (Chapter 8). Air can be drawn through a series of slits sequentially diminishing in size so that air velocity increases from one stage to the next. Particles of sufficient mass will be deposited on the upper stage, while smaller particles capable of remaining in the airstream at a lower velocity will leave the airstream at higher velocities to be impacted onto the collection surfaces. This type of sampler is known as the cascade impactor. May[19] described the first prototype that separated particles into four size fractions. Operating at 17.5 L min^{-1}, particles are deposited onto glass microscope slides coated with adhesive. This cascade impactor is no longer available commercially, but has been superceded by a seven-stage version, operated at 10 or 20 L min^{-1}, known as the "ultimate cascade impactor," that again collects particles onto glass microscope slides (Figure 9.3).[20]

Cascade impactors to collect particles by personal sampling in the breathing zone also are available.[21] Originally designed to measure workers' exposure to coal dust in mines, these samplers are small enough to be clipped to the lapel. Airborne particles are sampled at 2 L min^{-1} with suction provided by a battery operated pump (Figure 9.4). Three versions are available that collect particles in four, six or eight size fractions on collection substrates consisting of metal or Mylar plastic discs. Air passes through six radial slots in each stage and these are progressively narrower to facilitate size fractionation. The collection substrates have corresponding slots cut in them and are offset so that the collection surfaces between the slots are directly below the jets of the previous stage. These surfaces are coated with adhesive to ensure particle deposition.

Particles deposited on Mylar membranes can be observed directly under the microscope, viable spores recovered by rinsing to form a suspension for inoculating agar plates[22] or, in a modification of the original eight stage sampler, trays containing gelatin medium can replace the flat collection substrates to improve collection of viable microorganisms.[23] Although inlet efficiency is low for particles of aerodynamic diameter >20 μm, the eight stage version has been used successfully to collect aeroallergens, while giving data on particle size distribution.[24]

Sieve Samplers

Like slit samplers, sieve samplers are available for collection into a single size fraction or subdivision of the catch into several fractions. For aerobiological sampling, the most widely utilized are the stacked sieve samplers.

Stacked Sieve

The principle of collection by cascade impaction onto sticky surfaces can be employed also to draw air through a series of perforated plates or sieves with holes of a defined diameter. The principle of collection is the same as that of a slit sampler, but there are more deposition sites and the collection substrate can be a flat metal plate coated with adhesive, usually used only for non-viable dust collection, or agar media in petri dishes. The most widely employed example of this type of impactor and probably the most widely used of all aerobiological samplers is the Andersen viable impactor (Figure 9.5).[25] Suction is at 28.3 L min^{-1} (1 cfm). Particles of aerodynamic diameters from >15 to 1 μm are separated into six size fractions as the airstream passes through plates perforated by 400 holes of sequentially diminishing diameter and then impact onto the surface of agar plates positioned below each sieve plate. The sampler is designed for glass petri dishes with aluminum lids (supplied by the sampler manufacturer). The use of plastic petri dishes may give rise to electrostatic effects, but it can be argued that the small problems of wall losses can be offset by the practicalities of use. A two-stage version of the Andersen microbial impactor is available. Each stage has 200 orifices, sampling rate again is 28.3 L min^{-1} and particles are separated before impaction into "respirable" and "non-respirable" fractions. The upper stage corresponds roughly to deposition on stages 1 and 2 of the six stage version and the lower stage to deposition on stages 3 to 6. However, side-by-side comparisons with the six-stage version revealed a lower recovery rate except when sampling dilute concentrations (<1000 particles m^3) of larger particles (>1 μm).[26] These differences may be at least partially resolved by using correction factors (discussed later in this chapter)

Figure 9.3. May 'ultimate' cascade impactor. Slit-to-adhesive coated slides, with collection in seven stages of sequentially smaller size fractions. External mains operated vacuum supply required.

calculated specifically for the 200-hole two-stage impaction sampler,[27] but a recent study showed recovery of bacteria by the two-stage impaction to be approximately 40% less than that of the six-stage version.[28]

Single Stage Sieve Samplers

Collection by this principle has been used in the Surface Air Sampler (SAS), of which two models exist. One has a plate perforated by 220 holes and uses a 50 mm Rodac agar plate, the other has 260 holes and uses a 90 mm Rodac plate (Figure 9.6). Both sample at 180 L min^{-1} by means of a suction fan mounted behind the sampling chamber and can be run from a 12 V rechargeable power supply. Performance tests have shown the sampler to be efficient at collecting larger particle size aerosols, from 4 to 20 μm in diameter, but the efficiency of collection falls off rapidly below this figure, i.e., corresponding to smaller spores or viable cells made airborne singly[29] and its collection does not compare favorably with other bioaerosol samplers for airborne bacteria[28] or fungi.[18]

A battery operated single-stage sieve sampler (Portable Air Sampler, Model PASA/B, Burkard Mfg Co., Rickmansworth, UK) is available that collects at 10 L min^{-1} through a plate perforated by 100 holes (Figure 9.7). The collection substrate in this instance is a

Figure 9.4. Sierra Marple personal cascade impactor. Slit- and hole- to-adhesive coated impaction substrates (Mylar plastic membranes or metal coupons) in eight stages of sequentially smaller size fractions. The sampler can be lapel-mounted with suction provided by a battery-operated vacuum pump.

standard 90 mm petri dish filled with agar. The manufacturer's information suggests collection of particles 5 µm in diameter with an efficiency >95% (although efficiency drops to 40% with 2 µm particles), but there are at present no published reports of its performance in the field.

The six-stage Andersen viable impactor[25] has been adapted for single-stage use in two ways. The N6-Andersen sampler is a single-stage sieve sampler in which only the sixth stage is used. Air is drawn through the single sieve plate at 28.3 L min^{-1} and particles leave the airstream to impact onto the surface of an agar plate positioned below. Its performance has been found to be similar to the six-stage version[30] and to compare favorably with other samplers;[12,31] furthermore, it is more economical in terms of number of agar plates required and sampler processing time. However, it does not give size fractionation data.

Another adaptation of the Andersen sampler is known as the S6 sampler.[32] In this, an agar plate is loaded into only the sixth stage of the standard six-stage sampler. Airborne particles pass through the sieve plates of the first five stages before being impacted onto the agar plate below the sixth. Again, the advantage of this is the reduction in the number of agar plates required and it has been estimated that the time taken to collect a sample was reduced to about one third.[30] However, in a comparison between the N6 and S6 samplers, the S6 sampler consistently gave lower values, probably because of impaction and wall losses in the upper stages.[30]

Analysis

After collection, agar plates are incubated to cultivate microbial colonies from the deposited cells. For six-stage Andersen microbial impactors, a correction factor allows calculation of the numbers of particles deposited per unit volume of air sampled, taking into

Figure 9.5. Andersen microbial impactor. Hole-to-agar medium impaction in six stages of sequentially smaller size fractions. Also available as a two-stage impactor and as a single-stage impactor in two formats (N6 and S6). External mains operated vacuum supply required.

account coincident deposition of particles on the same deposition site in the lower stages when sampling from larger concentrations of airborne cells.[25,29,33] As described above, positive hole correction factors now are available for 200-hole microbial impactors.[27] Alternatively, after exposure, agar can be removed from the petri dishes, homogenized and diluted to inoculate further agar plates. Gelatin also can be used as a collection medium in petri dishes and can be melted in warm water to prepare an inoculum for agar plates. These methods have been successful for sampling from highly contaminated environments where incubation of agar plates from impactors to cultivate cells directly from deposits would have resulted in overgrowth.[10] It must be noted that, when compared to conventional use of the Andersen sampler, yields will be greater because clumps of cells are being disrupted.[22] Another possible use is to overlay plates with filters, from which deposits can be resuspended and provide an inoculum. However, this leads to particle bounce and re-entrainment into the airstream, causing apparent differences in size characteristics and possible losses.[22]

Figure 9.6. SAS surface air sampler. Single-stage hole-to-agar medium impaction, using 'Rodac' plates. Battery operated with internal vacuum source.

IMPINGERS

In an impinger, air is drawn by vacuum through a body of liquid and particles leave the airstream to be collected by impingement into the liquid. The liquid impinger was first described by Greenburg and Smith[34] as a dust sampler and was later adapted for use as a bacterial aerosol sampler.[35] Most impingers are constructed from glass with a single collection chamber and suction applied to a side arm. Air is drawn through a curving intake tube followed by an impinging jet that acts as a critical orifice to determine velocity and thus defines the flowrate. In the original designs, the distance between the impinging jet and the base of the sampler was 4 mm, thus the air jet displaced the body of impingement liquid and allowed particles to hit the base of the sampler, which may have affected viability of some collected microorganisms[36] (see also Chapter 6). This led to the development of alternatives, such as the Shipe sampler,[37] in which the air jet was introduced into the chamber at an angle, and the Porton raised all-glass impinger, or AGI-30 (Figure 9.8), in which the jet is raised to 30 mm above the base of the sampler. This increased sampling efficiency for viable cells by reducing the velocity at which particles were impinged and lessened the damage caused by contact with the base of the sampler.[36] The AGI-30, like the Andersen viable impactor, was adopted as a reference sampler for the collection of viable airborne microorganisms,[38] consequently, like the Andersen viable impactor, it is utilized widely. A sampling rate of 12.5 L min^{-1} is achieved by mains vacuum pump, usually collecting into 20 mL of an aqueous fluid. Suitable dilutions of the microbial suspension in the collection fluid can be spread onto the surface of agar media and incubated to reveal colonies developing from viable cells. The AGI-30 has been shown to collect bacterial aerosols with greater efficiency than a slit-to-agar sampler.[39] Recovery of airborne viruses with an AGI-30 was similar to that of a modified slit sampler,[16] while impingement

Figure 9.7. Burkard single-stage impactor. Single-stage hole-to-agar medium impaction, using standard disposable agar plates. Battery operated with internal vacuum source.

as a means of collecting airborne foot and mouth virus was found to be better than collection by filtration.[40] Because aerosols are collected by suspension in liquid, there is no strict limitation to the duration of sampling (other than owing to evaporation of collecting fluid) as the impinger collection fluid can be diluted before plating out. Consequently, this method of collection has an advantage over methods such as impaction where the deposition site may quickly become overloaded.[3] Similarly, Lembke et al.[41] found AGI-30 and Andersen samplers to be of equal use to examine bacterial aerosols in waste disposal sites, although the range of airborne microorganisms over which the AGI-30 could be used, from 10^3 to 10^7 cfu m^{-3}, was greater than that of the Andersen sampler (10^3 to 10^5 cfu m^{-3}). Laboratory tests using the AGI-30 as a reference sampler for collecting airborne bacteria showed it to have a similar collection efficiency to six-stage Andersen samplers and a better recovery than most other bioaerosol samplers tested.[28] For airborne fungal spores, viable recovery with impingers was less than that with filters,[42] but this probably reflected the improved collection characteristics of filters for larger, more robust, spores (see Chapter 8), rather than a reduced collection capacity of the impingement method.

Other single-stage liquid impingers also are available, such as midget and micro impingers (Figure 9.9), and are small enough for personal monitoring of exposure to microorganisms but do not allow for effects caused by vapor phase rehydration that can occur on inhalation (see Chapter 6). Both are made of Pyrex glass and are, in most respects, smaller versions of the AGI-30, except that the air inlet is straight instead of curved. Volume of the collecting chamber for midget impingers is 25 mL and for micro impingers 2.5 mL, containing 10 and 1 mL collecting fluid, respectively, under normal sampling conditions. Normal sampling rate for midget impingers is 1 L min^{-1} and for micro impingers is 0.1 L min^{-1}. Both can be employed with battery-operated vacuum pumps. Their advantages are similar to those of the AGI-30, with the additional one of their possible use for personal sampling if required (either type can be placed in a holster and clipped to the lapel). In practice, however, the volume of collection fluid in the micro impinger generally may be insufficient to allow the required analyses to be done. Midget impingers are deployed as standard samplers for measurement of some non-biological aerosols, such as acrylamide,[43] but their utilization for biological aerosols has been less well documented. However, Macher and First[44] found them to be as efficient as the AGI-30 in recovering aerosols of bacteria generated in a test chamber. In work environments they are reported to be better than filtration samplers for recovering airborne bacteria.[45] Their inlet

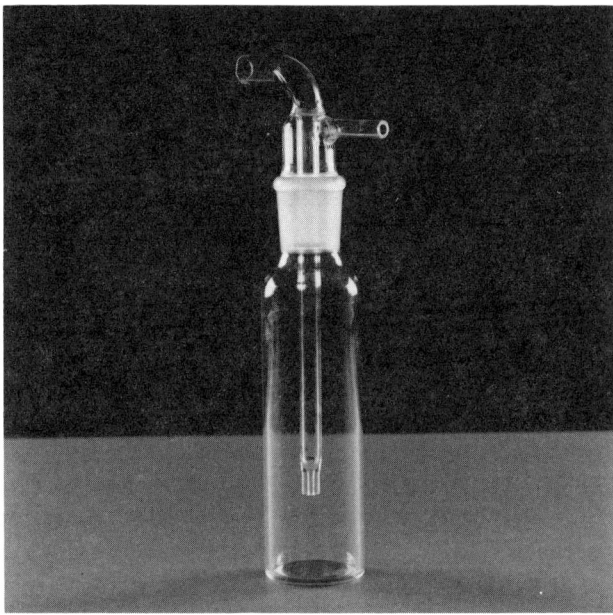

Figure 9.8. AGI-30 liquid impinger. External mains operated vacuum supply required.

efficiency has been found to decrease for particles above 10 μm, but midget impingers also have been used successfully to examine the microbiological content of aerosolized silage samples.[47]

Size Fractionating Impingers

A sampler combining the advantages of collection into different size fractions with collection by impingement into liquid was designed by May.[48] This sampler, known as the multistage all glass liquid impinger (MSLI, Figure 9.10), is available in three sizes collecting at 55, 20 and 10 L min^{-1} by means of a mains vacuum pump, and each separates collected airborne particles into three size fractions. In addition, it was designed to collect microorganisms from the air more gently than the single-stage impingers.

Air drawn into the intake tube of the first stage flows over a sintered glass disc constantly wetted by collection fluid held in the surrounding reservoir. The liquid level must be sufficient just to cover the glass sinter constantly during use, usually 10 mL for each stage. Particles of sufficient mass impact onto the disc and are washed into suspension by the agitated fluid in the reservoir. The airstream passes into a second chamber similar to the first except that the inlet diameter is smaller, thus increasing the air velocity. Further particles are trapped on another wetted sintered glass disc and washed into suspension. Finally, the airstream passes through a tapered jet into the third chamber. The jet curves so that the airstream enters the chamber tangentially, thus impinging the remaining particles into 10 mL of collection fluid swirling under the influence of the airstream around the base of the chamber. Each chamber has a plugged port for introduction and removal of collection fluid. For the 55 L min^{-1} version, the 50% size cutoff for particles in stages 1, 2 and 3 are 6, 3.3 and 0.5 μm, respectively, with an estimated size range for collection of >6 μm in stage 1, 3 to 6 μm in stage 2 and <3 μm in stage 3, which roughly corresponds to upper

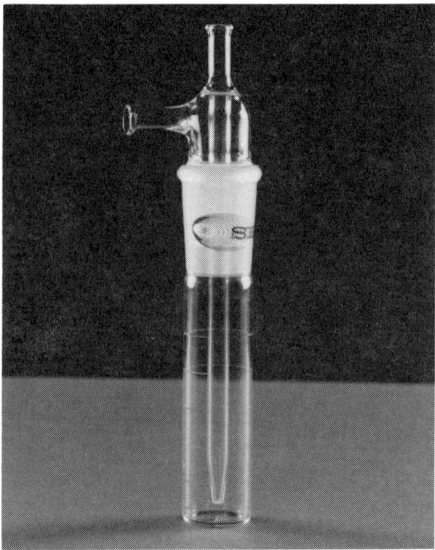

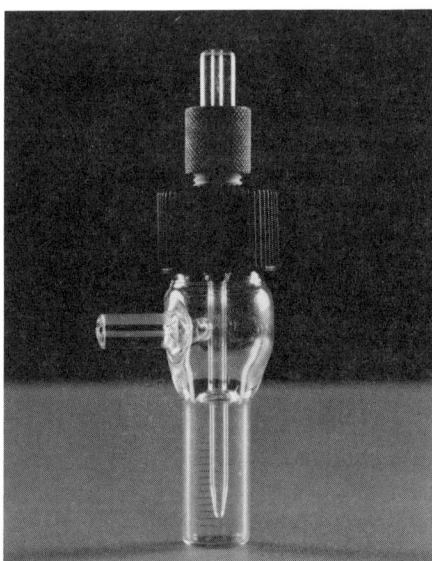

Figure 9.9. Midget (9a) and micro (9b) liquid impingers. Both can be lapel-mounted with suction provided by battery-operated vacuum pump.

respiratory tract, bronchial and alveolar deposition, respectively.[48] Airflow through the sampler is controlled by means of a critical orifice positioned in the outlet from the third stage, consequently collection fluid evaporation is less than in other impingers.

The MSLI is not used as widely as the AGI-30 when impingement is chosen as a method of collection, nor is it often used as an alternative to the Andersen sampler, despite its size fractionating capabilities. However, it has been found to collect with an efficiency similar to the Andersen impactor airborne bacteria in a wastewater treatment plant[49] and airborne bacteria and fungi in waste disposal sites.[50] The major drawback with the MSLI is its construction. It is made out of pyrex glass, so can be sterilized by autoclaving, but

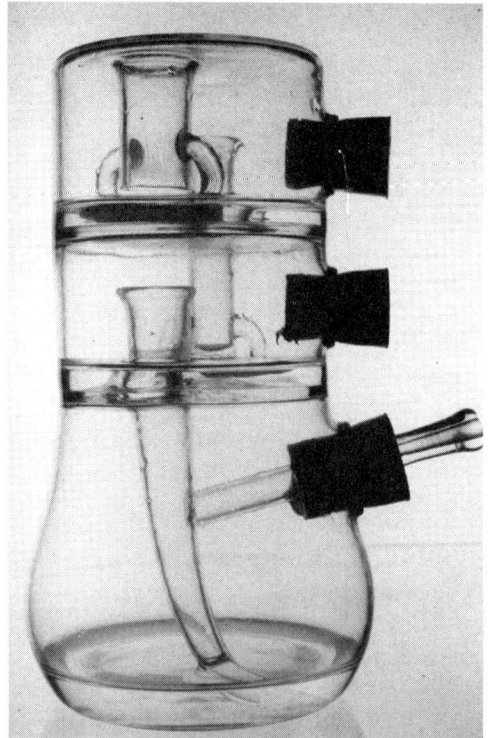

Figure 9.10. Multistage all glass liquid impinger. Collection of bioaerosols into three size fractions. External mains operated vacuum supply required, connected to third stage through a critical flow orifice.

because of its complicated design is difficult to construct accurately. Consequently, there can be much sampler-to-sampler variation in inlet characteristics, size fractionation and therefore collection efficiency. Also, although not expensive, MSLIs are fragile, possibly more so than the AGI-30. This design of sampler may soon be available made from a plastic material, but at this early stage no information of its collection characteristics, compared to the glass version, is available. Attempts have been made to adapt the MSLI so that it can be manufactured from steel or aluminum, to ensure precise construction and to modify it so that the collected size fractions correspond to the ISO/CEN definition of the inhalable, thoracic and respirable fractions.[51] A prototype has been made and characterized in laboratory tests using particles of known size, but is yet to be tested in the field. It will be some time before this sampler would be widely available. Similarly, at present there are no size-fractionating personal sampling liquid impingers available, although a prototype has been designed and currently is being tested in the laboratory.[52]

CENTRIFUGAL SAMPLERS (CYCLONES, CENTRIFUGES)

The principle of collection by centrifugation involves the creation of a vortex in which particles with sufficient inertia leave the airstream to be impacted upon a collection surface. Two examples of this principle are used to collect bioaerosols, the cyclone sampler and the centrifugal air sampler, collecting into liquid and onto semi-solid medium, respectively.

Cyclone Samplers

In this type of sampler, air is introduced tangentially near the top of a cylindrical or inverted conical chamber and withdrawn through an exit port situated centrally at the top. The airstream flows spirally down the chamber following the inner wall, then reverses and spirals up the center of the chamber and out through the exit. Particles of sufficient inertia are deposited on the wall of the cyclone chamber or move to the bottom of the chamber to be retained in a collection vessel.[53] This principle is used frequently in industry on a large scale to remove airborne particles from process air, for example, in chimney stacks or as a form of air conditioning. However, it can also be employed to collect bioaerosols for subsequent analysis. Errington and Powell[54] tested two sizes of cyclone samplers, with flow rates of 75 and 338 L min^{-1} and made of steel or perspex to the above design, and found them to be effective in collecting airborne microorganisms.

Cyclone samplers available commercially for bioaerosol sampling are typically made from pyrex glass with a ground glass Quickfit port at the base of the inverted conical chamber, to which can be attached a collection flask (Figure 9.11). Using a peristaltic pump, liquid is fed into the air inlet to wash particles deposited on the inner walls of the chamber into the collection vessel at the bottom. To ensure adequate wetting of the walls, a detergent such as Tween may be added to the liquid feed. Air sampling rates may range from 75 to 1000 L min^{-1}, depending on the size of cyclone.

There are many advantages to this method of sampling. Viable particles are collected into liquid, which may help to preserve their integrity. They are moistened while still in the airstream as a result of the liquid feed and this may aid preservation further, and to some extent mimic the circumstances for bioaerosols entering the mammalian respiratory tract (see also Chapter 6). Suitable dilution prior to analysis of the particles suspended in collection fluid means that there is no theoretical restriction to sampling time, and a fraction collector can be attached to the collection port to allow stepwise analysis throughout a long sampling period. There is little pressure drop across this type of sampler, therefore a powerful vacuum source is not required, their design is simple and therefore they are cheap to make. However, the theory behind their operation is complex and not well described[55] and in practice the positioning of the liquid feed inlet is critical to ensure complete washdown; also evaporation losses of up to two thirds of the collection fluid can occur.[56] Despite this, they have been used successfully to collect airborne microorganisms, including aerosolized bacteriophages[57] and have been found to collect airborne bacteria with an efficiency similar to that of the AGI-30.[56] They have also proved to be useful when sampling bioaerosols of a non-microbiological, proteinaceous origin.[58,59] However, when compared to slit-to-agar samplers they appeared to decrease in collection efficiency when sampling airborne fungal spores at high (10^5 to 10^6 cfu m^{-3}) concentrations in work environments.[10]

Another application of cyclone devices is for size separation of particles prior to sampling. Personal filtration samplers used for gravimetric dust sampling or for bioaerosols sampling (see Chapter 10), can be fitted with a presampling cyclone so that particles larger than respirable size (>4.25 μm) collect in a trap at the bottom of the cyclone and those of a respirable size collect on the filter. Also, cyclone samplers can be cascaded to provide for particle size fractionation (Chapter 8).

Centrifugal Samplers

The most frequent example of this type is the Reuter centrifugal air sampler, or Biotest RCS biocollector (Figure 9.12). Air is drawn into the sampler by an impeller housed inside

Figure 9.11. 'Aerojet' cyclone sampler. Airborne particles entering tangentially are mixed in the cyclone with liquid fed by peristaltic pump via a needle in the inlet. Liquid collects in a chamber at the base. External mains operated vacuum supply required, connected to the outlet at the top.

an open shallow drum, then accelerated by centrifugal force toward the inner wall of the drum. Lining the inner wall is a plastic strip supporting a thin layer of agar medium, onto which airborne particles are impacted. These strips are subsequently removed and incubated to reveal microbial colonies from deposits. The motor for the impeller is battery operated and the whole sampler unit is small enough to be hand-held. Consequently, this type of sampler has become very popular, especially for testing microbiological air quality in hospitals,[60] but it has also been employed in office and factory settings.[61] However, there has been some concern expressed about its effective flow rate and collection efficiency. Air enters and leaves the sampler through a single opening, therefore it is difficult to quantify the flow rate. The manufacturer's measured sampling rate is calculated from the rate of impeller revolution (4092 rpm), which, it is stated, gives an effective sampling rate of 40 L min^{-1}, but studies have shown that the effective sampling rate may be nearer 100 L min^{-1} [62] and that while it collects larger particles (15 µm) efficiently, less than 10% of particles 2 µm in diameter (therefore approximating to the size of single bacterial cells) were deposited.[63] This therefore must place in question the conclusions from other studies that this sampler performed better than slit-to-agar samplers[64–66] and liquid impingers[65] in recovering airborne bacteria. A more recent study[28] suggested that its efficiency in collecting viable airborne *E. coli* cells was less than 1% of a reference AGI 30 impinger. However, the sampler has recently been redesigned to allow separate entry and exit air flow (RCS + sampler) and this may improve its performance.

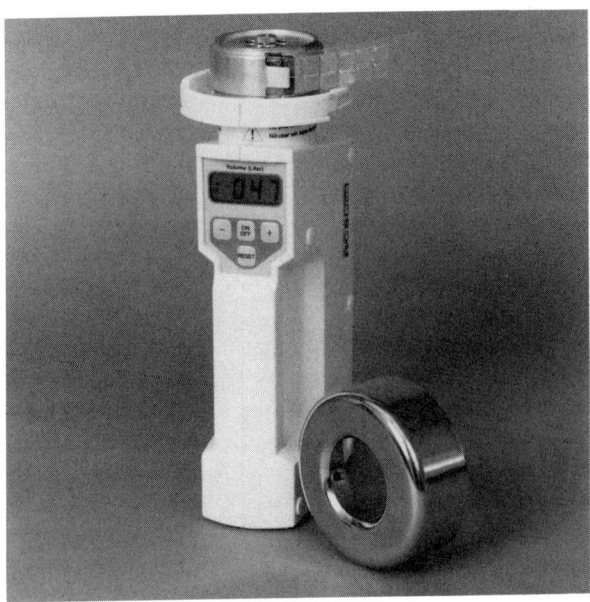

Figure 9.12. RCS Biotest centrifugal sampler. Impaction of airborne particles onto agar media on plastic strips following centrifugal acceleration. Battery operated with internal vacuum source.

PRACTICAL ASPECTS

High Volume Sampling

Most samplers described in this chapter have low to medium air flow rates, ranging from 1 L min^{-1} for personal samplers to 40 L min^{-1} for some of the area samplers. However, in some instances there may be a need to sample at higher flow rates, usually when the bioaerosol is dilute, to overcome the need to sample for extended periods. The drawback is that high-volume sampling is more likely to compromise the integrity of the bioaerosol, thereby leading to under-representation. In many instances high-volume sampling may be inappropriate, especially in enclosed environments, because it may 'strip' the atmosphere of biological particles faster than the rate of bioaerosol generation, again leading to a misrepresentation of the true bioaerosol concentration.

If high-volume sampling is required, then the recommended samplers are the 55 L min^{-1} version of the multistage liquid impinger and cyclone samplers, which have been used at rates up to 1000 L min^{-1} to collect airborne microorganisms. Cyclones are probably the most appropriate for collection of dilute aerosols as, together with the large throughput of air that is possible, the liquid collected can be recycled to concentrate further the suspended particles.

A high-volume cascade impactor collecting airborne particles in five size fractions from >7 μm to <0.5 μm usually onto filters operates at a flow rate of 1.13 m^3/min (40 cfm). This has been used mostly for gravimetric analysis of airborne dust, but potentially could have applications for collecting and characterizing dilute aerosols of allergens if they were sufficiently robust to withstand effects of dehydration caused by high air throughput.

Sampling of Large Particles

Large airborne particles of biological origin usually are pollen grains, large-spored fungi or clumps of spores. Because of their size (>10 μm aerodynamic diameter) they are more easily impacted than small particles and the methods of choice are usually impaction. Pollens and plant pathogenic fungi often are counted and identified by direct microscopy, therefore particles are impacted onto adhesive surfaces for subsequent mounting on microscope slides. Samplers most suited to the collection of large airborne particles therefore are spore traps[6] or rotorod samplers.[8,17]

Other Practical Aspects

It is difficult to compare the "sampling efficiencies" of the various aerobiological samplers available, not only because of the different modes of collection employed, but also because of differences in the nature of bioaerosols. Therefore, a single most efficient sampler is unlikely to be identified. For example, Henningson et al.[67] noted that the relative efficiency of collection of microorganisms differed according to the environment being sampled, especially when Andersen microbial impactors were compared with cyclone samplers collecting into liquid. Particles supporting clumps of cells, which remained intact on agar plates in the Andersen sampler to yield one colony, were disrupted to yield several from cyclone samplers. However, differences between the two types of sampler were greater when sampling in animal barns than in urban environments, suggesting differences in the number of microorganisms supported by a single particle. Therefore, the practical applications of the samplers described in this chapter and in Chapter 10 depend entirely on the bioaerosol to be sampled and an awareness of the advantages or limitations of the samplers employed. For example, a spore trap may allow monitoring of airborne microorganisms continuously over a period of days, but is inappropriate if subsequent cultivation is required, or if the microorganisms of interest have a small aerodynamic diameter. Probably the most crucial factor is the concentration of the bioaerosol, which may dictate the usefulness of a sampling device. For example, impactors such as the Andersen sampler deposit airborne particles onto a finite area; therefore, if this type of sampler is utilized in highly contaminated environments a very short sampling time may be necessary. This can lead to large errors when attempting to collect a sample representative of conditions in the environment over a long period of time. There are similar drawbacks to the use of the centrifugal (RCS) bioaerosol sampler, which has a deposition site much smaller than that of the Andersen viable impactor. Another drawback with these samplers is that if different groups of microorganisms are to be studied, which necessitates the inoculation of different agar media, repeat samples must be taken at each sample point, which increases the sampling time and may result in errors if rapid and large variations occur quantitatively or qualitatively in the airspora. However, size-fractionating impactors such as the Andersen microbial impactor provide valuable information about the size of airborne particles supporting microbial cells, and therefore about their potential for penetration into the human respiratory tract and thus to cause disease, for example, when determining the risks of bacterial aerosols released from domestic waste landfill sites.[68] Also, there are considerable advantages to the collection of airborne microorganisms of interest directly onto the medium upon which it will subsequently be grown, both in terms of minimizing subsequent storage and handling losses and of enhancing survival. In addition, some species of microorganisms, such as actinomycetes, may form colonies from spores collected on plates exposed in the Andersen sampler but may fail to do so on spread plates.[4] Air diluters are available commercially for fitting onto the inlet of a sampler to mix sample air with sterile air in proportions of 1:10 up to 1:1000, but there are no reports of their use in conjunction with microbiological air samplers. Alternatively, as described above, the contents of petri dishes can be homogenized and diluted to use as an inoculum for further agar plates to overcome overloading problems.

For sampling over longer periods and largely independent of airborne concentrations, collection into liquid by impingement or by cyclone should be the methods of choice. However, care must be taken to store samples adequately and to process them as quickly as possible after collection to prevent changes, quantitatively or qualitatively, in the microbial sample. Some sampling devices, such as pyrex glass MSLI or AGI-30, are inexpensive by comparison to many other devices, but their use may be restricted in some instances, e.g., to monitor airborne contaminants in food factories, because of the risk of breakage, as glass contamination of foods is difficult to detect.

To sample sequentially, for known periods over a longer timespan, spore traps are appropriate for larger particles and non-viable recovery, or cyclones can be used with liquid separated in a fraction collector; otherwise, sampling can be done with a series of samplers.

SUMMARY

Most of the samplers designed specifically for the collection of bioaerosols have been described in this chapter. Many different types of sampler exist, but not all are suitable for use in every environment or for collecting all bioaerosols. When choosing a sampler, many factors must be taken into consideration. The ideal sampler is that which is most efficient in recovering all bioaerosols from the environment being sampled and which allows all the required analyses to be performed. In practice, it may not be possible to satisfy all these criteria and a balance must be obtained. No single sampler is appropriate for all sampling conditions and the use of more than one type of sampler may provide complementary results. An awareness of the environment to be sampled and of the types and limitations of the samplers available is required.

REFERENCES

1. Johnston, J.R., A.M. Butchart and S.J. Kgamphe. "A comparison of sampling methods for airborne bacteria," *Environ. Res.* 16:279–284 (1978).
2. Boutin, P., M. Torre, R. Serceau and P.J. Rideau. "Atmospheric bacterial contamination from landspreading of animal wastes: evaluation of the respiratory risk for people nearby," *J. Agric. Eng. Res.* 39:149–100 (1988).
3. Buttner, M.P. and L.D. Stetzenbach. "Evaluation of four aerobiological sampling methods for the retrieval of aerosolized *Pseudomonas syringae*," *Appl. Env. Microbiol.* 57:1268–1270 (1991).
4. Lacey, J. and J. Dutkiewicz. "Isolation of actinomycetes and fungi from mouldy hay using a sedimentation chamber," *J. Appl. Bacteriol.* 41:315–319 (1976).
5. Sayer, W.J., N.M. MacKnight and H.W. Wilson. "Hospital airborne bacteria as estimated by the Andersen sampler versus the gravity settling culture plate," *Am. J. Clin. Pathol.* 58:558–566 (1972).
6. Hirst, J.M. "An automatic volumetric spore trap," *Ann. Appl. Biol.* 39:257–265 (1952).

7. Clarke, A.F. and T. Madelin. "Technique for assessing respiratory health hazards from hay and other source materials," *Equine Veter. J.* 19:442–447 (1987).
8. Solomon, W.R., H.A. Burge and J.R. Boise. "Performance of adhesives for rotating arm impactors," *J. Appl. Clin. Immunol.* 65:467–470 (1980).
9. Bourdillon, R.B., O.M. Lidwell and J.C. Thomas. "A slit sampler for collecting and counting airborne bacteria," *J. Hyg. (Camb)* 14:197–224 (1941).
10. Blomquist, G., G. Strom and L.H. Stromquist. "Sampling of high concentrations of airborne fungi," *Scand. J. Work Environ. Health* 10:109–113 (1984).
11. Lundholm, I.M. "Comparison of methods for quantitative determinations of airborne bacteria and evaluation of total viable counts," *Appl. Env. Microbiol.* 44:179–183 (1982).
12. Verhoeff, A.P., J.H. van Wijnen, J.S.M. Boleij, B. Brunekreef, E.S. van Reenen Hoekstra and R.A. Samson. "Enumeration and identification of airborne viable mould propagules in houses," *Allergy* 45:275–284 (1990)
13. Verhoeff, A.P., J.H. van Wijnen, P. Fischer, B. Brunekreef, J.S.M. Boleij, E.S. van Reenen Hoekstra and R.A. Samson. "Presence of viable mould propagules in the indoor air of houses," *Toxicol. Ind. Health* 6:133–145 (1990).
14. Decker, H.M., R.W. Kuehne, L.M. Buchanan and R. Porter. "Design and evaluation of a slit-incubator sampler," *Appl. Microbiol.* 6:398–400 (1958).
15. Dahlgren, C.M., H.M. Decker and J.B. Harstad. "A slit sampler for collecting T-3 Bacteriophage and Venezuelan Equine Encephalomyelitis virus. I. studies with T-3 bacteriophage," *Appl. Microbiol.* 9:103–105 (1961).
16. Kuehne, R.W. and W.S. Gochenour. "A slit sampler for collecting T-3 Bacteriophage and Venezuelan Equine Encephalomyelitis virus. II. studies with Venezuelan Equine Encephalomyelitis virus," *Appl. Microbiol.* 9:106–107 (1961).
17. Kozak, P.P., J. Gallup, L.M. Cummins and S.A. Gillman. "Currently available methods for home mould surveys; description of techniques," *Ann. Allergy* 45:85–89 (1980).
18. Buttner, M.P. and L.D. Stetzenback. "Monitoring airborne fungal spores in an experimental indoor environment to evaluate sampling methods and the effects of human activity on air sampling," *Appl. Environ. Microbiol.* 59:219–226 (1993).
19. May, K.R. "The cascade impactor: an instrument for sampling coarse aerosols," *J. Phys. Eng. J. Sci. Instr.* 22:187–195 (1945).
20. May, K.R. "An ultimate cascade impactor for aerosol assessment," *J. Aerosol Sci.* 6:413–419 (1975).
21. Rubow, K., V. Marple, J. Odin and M.A. McCawley. "A personal cascade impactor: design, evaluation and calibration," *Am. Ind. Hyg. Assoc. J.* 48:532–538 (1987).
22. Crook, B., P. Griffin, M.D. Topping and J. Lacey. "An appraisal of methods for sampling aerosols implicated as causes of work-related respiratory symptoms," in *Aerosols, Their Generation, Behaviour and Applications,* (London, The Aerosol Society: 1988), pp 327–333.
23. Macher, J.M. and H.C. Hansson. "Personal size separating impactor for sampling microbiological aerosols," *Am. Ind. Hyg. Assoc. J.* 48:652–655 (1987).
24. Gordon, S., R.D. Tee, D. Lowson, J. Wallace and A.J. Newman-Taylor. "Reduction of airborne allergenic urinary protein from laboratory rats," *Brit. J. Ind. Med.* 49:416–422 (1992)
25. Andersen, A.A. "New sampler for the collection, sizing and enumeration of viable airborne particles," *J. Bacteriol.* 76:471–484 (1958).
26. Gillespie, V.L., C.S. Clark, H.S. Bjornson, S.J. Samuels and J.W. Holland. "A comparison of 2-stage and 6-stage Andersen Impaction for viable aerosols," *Am. Ind. Hyg. Assoc. J.* 42:858–864 (1981).
27. Macher, J.M. "Positive hole correction of multiple jet impactors for collecting viable microorganisms," *Am. Ind. Hyg. Assoc. J.* 50:561–568 (1989).

28. Jensen, P.A., W.F. Todd, G.N. Davis and P.V. Scarpino. "Evaluation of eight bioaerosol samplers challenged with aerosols of free bacteria," *Am. Ind. Hyg. Assoc. J.* 53:660–667 (1992)
29. Lach, V. "Performance of the surface air system air samplers," *J. Hosp. Inf.* 6:102–107 (1985).
30. Jones, W., K. Morring, P. Morey and W. Sorensen. "Evaluation of the Andersen viable impactor for single stage sampling," *Am. Ind. Hyg. Assoc. J.* 46:294–298 (1985).
31. Smid, T., E. Schokkin, J.S.M. Boleij and D. Heoderik. "Enumeration of viable fungi in occupational environments: a comparison of samplers and media" *Am. Ind. Hyg. Assoc. J.* 50:235–239 (1989).
32. Solomon, W.R. and J.A. Gilliam. "A simplified application of the Andersen sampler to the study of airborne fungus particles," *J. Allergy* 45:1–13 (1970).
33. Peto, S.A. and E.O. Powell. "The assessment of aerosol concentration by means of the Andersen sampler," *J. Appl. Bacteriol.* 33:582–598 (1970).
34. Greenberg, L. and G.W. Smith. "A new instrument for sampling aerial dust," U.S. Bureau of Mines RI 2392 (1922).
35. Henderson, D.W. "An apparatus for the study of airborne infection," *J. Hyg. Camb.* 50:53–68 (1952).
36. May, K.R. and G.J. Harper. "The efficiency of various liquid impinger samples in bacterial aerosols," *Brit. J. Ind. Med.* 14:287–297 (1957).
37. Shipe, E.L., M.E. Tyler, D.M. Chapman. "Bacterial aerosol samplers. Development and evaluation of the Shipe sampler," *Appl. J. Hyg.* 53:337–354 (1959).
38. Brachmann, P.S., R. Ehrlich, H.F. Eichenwald, V.J. Gabelli, T.W. Kethley, S.H. Madin, J.R. Maltman, G. Middlebrook, J.D. Morton, I.H. Silver and E.K. Wolfe. "Standard sampler for assay of airborne microorganisms," *Science* 14, 1295 (1964).
39. Ehrlich, R., S. Metler and L.S. Idoine. "Evaluation of slit sampler in quantitative studies of bacterial aerosols," *Appl. Microbiol.* 14:328–330 (1966).
40. Thorne, H.V., and T.M. Burrows. "Aerosol sampling methods for the investigation of foot and mouth disease and the measurement of virus penetration through aerosol filters," *J. Hyg. Camb.* 58:409–417 (1960).
41. Lembke, L.L., R.N. Kniseley, R.G. van Nostrand and M.D. Hale. "Precision of the all glass impinger and the Andersen microbial impactor for air sampling in solid waste handling facilities," *Appl. Environ. Microbiol.* 42:222–225 (1981).
42. Silas, J.C., M.A. Harrison, J.A. Carpenter and J.B. Floyd. "Comparison of particulate air samplers for detection of airborne *Aspergillus flavus* spores," *J. Food Prot.* 49:236–238 (1986).
43. Health and Safety Executive. "MDHS 57: Acrylamide in air - laboratory method using high performance liquid chromatography after collection in an impinger containing water," HSE Publications, Sudbury, UK (1987).
44. Macher, J.M. and M.W. First. "Personal air samplers for measuring occupational exposure to biological hazards," *Am. Ind. Hyg. Assoc. J.,* 45:76–83 (1984).
45. Travers, S.A., B. Crook and P. Griffin. "Microbially contaminated oil mists in engineering works," in *Airborne Deteriogens and Pathogens* (London, Biodeterioration Society, Occasional Publication 6: 1989) pp 69–73.
46. Lyons, C.P. "Sampling efficiencies of all-glass midget impingers," *J. Aerosol Sci.* 23, Suppl. 1:s599–s602 (1992).
47. Dutkiewicz J., S.A. Olenchock, W.G. Sorenson, V.F. Gerencser, J.J. May, D.S. Pratt, V.A. Robinson. "Levels of bacteria, fungi and endotoxin in bulk and aerosolised corn silage," *Appl. Environ. Microbiol.* 55:1093–1099 (1989).
48. May, K.R. "Multistage liquid impinger," *Bact. Rev.* 30:559–570 (1966).
49. Zimmerman, N.J., P.C. Reist and A.G. Turner. "Comparison of two biological aerosol sampling methods," *Appl. Environ. Microbiol.* 53:99–104 (1987).

50. Crook, B., S. Higgins and J. Lacey. "Methods for sampling airborne microorganisms at solid waste disposal sites" in *Biodeterioration 7*. D.R. Houghton, R.N. Smith and H.O.W. Eggins, Eds., (London, Elsevier: 1987) pp 791–797.
51. Soderholm, S.C. "Proposed International Conventions of particle size - selective sampling," *Ann. Occup. Hyg.* 33:301–320 (1989).
52. Bradley, D., G.D. Burdett, W.D. Griffiths and E.P. Lyons. "Design and performance of size selective microbiological samplers," *J. Aerosol Sci.* 23 Suppl.1:s659–s662 (1992).
53. Ranz, W.E. "Wall flows in a cyclone separator: a description of internal phenomena," *Aerosol Sci. and Technol.* 4:417–432 (1985).
54. Errington, F.P. and E.O. Powell. "A cyclone separator for aerosol sampling in the field," *J. Hyg. (Camb.)* 67:387–399 (1969).
55. Hering, S.V. "Inertial and gravitational collectors," in *Air Sampling Instruments for Evaluation of Atmospheric Contaminants*. (ACGIH: 1989, 7th ed.) pp 339–385.
56. White, L.A., D.J. Hadley, D.E. Davids and R. Naylor. "Improved large volume sampler for the collection of bacterial cells from aerosol," *Appl. Microbiol.* 229:335–339 (1975).
57. Wheeler, D., H.E. Skilton and R.F. Carroll. "The use of bacteriophage as tracers of aerosols liberated by sludge suction appliances," *J. Appl. Bacteriol.* 65:377–386 (1988).
58. Behizad, M., R.R. Cumming, and F.J. Powell. "An on-line detection system for aerosols of proteolytic enzymes," *Ann. Occup. Hyg.* 32:499–508 (1989).
59. Griffin, P., F.P. Roberts and M.D. Topping. "Measurement of airborne antigens in a crab processing factory," in Proc. 7th Int. Symp. on Inhaled Particles, Edinburgh (1994), in print.
60. Casewell, M.W., N. Desai and E.J. Lease. "The use of the Reuter centrifugal air sampler for the estimation of bacterial air counts in different hospital locations," *J. Hosp. Inf.* 7:250–260 (1986).
61. Grillot, R., S. Parat, A. Perdrix and J. Croize. "Contamination of air systems: 1987–1988 assessment and prospects of the Grenoble intervention group," *Aerobiologia* 6:58–65 (1990).
62. Clark, S., V. Lach and O.M Lidwell. "The performance of the Biotest RCS centrifugal air sampler," *J. Hosp. Inf.* 2:181–186 (1981).
63. Macher, J.M. and M.W. First. "Reuter centrifugal air sampler: measurement of effective airflow rate and collection efficiency," *Appl. Environ. Microbiol.* 45:1960–1962 (1983).
64. Placencia, A.M., J.T. Peeler, G.S. Oxborrow and J.W. Danielson. "Comparison of recovery by Reuter centrifugal air sampler and slit to agar sampler," *Appl. Environ. Microbiol.* 44: 512–513 (1982).
65. Delmore, R.P. 1981 and Thompson W.N. "A comparison of air-sampler efficiencies," *Med. Device Diagn. Instrum.* Feb:45–53 (1981).
66. Nakhla, L.S. and R.F. Cummings. "A comparative evaluation of a new centrifugal air sampler (RCS) with a slit air sampler (SAS) in a hospital environment," *J. Hosp. Infec.* 2: 261–266 (1981).
67. Henningson, E., R. Roffey and A. Bovallius. "A comparative study of apparatus for sampling airborne microorganisms," *Grana* 20:155–159 (1981).
68. Rahkonen, P., E. Matti, M. Laukkanen and M. Salkinoja-Salonen. "Airborne microbes and endotoxins in the work environment of two sanitary landfills in Finland," *Aerosol Sci. Technol.* 13:505–513 (1990).

CHAPTER **10**

Non-Inertial Samplers: Biological Perspectives

B. Crook

INTRODUCTION

This chapter is concerned with the collection of bioaerosols and describes those samplers that do not rely on the inertia of the airborne particle for their operation; consequently, the way in which particles are collected is less dependent on their size (see also Chapter 8). Methods for collecting bioaerosols that are included in this group are filtration, electrostatic precipitation and thermal precipitation.

FILTRATION

Sampling by filtration is by far the most widely used method for collecting non-biological airborne particles, such as inorganic dusts and fibers. Consequently, there is a wide range of sampling devices and collection media, some of which have been used or potentially could be used to collect bioaerosols. The greatest advantage of filtration as a sampling method is that the equipment needed is simple and relatively inexpensive compared to many other sampling devices; also, sample collection and handling is straightforward. The sampling equipment required often is less bulky and more robust than that needed for other methods and samples can be taken continuously for long periods either at fixed points or by personal sampling. However, for bioaerosols, filtration poses two major disadvantages. Firstly, dehydration effects (Chapter 6) are caused by the large volume of air that may pass over a particle after its deposition on an essentially dry collection medium. Secondly, for subsequent analysis, it may be necessary to remove deposited material from the filters, and for some types of filter consistent and efficient recovery of deposits may be difficult. However, the latter may be overcome by the correct choice of filter, and for more robust bioaerosols (such as fungal and actinomycete spores, or aeroallergens present as components of dry dusts) dehydration during sampling may not significantly affect viability or integrity. But, electrostatic factors remain. There follows a summary of the types of filtration media and sampling devices commonly available.

FILTRATION MEDIA

The ability of filters to trap electrically neutral airborne particles can be predicted from their pore size and the material from which the filter is made. Filtration media may be divided broadly into three categories: fibrous filters, with a mesh of material whose fibers are randomly oriented; membrane filters, comprising a gel with interconnected pores of uniform size; and flat or etched membranes in which there are defined holes or pores. Liu et al.[1] have comprehensively reviewed the types of filtration media available. In this chapter, those filters used, or most applicable to, bioaerosol sampling will be described.

Each type of filter will trap particles smaller than the actual pore size because particles may strike the edge of a pore, thereby being intercepted. This effect is taken into consideration when the pore sizes of depth type filters (i.e., those that trap particles within the filter matrix—fibrous and membrane filters) are defined.

Fibrous Filters

Filtration media of this type include paper/cellulose and glass fiber. The fibers form a random, criss-cross matrix through which the particle-laden airstream is directed. The pore size of fibrous filtration media have a range of sizes, of which the mean is the defined pore size. The route of the airstream through the filter is tortuous and particles may pass part of the way through a filter before entrapment. Thus, the trapping capability of this type of filter is enhanced by retention of particles within its depth.

Glass fiber is the most widely used filtration medium for gravimetric airborne dust sampling because it is robust and inert. However, to recover airborne microorganisms it has been used only infrequently because microbial cells may be trapped in the fibrous matrix, leading to possible inconsistencies in recovery. Non-microbial, proteinaceous aeroallergens may be present in the air in minute quantities and may therefore need concentration from a large volume of air before analysis. Large volume sampling therefore is appropriate. Glass fiber filters can sample at high air flow rates and have been used extensively to collect soluble proteinaceous aeroallergens.[2,3]

Membrane Filters

Membrane filters are formed from cellulose ester (cellulose acetate or cellulose nitrate), PVC, PTFE, nylon or gelatin gels mixed with solvents then formed into a thin film. The pores are created by evaporation of the solvent. Again, the airstream follows a tortuous route through the filter and particles may be trapped within the filter as well as on the surface. Cellulose ester filters are employed widely for microbiological analysis, especially of liquids, while porous gelatine filters have successfully collected airborne microorganisms.[4,5]

Nylon filters allow concentration of microbial cells from water samples for *Legionella* analysis, for example, because the filter material is mechanically strong, can be disinfected in boiling water and particles can be resuspended easily from the filter surface.[6] For these reasons, they would be appropriate also for collecting airborne microorganisms, but their use as such has not been documented.

Flat Filters

Flat filters are made of polymer or metal through which holes have been etched. Examples include silver, inorganic aluminum oxide (Anopore™), and polycarbonate (Nuclepore™) membrane filters. Polycarbonate filters are made by extruding polycarbonate as a thin continuous sheet, then bombarding it with neutrons. Tracks formed by neutrons passing through the material are then acid-etched to produce separate, parallel, straight-through holes of a uniform and defined size. In this type of filter, particles are collected on the surface rather than being trapped within it. For this reason, polycarbonate filters are the most common filtration media for sampling bioaerosols, especially microorganisms. Anopore filters also have a defined pore size and a flat collection surface.

Filter Holders

Filter holders are essentially a rigid support for the filtration medium, connected to a flow rate gauge and vacuum source. However, the physical configuration of filter holders can affect the ability of airborne particles to reach the filter and to be deposited. A wide range of designs of filter holders are available commercially for collection of airborne particles in general and therefore for collecting bioaerosols. For most well characterized filtration samplers, the inlet has been designed to collect particles within a defined size range. Open-faced holders, however, have less control over the airstream reaching the filter, consequently they collect particles over a wide range of sizes.

For general area sampling, simple medium to large flow rate samplers consist mostly of an open-faced filter holder, with a suitable filter, connected to a calibrated flow meter then to a vacuum pump or air mover, sometimes with adjustable flow rate. Mid-range samplers collect typically at 30 to 70 L min^{-1} onto a 37 to 50 mm diameter filter, while large volume samplers collect at rates up to 1 m^3 min^{-1} onto 200 × 250 mm filters. The last type of sampler has been employed extensively to sample bioaerosols such as proteinaceous materials, especially allergens,[2] as has a lower flow rate sampler,[3] operated at 180 L min^{-3}.

Smaller filtration samplers, with lower flow rates (1 to 4 L min^{-1}), are primarily for collecting aerosols from the breathing zone and are connected to battery operated portable vacuum pumps. For gravimetric dust sampling in the breathing zone, a 25 mm diameter filtration sampler with a single inlet is appropriate, but in the UK this is now largely superseded as a method of choice by the UKAEA seven-hole sampling head, in which the airstream passes through a screen with seven 4 mm diameter holes arranged symmetrically before reaching the 25 mm diameter filter. This method collects the inhalable fraction of the total airborne particulate matter and is recommended in the UK for gravimetric analysis of airborne dust.[7] Sampling is at a rate of 2 L min^{-1} onto a glass fiber filter (Figure 10.1). Another sampler (IOM) has been developed recently for gravimetric analysis of inhalable airborne dusts.[8] Here, a 25 mm diameter glass fiber filter is mounted in a removable cassette with a 15 mm diameter inlet. Both are appropriate for collecting bioaerosols (Figure 10.2).

For gravimetric analysis of airborne particles using personal samplers, size selective sampling can be done using a cyclone separator.[7] Airborne particles drawn through the inlet at 1.9 L min^{-1} pass through a cyclone (described in Chapter 9) which removes larger particles and allows only those of respirable size to reach the 25 mm diameter filter (Figure 10.3).

A disposable polypropylene filter holder, often referred to as an aerosol monitor, is used commonly to collect bioaerosols, particularly microorganisms. This is made up of three sections; a filter support with air outlet, a cylinder that acts as a cowl for the filter while forming the main chamber of the monitor, and a top. The sampler can be open-faced with the top removed or closed-face by opening a 4 mm plugged inlet in the center of the top. The sampler is available for two sizes of filters, 37 and 25 mm diameter, and with 12 mm or 50 mm length cowls. The 25 mm diameter version also is available made from black, carbon-filled polypropylene to improve electrical conductivity and overcome

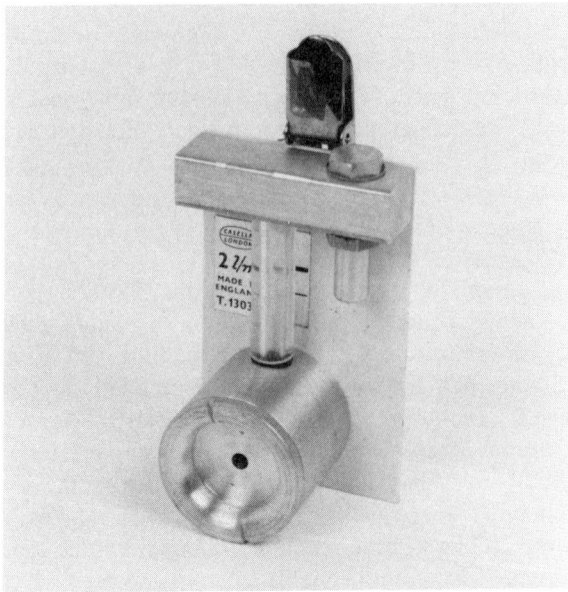

Figure 10.1. Single-hole (1a) and UKAEA seven-hole (1b) filtration samplers. Lapel mounted with suction provided by a battery operated vacuum pump.

problems, such as uneven deposition of particles on filters and on internal walls, caused by electrostatic effects (Figure 10.4). In some designs, the shape of the outlet is different which, according to the manufacturer, also gives a more even deposition of particles on the filter. Air monitor cassettes are approved by NIOSH and EPA in the United States for airborne particulate sampling, but are not recommended in the UK. Some concern has been expressed about the low inlet efficiencies and possible electrostatic effects. Used open-faced (i.e., with the top section removed) the sampler cassette was found to oversample when facing into the wind but to undersample when the inlet was angled away from the wind.[9] They have been shown in experimental trials to be more efficient when used closed-faced than open-faced, where their collection efficiency was found to drop for larger particles, but this was improved when the sampler was placed on a mannikin (i.e., when personal sampling was simulated[10]). To collect particles of respirable size, a cyclone separator similar to that described above is available for attachment to the open-faced aerosol monitor cassette.

For microbiological air sampling, aerosol monitor cassettes have the advantage that the filter and holder form an integral, disposable unit that can be assembled and kept sterile before use. After exposure the unit, with collected deposit, may be resealed for transport to the analytical laboratory. As will be described in greater detail later, liquid then can be introduced into the chamber to recover deposits from the filter, without the risk of losses

Figure 10.2. IOM filtration sampler. Lapel mounted with suction provided by a battery operated vacuum pump.

or cross-contamination that could arise if the filter were removed before analysis. If particles are recovered in this way, deposits also will be taken from inside the monitor chamber. Consequently, any airborne particle that passes through the sampling inlet will be recovered, irrespective of whether it has been deposited on the filter or whether electrostatic charges caused it to be deposited on the inner walls of the sampler. Therefore, the only question that remains is whether electrostatic charges affect inlet efficiency. It is likely that they may. However, the IOM sampler, which is currently used for gravimetric dust analysis and is well characterized, could also be used in a similar way as it has a removable cassette filter.

ANALYSIS

Particles collected onto filters may be analyzed *in situ* by microscopy, or may be removed from the filter and analyzed as described previously (Chapters 8 and 9).

Microscopy

Airborne microorganisms captured onto filters may be analyzed by light, fluorescence or scanning electron microscopy. Some filters can be rendered transparent for light microscopy, e.g., cellulose acetate filters, if mounted in glycerol triacetate, or polycarbonate filters if mounted in microscope immersion oil. Collected particles of biological origin, (e.g., some types of bacterial cells, fungal spores and pollen grains), then can be recognized by their size, shape, color and aggregation, and by reference to standard texts.[11-12] Values for total numbers of collected particles can be derived from direct counting, correcting for the area counted, the area of sample deposition and the volume sampled, to obtain a value that may be expressed as particles per unit volume of air sampled. To maximize counting precision, formulae have been derived to include allowances for aggregation of particles.[13]

Figure 10.3. Filtration sampler with cyclone preseparator. Removes thoracic fraction of inhalable particles to allow collection on filter of respirable fraction only. Lapel mounted with suction provided by a battery operated vacuum pump.

Direct light microscopy, as described above, enables a total value to be derived for deposited cells, but does not differentiate between viable and non-viable cells. Techniques are available, involving specific staining and light microscopy, that a differentiation.[14]

Microbial deposits, preferably on flat filters such as polycarbonate, can be stained with a fluorescent dye, then mounted in a suitable, non-fluorescent immersion oil such as Cargill oil type A, on a microscope slide. When viewed with an incident fluorescence (epifluorescence) microscope having an ultraviolet mercury vapor light source and equipped with suitable excitation filters, microbial cells thus treated produce a fluorescent image. Fluorescein diacetate permits staining, observation and enumeration of microorganisms,[15] but the fluorescence produced has a limited lifespan. A more consistent fluorescent stain, with a longer lifespan (stained cells will maintain their fluorescence for at least 12 months), is Acridine orange.[16] Cells collected from water samples onto black-dyed Nuclepore filters and stained with Acridine orange can be differentiated from the dark background, sized, typed according to cell shape, and classified as viable or non-viable according to their fluorescent color dependent on the relative proportion of DNA and RNA nucleic acids present in the cells (orange fluorescence indicates viable and green non-viable cells).[17] This technique has been developed as a rapid test for microbial contamination in food and dairy products[18] and has been adapted successfully as an analysis method for airborne microorganisms.[19-20] Another possible application of filtration sampling plus u.v. fluorescence microscopy is for specific detection of collected airborne cells conjugated to fluorescent-labeled antibodies.[21] More recently, aluminum oxide (Anopore) filters have been shown to yield bacterial counts one third greater than those on polycarbonate filters when utilized for direct epifluorescent counting of cells recovered from liquids,[22] but there is no record of their use for sampling airborne microorganisms.

Microbial deposits on polycarbonate filters can be mounted on aluminum stubs, sputter-coated with gold, then analyzed by scanning electron microscopy. Once again, as well as counting, some identification is possible from cell shape, size and structure and this technique successfully permitted the analysis of collected airborne microorganisms, especially fungal spores.[23-25]

Figure 10.4. Plastic 'aerosol monitor' samplers. Various configurations; 37 mm dia. filter with and without extended cowl, 25 mm dia. filter with holder made from carbon filled non-conducting polypropylene. Can be used open- or closed-face, lapel mounted with suction provided by a battery operated vacuum pump.

Cultivation

Microbial deposits on filters may be cultivated either by placing them onto the surface of agar plates or by washing deposits from them to form aqueous suspensions that, with appropriate dilution, can provide inocula. Cellulose acetate membrane filters are used extensively to collect particulates from water and food for microbiological analysis. Here, the microbial deposit is likely to be small and the filter can be placed directly onto an agar plate or a nutrient-soaked pad, then incubated to develop colonies from the deposited cells. For aerobiological sampling, especially in a contaminated environment, many more microorganisms are likely to be deposited and, to avoid overgrowth of colonies, it is better to remove cells from the filters for analysis (see below). With cellulose acetate filters, some microbial cells may be trapped within the pore matrix and recovery therefore may be inconsistent. Gelatine filters overcome this problem. After sampling, they can be placed directly onto the surface of agar plates, where they melt and are absorbed by the agar medium. Alternatively, they can be melted in an isotonic solution and the resulting suspension used as an inoculum.

Airborne particles may be washed from the surface of filters, especially flat filters such as polycarbonate membranes, by transferring them from the sampling device into a sealed container, adding an appropriate volume of wash liquid then agitating to resuspend the deposit. Alternatively, with sealable filter holders such as the aerosol monitors described earlier, the deposit can be recovered into suspension by washing from the filter surface while it is still in position in the filter holder. For microorganisms, an aqueous suspending fluid such as peptone water or mineral salts solution (quarter strength Ringers) with a simple sugar such as inositol added,[26] maximizes survival of the collected viable cells. A wetting agent, such as Tween 80, may be included to facilitate resuspension of deposits from filters. The resulting suspension then can be diluted and known volumes used to inoculate agar plates, or prepared for direct counting by fluorescence microscopy, as described above.

Airborne Allergens

Airborne allergens can be collected on filters, then recovered into an extracting solution, such as phosphate buffered saline in which the proteinaceous allergen material dissolves.[2] Recovery of material from the filter can be maximized either by elution or by soaking or macerating the filter in the extraction buffer.[2,27,28] The resulting solution can be concentrated, if necessary, by freeze-drying, then used in immunoassays with sera from exposed people to determine allergic response to the aeroallergen.[2] Most aeroallergen sampling uses high-volume samplers and glass fiber filters; however, PTFE filters are a successful alternative to glass fiber filters.[29] Also, where more sensitive immunoassays allow samples to be taken from smaller volumes of air, PTFE membranes in personal sampling devices can be operated at 2 L min^{-1}.[30] These filters are less bulky and proteins can be eluted by placing them with the buffer into the barrel of a syringe, then compressed after soaking to recover the material in solution.[31]

RELATIVE PERFORMANCE OF FILTRATION SAMPLING

Most comparative evaluations of filtration sampling for bioaerosols have concentrated on collecting microorganisms onto Nuclepore polycarbonate filters using disposable plastic aerosol monitor cassettes. Their performance was found to be comparable to that of slit-to-agar samplers (described in Chapter 9) when collecting low concentrations of airborne microorganisms in fairly clean environments.[32] When polycarbonate filter sampling was employed for direct counting of collected airborne fungal spores in agricultural environments, yields were greater than those for cultivation and viable counting, which reflected the difference between viable and non-viable cells. Both had greater yields than from impaction sampling directly onto agar by Andersen sampler, which reflected the difference between counts from intact particles impacted onto agar, in which a clump of cells may yield only one colony, and particles disrupted when washed from filter surfaces to yield more than one colony when subsequently cultivated.[24,25] These results were confirmed in comparisons between the performance of polycarbonate filters and Andersen samplers for collecting airborne microorganisms in domestic waste handling sites.[26] Further comparisons in this study showed collection characteristics for polycarbonate filters and multistage liquid impingers to be similar. In other studies, the filtration method was shown to be comparable to slit-to-agar samplers and cascade impactors for collecting airborne fungi in contaminated work environments. These studies also noted the ease of use of the filtration method from a practical point of view.[20,33]

When gelatine filters collected airborne microorganisms, their collection efficiency was found to be similar to that for collection onto cellulose acetate membrane filters.[34] However, they were considered to be unsatisfactory for collecting airborne bacteria because, during extended sampling, the gelatine dried out, which placed dehydration stresses on the collected microorganisms. On the other hand, if collecting aerosols where the air humidity is high, it is possible that the structure of the gelatine membrane becomes distorted and alters the collection characteristics.

ELECTROSTATIC PRECIPITATION

For this method of collection, particles are separated from the airstream by electrical rather than inertial forces (see Chapter 3), therefore collection is achieved with little pressure drop and a relatively small amount of power is needed. A vacuum source draws air into the sampler, housing a high voltage field, or corona discharge. As particles in the airstream pass through the corona they acquire an electrical charge, causing them to be attracted to a grounded or oppositely charged metal collection surface. This method of sampling collects bioaerosols into liquid or onto a thin flat material such as filter paper. Samplers of the first type are no longer available commercially, but are described here to illustrate the principles of this method of collection.

Collection into Liquid

In the Litton-type large volume electrostatic air sampler (LVAS), the collection surface consists of a rotating aluminum disc into the center of which liquid is delivered by a peristaltic pump. The liquid flows across the surface of the disc as a thin film, taking with it the deposited particles, is collected in a channel at the edge of the disc and delivered by a second peristaltic pump into a collection reservoir. This sampler is capable of collecting at an air flow rate of 1000 L min^{-1} (Figure 10.5). The LEAP sampler collects by electrostatic precipitation in the same way at flow rates ten times greater.[35]

Collection onto Solid Surface

Another type of electrostatic sampler (LVAS, Model M3-A, Sci. Med. Inc., Eden Prairie, MN) is divided into two sections, one where airborne particles acquire electrical charge and another where they are collected. Airborne particles drawn through the sampler at 4 to 10 L min^{-1}, by means of a separate vacuum supply, are subjected to positive ions generated by a corona discharge from a thin wire. The particles, now positively charged, pass into the collecting chamber, on the upper surface of which is a metal plate with positive voltage. This positive voltage pulses on and off, causing the particles to be driven to a collecting plate on the lower surface. The part of the cycle when the voltage is switched off allows the chamber to be filled once again with charged particles. Consequently, the sample volume is independent of the flow rate through the sampler and dependent on the number of precipitating cycles. On the collecting area, measuring 6 × 18 cm, can be placed paper or glass fiber filters, glass or metal slides, or coated EM grids.

Analysis

Bioaerosols collected into liquid using electrostatic precipitation can be treated in the same way as other aqueous suspensions from samplers, as described in Chapter 9. For microbiological analysis the sample can be diluted for inoculating agar plates, prepared for microscopic analysis or for biochemical assays. The suspension also can be prepared for immunoassays.[36] Airborne particles electrostatically precipitated onto solid surfaces can be observed microscopically, or they can be eluted to form a suspension analyzed as above.[37]

Figure 10.5. Large volume air sampler (LVAS) electrostatic precipitator. External (a) and internal (b) features. Air drawn through the inlet at the top, drawn by an integral mains operated vacuum source, passes through a corona discharge to deposit particles on a rotating disc in the upper chamber (seen in b). Liquid fed by peristaltic pump in lower chamber to center of disc takes suspended particles to channel at perimeter, from where it is pumped to a collection vessel positioned in the lower chamber.

Performance

The LVAS, collecting into liquid, was compared with all-glass impingers and Andersen impactors and found to be the only one of the three types capable of recovering measurable quantities of airborne rabies virus.[38] It has also been deployed successfully to recover airborne human respiratory disease viruses.[39] The combination of collection from high volumes of air and collection into liquid was obviously a significant factor. Other studies have shown the usefulness of the LVAS for recovering airborne bacterial cells,[40] bacteriophages[41] and fungal spores.[37] They also compared favorably with the more widely used filtration methods for collecting non-microbial, proteinaceous aerosols.[36,42] A possible drawback of electrostatic precipitation is that the corona discharge produces oxides of nitrogen and ozone that may adversely affect viability of some airborne microorganisms.[41,43]

Figure 10.5 (continued).

THERMAL PRECIPITATION

Particles can be collected by passing dust-laden air through a narrow channel in which there is a temperature gradient perpendicular to the direction of the airstream (see Chapter 3). Airborne particles entering this temperature gradient move in the direction of decreasing temperature (a phenomenon known as thermophoretic motion).[44] To collect airborne microorganisms, a device was designed using this principle in which air was drawn into a chamber between two metal plates. The upper plate was heated to 125°C while the lower plate was cooled by a circulating water heat exchanger. As the aerosol passed through the chamber, particles moved away from the heated plate and were deposited on the lower cooled plate. Microscopic examination of collected deposits was possible if a glass microscope coverslip provided the collection surface, or if a filter paper collected deposits this could be transferred to the surface of an agar plate to allow colonies to grow from deposited microbial cells.[45,46]

If sufficient thermal gradient is established in the sampling chamber, this type of sampler is very efficient at collecting small particles, i.e., 5 µm to <0.01 µm.[47] It proved effective in recovering airborne bacterial cells, operating at an efficiency similar to an AGI-30 impinger and greater than that of a midget impinger.[46] Like other non-inertial samplers, air flows freely through the sampler, therefore pressure drop is small and a powerful vacuum source is not needed, but this method allows for only a low collection rate, between 7 cm^3 min^{-1} and 1 L min^{-1}. It has a small collection area compared to other samplers and the temperatures generated may adversely affect the viability of some microorganisms.

Consequently, it has been mostly a research tool and has little serious application for bioaerosol collection.

PRACTICAL ASPECTS OF SAMPLING

High-Volume Sampling

Electrostatic precipitators such as the LVAS permit sampling at high flow rates into liquid and, if necessary, the collection fluid can be recycled to concentrate the sample into a small volume. With filtration, a limit to flow rate is posed only by the flow rate through the filter medium and the suction power of the vacuum source, after which the major limit is the physical loading capability of the filter. In practice, however, because of dehydration effects, high-volume sampling by filtration may adversely affect the integrity of some microbial cells and of labile proteinaceous aeroallergens. Thermal precipitators are not appropriate for high-volume sampling.

Sampling Large Particles

Filtration allows samples of the largest airborne particles to be collected, depending on the inlet characteristics of the filter holder. For this reason, it has been the method of choice for aeroallergen sampling,[2] where large particles such as pollen grains may be important.

Other Practical Aspects

Sequential sampling can be achieved readily with electrostatic precipitators collecting into liquid, plus a fraction collector in conjunction with the pumped collection fluid. For filtration, samples can be taken sequentially by changing the filter at regular intervals.

Long-term sampling can be done using electrostatic precipitators collecting into liquid, where appropriate dilutions can be made. For electrostatic precipitation onto a solid collection surface, filtration with direct microscopic observation of the deposit, or where a filter with deposit is to be placed directly onto an agar plate to culture microorganisms, overloaded samples can present a problem. Where deposits are to be recovered from the surface of filters, the upper limit to collection is dictated only by physical overloading and blockage of filters. However, depending on the nature of the bioaerosol, dehydration effects similar to those for high-volume sampling may adversely affect recovery. As discussed previously, high temperatures in thermal precipitators may adversely affect viability.

SUMMARY

Of the non-inertial samplers, filtration is the simplest and most common for collection of bioaerosols. Equipment is relatively inexpensive because it is non-specialized and a wide range of analyses can be performed on filter samples. However, despite its adaptability, filtration poses some limitations, especially for bioaerosols likely to be significantly affected by dehydration stresses. For this reason, it may be considered a method of choice for more robust bioaerosols, such as fungal spores or dust-borne aeroallergens, but if used to collect more sensitive materials, such as vegetative bacterial cells, results should be treated with caution. Other methods described in this chapter, such as electrostatic and thermal precipitation, may have potential uses in some situations, but the specialized equipment and experience needed limits their availability to that of a research tool.

REFERENCES

1. Marple, V.A. and J.E. McCormack. "Personal sampling impactors with respirable aerosol penetration characteristics," *Am. Ind. Hyg. Assoc. J.* 44:916–922 (1983).
2. Agarwal, M.K., J.W. Yunginger, M.C. Swanson and C.E. Reed. "An immunochemical method to measure atmospheric allergens," *J. Allergy Clin. Immunol.* 68:194–200 (1981).
3. Swanson, M.C., M.K. Agarwal and C.E. Reed. "An immunochemical approach to indoor aero allergen quantitation with a new volumetric air sampler: studies with mite, roach, cat, mouse and guinea-pig antigens," *Aller. Clin. Immunol.* 1976:1974–1979 (1985).
4. Rotter M. and W. Koller. "Sampling of airborne bacteria by gelatin foam filters," *Zbl. Bakt. Hyg. I. Abt. Orig. B* 157:257–270 (1973).
5. Koller, W. and M. Rotter. "Further investigations on the suitability of gelatine filters for the collection of airborne bacteria," *Zbl. Bakt. Hyg. I. Abt. Orig. B* 159:546–559 (1974).
6. Dennis, T.J.L. "Isolation of legionella from environmental specimens," in *A Laboratory Manual for Legionnella*, T.G. Harrison and A.G. Taylor, Eds., (Chichester, John Wiley: 1988) pp. 31–44.
7. Health and Safety Executive, "MDHS 14: General methods for the gravimetric determination of respirable and total inhalable dust," HSE Publications, Sudbury, Sulfolk.
8. Mark, D. and J.H. Vincent. "A new personal sampler for airborne total dust in workplaces," *Ann. Occup. Hyg.* 30:89–102 (1986).
9. Fairchild, C.I., M.I. Tillery, J.P. Smith and F.O. Valdez. "Collection efficiency for field sampling cassettes," Los Alamos Sci. Lab. Report LA8640-MS (1980) pp. 1–22.
10. Buchan, R.N., S.C. Soderholm and M.I. Tillery. "Aerosol sampling efficiency of 37mm filter cassettes," *Am. Ind. Hyg. Assoc. J.* 47:825–831 (1986).
11. Gregory, P.H. *The microbiology of the atmosphere.* 2nd ed. (London, Leonard Hill: 1973) p 377.
12. Nilsson, S. *Atlas of airborne fungal spores in Europe* (Berlin, Springer-Verlag: 1983) p 139.
13. Eduard, W. and O. Aalen. "The effect of aggregation on accounting precision of mould spores on filters," *Ann. Occup. Hyg.* 32:471–479 (1988).
14. Bets, R.P., P. Bankes and J.G. Banks. "Rapid enumeration of viable microorganisms by staining and direct microscopy," *Appl. Microbiol.* 9:199–202 (1989).
15. Paton, A.M. and S.M. Jones. "The observation and enumeration of microorganisms in fluids using membrane filtration and incident fluorescence microscopy," *J. Appl. Bact.* 38: 199–200 (1975).
16. Zimmerman, R. and L.A. Meier-Reil. "A new method for fluorescent staining of bacterial populations on membrane filters," *Kieler Meeresforschung* 30:24–27 (1974).
17. Hobbie, J.E., R.J. Daley and S. Jasper. "Use of nuclepore filters for counting bacteria by fluorescence microscopy," *Appl. Environ. Microbiol.* 33:1225–1228 (1977).
18. Pettipher G.L. and U.M. Rodrigues. "Rapid enumeration of microorganisms in foods by the direct epi-fluorescence filter technique," *Appl. Env. Micro.* 44:809–813 (1982).
19. Palmgren, U., G. Strom, G. Blomquist and P. Malmberg. "Collection of airborne microorganisms on Nuclepore filters, estimation and analysis - CAMNEA method," *J. Appl. Bacteriol.* 61:401–406 (1986).

20. Palmgren U., G. Strom, P. Malmberg and G. Blomquist. "The nuclepore filter method: A technique for enumeration of viable and non-viable airborne microorganisms," *Amer. J. Ind. Med.* 10:325–327 (1986).
21. Jost R. and H. Fey. "Rapid detection of small numbers of airborne bacteria by a membrane filter fluorescence antibody technique," *Appl. Micro.* 20:861–865 (1970).
22. Jones, S.E., S.A. Ditner, C. Freeman, C.J. Whitaker and M.A. Lock. "Comparison of a new inorganic membrane filter (Anopore) with a track etched polycarbonate membrane filter (Nuclepore) for direct counting of bacteria," *Appl. Environ. Microbiol.* 55:529–530 (1989).
24. Heikkila P., M. Kotimaa, T. Tuomi, T. Salmi and K. Lonhelainel. "Identification and counting of fungal spores with scanning electron microscopy," *Ann. Occup. Hyg.* 32: 241–248 (1988).
25. Heikkila P, T. Salmi and M. Kotimaa. "Identification and counting of fungal spores by scanning electron microscopy," *Scandinavian Journal Work Environment Health* 14:66–67 (1988).
26. Crook, B., S. Higgins and J. Lacey. "Methods for sampling airborne microorganisms at solid waste disposal sites" in *Biodeterioration 7*. D.R. Houghton, R.N. Smith and H.O.W. Eggins, eds, (London, Elsevier: 1987), pp 791–797.
27. Atkinson S., R.D. Tee, A.J. Newman and A.J. Newman Taylor. "Investigation of factors affecting rat urinary aeroallergen levels," *J. Allergy Clin. Immuno.* 1985 (supplement): 227 (1990).
28. Atkinson, S., R.D. Tee and A.J. Newman Taylor. "Measurement of rat urinary allergen; optimisation of filter elution techniques and investigation of factors affecting airborne levels," in *Proceedings 7th International Symposium on Inhaled Particles,* Edinburgh (1991), in print.
29. Gordon, S., R.D. Tee, D. Lowson and A.J. Newman Taylor. "Comparison and optimisation of filter elution methods for the measurement of airborne allergens," *Am. Occup. Hyg.* 36: 575–587 (1992).
30. Gordon, S., R.D. Tee, D. Lowson, J. Wallace and A.J. Newman Taylor. "Reduction of airborne allergenic urinary protein from laboratory rats," *Brit. J. Ind. Med.* 49:416–422 (1992).
31. Edwards, R.G., M.F. Beeson and J.M. Dewdney. "Laboratory animal allergy: the measurement of airborne urinary allergens and the effects of different environmental conditions," *Laboratory Animals* 17:235–239 (1983).
32. Fields, N.D., G.S. Oxborough, J.R. Puleo and C.M. Herring. "Evaluation of membrane filter field monitors for microbiological air sampling," *Appl. Micro.* 27:517–520 (1974).
33. Blomquist, G., G. Strom and L.H. Stromquist. "Sampling of high concentrations of airborne fungi," *Scand. J. Work Environ. Health* 10:109–113 (1984).
34. Macher, J.M. and M.W. First. "Personal air samplers for measuring occupational exposure to biological hazards," *Am. Ind. Hyg. Assoc. J.,* 45:76–83 (1984).
35. Decker, H.M., L.M. Buchanan and D.E. Frisque. "Advances in large volume air sampling," *Contamination Cont.* 8:13–20 (1969).
36. Griffin, P., B. Crook, J. Lacey and M.D. Topping. "Airborne scampi allergen and scampi peelers asthma," in *Aerosols, their Generation, Behaviour and Applications* (London, The Aerosol Society: 1988), pp 347–352.
37. Crook, B., P. Griffin, M.D. Topping and J. Lacey. "An appraisal of methods for sampling aerosols implicated as causes of work-related respiratory symptoms," in *Aerosols, their Generation, Behaviour and Applications,* (London, The Aerosol Society: 1988), pp 327–333.
38. Winkler, L.U.G. "Airborne rabies virus isolation," *Bulletin Wildlife Disease Assoc.* 4:37–40 (1968).
39. Gerone, P.J., R.B. Couch, G.V. Keefer, R.G. Douglas, E.B. Derrenbacher and V. Knight. "Assessment of experimental and natural viral aerosols," *Bact. Rev.* 30:576–584 (1966).

40. Houwink, E.H., and W. Rolrink. "The quantitative assay of bacterial aerosols by electrostatic precipitation," *J. Hyg. Camb.* 55:544–563 (1957).
41. Morris, E.J., H.M. Darlow, J.F.H. Peel and W.C. Wright. "The quantitative assay of monodispersed aerosols of bacteria and bacteriophage by electron prostatic precipitation," *J. Hyg. Camb.* 59:487–496 (1961).
42. Griffin, P., F.P. Roberts and M.D. Topping. "Measurement of airborne antigens in a crab processing factory" in Proceedings 7th International Symposium on Inhaled Particles, Edinburgh (1991), in print.
43. Cox, C.S. *The Aerobiological Pathway of Microorganisms.* (Wiley Interscience, Chichester, 1987), p. 293.
44. Waldmann, L. and K.H. Schmitt. "Thermophoresis and diffusiophoresis of aerosols," in *Aerosol Science*, (London, Academic Press: 1966) Chap. 4.
45. Kethley, T.W., M.T. Gordon and C. Orr. "A thermal precipitator for aerobacteriology," *Science* 116:368–369 (1952).
46. Orr, C., M.T. Gordon and M.C. Kordecki. "Thermal precipitation for sampling airborne microorganisms," 4:116–118 (1956).
47. Watson, H.H. "The sampling efficiency of the thermal precipitator," *Brit. J. Appl. Physics* 2:78–81 (1958).

CHAPTER 11

Modern Microscopic Methods of Bioaerosol Analysis

K.J. Morris

INTRODUCTION

The majority of viable particles within the atmosphere are spores of fungi, myxomycetes, bryophytes and pteridophytes, as well as pollen grains of flowering plants, moss gemmae, propagules of lichen, cells of algae, vegetative cells and endospores of bacteria, cysts of protozoans and virus particles.[1] Particles of biological origin usually vary in size from below 1 μm to approximately 50 μm or larger. Viruses typically range from 0.005 to 0.05 μm, while bacterial cells and spores typically range from 0.2 to 30 μm in length. Pollens and plant spores are generally larger, with diameters between 10 and 100 μm. Assuming a density of 1.0 g/cm^3, the settling rate of a 0.1 μm diameter sphere in still air is 0.3 cm/h, a 1.0 μm sphere is 13 cm/h, and a 40 μm sphere is 300 cm/min, according to Stokes law.[2] Thus, large airborne particles stay airborne in non-turbulent air for only a short period and are removed by sedimentation, although they may be resuspended by wind or physical disturbance. Sub-micron particles will stay airborne for days even in still air, and generally are removed by rain, diffusion to surfaces or by coagulation with other particles. Biological particles in the air may consist of single or unattached organisms or may occur in the form of clumps composed of a number of microbes. The organisms may also adhere to dust particles or exist as a free floating particle surrounded by a film of dried organic or inorganic material. Organisms can be associated with liquid droplets, e.g., as splash drops from sewage processing. Some microbes become airborne while in an actively metabolizing phase. Vegetative cells are important to health, as they include the primary etiological agents of communicable diseases. However, more commonly outdoor bioaerosols contain mostly spores, that are hardier, metabolically less active and often better adapted to dispersal. Many vegetative cells do not ordinarily survive very long in air unless other factors are favorable (see Chapter 6). However, some pathogens, such as staphylococci, streptococcus and the tubercle bacillus, will survive for relatively long periods and may be carried considerable distances while still viable, or they may settle on surfaces and be resuspended as an aerosol during activities such as sweeping and bedmaking.[3]

Main methods of bioaerosol sampling are based on impingement in liquids, impaction on solid surfaces, sedimentation, filtration, electrostatic precipitation and thermal precipitation (see Chapters 9 and 10). Frequently, particularly with bacteria or fungi, the collected organisms are isolated in culture medium or on plates. Other chapters within this book are concerned with these methods of bioaerosol sampling and subsequent culture of viable particles. This chapter, although mentioning them, will concentrate on physical methods of counting, morphometry classification (size and shape), and identification techniques presently available (see also Chapter 12).

In the origins of microbiology at the end of the last century and the first half of this century, sophisticated procedures were developed for the classification of microorganisms. This was largely as a result of recent discoveries demonstrating that microorganisms are an important cause of human disease. Identification of microorganisms, in particular bacteria, were based principally on the techniques of culturing using solid and streak plate or pour plate methods, from which the organisms were isolated and identified. Early taxonomists relied mainly on cell size, cell shape, the form of colonies, the growth in various types of broth, histological staining and pathogenicity. Hundreds of such identification methods now are described in the literature.[4] Examples of a few general texts are given in the references.[5-12] Over the last 40 years, it has become accepted generally that the air is an important route for disease transmission. This has led to many advances in the measurement of airborne viable organisms. The majority of sampling techniques still are concerned with counting and identifying bacteria, although methods have been developed specifically for viruses, fungi and yeast as well. More recently, problems such as hay fever and asthma have led to further interest in counting and classifying pollen grains, fungal spores and other biological allergenic materials. It is clearly important to sample airborne biogenic material in places such as hospitals, near sewage works, around or inside industrial plants and military establishments, in clean rooms, and in areas where people or their livestock and crops are concentrated (see also Chapters 18–21).

Recent improvements in electronics and immunoassay techniques have led to the development of fully computer software-driven image analysis systems, automated colony counters, and the introduction of fluorescent and specific antibody stains. The latter have proved valuable in the development of rapid bacterial detection systems, such as those using flow cytometers[13] (see also Chapters 12 and 13).

DIRECT VISUALIZATION AND MEASUREMENT OF SPORES AND POLLENS

The spore traps and filters described below and in Chapters 10, 14 and 16 are suitable for the sampling and direct visualization of bacterial cells, fungal hyphea, spores and pollen under a light or scanning electron microscope. However, viable vegetative cells in the air generally are desiccated by these sampling procedures and so cannot be identified easily.

Sedimentation and Impaction Devices

The number of biogenic particles in the air can be estimated directly using various non-culture sampling techniques. These are reserved typically for sampling material such as pollen grains or bryophyte spores, e.g., as part of hay fever research or assessing airborne fungal spores that produce rust diseases in crops. A spore trap generally relies on a sticky slide to collect the particles. Very simple devices, such as 'gravity slides', often are used in remote areas,[14] although the data from these are qualitative, as accurate estimation of sampled air volume is not possible and detection is determined by particle size and settling velocity.

More quantitative spore traps involve samplers through which air is drawn by pumps, fans or aspirators. The Hirst spore trap[15] is a power-driven sampler consisting of a single impactor slit, behind which is placed a sticky microscope slide moving at 2 mm/h. Over 24 h a trace is deposited in a band 48 mm long. This can be scanned under a light microscope to obtain a daily mean, or traversely every 4 mm to get a reading of the air spora content every two hours throughout the day and night. This and other bioaerosol samplers are discussed in detail by Gregory.[16] As the volume of air sampled is known, it is a simple matter to convert total spore counts to number of spores per m^3 of air.

A similar device occasionally used for spore counting is the Rotorod,[16,17] which consists of a U shaped square section brass rod. The two vertical arms are swept through 120 L of air per min, by a 12V motor, and particles are collected on the leading edge by impaction. The arms are removed after sampling and viewed directly under the microscope, and the biogenic particles counted by eye in the manner described below. Sticky tape may be attached to the arms for easier sample handling and storage. The Rotorod is mainly suitable for pollens and spores, and may be used to advantage when there is an absence of electrical power at the sampling site. This instrument is inefficient at sampling particles smaller than 7 μm, at which size the collection efficiency is down to 50%.

Even the best of these impaction samplers tend to have low and inconsistent sampling efficiencies,[16,17] and all are adversely affected by changes in wind speed.

After sampling, the collecting slide or plate generally is placed directly under a light microscope. Identification of the sampled spores and pollens normally is made by reference to prepared slides obtained from known species, or from photographs of the same. However, scanning samples under the microscope is very tedious, unless the spore density is dense. Collected samples also may be washed off the collecting surfaces and plated onto Petri dishes of solid media, incubated, and the formed colonies counted, to estimate the total number of viable organisms. However, it is rare for spore trap samples to be cultured, as vegetative cells generally are made non-viable owing to desiccation during sampling.

Filters

Membrane and cellulose filters have long been used for sampling atmospheric aerosols. These are normally simply placed into a suitable filter holder and attached to a pump and flow meter. However, the desiccating environment on these filter surfaces kills most vegetative cells if sampling periods are long or relative humidity is low, though the more robust bacterial spores do survive. After exposure, filters can be mounted as a transparency on a glass slide using various filter clearing techniques, and scored under a light microscope. Filters with black backgrounds may be used, and the biological material stained with fluorescent dyes for ease of counting. Alternatively, after sampling the filter can be placed directly onto the surface a Petri dish for culturing or shaken in a suitable medium and the suspension plated out. After incubation, colonies are counted by eye, by using an image analysis system or by using dedicated colony counting equipment.

Filtration has an advantage of being able to sample a known volume of air. Scoring bio-particles on the sampling medium generally involves counting manually by eye, under a light microscope, to distinguish biogenic particles from inert particles where possible. The total mass of particles on the filter can be estimated by weighing the filter before and after sampling, but care must be taken to avoid relative humidity (RH) effects on the filter and entrained particles, so drying the filter in a desiccator, or in air of a consistent %RH, before each weighing is required. By dividing the particle mass by the sampling volume, the mass of particles per unit volume of air can be calculated.

MEASUREMENT OF VIABLE MICROORGANISMS IN THE AIR

These bioaerosol aerosol sampling techniques are more suitable for measuring viable vegetative cells in the air, and may be used to culture these, as well as bacterial and fungal spores, provided the collecting medium is suitable (see also Chapter 6). Distinction should be made between sampling methods that tend to disaggregate bacterial clumps and thus measure total the viable number of viable organisms in the air, e.g., impingers and bubblers, and those with which one particle made up of many bacteria may form a single colony, e.g., gravity Petri dishes and slit samplers. However, problems such as selectivity of culture media, temperature, aeration and competitive or antibiotic interactions, make it difficult to develop a completely non-selective culture method for recording concentrations of viable organisms in the air. For this reason, all bioaerosol samplers are likely to underestimate numbers of airborne viable organisms.

Sedimentation and Impaction Devices

The 'gravity Petri dish' has been in common use for qualitative bioaerosol measurements, where a dish of sterile medium is left open at the sampling site for periods of 1 to 10 min to investigate the cultivable bacterial or mould flora of the atmosphere. Indoors, the method is subject to distortion owing to the sedimentation rate, as it preferentially selects larger particles. Outdoors, the dish also is subject to aerodynamic effects from the edge of the dish. Apart from convenience and economy, the method is valued for precision in identifying captured organisms and for selectivity when sampling is aimed at a group of organisms. The number of colonies can be counted on the dish, or cells may be removed for further selection using specialized media, or for direct visualization under a light, or electron microscope, after suitable mounting and staining.

Forced air flow impactors that collect the particles on or in nutrient media are some of the most popular bioaerosol samplers now in use.

Impingers and liquid scrubbing (bubblers) devices draw known volumes of air through a selective or general liquid nutrient media. This minimizes desiccation damage to microbes during sampling, although bacterial clumps may be broken up, and microorganisms may grow and divide in the media. The particles in the media may be counted directly using a microscope cell slide such as a hemocytometer. More frequently, the culture is plated out on media in a Petri dish, is incubated and the formed colonies are counted. Provided the volume of media remaining after sampling and the volume of sampled air are known, the counts can be converted to airborne concentrations. Individual organisms can be identified using standard microbiological techniques.

Slit samplers, such as the Casella, are designed primarily for indoor bioaerosol sampling. A stream of air is drawn through a narrow slit placed just above the surface of slowly rotating Petri dish containing sterile media. After a few minutes sampling, the Petri dish is removed and incubated so that colonies may develop and be counted.

Cascade impactors separate the particles by size during sampling. The Anderson cascade[18] sampler draws air, at a rate of 28.3 L per min, through a series of identically sized circular plates, each perforated with 400 holes through which particles are deposited onto sterile medium in Petri dishes. Succeeding plates have progressively smaller holes, so that the largest particles (> 11 μm aerodynamic diameter) are deposited in the first dish and the smallest particles (> 0.65 μm) are deposited in the final dish. After sampling, the Petri dishes are removed and incubated and the colonies counted. This impactor does suffer from desiccation-related problems because as the airstream passes over the agar plate it removes surface moisture, and simultaneously reduces the plate's ability to impact more particles owing to a loss of surface stickiness. Drying of the agar also affects viability of organisms already trapped on the agar plate, reducing the count obtained.

Gelatin Filters

More recently, gelatin filters have been developed that reduces the problem of desiccation when using conventional filters. The filters are supplied pre-sterilized by gamma irradiation, and are used in a pumped system such as the Sartorius MD8 (Sartorius Separation Technology, Epsom, Surrey, UK). The high level of associated moisture in the filters helps retain the viability of organisms long after a standard membrane would have allowed adverse desiccation effects to reduce the viable count. After sampling, the filter is normally placed directly onto solid culture medium, where it dissolves, and colonies grow that are subsequently counted. Alternatively, the filter can be dissolved and microorganisms re-filtered, if inhibitors such as antibiotics or antiseptics have to be removed. This indirect method of detection has proved successful when dealing with phages and other viruses.

LIGHT MICROSCOPY

Many bacterial classification techniques still rely on traditional smears on glass slides, histological staining, with subsequent viewing under a standard bright field light microscope. Such smears are obtained from liquid cultures or plates of bioaerosol samples. Under the light or electron microscope, bacterial cells often are classified according to their morphometry, as cocci for spherical cells, bacilli for rods, and spirilla for a helical outline.

Visual inspection of impacted bioaerosol samples from spore traps and filters using a microscope obviously is limited to recognizable particles, and one must be aware of this restriction when dealing with anything other than pollen or complex fungal spores.[19] The light microscope may be used to count or inspect colonies cultured on solid media, especially where some could be too small to be resolved by the naked eye.

The Mono-Objective Bright Field Light Microscope

The standard laboratory light microscope is the mono-objective form that uses an ordinary objective and regular paired eyepieces located in a 'tube' above the specimen. By the use of prisms an identical image is provided to each eyepiece, giving a non-stereoscopic effect. Objectives for these microscopes range from 1 to 100 × magnification. The eyepiece adds a further 7 to 15 × magnification, giving a practical total magnification factor ranging from 7 × to about 1200 ×. The specimen normally is illuminated from below by transmission from a tungsten or quartz white light source. Epi-illumination from above may be used for opaque objects or for fluorescence microscopy. Set up for standard bright field work, the specimen is viewed as a darker object against a white background.

Inverted mono-objective light microscopes, often with heated stages, also are available. These may be used for viewing sediments, living cells in culture (e.g., wells), or anything that is present at the bottom of a clear container. Although the objectives are placed below the specimen stage and illumination is from above, i.e., the inverse of a normal light microscope, these devices work on the same principle as the mono-objective light microscope.

Another form of light microscope also is available, namely the wide field stereoscopic binocular microscope, often called the dissecting microscope. With this microscope the body tubes are fitted with two matched objectives and eyepieces that are focused simultaneously on to the same area of the object, each eye viewing the field from a different angle, giving a full stereoscopic image. This microscope functions best in the range of 5 × to 45

× magnification, but magnification up to 300 × is achievable. Illumination may be transmission or incident. Although the stereoscopic microscope is useful for low power applications such as observation of colonies on solid media, the main microscope used for visualizing bioaerosol samples is the mono-objective high-power microscope, and discussion in this chapter will be confined to this type of light microscope.

The resolving power of the unaided eye is 0.1 mm, while the maximum resolving power of the mono-objective light microscope is about 0.2 μm at 1200 × magnification. In general, a range of objectives are required for the microscope, say 4 ×, 20 ×, 40 × and 100 ×. An additional variable set of magnification lenses above the turret, normally 1.0 to 2.0 ×, is also extremely useful. Eyepieces generally add a further 10 × magnification. Fluorescence and phase contrast objectives are normally 20 × to 100 ×. Objectives should have high-quality optics, and cost around £500 to £1,500 each. High-power objectives have short working distances, e.g., the distance between the bottom of a 100 × objective and the object in view is about 0.15 mm, which may cause problems with some samples. Special dry long working distance objectives are available, with the working distance of these 100 × objectives being around 0.7 mm (nearer that of a 40 × objective). Special dry objectives also are available for slides viewed without coverslips.

Oil immersion often is used with high power objectives, particularly 100 ×, as it offers resolving power half as much again over dry objectives. It is standard for high-power fluorescence work. However, oil immersion is messy, particularly where dry and oil objectives are mixed on the same turret, and it may be impossible to maintain oil contact when using motorized multi-slide stages controlled by automated image analyzers.

Image analysis systems typically will be connected to a microscope system capturing images via a black-and-white camera with a C-Mount thread or a color camera with EMG bayonet fitting. Image analysis systems will be described in detail later as they may be employed for measurements using light or electron microscopy, and for counting bacterial colonies using an epidiascope. If it is intended to attach an image analysis system to the microscope in the future, it is advisable to contact the manufacturer of the system to receive advice on which microscopes they preferentially adapt for use with their system. In general, a large dedicated photomicroscope, such as the Nikon Microphot FXA shown in Figure 11.1, is ideal, as they are sturdy, have linear optics, and dedicated ports for adding 35 mm, Polaroid and video cameras.

Cells and spores may be viewed using a standard bright field optical microscope. However, except for some pigments, most cellular components absorb little light in the visible region, hence the reliance on fixation and special histological stains. The detailed visualization of viruses and the ultrastructure of all microorganisms is only possible using transmission electron microscopy. Surface details are generally not well resolved by the light microscope owing to its limited depth of field, and for this purpose the scanning electron microscope is preferred. The recently developed laser confocal microscope offers better depth of focus than the conventional microscope, comparable resolution, and it can function as a fluorescence microscope for fluorochrome dyes.

A typical bright field laboratory binocular mono-objective microscope will cost, at 1994 prices in the UK, between £4,000 and £7,000. A dedicated photomicroscope such as the Nikon Microphot FXA will cost nearer £20,000, with an extra £1,000 for phase contrast and £200 to £1,500 for each objective. A fluorescence system will cost a further £5,000 excluding objectives. The phase contrast, polarizing, interference and fluorescence microscope all require extra hardware and specialized objectives that must be used in addition to the basic bright field microscope.

It is advisable to purchase the microscope and accessories with a view to upgrading the microscope in future. For example, buy Plan objectives where available, as these are

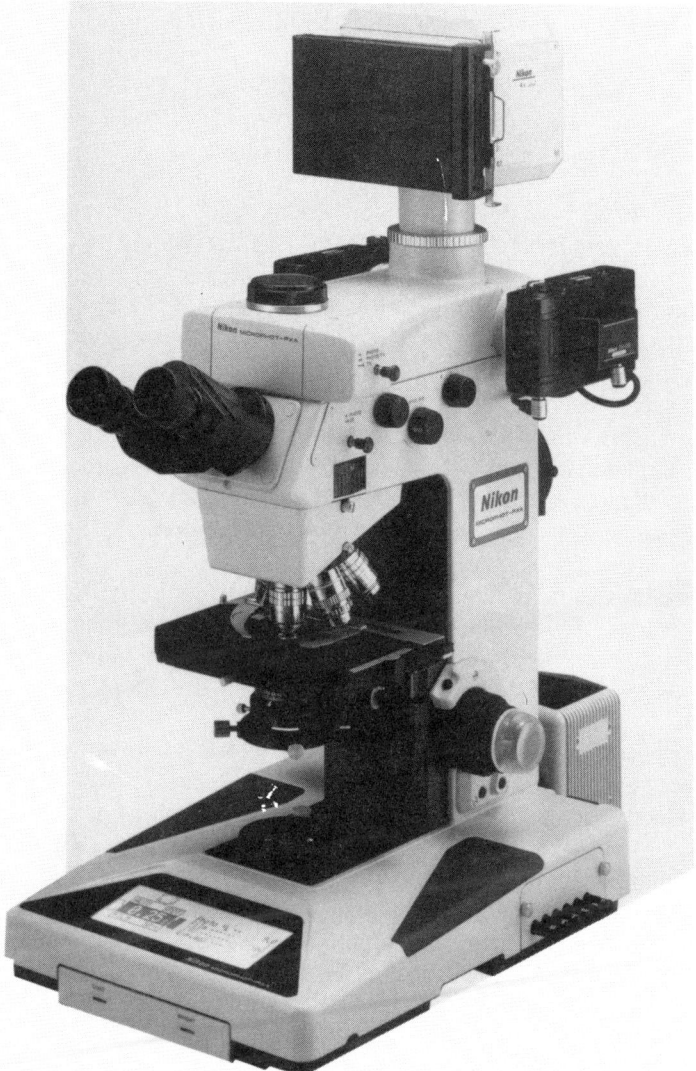

Figure 11.1. The mono-objective light microscope, in this case a Nikon FXA photomicroscope (Nikon UK Ltd., Halesfield 9, Telford, Shropshire, UK).

required for image analysis systems and photography, having linear optics and being in focus over the whole field of view. Generally, very good advice can be obtained from the technical support section of the major microscope manufacturers and distributors.

Visualization of Cell Structures Using Traditional Histological Stains

The use of selective stains to enhance different cellular components now is well established, and generally these are viewed under a clean standard bright field light microscope correctly set up with either Nelson or Kohler illumination.[20] Microorganisms must be fixed and dehydrated prior to staining; normally these are present as a smear on a glass slide, taken from a colony grown on solid culture medium. Larger microbes must also be embedded and sectioned. Such procedures introduce many changes to the biochemical and morphometric nature of the material, even reducing the cell size by 15% or so. Histological staining is generally applied to microorganisms cultured from aerosol samples. This is the most efficient way of identifying viable airborne microorganisms. However, this method destroys all evidence of the original size and number of the airborne bio-particles. Survival of vegetative cells for good histology on the collection surface is likely to be difficult if they have been subjected to dehydration.

Of the many bacterial stains in common use, methylene blue serves for simple examination. The Gram stain is the most important differential staining procedure, where Gram-negative bacteria lose the violet stain and take a counter stain. The staining response reflects important differences in the cell wall structure of the two cellular classes, for example, the bactericidal action of Penicillin and the enzyme lysozyme, found in tears and egg white, is much more effective against Gram-positive cells. Very few species have a variable Gram stain response.

Bacteria with high cellular lipid concentrations are difficult to stain, but may be stained with hot fuschin-containing phenol. They retain the dye after treatment with acidified ethanol, and are termed acid-fast. This Ziehl-Neelsen carbol fuschin stain may be used for non acid-fast bacteria which are difficult to stain otherwise, such as spirochaetes and legionellae.

The presence of intracellular bodies of cytoplasmic inclusions of reserve material is also used as an aid in the identification of certain types of bacteria, such as the presence of metachromatic granules being characteristic of most Corynebacteria (e.g., diptheriae). These granules are stained dark violet, with the cytoplasm counterstained yellow, using Neissers stain.

Bacterial spores may be stained with malachite green and safranine solution. This stains the spores green, sporangia and vegetative cells pink.

More detailed descriptions of traditional stain methods for microorganisms are to be found in standard microbiology[5,9,12] and histological[21,22] texts, and discussion of requirements with experienced histologists can be invaluable. More recently, various fluorochrome[23] and specific enzyme or monoclonal antibody stains[24,25] have been developed and are in general use for identifying microorganisms. Monoclonal antibodies may be conjugated with fluorescent labels, heavy metals or radioisotopes, for detection by fluorescence microscopy, electron microscopy and autoradiography, respectively.[10,24,26,27]

Counting Objects on a Slide

Generally, sticky slides or plates from spore traps are viewed directly under the microscope. As well as identifying the main organisms from morphometry and staining, often an estimate of the number present on the slide is required, to provide airborne numbers. Normally the slides are scored manually under a light microscope. Spores and pollens are most easily discerned. Vegetative cells may be difficult to score, even with suitable staining techniques. Many other objects of biogenic origin often will be seen, such as dried plant parts, animal hair and skin, algal fragments, fern sporangia, insects and their parts, and fungal hyphae. Occasionally talc particle contamination may be present, owing to handling slides with unwashed surgical gloves.

Prior to viewing the slide, the microscope must be clean and set up correctly according to the manual supplied from the manufacturer. This is important as it reduces operator fatigue due to eyestrain as well as ensuring the microscope is optimized for resolution and detection. For accurate estimates of the number of spores, pollens, etc., per unit volume of air, the total area of the sample on the slide and the volume of air sampled must be known, if a forced airflow sampler was used. 'Gravity slides' sample the column of air directly above the field measured, so for these only the area of the counted field, and the sampling time, is required.

Specific areas of the slide (fields) are sampled randomly across the slide, generally in a raster or snake pattern scan. An acceptable method of selecting a random field is to look away from the microscope, move the stage, and refocus on the new field. A suitable graticule must be inserted into the eyepiece. For counting, a simple squared grid graticule is used. When viewing through the eyepiece, the grid is superimposed onto the sample under the microscope. It is then a simple matter to score the number of items of interest within the complete grid. Various types of square grid can be obtained. Some are indexed to aid identification of areas of interest. Others have a checkerboard pattern to help distinguish the position of interest, where the darker squares are translucent and the lighter squares are transparent, thereby avoiding eyestrain during prolonged counting. If it is required to compare the proportion of large to small particles in a sample, the Miller graticule may be used, where the small particles are only counted in a small square on one side of the larger square graticule, the result being multiplied by ten for comparison with the number of larger particles in the large square. Graticules and calibration slides may be purchased from Graticules Ltd., Morley Rd., Tonbridge, Kent, UK.

The area sampled with the graticule depends on the magnification of the objective, eyepiece and intermediate lenses. This can be determined by placing a calibration slide on the stage prior to counting. Calibration slides contain a simple etched calibration scale. These range from 0.1 to 5 mm with divisions ranging from every 2 μm to every 0.1 mm, and are selected depending on the magnification used. This calibration slide should have a calibration certificate, traceable to national standards (e.g., the National Physical Laboratory, Teddington, Middlesex, UK). The square graticule simply is lined up with the calibration scale at the correct sampling magnification and the length and height (obtained by rotating the stage or eyepiece) of the complete square graticule field is measured. From this, the area of the sampling grid can be determined (length × height). Repeated random fields are then scored across the slide, and scores manually logged using a mechanical digital counter or memory if very few fields contain objects of interest. After the field has been scored the numbers are written onto a paper score sheet and the next field is selected. In general, the required number of fields to be scanned will depend on the density of the items of interest on the slide. The more fields sampled, the more likely is it that very rare objects are located, as well as a better sampling accuracy. In general, at least 200 fields per sample would be scored, or a predetermined large number of objects of interest (say 500). The field sampling regime should be set up before any scoring begins, after assessment of a few slides. To keep consistency, it must be rigorously adhered to, unless the regime proves unsatisfactory on new samples.

Sampling rules also must be defined to prevent repeat counting of the same object. For spherical objects it is generally considered easiest to count only objects that are cut by the top and left-hand side of the grid. Any objects cut by the right side or bottom of the grid are ignored. If these rules are adhered to, a slide can be completely raster scanned with no repeat counting of any bio-particles. Automatic image analyzers apply similar rules, although these may count all objects whose center of gravity is within the field frame.

Rod-shaped objects, such as animal hair and fungal hyphae, that may have large aspect ratio (length/diameter > 3), pose more of a problem, as if the magnification is high there may be sampling bias toward the longer objects. If all the lengths of such objects are less than 20% of the length of the field grid, then there should be no problem, and the sampling regime given above will suffice. If the length of the objects exceeds 50% of the length of the field grid, then either the magnification should be reduced or World Health Organization fiber-counting rules[28] applied. For counting, the rules are quite simple, and are based on counting fiber ends. All fibers that are entirely within the grid are counted as one object. All fibers cut by the grid and having one end in the field are counted as half an object. All fibers cut by the graticule grid, such that neither end is within the field, are ignored. Bundles of fibers are more of a problem. These may be counted as one fiber, subject to the end rules described, or if the fibers are easily distinguishable, e.g., overlaid on each other, they may all be counted.

To suit modern requirements for Good Laboratory Practice, as defined by the UK Department of Health,[29] a pre-prepared sheet for logging scores can be useful, where the operator enters his name, the date, slide details, experiment number, magnification, and other relevant details, prior to scoring. The sheet will be signed when the scan has been completed, once the operator has ensured that the information entered is correct. It is important to ensure that the magnification is recorded correctly, as many microscopes have additional magnification lenses above the turret, typically 1.0 ×, 1.2 ×, 1.6 × and 2.0 ×, and the selection of the incorrect value is very easy and difficult to perceive when viewing. Care also must be taken when switching magnification to identify objects at higher power. Normally the data from the score sheet will be entered into a personal computer spreadsheet, such as Microsoft Excel or Lotus 123. The name of the spreadsheet data file also should be recorded on the score sheet for cross-referencing purposes.

Once the slide has been scored completely in the manner defined by the sampling regime protocol, the total score can be converted into number of bio-particles per unit volume of air by the following simple equation, provided that the sampler is a forced airflow device:

$$N = (T \times (A \div F)) \div V$$

where N is the mean number of bioaerosol particles in 1 m^3 of air sampled T is the total number of bioaerosol particles counted in all fields. F is the total area in m^2 of all fields scored (field area × number of fields scored). A is the total area in m^2 of the sample on the slide. V is the volume of air sampled in m^3. It is likely that the slides will be differentially counted, and that each particle will be categorized and each category scored separately. For example, biogenic particles could be broadly categorized into pollen grains, fungal spores and miscellaneous fragments. Thus, the relative numbers of bioaerosol particles falling into each category also will be known, and often these are expressed as total counts per category, counts per cm^2 of slide, and each categories percentage of the total counts. Counts from 'gravity slides' normally are only expressed in this way, as the volume of air sampled is not known (although more complex calculations based on the particles size, density, and settling velocity may be used to estimate the airborne concentration).

When counting a slide using an image analyzer, the method is the same as described above. Viewing the field on a computer monitor has advantages even with manual counting, as it causes less eyestrain than viewing down the microscope, so counting times can be extended. An image analyzer also can provide a frame or grid overlaid on the screen to aid manual counting.

Manually Measuring the Size of Objects on a Slide

Although image analysis devices, described in detail later, now have superseded the use of eyepiece graticules for measuring object size, graticules still are available and are simple to use. Excluding the cost of the microscope, they are also very cheap compared to an image analysis system and still are very useful when only a limited number of particles require measurement. The basic principle is to compare the particles to circles, rectangles or shapes present on the graticule. These graticules will estimate both area and diameter, if the object measured approximates a disk or sphere, using the simple equation for the area of a circle (circle area = $\pi \times$ radius^2). The graticules are calibrated using calibration slides placed under the microscope, at the magnification to be used for measurement. For mostly circular objects, suitable graticules are the Patterson globe graticule, the British Standard Graticule, the Porton graticules, or the Fairs graticule (Graticules Ltd., Tonbridge, Kent, UK). Accurately sized polymer particles, e.g., Dynospheres, may be used for size and area calibration checks. Image distortion across a field can be determined by viewing silicon test specimens marked with a square grid (Agar Aids, 66a Cambridge Rd., Stansted, Essex, UK).

For irregular or rod-shaped objects, horizontal, vertical or crossed micrometer graticules may be more suitable. These have a divided scale, normally 0.0 to 100.0 with 1.0 divisions. The eyepiece scale is calibrated by comparison with a calibration slide under the microscope. This is achieved by superimposing the eyepiece over the calibration slide scale, looking for points of coincidence between the two scales, and dividing the length between the coincidence points into the number of eyepiece micrometer divisions along this length. This factor will be the length of each division on the eyepiece graticule. The eyepiece graticules now can be used to measure the length and diameter of each object by direct superimposition of the scale over that object, provided the magnification of the microscope is not changed.

A more sophisticated device for measuring object lengths is the filar micrometer eyepiece, which contains a movable hair in the eyepiece that is moved relative to a fixed hair by turning a calibrated micrometer screw. Calibration is carried out using a stage micrometer.

The sampling regime for selecting approximately spherical objects within fields is the same as that used in the previous section for object counting, e.g., all objects fully within the field are sized as well as those cut by the top and left-hand side of the field frame. For long objects such as fibers, special object sampling rules must be applied to prevent oversampling of the longer bioparticles, which would give rise to biased bioaerosol size distributions. One method of overcoming selection bias in these cases is to sample only objects that have a downward end visible within the field frame. All objects that meet this criteria must be measured, even if they extend well beyond the field. Another fiber sampling method is to weight the distribution, by measuring all objects fully within the field twice (both ends visible) while measuring objects cut by the field frame only once (one end visible).

Eyepiece graticules also have been developed for assessing the morphometry of objects, in particular the histo-morphometry of tissue sections, e.g., the Merz and Weibel graticules. However, such graticules now have largely been rendered obsolete by the introduction of software-driven image analysis systems.

Counting Objects in a Liquid Suspension

The light microscope together with a hemocytometer can be used to estimate accurately the number of particles present in a sample of liquid suspension.[20] Originally these were designed for counting blood cells, and are commonly available from laboratory suppliers. The hemocytometer consists of a glass slide with a central flat well upon which a gridded ruling of 9 squares, each of 1 mm^2, is engraved. The central square of these is subdivided into 400 small squares, marked as a grid of 5 × 5 squares each subdivided into the 4 × 4 small squares, with each small square being 1/400 mm^2. The depth of the well is fixed, normally to 0.1 or 0.2 mm, by covering the well with thick coverslips manufactured for the purpose. Thus, when viewed under the microscope the volume of suspension above the small squares can be calculated, i.e., 0.00625 mm^3 within each group of 25 (5 × 5) small squares.

The coverslip is placed over the ruled surface and pressed into place until Newton's rings are seen between the glass slide and coverslip. After discarding the first few drops, a drop of the suspension of particles then is allowed to flow by capillary action between the cover slip and the upper ruled area of the chamber. If the cells are motile they may be killed by the addition of formalin to the suspension. The drop should fill the center marked area but not be allowed to overflow into the moat. After standing for 10 to 15 minutes, the number of bacteria, pollen or spores are counted in the 25 small squares within the central square mm. These small squares should be taken from the four corners and the center in horizontal strips of 5, at a magnification of 400 to 500 ×. The area counted in 25 small squares is 0.0625 mm^2, thus the number of biogenic particles per cm^3 of suspension is:

$$\text{Number counted} \times 16 \times 10 \times \text{Any dilution factor} \times 1000$$

In the case of bacterial cultures the counts may not relate to the bacterial count in the original aerosol sample.

The Sedgewick Rafter Cell is similar in construction and use to the hemocytometer, although the latter is more commonly available and generally one is to be found in a drawer in most microscopy laboratories.

Another technique that may be used for counting microorganisms in suspension involves drying a small volume on a special slide engraved with circles,[30] and counting the objects in a manner similar to that described in the section "Counting Objects on Slide."

Image analyzers also can count (and size) bio-particles in suspension using the same methods as described, although the slide markings may interfere with object recognition.

The Dark Field Illuminated Microscope

In dark field illumination the light beam strikes the object obliquely and no direct light enters the objective and instead only light scattered or reflected by the object reaches the objective. This makes the object appear brightly on a dark background. Colored effects also can be obtained with Rheinberg differential color illumination, in which the black stops in the condenser used for dark field illumination are replaced with colored central stops and rings. Objects suitable for dark field and Rheinberg differential color illumination are living unstained bacteria, yeast and molds in aqueous suspension, and plant and animal hairs.[20] They are appropriate also for organisms that do not readily stain. Although the resolving power is the same as the standard microscope, the system can indicate the presence of sub-micron particles, such as the larger viruses, which are seen as dots of light.

The Phase Contrast Microscope

Cell morphology may be determined with unstained wet preparations using phase contrast microscopy, with a coverslip, and viewing at high magnification. As different regions of the cells have regions of varying refractive indices, light from these regions can form patterns of destructive interference causing sharp contrasts. Unstained living cells such as yeast, bacteria and fungi, viewed under phase contrast microscopy, reveal structures previously seen only in fixed and stained preparations. A National Physical Laboratory calibration slide is available for assessing the resolving power of the phase contrast system, at all magnifications (NPL, Teddington, Middlesex, UK).

The phase-contrast method is very important for viewing living cells, being quite transparent under the standard microscope, and has a considerable advantage over standard light and electron microscopes in this respect. However, fixing and staining still provides the best method of contrasting cellular structures, and unfortunately the resolving power of the light microscope is not improved by phase-contrast.

The phase-contrast microscope is used for bacteria motility tests, although the arrangement of the flagella cannot be determined. For example, polar flagellated bacteria show an apparent rapid darting movement, whereas peritrichous cells usually move at a lower speed. True motility must be distinguished from movement owing to Brownian motion. Phase-contrast can be used also to visualize spores that appear as brilliant refractile objects. Bacteriophages and some viruses also can be seen with the phase-contrast microscope. The microscope is also useful as well for observing natural fibers, animal and plant hairs.

Viewing bacterial capsules, a sticky layer that forms around the cells, is made possible by negative staining of live bacteria with India ink and subsequent viewing under a phase-contrast microscope. The capsule appears as a bright zone between the dark cells and grey background.[4]

The Polarizing Microscope

In this microscope a polarizing disc is fitted to the substage and a further rotating polarizing disc is fitted above the objective, termed the analyzer. This enables light still vibrating in one plane from the lower polarizer to be completely extinguished, while any diffraction exhibited by the object can be observed. Some ultrastructural cell components are bifringent and the polarizing light microscope was used (before the development of the transmission electron microscope) to analyze them indirectly, since bifringence is dependent on structural properties smaller than the wavelength of light.

The microscope is particularly useful in the study of fibrillar structures such as the mitotic apparatus. It is suitable also for the identification of natural fibers, animal and plant hairs.[20]

The Interference Microscope

The interference microscope is a polarizing microscope combined with an interferometer. The interferometer works by passing one light wave front through the object, while another wave bypasses it, and interference then takes place between these two wave fronts. A rotating polarizer and analyzer also is built in. These microscopes have the advantage of continuously variable phase contrast, without any associated halo around the objects. They have been used for estimating the dry mass, water content, lipid, nucleic acid and total protein content of cells, as well as determining the cell thickness and cell volume.[20]

The Fluorescence Light Microscope

Introduction

Fluorescence is a type of luminescence in which light is emitted from molecules for a short period of time following the absorption of light, and where the delay between absorption and emission is about 10^{-8} s. Various cellular constituents do exhibit autofluorescence, particularly when excited in the UV and blue regions of the spectrum, although this has limited value compared to the use of modern exogenous fluorescent probes and stains.

Probably the most commonly used fluorochrome for staining bioaerosol microorganisms for counting is acridine orange. This is a non-specific stain for DNA and RNA, and fluoresces green or red depending on uptake and other factors. Others such as ethidium bromide (staining DNA and RNA) and fluorescein isothiocynate (staining proteins) also are employed. These molecules absorb and emit light at precise wavelengths, so can be excited and detected in complex mixes of molecules. It is possible to detect as few as 50 fluorescent molecules in 1 μm^2 of cell using a fluorescence microscope.[31]

Many fluorescent dyes have been developed for use as *in vivo* probes studying the metabolism of sodium, chloride and calcium in isolated mammalian cells in culture, as well as intracellular pH and changes in membrane potential.[32,33] Specialized image analysis systems have been developed to exploit this technology, such as the MagiCal dynamic video imaging systems marketed by Applied Imaging (Southshields, UK).

With the development of monoclonal antibodies, specific immunological detection of individual microbe species and even strains is possible.[24] These antibodies are labelled with fluorochromes such as fluorescein isothiocynate (FITC) or rhodamine B, and are thus able to be visualized under a fluorescence microscope. However, owing to the high specificity of the antigen-antibody reaction, the technique is not suitable at present for general bioaerosol investigations as one must decide in advance which antigen will be assayed.

Using the Fluorescence Microscope

The fluorescence microscope has similar optical paths to the standard bright field microscope, and is designed to maximize the collection of fluorescent light while minimizing the collection of excitation light. Although transmitted fluorescence systems are available, epi-illumination (incident) is the preferred system. The light source is either a mercury arc, xenon arc, quartz halogen, or tungsten lamp. The choice of lamp depends on the spectral output and radiance stability required. With epi-illumination, a chromatic beam splitter or dichroic mirror is placed above the objective, and light is supplied to it though an excitation filter that has a single large transmission peak at the wavelength of interest. The beam splitter or dichroic mirror is designed to reflect a light below a cut-off wavelength down through the objective, which functions as a condenser, onto the specimen. This excites the fluorochrome in the specimen, and light emitted from the specimen passes back through the objective, and that above the cut-off wavelength of the beam splitter is not reflected but passes through the splitter and on to the eyepiece or detector. A blocking or barrier filter is mounted between the eyepiece and the beam splitter, and this generally is incorporated into the beam splitter. The excitation filter, chromatic beam splitter and blocking filter set usually are mounted in a rotatable filter holder above the objective, as different fluorochromes have different excitation and emission spectra.

Compared to the exciting light, the fluorescent radiation emitted will have lost energy and its wavelength will be longer than that of the exciting light. Consequently a fluorescing substance can be excited by light in the near-UV invisible range and be seen in the visible range.

As the excited light passes through the objective, the material of the lens or lens cement can emit fluorescence under UV and violet light excitation. Hence, special objectives made of non-fluorescent materials are required. Fluorochromes generally are bleached after a few minutes' exposure to the exciting light, although this effect can be reduced by the use of special immersion oils and mounting media developed for fluorescence use. One advantage in utilizing an epi-illumination fluorescence system is that it can be combined with conventional microscope optics such as phase contrast, polarization and interference, provided special fluorescence compatible objectives are used. Thus, these conventional optics may be employed to locate fields of interest prior to switching to fluorescence, thereby minimizing the exposure time of the fluorochrome to exciting light.

The microscope manufacturer will supply suitable hardware, filter sets and objectives for the fluorochromes to be used, and advise on the system required for the fluorochrome of choice (suitable fluorescence attachments only can be purchased from the manufacturer of the microscope base).

Direct counting of microbial cells stained with fluorescent dyes such as acridine orange and fluorescein isothiocyanate under a fluorescence microscope is a well established method of locating microorganisms on slides or filters.

The acridine orange method described by Bjornson[34] may be used for suspensions of microorganisms. Suspensions of microorganisms are killed with formalin at a concentration of 1% and then sub-samples of 1 mL are stained by addition acridine orange, at a concentration in the sample of 0.02%, for 2 to 5 minutes. The 1 mL suspension is filtered through a 0.2 μm pore size Nucleopore filter, prestained with Iragan black to minimize background interference. The filter is rinsed with distilled water, dried and mounted on glass slides using Cargille B (fluorescence compatible) immersion oil.

For microorganisms present as smears on slides the acridine orange method given by Bancroft and Cook[21] is typical. The alcohol-fixed smear is rinsed in distilled water, then 1% acetic acid for a few seconds, and then in two changes of distilled water for 2 min. The slide is dipped in a 0.01% solution acridine orange stain for 3 min. The stain is made by diluting 0.1% aqueous acridine orange by 1:10 with 0.06 M phosphate buffer (pH 6.0). The slide is rinsed in pH 6.0 buffer for 1 min, and differentiated with 0.1 M calcium chloride for 0.5 to 1 min, and then washed and mounted in phosphate buffer. Alternatively the slide may be washed and mounted with water soluble Aqua-poly Mount (Polysciences, Warrington, PA), a fluorescence compatible mounting medium, for a more permanent specimen.

The acridine orange specimens are viewed under a fluorescence microscope. The wavelength of the excitation light is at 500 nm, while the emitted light from the molecule is at 530 nm. The DNA fluoresces with a yellow green color, while RNA and some mucins fluoresce red (red indicates a relatively high uptake of the dye). Cells therefore may appear apple green or red. The fluorochrome stained microorganisms then are counted under fluorescent light in the manner described previously. Types of organisms may be distinguished by the use of non-specific fluorochromes, for example in acridine orange stained soil microflora, eukaryotic algae autofluoresce red, cyanobacteria autofluoresce gold and acridine orange (AO) stained bacteria fluoresce green.[35] High concentrations of AO may cause non-specific orange staining of cells. The color of AO fluorescence is not always an indication of the cell physiological state, as active cells (and those in which DNA and RNA have degenerated) fluoresce orange. Inactive bacteria contain predominantly DNA to which AO attaches as a green fluorescing monomer. Highly active cells contain large amounts of RNA to which AO attaches as a red fluorescing dimer, which masks the green fluorescence of DNA.[35] Occasionally non-biogenic particles or the background may autofluoresce or fluoresce green from uptake of the AO stain. This is a common problem with fluorescence microscopes. The operator normally must judge whether each fluorescent object is an a bioaerosol particle or an artefact. The fluorescence dye is bleached out of the sample by the exciting light. This bleaching is a permanent destruction of the fluorochrome by light-induced conversion of the fluorochrome to a chemically non-fluorescent compound. This process requires light and oxygen, so mounting the specimen with a coverslip and using Aqua-poly, Citifluor or a similar photoretardent mounting medium will provide some protection, reducing the rate of bleaching by up to a half.[35] To avoid bleaching, the fluorescence system normally is not switched in until the field is selected and ready to count. Normal white light transmission illumination with phase contrast, bright field etc. may be used instead for field selection.

Similar techniques are employed with other non-specific fluorescent biological stains such as ethidium bromide,[36] fluorescein isothiocyanate[37] and DAPI[38] (4'6-diamidino-2-phenylindole dihydrochloride). The ethidium bromide and DAPI is taken up by DNA, while fluorescein isothiocynate is a standard label for proteins. Ethiduim bromide, DAPI and fluorescein isothiocyanate require excitation at 520 nm, 360 nm and 490 nm respectively, with emission at 610 nm, 480 nm and 525 nm, respectively. Fluorescein may be viewed with the same filter set as used for acridine orange, but the other two fluorochromes require their own filter sets. As filter sets may be changed rapidly in the head of the microscope, the same sample may be stained with different fluorochromes, and these can be visualized distinctly and independently if their excitation and emission characteristics are suitable.

Fluorochromes may be utilized to stain bioaerosol samples impacted onto dry surfaces by spore traps, and will stain bacteria, fungi, yeast and spores. Incubation with nalidixic acid can distinguish between living and dead cells, where 'viable' cells become elongated and swollen.[39,40] Nalidixic acid is an antibiotic that inhibits DNA gyrase and hence prevents cell division, so cells growing in CO_2/air will elongate instead, while dead or dormant cells stay the same size and shape. This method is fairly subjective though, and it is difficult to say whether the uncounted cells actually are dead. Another method for detecting viable cells involves acridine orange combined with INT (2-(p-iodophenyl)-3-(p-nitrophenyl)-5-phenyl tetrazolium chloride), which indicates the number of respiring organisms.[41] Matcham et al.[42] discuss similar fluorescence methods for assessing fungal viability and biomass. Because only very small amounts of fluorochrome are required to be taken up by cells for fluorescence microscopy, they can be used to stain and view living microorganisms in culture.

Fluorochromes also have been conjugated to antibody proteins as a method for visualizing these antibodies as they attach to cell markers in antigen-antibody reactions. As the antibody is produced in animals immunized with the specific antigen, this technique is termed immunofluorescence. It is useful for the identification of medically important microorganisms but more recently this fluorescent antiserum technique has been applied to microbial ecology,[43] although its high specificity precludes its use as a general technique for identifying bioaerosol particles. Immunofluorescence methods for yeast are discussed by Pringle et al.[26]

Fluorescein diacetate (FDA), a nonfluorescent nonpolar derivative of fluorescein that readily penetrates cell membranes, has been used with mixed success to detect viable microorganisms. Intracellular FDA is hydrolyzed by non-specific esterases, resulting in release of the fluorochrome fluorescein. As fluorescein accumulation depends on intact membranes and active metabolism, only cells that were active during FDA exposure should fluoresce.[44]

Some species of microorganisms may contain naturally fluorescent pigments, and can be made to autofluoresce given the correct exciting light wavelength, e.g., cyanobacteria have the phycocyanin accessory pigment[45] that has a fluorescent peak at 655 nm. The primary photosynthetic pigment chlorophyll a in autotrophs (plants), such as algae, also autofluoresces, with a peak at 665 nm.

The Confocal Microscope

The confocal microscope is a scanning optical microscope that can build up a complete picture of a small field by sequentially scanning through the specimen. A laser or bright arc lamp is brought into sharp focus within the specimen by a microscope objective, and light from this focused probe is collected and brought to a second focus equivalent to or confocal with the light source. An aperture is placed in the image plane, which strongly selects the confocal rays, removing out-of-focus light. There are three light scanning methods used: stage scanning, beam scanning and tandem scanning.[46]

The beam-scanning confocal microscopes generally are utilized for biological imaging. In the beam-scanning (point-scanning) confocal microscope, illumination usually is scanned by a computer-controlled moving mirror, and the collected light from the specimen focused through an aperture on to a sensitive detector, such as a photomultiplier or silicon photodiode. The signal from this point detector then is fed into a framestore for image display, processing and storage. By the use of computers, the confocal microscope builds up a focused black-and-white digitized image of a slice through the specimen. As the video system is a digital image processor, the confocal microscope often contains limited image analysis software as well, such as that for measuring object lengths, as well as for image enhancement and pseudocoloring.

Most beam-scanning confocal microscopes are based on a standard light microscope and use epi-illumination in the same manner as a fluorescence microscope. Standard dichroic mirrors and filters select the imaging wavelengths. Thus, the confocal microscope may be used to detect fluorochromes. The wavelength of the excitation is limited by the type of laser employed. For example, the BioRad MRC1000, shown in Figure 11.2, has a krypton-argon laser that can detect three fluorescent labels in the blue, yellow and red wavelengths, allowing triple labeling experiments to be carried out. Confocal UV microscopes also are available for fluorochromes such as DAPI, using, for example, a 325 nm helium-cadmium laser.

The confocal microscope is useful for investigating cells in culture, e.g., with *in vivo* probes studying intracellular pH and the metabolism of sodium chloride and calcium in mammalian cells. It can provide useful morphological information on, for example, colony growth on solid media, or *in vivo* cell morphology. If required, the confocal system may be combined with an inverted light microscope. Compared to a conventional light microscope, the confocal microscope has a much larger depth of focus, being able to build up an image of a vertical 'slice' through the specimen. However, it has no better resolution than a standard light microscope.

The confocal microscope can create 3D stereoscopic images using appropriate software. The confocal microscope's advantage over the scanning electron microscope (SEM) is that it can be used on living cells, although it has much poorer resolution than the SEM. A typical beam-scanning confocal system such as the BioRad MRC1000, shown in Figure 11.2, in the UK costs around £60,000 to 80,000, excluding microscope.

The Scanning Electron Microscope

The scanning electron microscope (SEM) is also a scanning system like the confocal microscope. However, in this case a fine beam of electrons of 5 to 50 keV is scanned across a sample in a series of tracks. These electrons interact with the sample, producing back scattered electrons (BSE), secondary electron emission (SEE), light or cathodoluminescence and X-rays, which may be detected and displayed on the screen of a cathode ray tube.

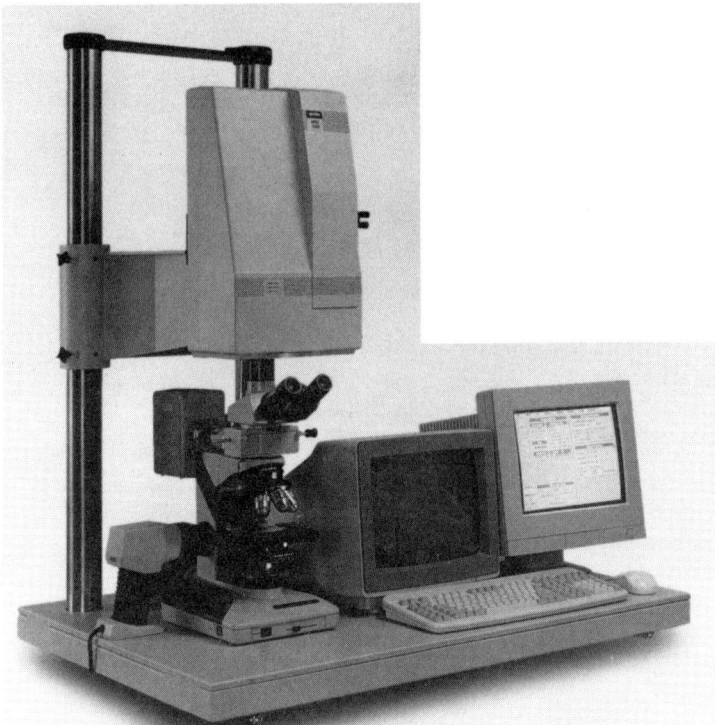

Figure 11.2. The MRC1000 confocal microscope (BioRad Microscience Ltd., Hemel Hemstead, Herts, UK).

SEE is the normal mode of SEM operation. Using a scanning generator, the electron beam traverses the specimen in a raster pattern. The secondary electrons emitted are collected onto a positively charged plate, the anode, creating a signal proportional to the number of electrons hitting the anode. This signal is amplified and used to modulate the intensity of a spot-scanning cathode ray tube. On the screen the specimen appears to be viewed from above, although the motorized sample stage can often be tilted. In BSE mode the particles appear to have been illuminated from a point source and give the impression of height due to shadows. Typically, samples are mounted on 25 mm diameter stubs, made from a conducting material such as aluminum or carbon. The surface of the sample normally must be coated with a conducting material, and typically the specimen will be sputter coated with carbon, for X-ray probe micro-analysis, or a metal such as gold, for high resolution viewing. Some SEM (and TEM) are equipped with special probes to detect semi-quantitatively the presence of the heavier elements by X-ray microprobe analysis.[47,48]

The SEM can view samples at magnifications varying from 10 to 400,000 × at a resolution of about 5 to 20 nm (which varies with accelerating voltage). Its depth of focus is several millimeters, about 300 times that of the light microscope. A SEM system in the UK costs from £80,000 to £140,000, depending on whether X-ray microanalysis is required. An example SEM is shown in Figure 11.3. Some SEMs are available with quoted resolutions nearer 1 nm, but these are expensive at £200,000.

All objects are subjected to a high vacuum within the microscope, and to withstand it, and to allow them to be metal coated, all specimens must be anhydrous. Specimens containing appreciable amounts of water will be distorted badly, which is clearly a problem with biological samples such as microorganisms. Hardening of the cells in fixative such

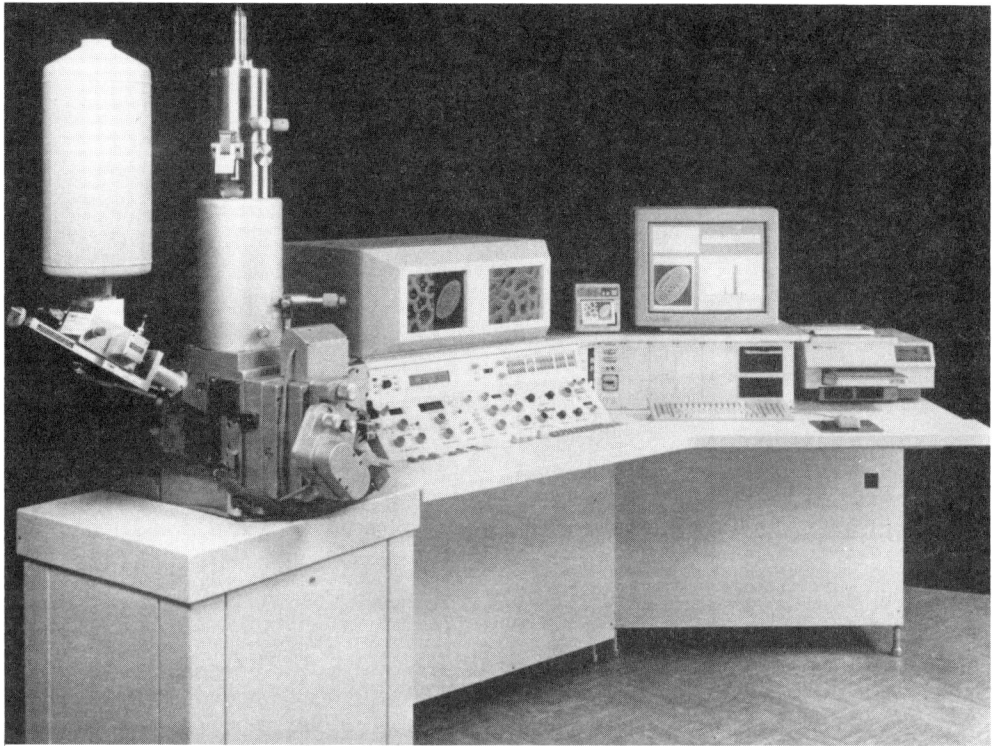

Figure 11.3. The scanning electron microscope, a CamScan CS44 (Cambridge Scanning Co Ltd., Bar Hill, Cambridge, UK).

as glutaraldehyde does help. Damage is further reduced if the water is replaced with solvents such as acetone or ether. Alternatively, freeze-drying or critical point drying in CO_2 may give better preservation.[49,50]

The SEM only reveals surface detail, although it does this with clarity and good depth of focus. The SEM can provide information on morphology of bacteria in soil, and can visualize cells on particle surfaces or in aggregates. It also can give excellent surface details of spores and pollens, and these may be viewed on the surface of a collection filter with the minimum of preparation. SEMs are able also to provide 3D red/green stereoscopic images of specimens, if required.

Modern SEMs (and TEMs) have dedicated computers built in. It is possible therefore to directly convert the SEM (or TEM) image to digital form and store it on a floppy disk (generally in TIF or IMG format) for transfer to an image analysis computer. Alternatively, the SEM may have its own image processing and analysis software, as well as optical disk storage for image archiving. However, there may be difficulties when using an older machine, and with these, photographs are taken of the SEM screen for subsequent size measurements of the objects using an epidiascope attached to an image analyzer. The SEM may be employed for estimating airborne concentrations and aerosol size distributions of particles, e.g., by counting particles on filters. However, sampling regimes may be complicated by the requirement to sample fields at different magnifications. SEM calibration is normally by the use of etched silicon crystal grids, accurately manufactured spherical latex particles or various traceable standard specimens (Agar Aids, Stanstead, Essex, UK).

The Transmission Electron Microscope

The transmission electron microscope (TEM) also uses electrons instead of light rays. With acceleration voltages of 100 keV an electron has the properties of an electromagnetic wave of about 0.04 nm, 10,000 times shorter than visible light. The resolving power of a TEM is about 1 nm, and some manufacturers quote resolutions of 0.2 nm when operating at 200 KeV. A beam of electrons is projected from an electron gun and passed through a series of electromagnetic lenses. A condenser lens collimates the beam onto the specimen, and an enlarged image is produced by magnifying electromagnetic lenses. As in the SEM, the image is visualized on a cathode ray tube. The penetrating power of the electrons is very weak, so only very thin specimens of less than 0.1 μm thickness generally are examined with the conventional TEM.

As with the SEM the specimen must be dehydrated, as the TEM also operates under high vacuum. Contrast in the specimen is due to the scattering of electrons within the specimen. Low mass elements that make up biological tissue and cells do not scatter the electrons well, and hence have poor contrast. The contrast in such specimens normally is enhanced by staining with salts of various heavy metals, such as uranium, osmium, cobalt and lead.[51] Single-cell microorganisms usually are fixed, dehydrated, embedded in resin and sectioned using an ultra-microtome fitted with a glass or diamond knife. The section then is mounted on special metal or carbon grids and the section is stained. Very small particles such as viruses, protein molecules and bacterial flagella normally are placed directly onto the support grid and negatively stained instead (where the stain is used to increase opacity of the field surrounding the particle). Other sophisticated techniques such as metal shadowing and freeze-fracturing have been developed also for TEM. The interested reader should consult specialist literature for detailed information on the subject of TEM and specimen preparation.[52]

Over the last 30 years, TEM has provided enormous insights into the ultra-structure of cells and viruses. More recently metal (e.g., gold and platinum) conjugated antibodies have been used to pinpoint the location of enzymes within the cell.[10,27]

TEM has been used also to visualize particles that can be photographed for morphometric analysis,[17] although modern machines have video cameras, TV monitors, computers and image analysis/enhancement software built in. Calibration usually is carried out with narrowly defined metal meshes, shadowcast diffraction grating replicas, or accurately manufactured particles (Agar Aids, Stansted, Essex, UK). Costs for a TEM system are comparable to SEM, although TEM operating costs may be higher, as it is more labor intensive, and a precision ultramicrotome also is required for its use. A typical TEM system is shown in Figure 11.4.

Another electron microscope, the scanning transmission electron microscope (STEM) can provide quantitative information about biological materials, such as the total ion concentrations in small organelles. The STEM is discussed by Andrews and Leapman.[53]

THE IMAGE ANALYZER

Introduction

Image analysis is concerned with the counting, quantification and classification of objects within images. Before the advent of cheap computers nearly all image analysis was carried out manually by eye, which was slow and labor intensive. However, now there are a multitude of computer software-driven image analysis systems on the market. These

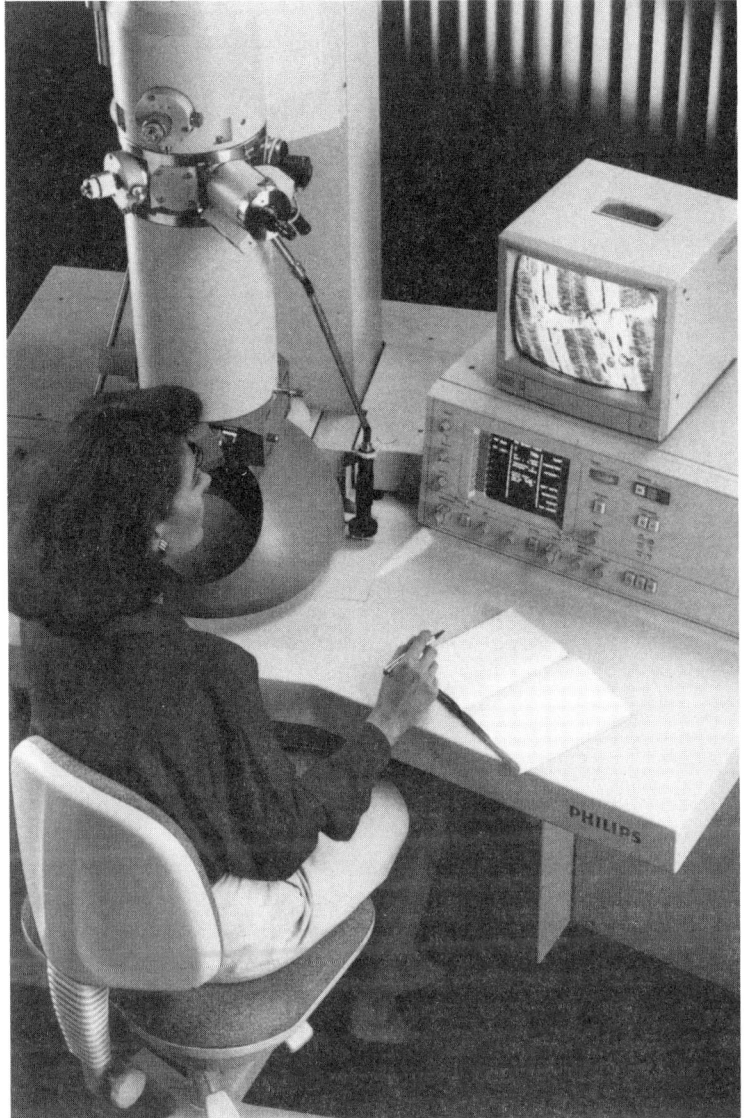

Figure 11.4. The transmission electron microscope, a Phillips CM100 (Philips Electron Optics, Cambridge, UK).

range from very small PC-based systems with a video card to expensive versatile machines that can control a microscope motorized stage, autofocus the image, and analyze samples automatically. Non-technical reference books discussing the principles of computer image analysis are available.[54,55]

A typical image analysis system comprises a video input, normally a video camera, an analog-to-digital converter, a PC-based computer with special video processing hardware, an additional monitor for visualizing the image while running the image analysis software, and output devices such as a text printer. The UK has many specialized manufacturers producing general purpose systems, such as Applied Imaging's Magiscan (Applied Imaging, formally Joyce Loebl, Hylton Park, Sunderland, Tyne and Wear, UK), Seescan Ltd's Sonata/Symphony (Cambridge, UK), Leica-Cambridge's Quantimet (formerly Cambridge

Instruments, Cambridge, UK). The price of a full system varies from around £10,000 to £80,000, excluding microscope. A typical image analysis system is shown in Figure 11.5. No programming knowledge is required to run a modern image analysis system, and anyone used to a microscope, and a personal computer running DOS and Microsoft Windows, will find any system relatively straightforward for simple tasks.

Video Cameras

The type of video camera required for image capture depends on the image analyzer use. Black-and-white (B&W) systems were often supplied with a Newvicon or Chalnicon thermionic tube camera with unity gamma (for densimetric measurements). These cameras have good light sensitivity, with the peak around 700 and 600 nm, respectively. They can detect normal and brightly fluorescing objects down a microscope (with the Newvicon more sensitive in the red and the Chalnicon more sensitive in green region). Weakly fluorescing microscopic objects have to be detected with an image intensified camera, such a silicon intensified target camera (SIT, peak sensitivity 440 nm) or the extremely sensitive intensified silicon target camera (ISIT, peak sensitivity 480 nm). The ISIT camera in particular is very noisy and requires image enhancement processing to improve picture quality. Solid state cameras based on charged coupled devices (CCD) are now more common than thermionic tube cameras, having the advantages of robustness and linear optics. Color cameras are only required for full color image processing systems. Frequently these are triple CCD or tube cameras, where there is one detector for each of the primary colors (red, green, blue). In the UK a good B&W camera costs around £2,000, a good color camera costs £5,000 to £8,000, whereas a good ISIT B&W camera is over £20,000. Despite the use of colored stains, often a B&W system is suitable for most applications when combined with optical filters, and is considerably cheaper.

The video camera normally will be connected to a port on the top of the microscope, and receive images from the sample simultaneously with the eyepiece. Alternatively, the camera may be fitted with a lens and attached to an epidiascope to view photographs, negatives or large objects. Occasionally, input images are captured from video recorders, via special interfaces from electron microscopes, from digitized images stored on floppy disks or traced with a stylus via a graphics tablet.

Image Analysis

The output from the B&W video camera is fed into an analog-to-digital converter circuit where the TV image is converted into a digitized image made up of square picture elements (pixels). The favored number of pixels is 512×512 (262,144 pixels) in most machines, although 256×256 or 1024×1024 may be used. Low resolution (256×256) has the advantage of fast processing. The high resolution 1024×1024 generally is not used, as it increases image processing time, requires the use of large image monitors so that objects made of small clusters of pixels can be seen, and the resolution of many cameras is not good enough. For a digitized B&W image each pixel must be assigned a grey level corresponding to how white or black that part of the picture was. A 4-bit, 6-bit or 8-bit pixel has 16, 64, or 256 possible grey levels per pixel (where 0 is black, 256 white, and 1 to 255 is increasingly brighter shades of grey, with an 8-bit converter). Again, more grey levels mean more computing power is required, so 64 grey levels per pixel is usual at present. The capturing of images and all subsequent image processing is under the control a computer that interrogates the video processing board.

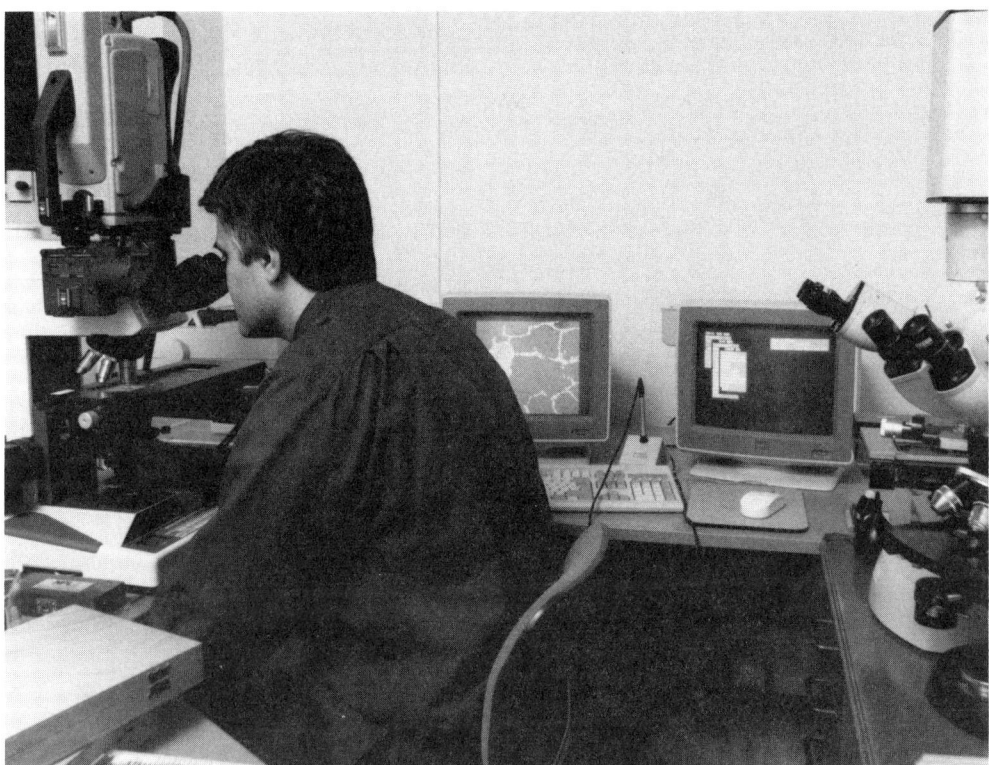

Figure 11.5. A typical general purpose image analysis system, in this case an Applied Imaging Magiscan full color system attached to a Nikon FX photomicroscope with computer-controlled motorized stage and focus (far left), an epidiascope (not shown) and a Zeiss Universal microscope with manual motorized stage (rear right). The image monitor is shown on the left, and is edited by a light pen. The computer monitor is shown on the right, and is controlled by a keyboard and mouse.

The computer runs a "user friendly" menu driven program that lists all the image process options available. This normally is shown on a separate color VDU screen from that used to display the image. Once captured by the software, the digitized image is put into video memory where it can be processed (and the image screen no longer shows a live digital image).

A captured B&W digitized image is now in a form where it can be manipulated by the computer. For example, camera images with a lot of noise can be captured many times and the images averaged. All constant information, i.e., the image of interest, will be retained but randomly moving interference will be removed. This is an important technique used with image intensified cameras. Geometric manipulation may be used also by single pixel transformations, e.g., to rotate the image or reverse black and white (invert) in the image to change negative film images to positive photographs. Where shading problems exist, e.g., low magnification condenser effects, a background image of a clear field may be subtracted from one with the sample present, to remove this shading effect.

Special transform filters or convolutions may be applied to groups of pixels in the image to provide image enhancement, increased contrast or edge detection (often used only when counting objects as such transforms may distort the image slightly). However, it must be emphasized that these operations are slow, and generally never as effective as ensuring that the sample itself is prepared in a manner ideal for subsequent image analysis, e.g., by modifying staining techniques or ensuring that the microscope is set up correctly with linear illumination across the field, and its optics are free from dirt.

After any B&W image transformation, a special 1-bit (binary) image, where all pixels are either 'on' or 'off,' is used then for object detection. This binary image is overlaid onto the captured B&W digital image displayed on the image monitor. Normally this binary image is set such that all the pixels are 'off,' and the binary image is transparent. The binary image is grey level thresholded using the B&W digital image, by selecting a menu option in the image analysis program. To select all black to mid-grey pixels, the operator enters thresholding values between 0 and 32 (in a 6-bit image). The computer then switches 'on' every pixel in the binary image that overlays a pixel in the B&W image that has a grey level between 0 and 32. The switched 'on' pixel is given a color (say red) that masks the underlying B&W image (switched off pixels still allow the B&W image pixels between 33 and 64 to be seen). In this way the threshold is adjusted to detect all the objects of interest in the image.

The detected objects now are visible as a two-dimensional group of colored shapes on the screen. This binary image may be subjected also to other operations. For example, all objects not fitting a defined parameter, e.g., length smaller than 2 μm or area greater than 200 μm^2, could be deleted from the binary image using an object delete option. The objects can be eroded (removing a layer of pixels from the surface), dilated (adding a layer of pixels) or converted to an outline. Touching objects can be separated automatically, and fibrous objects can be skeletonized to a single arc (for fiber length). Most important of all is a binary editor, where the operator can manually edit the binary image to remove, select or modify objects based on that seen under the microscope and in the underlying B&W grey level image. All these binary operation routines, and many more, simply are selected as options on the software menu.

Once all the objects of interest have been selected, the binary image is then 'measured.' The appropriate measurements required will be selected from options given in the measurement menu. The computer will measure each object (groups of touching pixels), in a fixed raster scan starting from one corner of the image screen. Some typical object measurements are:

Object area. The area of all the pixels that reside within the boundary of the object. Using the object area measurement, the diameter of a circle of equivalent area may be calculated.

Detected area. The area of the thresholded pixels within the boundary of the object, excluding unthresholded holes within the object.

Perimeter. The perimeter of the object. The boundary of the object is determined by computer algorithms smoothing the pixels around the object perimeter (this may or may not be done in area measurements).

Length. The maximum Feret length (equivalent to a caliper measurement of the distance between the two pixels farthest apart on the object).

Breadth. The maximum breadth Feret of the object normal to the length measurement.

Aspect ratio. A shape factor (length/breadth).

Orientation. The angle of the length Feret relative to the screen horizontal.

Equivalent circle diameter. The diameter of a circle of equivalent area to the object (equivalent circle diameter = $2 \times$ (object area $\div \pi)^{0.5}$).

Width. The length of the widest part of the object in a horizontal plane.

Height. The length of the widest part of the object in a perpendicular plane.

Circularity. A shape factor that determines how circular the object is, where for a circle the value is 1.0 (circularity = perimeter$^2 \div (4\pi \times$ object area)).

Optical density. A measurement of the grey level pixel under the threshold detected pixel in the binary image, that can be used to estimate the mass or concentration of a material if viewed by transmission microscopy; see the literature for further details.[54,55]

Center of gravity. Determines the X,Y co-ordinates of the center of gravity of the object in the field, or on the slide, which is useful for locating the object relative to its neighbors. This location may be the pixel location within the 512 × 512 image on the screen, or the object's location on a slide that had the zero co-ordinate set before the slide is scanned (using a motorized stage).

Other object measurements will be available, such as arc measurements, or the measurement of objects within objects (nesting). The analyzer will report also various field measurements of all the objects in the binary image, such as the total detected area and percentage of the field thresholded. Other field and object measurement routines generally are available for morphometric analyses, such as overlaying the image with a grid for detecting X,Y intercepts, or the analysis of nodes within the image.

The field and object measurements will be written to disk normally, from where the data can be combined, modified (e.g., by applying size or field number criteria), statistically analyzed, and expressed graphically, say as frequency distributions or scatter plots. This is done by dedicated results packages written for the image analyzer. Alternatively, the data may be converted to ASCII format and loaded into spreadsheets or scientific graphic packages, such as FIG.P (Biosoft, Cambridge, UK).

Image analysis machines also allow interactive manual measurement of lengths on the screen using a light pen or mouse, where the two points (pixels) of interest are selected, and algorithms calculate and display the screen the distance between them.

Prior to field sampling, the image analyzer is calibrated using a standard glass micrometer slide under the microscope (or using a ruler, calipers or a scale on a photograph, under the epidiascope). Normally a light pen or a mouse provides a line across a known length of the scale, the length is typed in, and the computer calculates the length of each square pixel. From now on all measurements will be in the unit of calibration.

By measuring the detected area of sampled bioaerosol particles, it is possible to estimate the particle mass by assuming a unit density. The simplest method for this is to calculate the equivalent circle diameter of the object from its area, and to use this to calculate the volume of a sphere with this diameter. By simply multiplying the volume by the known or estimated density (g/cm^3), the mass of the particle can be estimated:

Equivalent circle radius = (object area ÷ π)$^{0.5}$

Volume of a sphere = 4/3 × (π × circle radius)^3

Estimated mass of sphere = Volume of sphere × density

Similar calculations can be applied to rod-shaped objects to estimate their volume, using the volume of a cylinder ($\pi \times r^2 \times h$) or possibly a cone ($1/3 \times [\pi \times r^2 \times h]$), but for more irregular-shaped objects the above sphere method generally is used, and the errors accepted.

These equations can be applied to all the objects measured, to estimate the total biomass present in the sample (the sum of all estimated bio-particle masses), or to estimate the total mass of bio-particles in a particular size range. Normally this is done simply by inputting the equations into the image analyzer's results software package. Methods for estimating mycelial biomass are to be found in the literature.[42,56]

Full-color image analyzers also are available. These create a digitized color image from three B&W images, one for the red component, one for the green, and one for the blue. For thresholding, the grey levels of the pixels in all three images have to be selected. As before, a single binary image is produced by thresholding, which is edited and measured in the same way as for those created from B&W images. Full-color systems may be better at detecting objects within difficult images having little contrast in B&W, but surprisingly many applications are adequately met by cheaper and faster B&W systems using optical filters. Color systems also may be run in B&W mode, with B&W cameras, if preferred.

Sophisticated image analyzers also can control motorized XY stages on microscopes, and adjust the focus (and even change optical filters or magnification). All image analyzers can be used manually by direct interaction with the software menu. Generally they include also an option to save a sequence of commands from the menu, and replay them either fully automatically or semi-automatically (when the analyzer will stop at certain places for user control, e.g., for thresholding or changing fields). These are termed task lists, and are saved in editable PC files, to be run when required. Using these task list files, an image analyzer can automatically raster, snake or randomly scan across an entire slide and measure objects in every field, when using a microscope fitted with autofocus and an XY motorized stage. However, often an operator is required to verify that artifacts are not being measured in error or that objects are not being missed, so semi-automatic task lists with operator interaction are the norm.

Although an image analysis system can count and measure objects very conveniently, they are less good at only counting object numbers, particularly in semi-automatic mode. This is because in the time taken to threshold and edit an image, an operator could have counted the objects by eye, particularly if the number of objects in the field is quite low. However by operating automatically with a motorized 16 slide XY microscope stage, an image analysis system can run continuously 24 h a day.

Although some image analyzers can measure properties like object texture, they are not yet anywhere near capable of identifying microorganisms or spores with the same precision as can an experienced operator by eye. Image analyzers apply only simple parameters such as object length or area to discriminate between bioaerosol particles. The introduction of specific monoclonal fluorescent dyes has changed this a little, as the analyzer can count and thus 'identify' stained organisms under a fluorescence microscope.

As the human brain poorly estimates parameters such as area or perimeter, for cell morphometric analyses and bio-particle size distributions, the image analyzer is clearly a valuable instrument. More recently, dedicated image analysis systems have been developed for automatic measurement, e.g., colony counting, ELISA plate reading,[57] cell metabolism[32,33] (e.g., calcium), metaphase finding and chromosome karyotyping. Image analysis systems have been developed or modified on a small scale specifically for microbiological detection,[58] but advances in image analysis are so rapid that these can become obsolete quite quickly.

Colony Counting

Bioaerosol samples frequently are cultured on solid medium within Petri dishes to estimate the number of airborne viable cells or particles (subject to the effects on viability of the rigors of sampling and the selectivity of the culture medium). The number of colonies present after incubation are counted and related to the volume of the air sampled, thereby estimating airborne concentrations. Traditionally, the number of colonies present after incubation were counted by eye, using trained technicians, often with the aid of a marker pen to mark off colonies scored. Today digital counter marker pens are available that count a colony every time the tip is pressed onto the agar surface, and the total counts are shown on an LCD display integrated into the pen.

A general purpose image analyzer also can count colonies, with a circular pointset traced around the edge of the Petri dish. The Petri dish would be detected using a video camera, with a focusing macro-zoom lens, attached to a epidiascope fitted with transmission and epi-illumination. Standard thresholding and object separation routines are applied to count the colonies. However, contiguous colonies present problems, even to an experienced human operator, and general purpose image analyzers can fall down in this respect. Also, colonies around the edge of the Petri dish tend to merge into the dish wall, and be difficult to count. So, in many applications the image analyzer gives disappointing colony count accuracy compared to that obtained by eye. Some analyzers permit the light pen or mouse to score off colonies on the image screen and give total counts in the same manner as counting by eye and pen.

Far more successful are dedicated image analyzer software specifically written for colony counting. These are run on dedicated PC systems (which may be based on the manufacturer's general purpose image analysis machines), and recent software-driven colony counters are far more successful than the older hardware-based systems around in the early 1970s. A typical modern colony counter is the Scan 500 produced by Seescan Ltd (Cambridge, UK), which costs £7,500, see Figure 11.5. The electronics is housed inside the lighting box, along with a B&W CCD camera, and the system is operated by a keyboard and monitor. The Scan 500 allows counting of over 600 plates per hour, and has options for different plate types and for work using microscopes. Using a password protected menu system, an authorized user sets up the counter for the plate type in use. Once set up, the counter can be operated by non-technical personnel, who repeatedly places plates under the camera and runs the automatic counting software. Each sample can be identified by free text or using a bar-code reader. The system has, for example, algorithms for detecting colonies at the edge of dishes, to identify and deal with 'spreaders' that may inhibit the growth of colonies, as well as automatic recalibration to compensate for variations in ambient lighting. Many experienced scientific users are critical of automatic colony counters on the grounds that they rarely give exactly the same counts per plate as they obtain by counting manually by eye. However, distinguishing colonies often can be subjective, and any errors in counting generally are below differences owing to random variation during sampling. A dedicated image analysis colony counter has the real advantage of being highly consistent and repeatable, very fast, with good record keeping of data, and suitable for applications having large throughputs of colony plates on a frequent basis.

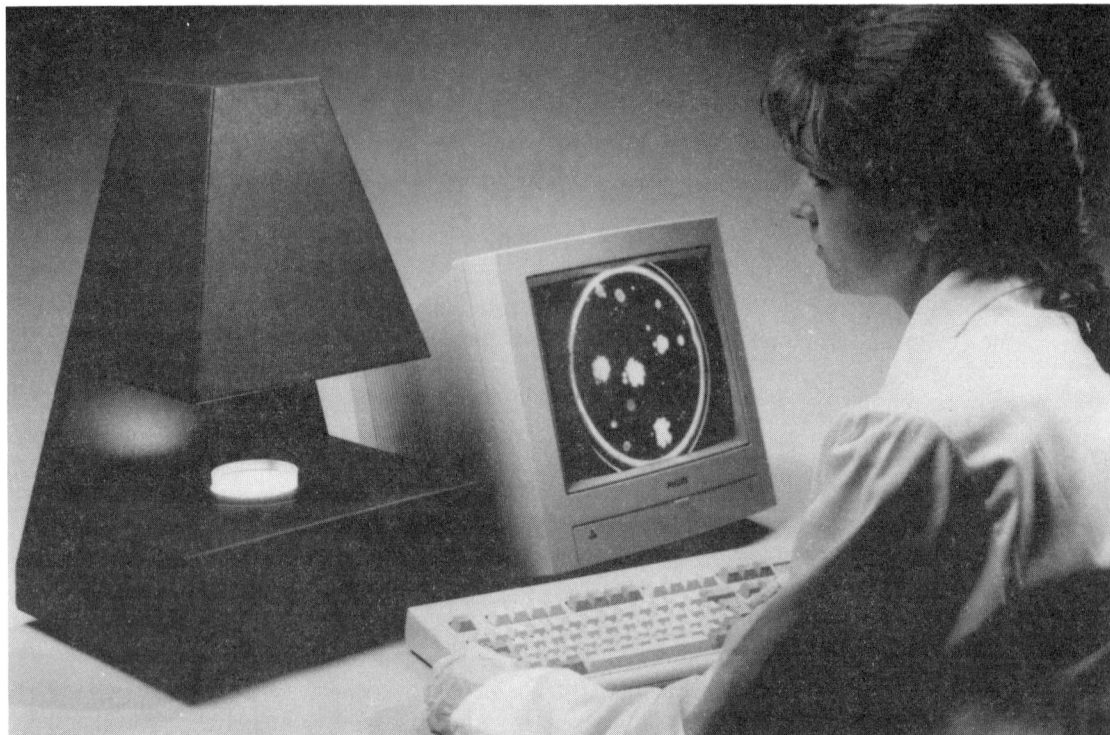

Figure 11.6. The Seescan Scan 500 colony counter (Seescan Ltd., Cambridge, UK).

OTHER PHYSICAL METHODS FOR DETECTING MICROORGANISMS

Owing to the time and expertise required to visualize, count and identify microorganisms using the microscopy techniques described, there is considerable interest in developing alternative methods for rapid microbial analysis (see also Chapter 12).

The use of fluorochromes for microbial detection with dual beam flow cytometers is discussed by Hadley et al.[23] The flow cytometer analyses cells in aqueous suspension, that are separated into a stream of droplets, one cell per drop, and passed at high speed through a sensing region of focused laser light plus detector. The amount of scattered light at right angles to the incident laser beam and liquid jet is dependent on cellular structure. Fluorochromes with distinguishable emission spectra can be selected to measure simultaneous binding of different antibodies. Some flow cytometers also may be used to sort cells of preselected size and/or fluorescence criteria, by deflection between two high-voltage plates.[40,60]

Other devices such as the Coulter Counter model MZ (Coulter Electronics Ltd., Luton, Bedfordshire, UK) and the Elzone 180 system (Particle Data, Elmhurst, IL) count cells and provide particle size analysis over the range 0.4 to 800 μm, although these systems cannot distinguish between living cells, spores or inert particles.

Various other techniques exist for the identification of organisms, such as enzyme linked immunosorbent assays (ELISA) used as diagnostics test for bacteria and viruses,[57] and pyrolysis mass spectroscopy, where samples from culture are pyrolyzed under vacuum and the evolved gas is recorded by a fast scanning mass spectrometer; the spectrum obtained is compared and matched to library data (model PYMS-200X, Horizon Instrument

Ltd., Heathfield, East Sussex, UK). Further information on alternative methods is given by Nelson,[13] Fox,[61] standard microbiology textbooks,[10,11] and in Chapter 13.

SUMMARY

The light microscope still has an important role in the identification of particles of biological origin, as it is relatively cheap and, most importantly, allows a direct visualization of the bioaerosol particles, and it can be used with living cells. The use of fluorescent dyes and specific cellular antibody stains has made the identification of viable cells more precise. The main failings of the light microscope are the resolution limit of about 0.2 μm and the limited depth of focus. The more expensive confocal light microscope offers an advantage of better depth of field, and is also ideal for use in visualizing fluorescent markers.

The scanning electron microscope has far better depth of field than even the confocal light microscope, and much better resolution (typically around 10 nm). However, the system is expensive, and can only provide surface details of objects, although it does this with remarkable clarity. As it operates under a vacuum, all samples must be processed to remove water, which may cause distortion of structure, and precludes the visualization of living material.

The transmission electron microscope has provided a wealth of information on the ultrastructure of cells. It has a resolution limit of about 1 nm. Ultrathin sections must be prepared, which are stained with metals to provide adequate contrast. These metals may be conjugated onto antibodies for specimen identification. Again, owing to the use of a vacuum, living material cannot be examined. The system is expensive. Both scanning and transmission electron microscopes can be used for semi-quantitative elemental analysis when an X-ray microprobe analyzer is incorporated.

The advent of cheap computer hardware and video systems has led to a growth in the number of image analysis systems available. These can cheaper than a laboratory microscope. However, sophisticated cameras and image analyzers are expensive. Although such systems have nowhere near the visual processing skills of a human operator, they can provide very useful information on such things as specimen morphometry or bioaerosol size distributions. They are also able to scan samples totally automatically, although with varying degrees of success. Dedicated image analysis systems have also been developed for the counting of colonies on plates.

The present rapid development of computers, image processing hardware and software, and various specific histological stains will undoubtedly lead to further advances in the microscopic methods used for the identification of bioaerosol particles.

REFERENCES

1. Grainger, J.M., and Lynch, J.M., Eds. *Microbiological method for environmental biotechnology.* The Soc. for Appl. Bacteriology Tech. Series No. 19. (Academic Press, London, 1984).
2. Corn, M. "Non-viable particles in the air," in *Air pollution,* 2nd ed. Stern, A.C., Ed., (Academic Press, New York, 1968), pp 47–94.
3. Wolf, H.W., Saliy, P.S., Hall, L.B., Harris, M.M., Decker, H.M., Buchanan, L.M., and Dahlgren, B.S. *Sampling microbial aerosols.* Public Health Monograph 60. (Public Health Service Publication 686. Library of Congress No. 59-60091, 1959).

4. Hawker, L.E., and Linton, A.H. *Micro-organisms: function, form and environment.* (Edward Arnold, London, 1971), p. 529.
5. Laynyi, B. "Classical and rapid identification methods for medically important bacteria." in *Methods in microbiology, vol. 19,* R.R. Colwell and R. Grigorova, Eds. (Academic Press, London, 1987), pp. 1–68.
6. Colwell, R.R., and Grigorova, R., Eds. "Current methods for the classification and identification of microorganisms" in *Methods in Microbiology,* vol. 19. (Academic Press, London, 1987).
7. Guthrie C., and Fink, G.R., Eds. "Guide to yeast genetics and molecular biology" in *Methods in enzymology,* vol. 194. (Academic Press, London, 1991).
8. Hewitt, W., and Vincent, S., Eds. *Theory and application of microbiological assay.* (Academic Press, London, 1989).
9. Salle, A.J. *Fundamental principles of bacteriology,* 6th ed. (McGraw-Hill, New York, 1967).
10. Brock T.D., and Madigan M.T. *Biology of microorganisms,* 6th ed. (Prentice Hall, Englewood Cliffs, NJ, 1991).
11. Stanier R.Y., Ingraham J.L., Wheelis M.L., and Painter P.R. *General microbiology,* 5th ed. (Prentice Hall, 1987).
12. Boyd R.F., and Hoerl B.G. *Basic medical microbiology.* (Little, Brown, Boston, 1991).
13. Nelson, W.H., Ed. *Instrumental methods for rapid microbiological analysis.* (VCH Inc., USA, 1985).
14. Gaur, R.D, and Kala, S.P. "Studies on the aerobiology of a Himalayan alpine zone, Rudranath, India," *Artic Alpine Res.* 16:173–183 (1984).
15. Hirst, J.M. "An automatic volumetric spore trap," *Ann. Appl. Biol.* 39:257–265 (1952).
16. Gregory, P.H. *The microbiology of the atmosphere,* 2nd ed. (Leonard Hill, Bucks, UK, 1973).
17. Allen, Y. *Particle size measurement,* 3rd ed. (Chapman and Hall, London, 1981).
18. Anderson, A.A. "New sampler for the collection, sizing, and enumeration of viable airborne particles," *J. Bacteriol.* 76:471–484 (1958).
19. Burge, H.A. (1985) "Indoor sources for airborne microbes," in *Indoor air and human health,* Gammage R.B., and Kaye S.V., Eds. (Lewis Publishers, Ann Arbor, MI, 1985) pp 139–170.
20. Needham, G.H. *The practical use of the microscope* (Charles C. Thomas, Springfield, IL, 1977).
21. Bancroft, J.D., and Stevens, A., eds. *Theory and practice of histological techniques,* 3rd ed. (Churchill Livingstone, Edinburgh, UK, 1990).
22. Bancroft J.D., and Cook H.C. *Manual of histological techniques* (Churchill Livingstone, Edinburgh, UK, 1984).
23. Hadley, W.K., Waldman, F., and Fulwyler, M. "Rapid microbiological analysis by flow cytometry," in *Instrumental methods for rapid microbiological analysis.* Nelson W.H., Ed. (VCH Inc., USA, 1985) pp. 67–89.
24. Chantler, S.M., and McIllmurray, M.B. "Labelled-antibody methods for detection and identification of microorganisms," in *Methods in microbiology, vol. 19.* Colwell, R.R., Grigorova, R., Eds. (Academic Press, San Diego, CA, 1987) pp. 273–331.
25. Newel, D.G., McBride, B.W., and Clarke, S.A. *Making monclonals - a practical guide to the production and characterization of monclonal antibodies against bacteria and viruses.* (Cambridge University Press, Cambridge, UK, 1991).
26. Pringle, P.R., Adams, A.E.M., Drubin, D.G., and Haarer, B.K. "Immunofluorescence methods for yeast," in *Methods in enzymology, vol. 194* (Academic Press, San Diego, CA, 1991) pp. 565–602.
27. Clark, M.W. "Immunogold labelling of yeast ultrathin sections. *Methods in enzymology, vol. 194* (Academic Press, San Diego, CA, 1991) pp. 608–626.

28. World Health Organisation. *Reference methods for measuring airborne man-made mineral fibers* (World Health Organisation, European Office, Copenhagen, 1985).
29. Department of Health. *Good Laboratory Practice, the United Kingdom compliance programme* (Department of Health, Crown Publications, London, 1989).
30. Trolldenier, G. "The use of fluorescence microscopy for counting soil microorganisms," *Bull. Ecol. Res. Comm. (Stockholm)* 17:53–59 (1973).
31. Taylor, D.L., and Salmon, E.D. (1989) "Basic fluorescent microscopy," in *Fluorescence microscopy of living cells in culture. Part B: Quantitative fluorescence microscopy - imaging and spectroscopy,* Wang Y., Taylor D.L., Eds. Methods in cell biology, vol. 30 (Academic Press, San Diego, CA, 1989) pp 207–237.
32. Wang, Y., Taylor, D.L., eds. *Fluorescence microscopy of living cells in culture. Part A: Fluorescent analogs, labelling cells and basic microscopy.* Methods in cell biology, vol. 29 (Academic Press, San Diego, CA, 1989).
33. Wang Y., and Taylor, D.L., eds. *Fluorescence microscopy of living cells in culture. Part B: Quantitative fluorescence microscopy - imaging and spectroscopy.* Methods in cell biology, vol. 30 (Academic Press, San Diego, CA, 1989).
34. Bjorson, P.K. (1986) "Automatic determination of bacterioplankton biomass image analysis," *Appl. Environ. Micriobiol.* 51:1199–1204 (1986).
35. Wynn-Williams, D.D. "Photofading retardent for epifluorescence microscopy in soil micro-ecological studies," *Soil Bio. Biochem.* 17:739–746 (1985).
36. Fisar, Z., Hysek, J., and Binek, B. "Quantification of airborne microorganisms and investigation of their interactions with non-living particles," *Int. J. Biometeorol.* 34:189–193 (1990).
37. Babiuk, L.A., and Paul, E.A. "The use of fluorescein isothiocyanate in the determination of the bacterial biomass of grassland soil," *Can. J. Microbiol.* 16:57–62 (1970).
38. Sieracki, M.E., Johnson, P.W., and Siebeth, J.M. "Detection, enumeration and sizing of planktonic bacteria by image analysed epifluorescence microscopy," *App. Environ. Microbiol.* 49:799–810 (1985).
39. Commission of the European Communities. *Methods for the detection of microorganisms in the environment* (Report EUR 14158 EN, ISBN 92-826-3858-8, Brussels, Luxembourg, 1992) pp 18–21.
40. Sayler, G.S., Burns, R., Cooper, J.E., Gray, T., Pedersen, J., Prosser, J., and Selenska, S. "Detection methods for modified organisms in the environment," in *The release of genetically modified microorganisms - Regem 2.* Stewart-Tull, D.E.S., Sussman M., Eds. (Plenum Press, New York, 1992) pp. 119–122..
41. Kogure, K., Simidu, U., and Taga, N. "A tentative direct microscopic method for counting living marine bacteria," *Can. J. Microbiol.* 25:415–420 (1976).
42. Matcham, S.E., et al. "Assessing fungal growth on solid media," in *Microbial methods for environmental biotechnology.* Grainger J.M., Lynch J.M., Eds. (Academic Press, London, 1984) p. 12.
43. Bohlool, B.B, and Schmidt, E.L. "The immunofluorescence approach to microbial ecology," *Advances in microbial ecology* 4:203–241 (1980).
44. Chrzanowski, T.H., Crotty, R.D., Hubbard, J.G., and Welch, R.P. "Applicability of the fluorescein diacetate method of detecting active bacteria in freshwater," *Microb. Ecol.* 10:179–185 (1984).
45. Schrieber, U. "Reversable uncoupling of energy transfer between phycobilins and chlorophyll in Anancystis nidulans. Light stimulation of the cold-induced phycobilisome detachment," *Biochim. Biophys. Acta.* 591:361–371 (1980).
46. White, N. "Confocal scanning optical microscopy," *Soc. Gen. Microbiolo. Quart.* 18:70–74 (1991).
47. Chandler, J.A. "X-Ray micro-analysis," in *Practical methods in electron microscopy, vol. 5, part II,* Glauert, A., Ed. (North-Holland, Amsterdam, 1977).

48. Reed, S.J.B. *Electron micro-probe analysis,* 2nd ed. (Cambridge University Press, Cambridge, UK, 1993).
49. Mercer, E.H., and Birbeck, M.S.C. *Electron microscope, a handbook for biologists,* 3rd ed. (Blackwell Scientific Pub., Oxford, UK, 1972).
50. Mir, J., and Esteve, I. (1992) The SEM in microbial ecology. *Microscopy and analysis, Issue 29* (ISBN 0958-1952, Rolston Gordon Communications, Bookham, UK, 1992) pp 31–33.
51. Lewis, P.R., and Knight, D.P. "Staining methods for sectioned material," in *Practical methods in electron microscopy, vol. 5,* Glauert, A., Ed. (North-Holland, Amsterdam, 1977).
52. Glauert, A., ed. *Practical methods in electron microscopy* (A series of 12 volumes discussing electron microscope techniques, Series ISBN 0 7204 4250 8, North-Holland, Amsterdam, 1977).
53. Andrews, S.B., and Leapman, R.D. "Biological scanning transmission electron microscopy," *Microscopy and Analysis, Issue 36,* (ISBN 0958-1952, Rolston Gordon Communications, Bookham, Surrey, UK, 1993) pp 19–22.
54. Joyce Loebl Ltd. *Image analysis: principles and practice* (ISBN 0 9510708 0 0, Applied Imaging Int. Ltd., formerly Joyce Loebl Ltd., Sunderland, Tyne and Wear, UK, 1985).
55. Misel, D.L. (1978) "Image analysis, enhancement and interpretation," in *Practical methods in electron microscopy, vol. 7,* Glauert, A., Ed. (North-Holland, Amsterdam, 1978).
56. Frankland, J.C., and Lindley, D.K. "A comparison of two methods for the estimation of mycelial biomass in leaf litter," *Soil Biol. Biochem.* 10:323–333 (1978).
57. Wreghitt, T.G., and Morgan-Capner, P. *ELISA in the clinical microbiology laboratory.* (Cambridge University Press, Cambridge, UK, 1991).
58. Sieracki, M.E., Johnson, P.W., and Sieburth, J.M. "Detection, Enumeration and sizing of planktonic bacteria by image analysed epifluoresence microscopy," *Appl. Environ. Microbiol.* 49:799–810 (1985).
59. Estep, K.W., MacIntyre, F., Hjorleifsson, E., and Siebeth, J.N. "MacImage, a user friendly image analysis system for the accurate measuration of marine organisms," *Mar. Ecol. Prog. Ser.* 33:243–253 (1986).
60. Shapira, H.M. *Practical flow cytometry,* 2nd ed. (Alan R. Liss, New York, 1988).
61. Fox, A., and Morgan, S.L., Eds. *Analytical microbiology methods: Chromatography and mass spectroscopy.* (Plenum Press, New York, 1991).

CHAPTER 12

Chemical Analysis of Bioaerosols

K.R. Spurny

INTRODUCTION

Progress in analytical and colloid chemistry, based on theoretical and instrumental advances, has had important and positive influences on the development of modern microbiological analysis. The last quarter century has seen a revolution in application of sensitive and increasingly informative methods of chemical analysis. This has happened to a large extent because advances in electronics, computers and laser technology have allowed practical applications of these methods that previously had been understood in theory but were cumbersome in use.[1] Detection sensitivity is now high and therefore analyses of very small samples and even of 1 μm single particles are possible.[2]

A list of these methods is a long one, ranging from atomic and molecular spectroscopy, on the one hand, to separation and surface science on the other. In addition to infrared and nuclear resonance spectroscopy, which began the revolution, we must add the various fluorescence spectroscopies, Raman and resonance Raman spectroscopy, and associated techniques, such as coherent anti-Stokes Raman spectroscopy. Separation methods, e.g., gas chromatography and high pressure liquid chromatography, when coupled with mass spectrometry or alone, have enhanced the revolution. These and many other methods have increased vastly our capacity to do basic research.

One fruitful area of application is the development of rapid methods for detecting and identifying microorganisms in general and of bioaerosols in particular. Physical methods for detection of microorganisms in air (Chapter 11) depend on measuring the number, size and shape of aerosol particles, but these measurements indicate only the presence of aerosol particles in a size range. For example, a sudden increase in the ratio of the number of particles in the size range of 0.5 to 5.0 μm could be indicative of artificially generated biological aerosols. Physical measurements, being non-specific, make no distinction between a microbial particle and particles of other origin.

Chemical methods being more specific than physical methods, are able to distinguish between bioaerosol and other aerosol particles; in some cases even different microbial species can be identified. Nevertheless, they still cannot fully replace classical methods that enable concentrations of viable microbial particles to be established as well as their specification.

Classical bacterial detection and identification are usually based on morphological evaluation of organisms as well as viability tests in or on various media, susceptibility to various phages and antibiotics, and ability to metabolize various compounds, etc. Generally, no single test provides a definitive identification of an unknown bacterium. Consequently classical methods are time consuming and cannot be developed as on-line and in-situ measurement and monitoring procedures. Modern physical methods of chemical analysis are exciting and promising alternatives to classical identification procedures but the future will surely confirm this promise.

Modern analytical instruments and procedures are characterized by rapid acquisition and high reproducibility of data, as well as frequently featuring computer-aided data recording and interpretation. They can be developed and used for in-situ and on-line monitoring.

Chemical and physico-chemical methods for detecting and identifying microorganisms depend on chemical measurements of mainly biochemical components of unique chemical structures in microorganisms and consequently can distinguish biological from non-biological materials. Commonly used for this purpose are tests for the presence of nuclei acids, adenosine, triphosphate, and proteins. For example, biological compounds (e.g., proteins, nucleic acids, peptides, carbohydrates) react with certain dyes to produce shifts in light absorption characteristics of the latter. Measuring this shift by conventional spectroscopy can provide the basis for detecting biological compounds in an unknown sample. For example, acridine orange, which becomes bound to nucleic acids, produces fluorescence in the microbial particles under ultraviolet light.

Such methods are fairly rapid and provide information about the biological nature of the sample but have several limitations. Notably, none of the methods depends on viability (Chapter 6) of the sample and consequently, these methods provide no information on the living state of the microorganisms and provide little information for differentiating among them.

Laser spectroscopical methods, chromatography, mass spectrometry and various combinations of these methods have attracted more attention for rapid analysis of microorganisms. The composition of cell walls and DNA content differ markedly between bacterial species and therefore can be used for detection and determination of microorganisms in different milieus, including in air.

The following review, while although not fully exhaustive, deals with the most common or promising methods that already have been applied for detection of microorganisms in liquids and in some cases also aerosols. Some of them are more or less of historical importance only, but the majority of them promise to be successful for future monitoring of bioaerosols.

COLLECTION AND SPECIMEN PREPARATION

The majority of successful applications of chemical detection methods have been with liquids. The first task in sampling bioaerosols for chemical detection therefore is to obtain a suitable quantity of preconcentrated microorganisms in a relatively small volume of liquid, e.g., aqueous suspension. Microbes dispersed in air are subjected to different physical and toxic stresses[3–5] (Chapter 6). Desiccation is experienced by all airborne microbes. Radiation, oxygen, ozone and its reaction products and various pollutants also decrease viability and infectivity through chemical, physical and biological modifications to predominantly phospholipids, protein and nucleic acid moieties. Changes in water content therefore occur in all microorganisms while airborne and during sampling and represent the most fundamental potential stress. The act of sampling itself can induce significant viability losses as can prehumidication before collection as lost water is slowly replaced. Viruses without structural lipids also are destabilized by desiccation and viruses with structural lipids are least stable at high relative humidities. Different bacteria species have different optimum aerosol survival temperatures, which in most cases is about +10°C:

Nonviable microbes may not be considered as dead because under certain conditions damaged moieties may be repaired (Chapter 6).

Bioaerosol particles have a broad range of particle sizes and selective sampling methods (Chapters 9 and 10) sometimes have advantages. Microbial particles often occur in the atmosphere as agglomerates, i.e., coagulates with other material such as dust particles or fog droplets, etc., and in dusty atmospheric environment microorganisms may be carried on particles larger than 3 μm in diameter.[6]

As mentioned above, when chemical detection and evaluation methods are applied, the viability of the sampled biological particles is not necessarily important. Therefore, several methods used for general aerosol sampling can be applied also for the collection of bioaerosols.[7] As we will see later, indirect methods for the determination and identification of biological aerosols can be divided into two categories, viz., analysis of bulk samples and analysis of individual particles.[2]

Bulk samples of bioaerosol particles can be collected onto a solid or into a liquid. Such samples then can be used directly or after further processing. Membrane and nucleopore filters are the most useful tools for filtration sampling[8,9] while cascade and virtual impactors, aerosol centrifuges, and thermal and electrostatic precipitators enable sample collection onto thin organic films.[2]

Sampled particles then can be analyzed and evaluated either in bulk or in some cases as a collection of individual single particles.[2] In the second case the analytical equipment (e.g., mass or Micro-Raman spectrometer) has to be combined with a microscopical tool (LM, SEM, etc). For several methods, e.g., fluorescence, luminescence, infrared and Raman-spectroscopies, as well as for flow cytometry and electrochemical methods, samples suspended in liquid are desirable and for these the impinger methods as well as "wet microcyclones" provide most suitable samplers.[7]

"Aerosol beam procedures" are suitable for on-line and in-situ detection of single particles. These methods are combined mainly with laser mass spectrometers and therefore will be described later.

Preparation and optimization of liquid suspension or solid samples for different analytical procedures follows collection of airborne material. For the application of gas chromatographic and mass spectrometric methods, the analyte first must be converted into the vapor phase.[1] In addition, vaporization of biological materials must be preceded by fragmentation into smaller units, and two methods for the vaporization and fragmentation of large biological compounds have been used. In the first method, macromolecules are converted into smaller units by hydrolysis and subsequently reaction with suitable compounds to give relatively volatile derivatives.

The second method involves heating the sample in a non-oxidizing atmosphere (pyrolysis) whereby large molecules are fragmented to produce small units in the vapor phase. Fragmentation takes place at preferred junctions in the molecules and the fragments contain information about their parent molecules.

Several designs of pyrolyzers have been reported in the literature[8-12] and are classified in terms of their mode of operation, either the continuous (e.g., furnace) or pulse. The pulse mode pyrolyzers are used commonly and include resistively heated pyrolyzers, Curie-point pyrolyzers, and laser pyrolyzers[13] and encouraging results have been obtained with Curie-point pyrolyzers.[12-15] In this method a small sample (5 to 20 μg) is deposited directly onto a ferromagnetic wire and high frequency inductive heating is employed to raise the temperature of the wire to its Curie temperature. This takes about 100 ms to reach equilibrium temperature and typically the sample is pyrolyzed in 1 to 10 s. Fully automated systems for Curie-point pyrolysis by GC and MS are available.[13]

Laser pyrolysis possesses many desirable features[2,13] including high temporal and spatial resolution, very fast heating rate, and ability to pyrolyze extremely small amounts of sample (e.g., 10^{-12} g or less), enabling single-cell analysis. More detail of methods for sample and analyte preparation for individual analytical procedures will be described later.

MICROSCOPICAL DETECTION AND IDENTIFICATION

Light microscopy (LM), scanning electron microscopy (SEM), transmission electron microscopy (TEM) and analytical electron microscopies (ASEM and ATEM) have been used for counting and identifying single bacterial particles (or colonies) sampled from the atmospheric environment by impaction and filtration.[16] Mainly nonspecific identifications or differentiations then are possible. Air particle sizes and forms as well as their staining by different dyes can be observed and measured. Specifically stained biological particles can be examined with an automated microscope. Fluorescence light microscopy (FLM) and fluorescence scanning microscopy (FSM) provide the most sophisticated methods for microscopical evaluation of airborne microorganisms. FLM and ethidium bromide as a selective fluorescence dye has provided a useful method for ambient air measurement and for CFU (colony-forming units) estimation.[17,18] Applications of FSM are also promising because of high resolution and better automation facilities.[19] In the FSM the higher energy laser light interacts with the specimen (e.g., nucleopore filter with bacterial particles), which then emits lower energy fluorescent light that passes back through the microscope optics, scanner, and dichroic to a simple achromatic lens that focuses it on an aperture. A detector (usually a photomultiplier tube) picks up this fluorescence and the output signal is amplified and digitized. An image then is formed by electronically storing and processing the single picture elements obtained from the scanned area.

SEM, ASEM, TEM and ATEM also have been successfully applied for the identification and quantification of airborne microorganisms. Millipore and nucleopore filters are very suitable for sampling and individual microbial particles may be counted on the filter specimen using magnifications of about 3000×.[20-25] ASEM and ATEM enables single microbial particles to be analyzed by the EDXA (energy dispersive x-ray analysis) procedures to give elemental spectra from each individual particle.[2] Newer analytical electron microscopes also include analysis of very light elements (e.g., C) and by means of such fingerprinting methods biological and mineral particles easily can be distinguished. In other investigations, combinations of LM and FM (fluorescence microscopy) with SEM were successfully used.[20]

CHROMATOGRAPHIC METHODS

Capillary gas chromatography (GC) or GC coupled with mass spectrometry (GCMS) are fast, specific and sensitive techniques for directly characterizing microorganisms. Furthermore, analyzing structural components rather than metabolic products does not require organisms to be viable, and consequently lengthy secondary (and possibly primary) culturing may be avoidable.

Pyrolysis-gas chromatography (PGC) can be used for detection and identification of microorganisms as a bulk method and pyrograms of sampled microorganisms predominantly contain amino acids, sugars, carbohydrates, purine and pyrimidine bases, etc. PGC was first applied in the 1960s and demonstrated that pyrograms of microorganisms and cell fractions generally are very complex (Figure 12.1). Consequently microorganisms have to

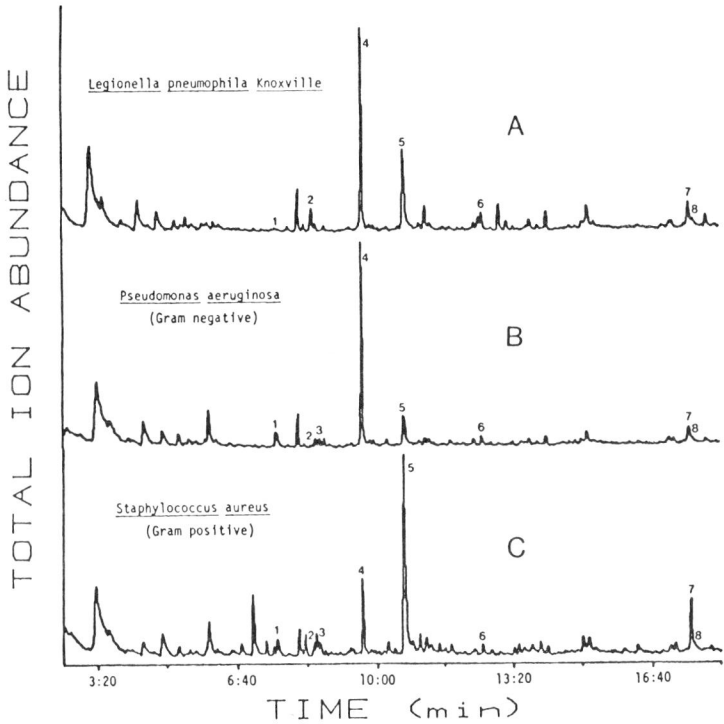

Figure 12.1 Example of three high-resolution capillary pyrograms of Legionella pneumophila (A), Pseudomonas aeruginosa (B) and Staphylococcus aureus (C).[26]

be differentiated by two or three characteristic peaks in a pyrogram containing perhaps two hundred or more components. In the past this meant tedious visual comparisons of many complex pyrograms, but since the beginning of the 1970s computerized pattern recognition techniques have been used to speed up the data analysis.[26,27]

Mass spectrometry (MS) is a critical component of this chemotaxonomic analytical approach because GC has non-selective detection (e.g., flame ionization) that does not permit unequivocal identification of chromatographic peaks. MS improves selectivity as well as allowing identifications or at least partial characterization of the chemical nature of the constituents of interest. The combined technique of GCMS can separate and identify only volatile components, but unfortunately most macromolecular components of bacterial cells are not volatile. Therefore, bacterial samples have to be chemically treated before GC, MS and GCMS analysis with the most suitable treatment being thermal fragmentation, i.e., pyrolysis.

LASER SPECTROSCOPICAL METHODS

Combining lasers with classical spectrometric methods like atomic emission and absorption spectroscopy, molecular infrared and Raman spectroscopy, fluorescence and luminescence spectroscopy, and particularly mass spectroscopy, have greatly improved their performance such that the latter now can provide very fast and sensitive analysis and identifications of very small amounts of elements or compounds in small volumes of samples.

Fluorescence and Luminescence Spectroscopy

Fluorescence- and luminescence-based identification techniques generate data from only some molecular components of the microorganism. These are therefore more specific.[28]

We may distinguish primary and secondary fluorescence spectroscopical methods with each one being subdivided into direct and indirect methods. Primary fluorescence methods are those in which naturally fluorescent components of bacteria are examined. Consequently, only bacteria containing or producing some fluorescent pigment may be examined by primary fluorescence methods, but these become less specific when applied to a mixture of bacteria.

Secondary fluorescence methods, in contrast, involve introducing a foreign fluorophore to the bacteria before identification. Such methods therefore do not depend on any natural fluorescence and can be applied more generally. Immunofluorescence assays are of particular promise. Microbes that are inhaled by an animal host provoke an allergic response[29] and the ensuing antibodies then can be identified by specific fluorescence measurements, usually by addition of fluorescein.

Both fluorescence procedures may be used as direct and indirect methods. Direct analyses are those in which the bacterial cells are present during the analysis, unlike indirect methods in which the cells are absent.[30] Direct immunofluorescence assays are highly specific, and often the bacterium of interest can be detected in the presence of many similar bacteria.

Chemiluminescence can be defined as the light, usually visible, that is emitted as part of the energy released during certain exothermic reactions.[31-32] A variety of substances exhibit chemiluminescence, with the most suitable for bacteria being luminol. When bacteria are placed in a highly alkaline media (pH 13) they lyse and the haem moiety is released into the reaction mixture. These luminescence techniques can be used for the detection but not for the identification of microorganisms, but nevertheless are very sensitive and permit the detection of about 10^3 microorganisms per milliliter.

Bioluminescence is caused by a sequence of reactions involving firefly luciferin, adenosine triphosphate, luciferase, magnesium and oxygen but the application of bioluminescence to the detection of microorganisms in the atmosphere is difficult because the luciferase becomes denaturated.[33] On the other hand, the chemiluminescent techniques have been applied successfully in detecting bioaerosols.[34,35] (see also Chapter 13) The bioaerosol is sampled by some suitable collection method (e.g., a wet cyclone air sampler) and the bacteria are collected into a solution. Iron porphyrins (e.g., haem) from bacteria catalyse the reaction between luminol and an oxidizing radical (e.g., perborate). Sodium hydroxide to causes bacterial lysis thereby releasing haem and is also necessary for the luminol reaction:

$$\text{Luminol} + \text{ROOH} \xrightarrow[\text{hematin}]{\text{NaOH}} \text{aminophthalate} + \text{ROH} + N_2 + h\nu$$

The emitted light ($h\nu$) is proportional to the amount of haem released from the bacteria provided the reaction takes place under fixed conditions. Light emission is measured with a photomultiplier tube and the signal is registered by a recorder or a display. Standard calibration curves for different bacteria can be constructed and a mean value curve from these curves used to determine amounts of bacteria in an unknown sample.

Infrared and Raman Spectroscopy

Infrared (IR) spectra applied to identifying bacteria was first reported in the 1950s.[28] Sampled bacteria are smeared onto an IR cell and IR absorbance spectra are acquired using conventional instrumentation. The resultant spectra represent the chemical composition of the bacterium under investigation and, hence, are generally complex and broad band because of the large number of components in the spectra. These constraints limit the application of this technique, though Fourier Transform Techniques (FTIR) are more promising.[36]

Raman spectra are obtained by placing a sample in a beam of monochromatic radiation and measuring the intensity of the light scattered at longer wavelengths (or smaller values of frequency). The Raman spectrum of a biological molecule is in effect a vibrational spectrum and modern laser Raman spectrometers (e.g., the Micro-Raman Laser-Spectrometry) permit routine examination of solid or liquid microparticles.[37]

High quality Raman spectra in the visible region have been obtained from very small aggregates of chromobacteria and useful spectra obtained from individual cells in bacterial mixtures. Argon laser excited resonance Raman spectrometry has been successfully applied to such measurements and organisms present in the vicinity of the laser beam and in the beam itself (Figure 12.2) were observed and counted with ease by means of a television image via a vidicon tube attached to the microscope. Figure 12.3 shows spectra of *R. palustris*. These results are most promising for the successful detection of bioaerosols, as the method permits analysis of airborne bulk samples as well as detection and identification of single microbial particles.[38–40]

Mass Spectrometry

Mass spectrometry (MS) has proved the most promising modern method for chemical detection and identification of microorganisms, including bioaerosols.[41–45]

For bioaerosols the most suitable mass spectrometer is the pyrolysis mass spectrometer (PyMS). For mass spectrometric analysis of bioaerosols, the laser pyrolysis has many advantages including high temporal and spatial resolution, very fast heating rates, and ability to pyrolyze extremely small amounts of sample (e.g., 10^{-12} g or less) so that single-cell analysis is possible. Laser pyrolysis mass spectrometers are available commercially and a frequency quadruplet Neodymium-YAG laser[46–47] is used for the Laser Microprobe Mass Analyzer (LAMMA). The laser is focused to micron size and has a spatial resolution of this order. Special cascade impactors[48] or filters (especially nucleopore filters[49]) are used for sampling atmospheric particles, including microorganisms.[50,51] The specimen (Figure 12.4) is observed by light microscopy and single particles are irradiated by the high energy laser beam, pyrolyzed and analyzed by the time-of-flight (TOF) mass spectrometer. Particle concentrations as well as mass spectra of individual single bacterial particles are obtained. However, sample evaluation is time consuming by this method. On the other hand, Sinha et al.[13,41–45] have developed a useful modification of a PyMS. His particle analysis by mass spectrometry (PAMS) is an *in-situ* and on-line procedure in which aerosol particles are introduced directly from the atmosphere in the form of a particle beam into the ion source of the mass spectrometer. This method therefore enables analyses of single particles on a continuous, real-time basis. The main components of the PAMS system are the particle beam generator and a quadruple mass spectrometer.

Particle beams (analogous to molecular beams) are produced when aerosols expand through a fine nozzle into a vacuum.[13,52] The particle beam enters the mass spectrometer chamber and individual particles are pyrolyzed. Two different pyrolysis systems have been

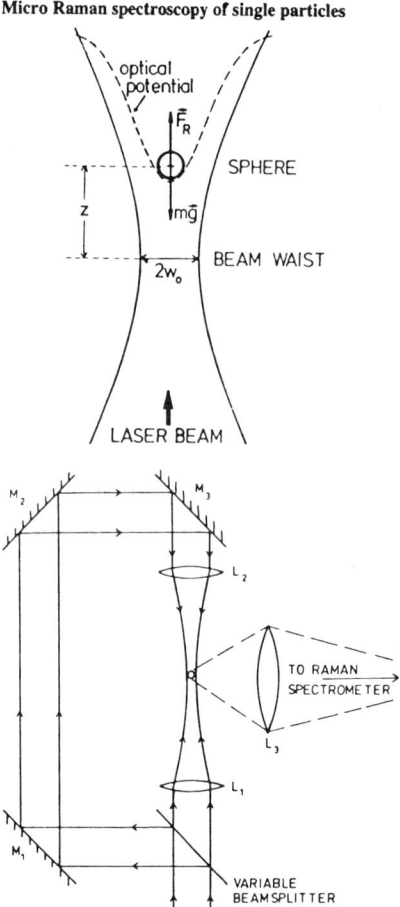

Figure 12.2 Schematic diagrams of spherical particles sitting in one and in two (lower picture) laser beams.[37]

examined and used. In the first, the accelerated aerosol particles impact onto a V-shaped zone refined rhenium filament (Figure 12.5). The filament is heated resistively and maintained at a constant temperature in the range of 200–1400°C. A particle striking the filament produces a plume of vapor molecules; these are ionized by electron impaction *in-situ*. The second pyrolysis system uses multiple lasers (Figure 12.6[42]) with the particles being pyrolyzed and ionized by a laser pulse while in flight in the aerosol beam. The Nd-YAG laser used for this pyrolysis, volatilization and ionization of the aerosol particles delivers 1 J of energy per pulse. In the latest development a "three laser system" is used: one for measuring particle size, the second for the volatization and pyrolysis, and the third for additional ionization.[53]

Bacterial particles in the aerosol beam are pyrolyzed into small mass fragments and a burst of ions is produced from individual particles after volatilization and ionization by electron impaction in the ion source of the mass spectrometer (or also in the additional laser). The mass spectrometer then measures the intensities of the different masses by scanning them in time. Intensities of the different mass peaks are normalized to the most intense peaks in their respective spectrum (Figure 12.7). The similarity between spectra is because the microorganisms have essentially the same major chemical building blocks. However, some differences are apparent by visual inspection while more objective comparisons of mass spectra and their reproducibility can be made by applying statistical proce-

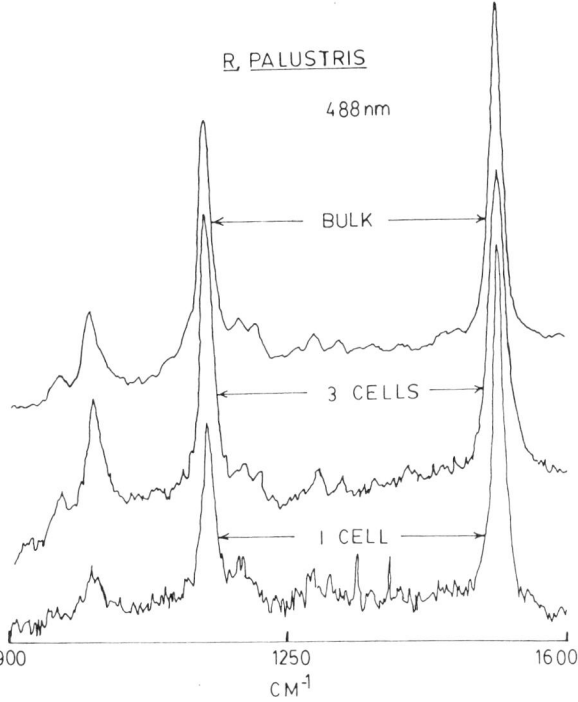

Figure 12.3 A comparison of bulk spectra from pure cultures and single-cell detection spectra.[40]

dures.[13,54] A library of mass spectral data of different microorganisms grown under different conditions should be compiled for identification by comparison.

Further improvement in bacterial chemotaxonomic characterization and fingerprinting has been achieved by applying Fast Atom Bombardment Mass Spectroscopy (FABMS). Positive and negative ion fast bombardment mass spectra may be obtained from lysed bacteria without extraction as phospholipids and other polar lipids are selectively desorbed to provide molecular ions. FABMS may serve also as an aid in differentiating between microorganisms.[55,56]

FLOW CYTOMETRY

In contrast to immunofluorescence assay that uses one highly specific reaction for identification of a bacterium, flow cytometry employs a condition of much less specific reactions to form a characteristic "response pattern."[57,58] In flow cytometry a sample of bacteria is stained with one or more fluorescent dyes, after which the pattern of dye uptake for each bacterium is determined with a flow cytometer. In the case of bioaerosol analyses, airborne microbial particles first have to be collected in liquids, e.g., by a wet microcyclone.

Most flow-cytometric systems allow simultaneous measurement of cell volume and two or more fluorescence signals (Figure 12.8) and sophisticated flow cytometers employing mercury-arc or laser light sources are commercially available.[58] Bacterial numbers, sizes and identification patterns can be provided simultaneously. The derived histograms provide useful fingerprints for distinguishing species and this method, named Mixed Dye-Fluorometry, is highly specific and may be used for multicomponent bacterial detection.

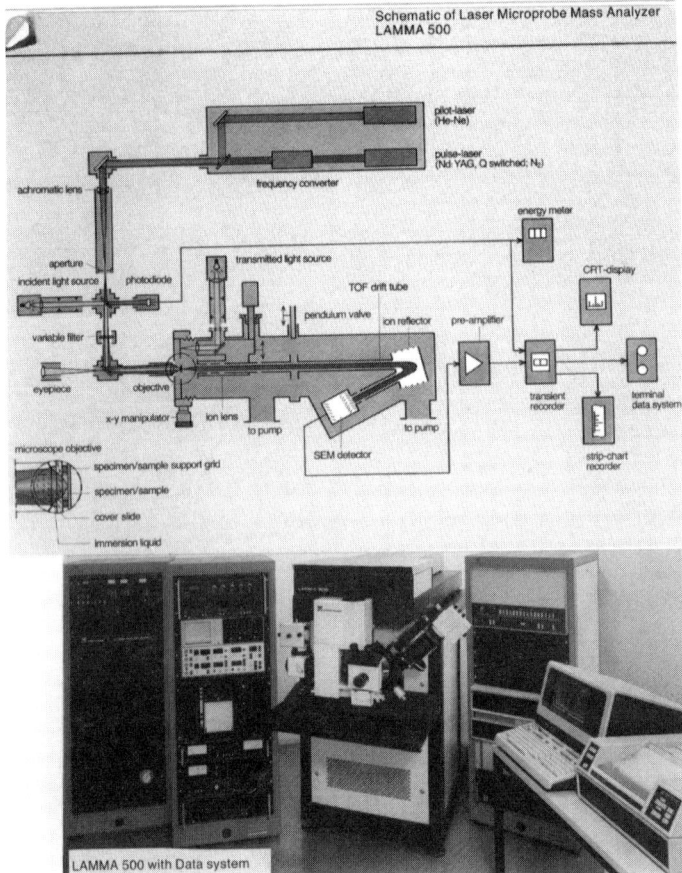

Figure 12.4 Schematic representation and a picture (below) of the LAMMA 500 commercial equipment.[46]

ELECTROCHEMICAL METHODS AND SENSORS

Decreased electrical impedance of a culture occurs when actively metabolizing bacteria utilize large molecules in a culture medium and form smaller ion pairs. The sensitive instruments now available usually can detect actively metabolizing bacteria when 10^6 to 10^7 bacteria per milliliter are present in the culture liquid.[59]

Impedance (Z) is a measure of the total opposition to the flow of a sinusoidal altering current in a circuit containing resistance (R), inductance (Λ) and capacitance (C). The relationship is:

$$Z^2 = R^2 + \frac{\Lambda}{2\pi \, f \, C^2}$$

where f is the frequency.

In electrochemical methods, bacteria must be viable and their concentration in the liquid milieu relatively high (> 10^6/mL). Applications of these procedures to bioaerosols are therefore not very promising, unless concentrated samples are available.

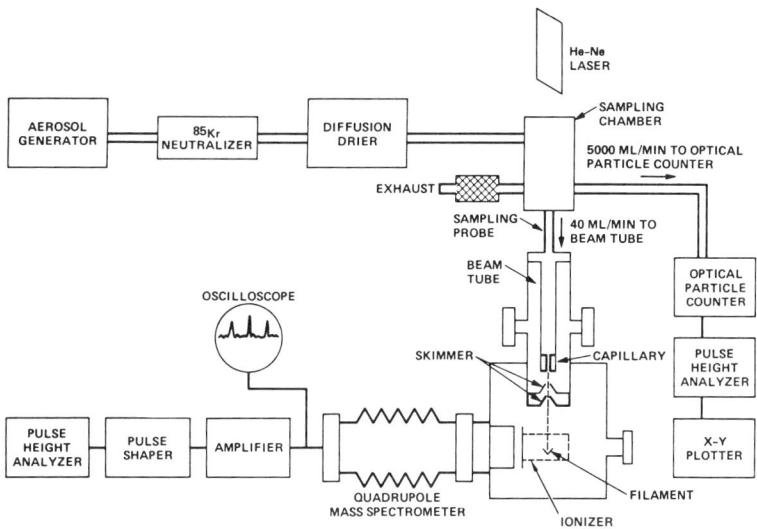

Figure 12.5 Schematic diagram of particle beam MS apparatus.[44]

The recent development of sensors using immunobiological techniques offers excellent selectivity through the process of antibody-antigen recognition and has revolutionized many aspects of chemical and biological sensor technologies.[60–62] Immunosensors use different detection techniques, e.g., electrochemistry, radioactive labelling, piezoelectricity, spectroscopic methods, etc., while rapid and sensitive detection of toxins and pathogens has been accomplished recently using a Light Addressable Potentiometric Sensor (LAPS[61]). The detection step is preceded by the formation of a streptavidin-antibody/antigen/antibody-urease complex followed by its capture onto a biotinylated nitrocellulose membrane. The LAPS is a semiconductor-based detector that monitors changes in pH over time due to conversion of urea to ammonia by the urease in the nitrocellulose bound toxin/pathogen antibody complex. Automated LAPSs have a continuous tape of nitrocellulose onto which analytes drawn in by an air sampler and reacted with the appropriate antibody mix are captured. The tape is automatically positioned under both the filtration unit and the LAPS unit. The detection limit is approximately 6000 organisms per sample.[62]

OTHER CHEMICAL METHODS

Modern instrumental chemical analysis is in continuous and fast development, and therefore it may be expected that already proven procedures will be improved and new principles discovered and developed. For similar reasons, new applications for microbiological detection and identification including bioaerosols can be expected. Many known methods for physical and chemical characterization of individual airborne particles[2] probably could be also extended for the fast and non-specific detection of bioaerosols.

The well-known SIMS techniques has similarities with MS and with LAMMA. In principle SIMS (Secondary-Ion-Mass-Spectroscopy) is a surface-sensitive technique in which a solid surface under bombardment by so-called primary particles (i.e., atoms or molecules with energies in the keV range) emits a number of secondary particles from the outermost molecular layers of the specimen. Emitted secondary particles comprise neutral atoms or molecules, positively or negatively charged atomic or molecular ions, excited particles, electrons or protons. An analysis of the positive and negative secondary ions

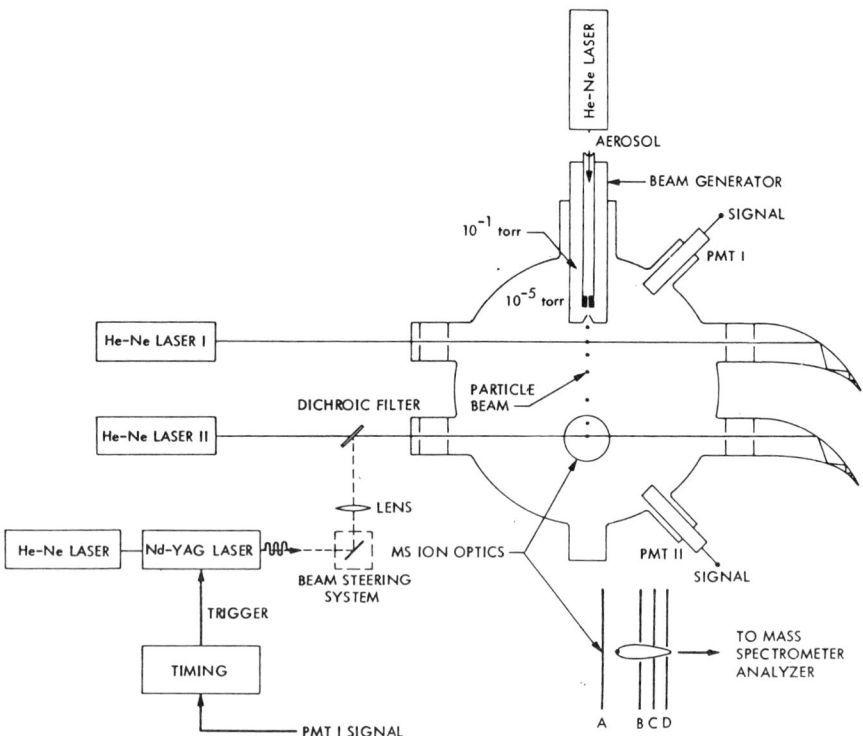

Figure 12.6 Schematic diagram of the laser-induced volatilization and ionization of particles in the aerosol beam.[42]

yields a mass spectrum, and this method is sensitive (detection limit of about 10^{-15} g or less). Bioaerosols collected by filters or cascade impactors may be directly analyzed as a bulk sample to give mass spectra of the organic parts of microorganisms, thereby providing fingerprints for different species or alternatively providing information on the presence of airborne microorganisms.[63] One other modern analytical technique that can be exploited for non-specific detection of bioaerosols is atomic spectroscopy, particularly atomic emission spectroscopy (AES). The classical flame photometry,[64] as well as the newer equipment of the AAS and ICP provide, specific element analysis of single aerosol particles.[65,66]

Applications of AES to atmospheric aerosol analysis started in the 1950s and much improved in the 1980s, enabling elements such as K, Na, Li, Ca, Ba, etc., to be detected in particles smaller than 1 μm. Organic particles (e.g., carbon, phosphorus and some pyrolysis products) can be detected in the flame or in the gas plasma. Some new combinations (e.g., AES with GC and MS) further promote applications of these techniques in microbiology as well as in bioaerosol detection.

FUTURE DEVELOPMENTS

In previous descriptions and considerations, we have emphasized that applications of modern methods of chemical analysis in the fields of microbiology and bioaerosols are still under development.[67,68] In general, chemical procedures are not yet completely satisfactory. Results to date are very encouraging but are not yet conclusive. Procedures still need to be faster, more sensitive and more specific. For example, fluorescence and luminescence

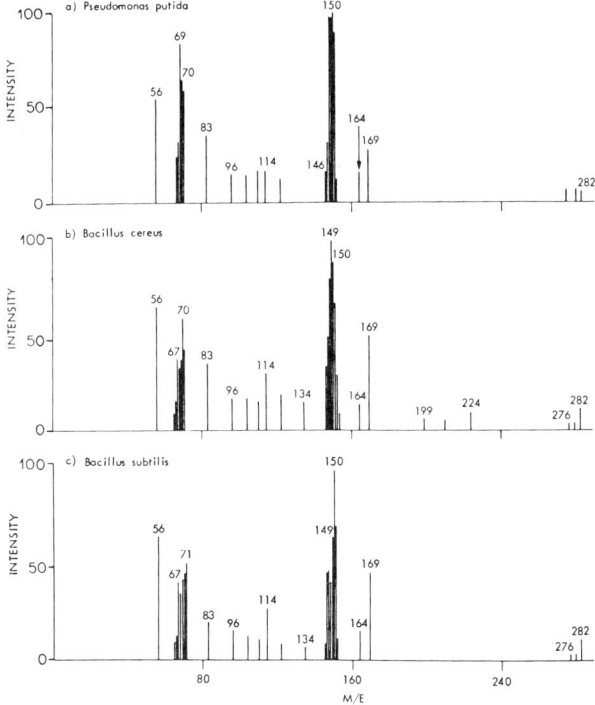

Figure 12.7 Mass spectra of three different bacterial particles.[44]

techniques may be used to characterize bacteria on the basis of the unique mixture of enzymes inherent in each bacterial type. Highly successful monitoring is then possible by means of immunofluorescence assays[69] but in the future bacterial identifications by fluorescence techniques will also require a good technique for analysis of mixtures of fluorophores. One very promising technique for this is Total Luminescence Spectroscopy (TLS).

Chemiluminescence probably is the only one that provides a standardized practical method for monitoring bioaerosols. Nevertheless, this procedure also needs further improvement especially for distinguishing between background material and living organisms.

Flow cytometry based on measurement either of scattered light or fluorescence has the potential for rapid detection and analysis of microorganisms. A remarkable range of new fluorochromes with improved fluorescent intensities is now available. The combination of new fluorochromes and equipment provides rapid methods of flow cytometry and can offer remarkable capacity for analysis, detection and identification of microorganisms, including bioaerosols.

Raman spectroscopy, particularly the Laser-Micro-Raman procedure for analysis of single particles, is an attractive and promising method, having applications for bioaerosol analysis but still needs more basic as well as applied investigation. UV should be more effective in exciting species-specific Raman resonances since most chemotaxonomic markers for bacteria absorb in the UV only. UV-excited resonance Raman spectra of bacteria may be expected to be much richer in detail than equivalent visible spectra and should be obtained in general with even better sensitivity as background fluorescence

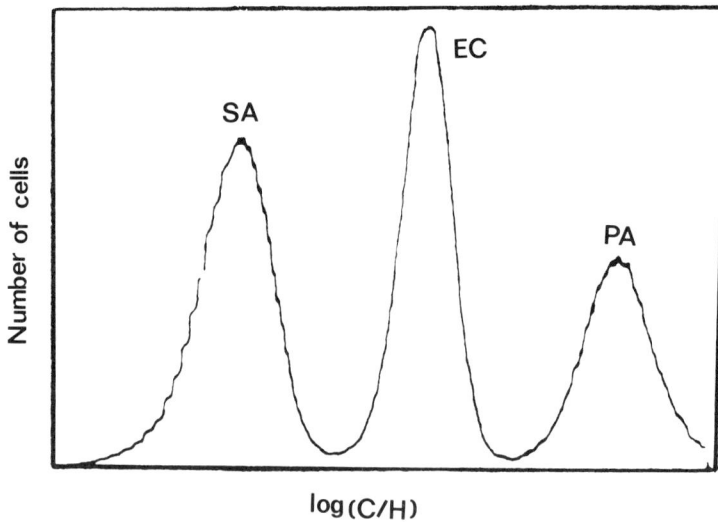

Figure 12.8 Distribution of the ratio of two fluorescence signals (C,H) over the cell population of *Staphylococcus aureus* (SA), *Escherichia colie* (EC), and *Pseudomonas aeruginosa* (PA) after double DNA staining.[57]

should be nearly zero. A resonance Raman microprobe instrument with UV laser excitation that enables analysis of single particles should be able to detect and identify individual bacterial, cells probably at the species level.[40]

Chromatographic methods (mainly Py-GCMS) are useful for direct characterization of microorganisms.[70–71] They require manual sample handling and relatively large samples and consequently applications in the field of bioaerosol analysis are not very hopeful.

Laser-microprobe MS (e.g., LAMMA, PAMS and FABMS) are methods approaching the point of practical applications of bioaerosol analysis. The miniaturized mass spectrographs combined with a three laser system together with a well-designed and miniaturized aerosol beam-sampling system will be able to count, size and identify single airborne particles, including airborne microorganisms *in-situ,* on-line and in real time. Such a system will also generate a complete mass spectrum from individual particles that, together with computerized pattern-recognition techniques, will help differentiate the mass spectra of different microorganisms.[13] It is probable that other interesting and exciting ideas may be expected in this topic during their further future development. One idea could be to combine two or three procedures mentioned in this chapter into one sophisticated apparatus.

SUMMARY

This chapter briefly describes some of the more important and promising chemical methods that are applicable to detection and identification of microorganisms in general and bioaerosols in special cases. Microscopical, chromatographical, laser spectroscopical, cytometric, electrochemical, as well as other methods of modern chemical analysis, have been shown to be capable and useful tools for these purposes. Nevertheless, more research and investigations are required, together with development of new apparatus if practical bioaerosol monitoring systems are to be realized.

REFERENCES

1. Nelson, W.H., Ed. "Instrumental Methods for Rapid Microbiological Analysis" (VCH Publishers, 1985) pp 219.
2. Spurny, K.R. (Ed.) "Physical and Chemical Characterization of Individual Airborne Particles,"Ellis Horwood Series in Analytical Chemistry (John Wiley and Sons, Chichester, 1986) pp. 418.
3. Cox, C.S. "Airborne Bacteria and Viruses," *Sci. Prog. Oxf.* 73:469–500 (1989).
4. Marthi, B., V.P. Fieland, M. Walter, and R.J. Seidler. "Survival of Bacteria during Aerosolization," *Appl. Environ. Microbiol.* 56:3463–3467 (1990).
5. Walter, M., B. Marthi, V.P. Fieland, and L.M. Ganio. "Effect of Aerosolization on subsequent Bacterial Survival," *Appl. Environ. Microbiol.* 56:3468–3472 (1990).
6. Simecek, J., J. Kneifova, and V. Stochl. "Investigation of the Microbial and Dust Air Pollution," *Staub-Reinhalt. Luft* 46(6):285–289 (1986).
7. Hochrainer, D., K.R. Spurny, and W. Fabig. "Sampling of Microbial Aerosols," Final Report (Fraunhofer-Society, Grafschaft, F.R.G., 1987) pp. 89.
8. Hadley, W.K. In "Second International Symposium of Rapid Methods and Automation in Microbiology." (Learned Information Ltd., Oxford, 1976) pp. 14–15.
9. Irwin, W.J., Ed. "Analytical Pyrolysis." (Marcel Dekker, New York, 1982) pp 578.
10. Lattimer, R.P. "Analytical Pyrolysis," *J. Annal. Appl. Pyrolysis* 17:1–3 (1989).
11. Ericsson, I., and R.P. Lattimer. "Pyrolysis Nomenclature," *J. Anal. Pyrolysis* 14:219–221 (1989).
12. Kim, M.G., H. Inove, and T. Shirai. "Development of a Curie-Point Thermal Desorption System and Its Application to the Analysis of Atmospheric Dusts," *J. Anal. Appl. Pyrolysis* 15:217–226 (1989).
13. Sinha, M.P. "Analysis of Individual Biological Particles in Air." In "Instrumental Methods for Rapid Microbiological Analysis," by W.H. Nelson, Ed. (VCH Publishers, 1985) pp. 165–192.
14. Snyder, A.P., J.H. Kremer, H.L.C. Meuzelaar, W. Winding, and K. Taghizadeh. "Curie-Point Pyrolysis Atmospheric Pressure Chemical Ionization MS," *Anal. Chem.* 59:1945–1951 (1987).
15. Chan, W.R., M. Kelbon, and B. Krieger-Brockett. "Single-Particle Biomass Pyrolysis," *Ind. Eng. Chem. Res.* 27:2261–2275 (1988).
16. Leahy, T.J. and M.J. Sullivan. "Validation of Bacterial-Retention Capabilities of Membrane Filters," *Pharmaceutical Technol.* (1978) pp. 7.
17. Fiser, Z., J. Hysek, and B. Binek. "Quantification of Airborne Microorganisms and Investigation of their Interactions with Non-Living Particles," *Int. J. Biometeorol.* 34:189–193 (1990).
18. Palmgren, U., G. Ström, P. Malmberg, and G. Blomquist. "The Nucleopore Filter Method: A Technique for Enumeration of Viable and Nonviable Microorganisms," *Am. J. Ind. Med.* 10:325–327 (1986).
19. Miller, W.I. and B. Forster. "Fluorescence and Confocal Laser Scanning Microscopy: Applications in Biotechnology," *Int. Laboratory,* March (1991) pp. 18–26.
20. Eduard, W., J. Lacey, K. Karlsson, U. Palmgren, G. Ström, and G. Blomquist. "Evaluation of Methods Enumerating Microorganisms in Filter Samples for Highly Contaminated Environments," *Am. Ind. Hyg. Assoc. J.* 51(8):427–436 (1990).
21. Lacey, J. et al. "Harmonization of Sampling and Analysis of Mould Spores," *Nordisk Ministerrad,* Copenhagen (1988) pp 70.
22. Eduard, W.P. Sandren, B.V. Johansen, and R. Bruun. "Identification and Quantification of Mold Spores by SEM," *Ann. Occup. Hyg.* 32:447–455 (1988).

23. Eduard, W., and O. Aalen. "The Effect of Aggregation on the Counting Precision of Mold Spores on Filter," *Ann. Occup. Hyg.* 32:471–479 (1988).
24. Palmgren, U., G. Ström, G. Blomquist, and P. Malmberg. "Collection of Airborne Microorganisms on Nucleopore Filters: Estimation and Analysis," *J. Appl. Bacteriology* 61:401–406 (1986).
25. Greene, V.W. "Apartichrome Analyser," *Environ. Sci. Technol.* 2:104–109 (1968).
26. Fox, A. and S.L. Morgan. "The Chemotaxonomic Characterization of Microorganisms by Capillary Gas Chromatography and Gas Chromatography-Mass Spectrometry," pp. 135–164. In "Instrumental Methods of Rapid Microbiological Analysis," by W.H. Nelson, Ed. (VCH Publishers, 1985).
27. D'Agostino, P.A., L.R. Provost, J.F. Anacleto, and P.W. Brooks. "Capillary Column Gas Chromatography-Mass Spectrometry and Gas Chromatography-Tandem Mass Spectrometry Detection of Chemical Warfare Agents in a Complex Airborne Matrix," *J. Chromatography* 504:259–268 (1990).
28. Rossi, T.M., and I.M. Warner. "Bacterial Identification Using Fluorescence Spectroscopy," pp. 1–50 In "Instrumental Methods of Rapid Microbial Analysis," by W.H. Nelson, Ed. (VCH Publishers, 1985)
29. Alan, R. "Immunoassays: Chemical and Laboratory Techniques for the 1980s," (Alan R. Liss, New York, 1980).
30. McElroy, W.D., and H.H. Seliger. "Bioluminescence in Progress," (Princeton University Press, Princeton, NJ, 1966) pp 432
31. Cormier, M.J., P.M. Hercules, and J. Lee. "Chemiluminescence and Bioluminescence," (Plenum Press, New York, 1913).
32. Baeyens, W.R.G., D. Dekeukeleire, and K. Korkidis. "Luminescence Techniques in Chemical and Biochemical Analysis," (Marcel Dekker, New York, 1991). Practical Spectroscopy Series Vol. 12, pp. 654.
33. Neufeld, H.A.. J.G. Pace, and R.W. Hutchison. "Detection of Microorganisms by Bio- and Chemiluminescence Techniques." In "Instrumental Methods for Rapid Microbiological Analysis," by W.H. Nelson, Ed. (VCH Publishers, 1985) pp. 51–65.
34. Hallin, P., G. Linfors, and G. Sandström. "A Device for Rapid Detection of Bacteria in Air by a Chemiluminiscent Technique," 5th Nord. Symp. in Aerobiology (1983) pp. 47–50
35. Hallin, P., G. Linfors, and G. Sandström. "Investigation of Variation in the Concentration of Bacteria at Outdoor Testing with the Use of Detector for Aerosols and Bacteria," Natl. Defence Res. Institute in Umea, Sweden (FAO Report C 40201–22(1984) pp. 18.
36. Grader, G.S., R.C. Flagan, J.H. Seinfeld, and S. Arnold. "Fourier Transform Infrared Spectrometer for a Single Aerosol Particle," *Rev. Sci. Instrum.* 58(4):584–588 (1987).
37. Schrader, B. "Micro-Raman Fluorescence and Scattering Spectroscopy of Single Particles." In "Physical and Chemical Characterization of Individual Airborne Particles," by K.R. Spurny, Ed. (John Wiley and Sons, Chichester, 1986) pp. 358–379.
38. Hartman, K.A., and G.J. Thomas. "The Identification, Interactions and Structure of Viruses by Raman Spectroscopy." In "Instrumental Methods for Rapid Microbiological Analysis," by W.H. Nelson, Ed. (VCH Publishers, 1985) pp. 91–134.
39. Schweiger, G. "Single Microparticle Analysis by Raman Spectroscopy," *J. Aerosol Sci.* 20(8):1621–1624 (1989).
40. Dalterio, R.A., M. Baek, W.H. Nelson, D. Britt, J.F. Sperry, and F.J. Purcell. "The Resonance Raman Microprobe Detection of Single Bacterial Cells from Chromobacterial mixture," *Appl. Spectroscopy* 41(2):241–244 (1987).
41. Sinha, M.P., C.E. Griffin, D.D. Norris, T.J. Estes, V.L. Vilker, and S.K. Friedlander. "Particle Analysis by Mass Spectroscopy," *J. Colloid Interface Sci.* 87(1):140–153 (1982).
42. Sinha, M.P. "Laser-Induced Volatilization and Ionization of Microparticles," *Rev. Sci. Instruments* 55(6):886–891 (1984).

43. Sinha, M.P., and S.K. Friedlander. "Real Time Measurements of NaCl in Individual Aerosol Particles by MS," *Anal. Chem.* 57(9):1880–1883 (1985).
44. Sinha, M.P., R.M. Platz, S.K. Friedlander, and V.L. Vilker. "Characterization of Bacteria by Particle Beam MS," *Appl. Environ. Microbiol.* 49(6):1366–1373 (1985).
45. Sinha, M.P., and S.K. Friedlander. "Mass Distribution of Chemical Species in Polydisperse Aerosol," *J. Colloid Interface Sci.* 112(2):573–582 (1986).
46. Hillenkamp, F., and R. Kaufmann. "Laser Microprobe Mass Analysis (LAMMA)." In "Laser Applications in Medicine and Biology," by M.L. Wolbarsh, Ed. Vol. 4 (Plenum Press, New York, 1982).
47. Kaufmann, R. "Laser Microprobe Mass Spectroscopy of Particulate Matter," in "Physical and Chemical Characterization of Individual Airborne Particles," by K.R. Spurny, Ed. (John Wiley and Sons, Chichester, 1986) pp. 226–250.
48. Wieser, P., and R. Wurster. "Application of Laser Microprobe Analysis to Particle Collectives." In "Physical and Chemical Characterization of Individual Airborne Particles," by K.R. Spurny, Ed. (John Wiley and Sons, Chichester, 1986) pp 251–270.
49. Spurny, K.R., and J.W. Gentry. "Aerosol Fractionation by Graded Nucleopore Filters," *Powder Technology* 24:129–142 (1979).
50. Böhm, R. "Sample Preparation Technique for the Analysis of Bacteria by LAMMA," *Fresenius Z. Anal. Chem.* 308:258–259 (1981).
51. Seydel, U., and B. Lindner. "Qualitative and Quantitative Investigations of Mycobacteria with LAMMA," *Fresenius Z. Anal. Chem.* 308:253–257 (1981).
52. Kievit, O.J.C.M. Marijnissen, P.T.J. Verheijen, and C.M.B. Scarlett. "Some Improvements on the Particle Beam Generator (PBG)," *J. Aerosol Sci.* 21(1):685–688 (1990).
53. Marijnissen, J.C.M., C.M.B. Scarlett, and P. Verheijen. "Proposed On-line Aerosol Analysis Combining Size Determination and Time-of-flight Mass Spectrometry," *J. Aerosol Sci.* 19(7):1307–1310 (1988).
54. Kistemaker, P.G., H.L.C. Meuzelaar, and M.A. Posthumus. In "New Approaches of the Identification of Microorganisms," by C.G. Heden, and T. Illeni (John Wiley and Sons, London, 1975) pp. 179–191.
55. Heller, D.N., R.J. Cotter, C. Fenselau, and O.M. Uy. "Profiling of Bacteria by Fast Atom Bombardment Mass Spectroscopy," *Anal. Chem.* 59:2806–2809 (1987).
56. Kissel, J. "The Giotto Particulate Impact Analyzer," *Phys. Bull.* 43(5):131–133 (1987).
57. Hadley, W.K., F. Waldman, and M. Fulwyter. "Rapid Microbial Analysis by Flow Cytometry." In "Instrumental Methods for Rapid Microbiological Analysis," by W.H. Nelson, Ed. (VCH Publishers, 1985) pp 67–89.
58. Melamed, M.R., P.F. Mullaney, and M.L. Mendelson, Ed. "Flow Cytometry and Sorting," (John Wiley-Liss, New York 1990) pp 407.
59. Hadley, W.K. and D.M. Yajko. "Determination of Microorganisms and their Metabolisms by Measurement of Electrical Impendence." In "Instrumental Methods for Rapid Microbiological Analysis," by W.H. Nelson, Ed. (VCH Publishers, 1985) pp. 193–209.
60. Smith, D.S., M. Harsan, and R.D. Nargessi. "Principles of Fluoroimmuno-assay Procedures." In "Modern Fluorescence Spectroscopy," by E.L. Wehry, Ed. (Plenum Press, New York, 1982) Vol. 3. Chap. 4.
61. Mackay, R.A., M.T. Godde, P.J. Stopa and A.W. Zulich. "Light Addressable Potentiometric Sensor Based Detection of Toxins and Pathogens," ACS, 201st Natl. Meeting, Atlanta, Div. of Environ. Chem. Preprint Papers 31(1):351–354 (1991).
62. Vo-Dinh, T., G.D. Griffin, M.J. Sepaniak, and J.P. Alarie. "Development of Fibreoptic Fluoroimmunosensor for Chemical and Biological Analysis," ACS, 201st Natl. Meeting, Atlanta, Div. of Environ. Chem. Preprint Papers 31(1):347–350 (1991).

63. Klaus, N. "Aerosol Analysis by Secondary Ion Mass Spectroscopy." In "Physical and Chemical Characterization of Individual Airborne Particles," by K.R. Spurny, Ed. (John Wiley and Sons, Chichester, 1986) pp. 331–339.
64. Clark, N.J. "A Scitilation-Spectrometer with Coincidence Counting for Aerosol Analysis and Source Attribution," *Atm. Environ.* 19(1):189–194 (1985).
65. Kawaguchi, H., N. Fukasawa, and A. Mizuike. "Investigation of Airborne Particles by Inductively Coupled Plasma Emission Spectroscopy Calibrated with Monodisperse Aerosols," *Spectrochimica Acta* 41B(12):1277–1286 (1986).
66. Sneddon, J. "Impaction-Electrothermal AAS for the Direct and Near Real-Time Detection of Metals in Aerosols," *Appl. Spectroscopy* 43(6):1100–1102 (1989).
67. Nelson, W.H. (Ed.) "Modern Techniques for Rapid Microbiological Analysis" (VCH Publishers, New York, 1991) pp 263.
68. Nevalainen, A., J. Pastuszka, F. Liebhaber and K. Willeke. "Performance of Bioaerosols Samplers: Collection Characteristics and Sampler Design Consideration," *Atm. Environ.* 24A(4):531–540 (1992).
69. Van Emon, J.M. and R.O. Mumma. "Immunochemical Methods for Environmental Analysis," ACS Symposium Series 442 (Am. Chem. Soc., Washington, DC, 1990) pp 229.
70. Fox, A., S.L. Morgan, L. Larsson and G.O. Odham, Eds. "Analytical Microbiology Methods, Chromatography and Mass Spectrometry" (Plenum Press, New York, 1990) pp 280.
71. Matney, M.L., T.F. Limero and J.T. James. "Pyrolysis-Gas Chromatography/Mass Spectrometry Analyses of Biological Particulates During Recent Space Shuttle Missions," *Anal. Chem.* 66:2820–2828 (1994).

CHAPTER **13**

Biological and Biochemical Analysis of Bacteria and Viruses

Andreas Hensel and the late Klaus Petzoldt

INTRODUCTION

Bacteria and viruses show the same aerodynamic behavior as other non-viable organic or anorganic particles when airborne (see also Chapter 3).[1,2] Due to their small mass and widespread occurrence at the interface between aqueous and gaseous phases, microbial air contaminants frequently can be isolated from air samples.[3] However, based on the properties of procaryotes as autonomous, unicellular life forms and due to the reproduction-cycle of viruses that infect viable cells to replicate their genes, their biological impact as potential biohazards is high. The consequences of spread by the airborne route requires adequate methods for identification and quantification to evaluate the risks of microbial air contamination.

In 1897, when the first basic review by Flügge was published focusing on airborne microorganisms as an ubiquitous source of infectious diseases, it was known already that microorganisms recoverable from the aerosol state belong to different taxonomic groups.[4] The broad spectrum includes harmless saprophytic species as well as pathogenic microbes that can spread disease among humans, animals and plants.[5,6] Because airborne microorganisms always are present in ambient air, they can be inhaled and deposited in the different compartments of the airways. The respiratory tract continuously has to provide a large variety of active and efficient defense mechanisms in order to maintain health.[7] These mechanisms include physical clearance of deposited particles by mucociliary action as well as mucus secretion, cough, bronchoconstriction, reflex changes in breathing pattern, and inactivation and exclusion of viable organisms by mucus, specific antibodies, and by pulmonary macrophages.[8–10]

Some species are of medical importance as infectious agents, as allergens or in increasing the endotoxin lung burden. Depending on deposition patterns, they and their metabolic products can cause adverse pathophysiological effects in the respiratory tract. During the initial phases of infection these pathogenic microorganisms are able to overcome the non-specific and local immune defense mechanisms using adhesions or invasive strategies. For these microbes the airborne state is the natural way of spreading their population by finding a natural host to maintain the species' survival, e.g., *Influenza* virus infection in humans and animals. The pathogenic state is the product of dynamic selective pressures on microbial populations.

The number of microorganisms present in ambient air is used as a parameter to assess hazards to public health.[11] High levels of microbial air contaminants increase the risk of lung infection by depressing the clearance capacity of the respiratory tract.[12,13] The taxonomic diversity of potentially aerosolized bacteria and viruses means that several prerequisites for sampling and identification facilities are necessary.

This chapter elucidates methods for isolation and identification of airborne bacteria and viruses. Studying airborne bacteria and viruses is a distinctive behavior compared to their non-aerosolized counterparts and this has to be taken into consideration, otherwise artifacts and experimental setbacks may occur. Nevertheless, methods and procedures of conventional bacteriology and virology can be used if adapted first for the purposes of aerobiology.[14] Thus, standard techniques are described only briefly. For readers requiring more detailed information, references are provided. We have tried to indicate general courses of action and to point out peculiarities of analyzing the biological characteristics of bacteria and viruses collected from air samples.

APPROACH TO A MICROBIOLOGICAL STUDY

General Limitations

Because of the biological diversity of airborne microbes, there is a lack of a convincing standard schedule for their qualitative and quantitative determination. Nevertheless, sensitive detection methods have been established for most species, but the aerosol state can substantially change surface structures, antigenicity, and reproduction of the microbes. Furthermore, the type of sampling device or collecting media chosen may unintentionally have adverse effects on quantitative results (see also Chapter 6). However, one must keep in mind that the detectability of microbes is restricted by the properties of the methods used.

The ability to estimate accurately and reliably the total number of living bacteria in environmental samples is under discussion and thus its value in indicating public health safety is in question. Comparisons between results of plate counting, direct microscopic enumeration, and indirect activity measurements indicate that the number of bacteria capable of forming colonies on a solid medium is less than the number actually present and metabolically active, often by several orders of magnitude. These findings from quantitative studies on undamaged viable microorganisms in their natural environment suggest that currently employed standard methods may be inadequate to protect human health.[15] These findings should be taken into consideration when planning a study on airborne microorganisms.

Choice of Methods

Questions immediately arising when considering the design of an experimental study are which methods are available and what are their limitations. Therefore, the aims of the microbiological study must be clear.[16] Basically, a distinction can be made between naturally occurring bioaerosols of microorganisms, with heterogeneous flora of microbes of all shapes and sizes, e.g., hospital, industrial or agricultural environments, and experimentally generated bioaerosols with definite regimens of particle sizes consisting of characterized strains of microorganisms.[17]

Various techniques can be employed to maintain maximum recovery or detection of microorganisms from an air sample, and in this regard, qualitative methods can be distinguished from quantitative methods. Furthermore, viable or infectivity assays are used to determine ability of bacteria to multiply and of viruses to replicate, while non-cultural methods are needed to detect microbial products, antigens, or nucleic acids.

The useful criteria for deciding on the general approach are: (i) relevance of the results, (ii) technical effort needed to establish the method, (iii) rapidity of test performance, (iv) reproducibility, (v) sensitivity and specificity, and (vi) costs.

In some cases, studies may be performed without emphasis on specific identification of a particular microorganism. One example is rapid routine monitoring to evaluate total numbers of microbial air-contaminants for public health in general. This usually requires conventional culturing methods with no further differentiation of the microbes sampled.[16] Another approach may be real-time control for sterile conditions in surgery rooms, where a rapid response is required. To detect microbial air-contaminants, rapid and automatic systems, (e.g., based on luminescence), should be given preference over culture methods.[18,19]

For a wide range of studies, characterization of the biological properties of sampled species cannot be ignored. Identification of species sampled, isolation of these strains, and data on their viability or infectivity may be required in order to evaluate their ability to grow, metabolize, respire, or divide in natural environments or in the host. In these circumstances "universal" collecting and culturing media should be used, e.g., buffered impinger fluids or blood agar plates, appropriate for a broad spectrum of microorganisms. In addition, if quantitative information on an obligate infectious agent is desired, for example to study clearance mechanisms of inhaled *Haemophilus* bacteria in laboratory animals, it may be necessary to perform extensive preliminary studies on aerosolization and sampling conditions, to provide a maximum of viable bacteria that can be offered for inhalation (Chapter 6). Therefore, the question of the repair mechanisms of the chosen strain operational under experimental conditions has to be clarified in detail, so that reproducibility may be obtained. For that reason, both construction and location of the sampler have to be taken into consideration.[20]

If it is necessary to establish a highly specific test system for a pathogen, time and costs for the necessary development should be anticipated. But, in general, the more information required about a particular microorganism the more preliminary work is involved. The main reason is that for the majority of airborne microbes little is known about the kinetics of biological decay and repair mechanisms and how to overcome sublethal damage after collection from the aerosol state. If the validation of the employed method is not previously carried out, the incidence of inaccurate results may increase. This is a problem also with commercially available sampling systems, while adverse effects on microorganisms caused by the sampling device may occur (Chapter 6).

Another methodological factor is the specificity of the chosen test system. For example, quantitative detection of a particular species of virus may be performed unsuccessfully using an infectivity assay, (e.g., a plaque test in tissue cultures,) particularly when large volumes of air are processed. A more sensitive method detecting specific viral antigens (e.g., an enzyme immunoassay) may result in a positive diagnosis when analyzing the same sample.

Some non-cultural methods supplement the common panel of viable and infectivity methods. For instance, several biological effects can be observed after inhalation of bacterial endotoxin in man and animals.[21,22] Because viable bacteria are not required for such pathological reactions, this demonstrates that medical and working environments cannot be exclusively monitored only for viable and infective properties of microorganisms. In these special cases, both the total number of airborne microbes as well as their capability to multiply their genomes is of medical interest.[23–25]

Microbiological Aspects

Aerial transport and aerodynamic behavior of airborne microorganisms are well documented and the subject of many comprehensive studies (see also Chapter 3).[26,27] The aerosolized state poses multiple stresses on a microorganism, which can lead to time-dependent inactivation. The major factors responsible for biological decay are dehydration, rehydration, radiation, the open-air factor, oxygen concentration, and temperature (see also Chapter 6).[28,29]

Indoors probably the most important of these factors influencing the results of aerobiological studies is relative humidity (RH).[30,31] Multiplication or replication cycles of microbes are strictly confined to aqueous conditions, and the lower the RH the more intensively microorganisms lose cell water.[30] RH under 50% is known to be usually lethal for most vegetative bacteria. When bacteria are aerosolized into an atmosphere with an RH of 85% the cytoplasma of these bacteria show approximately the same RH as a saturated solution of sugar.[31] This example is from everyday practice of using sugar solutions and highly concentrated brines for conservation purposes to prevent microbial decomposition of food-stuffs.

Within physiological limits, effects of dehydration on microbes initially are reversible. A single microbe inactivated by the loss of disposable water may show metabolic activity when rehydrated. Subsequently, if the organism has the necessary repair mechanisms and can activate them, then normal physiological properties can be recovered.[32] Because in non-anhydrobiotic organisms utilization of disposable water is one important factor limiting microbial survival in natural environments, the activation of these repair mechanisms has to be discussed as physiological reactions to maintain survival of the species.[30] For example, approximately only 15% of the original cell water but the same amount of proteins can be detected in spore-bearing bacteria.[33] Also, vegetative cells are able to compensate effects of short dehydration conditions by an increase in the inner pressure of the cell with no extensive damage of the enzymatic functions. Under suitable conditions some bacteria species are even able to divide during the aerosol state.[34]

Some naturally occurring substances, like glycerol or KCl, have important protective effects on survival of microorganisms in extreme habitats. Some bacteria species that are metabolically active in brines are known to use such substances for survival.[35] Similar effects can be demonstrated with some alcohols, carbohydrates, or amino acids. These desirable effects also may be used to increase the recovery rates of airborne microbes; when viable assays have to be performed and the portion of lethally damaged microbes can be reduced significantly by such additives, this may also enhance the sensitivity of the assay. When experimental microbial aerosols are generated, a preliminary controlled loss of cell water by means of lyophilization may lead to more viable microbes being recovered from the aerosol state.[36,37]

Within this context it is worth noting that physiological breathing in man and animals leads to an obligatory (warming and) vapor phase rehydration of inhaled microorganisms due to the water-saturated atmosphere in the respiratory system (see also Chapter 6).

When studying aerosols containing more than one organism in a single aerosol particle, it should be kept in mind that these organisms will show only a single colony forming unit (CFU) when sampled onto agar, but the biological effects (e.g., from lipopolysaccharides = LPS) that may be induced by inhalation can be many times higher than those of a single cell. In addition there can be effects owing to the action of vapor phase rehydration on the cells.

To get reliable results from microbiological studies, one needs to bear in mind that airborne microbes behave differently to their undamaged counterparts. In a population of aerosolized microbes, usually most of the individuals have damaged membranes. Thus, when performing viability assays, the choice of sampling and culturing media as well as incubation conditions are of basic importance for the evaluation of the microbe's ability to multiply. Survival of microorganisms is also a prerequisite for infectivity. All factors known to influence survival must affect infectivity as well. Furthermore, the aerosol state can make microbes susceptible to chemicals ordinarily ineffective for undamaged microorganisms of the same strain (e.g., the effects of chemical additives that make media selective). Because most pathogens isolated from natural clouds are normally present only in small numbers, any adverse effect on growth behavior or infectivity should be avoided.

Sub-lethally stressed bacteria populations are believed to be more susceptible than undamaged counterparts, in general, owing to changes in their physicochemical microenvironment, and therefore may exhibit greater variability in response to added nutrient.[15] During media preparation this can be especially an effect of heat formation of peroxides and free radicals, and the Maillard Reactions also may occur.[29]

A failure of various biochemical functions and damaged membranes (e.g., changes in protein, lipoprotein, and carbohydrate structures of the coat of viruses, or bacterial membranes) can cause lethal damage followed by plasmolysis (see also Chapter 6). Culturing this part of the population, which still is able to activate repair mechanisms, results in greatly extended lag-phases in growth behavior of the microorganisms. Thus, in aerobiological practice, it is necessary to lengthen incubation times for viable assays to activate the repair mechanisms and to restore ability to multiply. All other incubation conditions, like incubation temperature, oxygen tension and ionic strength, also need detailed consideration. If non-optimized culturing conditions are used, numbers of viable microorganisms, (i.e., those that are potentially able to survive) routinely are underestimated.

Generalities that can be made about aerosol inactivation of viruses are that they are more resistant to effects of dehydration and oxygen because of the relative simplicity of their structures. For viral survival, the content of lipids in their outer coat is very important. Higher lipid containing viruses are more aerostable at low RH; viruses with little or no lipid content are more stable at higher RH.[38] Under suitable conditions the infectivity of some virus species in natural environments can be preserved over long periods. Virusal denaturation of surface structures prevents cell invasion, thereby causing infectivity loss.[39] The initial effect leads to loss of infectivity is denaturation of capsid proteins or other surface structures like lipid membranes.[40] Nucleic acids are inactivated more slowly.[41,42]

Factors mainly influencing viability and infectivity of experimental microbial aerosols are growth conditions and strain selection. The fate of these microorganisms depends on the medium chosen for cultivation, length of incubation, ionic strength, oxygen tension, temperature, harvesting conditions, and storage. There are also differences in recovery between aged microbial aerosols and aerosols immediately collected from air samples.[43]

Because a number of variables affect bacteria and viruses in the aerosol state, it may be impossible to generalize the most appropriate experimental sampling conditions to be employed in experimental aerobiological studies.[44]

Effects of Collecting Devices on Microbiological Results

Maximal microbial recovery rates from air samples are dependent on the construction principles of the sampler used. Considerable amounts of microbes can become inactivated under normal sampling conditions (see also Chapter 6).

In general, there are two types of samplers used to collect material for viable assays. Microorganisms can be collected on to solid surfaces, (e.g., agar plates, glass slides, or filters), and also into liquids. The total loss caused by the design as well as the loss of viability and infectivity is due not only to aerosol conditions but also can be caused by the sampler, as well as through sedimentation and particle inertia that mainly depends on the design of the apparatus (see also Chapter 3).[45]

The determination of microbial viability using air-samplers to collect the microbes suffers from one major weakness—namely, that until recently it has been difficult to evaluate the total efficacy of a sampler. However, for standardization procedures two methods now are in routine use.[46,47]

First, a defined number of microorganisms is aerosolized and then the number of sampled bacteria or viruses is determined and compared to the primary number of microbes. However, unfortunately, even then some microbes may be inactivated and damaged lethally by the generating procedure, and by the aerosol state. Second, to compare the efficiency of a particular air-sampler with that of another construction type, one has to run the devices simultaneously in an atmosphere containing aerosolized microorganisms. Following a defined time schedule, the number of microbes sampled by each single apparatus is compared. However, this method is unsuitable for determining how many microorganisms are aerosolized, because the sampling efficiency of air-samplers may differ, and be a function of particle size.

Differences may arise also through differences in media composition, incubation temperature, oxygen tension and ionic strength. Comparing experimental results than can be difficult when the biological and technical principles of the sampling devices differ through construction, (e.g., collecting airborne bacteria with gelatin filters or using agar plates as impaction disks for solid-medium cascade impactors) (see also Chapters 9 and 10).

Major studies of airborne microorganisms in natural environments have been performed using solid-media samplers, like cascade or slit samplers.[48-52] Once exposed to an aerosol the petri dishes used for those types of samplers need no further manipulation other than incubation. When working with presterilized air in experimental set-ups, or when performing studies on a particular environmental airborne microorganism cultured on blood agar, then high dust levels of organic or anorganic debris co-sampled with the microbes often can be neglected. Additionally, not all contaminating (extra) colonies on the petri dishes play a role in the regular evaluation of the microbiological results.

Nevertheless, the main problem in high-volume air-sampling for microbiological analysis is interference caused by dust, or cross-reactions with other microbial background particles. This may influence the specificity of diagnostic test systems. Generally, when the degree of contamination is known, the counting range can be adjusted systematically by means of the air flow. Aerosol concentrations can be maintained so that the microbes in question can be distinguished from other concomitant air-contaminants. Even so, sampled contaminants may overgrow the microorganisms of interest. For that reason, solid media used for bacteriological determination, such as agar plates or gelatin filters, often require addition of substances to make these media selective—for example, acridine dyes or antibiotics.

When running samplers under continuous air flow conditions, (e.g., filters or cascade impactors), long sampling times lead to prolonged drying stress through dehydration for the collected microorganisms. This causes membrane damage and inactivation. This phenomenon also occurs on surfaces of solid-medium samplers using agar plates. To increase microbial recovery rates for viable assays, the use of protective additives (like glycerol) placed on the impaction surfaces of the sampler may be required. Rehumidification of the sampling air-stream prior to sampling enables an
equilibration period of several seconds for the trapped microorganism, so may also be advisable for improving survival.[29,53]

Because of high acceleration rates of the sampling air-stream that are probably enhanced at the lower stages of cascade impactors, it should be noted that mechanical impaction forces may lead to higher proportions of inactivated microbes. This limitation cannot be completely overcome using other particle-size-dependent methods. For example, liquid impingers are also known to develop shear forces that may influence the survival of predamaged microorganisms.[54] Adverse effects on viability and infectivity of airborne microbes caused by other sampler devices, e.g., thermal or electroprecipitation samplers, have been observed.[55-57]

A potential disadvantage of filter sampling methods for the detection of microbes basically is related to the continuous air-flow passing the filter membrane (see also Chapter 10). If the diameter of filter pores chosen for microbial analysis is too small, this results in filter obstruction, an increase in the filter's flow resistance, and a decrease in sampling efficacy. Thus, filters with pore sizes ranging from 0.2 to 0.5 µm should be employed. However, particles containing microorganisms like viruses include spirochaetes, rickettsia, or *Mycoplasma* species with small aerodynamic diameters, may pass the filter without being deposited.[58-60]

For routine microbiological analysis of aerosols with unknown microbial contents, media are required that can fulfill the nutritional necessities of the expected flora. When liquid impingers are used, it is possible to examine aliquots of the collecting fluid concurrently in various test systems, by utilizing dilution techniques.

Most of the impinger types commonly utilized are known to damage or kill some strains of vegetative cells because of the violent impaction. Sampling efficiency is size-dependent[61] and may lead to low deposition rates when collecting viruses from air samples. Usually the recovery rates are underestimated because of the small aerodynamic diameters of virus particles, if naked.

Based on their small mass, the transition of microorganisms from the gas to the liquid phase depends on composition and properties of their surface structures. For example, hydrophilic polysaccharide capsules of some particular bacteria species, like *Haemophilus* species or *Klebsiella* species, may enhance the transition. Representing the other extremes, the *Herpes* virus, for example, has hydrophobic lipid membranes, and the cell walls of *Mycobacterium* species consist of wax-like structures with high lipid content. In these cases wetting may be poor. To overcome such a problem, detergents (e.g., Tween-80), can be used to achieve higher deposition rates.[54,62] Also they can prevent foaming of protein-enriched collecting fluids in bubble-type samplers. Another application depends basically on the fastidious behavior of bacteria to adsorb at liquid surfaces quickly.[63] Using detergents, the distribution between bacteria present at the fluid boundary and within the original collecting fluid may be prevented.

ASSAYING AIRBORNE MICROORGANISMS AND THEIR COMPONENTS

Detection of a particular airborne microbe depends on the ability to assay air samples for species-specific components, or typical biological activities. For that reason an adequate sampling procedure is performed in order to deposit microorganisms into liquids or onto solid surfaces. Subsequently, the actual assay has to be carried out. Depending on the material provided by the sampling method, either non-cultural techniques, or assays of viability or infectivity, can be executed.[14] Although samples should be processed as soon as possible, appropriate refrigeration is possible without significantly altering the collected microbes for some species (see also Chapters 9 and 10).

When seeking organisms that are rare in the air, it may be useful to sample large volumes of air through a number of fluid samplers connected in parallel and to concentrate the catch by filtration, centrifugation, precipitation, or lyophilization. This may also work for non-cultural detection methods to increase sensitivity and specificity.[64,65]

Assaying Viability and Infectivity of Bacteria and Viruses

Even under experimental laboratory conditions it can be difficult to distinguish viable from inactivated microorganisms. Some methods are available that allow an estimation of the ratio of total number of microbes to the number of viable members of the species, (e.g., by radioactive labeling or by determination of the uptake of acridine dyes[15]). The criterion commonly used in microbiology to determine viability or infectivity is the ability of a microorganism to multiply in a suitable culturing media or in the natural host (Table 13.1).

Collecting Media

Aerosol samplers for collecting microorganisms should be cleaned and sterilized or disinfected before and after use, usually by ethylene oxide, alcohol, or autoclaving.

When liquid samplers are employed to collect airborne microbes, aliquots of collecting fluids normally are available for isolation and further differentiation of the microbes. Impinger fluids also can be used for appropriate serial dilution on plates or for titration in animals. If titration in tissue cultures or animals is necessary, an appropriate control medium should be used to demonstrate non-toxicity of the collecting fluid.

Collecting fluids in liquid impingers must be compatible with the organism or biological material being sampled. One basic requirement is that the microbes once deposited in the collecting fluids are unable to multiply. For that reason poor liquid media are usually taken for sampling purposes.[66] The fluid compositions may range from distilled water or buffered salt solutions for collecting vegetative bacteria cells and for spores, to egg yolk-enriched solutions for viral or rickettsial bioaerosols. Gelatin-phosphate solution, tryptose-saline solution, Eagle's minimum essential medium, or 5% (v/v) skim milk in distilled water also are employed for recovering vegetative bacteria from the aerosol state. Amino acids (e.g., cysteine), sugars (e.g., sucrose, or trehalose), betaine, pyruvate, catalase, or peptone incorporated into collection media can improve survival rates.[67–71] Hypo- and hypertonic shocks should be avoided as well as drastic changes in temperature. For re-isolation of lyophilized bacteria from the aerosol state, enriched media may be more effective than poor media in obtaining highest recoveries.[43,72]

Nutrient agar plates often are chosen for use in solid-medium samplers. For more stringent demands, blood agar or tryptose-soy-agar may be adequate solid media. Both are established as two of the most popular solid media in medical microbiology. The majority of naturally occurring airborne bacteria including presumptive pathogens are easy to differentiate and the colonies can be used for subculturing without serious problems.

For selective inhibition of contaminant organisms in mixed population clouds acridine dyes, antibiotics, etc., may be used if necessary.[73] Selective media are in common use in bacteriology to simplify differentiation procedures, (e.g., to distinguish *Enterococcus* species from other *Streptococcus* species). For qualitative analysis, selective enrichment can be performed (e.g., selenite broth for enrichment of *Salmonella* species). An enrichment medium allows enhancement of the growth of certain bacteria species, while inhibiting the development of unwanted microorganisms. This detection method can be successful even when the organism sought is rare within the sample, or the contamination level is extraordinarily high.

Table 13.1. Test Systems for Assaying Viability and/or Infectivity of Airborne Bacteria and Viruses

Test System No.	Principle[a]	General Comments for Test Systems No.
1. Plate counting	Viable procaryotic cells multiply on or within solid or liquid culturing media, either after direct deposition or after culturing of collecting fluids	1.- 4. Aliquots or concentration steps are recommended, if necessary; titration or end-point dilution are possible
2. Tissue culture	Viruses (and some bacterial species) taken from collecting fluids are able to infect susceptible cell lines and can be detected by the induction of plaques in the agarose covered cell monolayer	1.- 3. Reliable results highly dependant on the type of sampler, the sampling conditions and the chosen collecting/culturing medium
3. Embryonated chicken eggs	Inoculation of sampling fluids into the compartments of the embryonated egg has been used for a small spectrum of certain viruses (e.g., for propagation of Myxovirus) or bacteria (e.g., bacillus piliformis) which need higher demands for culturing	2.- 4. Also recommended for detection of bacterial toxins or toxigenic metabolites
4. Animal inoculation	Direct aerosol exposure or inoculation of sampling fluids into susceptible experimental animals is recommended in cases of highly contaminated air or if certain pathogenic bacteria and viruses in low aerosol concentrations (e.g., Francisella tularensis, coxsackie virus) have to be detected	2.- 3. When infectivity assays have to be established, these test systems should have preference for minimizing animal studies 3.- 4. Recommended under circumstances where tissue culture techniques are not available

[a] For further details see standard microbiological textbooks and reviews (e.g., references 33, 66, 75-79, 121).

The design of a selective medium for the purposes of air-sampling is difficult when taxonomic or physiologically closely related organisms have to be selected (e.g., to isolate staphylococci from an entire aerosol population mainly consisting of micrococci). Selection of media is dependent on knowledge of the normal flora present in the collecting environment being examined and the spectrum of microorganisms that may be found. It is common knowledge in routine microbiology that substances used to make media selective also can hinder the growth of that microbe to be determined. Aerosolized microbes usually undergo stresses leading to membrane damage and that is why use of selective media as liquid or solid collecting media cannot completely fulfill the demands for aerobiological purposes.

Evaluation of Bacterial Viability

For plating bacteria, 0.1 to 0.2 mL of an adequate dilution should be placed directly on the agar plate and with a spatula be spread uniformly onto the solid medium's surface. An average plate count of at least four plates for a minimum of each dilution step (or more if required for statistical analysis) should be utilized to determine the concentration of the microorganisms in the impinger fluid. The countable range for cultural assays should be 20 to 200 colonies per plates. Gelatin filters are popular for direct culturing on various agar plates.[58,59] Membrane filters also can be directly incubated on agar plates, but do not provide optimal growth conditions.[74]

Some microorganisms are difficult to isolate, e.g., *Legionella*, *Haemophilus*, *Chlamydia*, and *Mycoplasma*, or grow very slowly, e.g., some *Mycobacterium* species. The lack of information about the microbial content of an aerosol may limit the number of diagnoses because the composition of media, incubation time, incubation atmosphere (e.g., aerobic, anaerobic), and the temperature chosen for a proposed study may influence the microbiological results. Incubation times required for successful isolation of particular species may vary from hours (e.g., some *Micrococcus* species,) to several weeks (e.g., some *Mycobacterium* species). Thus, if bacteria of different species are cultured simultaneously, the differences in growth rate, depending on species generation times, may cause difficulties. If fast-growing microbial contaminants are expected, (e.g., micrococci, aerobic spore-forming organisms, fungi, or swarming bacteria species), long initial incubation times should be avoided. Similarly, those bacteria that regularly grow in small colony sizes (e.g., *Streptococcus*) are in danger of being disregarded. Further differentiation to determine genus, tribe, or species of a cultured microorganism is performed routinely on the basis of phenotype and typical metabolic activities. Typing by means of phages and specific antibodies may be carried out if necessary.[75]

Infectivity Assays

Viruses, the smallest infectious agents (20–300 nm), replicate only in living cells. The host range of some species is known to be extremely limited; other viruses infect a broad spectrum of cell types. For viruses definitions become more complicated because "viability" is measured conventionally in terms of infectivity of tissue cultures, the embryonated egg, or animals.[76,77] Infectivity assays such as embryonated chicken eggs or tissue cultures are used to detect some bacteria (e.g., *Bacillus piliformis*), and bacterial toxins (e.g., botulism toxin), that cause damage to or malfunction of host tissue. To detect such toxins in aerosol samples, adequate concentrations in the sample are required, otherwise non-cultural methods should be given preference.

Cultivation in Tissue Cultures

The ideal tissue culture for a study of virus infectivity is the susceptible host or target cells from the natural host. However, because this is not always possible, it is sometimes necessary to use cell cultures.[78]

Cell culture is the cultivation *in vitro* of dissociated single cells. There are three basic types. Primary cell cultures, consisting of a mixture of dispersed cells (usually by trypsinization), are obtained from minced organs, usually from kidney or lung. These cells can be cultured for a limited number of passages only. Once passaged *in vitro*, this primary cell culture becomes a cell line. The second type are diploid cell lines. The most commonly employed cell lines are composed of fibroblasts with complement chromosomal patterns, that have undergone a change that allows < 50–70 culture passages. The third type are heteroploid continuous cell lines that have been derived from malignant tissues or from diploid cell lines.

Although established cell lines (e.g., HeLa, Sirc, Vero, HEp-2) originated from defined precursors, in scientific usage there may be differences in growth behavior and virus susceptibility between different diagnostic laboratories.

Conditions and containers for growth and maintenance of cell cultures may vary and depend mainly on the diagnostic needs of laboratories. Multiplication of a virus in a sensitive cell line can be evaluated by visible reactions, but it may be necessary to subculture the infected cells before a viral effect can be found. The following morphologic changes related to virus infection may occur: a cytopathic effect in the cell culture that may have a typical appearance (e.g., *Poliovirus*, Foot-and-mouth-disease virus, *Herpes simplex* virus), an inhibition of metabolism in the cell (e.g., *Enterovirus*), an ability to bind erythrocytes due to virus-specific hemagglutinin in the cell membrane (e.g., *Influenza* virus, *Parainfluenza* virus), an appearance of complement-fixing antigen (*Adenovirus*, *Pseudorabies* virus), and a transformation or interference with a non-pathogenic virus (e.g., *BVD* virus) or an oncogenic virus (e.g., *SV 40*, *Rous sarcoma* virus). When such a cell culture is examined, the viral effects may be known, but the name of the virus is not. Hence, virus identification based on infectivity assays needs confirming by a second test system, e.g., immunofluorescence, enzyme immunoassay, or neutralization test.[76,78,79]

The most common test system to determine infectivity of a given virus is the plaque assay. Monolayers of host cells are inoculated with appropriate dilutions of virus and to prevent the spreading of virus throughout the layer, after a short adsorption period, the culture is overlaid with a medium containing agar. After an appropriate incubation period, cytopathic effects start, usually focally, producing small plaques indicating infected cells. Theoretically, a single infective virus can induce one plaque-forming unit (PFU), but in most cases the ratio between the number of infective virus particles and the host cell is about 1,000:1. The main reasons for this phenomenon are (i) inhibition of viral replication caused by the agar overlay, (ii) aggregation of virus particles, and (iii) the presence of defective viral particles in the sample. When an air sample with unknown virus content has to be screened without focusing on a particular species, primary or secondary cell cultures should be given preference, because they allow a broad spectrum of virus species to multiply.[79] Preferably, cell cultures derived from the host or related species should be utilized (see also Chapter 6). Rehumidification of the air sample is known to increase the number of PFU of some virus species.[53] Infectivity assays to detect virus can be performed with collecting fluids, but in most cases large amounts of air have to be sampled before detectable numbers of infective virus particles can be provided for test performance.[80] To prevent bacterial growth, 200 units of penicillin and 100 µg of streptomycin per mL may be added to the sample. Because collecting fluids are usually highly contaminated, the suspension has to be filtered through 0.45 µm and 0.22 µm pores. Aliquots are placed on cellular monolayers (limiting dilution of the sample may be done, if necessary) and incubated for 1 hour or more at 37°C to allow absorption of the virus particles. The collected specimens then may be removed or left on the cell culture, but fresh media should be added to the cultures. These should be observed for cytopathic effects following a 1 to 2 week incubation period. Most species of cytopathic viruses take 24 to 72 hours to induce the visible cytopathic effects (CPE), but this is a function also of the inoculated dose, while titration may be performed to evaluate the $TCID_{50}$ (i.e., the dose required to infect 50% of the tissue cultures). Furthermore, for low initial virus concentrations in a specimen, up to 5 blind cell passages are recommended to achieve detectable morphological changes.

Cultivation in Embryonated Eggs

To isolate and cultivate some particular microorganisms (e.g., *Chlamydia* species, *Influenza* virus, *Bacillus piliformis*) without inoculating animals, it is essential to use embryonated chicken eggs (ECE).[79] For successful isolation various routes of inoculation can be evoked to reach the various cell types of the embryo and titration of collection fluids may be performed to determine the EID_{50} of the sample (i.e., the dose required to infect 50% of the ECE). The multiplication of microorganisms within the ECE leads to typical pathological changes: e.g., direct replication of a virus (Newcastle disease virus), death of the embryo (*Encephalitis* virus), production of plaques (*Herpesviridae*) or pocks at the chorioallantoic membrane (*Vaccinia* virus, Small pox virus), or production of hemagglutinating antibodies in the embryonic fluids (e.g., *Influenza* virus). Repassage of homogenized ECE suspensions from primary infected chick embryos can be performed to detect low amounts of virus in a specimen.

Animal Inoculation

Although most diagnostic laboratories are restricted to *in-vitro* methods and do not have the facilities for housing laboratory animals (e.g., mice, rats, guinea pigs, hamsters, rabbits, chickens, or monkeys), the experimental inoculation of pathogens into animals is helpful as a tool to prove bacteriological diagnosis or for primary isolation in the case of viruses.[77]

In diagnostic microbiology, several microorganisms are difficult to isolate and to differentiate by *in-vitro* methods (e.g., *Brucella* species, *Clostridium* species, *Legionella* species, some *Mycobacterium* species, some *Enterovirus*, etc.) or hazardous to handle (e.g., *Bacillus anthracis*, *Francisella tularensis*, HIV, some *Arbovirus*). The particular inoculation route employed depends on ability of the local tissue to support the particular infection. For instance, many viruses introduced into the normally sheltered tissues of the central nervous system find a susceptible environment for multiplication.

Animal responses to inoculated microbial samples may be determined by immunological tests (e.g., antibody titres, skin test to demonstrate sensitivity to *Mycobacterium tuberculosis* infection) or by pathomorphological examination at necropsy. This is often performed dose dependently. Typical reactions or pathological changes furnish evidence for a specific infection or allergic reaction. Moreover, histochemical and immunohistochemical techniques can confirm the results.

If pathogenic microorganisms are present in a sample of air, animals can be deployed as living samplers to detect the infectious agents. Exposing a susceptible animal provides information on tissue tropism, virulence, pathogenesis, and transmissibility of microorganisms. This may be a way to overcome problems with low aerosol concentrations of an agent, or with highly contaminated air samples.

Doses used for animal exposure or inoculation are routinely expressed in terms of number of CFU or PFU. When animals are employed in the assay, the pathogenicity is expressed as the LD_{50} value (i.e., the lethal dose required to affect 50% of the host animals), or ED_{50} (i.e., the effective dose, required to induce a typically defined clinical sign of disease in 50% of the animals). Because several factors are known to affect inhaled microbes, these calculated values should be used with caution. ED_{50} also is a function of the host. To evaluate the pathogenicity of a microbial strain, characterization is needed of the deposition and translocation patterns, (i.e., the distribution and kinetics within the host, and the methods to determine the challenge level).

Non-Cultural Methods

Rapid detection and identification of microorganisms is of great importance in a diverse array of applied and research fields.[18] In routine diagnostic microbiology there are a multitude of techniques available to identify microbes by means of direct examination, by their antigenic structural components, by their nucleic acids, and by their metabolism products. The following non-cultural techniques have been selected for applicability to aerobiological analysis and underlying principles are described briefly.

Although until now most studies on airborne bacteria have used cultural methods to assess properties of microbial aerosols, some technological progress has been introduced gradually into diagnostic microbiology. In particular, antibody test systems, gas-liquid chromatography, chromogenic substrate tests, and molecular nucleic acid hybridization can be utilized for obtaining a specific and rapid microbiological diagnosis (see also Chapter 12). Most can be adapted easily for purposes of aerobiology to determine microbial bioaerosols. In the future, these promising new techniques may be utilized to initiate substantial changes in environmental microbiological research (Table 13.2).

Direct Examination

Gross examination of specimens is particularly valuable in collecting fluids from bioaerosols with known contents of microbes. A haemocytometer chamber permits determination of total numbers of bacteria. This technique can be applied as well when information about the arrangement or other distinctive morphological features (e.g., bacteria in chains or clusters) of the sampled microbes is needed. However, non-microbial contaminating particles may falsify the results. Furthermore, when impingers are employed aggregates of bacteria can become broken up by turbulence. Thus, total numbers of bacteria may be overestimated. If such a bioaerosol is monitored simultaneously by a solid-medium sampler, differences in CFU will result from this phenomenon. In the case of low pollution levels of interfering particles, it is possible to use automated detection systems to determine total cell contents, e.g., Coulter counter.[34,81]

Staining

Bacterial membranes, cell walls, and capsules can be stained with specific dyes to reveal valuable information about the microbial contents of samples (e.g., bacteria impacted on a glass slide or in liquid). Although the Gram stain or Giemsa stain are ordinarily satisfactory for determining cellular characteristics, the acridine orange stain sometimes is worthwhile for detecting viable bacteria within a collecting fluid.[15] For detection of *Mycobacterium* species, the Ziehl-Neelsen acid-fast stain may be performed. Because general interference with organic debris can complicate the diagnosis, and staining methods are mostly not species-specific, the results should be interpreted cautiously.

Electron Microscopy

By means of electron microscopic (EM) examination, the shape and other morphological characteristics (e.g., branching, filaments with spherical bodies, pointed ends, or granular forms) can be determined.[82] For examination, filter membranes or glass slides may be used directly, or homogenous collecting fluids may be processed.

Table 13.2. Non-Cultural Methods for Assaying Airborne Bacteria and Viruses from Sampling Media

Method	Principle	Reliability[a]	Sensitivity	Selected References[b]
Direct examination				
1. Staining and light microscopy	Demonstration of bacterial or viral or bacterial structural components with mono- or polychromatic dyes (for some species specific stains known)	Low	Low	15, 82
2. Electron microscopy	Demonstration of viral or bacterial particles (stained preparations)	Low	Low	84, 85
3. Immune electron microscopy	See 2., signal amplification or particle aggregation by specific antibodies	Medium	Medium	14, 83-85
Antibody-based detection of specific microbial antigens				
4. Immunofluorescence (Epifluorescence microscopy)	Detection of viral or bacterial structural components by fluorogenic labeled specific anti-bodies using an UV-microscope	Medium-High	Medium-High	60, 85-90
5. Enzyme immuno-assay/ Radioimmunoassay	Capture of viral or bacterial components (including toxins and metabolic substances) by specific antibodies, direct or indirect detection by using enzymatic activity and chromogenic substrates or radioactivity	Medium-High	High	85, 92-96
6. SDS-PAGE/Immunoelectroblot	Chromatographic separation of viral or bacterial structural components depending on their molecular weight, transfer to immobilizing matrices, detection by enzyme immunoassay	Medium-High	Medium	94, 97

Detection of specific microbial nucleic acids

7. DNA/RNA hybridization	Detection of species (genus)-specific nucleic acids by labeled DNA/RNA probes	High	High	98-103
8. Polymerase chain reaction	See 8., amplification of the target DNA/RNA	High	High	98, 104-107

Detection of specific microbial products

9. Chromogenic/fluorogenic enzyme substrate tests	Detection of genus (species)-specific bacterial enzymes visualized by chromogenic or fluorogenic substrates	High	Medium-High	108-110
10. Gas-liquid chromatography	Chromatographic identification of bacterial metabolites or components by their solubility between a gaseous and a liquid phase	High	High	111-113
11. Bioluminescence or Chemiluminescence	Measurement of bacterial, e.g., ATP present in liquid samples performed by a photometric analysis of a light-emitting substrate	Medium	Medium	18, 19, 114-116
12. Limulus test	Quantitative determination of LPS in liquid samples by using specific enzyme activity	Medium	High	117-120

[a] Assumes reproducible and reliable sampling conditions and experience in test performance. [b] For further details see standard microbiological textbooks (i.e., reference 121).

These techniques are not that valuable for direct diagnosis because sensitivity is poor (e.g., to find a single virus particle on a 200 mesh grid the minimum concentration of the microbes has to be $>10^5$ particles per mL) and further differentiation of morphologically and taxonomically related species is impossible.

Immune Electron Microscopy

The immune electron microscopic (IEM) examination of microorganisms is an improvement on the methods described above. Specific antibodies, preferably polyclonal antisera, can be incubated with the microbe-containing collecting fluid to obtain antigen--antibody complexes. These are transferred to precoated EM grids and subsequently stained. For the following EM examination, the identification rates can be enhanced by previous coupling of the antibodies to amplifying conjugates (e.g., gold particles).[83,84]

Antibody-Based Detection of Specific Microbial Antigens

Antibody-based detection systems require antigen-specific polyclonal or monoclonal antisera. Positive results are visualized using color reactions by means of microscopes, enzyme immunoassay readers, or the naked eye. In general, these techniques are highly specific and sensitive, but need some serological experience to be performed well.[85]

Immunofluorescence Tests

Immunofluorescence (IF) techniques are based on the combination of a specific antibody with unaltered immunologic reactivity covalently bound to a fluorochrome (e.g., rhodamine derivates, Texas red, FITC = fluoresceine isothiocyanate). These dyes can be excited with light of specific wavelengths. First, labelled antibodies are incubated with a collected sample. Fluorescence that indicates the presence of antigen-specific binding of the antibody to the microorganism then can be detected with a special IF microscope.

IF techniques are highly sensitive, simple, and quick. The presence of specific microbial antigen may be detected in impinger collecting fluids, cell cultures, tissue slides, and bacteria colonies from agar plates.[85-90] Examination of bioaerosol particles directly deposited on to solid surfaces may be a valuable method to determine the microbial content.[60,86-89]

For each specimen examined, adequate controls are necessary to distinguish autofluorescence, or nonspecific fluorescence from a positive control. Additionally, taxonomically related species have to be tested for antigenic cross-reactions. Normally, serological techniques do not determine microbial viability. Complementing these methods with viable stains (e.g., acridine orange) would be one way to overcome this limitation. To extend their application, IF may be complemented with a fluorescence-activated cell sorter (FACS) to achieve fast and specific information combined with automation.

Enzyme Immunoassays and Radioimmunoassays

Immunoassays are rapid, highly sensitive and specific methods of detecting microbial antigens. The underlying principle of enzyme (EIA) and radioimmunoassays (RIA) is the combination of a specific antibody either conjugated with an enzyme or labeled radioactively. The presence of microbial antigen in the specimen is detected subsequently by visualization of enzyme activities using colorful substrates or by measuring radioactivity.[91] Alkaline phosphatase and horseradish peroxidase are the enzymes commonly used. EIA or ELISA (enzyme-linked immunosorbent assay) developed out of a need to circumvent the disadvantages of using radioisotopes in diagnostic laboratories.[92]

EIA performed as a capture or sandwich assay or RIA used as a competitive-type can provide quantitative information about antigen content in a specimen.[85,93] Specificity of these techniques may be enhanced by the use of monoclonal antibodies developed against specific antigens. Sensitivity of EIA and RIA techniques can be increased in magnitude by amplification systems (e.g., biotin-streptavidin).[85] These techniques can be performed as well on solid-phase matrices. Not only polystyrene microtube wells, plastic beads, or ferrous metal beads may be utilized as solid surfaces, but also membranes or filters from air-sampling.[94–96] Thus, detection of airborne microbial antigen is possible, carrying out test performances either with directly deposited particles from the bioaerosol state, or by using prepared collecting fluids.

Immunoelectroblot Techniques

Immunoelectroblot techniques are primarily qualitative techniques. They are valuable methods for determining bacterial and viral antigens in an air sample and are as sensitive as EIAs, but potentially more specific.[97] Three basic techniques are involved. First, structural components of microbes, like proteins, polysaccharides, or LPS, are separated by sodium dodecyl-polyacrylamide gel electrophoresis (SDS-PAGE) depending on their molecular weights. This results in a specific pattern of bands. If necessary, coomassie blue or silver staining may be performed. Secondly, these electrophoresed ligands are electrotransferred to immobilizing matrices (e.g., nitrocellulose, or polyvinylidene fluoride). Subsequently, an immunoassay is used to identify reactivity with antigenetic bands by means of a precipitating chromogenic substrate. The high specificity of this technique is made possible by direct visualization of antigen-antibody reactions when genus-, species-, or strain-specific microbial agents are present in the specimen.[97] To achieve maximum sensitivity, preliminary concentration steps have to be performed where necessary.

Detection of Specific Microbial Nucleic Acids

Molecular nucleic acid hybridization-based assays allow direct culture confirmation of a potential pathogen in a specimen. Nucleic acid probes are able to bind their complementary references to form a novel duplex molecule of DNA or RNA. Probes can be tailored to be genus- or species-specific, depending on their nucleotide sequence. Once a nucleotide sequence is evaluated, the probes may be provided by an automated synthesizer.

Using species-specific oligospecific oligonucleotide probes, the genus-specific and species-specific nucleic acids are visualized by specific markers (e.g., radioactive-labeled probes detected by radiographic films, or biotinylated probes read colorimetrically).[98,99] Sometimes amplification to a large number of copies is necessary using the polymerase chain reaction (PCR).[100]

Nucleic acid hybridization allow rapid identification of a particular microorganism using fingerprint techniques, (i.e., detecting DNA patterns pretreated with restriction enzymes.)[101] The fragments of the digest are separated in an agarose gel by electrophoresis and stained with ethidium bromide. Furthermore, Southern blotting (i.e., electrotransfer to a solid matrix, like nitrocellulose or nylon membranes) generates a hybridization pattern with specific DNA probes. The visualization of these probes using chromogenic or radioactive labeled markers leads to a final diagnosis.[102]

Antimicrobial susceptibility testing is not possible using these methods, while the inability of strain-typing performed concurrently with the identification of the species may be a disadvantage. The major breakthroughs of this technology are the rapidity of hybridization reactions and high specificity of probes, making this technology valuable for microbiological diagnostics.[103,104] Moreover, with new signal and target amplification techniques, like PCR, very low quantities of an infectious agent can be detected in a sample even for those microbes that are not culturable under normal laboratory conditions. Additionally, when probes become more routinely available, such methods will enhance the identification of nucleid acids of environmental microorganisms.[105–107]

Detection of Specific Microbial Products

In contrast to the use of conventional differential media, often requiring many days to achieve an observable end-point reaction, several new methods have been developed to shorten the time for identifying unknown bacterial species through their biochemical characteristics when recovered from solid-medium cultures or directly from collecting fluids.

Chromogenic and Fluorogenic Enzyme Substrate Tests

For direct and rapid identification of unknown bacteria, reaction of a genus- or species-specific enzyme is visualized with chromogenic or fluorogenic substrates. Hydrolysis either results directly in a colored product or produces a by-product that can be detected with color-developing reagents, e.g., hydrolysis of the amino-acid para-nitroanilide releases a free amino-acid and the yellow para-nitroaniline chromophore. Systems utilizing chromogenic substrates allow identification of bacteria colonies recovered on a solid medium within 4 hours or less. These test reactions, particularly the various commercially packed kit systems, are based on preformed enzymes and are not growth-dependent. The biochemical reactions are well-defined and end points are easy to determine. The influence of experimental uncertainties usually is small. Providing accuracy in test performance, the methods generally are sensitive with a high level of reliability.[108–110]

Gas-Liquid Chromatography

Microbial metabolites and components such as carbohydrates, quinones, aliphatic and aromatic amines, and short-chain acids, extracted from growth media or from collecting fluids, may be used for taxonomic identification of microorganisms. The composition of these traits are genetically determined and a reflection of the microbe's genome. Several chromatography methods are used in diagnostic microbiology, (e.g., thin-layer chromatography, liquid solid chromatography, high-performance liquid chromatography). The most popular for chemical analysis of microorganisms is gas-liquid chromatography (GLC).[111]

The basic components of a gas-liquid chromatograph are a sample injection device, the analytical column, and a detection system. The underlying principle of this method is separation of sample components (by their chemical and physical properties) between two phases, one mobile and the other stationary. The stationary phase is liquid at the operating temperature of the chromatograph. The analytical glass column is packed with an inert substance (celite, silica) coated with an inert, volatile liquid. Methyl silicone or carbowax commonly is used. In GLC the mobile phase is an inert gas, e.g., helium, argon, or nitrogen. When the volatilized sample is injected into the stream of the carrier gas, the different compounds are distributed between the two phases. Compounds pass through the column depending on their affinity for the stationary phase. A detector system measures components as they emerge from the column. Column retention times are used to identify specific compounds when compared to retention times of known standards. Microbial metabolites extracted from growth media with organic or liquid solvents can be distinguished by their typical patterns.[112] For some particular bacteria, direct examination of liquid specimens also provides valuable results.[113] This technique is highly sensitive and the results are objective. Organisms may be processed that are otherwise to distinguish (e.g., *Clostridium* species, *Shigella* species) or are time-consuming using conventional identification tests. The disadvantages are high capital investment for equipment and the need to establish references with which unknown samples can be compared.[111]

Luminescence

The detection of bacterial adenosin triphosphate (ATP) present in a liquid sample provides an index for biomass (see also Chapter 12). The underlying principle of the bioluminescence test is the emission of a photon of light energy following the oxidation of a reduced substrate (luciferin) by an enzyme (luciferase). The non-bacterial, somatic cell ATP has to be degraded before a second lysing agent is added that releases ATP. The amount of bacterial ATP present in the mixture is quantified by adding luciferin-luciferase and the amount of emitted light is measured with a special bioluminescence photometer. Intensity of light emission is proportional to content of bacteria in the sample.[114] In microbiological practice, this method is not very sensitive and requires detailed standardization of the non-specific background.[18,115,116]

Chemiluminescence methods are based on bacterial cytochromes that catalyze the reaction of luminol and an oxidizing agent. These methods can be used for continuous and automated bioaerosol control.[18,19]

Limulus Test for LPS Detection

Lipopolysaccharides (LPS) are components of the outer membrane of all Gram-negative bacteria species (Gram-positive bacteria cannot be detected with this test). The lipid portion of the molecule, which is anchored in the cell membrane, is the strong, biologically active pyrogen. The pathophysiological effects of endotoxin, the whole LPS molecule, do not require living bacteria.[117]

To evaluate the content of LPS in a sample, there are two common methods. One is the pyrogen test in rabbits by means of *in-vivo* application of the sample. The other used for *in-vitro* testing is the Limulus test.[118] In the presence of calcium amebocytes (blood cells), lysates (LAL) from *Limulus polyphemus*, the horseshoe crab, can be activated by LPS to an enzyme converting soluble protein to an unsoluble complex.[119,120] Performed with collecting fluids, a well-functioning Limulus test should be able to detect approximately 10^3 to 10^4 Gram-negative bacteria per mL.

SUMMARY

A great taxonomic diversity of microorganisms is distributed throughout atmospheric and human environments. Although most species are harmless, some are potential pathogens. For that reason airborne microbes need to be considered in terms of their biological and biochemical properties in order to evaluate biohazards for humans, animals, and plants. The bioaerosol state can substantially change surface structures, antigenicity, and reproduction of the microbes resulting in adverse effects on viability and infectivity. Dehydration is known to be the most important indoor factor affecting survival capacity of microorganisms. Furthermore, the type of sampling device and collecting media chosen, influence the accuracy and reliability of detection methods. Because of the broad spectrum of potentially bioaerosolized bacteria and virus species no convincing standard schedule is available for their qualitative and quantitative evaluation.

Methods and procedures of conventional bacteriology and virology have been adapted for the purposes of aerobiology. Depending on the nature of the bioaerosol to be analyzed and the requirements of the study, viable assays or non-cultural techniques may be performed for monitoring or screening airborne microbes.

Evaluation of bacterial viability usually is performed with plate-counting methods. Infectivity assays for detection of microbes, and especially viruses, include cultivation in tissue cultures, in embryonated eggs, or inoculation of animals. For non-cultural testing there are a multitude of techniques available to identify microbes by means of direct examination, antigenic structural components, nucleic acids, and metabolic products. In particular, staining methods, antibody-based test systems, chromatography, chromogenic substrate tests, and molecular-nucleic acid hybridization provide specific and rapid microbiological diagnosis. Some newer technologies have been introduced gradually into diagnostic microbiology and have provided substantial improvements for microbiological diagnostics in aerobiology.

REFERENCES

1. Zajic, J.E., I. Inculet, and P. Martin. "Basic concepts in microbial aerosols," *Advan. Biochem. Engineer. Biotechnol.* 26:51–91 (1983).
2. Adams, A.J., D.E. Wennerstrom, and K. Mazumder. "Use of bacteria as model nonspherical aerosol particles," *J. Aerosol Sci.* 16:193–200 (1985).
3. Loosdrecht van, M.C.M., J. Lyklema, W. Norde, and A.J.B. Zehnder. "Influence of interfaces on microbial activity," *Microbiol. Rev.* 54:75–87 (1990).
4. Flügge, C. "Über Luftinfektion," *Z. Hyg. Infekt.-Kr.* 25:179–193 (1897).
5. Lidwell, O.M. "The microbiology of air," in *Topley and Wilson's principles of bacteriology, virology and immunity, Vol.1, General bacteriology and immunity*, A.H. Linton, and H.M. Dick, Eds. (London, Melbourne: Edward Arnold, Hodder & Stoughton, 1990), pp. 226–241.
6. Wathes, C.M. "Airborne microorganisms in pig and poultry houses," in *Environmental aspects of respiratory disease in intensive pig and poultry houses, including the implications for human health*, J.M. Bruce, and M. Sommer, Eds. (Commission of the European Community, Report EUR 10820 EN, 1987), pp. 57–71.
7. Phalen, R.F. *Inhalation studies: Foundations and techniques.* (Boca Raton, FL: CRC Press, 1984), pp. 151–172.
8. Naumann, H.H. "On the defense mechanisms of the respiratory mucosa towards infections," *Acta Otolaryngol.* 89:165–176 (1980).

9. Robertson, B. "Basic morphology of the pulmonary defense system," *Eur. J. Respir. Dis.* 61(Suppl.107):21–40 (1980).
10. Wilkie, B.N. "Respiratory tract immune response to microbial pathogens," *J. Am. Vet. Med. Assc.* 181:1074–1079 (1982).
11. Spendlove J.C. "Industrial, agricultural and municipal microbial aerosol problems," *Develop. Ind. Microbiol.* 15:20–25 (1974).
12. Edmondson E.B., J.A. Reinarz, A.K. Pierce, and J.P. Sanford. "Nebulization equipment: A potential source of infection in gram-negative pneumonias," *Am. J. Dis. Child.* 111:357 (1966).
13. Pierce, A.K., and J.P. Sanford. "Bacterial contaminations of aerosols," *Arch. Intern. Med.* 131:156–159 (1973).
14. Noble, W.C. "Sampling airborne microbes - Handling the catch," *Symp. Soc. Gen. Microbiol.* 17:81–101 (1967).
15. Roszak, D.B., and R.R. Colwell. "Survival strategies of bacteria in the natural environment," *Microbiol. Rev.* 51:365–379 (1987).
16. Fannin, K.F. "An approach to the study of environmental microbial aerosols," *Water Sci. Techn.* 13:1103–1114 (1981).
17. Rotter, M., W. Koller, H. Flamm, W. Resch, and J. Schedling. "Sampling of airborne bacteria by gelatine filters in an automatic sampler," in *Airborne Transmission and Airborne Infection: IV. International Symposium on Aerobiology in Enschede*, J.F.P. Hers, and K.C. Winkler, Eds. (Utrecht: Oosthoek Publishing Co., 1973), pp. 47–50.
18. Nelson W.H. *Instrumental methods for rapid microbiological analysis* (Weinheim, New York: Verlag Chemie, 1985), pp. 1–219.
19. Oleniacz, W.S. "Chemiluminescence method for detecting microorganisms in water," *Environ. Sci. Technol.* 2:1030–1033 (1968).
20. Breiman, R.F., W. Cozen, B.S. Fields, T.D. Mastro, S.J. Carr, J.S. Spika, and L. Mascola. "Role of air sampling in investigation of an outbreak of Legionnaires disease associated with exposure to aerosols from an evaporative condenser," *J. Inf. Dis.* 161:1257–1261 (1990).
21. Burrell, R., and S.H. Yeh. "Toxic risks from inhalation of bacterial endotoxin," *Br. J. Ind. Med.* 47:688–691 (1990).
22. Delucca, A.J., and M.S. Palmgren. "Mesophilic microorganisms and endotoxin levels on developing cotton plants," *Am. Ind. Hyg. Assoc. J.* 47:437–442 (1986).
23. Clark, C.S., R. Rylander, and L. Larsson. "Levels of gram-negative bacteria, Aspergillus fumigatus, dust, and endotoxin at compost plants," *Appl. Environ. Microbiol.* 45:1501–1505 (1983).
24. Burrell, R., R.C. Lantz, and D.E. Hinton. "Mediators of pulmonary injury induced by inhalation of bacterial endotoxin," *Am. Rev. Respir. Dis.* 137:100–105 (1988).
25. Olenchock, S.A. "Quantitation of airborne endotoxin levels in various occupational environments," *Scand. J. Work Environ. Health* 14:72–73 (1988).
26. Bovallius A., R. Roffey, and E. Henningson. "Long-range transmission of bacteria," *Ann. NY Acad. Sci.* 353:186–200 (1980).
27. Cox, C.S. "Airborne bacteria and viruses," *Sci. Prog.* 73:469–500 (1989).
28. Dimmick, R.L., and A.B. Akers *An introduction to experimental aerobiology* (Chichester, New York: John Wiley and Sons, Inc., 1969).
29. Cox, C.S. *The aerobiological pathway of microorganisms* (Chichester, New York: John Wiley and Sons, Inc., 1987), pp. 172–205.
30. Brown, A.D. "Physiological problem of water stress," in *Strategies of microbial life in extreme environments*, M. Shilo, Ed. (Weinheim, New York: Verlag Chemie, 1979), pp. 65–81.

31. Horowitz, N.H. "Biological water requirements," in *Strategies of microbial life in extreme environments*, M. Shilo, Ed. (Weinheim, New York: Verlag Chemie, 1979), pp. 11–13.
32. Andrew, M.H.E., and A.D. Russell. *The revival of injured microbes* (New York: Academic Press, Inc., 1984), pp. 1–384.
33. Schlegel, H.G. *Allgemeine Mikrobiologie* (Stuttgart, New York: G. Thieme Verlag, 1992), pp. 79–84.
34. Dimmick, R.L., H. Wolchow, and M.A. Chatigny. "Evidence that bacteria can form new cells in airborne particles," *Appl. Environ. Microbiol.* 37:924–927 (1979).
35. Brown, A.D. "Microbial water stress," *Bacteriol. Rev.* 40:803–846 (1979).
36. Cox, C.S. "Aerosol survival of Pasteurella tularensis disseminated from the wet and dry states," *Appl. Microbiol.* 21:482–486 (1971).
37. Strange, R.E. and C.S. Cox. "Survival of dried and airborne bacteria," *Symp. Soc. Gen. Microbiol.* 26:111–154 (1976).
38. Mohr, A.J. "Development of models to explain the survival of viruses and bacteria in aerosols," in *Modeling the environmental fate of microorganisms*, C.J. Hurst, Ed. (Washington, DC: American Society for Microbiology, 1991), pp. 160–190.
39. Harper, G.J. "The influence of environment on the survival of airborne virus particles in the laboratory," *Arch. Gesamte Virusforsch.* 13:64–71 (1963).
40. Harper, G.J. "Airborne microorganisms: survival tests with four viruses," *J. Hyg., Cambridge* 59:479–486 (1961).
41. Akers, T.G., and M.T. Hatch. "Survival of a picorna virus and its infectious ribonucleic acid after aerosolization," *Appl. Microbiol.* 16:1811–1813 (1968).
42. Adams, A.P., J.C. Spendlove, R.S. Spendlove, and B.B. Barnett. "Aerosol stability of infectious and potentially infectious reovirus particles," *Appl. Environ. Microbiol.* 44: 903–908 (1982).
43. Goldberg, L.J., and I. Ford. "The function of chemical additives in enhancing microbial survival in aerosols," in *Airborne transmission and infection*, J.F.P. Hers, and K.C. Winkler, Eds. (Chichester, New York: John Wiley and Sons, Inc., 1973), pp. 86–89.
44. Goodlow, R.J., and F.A. Leonard. "Viability and infectivity of microorganisms in experimental airborne infection," *Bacteriol. Rev.* 25:182–195 (1961).
45. Zebel, G. "Modellrechnungen über den Einfluß des Windes auf die Einsaugkoeffizienten von Staubteilchen für zwei verschiedene Probennahmeköpfe," *Staub-Reinh. Luft* 39: 349–356 (1979).
46. May, K.R., and G.J. Harper. "The efficiency of various liquid impinger samplers in bacterial aerosols," *Br. J. Ind. Med.* 14:287–297 (1957).
47. Tyler, M.E., and E.L. Shipe. "Bacterial aerosol samplers," *Appl. Microbiol.* 7:337–349 (1959).
48. Luckiesh, M., A.H. Taylor, and L.L. Holladay. "Sampling devices for airborne bacteria," *J. Bacteriol.* 52:55–65 (1946).
49. Anderson, A.A. "A new sampler for the collection, sizing, and enumeration of viable airborne bacteria," *J. Bacteriol.* 76:471–484 (1958).
50. Ehrlich, R., S. Miller, and L.S. Idoine. "Evaluation of slit sampler in quantitative studies of bacterial aerosols," *Appl. Microbiol.* 14:328–330 (1966).
51. Henningson, E., R. Roffey, and A. Bovallius. "A comparative study of apparatus for sampling airborne microorganisms," *Grana* 20:155–159 (1981).
52. Melvaer, K.L., and H. Ringstad. "Collection of bacterial aerosols by means of slit sampler: A face-mask study," *Acta Pathol. Microbiol. Immunol. Scand. B.* 94:325–328 (1986).
53. Warren, J.C., T.G. Akers, and E.J. Dubovi. "Effect of prehumidification on sampling of selected airborne viruses," *Appl. Microbiol.* 18:893–896 (1969).

54. Thorne, P.S., M.S. Kiekhaefer, P. Whitten, and K.J. Donham. "Comparison of bioaerosol sampling methods in barns housing swine," *Appl. Environ. Microbiol.* 58:2543–2551 (1992).
55. Kethley, T.W., M.R. Gordon, and C. Orr. "A thermal precipitator for aerobacteriology," *Science* 116:368 (1952).
56. Morris, E.J., H.M. Darlow, J.F.H. Peel, and W.C. Wright. "The quantitative assay of mono-dispersed aerosols of bacteria and bacteriophage by electrostatic precipitation," *J. Hyg. Cam.* 59:487–596 (1961).
57. Orr, C. "Thermal precipitation for sampling air-borne microorganisms," *Appl. Microbiol.* 4:116–118 (1966).
58. Rotter, M., and W. Koller. "Sammlung von Luftkeimen mit Gelatinefiltern," *Zbl. Bakt. Hyg. I. Abt. Orig. B* 157:257–270 (1973).
59. Koller, W., and M. Rotter. "Weitere Untersuchungen über die Eignung von Gelatinefiltern zur Sammlung von Luftkeimen," *Zbl. Bakt. Hyg. I. Abt. Orig. B.* 159:546–559 (1974).
60. Palmgren, U., G. Ström, G. Blomquist, and P. Malmberg. "Collection of airborne micro-organisms on Nuclepore filters, estimation and analysis - CAMNEA method," *J. Appl. Bacteriol.* 61:401–406 (1986).
61. May, K.R., and G.J. Harper. "The efficiency of various liquid impinger samplers in bacterial aerosols," *Brit. J. Ind. Med.* 14:287–297 (1957).
62. Davies, R.R. "Viable moulds in house dust," *Trans. Br. Mycol. Soc.* 43:617–622 (1960).
63. Marshall, K.C. *Interfaces in microbial ecology* (Cambridge, London: Harvard University Press, 1976).
64. Lamberg, R.E., R.F. Schell, and J.L. LeFrock. "Detection and quantitation of simulated anaerobic bacteremia by centrifugation and filtration," *J. Clin. Microbiol.* 17:856–859 (1983).
65. Bej, A.K., M.H. Mahbubani, J.L. Dicesare, and R.M. Atlas. "Polymerase chain reaction-geneprobe detection of microorganisms by using filter-concentrated samples," *Appl. Environ. Microbiol.* 57:3529–3534 (1991).
66. Wolf H.W., P. Skaliy, L.B. Hall, M.M. Harris, H.M. Decker, L.M. Buchanan, and C.M. Dahlgren. "Sampling microbiological results," in *Public Health monograph no. 60* (Washington, DC: U.S. Public Health Service, 1964), pp. 3–40.
67. Brewer, D.G., S.E. Martin, and Z.J. Ordal. "Beneficial effects of catalase or pyruvate in a most-probable-number technique or the detection of Staphylococcus aureus," *Appl. Environ. Microbiol.* 34:797–800 (1977).
68. Larsen, P.I., L.K. Sydnes, B. Landfald, and A.R. Strom. "Osmoregulation in Escherichia coli by accumulation of organic osmolytes: betaines, glutamic acid, and trehalose," *Arch. Microbiol.* 147:1–7 (1987).
69. Le Rudulier, D., A.R. Strom, A.M. Dandekar, L.T. Smith, and R.C. Valentine. "Molecular biology of osmoregulation," *Science* 224:1064–1068 (1987).
70. Marthi, B., and B. Lighthart. "Effects of betaine on enumeration of airborne bacteria," *Appl. Environ. Microbiol.* 56:1286–1289 (1990).
71. Marthi, B., B.T. Shaffer, B. Lighthardt, and L. Ganio. "Resuscitation effects of catalase on airborne bacteria," *Appl. Environ. Microbiol.* 57:2775–2776 (1991).
72. Baird-Parker, A., and E. Davenport. "The effect of recovery medium on the isolation of *Staphylococcus aureus* after heat-treatment and after storage of frozen or dried cells," *J. Appl. Bacteriol.* 28:390–396 (1965).
73. Kingston, D. "Selective media in air sampling: a review," *J. Appl. Bacteriol.* 34:221–232 (1971).
74. Muir, W., and G.R. Milne. "The use of membrane filters for the detection of airborne contamination," *J. Med. Lab. Technol.* 20:85–91 (1963).

75. Starr, M.P., H. Stolp, H.G. Trüper, A. Balows, and H.G. Schlegel *The procaryotes. A handbook on habitats, isolation, and identification of bacteria* (New York, Heidelberg: Springer-Verlag, Vol. 1 & 2, 1981).
76. Castro, A.E. "Laboratory diagnosis of virus infections," in *Review of veterinary microbiology*, E. L. Biberstein, and Y. C. Zee, Eds. (Boston, Oxford: Blackwell Scientific Publications, Inc., 1990), pp. 410–424.
77. Janda, J.M., E.P. Desmond, and S.L Abbott. "Animal and animal cell culture systems," in *Manual of clinical microbiology*, A. Balows, W.J. Hausler, K.L. Herrmann, H.D. Isenberg, and H.J. Shadomy, Eds. (Washington, DC: American Society for Microbiology, 5th ed., 1991), pp. 137–146.
78. Spendlove, J.C., and K.F. Fannin. "Methods of characterization of virus aerosols," in *Methods in environmental virology*, C.P. Gerba, and S.M. Goyal, Eds. (New York, Basel: Marcel Dekker, Inc., 1982), pp. 261–329.
79. Koneman, E.W., S.D. Allen, V.R., Jr. Dowell, W.M. Janda, H.M. Sommers, and W.C. Jr. Winn, Eds. "Diagnosis of infections caused by obligate intracellular pathogens: viruses, chlamydia, and rickettsia," in *Color atlas and textbook of diagnostic microbiology* (Philadelphia: J.B. Lippincott Company, 3rd ed., 1988), pp. 691–764.
80. Guerin, L.F., and C.A. Mitchell. "A method for determining the concentration of airborne virus and sizing droplet nuclei containing the agent," *Can. J. Comp. Med.* 28: 283–289 (1964).
81. Kubitschek, H.E. "Counting and sizing micro-organisms with the Coulter Counter," in *Methods in Microbiology*, J.R. Morris and D.W. Ribbons, Eds. (London: Academic Press, Vol.1, 1969), pp. 593–610.
82. Beveridge, T.J., and L.L. Graham. "Surface layers of bacteria," *Microbiol. Rev.* 55: 684–705 (1991).
83. Romano, E.L., and M. Romano. "Staphyloccocal Protein A bound to colloidal gold: a useful reagent to label antigen-antibody sites in electron microscopy," *Immunochem.* 14:711–714 (1977).
84. Boguslaski, R.C. "Immunoassays monitored by virus, particle, and metal labels," in *Clinical immunochemistry: principles of methods and applications*, R.C. Boguslaski, E.T. Maggio, and R.M. Nakmura, Eds. (Boston: Little, Brown & Co., 1987), pp. 211–219.
85. Rosebrock, J.A. "Labeled-antibody techniques: fluorescent, radioisotopic, immunochemical," in *Manual of clinical microbiology*, A. Balows, W.J. Hausler, K.L. Herrmann, H.D. Isenberg, and H.J. Shadomy, Eds. (Washington, DC: American Society for Microbiology, 5th ed., 1991), pp. 79–86.
86. Zimmermann, R., and L.A. Meier-Reil. "A new method for fluorescence staining of bacterial populations on membrane filters," *Kieler Meeresforschung* 30:24–27 (1974).
87. Jones, J.G., and B.M. Simon. "An investigation of errors in direct counts of aquatic bacteria by epifluorescence microscopy, with reference to a new method for dyeing membrane filters," *J. Appl. Bacteriol.* 39:317–329 (1975).
88. Pettipher, G.L., R. Mansell, C.H. McKinnon, and C.H. Cousins. "Rapid membrane filtration - epifluorescent microscopy technique for direct enumeration of bacteria in raw milk," *Appl. Environ. Microbiol.* 39:423–429 (1980).
89. Hobbie, J.E., R.J. Daley, and S. Jasper. "Use of Nucleopore filters for counting bacteria by fluorescence microscopy," *Appl. Environ. Microbiol.* 33:1225–1228 (1977).
90. Rodriguez, G.G., D. Phipps, K. Ishiguro, and H.F. Ridgway. "Use of a fluorescent redox probe for direct visualization of actively respiring bacteria," *Appl. Environ. Microbiol.* 58:1801–1808 (1992).
91. Strange R.E., J.E. Benbough, P. Hambleton, and K.L. Martin. "Methods for the assessment of microbiological populations recovered from enclosed aerosols," *J. Gen. Microbiol.* 72:117–125 (1972).

92. Forghani, B. "Radio-immunoassay," in *Diagnostic procedures for viral, rickettsial and chlamydial infections*, E.H. Lenette, and N.J. Schmidt, Eds. (Washington, D. C.: American Public Health Association, 5th ed., 1979), pp. 171–189.
93. Hamilton, R. G. "Antigen quantification: Measurement of multivalent antigens by solid-phase immunoassay," in *Immunochemistry of solid-phase immunoassay*, Ed. J.E. Butler (Boca Raton, FL: CRC Press, 1991), pp. 139–150.
94. Brown, W.R., S.E. Dierks, J.E. Butler, and J.M. Gershoni. "Immunoblotting: Membrane filters as the solid phase for immunoassays," in *Immunochemistry of solid-phase immunoassay*, Ed. J. E. Butler (Boca Raton, FL: CRC Press, 1991), pp. 151–172.
95. Ekins, R.P. "Competitive, noncompetitive, and multi-analyte microspot immunoassays," in *Immunochemistry of solid-phase immunoassay*, Ed. J.E. Butler (Boca Raton, FL: CRC Press, 1991), pp. 106–138.
96. Lund, A., A.L. Hellemann, and F. Vartdal. "Rapid isolation od $K88^+$ Escherichia coli by using immunomagnetic particles," *J. Clin. Microbiol.* 26:2572–2575 (1988).
97. Hechemy, K.E., R.W. Stevens, and J.M. Conroy. "Immunoelectroblot techniques," in *Manual of clinical microbiology*, A. Balows, W.J. Hausler, K. L. Herrmann, H. D. Isenberg, and H. J. Shadomy, Eds. (Washington, DC: American Society for Microbiology, 5th ed., 1991), pp. 93–98.
98. Kemp, D.J., D.B. Smith, S.J. Foote, N. Samaras, and M.G. Peterson. "Colorimetric detection of specific DNA segments amplified by polymerase chain reactions," *Proc. Natl. Acad. Sci. USA* 86:2423–2427 (1989).
99. Wahlberg, J., J. Lundeberg, T. Hultman, and M. Uhlén. "General colorimetric method for DNA diagnostics allowing direct solid-phase genomic sequencing of the positive samples," *Proc. Natl. Acad. Sci. USA* 87:6569–6573 (1990).
100. Saiki, R.K., P.S. Walsh, C.H. Levenson, and H.A. Erlich. "Genetic analysis of amplified DNA with immobilized sequence-specific oligonucleotide probes," *Proc. Natl. Acad. Sci. USA* 86:6230–6234 (1989).
101. Kühn, I., G. Allestam, T.A. Stenström, and R. Möllby. "Biochemical fingerprinting of water coliform bacteria, a new method for measuring phenotypic diversity and for comparing different bacterial populations," *Appl. Environ. Microbiol.* 57:3171–3177 (1991).
102. Sambrook, J., E.F. Fritsch, and T. Maniatis. *Molecular cloning: A laboratory manual* (Cold Spring Harbor, NY: Cold Spring Harbor Laboratory, 2nd ed., 1989).
103. Tenover, F.C. "Molecular methods for the clinical microbiological laboratory," in *Manual of clinical microbiology*, A. Balows, W.J. Hausler, K.L. Herrmann, H.D. Isenberg, and H.J. Shadomy, Eds. (Washington, DC: American Society for Microbiology, 5th ed., 1991), pp. 119–127.
104. Erlich, H.A. *PCR technology: principles and applications for DNA amplification* (New York: Stockton Press, 1989), pp. 1–246.
105. Tsai, Y.L., and B.H. Olson. "Detection of low numbers of bacterial cells in soils and sediments by polymerase chain reaction," *Appl. Environ. Microbiol.* 58:754–757 (1992).
106. Picard, C., C. Ponsonnet, E. Paget, X. Nesme, and P. Simonet. "Detection and enumeration of bacteria in soil by direct DNA extraction and polymerase chain reaction," *Appl. Environ. Microbiol.* 58:2717–2722 (1992).
107. Eisenstein, B.I. "New molecular techniques for microbial epidemiology and the diagnosis of infectious diseases," *J. Inf. Dis.* 161:595–602 (1990).
108. Manafi, M., W. Kneifel, and S. Bascomb. "Fluorogenic and chromogenic substrates used in bacterial diagnostics," *Microbiol. Rev.* 55:335–348 (1991).
109. Kämpfer, P. "Differentiation of Corynebacterium spp., Listeria spp., and related organisms by using fluorogenic substrates," *J. Clin. Microbiol.* 30:1067–1071 (1992).

110. Kämpfer, P., I. Kulies, and W. Dott. "Fluorogenic substrates for differentiation of Gram-negative nonfermentative and oxidase-positive fermentative bacteria," *J. Clin. Microbiol.* 30:1402–1406 (1992).

111. Sasser, M., and M.D. Wichman. "Identification of microorganisms through use of gas chromatography and high-performance liquid chromatography," in *Manual of clinical microbiology*, A. Balows, W.J. Hausler, K.L. Herrmann, H.D. Isenberg, and H.J. Shadomy, Eds. (Washington, DC: American Society for Microbiology, 5th ed., 1991), pp. 111–118.

112. Stoakes L., T. Kelly, B. Schieven, D. Harley, M. Ramos, R. Lannigan, D. Groves, and Z. Hussain. "Gas-liquid chromatographic analysis of cellular fatty acids for identification of gram-negative anaerobic bacilli," *J. Clin. Microbiol.* 29:2636–2638 (1991).

113. Taylor, A.J., and P.R. Skinner. "Gas liquid chromatography in medical microbiology," *Med. Lab. Sci.* 40:375–385 (1983).

114. Meighen, E.A. "Molecular biology of bacterial bioluminescence," *Microbiol. Rev.* 55:123–142 (1991).

115. Drow, D.L., C.H. Baum, and G. Hierschfield. "Comparison of the Lumac and Monolight systems for detection of bacteriuria by bioluminescence," *J. Clin. Microbiol.* 20:797–801 (1984).

116. Selan, L., F. Berlutti, C. Passariello, M.C. Thaller, and G. Renzini. "Reliability of a bioluminescence ATP assay for detection of bacteria," *J. Clin. Microbiol.* 30:1739–1742 (1992).

117. Rylander, R., and M.C. Snella. "Endotoxins and the lung: cellular reactions and risk for disease," *Prog. Allergy* 33:332–344 (1983).

118. Devleeschouwer, M.J., M.F. Cornil, and I. Dony. "Study on the sensitivity and specifity of the Limulus Amebocyte Lysate test and Rabbit Pyrogen Assays," *Appl. Environ. Microbiol.* 50:1509–1511 (1985).

119. Gardi, A., and G.R. Arpagaus. "Improved microtechnique for endotoxin assay by the Limulus amebocyte lysate test," *Analyt. Biochem.* 109:382–385 (1980).

120. Piotrowicz, B.I., S.E. Edlin, and A.C. McCartney. "A sensitive chromogenic Limulus lysate micro-assay for detection of endotoxin in human plasma and in water," *Zbl. Bakt. Hyg. A* 260:108–112 (1985).

121. Parker, M.T., and L.H. Collier, Eds., *Topley & Wilson's Principles of bacteriology, virology and immunity* (London, Melbourne: Edward Arnold, Hodder & Stoughton, Vol. 1.-5., 8th ed., 1990).

CHAPTER 14

Biological Analysis of Fungi and Associated Molds

T.M. Madelin and M.F. Madelin

INTRODUCTION

General Characteristics of Fungi and Associated Molds

Spores and hyphal fragments of fungi are ubiquitous in air, where they are sometimes the major pollutant and sources of infection or allergic reactions. For the great majority of dry-spored fungi, air is their natural dispersal medium, and accordingly they have evolved various mechanisms that enhance their effective dispersal and survival in air. From the aerobiological standpoint this is an important difference from bacteria and viruses. Fungal spores vary greatly in size, but most are in the range 2–50 µm; they are bigger than actinomycete and other bacterial spores and generally smaller than pollens (see Figures 14.1 and 14.2).

Airborne fungal spores cannot rise or move laterally in air except through wind and turbulence (see Chapter 4). Consequently they must first traverse the boundary layer of still air that adjoins the surface from which they have originated in order to reach a region of air movement, for otherwise they would not disperse far from their site of formation. Their introduction into the turbulent region is commonly achieved by mechanisms or structures that shoot them through the boundary layer (such as bursting asci) or drop them from a height into moving air (such as the fruit bodies of bracket fungi and toadstools), or simply raise them on microscopic stalks the millimeter or so required to expose them to wind or gusts. The dispersal of fungal spores is treated comprehensively by Ingold.[1] Not infrequently, fragments of mycelium and sporophores also become blown away. Some of these remain viable and capable of instituting new growth.[2]

Instead of introducing their spores directly into the wind, many fungi produce them in mucilage that prevents their release except when conditions are wet. Rain drops falling on slime-enmeshed spores may launch them into the air in tiny water droplets. These, if large enough, may be thrown considerable distances around their origin by "ballistic dispersal," while if small enough, may be carried passively by the wind. Droplets less than 100 µm in diameter swiftly evaporate unless the relative humidity is near to 100%, and will leave the contained spores to be dispersed dry upon the wind, possibly then to achieve long-distance dispersal.[3] However, for any airborne spores, high-altitude airborne transport is hazardous as a result of solar radiation, desiccation and cold.

Yeasts are fungi whose predominant form is unicellular, and usually multiply by budding. Under certain conditions some revert partially to mycelial or pseudomycelial form, while some are known to be alternative morphs of fungi that otherwise exist in mycelial form. Yeasts, especially *Sporobolomyces* species and *Aureobasidium pullulans*,

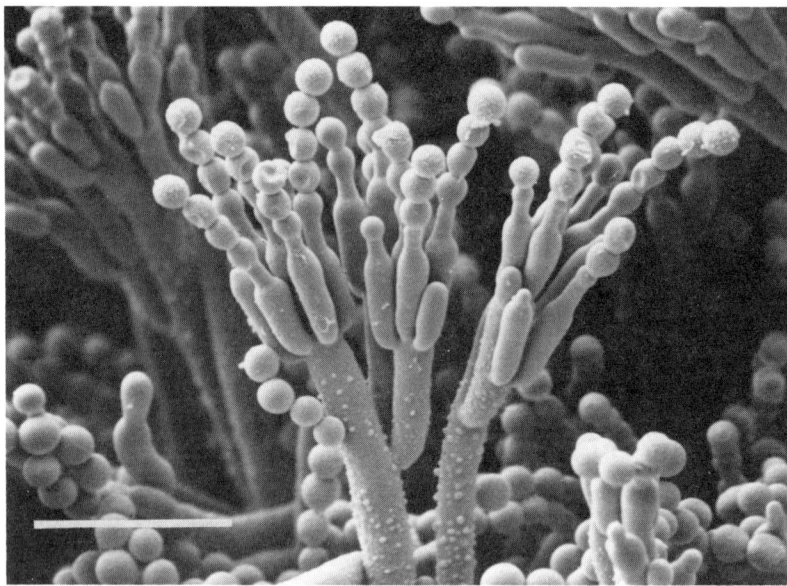

Figure 14.1. Scanning electron photomicrograph of conidiophores of *Penicillium camembertii* bearing chains of spores (courtesy of Dr. A. Beckett, University of Bristol). Scale bar 20 μm.

sometimes comprise the most abundant component of the outdoor air spora. The ballistospore-producing yeasts of the Sporobolomycetaceae clearly have specific provisions for wind-dispersal, but most yeasts lack active mechanisms for launching viable propagules, and probably depend on rain splash and wind-shaking. A few may have alternative morphs that are suited to wind-dispersal, such as the basidiomycetous *Filobasidiella neoformans* state of the pathogenic yeast *Cryptococcus neoformans*. Vegetation is a major source of yeasts in outdoor air. Yeasts and yeast-like imperfect fungi such as *Aureobasidium pullulans* are in fact the most abundant inhabitants of the surface of young green leaves (the phylloplane).[4] Yeasts in indoor air may originate from outdoor sources, but also may stem from such artifacts as cold-mist vaporizers used for indoor humidifying.[5]

Numbers of fungal propagules in outdoor air vary enormously with species, location, altitude, season, climate and time of day. For example, spores of *Sporobolomyces* are most abundant in air spora before dawn, spores of *Phytophthora infestans* after dawn, and spores of *Cladosporium, Alternaria* and *Ustilago* in early afternoon. After rain, near ground level there is often a "damp-air spora" dominated by various kinds of ascospore whose active release depends on the turgor of the asci in which they form. This damp-air spora replaces the "dry-air spora" consisting of spores of *Cladosporium, Alternaria,* smuts and rusts that is largely washed out of the air by the same rainfall giving rise to the damp-air spora. Counts of fungal spores commonly average thousands per cubic meter, but levels in excess of millions per cubic meter have been recorded. Although concurrent sampling at different outdoor sites, e.g., within a 10-kilometer radius, yields comparatively large differences in spore concentration, averaging sample data over increasingly long periods of time tends to reduce such differences till they are no longer significant.[6] Concentrations of airborne fungi are usually less in domestic buildings than outdoors but may be large if the former are damp and affected by mold growth, particularly during winter. Concentrations vary widely but may reach $4 \times 10^5/m^3$ air in extreme cases.[7] They can be comparably high in industrial and farm buildings.

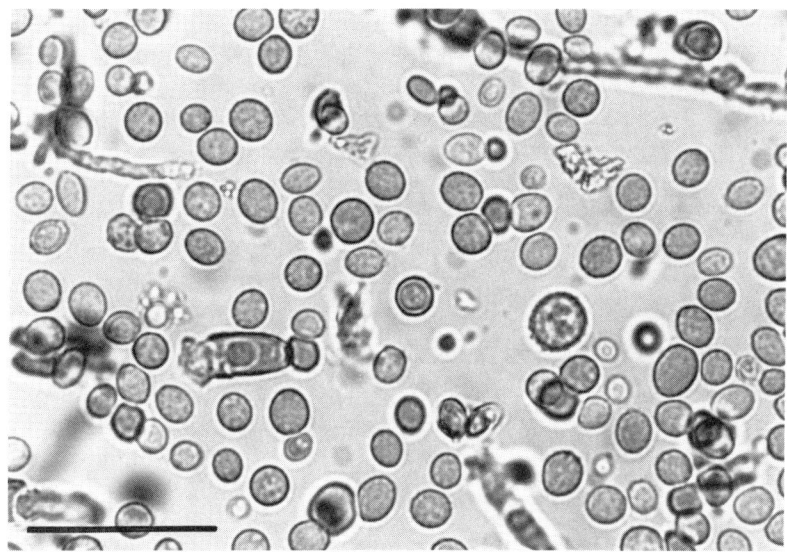

Figure 14.2. Airborne fungal spores released from disturbed straw, and trapped with May/RE Cascade Impactor. Scale bar 20 μm.

Actinomycetes are bacteria but like most fungi they are capable of branching filamentous growth and many produce dry airborne spores. For this reason, in aerobiological studies they often are assayed with the fungi. Thermophilic actinomycetes are common in composts that have heated during fermentation and are important as causative agents of occupation-related respiratory allergies, often occurring in high concentrations in air when compost is disturbed. Up to $2\times10^9/m^3$ have been recorded during farmwork with moldy grain, and nearly $1.6\times10^9/m^3$ air after shaking bales of hay.[8,9] Actinomycete spores are smaller than those of fungi, averaging about 1μm in diameter. Owing to their small size, they can reach the pulmonary region of the lung when inhaled. Their classification and biology has been comprehensively reviewed by Goodfellow, Mordarski and Williams.[10]

Myxomycetes (plasmodial or acellular slime molds) are a group of microorganisms traditionally associated with true fungi, but probably are more closely related to Protozoa. There are about 500 known species mostly inhabiting soils, litters, decaying wood and other vegetable materials. Their reproductive stages produce spores well-suited to airborne dispersal.[11] The spores are more or less spherical, single-celled bodies about 10 μm in diameter and are released into air from specialized sporangia. They have been detected in air samples[12] and have been implicated in human allergy.[13]

Principles of Aeromycological Assays

Different procedures for assaying air sporas need to be selected according to the environment and purpose for which information is sought. Reasons for conducting aerobiological surveys of fungi and associated molds include: measurements of air quality; detection of pathogenic organisms; epidemiological surveys and predictions; detection of mycotoxic fungi; detection and measurement of exposure to fungal and actinomycete aeroallergens. Some of these aims require counting and categorization of viable or infective propagules in air, but for others it may not be necessary to know the condition of trapped materials. For example, spores in indoor air, e.g., in animal houses, may do no more than add to particulate pollution of air, though in so doing additional stress may be added to defense systems of animal lungs, thereby increasing susceptibility to other airborne diseases. In relation to allergies, likewise it may be of no consequence whether airborne particles are alive or dead. Assay procedures described below relate to numbers, viability, infectivity, toxicity and allergenicity of sampled fungal aerosols.

TOTAL ASSAY OF AEROSOLS OF FUNGI AND ASSOCIATED MOLDS

Introduction

Since it is not always necessary to determine whether fungal and associated mold particles in air are viable, infective, allergenic or toxic, purely enumerative procedures often are sufficient, as for example when investigating the periodicity of fluctuations in air spora, or dustiness of materials such as hays and straws. A total survey is sometimes the first step in recognizing the existence of medical, veterinary or agricultural hazards. In practice, the distinction between total and other kinds of assay sometimes is obscured because many fungi need to be cultured for identification. Fungal identification is based almost wholly on morphology of reproductive states. Total assay procedures that do not rely on viability of catches will be considered first.

Non-Volumetric Total Assays

A number of techniques are employed that reveal rates of deposition of particles of fungi and associated molds on surfaces indoors or outdoors, but not their actual concentration in air, though they may provide a general indication. For some purposes, rates of deposition are more useful than airborne concentration. Because fungal and other mold particles generally are recognizable as such by light microscopy, samplers that trap particles in a way that permits such examination are widely used (Figure 14.2). Commonly this involves capture onto a sticky surface, with either hydrophilic or hydrophobic adhesives being used. The latter are particularly useful when it is important to avoid size or shape changes of captured particles through uptake of water. Commonly used adhesives include glycerine jelly, silicone grease, petroleum jelly (sometimes with added liquid paraffin) and polyvinyl alcohol. Use of transparent sticky tape also has been advocated.[14]

The simplest non-volumetric trap is the sticky slide. Being inexpensive and requiring no power, it is convenient for extensively replicated sampling in outdoor sites. If mounted horizontally, it primarily monitors gravitational sedimentation; if vertically, inertial impaction. However, it is not uniformly efficient for spores of different sizes as small spores will be under-represented in samples because they do not have the inertia required for impaction but will instead flow around slides in moving airstreams.

An agricultural method akin to use of sticky slides is to examine the population of fungal spores that has been deposited naturally on living leaves. Spores may be counted *in situ* or stripped off leaf surfaces in a transparent film of nail varnish or acrylic plastic, or be removed on the surface of sticky transparent tape.

It is sometimes desirable to assay the course of spore deposition over a period of time, especially because many fungi show diurnal periodicity of spore release. Several simple mechanical elaborations of the sticky slide principle that record sequential deposition have been described.[15,16]

A widely employed outdoors alternative to sticky slides is the vertical cylindrical rod. If made of glass, it may be made sticky and, after exposure, be examined directly; if made of steel, a band of polyethylene plastic coated with silicone grease may be wrapped around it, and be removed for subsequent examination.[17] Different rod diameters favor impactive capture of different sized spores; often rods of several sizes in the range 1 to 10 mm diameter are simultaneously employed. Factors influencing efficiency of impaction on vertical cylinders are discussed by Gregory.[18] Impaction efficiency of spores on vertical rods is greater than that on vertical slides, while varying less with windspeed, particularly with cylinders of relatively small diameter.[19] Larger trapping areas offered by slides are of significance only when spore concentrations are low.[20] Vertically oriented slides and rods sometimes are mounted on wind vanes so as always to face into the wind.

Volumetric Total Assays

Sampling a known volume of air has obvious advantages for quantitative analyses and comparisons. The aim of such volumetric assays must be to capture a representative aerosol in which all trapped spores and other particles of interest are present in the same proportions as in the air. Isokinetic sampling where airflow velocity at sampler intake is the same as that of air being sampled, ensures that a representative aerosol is collected (see also Chapter 4). This is most easily achieved under controlled conditions such as in a wind tunnel.[21] May[22] suggested that in outdoor conditions, where wind speed and direction are constantly changing, intake efficiency would be improved by placing a hemicylindrical baffle immediately behind a sampling nozzle, thereby forming a stagnation point where air is almost stationary at the point of sampling (see also Chapter 8).

Various volumetric samplers capture spores by their impaction from a high-velocity air stream onto sticky surfaces, air entering at a known rate through a slit, usually by the action of a suction pump. Because the composition of air spora varies with time, both short and long term, commercially available continuously recording samplers such as the Hirst-Burkard and Kramer-Collins that deposit the catch on slowly moving target surfaces find much use, especially in outdoor aerobiology. Constructional details of inexpensive vane-mounted[23-26] and non-vane-mounted[27] recording volumetric spore traps have been published.

Quantitative data can be obtained from continuous recording samplers by light microscopic examination of the trapping surfaces. Käpylä and Penttinen[28] evaluated different sampling strategies for deriving mean data from Hirst-Burkard volumetric spore traps and their conclusions are relevant to most traps of this general type. To measure airborne spore concentration at every hour, commonly a single traverse across the tape width is made. However, because transverse distribution patterns of trapped particles may be irregular, the whole width of tapes should be studied. To arrive at an estimate of daily (or other relatively long-period) mean, the common practice of making one or two scans along the length of the tape may prove unreliable because of these irregularities in transverse distribution. Instead, random microscope fields or selected traverses across the width of the tape are recommended. Twelve such traverses are enough to estimate a daily mean concentration. Systematic traverses are more effective than randomly spaced ones.

There are occasions when a spot reading of concentration and composition of prevailing air spora is adequate, as, for example, when assessing dustiness of hay samples.[29] The battery-driven Burkard portable hand-held sampler in which aerosols impact against a stationary sticky microscope slide can provide this information. Alternatively, aerosols may be drawn through filters of appropriate pore size, and the catch examined either microscopically or by cultural assay.[8] Where size-categorization is desirable, samplers such as the 7-stage May/RE cascade impactor (Research Engineers, Shoreditch, London) based on May's "Ultimate" cascade impactor,[30] are particularly valuable. Although the cascade impactor furnishes quantitative (numbers or mass/unit volume) and qualitative (microscopic) data on air spora, the time-consuming nature of particle-counting may limit its use. In such cases it may be feasible to relinquish visual qualitative data and instead use an automatic particle sizer and counter.

Light-based direct-reading particle sizers and counters fall into two categories according to method of sizing (see also Chapter 8). Either they measure light-scattering properties of a particle as it passes through a light beam, or they use laser Doppler velocimetry (LDV) to measure the size-dependent velocity of an accelerated particle in a fixed-flow field. The Aerodynamic Particle Sizer (APS) (TSI Inc., St. Paul, MN) is an example of the latter type of instrument; it has the advantage of measuring aerodynamic equivalent diameter which is a most useful size parameter for most purposes in aerosol science since it describes the behavior of a particle in air. Aerodynamic diameter is defined as the diameter of a unit-density sphere having the same settling velocity as the particle in question. Other advantages are the large number of size divisions (32 channels from 0.5 to 15 μm in the APS) and the wealth of numerical and graphical computer-generated data on particle numbers, surface area, mass, and respirable mass in a given volume of air. Most fungal spores prevalent in air lie within the APS size-range.

The light-scattering instruments (e.g., Climet Particle Analysis System: Redlands, California; Rion Particle Counter: Hawksley, Lancing, U.K.) have advantages of being portable, robust and easy to handle. Disadvantages common to both types of sizer are that intake nozzles may become blocked with dust during sampling and, in high particle concentrations, particle coincidence problems can occur when more than one particle enters the sensing volume at the same time, thus affecting both size and number measurements. The major disadvantage for aeromycologists, however, is that current particle sizers do not provide identification of sensed particles. Where the aim is to characterize a polydisperse dust that includes diverse fungal spora, e.g., in open air or in an animal house, then particle sizers may be used in conjunction with instruments, such as the May/RE cascade impactor, that allow visual analysis.[21,31] Visually characterizing a few representative fields on each slide is sufficient for qualitative interpretation of numerical data provided by particle sizers. Where an instrument such as the APS is used to size and count experimentally produced spore-clouds of a single species, it is useful nevertheless also to have a visual method of assessment to provide information on, for example, the amount of spore aggregation and purity of the aerosol.[31]

Samplers have been attached to aircraft for high level surveys of air spora,[18,32,33] including monitoring arrival or escape of fungal crop pathogens. Some have been designed specifically for mounting upon small radio-controlled model aircraft. The traps designed by Gottwald and Tedders[34] to be mounted one beneath each wing of a remotely piloted model plane, relied on airflow beneath the wings to cause aerosol impaction against the sticky surface of a drum rotatable by remote control to 20 different sampling positions. McCracken[35] described the design of a remote-piloted model aircraft carrying an aspirated sampler that trapped airborne particles on sticky rods positioned by radio control. Self-aspiration of the sampler obviated the need to maintain a high flying-speed while operating the aircraft in areas with frequent obstructions and poor visibility.

Volumetric samplers are not necessarily based on aspirated mechanisms. The rotorod[36] and rotoslide[37] impaction samplers trap particles on sticky surfaces of, respectively, narrow rods or edges of microscope slides that are rotated rapidly about an axis by an electric motor. They are volumetric insofar as they sample calculable volumes of air; their efficiency varies for different spore sizes, but generally is appropriate for work with fungi. Correction factors may be applied to adjust counts of spores of particular sizes.[37] The efficiency of the rotorod sampler is essentially independent of windspeed and direction.

It may be operated continuously for short sampling periods but its high efficiency causes rapid overloading. Hence, for sampling periods longer than about 2 hours, it should be operated intermittently. Variation in sampling rates between individual traps can be minimized by calibrating rotation rates with a stroboscope.[38]

Sutton and Jones[39] evaluated the relative merits of rotorod samplers, Burkard volumetric traps, and vertically oriented wind-vane-mounted glass rods and microscope slides for monitoring discharge of ascospores of *Venturia inaequalis* in orchards. Their consideration of scientific and practical aspects of different procedures is likely to prove helpful in making decisions about equipment to use in the field in other circumstances.

A problem arising when sampling air within buildings is that recoveries of particles indoors may reflect the pervasive abundance of air spora derived from outdoor sources. To minimize this enhancement of indoor recoveries, Solomon[40] scheduled his sampling program (in Michigan, USA) so that it was done on days with subfreezing temperatures, on 62% of which there was some snow cover.

Total Assay of Rain- and Splash-Dispersed Air Spora

Some fungal spores are specially adapted to dispersal by rain or rain splash (see also Chapter 4), and may be captured from the atmosphere simply by catching rain in a vessel. Precautions must be taken to avoid contamination by spores from sources close to the sampler.[18] However, captured water is liable to contain other materials that can render observation difficult. Bacterial growth may be prevented by addition of mercuric chloride.[41] Fungal spores can be separated from other particles and concentrated by collection at the interface between pairs of immiscible liquids, followed by filtration in oil through a membrane filter on which they may be microscopically examined, or by centrifuging them through a density gradient followed by deposition onto membrane filters.[41] To prevent loss of fungal spores from rain samples through germination, fungitoxic or fungistatic materials (such as phenol or glycerol respectively) may be added.[3]

Various samplers are suitable for capturing splash-borne plant pathogenic fungi.[3] Since spore-carrying droplets vary in size from large "ballistic" droplets down to small "airborne" ones, no single sampler is universally appropriate. For ballistic droplets, samplers with horizontal collecting surfaces are most suitable. Slides are simple, but lose spores through run-off of excess water. Funnels are effective but samples may require concentration (see above). Small airborne droplets and residual particles left after their evaporation are best sampled with high-volume samplers such as cyclones or Casella bacterial slit samplers. For volumetric samplers used above crops, attachment to a wind-vane will keep the orifice facing into the wind. For sampling within crop canopies, samplers having orifices facing upwards beneath rain shields and above the main splash zone are advisable. Operating samplers only during periods of rainfall avoids collecting large numbers of dry-dispersed fungal spores. Automatic control units have been developed that respond to moisture or sound of falling raindrops.[39,42]

Microscopic Identification of Air Spora Constituents

One may wish to identify different kinds of spore trapped in an air sampler or simply recognize spores of a single organism of particular interest. The latter is difficult if similar looking spores from different fungi are present, and can be particularly troublesome when studying organisms belonging to large taxa with relatively limited ranges of morphology such as rusts and mildews, and the small-spored storage fungi such as Aspergilli and Penicillia. An analysis of data from 14 different air-sampling stations found that fungal spores were counted unreliably.[43] If reliable spore data from multiple stations are to be obtained, personnel should be rigorously trained.

Although microscopic examination may serve to identify a spore to genus level, it seldom allows confident identification to the level of species. This can be a problem in most aerobiological surveys, but is particularly acute for fungal spores that are solitary, single-celled, rounded or ellipsoidal bodies; spores consisting of two or more cells have greater potential for structural diversity and hence for distinctiveness. Unfortunately, it is often those spores in which aeromycologists are most interested that present special difficulty in identification by microscopic examination. Sometimes circumstantial evidence points to the likely identity of spores, e.g., the presence of nearby sources of morphologically identical spores of identified species.

There are many well illustrated publications that are an aid in spore identification.[44-54]

VIABILITY

Introduction

Viability is demonstrated by spore germination, with or without culture for subsequent growth, or by vital staining.

Culture Techniques

If a spore can form a germ tube, it is viable. Incubating spores in a moist chamber usually allows germ tubes to form. However, not all fungal spores can germinate in the absence of nutrients.

More common methods of enumerating viable fungal and mold propagules in air are by trapping directly onto growth medium allowing colonies to form that subsequently may be counted and identified, or alternatively by capturing into fluid that then is used to inoculate culture media or even live potential hosts. In situations where plates of exposed culture media are likely to become overloaded, solid trapping medium may be used to prepare inocula for subsequent dilution plating. Depending on composition of media, e.g., agar or glycerol/gelatine gels, this may be achieved by either homogenizing or melting.[55]

The Hirst continuous volumetric spore trap has been used to yield viable spore counts.[56] The sampling slide was tightly covered with polythene film smeared lightly with pectic jelly, whence after exposure the coated polythene was used to prepare suspensions that were diluted serially and plated out onto malt extract agar.

Obligate parasites, or those with exacting requirements need living substrates for their detection. However, detection of some species culturable on agar sometimes may be facilitated by using natural substrata. For example, workers studying airborne dispersal in forests and sawmills of fungi causing bluestain in timber have done so by exposing living softwood discs.[57] Fungi may be identified on discs or after transfer to agar culture medium. There is a risk, though, of introducing fungal contamination by insects when surfaces of growth media or substrata are passively exposed in this way.[57]

Gravity-settling culture plates do not measure spore concentration in air, nor do they trap representative populations of viable propagules since they suffer from the fundamental limitation of underestimating small spores of low terminal velocity. They may even fail to detect species revealed by more sensitive samplers, such as Andersen viable samplers or total-recovery impingers. Andersen samplers and multistage liquid impingers (MSLIs) have the additional advantage of separating catches into size-fractions. This is particularly important in human and animal respiratory allergy studies since sites of deposition in lungs are determined by particle size. The three stages of the May liquid impinger[58] are intended to correspond with the principal deposition sites in the human respiratory system, viz., naso-pharyngeal, tracheobronchial, and pulmonary region, and they collect particles with aerodynamic diameters of >10, 4 to 10, and less than 4 μm, respectively. Colony counts in Petri dish cultures exposed in Andersen samplers should be corrected for effects of multiple impactions at one point.[59,60] Published reports should always state when this has been done.

Slit samplers such as the Casella Airborne Bacteria Sampler, in which aerosols are drawn through a slit and impact onto a slowly rotating Petri dish of agar culture medium, do not fractionate by size but serve to spread and sequentially distribute the inoculum.

Volumetric impactors, such as the Andersen and Casella samplers, in which particles are impacted directly onto growth medium, provide numbers of colony-forming units (CFUs). Fungal spores often are present in air as chains (Figure 14.1) or clumps of spores and may also be carried on rafts of plant or other material. Each impacted particle can give rise to just one CFU when incubated although this may consist of several viable propagules. Impingement into liquid leads to aggregates being dispersed into individual viable cells so that total numbers of microorganisms present may be estimated. Depending on the aims of an aerobiological study, either or both methods may be appropriate. Liquid impingers suitable for fungal assays include the All Glass impingers (AGI) and MSLIs. Effects on viability of plunging dry fungal spores into liquid at very high velocity have not been directly assessed, but a study comparing recoveries of bacteria from MSLIs containing different aqueous collecting solutions found quarter-strength Ringer's solution plus 20 g inositol per liter was best[61]; whether it was best for fungi was not established. It is possible to sample with an impinger for long periods before evaporation of the collection fluid imposes limits. A solution of glycerol, polyglycol and tris(hydroxymethyl)aminomethane has been used to reduce evaporation.[62] Centrifugal "Cyclone" samplers[63,64] have more gentle collection characteristics and usually incorporate fluid injection at the air inlet to form a thin layer on the inside wall.

Viable airborne particles also may be quantified by drawing air through a membrane filter of appropriate pore size that may then be placed directly on an appropriate nutrient agar plate if clusters of spores are not to be dispersed, or homogenized in water and the homogenate plated out if clusters are to be dispersed.[62] If polycarbonate filters are used, trapped fungal spores may be washed off for serial dilution. A good correlation was found between counts of viable airborne spores made by this method and by means of a cascade impactor.[55]

During extended sampling periods, agar plates or collecting fluids lose moisture. This can affect collection efficiency of samplers in a number of ways, for example, by altering the critical distance between the agar surface and the impaction jet in an Andersen sampler. Sterile 0.2% emulsion of oxyethylene docosanol swilled over the dried surface of agar plates and then poured off leaves a monomolecular layer of wax that prevents evaporation loss and allows very substantial increases in sampling time. It is ineffective in high-velocity liquid impingers that obstruct re-formation of the monomolecular layer, but is nevertheless useful in the upper stages of May multistage impingers.[65] Similarly, replacing agar with plates of gelatine (7.1% w/w) plus glycerol (50% w/w) greatly reduces drying in a slit sampler. The gel then is melted at 40°C and serially diluted for dilution-plate counts.[55]

An Andersen sampler attached to the upper surface of a Burkard volumetric spore trap has been used for the specific purpose of simultaneously comparing "viable" and "non-viable" volumetric collectors.[66] Results revealed that although viable (i.e., Andersen) recoveries of *Cladosporium* spores varied directly with non-viable (i.e., Burkard) counts, the viable:non-viable ratio fell as the spore levels rose.

Cultural recoveries are strongly influenced by selectivity of media employed and by incubation temperatures. Fungi mostly have wide cultural tolerances but prefer an acid medium with pH of 4 to 7. In our experience 2% malt extract agar and potato-dextrose agar, both containing 20 units/mL penicillin and 40 units/mL streptomycin, are good all-round media for collecting a wide range of airborne fungi. The antibiotics suppress bacteria. Addition of non-ionic surfactant Triton N-101 (Rohm and Haas) at 0.05% is effective in reducing radial growth of fast-growing species that otherwise could overwhelm slow-growing and late-appearing species.[67] Rose-bengal is another commonly used growth inhibitor but has been found to perform unsatisfactorily by several authors.[67–69] Comparisons of recoveries of airborne fungi on various media have been reported[68,70]

Non-selective air-sampling for fungi on a general purpose medium and incubation at 20–25°C produces a very mixed collection of species. Some (e.g., the Aspergilli) will grow and mature within a few days and start producing secondary colonies; these confuse colony counts and also tend to obscure small, more slowly developing species such as *Wallemia sebi*. Radial growth inhibitors are useful here, but it is important to count emerging colonies frequently for up to three weeks and to mark them on the reverse of the plate as they emerge. We have found that some basidiomycetes have become apparent only after several months storage in a cold room at 4°C. Ahlström and Käärik[71] reported that when 4–10 ppm of the systemic fungicide benomyl ("Benlate") was added to 1.5% malt extract agar, previously predominant deuteromycetes and ascomycetes were inhibited and the proportion of basidiomycetes developing was noticeably higher. Airborne propagules of basidiomycetes cannot as a rule be identified to species without being cultured. Colonies can be identified as basidiomycetes if they have clamp connections or give a positive laccase or tyrosinase reaction that indicates ability to produce ligninolytic enzymes.[71] Cultural and biochemical methods for identification of mycelia of wood-decay fungi have been described.[72,73] Fungi able to grow at low water activities such as *Wallemia sebi* and *Aspergillus glaucus* group grow preferentially on media supplemented with 20% sucrose, 7.5% NaCl or with dichloran-glycerol.[74] Some species require daylight in order to sporulate, and thus become identifiable, while other fungi sporulate better in near-ultraviolet "black" light.

Optimum growth temperatures for sporulation vary according to species, but most grow well at 20–25°C, or at room temperature. Some thermotolerant species such as *Aspergillus fumigatus* are more easily isolated at higher temperatures such as 37°C where their presence is not obscured by mesophiles. All medically important fungal pathogens grow well at 30°C including dermatophytes, but systemic pathogens also grow at 37°C. Thermophilic actinomycetes implicated in occupational respiratory allergies grow at 55°C although *Saccharomonospora viridis* performs better at 45–50°C. *Streptomyces* species common in straw may be allergenic and grow well at 37°C.

Culture media suitable for these actinomycetes are nutrient agar or tryptone-soy agar, both at half-strength, supplemented with 2% casein hydrolysate. Cycloheximide at 50 µg/mL suppresses fungal growth. Rifampicin at 5 µg/mL inhibits unwanted bacteria and some actinomycetes while encouraging others, notably *Saccharomonospora viridis*. A medium containing tyrosine distinguishes the highly allergenic *Thermoactinomyces vulgaris* from *T. thalpophilus*.[75] These species are visually similar and occur together in fermented composts. Actinomycetes, like most bacteria, prefer an alkaline pH of about 7.5.

It is worth emphasizing differences in mycological culture techniques from those employed by bacteriologists. Because of the relatively slow growth of fungi, agar plates should be incubated loosely enclosed in plastic bags to prevent drying, although conditions should always remain aerobic; this is particularly necessary at high incubation temperatures. Petri dishes of media should not be left open for their surfaces to dry before use because of possible contamination from the many spores common in air. During incubation, Petri dishes should not be inverted since spores will drop onto the lid and may escape to colonize other plates. Fungal cultures should never be opened and smelled because of the risk of inhaling spores. Mycological culture techniques are described by Onions et al.[51]

Certain fungi have specific characteristics that allow them to be detected by their growth on highly selective culture media. The "kerosene fungus," *Cladosporium resinae*, is a good example. It is best known as a contaminant of aviation jet fuel, though widely distributed in soil. A medium selective for this mold, consisting of V-8 Vegetable Juice agar with 0.1% creosote, has been used to isolate it from the air.[76] Media may also be diagnostic. Bothast and Fennell[77] devised an *Aspergillus* differential medium to distinguish members of the *Aspergillus flavus* group from other species of *Aspergillus* and common storage molds by their capacity to produce a yellow-orange pigment on the reverse side of the culture.

The possibility that trapped airborne spores are viable yet able to germinate only under special conditions has received little experimental attention. Basidiospores of the Bolbitiaceae from dried herbarium material were rarely germinable if placed directly on agar media, yet if incubated in a saturated atmosphere overnight before plating out, they germinated.[78] Such sensitivity to rapid rehydration might partly explain the commonly observed paucity of basidiomycetous colonies in agar plates exposed in air samplers. However, basidiospores are not the only spores responsive to vapor-phase rehydration. Conidia of the mold *Botryodiplodia ricinicola* rendered ungerminable by chilling at 5°C became germinable after this treatment.[79] It is possible that the dryness and cold to which fungal spores are commonly exposed while airborne, especially outdoors at high altitudes, renders some susceptible to damage, probably to their cell membranes, by rapid rewetting through contact with raindrops, dew, or wet surfaces.

Vital Staining

In spite of the attractive simplicity of vital staining techniques as a means of recognizing viable airborne propagules, they are not without disadvantages and limitations.

Fluorescence microscopy with fluorescein diacetate (FDA) has been used to recognize viable organisms in which it causes yellowish-green fluorescence.[80] Damaged and dead cells fail to accumulate it. The FDA technique has been used successfully on fungi in pure culture and in soil.[81] FDA also has been used in conjunction with propidium iodide (PI) that supposedly penetrates only dead cells, the two stains fluorescing green and red, respectively, when excited by blue light. However, only PI fluoresces when excited by green light.[82] Dittmer and Weltzien[83] who used FDA successfully to determine viability of cells of sclerotia of the fungus *Sclerotinia sclerotiorum*, give full details of the staining protocol. They also evaluated other vital staining methods with the same fungal material and found that the acridine orange (AO) fluorescence microscopy technique (in which AO is supposed to accumulate in dead or damaged cells more than in living ones) did not consistently distinguish live from dead cells. Mixed results have been reported with this technique. Calcofluor White M2R (CW), also known as "Cellufluor" (Polysciences, Inc., Warrington, Pennsylvania, USA), that stains cell polysaccharides, has been recommended for visualizing viable microbial cells on plant roots,[84] but was found to give varying results with sclerotial material.[83] For vital staining with CW, it is necessary to determine the appropriate non-toxic but reactive concentration of dye for the organism or cells under study. Europium (III) thenoyltrifluoroacetonate, 3-hydrate (EU[TTA]$_3$) that fluoresces bright red[85] usually stains nucleic acids only in living cells, but dead cells of *Sclerotium sclerotiorum* have been found to show the same fluorescence.[83] Combinations of EU(TTA)$_3$ and CW have been used successfully,[86] but proved unsatisfactory with *S. sclerotiorum*.[83]

Tetrazolium stains have been used as vital stains with variable success, for example, 2,3,5-triphenyl tetrazolium chloride[87] and nitroblue tetrazolium chloride.[88] Phloxine B (in 0.01 or 0.05% aqueous solution) has been reported effective as a means of distinguishing viable cells of yeast as well as thick-walled oospores; the stain penetrated the fungal cell only if the semipermeability of the plasma membrane had been lost, as happens in dead cells.[88]

Because natural autofluorescence has been found to relate to loss of viability, it has been suggested that it could serve as an inverse indicator of viability.[89]

The use of vital staining techniques to recognize viable cells among trapped air spora thus appears to be unreliable unless the particular technique chosen has been experimentally tested and proved effective for the particular kind of material being studied.

INFECTIVITY

Plant Pathogens

Assay of infectivity of plant pathogenic fungi requires the production of parasitic colonies within a live host that functions as a biological sampler.

The simplest biological sampler and most sensitive, because of its great size, is a commercially planted crop, though it provides only an index of deposition of infective units rather than their concentration in air. Deposition of infective units is often measured on a smaller scale by exposure of seedlings of susceptible species that have been raised under pathogen-free conditions, usually in specially screened greenhouses. After exposure for a specified period that may range from minutes to days, seedlings are incubated in a dew-chamber overnight to allow any infective propagules that are present to germinate and infect. Subsequent maintenance of seedlings in a pathogen-free greenhouse or illuminated growth-chamber allows visible lesions, colonies or sporing pustules to appear and be counted. Alternatively, the trap plants may be enveloped within polyethylene plastic bags immediately after exposure and thereafter kept in an ordinary greenhouse, taking precautions to avoid overheating by direct sunlight. It is usually assumed that numbers of pustules can be equated with numbers of effective dispersal units deposited on the plant.[90] However, "lesion-producing dispersal units" may consist of single spores or clusters of various sizes.[91] Numbers of lesions can be converted to numbers per unit area of exposed leaf per hour of exposure in the field. The concurrent incubation of pots or trays of seedlings that have not been exposed to the airborne inoculum is a prudent check on possible occurrences of cross-infections during incubation.

An alternative to exposing and incubating pots of pathogen-free seedlings is to expose detached leaves rather than whole plants, or alternatively, to detach leaves after exposure of seedlings. Detached leaves of wheat have been kept alive for more than six weeks while the obligate parasite *Erysiphe graminis* developed, by floating them on dilute solutions of benzimidazole in water (100 ppm) during incubation at 5°C with diurnal illumination for 20 hours. If incubated at 15°C, initial symptoms of infection were seen after only 6 days.[92] This technique is likely to provide greater security against the risk of chance infections in greenhouses. Airborne spores of powdery mildews and rusts have been enumerated by drawing air through a tube packed with detached leaves of barley seedlings that served both as a filter and host tissue. Leaves then were incubated in 13 ppm benzimidazole solution at 15°C under fluorescent lighting while infections developed.[93]

Animal Pathogens

Apart from dermatophytes and certain yeasts, fungi only rarely infect man or other animals and when they do, such infections are opportunistic and simply reflect the ability of fungi to exist under a variety of non-ideal conditions. For their animal host, however, consequences can be serious or even fatal since deep-seated mycoses are difficult to treat without harming the host. Fungal infections of interest to aeromycologists are those that are disseminated by the aerial route, and result in diseases acquired through inhalation of infective material. Such human mycoses include blastomycosis, coccidioidomycosis, histoplasmosis, and opportunistic fungal infections such as aspergillosis. Microorganisms must be able to survive and grow at body temperatures of around 37°C and under anaerobic or low-oxygen conditions.

Infectivity may be assayed by several methods. (a) Microscopic examinations of air samples from environments can provide evidence of an infective organism. Mounts of clinical materials such as sputum, body tissues and exudates can show the presence of proliferating fungal growth; this may be morphologically different from the form encountered outside the body, e.g., mycelial growth as a saprobe changes to a yeast form in body tissue in *Histoplasma capsulatum*, *Blastomyces dermatitidis* and *Paracoccidioides brasiliensis*. (b) Culture from clinical materials can reveal the characteristic aerobic form of fungi by which they may be identified. Selecting the best growth medium is important. (c) Pathogenicity tests can be carried out using infectious clinical material that is inoculated into laboratory animals to determine if the disease can be reproduced. This method has the advantage of avoiding handling the mycelial form with its inherent danger of possible dissemination of infectious particles into air. Alternatively, a mycelial culture of the suspect organism may be inoculated into an animal model to see if it converts to the tissue form, thereby confirming a tentative diagnosis based on morphological characteristics of the culture. (d) Live animal hosts can be used as biological air samplers to both capture and manifest pathogenic fungi difficult to isolate by means of artificial samplers.[94] (e) Immunologic tests have become increasingly important in assaying infectivity. Circulating antibodies produced by the host in response to an infective fungus can be detected in blood serum by standard agglutination, precipitin and immunofluorescence tests. Such serological procedures have proved most effective in coccidioidomycosis, aspergillosis and mycetoma.[95] Dermal sensitivity tests can be used to determine exposure to a particular pathogen and are useful in epidemiological studies of geographic spread.

Airborne Mycoses

Blastomycosis affects man, dogs and occasionally other animals and is caused by the dimorphic fungus *Blastomyces dermatitidis*. It is generally accepted that infection results from inhalation of conidia from colonies growing as saprophytes in the soil. It was originally thought to be confined to North America but a much wider geographic distribution is now recognized. The primary pulmonary infection may become disseminated throughout the body and frequently occurs in a cutaneous form. In man the disease is usually chronic, in dogs it tends to be fatal. Diagnosis is by methods of microscopy, culture, animal inoculation and serology. Blastomycin is commercially available for skin tests.

Paracoccidioidomycosis is found principally in South America and is similar in many respects to the North American blastomycosis described above. The aetiologic agent, however, is the dimorphic fungus *Paracoccidioides brasiliensis*. Infection is by the respiratory route; early lesions occur in the lung with possible later, general dissemination. Diagnostic methods are similar to those for *Blastomyces dermatitidis*. The yeast phases of the two fungi may be differentiated by the form of attachment of budding cells to the mother cell.[96]

Coccidioidomycosis is a highly infectious disease caused by inhalation of airborne arthrospores of the dimorphic fungus *Coccidioides immitis*. The disease is endemic in certain regions, principally in desert soils in North America, and reportedly affects man, monkeys, dogs, cattle, rodents and rarely, cats, horses, sheep and swine. It causes initial respiratory infection that occasionally progresses to the disseminated, highly fatal form. Evidence of infection by the airborne route has been demonstrated at an endemic site in Southern Arizona, where the disease was acquired by 5 out of 34 monkeys confined in open cages suspended above ground level for one year, although the fungus was not isolated on exposed plates of culture medium.[94] Large-scale skin-test programs using extracts of the fungus have been used to assay geographic boundaries of the disease. These also revealed the extensive incidence of the mild, self-limiting form. IgG and IgM precipitating antibodies to coccidioidin develop after infection, and complement-fixing antibodies also arise. Clinical material should be inoculated onto cycloheximide-chloramphenicol agar that selectively isolates *C. immitis* from almost all other bacteria and saprophytic fungi. The dangers of inhaling airborne arthrospores of the mold form of the fungus can be eliminated by inoculating suspected clinical material into mice for subsequent histological examination.

Cryptococcosis, caused by *Cryptococcus neoformans*, a yeast pathogenic to man and animals, has been isolated from soil in many parts of the world but is particularly associated with excreta of pigeons. In man it usually develops as an infection of the central nervous system. However, it is presumed that this results from hematogenous spread of the organism from primary lung lesions. Very small basidiospores of the sexual (teleomorphic) state of this fungus (viz., *Filobasidiella neoformans*) are thought to be the means of aerial dispersal. Slimy capsules of the pathogenic yeast phase may be difficult to see in microscopic examinations of body fluids. In culture it is slow-growing and sensitive to cycloheximide. Blood agar is a recommended medium for growth at 37°C. Intracerebrally inoculated mice usually die in 4 days to 2 weeks. Brain material should be examined for the encapsulated, yeast-like cells. Serological diagnostic methods are not very satisfactory.

Histoplasmosis, caused by the dimorphic fungus *Histoplasma capsulatum*, is sometimes fatal to man and animals, and is apparently acquired by the airborne route. A benign form of this disease also occurs. It has been isolated from air inside and outside chicken houses by means of a novel and highly selective technique in which particles were washed from the air by "venturi scrubbers" (intended for use industrially for cleaning air), the solid material collected by filtration, resuspended in saline with bactericidal antibiotics and injected into mice intraperitoneally. Four weeks after inoculation, the mice were sacrificed and liver, spleen and adrenals, after sterile maceration, were plated out onto antibiotic-containing culture media suitable for *H. capsulatum* on which colonies grew.[97] Microscopic examinations of clinical material are unrewarding. A recommended culture medium is brain heart infusion-cycloheximide-chloramphenicol(BHI-CC) and BHI-CC fortified with 6% blood.[96] Skin tests with histoplasmin may be inconclusive.

Pulmonary aspergillosis is caused by inhalation of conidia of *Aspergillus fumigatus*, *A. flavus* and other *Aspergillus* species able to survive at lung temperatures. Dissemination to other organs may occur. It has long been recognized as a serious disease of housed birds. The conidia are ubiquitous in the air of both town and country but do not invade or colonize healthy lungs. Incidence of invasive aspergillosis and infection by other opportunistic genera in man has increased in recent years with advances in transplant surgery and the accompanying use of immunosuppressive drugs, and also with the rise of immune-deficiency diseases. In microscopic examinations complete conidial heads may be seen. The aspergilli grow readily on standard fungal media but it must be borne in mind that they are very common contaminants on culture plates and also are routinely present in the upper respiratory tract. Atopic individuals may develop allergic aspergillosis diagnosed by an immediate response to a skin test with an extract of the fungus, sometimes followed by a later delayed response. Fungus balls or mycetomas may form in lung cavities, sometimes in post-operative, healed lesions. Aspergillomas are the most common type of mycetoma, with *Aspergillus fumigatus* the most frequent species. They do not usually invade healthy tissues and may resolve spontaneously.

TOXICITY

Many fungi produce secondary metabolites in nature or in artificial culture or both. Many of these are toxic to particular organisms, including farm animals and man, and thus have acquired economic and social importance.[98,99] Ingestion of molded foodstuffs is the usual means of conveying the toxin into the body, but there is increasing evidence that inhaled aerosols of fragments of mycelium, spores and molded substrate may be sources of mycotoxins of veterinary or medical significance. Both spores and dust pose a health hazard to handlers of grain, for both may contain mycotoxins. With different fungi, the relative distribution of mycotoxin between the mycelium-substrate (MS) matrix and spores differs. Among the toxins of certain *Aspergillus* and *Penicillium* species, aflatoxin B1, norsolorinic acid and secalonic acid D predominate in the MS-matrix, whereas aurasperone C and fumigaclavine C predominate in the spores. Methods for separating rice grain material into MS-matrix and spores for analysis have been described.[100] The term "toxomycoses" has been applied to diseases produced by inhalation of fungal spores, mycelia or fungally contaminated material.[101] Deleterious effects are caused without fungal growth in the host. Mycotoxins act as chemical agents rather than as allergens, but this distinction is not always easily made in practice.

Evidence that inhalation of mycotoxins is involved in human and animal disease is often circumstantial. Increased incidence of liver and biliary tract cancer in workers in livestock feed processing companies has been reported.[102] Imported raw materials used are prone to molding by *Aspergillus flavus* that produces aflatoxins known to cause liver carcinomas in diverse animals. From data on dust content of air, and average aflatoxin content of dust, likely inhalation doses of aflatoxin by workers were estimated. The study, however, concluded that the lung did not seem to be a target for carcinogenic effect of inhaled aflatoxin in humans. Nevertheless, the deaths from pulmonary adenomatosis of two men who had been working on a method for sterilizing peanut meal contaminated by

Aspergillus flavus have been reported.[103] Chemical tests on lung tissue of one of the men indicated the presence of aflatoxin.

Illness suggestive of trichothecene toxicosis experienced by occupants of a household in Illinois was attributed to a heavy infestation of *Stachybotrys atra* in air ducts and on damp fiberboard. Air was sampled by an electrostatic precipitator (Sci-Med, Eden Prairie, MN) that collected and sampled particulate matter in an aqueous solution. Tests for trichothecenes in samples proved positive. Trichothecenes also were isolated from *Stachybotrys*-molded materials in the house. After the house was cleansed of such materials, the residents no longer suffered from the illness. However, dust and debris proved to be highly irritating to the skin and respiratory system of those doing the cleaning. It was recommended that personnel handling such contaminated material should wear respirators and protective clothing.[104] Presence of trichothecene mycotoxins in artificially aerosolized conidia of *Stachybotrys atra* has been demonstrated.[105] Dust for analysis was trapped on glass-fiber filters through which the sampled air was drawn. The procedure for extraction of mycotoxins from the filters was described.

Organic dust toxic syndrome (ODTS) is an acute febrile, non-allergic, non-infectious respiratory illness caused by inhalation of organic dust from moldy silage, hay or corn.[106-107] It may affect farm workers when removing aerobic, fungally contaminated top layers or "caps" from silos before unloading cured feed,[107] but also has been reported when hay was strewn on the floor of a badly ventilated room used for a student party.[108] Involvement of fungal toxins is not proven, however, and May et al.[107] reported failure to demonstrate mycotoxins in silage associated with ODTS by either gas chromatography or mass spectroscopy.

Aerobiological techniques can be an aid to diagnosis of mycotoxicoses caused by ingestion of mouldy foodstuffs. Of the many mycotoxins that can cause disease in animals, only a few produce signs or lesions that permit the unequivocal diagnosis or even diagnosis with a high degree of probability. These include aflatoxin (aflatoxicosis), ergot alkaloids (ergotism), ochratoxin (ochratoxicosis), slaframine (slobber syndrome), sporidesmin (facial eczema), stachybotryotoxin (stachybotryotoxicosis) and zearalenone (estrogenic syndrome). Accordingly, where mycotoxin poisoning is suspected in housed or grazing animals, the fungal source may be sought by aerobiological means simply by shaking some suspect feed or bedding material, preferably contained within a bag, and exposing plates of suitable agar culture medium to the aerosol. An Andersen sampler is ideal but far from essential for this purpose. The majority of toxin-producing fungi will grow adequately on non-specific media such as malt extract, although the more xerophilic storage fungi such as the *Aspergillus glaucus* group prefer media with high osmotic pressure such as is produced by inclusion of 20% sucrose. Microscopic examination of airborne spores collected with samplers such as the May/RE cascade impactor[30] or the battery-driven single-stage Burkard portable hand-held sampler can be sufficient to identify morphologically distinctive spores[29] such as those of *Stachybotrys* and *Fusarium* spp. If the presence of mycotoxin-producing fungi is demonstrated by cultural or microscopic techniques, it is then probably prudent to subject feeds or bedding to chemical tests for the presence of mycotoxins in hazardous amounts. Extraction and analytical procedures vary for different mycotoxins,[98,99] and generally are best carried out by specialist laboratories. A comprehensive review of mycotoxic fungi, mycotoxins and mycotoxicoses is presented in the three-volume series edited by Wyllie and Morehouse.[99]

Antibiosis is a special kind of toxicity. Antibiotic effects of airborne fungi on other fungi and bacteria may affect results of aerobiological sampling by inhibiting germination and growth of some species on culture plates. This inhibition is sometimes visible as a clear untenanted zone around certain colonies. Where the object of sampling is to determine numbers of fungal propagules and species present in air it may be necessary to circumvent such growth inhibition by aseptically transferring each emergent colony onto a fresh agar plate.

ALLERGENICITY ASSAYS

Introduction

An allergy is a deleterious physiological consequence of an excessive immunological response to contact with a foreign substance. Airborne fungal and actinomycete spores and fragments together with pollens are major causes of respiratory allergy worldwide. Assaying for fungal allergens involves several stages. Analysis of the case history and clinical symptoms of a presenting subject is the first stage toward a presumptive diagnosis of an allergic reaction. Ideally this should be followed by an investigation of the aerial environment of the home or workplace in order to find possible causative agents. Suspect organisms or, more usually, antigens derived from them, are tested to discover if they elicit a reaction in the host. A positive reaction supports, but does not prove, the diagnosis of mold allergy. The diagnosis is further supported if the host ceases to exhibit symptoms when removed from the presumptive allergenic source. This is also the most effective therapeutic measure. Pharmaceutical treatments include anti-inflammatory, anti-histamine and broncho-dilatory drugs.

Symptoms of Respiratory Allergy

Clinical characteristics of fungal allergy are of several kinds and vary with the type of immunologic response, sensitivity of individuals and fungal species involved. Allergic rhinitis, bronchitis and asthma typically occur in atopic individuals and are characteristic of an immediate upper-airway response within minutes of exposure to the relevant allergen. This is predominantly a type 1 IgE-mediated immune response. *Alternaria* and *Cladosporium* spp. are two of the commonest genera causing seasonal mold allergies of this type. Allergic bronchopulmonary aspergillosis may present a spectrum of immunological reactions.[109] Localized allergic responses can occur in the bronchi, probably caused by antigen released when inhaled *Aspergillus* spores germinate *in situ*. Mycetomas in the lung may cause strong immunologic reactions.

Symptoms of fever, chills, dyspnoea and cough occurring 4–8 hours after exposure and taking up to 48 hours to resolve, are characteristic of hypersensitivity pneumonitis (HP), also known as extrinsic allergic alveolitis. This is usually an occupation-related disease that typically affects non-atopic individuals and requires a heavy initial sensitizing dose. Allergens penetrate to the gas-exchange tissue of lungs and numerous species of fungi and thermophilic actinomycetes have been implicated, each having spores small enough to penetrate to alveolar regions. Actinomycete spores are mostly about 1 μm in diameter. In man, names of diseases reflect the occupational source of causative organisms (e.g., farmer's lung, mushroom worker's lung, corkworker's lung)[110] and such bizarre conditions as "sauna taker's lung"[111] which is a reaction to *Aureobasidium pullulans*. A recent phenomenon is HP caused by inhalation of spores from molds growing in humidifiers and air-conditioning systems; these can contribute towards "sick building syndrome." Animals also suffer from HP; horses exhibit a form of chronic pulmonary disease that is characterized by a double expiratory effort (hence its colloquial name of 'heaves'), nasal catarrh, dyspnoea and reduced exercise tolerance. Like farmer's lung in man, it is associated with molds released from hay or straw that has been baled at high water content and subsequently has self-heated.[112,113] Major causative agents of HP in mouldy hay are the thermotolerant fungus *Aspergillus fumigatus* and certain thermophilic actinomycetes such as *Faenia rectivirgula*. Cattle also develop a similar condition when exposed to poor quality hays and their constituent molds.[114,115] The most characteristic immunologic feature of HP is the presence of precipitating IgG antibodies against the offending antigen.

Sampling the Environment

Any of the standard aerobiological techniques may be used for sampling home or work environments; strategically placed settle plates are a simple method in which the patient can cooperate. Personal samplers can be worn to characterize environments in the breathing zone of the individual to evaluate his specific exposure. Personal samplers may be passive, such as the slide sampler described by Leuschner and Boehm,[116] or battery driven such as the Miniature Continuous Monitor (MDA Scientific, Inc., Illinois). Kucharski[117] describes a personal sampler in which speed of air-sampling is related to respiratory rate of the wearer, using the correlation between pulmonary ventilation and pulse rate monitored by electrodes on the sampler wearer's chest. Since high concentrations of molds usually are required in order to elicit an immune reaction, the most prevalent molds are likely to be the causative agents, although some species are more antigenic than others. If symptoms indicate a hypersensitivity pneumonitis type of allergy, then particular attention must be given to the possible presence of thermophilic actinomycetes since these are frequently implicated. A range of incubation temperatures and selective and non-selective media should be employed in culture techniques to ensure that no culturable organisms remain undetected and choice of media should be based upon the nature of the environment. Cultural techniques do not necessarily reflect numbers of allergenic particles in the air since these do not have to be viable in order to elicit a reaction. Microscopic examinations of trapped spores can be helpful for quantitative assessments.

Identifying and Extracting the Antigens

Methods for production of fungal antigens have been comprehensively reviewed.[118] As yet, few attempts have been made to standardize and define cultural conditions for production of a particular antigen or allergen. This would be difficult since each species has different optimum growth conditions. Many pitfalls are associated with production of antigen extracts. The amount and kinds of specific antigens produced will often vary with cultural procedures and extraction methods. Commercially produced antigen extracts of more common mold allergens are available but may be seriously deficient in their antigenicity compared with freshly prepared extracts derived from the same molds cultured from the environment.[119,120] They also may vary widely from batch to batch and from one manufacturer to another since standardization and quality control have yet to be achieved. Whenever feasible, it is recommended that crude antigen extracts are prepared freshly from microorganisms found in the environment by growing in liquid culture in defined medium.[118,121] If undefined medium is used, antigens should be extracted by the double dialysis method that avoids contamination of extracts with medium-derived components.[122] Airborne exposure is mainly to spores and therefore cultures producing a sporogenous surface mat are preferable to shake-cultures that produce submerged mycelial growth. Levels of *Alternaria* antigens in the atmosphere have been quantified by exposing filter sheets in air samplers, eluting them, and assaying allergenic activity by the radioallergosorbent test (RAST) inhibition assay.[123]

Allergens are proteins or glycoproteins of small to medium size (2,000 to 100,000 daltons).[124] Most identified fungal antigens are glycopeptides; specific antigens are identified by using gel filtration, chromatography and chemical extraction methods.[118]

Testing for Allergy

The simplest method for determining allergenic activity of a mold extract is by direct skin testing using allergic subjects. Intradermal titration is more sensitive than skin prick tests. Immunological methods and their attendant problems have been comprehensively reviewed[110]; here we allude briefly to some of them. The characteristic immunologic feature of HP is the presence of IgG serum-precipitating antibodies against the specific antigen. Ouchterlony's gel diffusion technique,[125] as modified by later workers,[126] is a microtechnique of double diffusion in agar. Double diffusion tests are simple to perform, require little in the way of special equipment, and are reliable in practice. A large central well cut into a thin agar layer is filled with patient serum and is surrounded by smaller peripheral wells containing antigen extracts. Precipitating antibodies show as arcs between central and peripheral wells. Staining with Coomassie brilliant blue solution enhances clarity of arcs. Immunoelectrophoretic techniques of varying sophistication allow identification and characterization of complex fungal antigens into fractions of known molecular weights. Enzyme-linked immunosorbent assay (ELISA) techniques are becoming increasingly popular and may be used in antibody assay for IgE antibody as well as IgG and for detecting circulating antigen in serum. Radioimmunoassays include radioallergosorbent tests (RAST) in which antisera are radiolabelled for investigating allergic IgE-mediated reactions to fungi,[110] and for radiolabelling antigens to measure extent of antigen-antibody binding. IgE-binding capacity of extracts can be measured by means of enzyme allergosorbent tests (EAST),[127] while other tests for fungal allergy include agglutination tests, complement fixation and immunofluorescence tests.[110]

Positive results in skin tests and presence of precipitins are evidence of exposure and not necessarily of allergy. Up to 50% of exposed but asymptomatic individuals may react positively.[128] When antigen extracts from 18 species of fungi and thermophilic actinomycetes isolated from a horse stable were tested against sera of asymptomatic horses from the same stable, 50% of the apparently healthy horses exhibited serum precipitins to one or more of the antigens.[121] By contrast, symptomatic subjects do not always produce serum-precipitins and it has been suggested that in some individuals direct activation of the alternative pathway of complement eliminates the need for antibody to mediate the reaction.[129]

Reproduction of symptoms by inhalation challenge is considered confirmatory,[128] but this cannot be a recommended procedure since it is inherently dangerous. Cross-reactivity between species is a complicating factor; subjects sensitive to *Alternaria,* for example, may also react to *Curvularia, Stemphylium* and *Ulocladium.* Bronchiolar lavage can show changes in immunoglobulin levels indicative of different types of respiratory allergy.

Species Implicated in Allergies

The following list of predominantly conidial fungi reported as causative agents of human allergies is taken mainly from Cole and Samson[130] who give source references:

Alternaria alternata, Aspergillus clavatus, A. flavus, A. fumigatus, A. glaucus, A. nidulans, A. niger, A. ochraceus, A. penicilloides, A. umbrosus, A. versicolor, Aureobasidium pullulans, Botrytis aclada, B. cinerea, Cladosporium cladosporioides, C. herbarum, C. macrocarpum, C. sphaerospermum, Cryptostroma corticale, Drechslera sorokiniana, Epicoccum purpurascens, Fulvia fulva, Fusarium solani, Geotrichum candidum, Graphium sp., Penicillium brevicompactum, P. citrinum, P. decumbens, P. expansum, P. glabrum, P. herquei, P. implicatum, P. italicum, P. oxalicum, P. roquefortii, P. simplicissimum, P. verrucosum var. cyclopium, P. verrucosum var. verrucosum, P. waksmanii, Pleurotus ostreatus, Scopulariopsis brevicaulis, Trichophyton mentagrophytes, Trichothecium roseum, Verticillium lecanii, Wallemia sebi. Yeasts include *Candida* spp. and *Sporobolomyces*. Species of thermophilic actinomycetes associated with various forms of hypersensitivity pneumonitis include *Faenia rectivirgula, Saccharomonospora viridis, Thermoactinomyces thalpophilus, T. vulgaris* and *T. sacchari.* Other actinomycetes that have been implicated are *Streptomyces albus* and *S. olivaceus.*

SUMMARY

Principal characteristics of fungi, actinomycetes and myxomycetes relevant to aerobiology are described. Procedures for assaying aerosols of these organisms are reviewed. These comprise total assays that do not discriminate between viable and non-viable organisms, and assays that assess various aspects of viability, infectivity, toxicity and allergenicity.

REFERENCES

1. Ingold, C.T. *Fungal Spores, Their Liberation and Dispersal* (Oxford, UK: Clarendon Press, 1971).
2. Pady, S.M., and P.H. Gregory. "Numbers and Viability of Airborne Hyphal Fragments in England," *Trans. Brit. Mycol. Soc.* 46:609–613 (1963).
3. Fitt, B.D.L., H.A. McCartney, and P.J. Walklate. "The Role of Rain in Dispersal of Pathogen Inoculum," *Ann. Rev. Phytopathol.* 27:241–270 (1989).
4. Dickinson, C.H. "Fungi on the Aerial Surfaces of Higher Plants," in *Microbiology of Aerial Plant Surfaces,* C.H. Dickinson, and T.F. Preece, Eds. (New York: Academic Press. 1976) pp. 293–324.
5. Solomon, W.R. "Fungus Aerosols Arising from Cold-mist Vaporizers," *J. Allergy Clin. Immunol.* 54:222 (1974).
6. Kramer, C.L., and M.G. Eversmeyer. "Comparisons of Airspora Concentrations at Various Sites Within a Ten Kilometer Radius of Manhattan, Kansas, USA," *Grana* 23: 117–122 (1984).
7. Flannigan, B. "Mould Growth and Airborne Mould Spores as Potential Health Hazards in British Housing," Microbiology Congress, International Union of Microbiological Societies: Bacteriology and Mycology, vol. 111, Abstracts of Papers, Osaka, Japan (1990), p. 30.
8. Malmberg, P., A. Rask-Andersen, U. Palmgren, S. Höglund, B. Kolmodin-Hedman, and G. Stalenheim. "Exposure to Microorganisms, Febrile and Airway-Obstructive Symptoms, Immune Status and Lung Function of Swedish Farmers," *Scand. J. Work Environ. Health* 11:287–293 (1985).

9. Lacey, J, and M.E. Lacey. "Spore Concentrations in the Air of Farm Buildings," *Trans. Brit. Mycol. Soc.* 47:547–552 (1964).
10. Goodfellow, M., M. Mordarski, and S.T. Williams, Eds. *The Biology of the Actinomycetes* (London: Academic Press, 1984), 544 pp.
11. Madelin, M.F. "Myxomycetes, Microorganisms and Animals: a Model of Diversity in Animal-MicrobialInteractions," in *Invertebrate-Microbial Interactions*, J.M. Anderson, A.D.M. Rayner, and D.W.H. Walton, Eds. (Cambridge, U.K.: Cambridge University Press, 1984), pp. 1–33.
12. Lacey, J. "The Aerobiology of Conidial Fungi," in *Biology of Conidial Fungi*, vol. 1, G.T. Cole, and B. Kendrick, Eds. (New York, NY: Academic Press, 1981), pp. 373–416.
13. Santilli, J., Jr., W.J. Rockwell, R.P. Collins, and D.G. Marsh. "*Coprinus micaceus, Fuligo septica,* and *Alternaria alternata* Spore Allergy in Highly Ragweed-Sensitive Subjects," *J. Allergy Clin. Immunol.* 75:118 (1985).
14. McCoy, R.E., and A.W. Dimock. "A Scotch Tape Method for the Trapping and Examination of Airborne Spores," *Plant Dis. Rep.* 55(9):832–834 (1971).
15. Wood, F.A., and R.A. Schmidt. "A Spore Trap for Studying Spore Release from Basidiocarps," *Phytopathology* 56:50–52 (1966).
16. Livingston, C.H., M.D. Harrison, and N. Oshima. "A New Type Spore Trap to Measure Numbers of Air-borne Fungus Spores and Their Periods of Deposition," *Plant Dis. Rep.* 47(5):340–341 (1963).
17. Roelfs, A.P., and L.B. Martell. "Uredospore Dispersal from a Point Source within a Wheat Canopy," *Phytopathology* 74:1262–1267 (1984).
18. Gregory, P.H. *The Microbiology of the Atmosphere*, 2nd ed., (Aylesbury, UK: Leonard Hill, 1973).
19. Gregory, P.H. "Deposition of Airborne *Lycopodium* Spores on Cylinders," *Ann. Appl. Biol.* 38: 357–376 (1951).
20. Roelfs, A.P., V.A. Dirks, and R.W. Romig. "A Comparison of Rod and Slide Samplers Used in Cereal Rust Epidemiology," *Phytopathology* 58:1150–1154 (1968).
21. Madelin, T.M. "The Role of Fungi and Organic Dust in Chronic Pulmonary Disease of the Horse," PhD Thesis, University of Bristol (1990).
22. May, K.R. "Physical Aspects of Sampling Airborne Microbes," *Symp. Soc. Gen. Microbiol.* 17:60–80 (1967).
23. Pady, S.M. "A Continuous Spore Sampler," *Phytopathology* 49:757–760 (1959).
24. Husain, S.M. "An Automatic Suction-Impaction Type Spore Trap and Its Use with Onion Blotch Alternaria," *Phytopathology* 53:382–384 (1963).
25. Schenck, N.C. "A Portable, Inexpensive, and Continuously Sampling Spore Trap," *Phytopathology* 54:613–614 (1964).
26. Gadoury, D.M., and W.E. MacHardy. "A 7-Day Recording Volumetric Spore Trap," *Phytopathology* 73:1526–1531 (1983).
27. Panzer, J.D., E.C. Tullis, and E.P. Van Arsdel. "A Simple 24-Hour Slide Spore Collector," *Phytopathology* 47:512–514 (1957).
28. Käpylä, M., and A. Penttinen. "An Evaluation of the Microscopical Counting Methods of the Tape in Hirst-Burkard Pollen and Spore Trap," *Grana* 20:131–141 (1981).
29. Clarke, A.F., and T.M. Madelin. "Technique for Assessing Respiratory Health Hazards from Hay and Other Source Materials," *Equine Vet. J.* 19(5):317–322 (1987).
30. May, K.R. "An "Ultimate" Cascade Impactor for Aerosol Assessment," *J. Aerosol Sci.* 6:413–419 (1975).
31. Madelin, T.M., and H.E. Johnson. "Fungal and Actinomycete Spore Aerosols Measured at Different Humidities with an Aerodynamic Particle Sizer," *J. Appl. Bacteriol.* 72:400–409 (1992).

32. Hermansen, J.E., H.B. Johansen, H.W. Hansen, and P. Carstensen. "Notes on the Trapping of Powdery Mildew Conidia and Urediospores by Aircraft in Denmark in 1964," *Kgl. Vetr. and Landbohojsk Aarsskr.* 1965:121–129 (1965).
33. Trägardh, C. "Sampling of Aerobiological Material from a Small Aircraft," *Grana* 16: 139–143 (1977).
34. Gottwald, T.R., and W.L. Tedders. "A Spore and Pollen Trap for Use on Aerial Remotely Piloted Vehicles," *Phytopathology* 75:801–807 (1985).
35. McCracken, F.I. "Design and Evaluation of a Remote-Piloted Aircraft Spore Sampler (REPASS) Used over a Cottonwood Plantation," *Can. J. Bot.* 67:822–826 (1989).
36. Perkins, W.A. "The Rotorod Sampler," Second Semiannual Report, CML 186, Aerosol Laboratory, Stanford University, Palo Alto, CA (1957) 66 pp.
37. Ogden, E.C., and G.S. Raynor. "A New Sampler for Airborne Pollen: the Rotoslide," *J. Allergy* 40:1–11 (1967).
38. Aylor, D.E., and G.S. Taylor. "Escape of *Peronospora tabacina* Spores from a Field of Diseased Tobacco Plants," *Phytopathology* 73:525–529 (1983).
39. Sutton, T.B., and A.L. Jones. "Evaluation of Four Spore Traps for Monitoring Discharge of Ascospores of *Venturia inaequalis*," *Phytopathology* 66:453–456 (1976).
40. Solomon, W.R. "A Volumetric Study of Winter Fungus Prevalence in the Air of Midwestern Homes," *J. Allergy Clin. Immunol.* 57(1):46–55 (1976).
41. Rowell, J.B., and R.W. Romig. "Detection of Urediospores of Wheat Rusts in Spring Rains," *Phytopathology* 56:807–811 (1966).
42. Fitt, B.D.L., C.J. Rawlinson, and C.B. Smith. "A Comparison of Two Rain-Activated Switches Used with Samplers for Spores Dispersed by Rain," *Phytopathol. Z.* 105:39–44 (1982).
43. Burge, H.A., M.L. Jelks, and J.A. Chapman. "Quality Control of Multisource Aeroallergen Data," *Grana* 25:247–250 (1986).
44. Ellis, M.B. *Dematiaceous Hyphomycetes* (Kew, Surrey, England: Commonwealth Mycological Institute, 1971).
45. Ellis, M.B. *More Dematiaceous Hyphomycetes* (Kew, Surrey, England: Commonwealth Mycological Institute, 1976).
46. Ellis, M.B., and J.P. Ellis. *Microfungi on Land Plants: an Identification Handbook* (London: Croom Helm, 1985).
47. Ellis, M.B., and J.P. Ellis. *Microfungi on Miscellaneous Substrates: an Identification Handbook* (London: Croom Helm, 1988).
48. Domsch, K.H., W. Gams, and T.-H. Anderson *Compendium of Soil Fungi*, vols. 1 and 2 (London: Academic Press, 1980).
49. Arx, J.A.von. *The Genera of Fungi Sporulating in Pure Culture*, 3rd ed. (Vaduz, Germany: J. Cramer, 1981).
50. Carmichael, J.W., W.B. Kendrick, I.L. Conners, and L. Sigler. *Genera of Hyphomycetes* (Edmonton, Alberta: University of Alberta Press, 1980).
51. Onions, A.H.S., D. Allsopp, and H.O.W. Eggins. *Smith's Introduction to Industrial Mycology*, 7th ed. (London: Edward Arnold, 1981).
52. Barron, G.L. *The Genera of Hyphomycetes from Soil* (Huntington, NY: Robert E. Krieger Publishing Co., 1968).
53. Cole, G.T., and R.A. Samson. *Patterns of Development in Conidial Fungi* (London: Pitman, 1979).
54. *CMI Descriptions of Pathogenic Fungi and Bacteria*, (Kew, Surrey, England: Commonwealth Mycological Institute, 1964 onwards).
55. Blomquist, G., U. Palmgren, and G. Ström. "Improved Techniques for Sampling Airborne Fungal Particles in Highly Contaminated Environments," *Scand. J. Work Environ. Health* 10:253–258 (1984).
56. Baruah, H.K. "The Air Spora of a Cowshed," *J. Gen. Microbiol.* 25:483–491 (1961).

57. Dowding, P. "The Dispersal and Survival of Spores of Fungi Causing Bluestain in Pine," *Trans. Brit. Mycol. Soc.* 52(1):125–137 (1969).
58. May, K.R. "A Multi-stage Liquid Impinger," *Bacteriol. Rev.* 30:559–570 (1966).
59. Andersen, A.A. "New Sampler for the Collection, Sizing, and Enumeration of Viable Airborne Particles," *J. Bacteriol.* 76:471–484 (1958).
60. Peto, S., and E.O. Powell. "The Assessment of Aerosol Concentration by Means of the Andersen Sampler," *J. Appl. Bacteriol.* 33:582–598 (1970).
61. Crook, B., S. Higgins, and J. Lacey. "Methods for Sampling Airborne Microorganisms at Solid Waste Disposal Sites," Proceedings of the 7th International Biodeterioration Symposium (1988).
62. Henningson, E., R. Roffey, and A. Bovallius. "A Comparative Study of Apparatus for Sampling Airborne Microorganisms," *Grana* 20:155–159 (1981).
63. Errington, F.P., and E.O. Powell. "A Cyclone Separator for Aerosol Sampling in the Field," *J. Hyg. England* 67:387–399 (1969).
64. Cox, C.S. *The Aerobiological Pathway of Microorganisms* (Chichester, John Wiley and Sons, 1987), pp.66–69.
65. May, K.R. "Prolongation of Microbiological Air Sampling by a Monolayer on Agar Gel," *Appl. Microbiol.* 18:513–514 (1969).
66. Burge, H.P., J.R. Boise, J.A. Rutherford, and W.R. Solomon. "Comparative Recoveries of Airborne Fungus Spores by Viable and Non-Viable Modes of Volumetric Collection," *Mycopathologia* 61:27–33 (1977).
67. Madelin, T.M. "The Effect of a Surfactant in Media for the Enumeration, Growth and Identification of Airborne Fungi," *J. Appl. Bacteriol.* 63:47–52 (1987).
68. Burge, H.P., W.R. Solomon, and J.R. Boise. "Comparative Merits of Eight Popular Media in Aerometric Studies of Fungi," *J. Allergy Clin. Immunol.* 60:199–203 (1977).
69. Pady, S.M., C.L. Kramer, and V.K. Pathak. "Suppression of Fungi by Light on Media Containing Rose Bengal," *Mycologia* 52:347–350 (1960).
70. Morring, K.L., W.G. Sorenson, and M.D. Attfield. "Sampling for Airborne Fungi: A Statistical Comparison of Media," *Am. Ind. Hyg. Assoc.* 44(9):662–664 (1983).
71. Ahlström, K., and A. Käärik. "A Study of Airborne Fungal Spores with the Aid of the FOA Slit-Sampler," *Grana* 16:133–137 (1977).
72. Käärik, A. "The Identification of the Mycelia of Wood-Decay Fungi by Their Oxidation Reactions with Phenolic Compounds," *Stud. For. Suec.* 31:1–80 (1965).
73. Nobles, M.K. "Identification of Cultures of Wood-Inhabiting Hymenomycetes," *Can. J. Bot.* 43:1097–1139 (1965).
74. Hocking, A.D., and J.I. Pitt. "Dichloran-glycerol Medium for Enumeration of Xerophilic Fungi from Low Moisture Foods," *Appl. Environ. Microbiol.* 39:488–492 (1980).
75. Cross, T. "The Monosporic Actinomycetes," in *The Prokaryotes: A Handbook of Habitats, Isolation and Identification of Bacteria,* vol. 2, M.P. Starr, H. Stolp, H.G. Truper, A. Balows, and H. Schlegel, Eds. (Berlin: Springer Verlag, 1981) pp. 2091–2101.
76. Sheridan, J.E., and J. Nelson. "The Selective Isolation of the "Kerosene Fungus" *Cladosporium resinae* from the Air," *Int. Biodeterior. Bull.* 7(4):161–162 (1971).
77. Bothast, R.J., and D.I. Fennell. "A Medium for Rapid Identification and Enumeration of *Aspergillus flavus* and related organisms," *Mycologia* 66:365–369 (1974).
78. Watling, R. "Germination of Basidiospores and Production of Fructifications of Members of the Agaric Family Bolbitiaceae Using Herbarium Material," *Nature,* London 197:717–718 (1963).
79. Ogunsanya, O.C., and M.F. Madelin. "Sensitivity of *Botryodiplodia ricinicola* Conidia to Mild Chilling," *Trans. Brit. Mycol. Soc.* 69(2):191–195 (1977).

80. Rotman, B., and B.W. Papermaster. "Membrane Properties of Living Mammalian Cells as Studied by Enzymatic Hydrolysis of Fluorogenic Esters," *Proc. Natl. Acad. Sci. USA* 55:134–141 (1966).

81. Soderstrom, B.E. "Vital Staining of Fungi in Pure Cultures and in Soil with Fluorescein Diacetate," *Soil Biol. Biochem.* 9:59–63 (1977).

82. Butt, T.M., H.C. Hoch, R.C. Staples, and R.J. St. Leger. "Use of Fluorochromes in the Study of Fungal Cytology and Differentiation," *Expl. Mycol.* 13(4):303–320 (1989).

83. Dittmer, U., and H.C. Weltzien. "A Rapid Viability Test for Sclerotia with Fluorescein Diacetate," *J. Phytopathol.* 130:59–64 (1990).

84. Johnen, B.G. "Rhizosphere Microorganisms and Roots Stained with Europium Chelate and Fluorescent Brightener," *Soil Biol. Biochem.* 10:495–502 (1978).

85. Scaff, W.L., D.L. Dyer, and K. Mori. "Fluorescent Europium Chelate Stains," *J. Bacteriol.* 98:246–248 (1969).

86. Anderson, J.R., and D. Westmoreland. "Direct Counts of Soil Organisms Using a Fluorescent Brightener and a Europium Chelate," *Soil Biol. Biochem.* 3:85–87 (1971).

87. Pathak, V.K., S.B. Mathur, and P. Neergaard. "Detection of *Peronospora manshurica* (Naum.) Syd. in Seeds of Soybean, *Glycine max*," *EPPO Bull.* 8(1):21–28 (1978).

88. Roongruangsree, U.-T., C. Kjerulf-Jensen, L.W. Olsen, and L. Lange. "Viability Tests for Thick Walled Fungal Spores (ex: Oospores of *Peronospora manshurica*," *J. Phytopathol.* 123:244–252 (1988).

89. Wu, C.H., and H.L. Warren. "Natural Autofluorescence in Fungi and Its Correlation with Viability," *Mycologia* 76:1049–1058 (1984).

90. Aylor, D.E. "Deposition Gradients of Urediniospores of *Puccinia recondita* near a Source," *Phytopathology* 77:1442–1448 (1987).

91. Ferrandino, F.J., and D.E. Aylor. "Relative Abundance and Deposition Gradients of Clusters of Urediniospores of *Uromyces phaseoli*," *Phytopathology* 77:107–111 (1987).

92. Wolfe, M.S. "Physiologic Specialization of *Erysiphe graminis* f. sp. *tritici* in the United Kingdom, 1964–5," *Trans. Brit. Mycol. Soc.* 50:631–640 (1967).

93. Hermansen, J.E., H.B. Johansen, and H.W. Hansen. "A Method of Trapping Live Powdery Mildew Conidia and Urediospores in the Upper Air," *Kgl. Vetr. and Landbohojsk Aarsskr.* 1967:77–81 (1967).

94. Converse, J.L., and R.E. Reed. "Experimental Epidemiology of Coccidioidomycosis," *Bacteriol. Rev.* 30(3):678–694 (1966).

95. English, M.P. *Medical Mycology: Studies in Biology no. 119* (London: Edward Arnold, 1980).

96. Jungerman, P.F. and R.M. Schwartzman. *Veterinary Medical Mycology* (Philadelphia, Lea and Febiger, 1972) pp. 115,131.

97. Ibach, M.J., H.W. Larsh, and M.L. Furcolow. "Epidemic Histoplasmosis and Airborne *Histoplasma capsulatum*," *Proc. Soc. Exp. Biol. Med.* 85:72–74 (1954).

98. Smith, J.E., and M.O. Moss. *Mycotoxins: Formation, Analysis and Significance* (Chichester, U.K.: John Wiley and Sons, 1985).

99. Wyllie, T.D., and L.G. Morehouse, Eds. *Mycotoxic Fungi, Mycotoxins, Mycotoxicoses* (New York: Marcel Dekker, Inc., Vol. 1 1977, Vols. 2 and 3 (1978).

100. Palmgren, M.S., and L.S. Lee. "Separation of Mycotoxin-Containing Sources in Grain Dust and Determination of Their Mycotoxin Potential," *Environ. Health Perspect.* 66:105–108 (1986).

101. Northrup, S.C., and K.H. Kilburn. "The Role of Mycotoxins in Human Pulmonary Disease," in *Mycotoxic Fungi, Mycotoxins, Mycotoxicoses*, Vol. 3, T.D. Wyllie, and L.G. Morehouse, Eds. (New York: Marcel Dekker, Inc., 1978), pp. 91–108.

102. Olsen, J.H., L. Dragsted, and H. Autrup. "Cancer Risk and Occupational Exposure to Aflatoxins in Denmark," *Brit. J. Cancer* 58:392–396 (1988).

103. Dvorackova, I. "Aflatoxin Inhalation and Alveolar Cell Carcinoma," *Brit. Med. J.* 1:691 (1976).
104. Crofts, W.A., B.B. Jarvis, and C.S. Yatawara. "Airborne Outbreak of Trichothecene Toxicosis," *Atmos. Environ.* 20:549–552 (1986).
105. Sorenson, W.G., D.G. Frazer, B.B. Jarvis, J. Simpson, and V.A. Robinson. "Trichothecene Mycotoxins in Aerosolized Conidia of *Stachybotrys atra*," *Appl. Environ. Microbiol.* 53:1370–1375 (1987).
106. Emanuel, D.A., F.J. Wenzel, and B.R. Lawton. "Pulmonary Mycotoxicosis," *Chest* 67:293–297 (1975).
107. May, J.K., L. Stallones, D. Darrow, and D.S. Pratt. "Organic Dust Toxicity (Pulmonary Mycotoxicosis) Associated with Silo Unloading," *Thorax* 41:919–923 (1986).
108. Brinton, W.T., E.A. Vastbinder, J.W. Greene, J.J. Marx, R.H. Hutcheson, and W. Schaffner. "An Outbreak of Organic Dust Syndrome in a College Fraternity," *J. Am. Med. Assoc.* 258:1210–1212 (1987).
109. Wagner, G.E. "Bronchopulmonary Aspergillosis and Aspergilloma," in *Mould Allergy*, Y. Al-Doory and J.F. Domson, Eds. (Philadelphia, PA: Lea and Febiger, 1984), pp. 202–215.
110. Longbottom, J.L. "Applications of Immunological Methods in Mycology," in *Handbook of Experimental Immunology, vol. 4: Applications of Immunological Methods in Biomedical Sciences,* 4th ed., D.M. Weir, Ed. (Oxford, U.K.: Blackwell Scientific Publications, 1986), pp. 121.1–121.30.
111. Kurup, V.P., J.J. Barboriak, and J.N. Fink. "Hypersensitivity Pneumonitis," in *Mould Allergy*. Y. Al-Doory and J.F. Domson, Eds. (Philadelphia, PA: Lea and Febiger, 1984), pp. 216–243.
112. Derksen, F.J., N.E. Robinson, and J.S. Scott. "Aerosolized *Micropolyspora faeni* Antigen as a Cause of Pulmonary Dysfunction in Ponies with Recurrent Airway Obstruction (Heaves)," *J. Vet. Res.* 49:933–938 (1988).
113. Halliwell, R.E.W., J.B. Fleishman, M. Mackay-Smith, J. Beech, and D.E. Gunson. "The Role of Allergy in CPD of Horses," *J. Am. Vet. Med. Assoc.* 174:277–281(1979).
114. Pirie, H.M., C.O. Dawson, F.G. Breeze, I.E. Selman, and A. Wiseman. "Precipitins to *Micropolyspora faeni* in the Adult Cattle of Selected Herds in Scotland and North-West England," *Clin. Allergy* 2:181–187 (1972).
115. Wilkie, B.N. "Hypersensitivity Pneumonitis. Experimental Production in Calves with Antigens of *Micropolyspora faeni*," *Can. J. Comp. Med.* 40:221–224 (1976).
116. Leuschner, R.M., and G. Boehm. "Individual Pollen Collector for Use of Hay-fever Patients in Comparison with the Burkard Trap," *Grana* 16:183–186 (1977).
117. Kucharski, R. "A Personal Dust Sampler Simulating Variable Human Lung Function," *Br. J. Ind. Med.* 37:194–196 (1980).
118. Longbottom, J.L., and P.K.C. Austwick. "Fungal Antigens," in *Handbook of Experimental Immunology, Vol. 4: Applications of Immunological Methods in Biomedical Sciences,* 4th ed., D.M. Weir, Ed. (Oxford, U.K.: Blackwell Scientific Publications, 1986), pp. 7.1–7.11.
119. Hoffman, D.R., P.P. Kozak, S.A. Gillman, L.H. Cummins, and J. Gallup. "Isolation of Spore Specific Allergens from *Alternaria*," *Ann. Allerg.* 46:310–316 (1981).
120. Kurup, V.P., A. Resnick, G.H. Scribner, M. Gunassekaran, and J.N. Fink. "Enzyme Profile and Immunochemical Characterization of *Aspergillus fumigatus* Antigens," *J. Allergy Clin. Immunol.* 78:1166–1173 (1986).
121. Madelin, T.M., A.F. Clarke, and T.M. Mair. "Prevalence of Serum-Precipitating Antibodies in Horses to Fungal and Thermophilic Actinomycete Antigens: Effects of Environmental Challenge," *Equine Vet. J.* 23(4):247–252 (1991).
122. Edwards, J.H. "The Double Dialysis Method for Producing Farmer's Lung Antigens," *J. Lab. Clin. Med.* 79:683–688 (1972).

123. Agarwal, M.K., M.C. Swanson, C.E. Reed, and J.W. Yunginger. "Immunochemical Quantitation of Airborne Short Ragweed, *Alternaria,* Antigen E and Alt-1 Allergens: a Two-Year Prospective Study," *J. Allergy Clin. Immunol.* 72:40–45 (1983).

124. Aas, K., and L. Aukrust. "Immediate Hypersensitivity Responses to Fungal Agents," in *Mould Allergy,* Y. Al-Doory and J.F. Domson, Eds. (Philadelphia, PA: Lea and Febiger, 1984), pp. 133–146.

125. Ouchterlony, O. "Antigen-Antibody Reaction in Gels: Types of Reaction in Co-ordinated Systems of Diffusion," *Acta Pathol. Microbiol. Scand.* 32:231–240 (1953).

126. Longbottom, J.L., and J. Pepys. "Pulmonary Aspergillosis: Diagnostic and Immuno-logical Significance of Antigens and C-substance in *Aspergillus fumigatus,*" *J. Path. Bacteriol.* 88:141–151 (1964).

127. Kauffman, H.F., S. van der Heide, S. van der Laan, H. Hovenga, F. Beaumont, and K. de Vries. "Standardization of Allergenic Extracts of *Aspergillus fumigatus:* Liberation of IgE-Binding Components during Cultivation," *Int. Arch. Allergy Appl. Immunol.* 76: 168–173 (1985).

128. Fink, J.N. "Hypersensitivity Pneumonitis," *J. Allergy Clin. Immunol.* 74(1):1–9 (1984).

129. Edwards, J.H., J.T. Baker, and B.H. Davies. "Precipitin Test Negative Farmer's Lung—Activation of the Alternative Pathway of Complement by Mouldy Hay Dusts," *Clin. Allergy* 4:379–388 (1974).

130. Cole, G.T., and R.A. Samson. "The Conidia," in *Mould Allergy,* Y. Al-Doory, and J.F. Domson, Eds. (Philadelphia, PA: Lea and Febiger, 1984), pp. 66–103.

CHAPTER **15**

Aerobiology of Pollen and Pollen Antigens

Auli Rantio-Lehtimäki

INTRODUCTION

General Characteristics of Pollen Grains

In number, pollens dispersed by wind and constitute a small portion of the viable airborne particles. However, their importance as particles provoking allergies on a worldwide scale is much greater.

Pollen is the male gametophyte in both angiosperms and gymnosperms. In size, pollen grains vary between 5 μm and more than 200 μm[22] in diameter (see Figures 15.1 through 15.4). Only about 1/10 of species out of over 250,000 pollen, producing species are anemophilous,[30] i.e., their pollen grains are dispersed by wind. Most others are entomophilous, most often insect pollinated, but the pollinators also may be other animals, such as birds and bats. Some intermediate forms with both pollination systems also occur (e.g., in the genus *Salix*). Only anemophilous pollen are studied aerobiologically, as only they include important allergenic pollen and are thought to produce significant amounts of antigenic aerosols in air.

The typical features of wind-pollinated species may be summarized as follows: 1) large quantities of dry pollen with a more or less smooth surface are produced; 2) the stamens are usually large, borne on long filaments, well exposed, often hanging or organized in catkins; 3) the pollen usually is released in warm, dry weather, and 4) they are dispersed rapidly and efficiently over long distances; 5) they have large and well-exposed stigmas; 6) color, scent and nectar are frequently totally absent from the flowers.[23] The frequency of anemophily increases with latitude and elevation, being relatively low in tropical environments and high in boreal forests.[75]

A ripe pollen grain is surrounded by a wall—the sporoderm. The outer hard layer is called the exine, and the softer inner layer the intine. The exine is subdivided into further layers given different names by different authors.[22] Pollen grains are classified and identified on the basis of their morphology. Pollen grains either have different kinds of apertures or lack apertures altogether. In most cases, the aperture's part of the outer wall is thinner. Pores also frequently occur, e.g., alder pollen has four or five pores and birch only three. Grasses have one pore with a "lid." Important surface features for identification purposes are spines, nets, trabeculae or cones. Only a few pollen are easily identifiable to species level, e.g., grass pollen cannot be separated to species.

Principles of Pollen Assays

The selected method of pollen assay depends on the purpose of the study. Most aerobiological studies of pollen are carried out for allergological purposes. For instance,

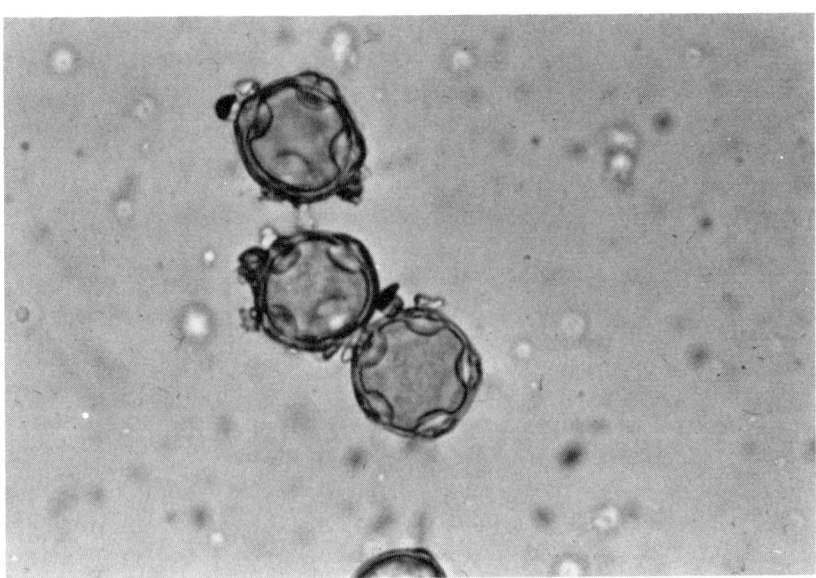

Figure 15.1 Alder (*Alnus*) pollen grains (18-21 × 23-30 μm).

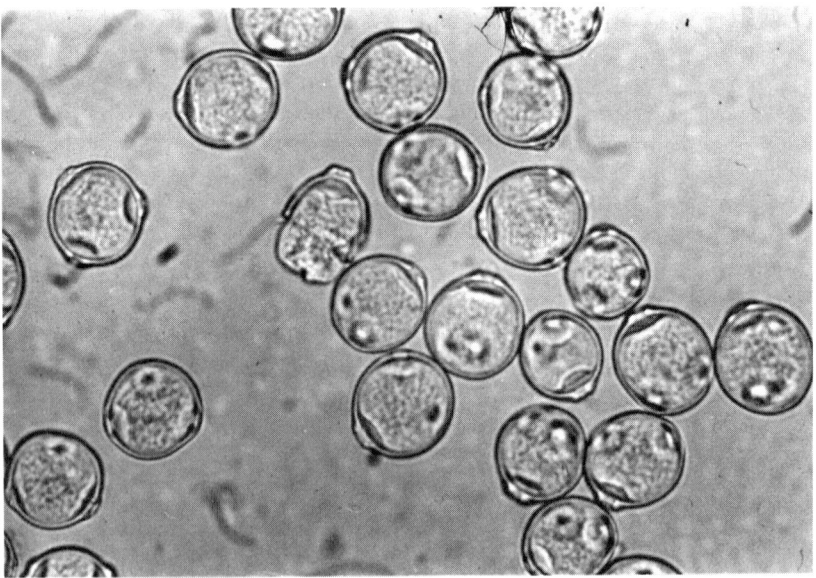

Figure 15.2 Birch (*Betula*) pollen grains (19-23 × 21-27 μm).

the pollen information services use only the daily mean frequencies for quite large areas, and so a non-viable method can be used. In some cases, it is important to analyze only the total pollen rain (e.g., studies of soil fertilization by pollen sediments); sometimes the focus is on the circadian variations in airborne pollen grains (e.g., in order to advise patients to stay indoors during peak pollen loads[70]).

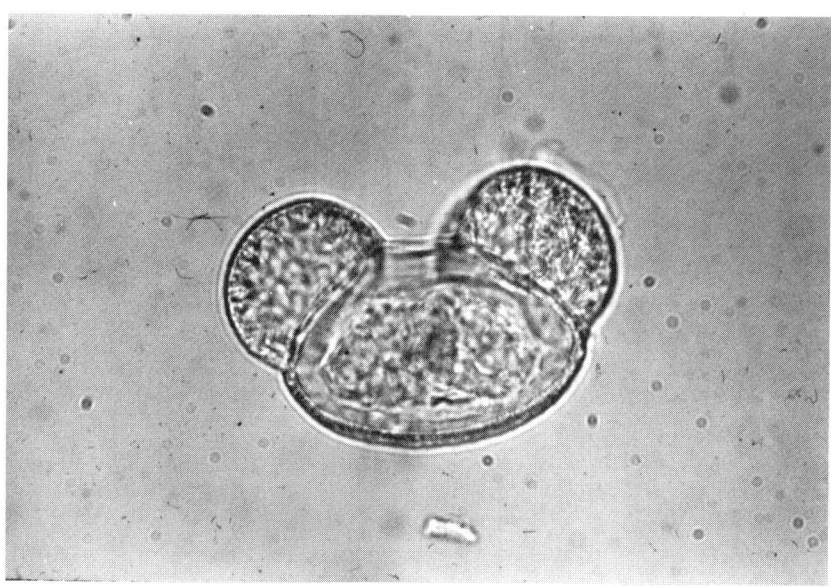

Figure 15.3 Pine (*Pinus*) pollen grains (45-50 × 65-80 μm).

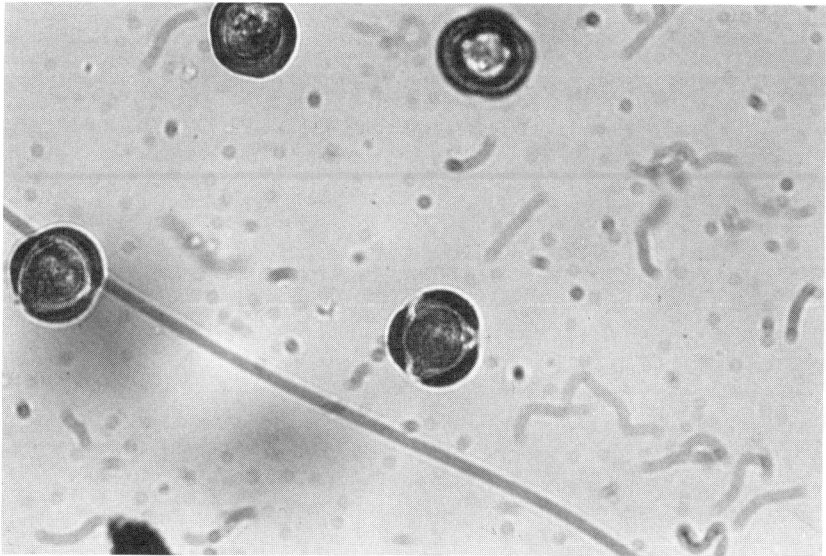

Figure 15.4 Mugwort (*Artemisia*) pollen grains (18-20 × 20-23 μm).

In order to obtain more precise data on the allergy-triggering factors present in the air, antigen concentrations are analyzed. Analyses of the viability of dispersed pollen are important when long-range pollination is under study, as in conifer seed orchards and when, for instance, additional pollination with stored pollen is needed, because the natural pollination process does not function efficiently (e.g., due to unfavorable weather in seed orchards).

TOTAL ASSAY OF POLLEN AEROSOLS

Introduction

For pollen information and forecasting purposes, the total pollen count is determined, and the most important allergenic pollen grains are analyzed at least to the genus level. The identification of pollen grains most often is based on light microscope analysis (see also Chapter 11), which utilizes easily observed pollen features such as ornamentation, germ pores and other characteristics of the surface of the pollen grain. The viability or allergenicity of the pollen grains in question usually is not determined.

Non-Volumetric Total Assays

Several sedimentation sampling methods have been developed for pollen analysis (cf. 45): Durham microscope slide,[20] hanging slides,[12] flags, passive impactor rods, sticky cylinders etc.[30] In forest pollen research, in particular, sedimentation methods are still in frequent use.[13] The impaction cylinder or rod with a small diameter is probably the best choice for non-volumetric assay but the catch is very much dependent on wind speed and direction.[73] For ecological studies where yearly total pollen deposition is analyzed, Tauber traps[92] are convenient. For individual sampling a non-volumetric pollen collector has been designed by Leuschner and Boehm[48] (cf. 45).

Pollen grains, on average, are rather heavy particles and therefore, sediment faster (1.5–39 cm/sec[30]) than most fungal spores, for instance. Moreover, it has been shown that different pollen types behave in different ways. Therefore, pollen quantity and the relative composition of the catch by sedimentation are not comparable with pollen obtained by volumetric methods.[73] Thus, the only acceptable pollen sampling methods in the field of aerobiology are volumetric assays.

Volumetric Total Assay

Intact Pollen Grains

At most aerobiological monitoring stations volumetric methods such as the Hirst, Burkard and Kramer-Collins traps are used (cf. 45). These traps are not isokinetic because wind speed variations have some effect on the catch, but the counts are sufficient for pollen information services. Airborne counts are expressed in particles per cubic meter of air. Threshold values for the airborne concentrations of different pollen grains which provoke symptoms in sensitized subjects have been reported[11,28,29,71,95] and the results from various stations all over the world are to a certain extent comparable with each other.

Several other volumetric, and even isokinetic, methods have been developed,[45] but most of them have several disadvantages, e.g., not being continuous in the operation, having overloading problems (e.g., whirling arm impactors), difficulties in the elution of particles (various filter samplers), complicated preparation, etc. When sampling nets are needed e.g., for pollen dispersal studies, a rotorod-net is very convenient, easy and cheap to build.

The University of Minnesota type inlet[49] provides a good solution to the problem of isokinetic sampling. The inlet is directed upwards and covered by a roof. Air is drawn in from each direction, thus eliminating any influence of wind direction. The suction volume remains constant and independent of wind speed (Table 15.1).

Table 15.1. Threshold Values in Use on the Basis of Birch as a Standard Pollen, When Sampling With the Burkard Trap.

Pollen	Threshold Values at Roof Level			Vol. Pollen (μm^3)
	Low	Moderate	Abundant	
Betula (used as a "standard")	< 10	10-100	>100	5600
Alnus	< 10	10-100	>100	6400
Artemisia	< 10	10-30	> 30	4200
Poaceae	< 10	10-30	> 30	17000 (Phleum pratense)

Threshold values of some important pollen grains which provoke allergic symptoms.

For individual sampling, several types of apparatus have been designed. One of the best functioning volumetric samplers with particle size selection is the Marple Personal Impactor Sampler[78] that may be used in the analyses of both airborne pollen counts and antigen concentrations. It is light to carry and thus is suitable even for small children.

Pollen Antigens

The total pollen antigen concentration in the air often provides more useful information for allergic subjects than the pollen count alone. One of the most important prerequisites in taking air samples for antigen analyses, though, is that the sampling apparatus should not let any particles pass through the system. All particles are retained when a cold trap follows the last filter or liquid sampling stage. Thus, even the smallest particles carried by small water droplets are trapped.

For antigen samples, a particle size selection is employed so that intact pollen grains are separated from smaller fragments and molecular sized particles. This is because one would like to know the antigenic concentration reaching the alveoli (the lowest parts of the respiratory tract) immediately after inhalation. The size separation may be carried out with a cascade impactor type apparatus on the top of a high-volume sampler[88] or with a virtual impactor[72]. A two-stage virtual impactor, SSBAS (size-selective bioaerosol sampler[42]), divides particles into three categories (over 7.2, 2.4–7.2, below 2.4 μm in diameter) in filter stages, followed by a liquid-cooled condensation unit, at -20°C, for sub-micron particles.

It has been shown that the airborne pollen count and the antigen concentration are not necessarily correlated. Sometimes pollen grains are empty of their allergenic content and, consequently, allergens often are found in high concentrations in small particle fractions even when there are no airborne pollen grains.[1-2,24,31,72,87-88] So far, no routine methods sensitive enough for the analysis of short-term changes in airborne antigen concentrations have been developed. The existing methods are rather time-consuming and tedious, RIA (radioimmunoassays, IgE methods[88]), or ELISA (enzyme-linked immunosorbent assay; IgG method[72]). An example of an ELISA procedure for pollen antigens is shown in Figure 15.5.

Schumacher et al.[82] have analyzed microscopically, using a semiquantitative fluorescent staining method, grass pollen antigens taken directly from Burkard tapes. They also found antigenic material outside of the pollen grains in small size fractions.

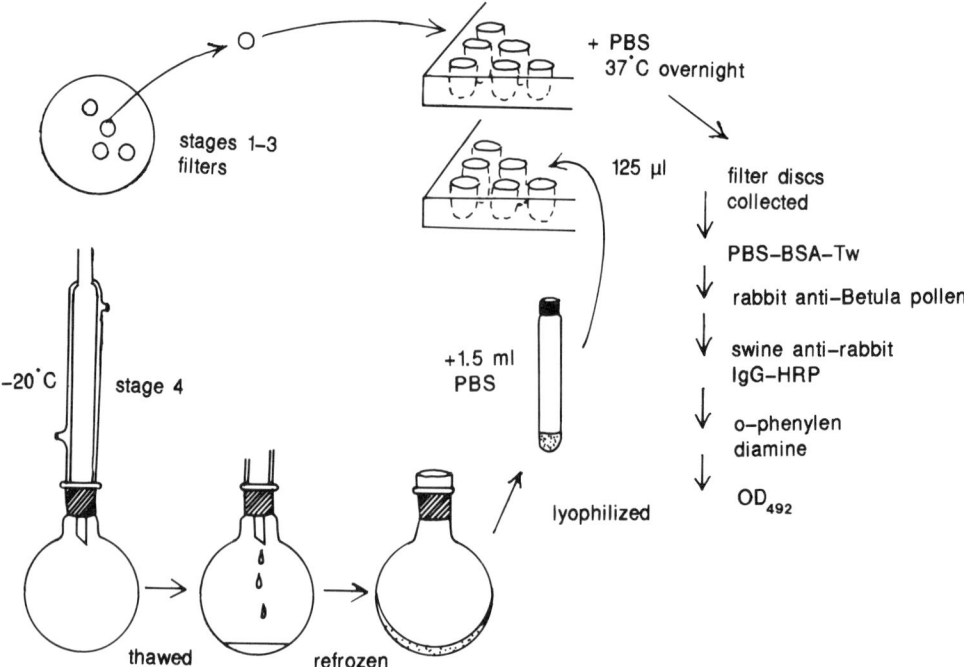

Figure 15.5 The ELISA (enzyme-linked immunosorbent assay) procedure for analyzing birch pollen antigenic activity in stages 1–4 of the SSBAS (size selective bioaerosol sampler; particle size classes: over 7.2 μm, 24–7.2 μm, below 2.4 μm in diameter, and sub-micron size class).

Pollen Grains and Pollen Antigens Indoors

Outdoor pollen can be transported indoors during ventilation, on people and their clothes and footwear, and on pets. Indoor air pollen counts may be high during the pollen season and under open window conditions concentrations of 600 pollen grains (p.g.) per cubic meter of air have been measured. In contrast, most indoor plants do not produce pollen or are not allergenic.[67]

Reports of indoor pollen concentrations in dust have ranged from 0 to 5.5 million per gram.[61] Finnish studies of birch pollen antigens in outdoor dust have shown that antigen activity remains at a high level for weeks after the pollen season.[64*] Sedimented dust outdoors is an important source of indoor dust.

POLLEN INFORMATION NETWORKS

Local Networks

Allergenic pollen grains have been shown to be so important that in tens of countries a national information service has been established to provide information about the pollen content of the air. A total of 264 European pollen monitoring stations were included in the *Traveller's Allergy Service Guide*.[93] Pollen information is given in most countries through the newspapers, radio and TV, and as a phone service.

All-Continent Networks

In North America, the pollen network (more than 50 gravity slide stations) is run by the American Academy of Allergy and Immunology's Pollen and Mold Committee. The Academy publishes an annual report. Daily reporting of volumetric pollen counts is carried out by the private sector, mainly the specialized weather networking channels.[15]

EAN, the European Aeroallergen Network, was established in 1986 for centralizing pollen data collection. All-European pollen reports are planned for the near future. Up until now, pollen counts from stations all over the continent have been sent weekly to the data bank at the University of Vienna, Austria.[38] These data have been made available for all contributing participants, for instance to provide information to travellers allergic to certain pollen grains.

The data accepted by the data bank have been based on volumetric sampling and this makes it possible to compare results for various sampling sites. The standardization of sampling sites is a difficult task. The sampling height, for instance, varies a great deal from site to site because of local conditions, but this cannot easily be changed for several practical reasons, e.g., for safety regulations. The vegetation around the sampling site also may vary considerably in both height and density and this gives a certain local influence on the whole picture. It has been accepted generally that the trap should be placed in an open site where the local vegetation does not have too much effect on the catch and the sampling height should be at least 10–15 m above ground level. The recommended maximum distance between samplings sites for a pollen sampling network is 170 km. However, nowadays a rather loose net, especially in the sparsely populated areas of Northern Europe, is the basis for the European pollen maps[39] and the information service.

It has been shown that not all pollen grains rise in a similar manner to the height of the trap. For example, 11-fold differences have been observed for *Artemisia* pollen at ground level at a distance of 200 m from the permanent Burkard station compared with the station itself at 15 m height.[71] In contrast, for tree pollen no such clear difference between these two heights was obtained because the source of tree pollen is high and dilution of the pollen also happens at a height greater than the height of the trap.

The analysis of pollen samples is not standard. A magnification of × 400–500 most frequently is employed for microscopic examination of pollen in Burkard samples. Several groups report using the sample analysis recommended by Käpylä and Penttinen,[41] while at some stations a random microscopic field method of analysis is used. The latter is a quicker method, and more sampling sites may be observed with the same effort as the former analysis system. It is possible also to analyze circadian variation from random field results.[70]

Microscope analysis by eye is always slow and tedious. Automatic image analysis would be of great help, but so far its suitability is limited. In future, we should have an extensive library of pollen grain characteristics needed for automated image analysis, and this should enable several more stations to operate and provide more local data on concentrations of allergenic pollen grains.

VIABILITY

Introduction

The rate of loss of viability of pollen when airborne differs from species to species. For example, rye pollen loses its viability within a few hours, whereas in oilseed rape pollen viability decreases gradually over 4–5 days. The loss of viability is mainly due to the egress of water through the pollen wall and its lipid seals[89] (see also Chapter 6). The viability of airborne pollen grains is of utmost importance for forest researchers and plant breeders. For instance, in Finnish Lapland, the climate is so harsh that the seed of Scots pine matures only a few times per century.[32] Therefore, pine clones have been brought to the south in an attempt to produce more and better seed for northern reforestation purposes. These plants must be of northern origin so as to retain the better winter-hardiness characteristics, etc., and they should be fertilized only with pollen of clones from the same latitudes.

The female flowers of forest trees may be receptive for several days before the start of local male flowering.[68,97] This allows the possibility of gene drift over considerable distances. Suitable air currents are known to carry high loads of foreign pollen over distances of even hundreds of kilometers.[3] Forest researchers, and especially breeders working with seed orchards, are eager to know, for instance, whether this long-range dispersed pollen is viable and capable of pollinating trees.

With reference to our own studies,[69] we have observed that the germination percentage of pine pollen remains high for much longer than the receptive female flowers exist.

Viability and allergenicity of pollen grains are not necessarily intercorrelated even though most of the allergenic proteins are found in the cytoplasm.[89] Franchi et al.[27] have shown that germination ability is higher when there are plenty of starch granules in the pollen grain. In long flowering plants, variations in starch content and germination percentage correlate negatively.

Methods of Studying the Viability of Pollen Grains

Various methods of analyzing the viability of pollen grains have been developed. Plant breeders, in particular, have used vital stains, inorganic acids, and *in vitro* germination. *In vivo* assays are difficult to perform because, for example, it takes up to two years to produce a result for conifers.

In vital staining, pollen is stained with 2,3,5-triphenylen-tetrazolium chloride (TTC) or related chemicals to test the oxidative enzyme activity. As long as the enzyme is active, pollen (not necessarily alive) will be stained red. Staining with a color dye often is used as an index of viability. Aniline blue in lactophenol stains callose with a high specificity, but most stains are not sufficiently accurate. Fluorochrome dyes can be adsorbed by protoplasmic organelles, e.g., acridine orange shows a characteristic green fluorescence in the nucleus and orange red in cytoplasm. Dead cells fluoresce yellow-green in the latter case.[89]

Treatment with inorganic acids, for instance 14.4% H_2SO_4 solution, tests the cell wall permeability of pollen grains and causes instant germ tube formation, called 'pseudogermination.'[89] This bursting of a germ tube may happen in seconds and it is assumed to correlate with the viability of the grain. However, this works only in apertured pollen grains. A 14.4% concentration of H_2SO_4 produces a maximum number of extruded tubes, while 19.2% H_2SO_4 causes total destruction of the grains.[89]

However, the best method for testing viability is to germinate the pollen grains. This may be achieved in liquid using several different techniques that have been described in detail by Stanley and Linskens,[89] e.g., a hanging drop in a moist chamber, spots or wells with germination liquid, on floating membranes, or dialyzing tubes in a moist environment.

Boron is an important component of germination media *in vitro*.[10] Stigmatic fluid has been shown to stimulate germination, and also contains high levels of boron. Pollen grains cultured in dense large populations germinate better and form longer pollen tubes than pollen germinated in small, evenly distributed populations.

Different pollen types demand different germination media (e.g., 10). A pollen grain is considered to have germinated when the length of the pollen tube exceeds half of the diameter of the pollen grain. A membrane floating on water, or a dialyzing tube hanging in a moist chamber, offers the possibility of transferring the samples in as many liquids as needed. But some pollens do not germinate well on a membrane while others do not even need a germination liquid.

Germination tests often are carried out on a solid agar surface. Lactose is one of the best sugar alternatives because it does not favor the growth of mold, which is often a problem in pollen germination assays.[89]

The percentage viability of grains often decreases very rapidly both in the air and during storage. On the other hand, some pollen grains are preserved rather well at freezing temperatures, while others survive better at room temperature.[62]

POLLEN ALLERGIES

Introduction

Charles Blackley[8] was the first to understand that grass pollen can cause seasonal symptoms in an allergic subject and affect the skin, eyes, nose and chest. He performed provocation tests on himself, reporting on both the immediate and late symptoms. The discovery of immunoglobulin E (IgE) in 1966 by the Ishizakas started a new era in immunological research. Allergy to airborne pollen was confirmed as being an IgE mediated allergy. However, no immunologically specific immunotherapy leading to the suppression of IgE antibodies so far has been developed. Pollen allergens probably have been studied more than any other allergen group.

In the temperate and boreal zones of the world, pollen allergies play an important role in human health. In tropical regions pollen allergies are far fewer because most plants are pollinated by insects or other animals.

During the peak pollen season a subject inhales approximately 1 μg pollen per day. Seasonal dosages of inhaled pollen allergens usually total no more than 0.06–1 μg (about 0.001–0.01 μg/kg) for adult humans.[51]

Clinical symptoms include both those of the upper respiratory tract (pollen grains are so big that they are retained in the nasopharyngeal part of the respiratory tract), and those of lower parts of the lungs,[56] i.e. ,asthmatic reactions (resulting from sub-micronic size particles and diffused pollen proteins).

Allergenic Pollen

Less than 100 of the allergologically well-documented pollen-producing plant species are known to be important causes of pollinosis.[52] Some of the plants (at least some grasses) have a world-wide distribution, but others occur only in limited geographical areas (like the birches in the boreal zones). Pollen has to become airborne in large amounts to become an allergen.[34]

Only very few Gymnosperm pollen are allergenic; of these *Cryptomeria* (sugi or Japanese cedar) is the most important but some reports of *Juniperus* (mountain cedar) and *Cupressus* have also been published. Despite the fact that it produces enormous quantities of pollen, the genus *Pinus* generally has been considered non-allergenic because of the large grain size, hydrophobic nature and low protein content. However, in New Zealand *Pinus radiata* D. Don has recently been shown to have allergenic properties, and cross-reactivity with ryegrass pollen allergens also was observed.[25]

In the Angiosperm families the following important allergenic genera are found: Hamamelidaceae (*Platanus*, plane tree, sycamore), Urticaceae (*Ulmus*, elm; *Morus*, mulberry; *Parietaria*, pellitory), Fagaceae (*Fagus*, beech; *Quercus*, oak; *Alnus*, alder; *Betula*, birch; *Carpinus*, hornbeam; *Corylus*, hazel), Salicaceae (*Populus*, cottonwood), Caryophyllaceae (*Chenopodium*, Lamb's quarters; *Kochia*, firebush; *Salsola*, Russian thistle; *Amaranthus*, pigweed), Aceraceae (*Acer*, maple), Oleaceae (*Fraxinus*, ash; *Olea*, olive), Plantaginaceae (*Plantago*, English plantain), Asteraceae (*Ambrosia*, ragweed; *Artemisia*, mugwort, sage; *Chrysanthemum*, ox-eye; *Taraxacum*, dandelion; *Xanthium*, cocklebur; *Parthenium*, feverfew.[83] Poaceae (*Agrostis, Alopecurus, Anthoxanthum, Avena, Bromus, Dactylis, Festuca, Holcus, Lolium, Phleum, Poa, Secale, Phragmites, Paspalum, Sorghum, Zea, Bouteloua, Cynodon*). The genus *Betula* is important in areas of the Northern Hemisphere, *Parietaria* in Southern Europe and *Ambrosia* in the Central parts of North America,[52] while *Parthenium* is especially important in India. In addition to being allergenic, this latter genus also is poisonous. Only a few cultivated plants are important allergologically in Europe, for example rye, which is the only cereal that is wind-pollinated. Only a few entomophilous plants have any allergological importance: the genus *Salix*, for example, has both pollination systems, anemophilous and entomophilous. Its pollen is sticky with pollenkitt to improve attachment to insects but it also is carried by strong winds when the pollen is mature and present in the flowers in huge amounts.

Allergic Reactions to Pollen Grains

Allergic reactions to pollen grains are traditionally called hayfever or pollinosis. Pollen primarily affects the mucous membrane of the upper respiratory tract. Clinically, the most frequent symptoms are allergic rhinitis or conjunctivitis, which is characterized by sneezing, watery eyes, nasal obstruction, itchy eyes and nose and frequently also coughing. Pollen is also known to cause asthmatic reactions that are localized in the lower respiratory tract. Skin reactions (urticaria) may occur, as well as reactions in the gastrointestinal tract, central or peripheral nervous system and the cardiovascular system.[89]

The most important allergenic antigens are proteins or polypeptides, although polysaccharides, glycoproteins, and lipoproteins also can act as allergens. The rapidity of the manifestation of the symptoms of hayfever suggests that allergens are readily available, and either they are present on the surface of the pollen grain or they diffuse out rapidly upon contact with the moist surfaces of the eyes and the upper respiratory tract. The pollen wall contains macromolecules including proteins and glycoproteins.[43] Some of the macromolecules have their sites of synthesis in the haploid pollen cytoplasm and are transported across the membrane to the wall. These proteins are strategically sited for immediate contact with the stigma surface, where they may be involved in recognition responses.

The main allergen of *Ambrosia artemisiifolia*, Amb a I, is held in the inner cellulosic intine in the pollen grain wall.[43] This layer also contains hydrolytic enzymes that are released immediately from mature pollen grains upon moistening. This release even may happen within 30 s. The allergens in the intine are released more slowly through the germinal aperture during germination. Similar observations have been made with respect to birch pollen allergens.[7]

Studies of Individual Pollen Allergens

Introduction

Pollen allergens have been studied for three decades. The only common characteristic of pollen allergens seems to be that they are water-soluble proteins with a molecular size ranging from 5,000 to 70,000 daltons.[52] The release of allergenic proteins from the pollen grain is a quick process that may happen in a few seconds—as also happens on human mucosa.[6] Under natural conditions this means that when pollen is impacted on a stigma, it immediately starts to release enzymes and form the germ tube.

Over 40 allergenic antigens have been purified and studied in detail. *Ambrosia artemisiifolia*, *Lolium perenne* and *Phleum pratense* are the best investigated plant species but amino acid sequences have been determined for only a few pollen allergens.[52]

Single allergens that have been studied by several different authors include: Cry j I, Cry j II (Japanese cedar, *Cryptomeria japonica*); Jun s I (mountain cedar, *Juniperus scopulorum*); Amb a I/II, Amb a III, Amb a V, Amb a VI (short ragweed, *Ambrosia artemisiifolia*); Amb t V (giant ragweed, *Ambrosia trifida*); Art v I, Art v II, Art v III (mugwort, *Artemisia vulgaris*); Par o I (wall pellitory, *Parietaria officinalis*); Par j I (*Parietaria judaica*); Bet v I, Bet v II (birch, *Betula verrucosa*); Cor a I (hazel, *Corylus avellana*); Aln g I (alder, *Alnus glutinosa*); Car b I (beech, *Carpinus betulus*); Que a I (oak, *Quercus alba*); Sal p I (Russian thistle, *Salsola pestifer*); Ole e I, Ole e II (olive, *Olea europaea*); Agr a I (redtop, *Agrostis alba*), Ant o I (sweet vernalgrass, *Anthoxanthum odoratum*); Lol p I, Lol p II, Lol p III, Lol p IV, Lol p V, Lol p IX, (rye grass, *Lolium perenne*); Phl p I, Phl p IV, Phl p V, Phl p VI, Phl p IX (timothy, *Phleum pratense*), Cyn d I (Bermuda grass, *Cynodon dactylon*), Sor h I (Johnson grass, *Sorghum halepense*), Dac g I, Dac g V (orchard grass, *Dactylis glomerata*), Fes e I (meadow fescue, *Festuca elatior*); Poa p I, Poa p V, Poa p IX (June grass, *Poa pratensis*) and Sec c I, (rye, *Secale cereale*).[4,5,21,33,35,36,46,51,53,54,57,74,76,77,79,80,84–86,91,94,96]

Methods of Studying Pollen Antigens

Several immunochemical techniques have been utilized in the study of pollen allergens. Proteinaceous allergens may be separated with crossed immunoelectrophoresis (CIE), isoelectric focusing (IEF), and electrophoresis in polyacrylamide gel (sodium dodecyl polyacrylamide gel electophoresis, SDS-PAGE). Allergens become visible through staining or using allergen-specific antibodies. Materials can be proved to be allergens by testing them against the sera of sensitized people, i.e., their IgE antibodies.

Only limited areas on the antigen take part in forming the bindings of immunocomplexes. These sites are called antigenic determinants or epitopes. Small molecules may become immunogenic when bound in a bigger molecule. These small molecules are called haptens.[56]

Allergens are named following the International Union of Immunological Societies (IUIS) recommendations. The name expresses the organism from which it originates. It has both an alphabetic and a numerical identification code. The organism is designated by taking the first three letters of the Latin name of the genus and one letter from the species name. The number refers to the order in which the allergens move furthest in the direction of the anode. If the separation is done with IEF, isoelectric points are determined. In SDS-PAGE, molecular weight is determined. These facts also can be used in the code. Purified allergens are named such that only Roman numerals are employed. Number I is always the main allergen. When the allergen also has a chemical name, both should be used together.[50]

Individual Pollen Allergens

The biological functions of plant allergens are poorly understood. In the case of the main allergen of birch pollen, Bet v I, homology with plant defense proteins has been shown.[9] Valenta et al.[94] identified birch pollen allergen Bet v II as profilin. Profilins can be found in almost all plants, including important food allergens. Profilins are known to regulate the eucaryotic cytoskleleton by sequestering monomeric actin and also participate in signal transduction via the phosphoinositide pathway.

In many cases allergens from related plants also are often similar in their immuno-chemical reactivity and in various physico-chemical aspects; for instance, the main allergens of birch, alder and hazel (Betulaceae) are very alike. Cross-reactions between these genera also are frequent. Each extract of pollen from these genera contains approximately 40 protein antigens but only a limited number of these are allergenic. Only one or two proteins are assessed as major allergens, i.e., most people sensitized to these plants get allergic reactions from just these proteins or, *in vitro*, the majority of patients' sera contain IgE that reacts strongly with this or these particular protein(s). Only a few proteins are highly species specific.[52]

The amino acid composition of Betulaceae-allergens is highly correlated and the NH_2-terminal amino acid sequences (ca. 25–30% of the complete sequence) of the major allergens exhibit a high degree of sequence identity, approximately 80–85%;[36] both hybridization experiments with cloned genes and restriction enzyme maps of cloned genes (Bet v I and Car b I)[52] indicate that the complete amino acid sequence of each main allergen is very similar.

Grass pollen allergens are of world-wide importance and were also some of the first to be studied. They constitute the largest group of related allergens to have been studied. An aequous extract of grass pollen from a single grass species will reveal approximately 20–30 antigenic components, of which at least one third binds IgE in sera from grass pollen-allergic individuals. However, only a few of the grass pollen allergens (usually two or three) will bind IgE from a major part of the sera.[53]

The Group I allergens have for many years been recorded as the major allergens in grass pollen. However, it now appears that the Group IX allergens are equally important. Group I, Group V and Group IX allergens are not allergologically cross-reactive.[44]

Allergens in Micron Size Fractions in the Air

Because of their large size, pollen grains should not be respirable. However, people who are allergic to pollen often show symptoms of asthma, indicating the presence of antigens similar to pollen antigens in size fractions smaller than those of intact pollen grains. Several authors[1,2,24,31,87,88] have documented antigenic activity in small particle size classes, as small as 0.1 μm and less. The existence of allergen-containing micronic particles was first demonstrated for ragweed and later for grass, birch and oak pollen. Several possibilities for the origin of small particle fractions have been suggested (e.g., 26). Pollen fragments produced by physical degradation in the environment are one likely source, especially when it is known that the protoplasm easily bursts out of the pollen grain. It is known also that the anther lining in ragweed and grasses is coated with orbicules that could be released into the environment.[44] Further possibilities include allergen-containing aerosols bound to various sizes of physical particulate matter in the environment. Fountain et al.[26] have detected the same antigens as found in birch pollen in various other parts of birch trees.

Recently Knox et al.[44] have shown that Lol p IX is present in starch granules in pollen grains of *Lolium perenne*, rye-grass. This is one of the few major respirable allergens known. Rye-grass pollen has been seen to rupture by osmotic shock on wetting. The frequency of micronic particles was observed to be especially high on days following rainfall (an increase of 50-fold was detected) during the grass pollen season, but airborne micronic particles are known to occur in fair weather, too. This phenomenon of increased antigen concentrations after rain was reported earlier by Stewart and Holt[90] and by Pehkonen and Rantio-Lehtimäki.[63] Increased antigen concentrations also have been shown experimentally to trigger asthma reactions.[44] However, the starch granules alone do not explain the high antigen concentrations recorded before the pollen season.[63]

Pharmacology

Detection of Pollen Allergy

Hay fever is a seasonal phenomenon. As more information about pollens is becoming available to the public, so more people are becoming aware of the possibility of allergies to pollen.

Several different methods exist for proving a specific allergy to pollen grains and checking pollen types that are the causative agents. Total IgE usually is determined by an immunochemical method. In normal individuals IgE comprises less than 0.001% of the total-Ig. People with hereditary high IgE are called atopic. In hayfever cases it has been shown that 25–50% of the total IgE is made up by IgE pollen antibodies. Thus, an increased total IgE level reflects the presence of specific IgE antibody and is an indicator of active allergy.[56]

Skin tests routinely are carried out in allergy laboratories. A pollen extract, when brought in contact with the skin, produces skin weal and flare; an immediate reaction follows after 15–20 minutes and a late reaction after 4–8 hours. The two basic skin test methods are the intradermal and prick tests. In the intradermal test an aqueous allergen extract is injected superficially in the skin, whereas in the prick test, a drop of glycerinated allergen extract is placed on the skin. The superficial layer is lifted with a lancet or needle. The positive control is histamine and the negative one is the diluent in use.[56]

A bronchial allergen provocation test may be utilized to confirm a positive skin test before the start of immunotherapy, for instance. It is, however, a more tedious test and as skin test extracts have improved and false positive test results decreased, bronchial provocation, at least in pollen allergies, is seldom necessary.

Provocations of eye conjunctiva and nasal mucosa with pollen extract also are used to some extent.[56]

Immunotherapy and Drugs

Immunotherapy (desensitization, hyposensitization) consists of a series of subcutaneous injections of allergen extract. This may be expected to reduce the patient's reactivity to the allergen in question. In the first phase, increasing dosages of the allergen are given at regular intervals. In the second phase, the highest tolerated dose is given as a maintenance dose. Hyposensitization only is effective if the patient is treated with extracts representing the actual sensitization source.[56]

Drugs against the symptoms of allergy have been improved to such an extent that the percentage of patients receiving immunotherapy lately has decreased. Antihistamines help in treating itching eyes and rhinitis, while sympatomimets are effective especially for nasal obstructions. Often an antihistamine and a sympatomimete are combined in the treatment. Chromoglicate is a safe drug even for children, but it has to be taken several times a day. Local steroids improve pollen rhinitis in 70–90% of cases.[56]

TOXINS

Toxic Pollen Grains

Toxic (to humans or other animals) pollen grains are very rare, but in India, honey from the species *Lasiosiphon eriocephalus* with a high pollen count has sometimes been found to cause nausea and vomiting.[89]

Pollen and Acid Rain

Pollen grains are very effective in catching airborne pollutants, especially since the size of pollen grains is suitable for 'cleaning' the air of very small dust and soot particles.[16]

Noll and Khalili[59,60] have shown the load of sulphate on tree pollen from different sources (both commercial and non-commercial) to be 0.2–0.5% of the pollen weight. Sulphate influx has been estimated at 0.5 kg per ha per year in some areas of the U.S. Nitrate concentrations were a magnitude lower. This could be important fertilizer, especially in nutrient-poor areas such as tundra ecosystems.

Pollen caught from the air has been shown to release high amounts of hydrogen ions when introduced into water; for instance, pollen of red pine produced 98 μg H$^+$ per gram of pollen added.[60]

Changes in Pollen Chemistry Due to Atmospheric Pollutants

Only a few investigations have been made of changes in the chemistry and morphology of pollen grains due to contact with pollutants compared to the many speculations made about 'angry town pollen' and 'nice country pollen.' In some experiments the influence of different pollutants has been studied, most often using SEM (scanning electron microscopy) and EDX (energy dispersive X-ray) analysis (e.g., 58, 14, 65). Experimentally, birch pollen has been exposed to an atmosphere of SO_2 (1%) and N_2 (99%) for 2 hours. This resulted in serious damage to the exine. Being subjected to a CO_2 atmosphere for 2 hours resulted in the collapse of the birch pollen grains.[14]

When *Dactylis glomerata* pollen grains from a non-polluted area were exposed to different treatments, e.g., car exhaust gas, cigarette smoke, UV irradiation and ozone, K$^+$ was the most abundant ion in the untreated pollen and Cl$^-$ in the treated pollen while UV induced no content modifications (see also Chapter 6). Some decrease in the amount of extracted allergen was recorded mainly after car exhaust gas and cigarette smoke treatments. Intense UV irradiation induced a marked modification of the major allergen, Dac g I, electrophoretic mobility.[66] It is possible, therefore, that the content and structure of pollen allergens are modified by environmental factors.

It has been suggested that air pollutants have played an important role in the increase in pollen allergies. Pollutants and bioaerosols may certainly interact, additively, synergistically, or inhibitively (see also Chapter 6). It seems clear that the most frequent interaction is the synergistic effect that aerobiological and chemical pollutants have on health.[47]

Pollen Grains in Toxin Tests

Trace amounts of various air pollutants and toxins have been proved to inhibit pollen germination and pollen tube formation. This is one of the most sensitive botanical indicators of atmospheric pollution.[40] Germination has been shown to be reduced by SO_2,[19,55,81] heavy metals such as lead and cadmium, and agrochemicals.[17,18]

SUMMARY

Only anemophilous pollens, i.e., these species where pollen grains are dispersed by wind, are aerobiologically important. Pollen allergens are significant agents in provoking allergic symptoms, and almost impossible to be avoided totally. However, pollen grains of less than 100 plant species are known to be important causes of pollinosis. At northern latitudes, pollen of the family Betulaceae is the most important group of outdoor allergens: but in the U.S., for example, *Ambrosia* pollen is far more important. Pollen information services operate in many areas and are well organized, at least in Europe, where the data bank in Vienna is continuously updated during the pollen season.

Recently, it has been shown that intact pollen grains are not the only sources of pollen allergens, but are found also in sub-micron fragments, and these may be especially important as allergy triggers. Airborne pollen counts and simultaneous antigen concentrations are not always correlated. Thus, instead of pollen information services, antigen reports may be more useful for clinicians and allergic subjects. In indoor environments, pollen antigens are more important than previously thought. Antigenic activity in sedimented dust remains high for weeks after the pollen shedding period and may be a source of resuspension into the air in indoor environments as well as out of doors.

Viability and antigenicity of pollen are not necessarily intercorrelated. A subject has been calculated to inhale approximately 1 μg pollen per day during the season and therefore the allergen concentration inhaled is unlikely to total more than 0.06–1 μg for an adult. Clinical symptoms include those of the upper respiratory tract and those of lower parts of the lungs, e.g., asthmatic reactions.

Pollen allergens have been studied extensively for more than 20 years, and many allergens (most of them proteins or polypeptides) are known in structure. Cross-reactivity between different pollen, food, fungal and animal allergens is common, and this complicates the identification of an allergy. Environmental changes may cause alterations in the chemistry of pollen grains, both as fertilizing male gametophytes and as allergenic agents.

REFERENCES

1. Agarwal, M.K., Swanson, M.C., Reed, C.E., and Yunginger, J.W. "Immunochemical quantitation of airborne short ragweed, *Alternaria* antigen E and Alt-I allergens. A 2 year prospective study," *J. Allergy Clin. Immunol.* 72:40–45 (1983).

2. Agarwal, M.K., Swanson, M.C., and Reed, C.E. "Airborne ragweed allergens: association with various particle sizes and short ragweed plant parts," *J. Allergy Clin. Immunol.* 74: 687–693 (1984).
3. Andersson, E. "Seed stands and seed orchards in the breeding of conifers," *Proceedings of the World Consultation on Forest Genetics and Tree Improvement,* FAO/FORGEN, Paper 8/1, 18 p (1963).
4. Avjioglu, A., Singh, M.B., Kenrick, J., and Knox, R.B. "Monoclonal antibodies to a recombinant allergenic peptide of Lol p IX and the identification of corresponding epitopes in different grasses," in *Molecular biology and immunology of allergens,* D. Kraft and A. Sehon, Eds. (Boca Raton, FL, CRC Press, Inc. 1993), pp. 149–151.
5. Ayuso, R., Polo, F., and Carreira, J. "Purification of Par j I, the major allergen of *Parietaria judaica* pollen," *Mol. Immunol.* 25:49–56 (1988).
6. Baraniuk, J.N., Bolick, M., Esch, R., and Buckley, C.E. III. "Quantification of pollen solute release using pollen grain column chromatography," *Allergy* 47:411–417 (1992).
7. Belin, L., and Rowley, J.R. "Demonstration of birch pollen allergen from isolated pollen grains using immunofluorescence and single radial immunodiffusion technique," *Int. Arch. Allergy* 40:754–769 (1971).
8. Blackley, C.H. "On the quantity of pollen found floating in the atmosphere during the prevalence of hay-fever, and on its relation to the intensity of symptoms," *Experimental researches on the causes and nature of Catarrhus aestivus,* 4:115–153 (1873).
9. Breiteneder, H., Pettenburger, K., Bito, A., Valenta, R., Kraft, D., Rumpold, H., Scheiner, O., and Breitenbach, M. "The gene coding for the major birch pollen allergen, Bet v I, is highly homologous to a pea disease resistance response gene," *EMBO J.* 8:1935–1938 (1989).
10. Brewbaker, J.L., and Kwack, B.H. "The essential role of calcium ion in pollen germination and pollen tube growth," *Am. J. Bot.* 50:859–865 (1963).
11. Brown, E.B., and Ipsen, J. "Changes in severity of symptoms of asthma and allergic rhinitis due to air pollutants," *J. Allergy,* 41:254–268 (1968).
12. Bryant, R.H., Emberlin, J.C., and Norris-Hill, J. "Vertical variation in pollen abundance in North-Central London," *Aerobiologia* 5:123–137 (1989).
13. Caron, G.E., and Leblanc, R. "Pollen contamination in a small black spruce seedling seed orchard for three consecutive years," *For. Ecol. Manage.* 53: 245–261 (1992).
14. Cerceau-Larrival M.-T., and Derouet, L. "Relation possible entre les elements inorganiques detectes par Spectrometrie X a Selection d'Energie et l'allergenicite des pollens," *Ann. Sci. Nat. Bot., Paris* 13:133–152 (1988).
15. Comtois, P. "Networking and forecasting in North America: how to reach a real time status," *Aerobiologia* 8:42–43 (1992).
16. Cox, R.M. "Sensitivity of forest plant reproduction to long range transported air pollutants: in vitro sensitivity of pollen to simulated acid rain," *New Phytol.* 95:269–276 (1983).
17. Cox, R.M. "Contamination and effects of cadmium in native plants," in *Cadmium in the environment,* H. Mislin, and O. Ravera, Eds. (*Experientia* Suppl. 50, 1986), pp. 101–109.
18. Cox, R.M. "The sensitivity of pollen from various coniferous and broad-leaved trees to combinations of acidity and trace metals," *New Phytol.* 109:193–201 (1988).
19. DuBay, D.T., and Murdy, W.H. "Direct adverse effects of SO_2 on seed set in *Geranium carolinianum* L.: a consequence of reduced pollen germination on the stigma," *Bot. Gaz.* 144:376–381 (1983).
20. Durham, O.C. "The volumetric incidence of atmospheric allergens; IV. A proposed standard method of gravity sampling, counting and volumetric interpolation of results," *J. Allergy* 17:79–86 (1944).
21. Esch, R.E., and Klapper, D.G. "Isolation and charcterization of a major cross-reactive grass group I allergenic determinant," *Mol. Immunol.* 26:557–561 (1989).

22. Faegri, K., and Iversen J. *Textbook of pollen analysis,* 4th ed., K. Faegri, K., P.E. Kaland, and K. Krzywinski, Eds. (New York, John Wiley and Sons, Inc., 1989).
23. Faegri, K., and van der Pijl, L. *The principles of pollination ecology* (Oxford, Pergamon, 1979).
24. Fernandez-Caldas, E., Swanson, M.C., Pravda, J., Welsh, P., Yunginger, J.W., and Reed, C.E. "Immunochemical demonstration of red oak pollen aeroallergens outside the oak pollination season," *Grana* 28:205–209 (1989).
25. Fountain, D.W., and Cornford, C.A. "Aerobiology and allergenicity of *Pinus radiata* pollen in New Zealand," *Grana* 30:71–75 (1991).
26. Fountain, D.W., Berggren, B., Nilsson, S., and Einarsson, R. "Expression of birch pollen-specific IgE-binding activity in seeds and other plant parts of birch trees (*Betula verrucosa* Ehrh)," *Int. Arch. Allergy Immunol.* 98:370–376 (1992).
27. Franchi, G.G., Pacini, E., and Rottoli, P. "Pollen grain viability in *Parietaria judaica* L," *Giornale Botanico Italiano* 118:163–178 (1984).
28. Frankland, A.W. "The pollen count and the patient," *Ind. J. Aerobiol.* 3:1–6 (1990).
29. Freedman, B. "Ragweed pollinosis control," *J. Environ. Health* 30:151–156 (1967).
30. Gregory, P.H. *The microbiology of the atmosphere* (Aylesbury, Leonard Hill, 1973).
31. Habenicht, H.A., Burge, H.A., Muilenberg, M.L., and Solomon, W.R. "Allergen carriage by atmospheric aerosol. II. Ragweed-pollen determinants in submicronic atmospheric fractions," *J. Allergy Clin. Immunol.* 74:64–67 (1984).
32. Henttonen, H., Kanninen, M., Nygren, M., and Ojansuu, R. "The maturation of *Pinus sylvestris* seeds in relation to temperature climate in Northern Finland," *Scand. J. For. Res.* 1:243–249 (1986).
33. Hoz, F. de la, Polo, F., Prado, J.M., Selles, J.G., Lombardero, M., and Carreira J. "Purification of Art v I, a relevant allergen of *Artemisia vulgaris* pollen," *Mol. Immunol.* 27:1047–1056 (1990).
34. Hyde, H.A. "Atmospheric pollen and spores in relation to allergy 1," *Clin. Allergy* 2:153–179 (1972).
35. Ipsen, H., and Løwenstein, H. "Isolation and immunochemical characterization of the major allergen of birch pollen (*Betula verrucosa*)," *J. Allergy Clin. Immunol.* 72:150–159 (1983).
36. Ipsen, H., and Hansen, O.C. "The NH_2-terminal amino acid sequence of the immunochemically partial identical major allergens of alder (*Alnus glutinosa*) Aln g I, birch (*Betula verrucosa*) Bet v I, hornbeam (*Carpinus betulus*) Car b I, and oak (*Quercus alba*) Que a I pollens," *Mol. Immunol.* 28:1279–1288 (1991).
37. Ishizaka, K., Ishizaka, T., and Hornbrook, M.M. "Physico-chemical properties of reaginic antibody. V. Correlation of reaginic activity with gamma-E-globulin antibody," *J. Immunol.* 98:840–852 (1966).
38. Jäger, S. "EANS—European aeroallergen network server," *Aerobiologia* 4:16–19 (1988).
39. Jäger, S., and Mandrioli, P. "Airborne grass pollen distribution in Europe 1992," *Aerobiologia* 8:3–39 (1992).
40. Kappler, R., and Kristen, U. "Photometric quantification of *in vitro* pollen tube growth: A new method suited to determine the cytotoxicity of various environmental substrates," *Env. Exp. Bot.* 27:305–309 (1987).
41. Käpylä, M., and Penttinen, A. "An evaluation of the microscopical counting methods of the tape in Hirst-Burkard pollen and spore trap," *Grana* 20:131–141 (1981).
42. Kauppinen, E.I., Jäppinen, A.V.K., Hillamo, R.E., Rantio-Lehtimäki, A.H., and Koivikko, A.S. "A static particle size selective bioaerosol sampler for the ambient atmosphere," *J. Aerosol Sci.* 20:829–838 (1989).
43. Knox, R.B. "The pollen grain" in *Embryology of angiosperms,* B.M. Johri, Ed. (Berlin, Springer-Verlag, 1984).

44. Knox, R.B., Taylor, P., Smith, P., Hough, T., Ong, E.K., Suphioglu, C., Lavithis, M., Davies, S., Avjioglu, A., and Singh, M. "Pollen allergens: Botanical aspects," in *Molecular biology and immunology of allergens*, D. Kraft and A. Sehon, Eds. (Boca Raton, FL, CRC Press, Inc. 1993), pp. 31–38.
45. Lacey J., and Venette, J. "Outdoor air sampling techniques," in *Bioaerosols Handbook*, C.S. Cox, and C.M. Wathes, Eds. (Chelsea, MI: Lewis Publishers, Inc., 1994).
46. Lauzurica, P., Marubi, N., Galocha, B., Gonzalez, J., Diaz, R., Palomino, P., Hernandez, R., Garcia, R., and Lahoz, C. "Olive (*Olea europaea*) pollen allergens. II. Isolation and characterization of two major antigens," *Mol. Immunol.* 25:337–344 (1988).
47. Lebowitz, M.D., and O'Rourke, M.K. "The significance of air pollution in aerobiology," *Grana* 30:31–43 (1991).
48. Leuschner, R.M., and Boehm, G. "Individual pollen collector for use of hay-fever patients in comparison with the Burkard trap," *Grana* 16:183–186 (1977).
49. Liu, B.Y.H., and Pui, D.Y.H. "Aerosol sampling inlets and inhalable particles," *Atmos. Environ.* 15:589–600 (1981).
50. Marsh, D.G., Goodfriend, L., King T.P., Løwenstein, H., and Platts-Mills, T.A.E. "Allergen nomenclature," *Allergy* 43:161–168 (1988).
51. Marsh, D.G., and Freidhoff, L.R. "HLA genes determining susceptibility to allergy," in *Molecular biology and immunology of allergens*, D. Kraft and A. Sehon, Eds. (Boca Raton, FL, CRC Press, Inc. 1993), pp. 1–9.
52. Matthiesen, F., Ipsen, H., and Løwenstein, H. "Pollen allergens," in *Allergenic pollen and pollinosis in Europe,* G. D'Amato, F. Th. M. Spieksma, and S. Bonini, Eds. (Oxford, Blackwell Scientific Publ., 1991a), pp. 36–44.
53. Matthiesen, F., and Løwenstein, H. "Group V allergens in grass pollens. I. Purification and characterization of the group V allergen from *Phleum pratense* pollen, Phl p V," *Clin. Exp. Allergy* 21:309–320 (1991b).
54. Mecheri, S., Peltre, G., and David, B. "Purification and characterization of a major allergen from *Dactylis glomerata* pollen, the Ag Dg 1," *Int. Arch. Allergy Appl. Immunol.* 78:283–289 (1985).
55. Mejnartowicz, L., and Lewandowski, A. "Effects of fluorides and sulphur dioxide on pollen germination and growth of the pollen tube," *Acta Soc. Bot. Pol.* 54:125–129 (1985).
56. Mygind, N. *Essential allergy*, (Oxford, Blackwell Scientific Publ., 1986).
57. Nielsen, B.M., and Paulsen, B.S. "Isolation and characterization of a glucoprotein allergen, Art v II, from pollen of mugwort (*Artemisia vulgaris* L)," *Mol. Immunol.* 27:1047–1056 (1990).
58. Nilsson S. "Poursuite des analyses de la pollution particulaire des pollens," *Ann. Sci. Nat., Bot.,* Paris 9:125–132 (1988).
59. Noll, K.E., and Khalili, E. "Dry deposition of sulfate associated with pollen," *Atmos. Environ.* 22:601–604 (1988a).
60. Noll, K.E., and Khalili, E.K. "Hydrogen ions associated with the dry deposition of pollen." 81st Annual Meeting of APCA, Dallas, (1988b), 23 pp.
61. O'Rourke, M.K., and Lebowitz, M.D. A comparison of regional atmospheric pollen with pollen collected at and near homes, *Grana* 23:55–64 (1984).
62. Owens, J.N., and Blake, M.D. "Forest tree seed production. A review of the literature and recommendations for future researcch," Information Report PI-X-53, University of Victoria, Department of Biology, Victoria, B.C., Canada (1985).
63. Pehkonen, E., and Rantio-Lehtimäki, A. "Variations in airborne pollen antigenic particles caused by meteorologic factors," *Allergy* 49:472–477 (1994).
64. Pehkonen, E., Rantio-Lehtimäki, A., and Yli-Panula, E. "Submicroscopic antigenic particles in outdoor air: a contamination source of indoor air," Proc. Indoor Air '93, the 6th Int. Conf. on Indoor Air Quality and Climate, Helsinki, Finland, 4:231–236 (1993).

65. Peltre, G., Panheleux, D., and Davis, B. "Environmental effect on grass pollen allergens," *Ann. Sci. Nat., Bot., Paris* 9:225–229 (1988).
66. Peltre, G., Derouet, L., and Cerceau-Larrival, M.-T. "Model treatments simulating environmental action on allergenic *Dactylis glomerata* pollen," Grana 30:59–61 (1991).
67. Pope, A.M., Patterson, R., and Burge, H., Eds., *Indoor Allergens,* (Washington, DC, National Academy Press, 1993).
68. Pulkkinen, P. "Time isolation in flowering of Scots pine seed orchards with northern clones; differences in background pollen sources," *Can. J. For.* (1994).
69. Pulkkinen, P., and Rantio-Lehtimäki, A. "Pine pollen viability and development of male flowering," *Abstracts. IUFRO Symposium, Victoria, BC, Canada* (1993).
70. Rantio-Lehtimäki, A., Helander, M.L., and Pessi, A.-M. "Circadian periodicity of airborne pollen and spores: significance of sampling height," *Aerobiologia* 7:129–135 (1991a).
71. Rantio-Lehtimäki, A., Koivikko, A., Kupias, R., Mäkinen, Y., and Pohjola, A. "Significance of sampling height of airborne particles for aerobiological information," *Allergy* 46:68–76 (1991b).
72. Rantio-Lehtimäki, A., Viander, M., and Koivikko, A. "Airborne birch pollen antigens in different particle sizes," *Clin. Exp. Allergy* 24:23–28 (1994).
73. Raynor, G.S. "Sampling techniques in aerobiology," in *Aerobiology. The ecological system approach,* R.L. Edmonds, Ed. (Stroudsburg, PA, Dowden, Hutchinson and Ross, Inc., 1979).
74. van Ree, R. van, Clemens, J.G.J., Aalbers, M., Stapel, S.O., and Aalberse R.C. "Characterization with monoclonal and polyclonal antibodies of a new major allergen from grass pollen in the group I molecular weight range," *J. Allergy Clin. Immunol.* 83: 144–151 (1989).
75. Regal, P.J. "Pollination by wind and animals: Ecology of geographic patterns," *Annu. Rev. Ecol. Syst.* 13:497–594 (1982).
76. Roebber, M., Hussain, R., Kapper, D.G., and Marsh, D.G. "Isolation and properties of a new short ragweed pollen allergen Ra6," *J. Immunol.* 131: 706–711 (1983).
77. Roebber, M., Klapper, D.G., Goodfriend, L., Bias, W.B., Hsu, S.H., and Marsh, D.G. "Immunochemical and genetic studies of Amb t V (Ra5G), an Ra5 homologue from giant ragweed pollen," *J. Immunol.* 134: 3062–3069 (1985).
78. Rubow, K.L., Marple, V.A., Olin, J., and McCawley, M.A. "A personal Cascade impactor: Design, evaluation and calibration," *Am. Ind. Hyg. Assoc. J.* 48:532–538 (1987).
79. Sakaguchi, M., Inouye, S., Taniai, M., Ando, S., Usui, M., and Matuhasi, T. "Identification of the second major allergen of Japanese cedar pollen," *Allergy* 45:309–312 (1990).
80. Scheiner, O., Vrtala, S., Laffer, S., Duchene, M., Kraft, D., and Valenta, R. "cRNA cloning of the major pollen allergens from Timothy grass (*Phleum pratense*) and rye (*Secale cereale*). Diagnosis of grass pollen allergy with recombinant allergens," in *Molecular biology and immunology of allergens*, D. Kraft and A. Sehon, Eds. (Boca Raton, FL, CRC Press, Inc., 1993), pp. 153–156.
81. Scholz, F., Vornweg, A., and Stephan, B.R. "Wirkungen von Luftverunreinigungen auf die Pollenkeimung von Waldbäumen," *Forstarchiv* 56:121–124 (1985).
82. Schumacher, M.J., Griffith, R.D., and O'Rourke, M.K. "Recognition of pollen and other particulate aeroantigens by immunoblot microscopy," *J. Allergy Clin. Immunol.* 82:608–616 (1988).
83. Seetharamiah, A.M., Viswanath, B., and Subba Rao, P.V. "Atmospheric survey of pollens of *Parthenium hysterophorus*," *Ann. Allergy* 47:192–196 (1981).
84. Shafiee, A., Yunginger, J.W., and Gleich, G.J. "Isolation and characterization of Russian thistle *(Salsola pestifer)* pollen allergens," *J. Allergy Clin. Immunol.* 67:472–481 (1981).
85. Smith, J.J., Olson, J.R., and Klapper, D.G. "Monoclonal antibodies to denatured ragweed pollen allergen Amb a I: Characterization, specificity for the denatured allergen and

utilization for the isolation of immunogenic peptides of Amb a I," *Mol. Immunol.* 25:355–365 (1988).

86. Smith, P.M., Singh, M.B., and Knox, R.B. "Characterisation and cloning of the major allergen of Bermuda grass, Cyn d I" in *Molecular biology and immunology of allergens,* D. Kraft and A. Sehon, Eds. (Boca Raton, FL, CRC Press, Inc. 1993), pp. 157–160.
87. Solomon, W.R., Burge, H.A., and Muilenberg, M.L. "Allergen carriage by atmospheric aerosol. I. Ragweed pollen determinants in smaller micronic fractions," *J. Allergy Clin. Immunol.* 72:443–447 (1983).
88. Spieksma, F.Th.M., Kramps, J.A., van der Linden, A.C., Nikkels, B.H., Plomp, A., Koerten, H.K., and Dijkman, J.H. "Evidence of grass-pollen allergenic activity in the smaller micronic atmospheric aerosol fraction," *Clin. Exp. Allergy* 20:273–280 (1990).
89. Stanley, R.G., and Linskens, H.F. *Pollen. Biology, Biochemistry, Management,* (Berlin, Springer-Verlag, 1974).
90. Stewart, G.A. and Holt, P.G. "Submicronic airborne allergens," *Med. J. Aust.* 143:426–427 (1985).
91. Taniai, M., Ando, S., Usui, M., Kurimoto, M., Sakaguchi, M., Inouye, S., and Matuhasi, T. "N-terminal amino acid sequences of a major allergen of Japanese cedar pollen (Cry j I)," *FEBS Lett.* 239:329–332 (1988).
92. Tauber, H. "A static non-overload pollen collector," *New Phytol.* 73: 359–369 (1974).
93. *Traveller's Allergy Service Guide,* S. Nilsson and F. Th. M. Spieksma, Eds., Palynol. Lab., Swedish Mus. Nat. Hist. and Fisons Sweden AB (1992).
94. Valenta, R., Duchene, M., Vrtala, S., Birkner, T., Ebner, C., Hirschwehr, R., Breitenbach, M., Rumpold, H., Scheiner, O., and Kraft, D., "Recombinant allergens for immunoblot diagnosis of tree-pollen allergy," *J. Allergy Clin. Immunol.* 88:889–894 (1991).
95. Viander, M., and Koivikko, A. "The seasonal symptoms of hyposensitized and untreated hay fever patients in relation to birch pollen counts: correlation with nasal sensitivity, prick tests and RAST," *Clin. Allergy* 8:387–396 (1978).
96. Villalba, M., Lopez-Otin, C., Martin-Orozco, E., Monsalve, R.I., Palomino, P., Lahoz, C., and Rodriques, R. "Isolation of three allergenic fractions of the major allergen for *Olea europaea* pollen and NH_2-terminal amino acid sequence," *Biochem. Biophys. Res. Commun.* 172:523–528 (1990).
97. Wheeler, N.C. "Variation in reproductive bud phenology among and within the Douglas-fir clones: significance for SMP-1983 results," Weyerhauser Forest Research Technical Report 050-3210/3. Centralia, WA, Weyerhauser Co. 25 (1983).

CHAPTER **16**

Outdoor Air Sampling Techniques

J. Lacey and J. Venette

INTRODUCTION

Bioaerosol particles are almost always present in outdoor air although their numbers, types and viability change with time of day, weather, season and geographical location. Particles from local sources can sometimes predominate. Most bioaerosol particles are pollen grains and fungal spores but some may be bacteria, algae, moss and fern spores, protozoa and even viruses. They come from many sources and their aerodynamic sizes can range from 1 to >100 μm and their numbers from <1 to >10^6 particles m^{-3} air. They may be found from ground level to high altitudes, but their presence above the troposphere has not been established beyond doubt.

Sampling instruments for particular components of the atmospheric bioaerosol[59,130,299] and their use[183–185,380] have been reviewed but no description covers all aspects of bioaerosol sampling out of doors (see also Chapters 9 and 10). Most often, bioaerosols have been sampled for pollens and fungal spores in the ambient environment or in agricultural crops or for bacteria associated with waste disposal. Many other organisms can be dispersed by wind at some stage of their life cycle, including weak flying insects (e.g., aphids), insects and arachnids that produce ballooning webs[277] and plant seeds that are so small or that have morphological adaptations (e.g., pappus, wings) enabling them to remain airborne for some distance.[219] However, these can hardly be classified as bioaerosols and are not treated here.

Aerobiology is a derived science. It cannot readily be separated from other biological disciplines but integrates the methodologies of mycology, virology, bacteriology, plant pathology, phycology and botany with aerosol science and meteorology. It enables the collection, growth, characterization and identification of the components of the bioaerosol so as to understand the biology of the organisms and how they become airborne; to provide the framework for the proper design and interpretation of studies; to derive the optimum designs for samplers and to understand their limitations; and to understand particle dispersion.

No components of the bioaerosol are known to complete their life cycles in the air although bacterial multiplication may sometimes occur.[85] Studies of outdoor bioaerosols have been described as outdoor anabiology since most particles have low metabolic rates (anabiosis or hypobiosis).[376] They only become important when the particles land and impact on other organisms or environments, resulting in pollination of plants, colonization of new substrates, the spread of genetically manipulated organisms, diseases in animals and plants or allergy but the extent of this impact may depend upon the environment providing the right conditions for infection or growth.

Environmental conditions can affect both sampling and interpretation of results since they can affect the numbers and viability of bioaerosol particles, their liberation, dispersal and deposition and contamination by other particulate matter, while gaseous pollutants that affect airborne microbes may sometimes make generalizations difficult.[328] Inevitably, there is much overlap between the methods used in different outdoor environments. Therefore, to minimize repetition, most methods are described in detail under ambient air sampling and only special methods or adaptions of these will be reported in other sections.

SOURCES OF BIOAEROSOLS

Sources of bioaerosols can be defined both aerobiologically and in terms of different biological particles or different substrates. Aerobiologically, sources are defined as point, line or area sources, depending on their size and shape in relation to the size of the sampling area. Point sources may be thought of as a single plant or tree or even a plot or field, if the area of interest is large enough. Line sources may be represented by a hedge or a row of inoculated plants in a plant disease experiment. Area sources may comprise a plot within a field or a field or larger area in a region (see also Chapter 14).

Most airborne pollens come from anemophilous plants but pollen from entomophilous plants sometimes may be airborne in sufficient numbers to be important causes of allergy.[5] Although on occasions modified by weather conditions, many pollens have clearly defined seasons of abundance corresponding to the flowering periods of the plants from which they originate. Grasses form the most numerous pollen types in many parts of the world but pollens from trees and other herbaceous plants can be numerous and often are implicated in allergy. Seasonal mean concentrations of different pollen types range from about 5–300 grains m^{-3} air but mean hourly concentrations, particularly of grasses and *Urtica*, can exceed 1500 m^{-3} with maxima up to 8000 grains m^{-3}.[198] Allergy to *Betula* (birch) is often most frequent in March-April, to grass pollen in May-June, to *Urtica* in June-August and to *Ambrosia* (ragweed) in August-September. Changing agricultural practice may affect the spectrum of airborne pollens. Thus, although oilseed rape (*Brassica napus*) is entomophilous, its increasing cultivation in Britain has increased both the amounts of airborne pollen and complaints of allergy.

Despite their widespread occurrence in the air, moss and fern spores have been studied much less than pollens. Spores of ferns are released explosively when sporangia spring open as they dry during the early morning. Both sporangia and spores can be found on trap slides but the latter are far more numerous. Both Hamilton[141] and Gregory and Hirst[133] record *Pteridium* spores to a maximum of 36 m^{-3} while Lacey[198] found an average of 1.7–1.9 m^{-3}, when there were only small sources close to the trap. However, within an area of bracken, concentrations exceeded 1800 spores m^{-3} on some days between August and October.[227] Maximum catches of sporangia occurred slightly earlier (8–9 am) than those of spores (9–10 am). Average concentrations of airborne moss spores may be only 0.1–0.3 m^{-3}.[198] Clubmosses and horsetails are much more restricted in their distribution than mosses and ferns but they have been trapped from air in Canada[24] although there are no volumetric data.

Algae may be dispersed from rocks, trees, soil and water by carriage on wind-blown soil particles, moss or fern spores, by rain splash of soil, by bursting bubbles or splashing from seas, rivers and sewage disposal plants, and by carriage in airborne foam.[317–318,363–364] At least 40 algae, including *Chlorococcum*, *Chlorella*, *Pleurococcus* and *Navicula*, were obtained from 20 m^3 air in the Netherlands[369] while *Gloeocapsa* or other members of the Chroococcaceae, occurred in concentrations up to 110 dispersal units, each composed of up to eight cells, m^{-3} air in southern England.[133] Smaller numbers were found in London,[141] and mean concentrations at two sites to the west of London were 7–11 m^{-3} air.[198] In Sweden, viable algae, mainly *Chlorella* and *Chlorococcum*, formed up to 47 dispersal units m^{-3} air.[363-364]

Fungi are much more numerous in the air than are pollen grains but their smaller size means that their volume contribution to the bioaerosol is often of the same order. Maximum concentrations are about 10^6 spores m^{-3} air but about 10^4–10^5 m^{-3} is more usual.[198] In most parts of the world, *Cladosporium* is usually the predominant type by day and *Sporobolomyces* by night.[195] Many species have characteristic diurnal periodicities governed by the interaction between spore liberation mechanisms and climatic conditions. Thus some species have spores that require water, as dew or rain, for their release, while others require drying conditions, or are released by mechanical disturbance, as when wind blows leaves, and some are liberated only by rainsplash. Seasonal periodicities are often governed by stages of crop growth or by climatic conditions. Fungi are ubiquitous in the environment but the major sources of the air spora undoubtedly are growing plants and decaying litter. *Cladosporium* and *Alternaria* are important colonizers of plant surfaces and fresh plant litter and overall are the most numerous airborne fungal spores through most of the world. However, they may be exceeded by other types in some regions at some seasons. Soil also contains many microbes, including fungi, but these can mostly only become airborne when winds are strong enough and soils dry enough to be blown. Soil particles, especially the organic fraction[83] 30–100 μm diameter, may be blown when wind speeds exceed ca 0.2 m s^{-1}. Wind blown soil, together with disturbance by vehicles and mechanical diggers, has been shown to be important in the epidemiology of coccidioidomycosis, a respiratory infection caused by the fungus *Coccidioides immitis*, which grows in desert soils in California,[51] and also in the spread of vesicular-arbuscular mycorrhizal fungi across a restored surface mine.[9,373]

Bacteria are less numerous in air than fungi but although sometimes they may be airborne as single cells, often they are carried on rafts of plant material, on skin squames or in droplet nuclei, protected by solutes or mucous. Bacteria tend to be more numerous in cities (up to 4000 m^{-3}, average 850) than in rural areas (up to 3400 m^{-3}, average 99).[34,176] Like fungi, they colonize aerial plant surfaces and are numerous in soils but, except for actinomycetes with spores easily dispersed mechanically, most have no specialized mechanisms for becoming airborne. Many bacteria in ambient air probably come from plants[206] whose leaves carry a diverse range of species,[29] some itinerants, some epiphytes or residents and some pathogens. They become airborne during rain and irrigation,[369] and during the mechanical handling and harvesting of crops.[76,280] Other sources include soil during cultivation, animal houses and farm wastes during disposal.[333] Most soil bacteria occur in the upper 5–45 cm and *Streptomyces*, especially, easily can become airborne in dust.[212] Many sources of airborne bacteria are man-made and perhaps the most important of these are sewage and animal waste disposal facilities. Disruption and digestion processes frequently produce bioaerosols containing bacteria, including fecal coliforms and fecal streptococci, and bacteriophages.[333] Such aerosols, together with those associated with farm animals, may contain endotoxins, lipopolysaccharide components of the cell walls of Gram-negative bacteria liberated by lysis.[265] On inhalation, these may cause febrile reactions. Large amounts of endotoxin (>50,000 EU/mg) are present also in pellets containing *Pseudomonas syringae*, an ice-nucleation-active bacterium used in the formation of artificial snow.[265] Bacteria, e.g., *Bacillus thuringiensis*, sometimes are aerosolized also as biological control agents against insect pests[152] and may drift downwind into non-target areas.[7]

Viruses and bacteriophage normally are released into the atmosphere from human and animal sources and from sewage, and only occasionally from plant material or through application as biocontrol agents. However, they are airborne rarely as individual particles. Their sampling and assay presented many problems before the availability of immunoassays and most evidence for their presence in bioaerosols comes from veterinary epidemiology, e.g., foot and mouth disease,[86,327,378] Newcastle disease[323] or through the exposure of susceptible animals.[47] However, rabies virus has been detected in caves inhabited by bats,[377] echoviruses, poliovirus and cocksackie virus downwind from effluent irrigation sites,[103,253,356] and Newcastle disease virus 64 m downwind from infected poultry houses.[165] Aerosols of *Heliothis* nuclear polyhedrosis virus, Douglas fir tussock moth virus, *Neodiprion sertifer* virus and gypsy moth virus, may be created artificially to control the host organisms[373] while plant viruses can become airborne when infected plants are damaged, as by high wind speeds (>6.9 m s^{-1}) in a tobacco crop infected with tobacco mosaic virus.[23] Viruses may be carried in fungal spores or nematodes dispersed with wind-blown soil.[266,362]

Airborne protozoa have been little studied for lack of convenient techniques. However, 0.1–0.2 protozoan cysts m^{-3} air were found in the air of Paris[251] and an average of 2.5 m^{-3} above a river bank near Heidelberg.[286] Their sources are mainly water and soil.

AMBIENT AIR

Usually ambient air is sampled to determine the incidence of airborne allergenic pollens, and fungus spores to aid diagnosis and treatment (see also Chapter 14). Many methods have been employed since Pasteur[274] first demonstrated the presence of microorganisms in the air and Blackley[30] and Cadham[49] showed, respectively, that airborne pollens and fungal spores can cause allergy. Although Pasteur and Miquel found deficiencies in sedimentation methods and abandoned their use,[130] and even Durham[89] found poor correlation between sedimentation and volumetric methods, the Durham sampler[90] was adopted by the American Academy of Allergy for pollen surveys. It has yielded much basic information on the seasonal occurrence of pollen and fungal allergens but its deficiencies became very apparent when automatic volumetric spore traps[153] became available. These revolutionized the accepted view of the air spora and showed that spore types poorly represented in sedimentation traps could be among the most abundant in the air. Subsequently, other samplers, employing different collection principles have been used (see also Chapters 9 and 10).

Sedimentation Samplers

Following Durham,[90] microscope slides (2.5 × 7.5 cm) coated with glycerine jelly, petroleum jelly or silicone grease are placed, adhesive side upwards, on a horizontal platform 2.5 cm above the lower of two circular discs 22.5 cm in diameter and 10 cm apart which act as rainshields. A similar device used in Britain, with discs 91.5 cm diameter and 28 cm apart, the lower supported 30.5 cm above ground or roof level, had the slide placed directly on the lower disc.[167] Slides are exposed usually for 24 h but sometimes have been changed every two hours to study the diurnal periodicities of grass and other pollens in the air.[167] After exposure, the deposit in a central 1.0 or 2.2 cm square is classified and counted microscopically to give the number deposited per square centimeter.

Despite their simplicity and the ease with which they may be used, the interpretation of results from horizontal slides is difficult because catches cannot be related to volume of air sampled. In still air, rate of deposition is proportional to terminal velocity (v_s) and, thus, to the square of particle radius. Consequently, large particles are deposited more rapidly than small. If *Lycoperdon* spores (v_s, 0.05 cm s^{-1}) were collected from a layer of air 0.5 mm thick, *Puccinia* urediniospores (v_s, 1.0 cm s^{-1}) would be collected from a column 10 mm thick and grass pollens (v_s, 3.1 cm s^{-1}) from a column 50 mm thick.[130] If both *Lycoperdon* spores and grass pollens were present in equal concentrations, 100 times more grass pollens would be collected than *Lycoperdon* spores. Correction factors, based on particle radius[63] or calculated by reference to differences between deposits on slides and in impactor samples,[90] have been suggested to enable results to be expressed volumetrically. Concentrations per cubic yard (0–765 m^3) obtained with an impactor were 3.6 times those per square centimeter of sedimentation slides. This factor was used by Bassett *et al.*[24] to give a 'volumetric count' of *Ambrosia artemisiifolia* pollen. However, Cocke[63] estimated the factor to be 7.3 from the calculated terminal velocity. Outdoor air seldom is still and even at low wind speeds, the upwind slide edge can produce a shadow with decreased deposition. At 1.1 m s^{-1}, no *Lycopodium* spores (v_s, 1.76 cm s^{-1}) were deposited close to this edge and overall deposition was decreased by 50% while at 1.7 m s^{-1} deposition was decreased almost to zero over the whole slide. At higher wind speeds deposition recovered but at 9.5 m s^{-1} as many spores were deposited by turbulence on both surfaces of slides held parallel to the wind, whether horizontal or vertical. Deposition on inclined slides, increased with wind speed along the leading edge but was affected less along the trailing edge until the presentation angle approached 90°.[135]

Open petri dishes containing agar medium can be substituted for horizontal slides in Durham or similar shelters or they can be exposed in the open to provide counts of viable microorganisms, including algae. However, settle plates are subject to similar limitations to horizontal slides. Plates are exposed usually for 10–30 min, but sometimes for only 15 s to 1 min or as long as 12 h. Open Petri dishes even have been held at arm's length, facing into the wind, from cars travelling at 72–100 km h^{-1} for 3–5 min and from airplanes for 1 min to trap algae and fern spores.[42–43] However, the turbulence created when the lid is taken off Petri dishes and replaced is responsible for much spore deposition. At all wind speeds, narrow edge drifts occur behind the rim of the leading edge and in front of the trailing edge.[136] Efficiency is low with wind speeds of 0.5 m s^{-1}, but high at 1.1 and 1.7 m s^{-1}, because of the large contribution from these edge drifts, and again low at 3.2 m s^{-1} because the agar surface is almost completely shadowed by the 1 cm high rim of the dish. Consequently, deposition velocity to Petri dishes apparently is less than particle terminal velocity.[56] Edge effects almost could be eliminated by placing the dish at the bottom of a cylinder, 13 cm deep and 11.5 cm diameter, mounted below a horizontal flat surface extending 11 cm up- and downwind.[136] The species that grow are determined by spore viability, agar medium and incubation temperature. Tauber[354–355] applied this principle in the design of a new deposition trap.

The Tauber trap, made from acrylic, consists of a cylindrical container 10 cm high × 10 cm diameter, closed at the top by a 15 cm diameter aerodynamically shaped collar with a 5 cm diameter central orifice. Particles entering the orifice are trapped in a thin layer of glycerol at the base of the container. A slit in the underside of the top-plate, or a hole closed with a rubber stopper, allows emptying of the sampler. The sampler either is exposed unprotected when wind blown and rain-borne particles will all be collected, or is protected by an aluminum rainshield, 45 cm diameter, held 12.5 cm above the opening of the trap. The trap may be left in place for a whole pollen season as it cannot be overloaded, unless the water-holding capacity of the contaniiner is exceeded. It is independent of wind direction but collection efficiency is again low and depends on particle terminal velocity and wind speed.[355] Efficiency may be increased by increasing the height: orifice diameter ratio.[275] Airborne algae may be detected by placing moistened sterile soil in a Petri dish in the base of a Tauber trap and exposing for standard periods.[39]

Inverted frisbees have been used as deposition samplers for particles larger than about 20 μm.[137] The inside is coated with liquid paraffin or a similar adhesive to improve retention and a collection bottle may be attached to a 80 cm spacer tube underneath to collect rainfall and any particles washed off the collecting surface. Alternatively, adhesive-coated tapes or slides could be attached to the inside.[380] The trap was designed to collect inorganic dusts but, perhaps, could be used to trap other large particles including pollens and fungus spores. In wind tunnel tests, mean collection efficiency for particles 50–183 μm diameter was 50%. It approached 100% for 50 μm particles at low wind speeds and was least for particles 80–100 μm diameter.

A pollen collector for individuals, attached to clothing near the breathing zone, was designed by Leuschner and Boehm[204] to determine personal exposure to airborne allergens over a 24 or 48 h period. This consisted of a plastic slide holder, with slits on all four sides to allow passage of air, and a round funnel, 8 mm high, mounted above the center of a slide coated with petroleum jelly. About 3.15 cm^2 of adhesive surface is exposed beneath the funnel and particles are deposited, probably mostly through turbulence as the patient moves about. Results obtained by microscopic examination of the deposit are purely qualitative and probably biased towards large particles. No extrapolation to concentration is possible.

Impaction on Static Rods

Both vertical and inclined slides have been used to determine the ambient air spora but these are relatively inefficient, especially at low wind speeds, and are extremely sensitive to wind direction.[130] Impaction on sticky cylinders is independent of wind direction, is more efficient than on flat surfaces and increases with increasing windspeed and decreasing cylinder diameter, except that large particles (e.g., *Lycopodium* spores, 32 μm) may be blown off small cylinders at high wind speeds and very small spores (e.g., *Calvatia gigantea*, 4 μm) may not be deposited under any conditions. However, no spores will be collected under still conditions regardless of their concentration. For practical use, a cylinder 0.53 cm diameter was adopted by Gregory.[128] This could be coated with a 1.6 cm square of transparent cellulose film dipped in glycerine jelly and mounted in a Durham-type rain shelter for exposure. Similar traps, exposed for 24 h, have been used in studies of the ambient air spora in India (Figure 16.1).[288,305]

Flag samplers are similar except that the tape is wrapped around a pin which is free to turn in a fused glass bearing. The trailing edges of the tape form a wind vane or flag.[146] Only the leading edge is coated with adhesive. Cylinder traps similarly may be mounted in front of a wind vane to decrease the trapping area that has to be scanned.[262]

After exposure, the cellophane squares are removed from the cylinders and mounted on microscope slides under a cover glass. Spores and pollens are counted microscopically along traverses 140–210 μm wide across the whole width of the square. Ramalingam[288] also isolated viable fungi and bacteria by coating the cellophane squares with a sugar-gelatine jelly. The spores then were suspended in 100 mL sterile water, diluted 1:5 or 1:10 and 2 mL was inoculated into agar media. Volumetric counts have been obtained from cylinder and flag samplers by multiplying the catch by the wind run during the period of exposure.[8,288]

Whirling Arm Impactors

The trapping efficiency of narrow vertical cylindrical or flat surfaces can be increased by whirling these rapidly around a circular path. Flat surfaces are easier to work with than

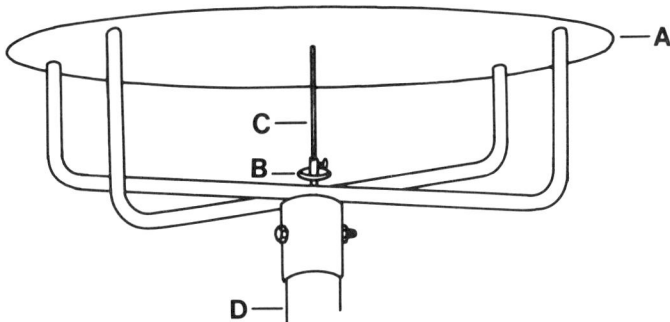

Figure 16.1. Vertical cylinder trap.[288] A, rain shield; B, cylinder support incorporating oil reservoir as a barrier to crawling insects; C, 4.8 mm glass rod bearing cellophane squares coated with petroleum jelly/paraffin wax or sucrose/gelatine gel; D, supporting pole.

are spherical ones and these have been incorporated into rotorod, rotobar and rotoslide samplers. Rotorod samplers are easily constructed using small battery operated electric motors fitted with vertical U-shaped collectors of 1.6 mm square brass wire with arms 6 cm long and 8 cm apart. Some commercial versions have a 1.3 mm wide lucite rod suspended from a horizontal support bar or a 0.38 mm wide H-shaped rod. The rod assemblies are rotated at 2000–3000 rpm, to give a linear speed of 10–15 m s^{-1}.[91,263,279] In rotoslide samples, the trapping surface is the long edge of a 7.5 × 2.5 cm microscope slide.[262] The samplers can be mounted on poles at any convenient height and their low cost enables large arrays to be set out to study vertical and horizontal dispersal gradients.[224–225,300,303–304]

Rotorod samplers are extremely adaptable, as illustrated by their portability when mounted in plastic lattice-work shopping bags.[134] Their use enabled the discovery of the source of *Pithomyces chartarum* spores in cut grass, for the first time in Britain, by following the concentration gradient. Rotorod samplers can be operated continuously for short periods, often up to 2–3 h, usually at the same time(s) each day, or intermittently, often for one minute in every ten, for up to 24 h and may collect both greater numbers and a wider range of pollen and fungal spore types than rotoslide samplers.[25] Overloading of the trapping surface may occur with large spore concentrations necessitating shorter sampling periods or intermittent sampling for ease of counting.[299] Time discrimination may be obtained by taking sequential samples.[301] Trapping surfaces may be protected when not sampling by the addition of swing shields that are drawn away by centrifugal force when the sampler is started.[297] Rotorod and rotoslide samplers are independent of wind direction although airflow around the rods in different versions is understood poorly. Trapping efficiency declines somewhat with increasing windspeed, with overloading and with decreasing particle size (Table 16.1).[91,231,263]

Rotorods, rotoslides and rotobars may be coated along their leading edges with a strip of cellulose tape trimmed, using a razor blade, to the same width as the rod followed by a thin layer of a petroleum jelly-paraffin wax mixture, silicone grease or glycerine jelly. A glucose-gelatine mixture was used to coat rotorods so that spores of *Histoplasma capsulatum*, (a pathogenic fungus commonly found in faecal accumulations beneath starling or grackle roosts or around bat colonies) were trapped and then detached by suspension in a suitable fluid that was injected into mice.[148] For microscopic examination, the exposed tape is removed from each arm and either mounted in one piece along a microscope slide under a 50 × 25 mm cover glass or cut into four pieces that are mounted side by side in gelvatol or glycerine jelly under two 26 × 26 mm cover glasses. Often the latter is easier for

Table 16.1. Trapping Efficiency of 0.154 cm Rotorods[91,231]

Species	Spore Size (μm)	Trapping Efficiency (%)	Reference
Bacterial endospores	0.5 × 1.0	0.5	231
Calvatia gigantea	4–5 diam.	19.3	231
Scleroderma lagerbergii ascospores	19.5 × 5 (aerodynamic diam.: 12)	30	91
Puccinia graminis urediniospores	14 × 24 (aerodynamic diam.: 20) 18.5 × 24 (aerodynamic diam.: 22)	89.5 95	231 91

counting. Lucite rods and rotoslides may be coated directly and, after exposure, mounted in an aqueous medium such as Calberla's solution (glycerol, 5 mL, 95% ethanol 10 mL, distilled water 15 mL) under a cover glass and then placed in a stage support before counting.[8] However, samples cannot be stored for later examination when many samples are collected over a short period. Culturing of fungal, moss and fern spores from rotorod catches also has been described either utilizing a modified stage adaptor to which liquid culture medium is added after inserting the exposed rod,[28] and of algae by streaking the rod across the surface of agar medium in Petri dishes.[43] The rod also was left on the surface of the agar after streaking so that any cells not removed could still grow.

The total volume sampled by the two arms of the sampler can be calculated using the formula:

$$V = 2\text{wld}\pi r \times 10^{-3}$$

where V = volume of air sampled in unit time (L min^{-1}), w = width of rod (cm), l = length of trapping surface (cm), d = diameter of circle swept (cm), r = speed of rotation (revolutions min^{-1}). An English version with vertical arms 0.16 × 6 cm, each 4 cm from the axis of rotation and turning at 2300 rpm, samples at approximately 110 L min^{-1}. For accurate determination of sample volumes though, frequent calibration of the rate of rotation using a stroboscopic lamp is necessary, especially if the voltage of the power supply is likely to fall during the sampling period.

A whirling arm trap for large aerosol particles (about 30–250 μm) has been constructed (Figure 16.2)[140] with an L-shaped sampling probe, about 2.8 cm internal diameter and 50 cm long, measured from the axis of rotation to the short elbow on the end, balanced by a similar dummy probe. The probe assembly rotates in a horizontal plane, the open end of the probe facing the direction of rotation and moving at a known tip speed that greatly exceeds usual ambient wind speeds and particle terminal velocities. Rotational speeds up to 1250 rpm and tip velocities up to 65 m s^{-1} were obtained with a 3/4 hp electric motor. Air is drawn in through the probe isokinetically at up to 1.3 m^3 min^{-1}. Particles are trapped by inertial deposition in the elbow near the probe entrance. This can be coated with vacuum grease dissolved in toluene to improve retention, and are recovered by redissolving the trapping surface and utilizing membrane filtration. A refinement that could be tested for biological use would be to cover the trapping surface with a coated cellulose or Melinex (ICI) film that afterwards could be mounted on microscope slides.

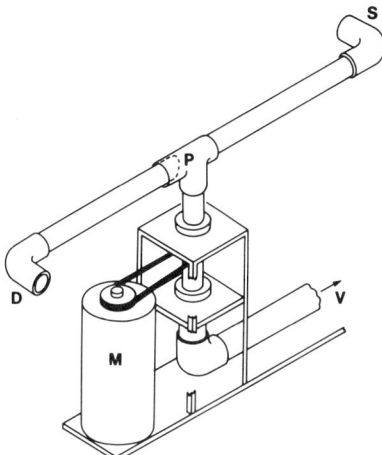

Figure 16.2. Simplified diagram of a rotating arm coarse particle sampler (after 140). Key: S, sample probe; D, dummy probe blocked with an aluminum plug at P; M, 3/4 hp motor giving 1250 rpm and tip velocity 65 m s^{-1}; V, to air mover and flow meter.

Impaction in Suction Traps

The cascade impactor[238] allows efficient sampling of airborne particles in four size fractions but can be overloaded easily. However, low pollen concentrations in the Shetland Islands in early spring enabled a cascade impactor, mounted on a windvane under a rainshield, to detect pollen grains blown from Scandinavia and Britain, over 250 km away.[368] Aspiration was at 17.5 L min^{-1} and slides, coated with petroleum jelly, were exposed for 24 h. Slides were counted at 350 times magnification either up to 100 grains of the predominant species, when grains were numerous, or along 20 traverses of the deposit, amounting to about half the area of the slide. The origins of the air masses carrying the pollen grains then were determined by plotting trajectories using synoptic weather maps.

The automatic volumetric spore trap,[153] based on the 14 × 2 mm second jet of a four stage cascade impactor,[238] revolutionized knowledge of the air spora, and revealed the importance of ascospores, basidiospores and the ballistospores of mirror yeasts (e.g., *Sporobolomyces*). For the first time, it allowed continuous monitoring of spore concentrations and a determination of the effects of time of day and weather on spore numbers and types.[154] The original version deposited particles onto a microscope slide that moved past the orifice at 2 mm h^{-1} but later versions (Burkard Manufacturing Company, Rickmansworth) used a transparent Melinex tape wound around the periphery of a rotating drum that moved past the orifice at the same rate. The inner end of the orifice is separated from the trapping surface by slightly less than 1 mm. Air is drawn through the sampler at 10 L min^{-1} and a wind vane rotates the sampler so that the inlet orifice faces into the wind. Similar samplers have been described with a 14 × 1 mm slit and an intermittently moving slide that produces 24 bands 2mm apart, each representing a 1-h sample.[192] An alternative incorporates a circular jet, 0.635 cm diameter, and a drum loaded with 24 slides that can be successively exposed for 15 min to 24 h when sampling at up to 26 L min^{-1}.[371] Another has a circular jet, 3 cm × 0.6 cm diameter, separated from the trapping surface by 0.5 cm and sampling at 5 L min^{-1} but with no wind vane to rotate the orifice to face into the wind.[365] Yet another has a vertically mounted orifice under a rainshield surrounded by a mesh screen to slow airflow and exclude large particles such as insects.[41] Smaller, more portable versions have been described by Gregory[129] and Ramalingam.[290] However, trapping efficiency has not been determined for most of these and because the Casella, Burkard and Lanzoni versions of the Hirst trap have been used far more widely and are available commercially, their use will be described in detail.

If contamination is to be avoided, a clean area is necessary for the preparation and mounting of the trapping surfaces. Slides for the 24 h sampler are coated directly along a 55 mm length with molten petroleum jelly containing 10% paraffin wax (melting point 54°C) using the end of a plastic plant label or glass microscope slide.[156] Preparation is aided by first coating the slide with 10% Solvar (a partially hydrolysed polyvinyl acetate preparation; Shawinigan Ltd.) and by gently heating the slide after preparation to remove stresses that cause splitting of the film. The drum of the 7-day model is mounted on a bracket supplied by the manufacturer and cleaned with hexane. Double-sided adhesive tape is placed across the circumference of the drum centered on a black line (B, Figure 16.3 a, b) on the side and one end of Melinex tape is attached to this, leaving half still exposed (Figure 16.3 c). The tape then is wound tightly around the drum, secured to the exposed adhesive tape and cut off flush with the first end using a razor blade or sharp scalpel, leaving no gap or overlap (Figure 16.3 d). The tape then can be cleaned lightly with hexane before coating. Often the coating adhesive has been applied with a flat 1.3 to 1.9 cm wide artist's acrylic brush or a plant label while rotating the drum but a small roller assembly (Figure 16.3 e) has proved extremely useful. The adhesive may be Dow Corning 280 Series silicone in xylene,[261] silicone grease,[60] Lubriseal stopcock grease (A.H. Thomas, Inc.),[46] petroleum jelly, usually containing 5–15% paraffin wax and sometimes also 8% phenol, dissolved in toluene or hexane.[62,227] It is essential that the tape is completely covered with a thin, uniform layer of adhesive.

The ends of a trace should be marked with a dissecting needle inserted through the inlet orifice or with a puff of *Lycopodium* spores before and after changing the drum to enable the beginning and end of the trace to be determined easily in the laboratory and any clock malfunctions identified. At the same time the flow rate should be checked using the flowmeter provided by the manufacturer, the clock wound and the intake orifice cleaned of any obstructions, such as insects or other large particles. In removing the drum, it is essential that the trapping surface is prevented from touching anything and that it is placed in the carrying box as rapidly as possible to prevent contamination. Drums should be labelled with the site, date and time of the start of the exposure and date and time of the end.

After exposure, glass slides can be warmed gently on a hot-plate to fix the spores. Slides then are inverted and gently lowered onto 50 × 26 mm cover glasses bearing sufficient mountant to cover the whole area. Once the cover glass has been picked up, the slides can be turned over and light pressure applied to spread the mountant. The slides should be left horizontal for at least 24 h while the mountant sets. Then they can be stored in any orientation. The mountant used originally was lacto-Solvar (20% aqueous Solvar, 50 parts; lactic acid, 25 parts; 6% aqueous phenol, 25 parts) but this has since been replaced by Gelvatol (Gelvatol, grade 40-20, 35 g; distilled water, 100 mL; glycerol, 50 mL; phenol, 2 g). Glycerine jelly (gelatin 10, g; distilled water, 60 mL; glycerol, 70 mL; phenol, 1.5 g) also may be used as a mountant. Some workers incorporate stains, e.g., basic fuchsin, trypan blue or aniline blue, into the mountant. This may be helpful where the primary interest is in pollens or hyaline fungal spores but for other spores, their natural color can be a useful aid to identification.

Melinex tapes are detached carefully from the drums of 7-day traps, ensuring that no adhesive tape is attached, and placed on the acrylic cutting bar provided by the manufacturer so that the start of the trace lines up with a time marker. The cutting bar is divided into

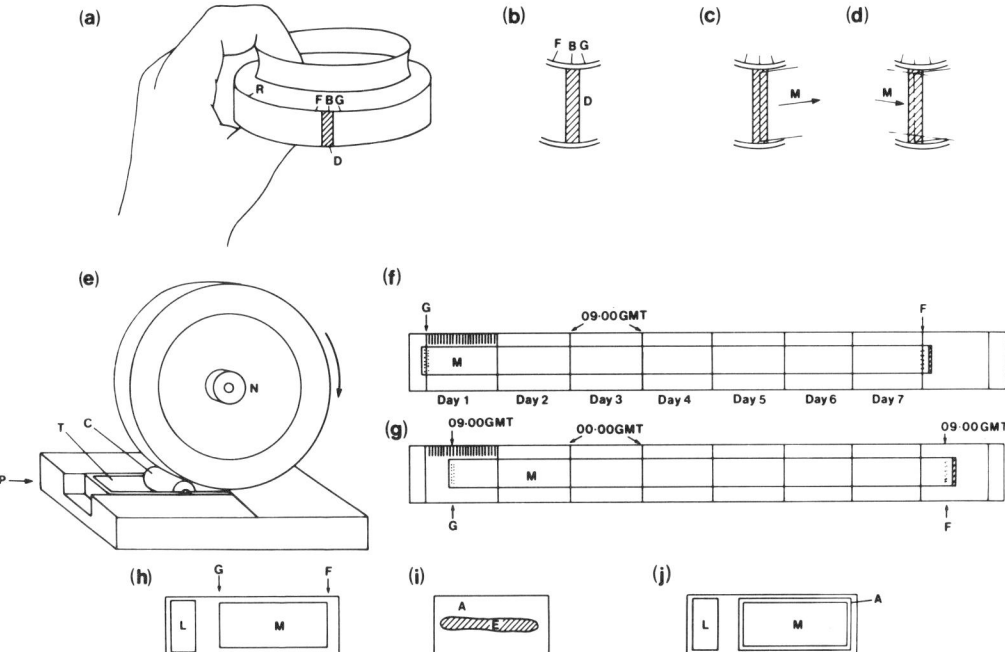

Figure 16.3. Preparation and mounting of automatic volumetric spore trap (Burkard) tapes (reproduced by permission of Mrs M.E. Lacey): (a) drum to be prepared; (b) double-sided sticky tape trimmed to width of depression; (c) Melinex tape attached to middle of tape; (d) Melinex tape attached to other half of tape after winding around drum and cut across line of contact with a sharp blade; (e) coating Melinex tape with petroleum jelly/paraffin wax in hexane by pushing trough against spring in holder until the cylinder makes contact with the rotating spore trap drum; (f) mounting block with exposed tape in place for cutting into 24 h segments beginning and ending at 0900 GMT each day; (g) mounting block with exposed tape in place for cutting into 24 h segments beginning and ending at midnight GMT each day; (h) 24 h segment of exposed tape on slide ready for mounting; (i) microscope cover glass bearing Gelvatol mountant onto which inverted slide is lowered; (j) prepared slide. Key: A, microscope cover glass; B, marker line on spore trap drum for alignment of double-sided tape; C, brass rotating cylinder; D, double-sided sticky tape; E, Gelvatol mountant; F, marker line on spore trap drum marking end of 7 day sampling period and corresponding position on exposed tapes; G, marker line on spore trap drum marking start of 7 day sampling period and corresponding position on exposed tapes; L, Label with serial number and start and finish dates and times; M, Melinex tape; N, nut securing drum to mounting bracket; P, trough pushed in until cylinder contacts drum; R, marker line on spore trap drum to align G with orifice; T, brass trough containing petroleum jelly/paraffin wax in hexane.

eight large segments, each 48 mm long, by lines across the full width and the first segment again is divided by short lines into 24 segments, each of 2 mm. For mounting, tapes have been divided in two ways. One is to line up the start of the tape with one of the main divisions on the cutter bar and then to cut it into 48 mm lengths, each starting at the time of day when the drum was changed (Figure 16.3 f). The other is to treat the main divisions as marking midnight each day and lining the start of the tape with the short marker corresponding to the time when the trap was changed (Figure 16.3 g). The tape then is cut into 48 mm lengths from midnight to midnight each day with shorter lengths at each end for the days when the drum was changed. The first method has the advantage that each segment can be timed to correspond with meteorological records while the second represents the full 24 h deposit for one day, avoiding any subsequent confusion over the time at which the trap was changed, but against that is the disadvantage that the deposit is split between two short segments from successive drums on days when the trap is changed. To measure contamination during handling, a useful precaution is to prepare blank slides with a 48 mm strip of Melinex tape coated at the same time as the tape for the trap. This slide is carried to the field in a slide box when the drum is changed, exposed during changing, then returned to the laboratory for mounting when the tape is mounted. For mounting, segments of tape are placed on a film of water, to exclude air bubbles, on separate suitably labelled microscope slides (Figure 16.3 h). Afterwards they are handled in the same way as described for slides (Figure 16.3 i, j).

Trapping efficiency depends on particle size, wind velocity and the quality of the trapping surface. Sampling at 10 L min^{-1} and using petroleum jelly to coat the slides, the trapping efficiency for *Lycopodium* spores in a wind tunnel was 62.4 to 93.8% with wind speeds from 1.5 to 9.3 m s^{-1}, with the minimum at a wind speed of about 5 m s^{-1}.[153] Both *Lycopodium* and *Ustilago perennans* spores were trapped with efficiencies >90% over the same range of wind speeds.[154] With wind speeds >2 m s^{-1}, *U. perennans* spores were caught more efficiently than *Lycopodium*. However, under field conditions with a mean wind speed of about 3 m s^{-1}, intake efficiency ranged from only 17.9% for 50 μm particles to 45.9% for 20-μm particles,[246] chiefly owing to the slow response time of the trap to changes in wind direction, the small size of the intake orifice and the low sampling rate. However, it is possible also that dry particles would stick less readily to the orifice than the aerosol used in these tests. Nevertheless, diurnal and seasonal changes in the air spora are sufficiently large for marked differences in spore numbers and types with time of day, weather and season to be apparent.

Impaction samplers also can be used to culture algae, fungi and bacteria from ambient outdoor air, often to complement results from samplers assessed microscopically. Both slit[33] and Andersen samplers[12] have been used. A Casella slit sampler, operating at 28 L min^{-1} was employed to assess the microflora at different locations around Ibadan, Nigeria[264] and a BIAP slit sampler, operating at 30 L min^{-1}, was deployed in Copenhagen, Denmark.[201] Slit samplers also were constructed by Brown[40] and Davies.[78] That constructed by Brown comprised a 62 cm diameter culture plate containing 400 mL of agar medium which was moved intermittently as it rotated past a 28.6 × 0.25 mm orifice to give a spiral deposit 2 m long. Every 30 min in a 24 h period, 20 L air was sampled and the plate moved so that each deposit was separated from the next by about 1 cm. Davies' sampler deposited spores through a 30 × 1 mm slit onto 10 cm Petri dishes in studies of the viability of airborne *Cladosporium* spores impacted at 20 L min^{-1}. A wind-oriented FOA slit sampler, with a wind vane and rainshield over a horizontal 50 × 0.2 mm inlet orifice, sampling at 20 L min^{-1} onto agar in a vertical 15 cm Petri dish has been used regularly to isolate bacteria in Sweden.[13–14,308]

Andersen samplers, operating at 28.3 L min^{-1}, have been employed to study particular groups of fungi[163-164,295] as well as in more general studies of the fungal air spora,[27,45-46,291] bacteria[14,34] and algae.[364] Different configurations for the sampler are available commercially with one, two or six (standard) stages while a seven-stage version with a modified hole-pattern to improve particle distribution on the upper stages was described by May.[240] The sampler is provided with an inlet placed centrally in a conical lid but deposition of particles on the first stage is more uniform without this cone.[240] The sampler can be deployed upright in still air but it should be oriented on its side facing into the wind in moving air. The broad multi-orifice suction area of stage one (jet velocity 1.08 m s^{-1}) can be considered a reasonable approach to isokinetic sampling.[240] The six stage sampler has been operated with a single Petri dish of agar medium placed below stage 6 in a wind oriented sampler with its entry cone,[331] but this gave inferior results to the full 6-stage standard Andersen sampler although results from a single-stage sampler, equivalent to stage 6 without the entry cone, correlated closely with those from the standard sampler.[177] Two-stage samplers also give smaller counts than six stage samplers.[123] Decreasing the number of stages loses size discrimination and increases the risk of overloading although the number of plates that must be prepared and counted is decreased. Andersen[12] claimed that plastic petri dishes collected 20% fewer particles than glass dishes but when many samples are to be collected, glass dishes become impractical. Poorer recovery could result if the depth of agar in the plates and its separation from the impaction jets above differs from that recommended. Optimal trapping efficiency with the size distribution between stages described by Andersen,[12] is obtained if the standard glass dishes supplied by the manufacturer are filled with 27 mL agar. The dimensions of plastic dishes differ from those of glass and Tiberg et al.[363] used 35 mL agar in a sampler modified for plastic dishes. When sampling at low temperatures (down to -20 to -25°C), the sampler may be mounted in an insulated box containing a heater unit to prevent the agar freezing.[14] Sampling period differs with the organisms of interest and with their abundance in the air. For algae, 2-h samples were taken during the afternoon.[362-363]

To estimate personal exposure to allergenic particles, two versions of suction impactors are commercially available. The Burkard personal sampler is a handheld cylindrical instrument with a rectangular bell-shaped orifice, 14 × 2 mm, in the top impacting onto a microscope slide inserted through the side and closed by a rotating ring seal. Suction (10 L min^{-1}) is provided by a small pump powered by a rechargeable battery that can operate for 8 h continuously. The slide is prepared and mounted in the same way as for a Hirst trap although only a small central area needs to be coated. The Sierra-Marple personal cascade impactor is an eight-stage sampler which impacts particles in different size fractions onto Mylar plastic discs at each stage. Adhesion to the discs can be improved by treating them with glycerol,[32] 1–5% agarose/50% sorbitol[71] or 2% agarose/15% glycerol.[283] Power is provided by a personal sampler pump operating at 1–4 L min^{-1} for up to 8 h. After exposure, particles are suspended in fluid (bacteriological peptone 0.1%, Tween 80 0.05%, inositol 2%, pH 7.0), diluted as necessary with quarter-strength Ringer's solution or 0.1% bacteriological peptone, before 0.1 mL aliquots are spread on plates of suitable agars. After incubating, the colonies that grow are classified and counted.

Virtual Impactors

Recently virtual impactors have been developed for bioaerosol sampling. The jet spore trap (Burkard Manufacturing Co. Ltd.) is essentially a virtual impactor sampling at 400 L min^{-1} and impacting particles onto a column of still air.[320] It was designed for the collection of *Erysiphe* spores in cereal crops, which then sedimented onto leaves in the base, but has been suggested also for the detection of airborne algae in Antarctica.[39,380] A two-stage virtual impactor has now been designed for the study of pollen allergens in different size fractions (Figure 16.4).[182,292] This consists of an inlet of the type designed at the University of Minnesota,[211] a preimpactor to remove large particles, such as insects and water droplets, two virtual impactors in series to collect coarse and fine particles which are deposited on filters, a backup filter, a condenser for vapor collection and a flow control unit. The minor flow of both virtual impactor stages is 10% of the total flow through that stage and particles are collected onto 47 mm diameter filters. The inlet flow rate is 18.5 L min^{-1} and the cutoff diameters for the preimpactor, first and second virtual impactor stages, are about 40, 7.2 and 2.4 μm, respectively. More *Betula* pollens but fewer *Pinus* pollens and *Cladosporium* spores were caught with the new sampler than with a Burkard trap.

Impingers

Liquid impingers have been used to collect infective dispersal units of *Coccidioides immitis* in desert areas where it is endemic. However, the spores are dispersed easily and should never be handled in Petri dishes in the open laboratory because of the risks of escape and serious infection. Samples from liquid impingers should be handled only in isolation cabinets and the catch, after dilution if necessary, cultured on plates or in roll tubes that can be adequately sealed with rubber caps.[80]

Because liquid impingers sampling at 28–57 L min^{-1} were unsuccessful, low velocity impingers ("air washers") have been used to sample air for algae and protozoa.[315,316] Air was drawn, at 14 L min^{-1}, through 4 mm diameter tubes into 50 mL soil water extract medium in assay bottles wrapped with heating tape and aluminum foil to prevent freezing when air temperatures were below 0°C. The samplers were operated for 2–8 h 5–150 m above ground level. However, the assays were not quantitative although seven species of algae and protozoa, as well as bacteria, moss, fern, and fungal spores, were collected 5 or 9 m above the ground in one study and 33 species of algae and seven species of protozoa at a height of ca. 50 m in another.[315,316] Evidence of distant dispersal was obtained by correlating counts with wind speeds and direction. Simultaneous sampling of airborne algae and water was achieved by mounting a small impinger on the mast of a radio-controlled boat that also carried a controllable plankton dip net.[317,318]

If multi-stage liquid impingers need to be used out of doors where there is air movement, a hemispherical baffle plate must be fitted behind the orifice to provide stagnation point sampling.[242]

Cyclone and High-Volume Samplers

Cyclones have rarely been used to determine the ambient air spora although they have been utilized for the mass collection of pollens and fungus spores for allergen preparation and for plant disease studies.[61,325,358,359] Cyclones differ greatly in their sizes, sampling rates and collection characteristics. Sample rates range from 1–850 L min^{-1} and collection is efficient for particles larger than 2–3 μm. Collection of microorganisms may be aided by the introduction of a water spray (1 mL min^{-1}) containing a wetting agent and perhaps a protectant such as inositol through the inlet.[69,95,353] Tate *et al.*[353] sterilized their cyclones with 70 or 90% ethyl alcohol before use and maintained spore viability after collection by placing water in the collection vial. A venturi scrubber sampling 0.45 m^3 air min^{-1} was used to collect spores of *Histoplasma capsulatum* although a 4-h sampling period was necessary for infectivity to be demonstrated in mice.[170]

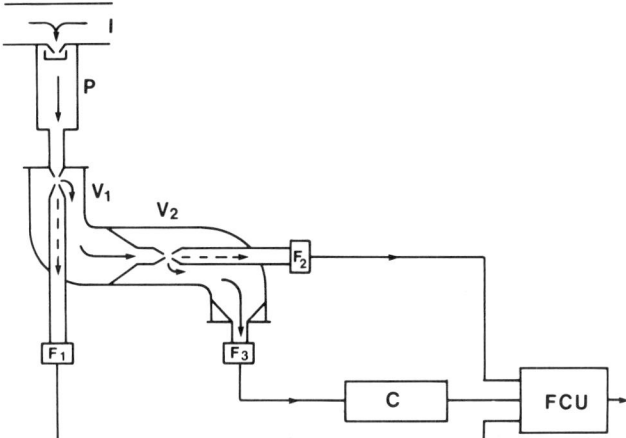

Figure 16.4. Simplified diagram of a two-stage virtual impactor with preimpactor and condenser unit (after 182, 292). Key: I, University of Minnesota inlet; P, preimpactor with 40 μm cutoff; V_1, V_2, virtual impactor stages with, respectively, 7.2 and 2.4 μm cutoffs; F_1, F_2, F_3, membrane filters; C, condenser unit; FCU, flow control unit;⟶, major air flow; ----→, minor air flows (10% of major air flow).

A high volume sampler for ambient bioaerosols incorporating cyclone collectors was designed by Fontanges and Isoard.[113] Air was sampled at 20 m^3 min^{-1}, screened to preclude debris, such as insects and leaves, and then passed through a cyclone that removed particles >10 μm. Smaller particles then were mixed with a fog of sterile distilled water, passed through a venturi jet causing instantaneous condensation of water onto particles which, after deceleration, were collected by another cyclone with efficient retention down to 1 μm. For sampling in arctic environments, samplers were fitted with 5–10 kW heaters.[14] The collected material was cultured to detect bacteria, actinomycetes and fungi.[113,172]

There are few studies of sampling for viruses in ambient air[313] but large-volume electrostatic air samplers (Litton-type) were superior to AGI-30 liquid impingers for detecting airborne rabies virus in caves housing bat colonies.[377] The large-volume sampler could concentrate the catch in small volumes of liquid, giving 100 times greater concentrations of particles than liquid impingers.

Filtration Samplers

Filtration has been little used to study the ambient air spora but membrane filters, sampling at 7 L min^{-1}, have been used to isolate algae and protozoa.[315] After exposure, filters were divided into two halves, one to be cultured in sterile water and one examined microscopically. A personal filter sampler with no power requirement was made from a respirator fitted with a paper filter trap.[77] This unit sampled air at a rate determined by the activity of the subject and tests with *Bacillus subtilis* spores gave retentions comparable with liquid impinger and high-volume samplers. Filtration has been also used to sample allergens in the ambient bioaerosol. Filters, 25 × 20 cm, have been mounted in the holder of a high volume sampler operated at about 600 l min^{-1} and exposed for up to 12 h. Allergenic material then was extracted with aqueous ammonium bicarbonate solution, dialyzed and freeze-dried before being reconstituted and assayed immunochemically using a radio-allergosorbent test (RAST).[4,196]

Frames, 20 × 20 cm, holding filters made of five layers of hydrophilic gauze impregnated with silicone oil and held perpendicular to the wind on a windvane have been employed to trap pollens and spores.[68,249] Similar filters were mounted vertically over a receptacle 1 m above the soil surface to measure simultaneously vertical deposition of pollens. Filters then were dissolved chemically and catches identified microscopically. However, the proportion of the airflow passing through these filters has never been determined and trapping efficiency for different sized particles at different wind speeds has never been tested. Much of the deposition could be due to turbulence. Similar wind-net samplers were used to trap marine algae >64 km from the nearest source.[247]

Precipitation Sampling

Rain falling through the atmosphere rapidly ends dispersal (see also Chapter 4). Microorganisms may impact on raindrops, be captured by cloud droplets or even form their nuclei. Efficiency of impaction is a function of raindrop radius and the terminal velocities of both microorganism and raindrop. Raindrops may be up to 5 mm diameter and have terminal velocities from 2–9 m s^{-1}. The optimum size for particle collection is about 2 mm[200,347] and is nearly the same for all particle sizes although collection efficiency declines as particle size decreases. Thus, 15% of *Lycoperdon* spores would be collected by 2 mm raindrops and none by 1 mm while 80% of *Puccinia* and *Erysiphe* spores would be collected by 2.8 mm drops.[130] By collecting rain through a 24-cm diameter funnel into a conical flask containing formalin to inhibit germination, Shenoi and Ramalingam[322] in India were able to identify 25 types of fungi, six algae and three angiosperm pollens with many more small '*Aspergillus*' type spores than was indicated by a Hirst trap. On average, 4700 fungal spores mL^{-1} rainwater were counted microscopically. Similarly Rosas et al.[310] isolated 600 to 6000 colonies mL^{-1} from rainwater in Mexico with seasonal changes in the occurrence of different taxa, although *Cladosporium* usually predominated. However, rain does not fall every day and in some regions may be limited to particular seasons. On average, rain falls in England for one tenth of each day while in India the weather is generally dry from November to April. Impaction on water droplets might be employed in a trap with no energy requirement, except that required to carry water to the top of a tower.[131] Drops would fall through a 2–3 m column of air into a collecting vessel and time discrimination could be obtained by using a mechanically operated laboratory fraction collector. However, the method is still untested.

Indirect but convincing evidence of high altitude transport of bioaerosols comes from studies of red snow in Sweden.[34-35] *Bacillus* species isolated from this snow and from Andersen sampler plates differed from resident types. Analysis of the trajectories of high altitude winds, elemental analysis of clay in the dust and fungal and pollen composition suggested that the air mass had originated north of the Black Sea, 1800 km away, where a sandstorm 36 h earlier had raised dust to high altitudes.

Biological Air Samplers

The term *biological air samplers* was used by Davies[80] to refer to the use of living plants and animals to detect pathogenic bioaerosols. Plant baits chiefly are employed for the study of fungi and bacteria pathogenic to plants (see Under Plant Canopies and in Crops) and animals. For instance, monkeys and dogs have been exposed in compounds in the Arizona desert to detect airborne spores of *Coccidioides immitis*.[65] Infection was detected in 15% of monkeys and 58% of dogs, although no colonies could be isolated in sedimentation plates. Similarly, animals were used to detect dermatophytic fungi released from dust in a cave.[214] The animals were killed and their lungs homogenized and either cultured or injected intraperitoneally into mice. After 3–6 weeks, *Trichophyton mentagrophytes* and *Microsporum gypseum* could be isolated from liver and spleen tissue although none were cultured from the lungs of exposed animals. Guinea pigs have been exposed in chambers through which air was passed from a hospital tuberculosis ward. Over 2 years, 156 guinea pigs were exposed and 45.5% developed tuberculosis, mostly in the chamber in which the air had not been pre-treated with ultraviolet irradiation. Since a guinea pig inhales 9.4 L h^{-1}, it was calculated that there were 0.003 infective particles m^{-3} in the first year and 0.009 m^{-3} in the second.

Choice of Sampler

Understanding the effects of environmental conditions on both the bioaerosol and the sampler are important in designing sampling strategies. Natural bioaerosols and environments often are complex and no one sampler or single-step sampling procedure can realistically be used to describe all the organisms in a given volume of air. Bioaerosols are usually mixed with gaseous pollutants and non-viable particulates, with radii ranging over 8 orders of magnitude and concentrations over 30[173] and include soot, dust and soil,[350] which may limit the value of some sampling systems.[98]

Outdoor bioaerosols are carried in air masses too large to be examined in their entirety so that sampling is necessary and the results extrapolated to the whole. Errors in sampling can thus lead to erroneous conclusions.[375] To collect a sample from which valid generalizations can be made can be difficult. All samples are representative but only of the specific conditions under which they were acquired. The better the sample and the environment are characterized, the better the generalization or extrapolation to other similar situations. A fundamental problem of representative sampling is that conditions for the generation, transport and deposition of epidemiologically important bioaerosols may occur only sporadically. In such situations, the volume of air required for a representative sample remains problematical. For pathogenic microorganisms, this may need to be related to the minimum infectious dose. Even though a single infectious particle might initiate an infection when placed in a favorable environment, in practice, larger doses are needed because not all cells are infective. When the number of target organisms is small, because the source is weak or distant and the spore cloud widely dispersed, or when trapping efficiency is small, detection in small samples is unlikely and negative results should be interpreted with caution.[99]

The choice of sampler for a given purpose is dependent on the organism of interest, whether it can be cultured or recognized microscopically, the precision of identification needed, the availability of selective culture media, the size of the airborne dispersion unit, their concentration, whether quantitative or qualitative information and size or time discrimination are required, the availability of a source of power, the need for portability and the resources, facilities and levels of skill available. Wherever possible, inertial traps, either suction or whirling arm, should be used because of their greater efficiency. Different spore traps yield different types of information and none gives a complete picture of the air spora. Microscopic assessment is least selective, allowing all particles to be counted and classified, but identification to species is seldom possible and often genera cannot be distinguished. Illustrations, such as those in Gregory,[130] Ogden *et al.*,[263] Smith,[324] Nilsson[258,259] and other spore and pollen atlases and standard mycological and palynological texts may help but these need to be supplemented by reference materials from local sources. Isolation in culture gives greater precision of identification of algae, fungi and bacteria but numerical estimates seldom coincide with microscopic counts,[25,45] because of low spore viability and the inability of organisms to grow on the chosen medium under the incubation conditions used. With fungi, the spore form obtained in culture may not be the same as that present in the bioaerosol and which is recognized microscopically. Thus, colonies of *Phoma* or *Fusarium* may have grown from ascospores of *Leptosphaeria, Nectria* or other genera.[45,122]

Lacey[195] compared published results for fungal spores from sedimentation plates and slides and microscopically assessed inertial traps from about 200 studies. Microscopic studies, especially those utilizing inertial traps, recorded many more ascomycotina, basidiomycotina, rusts, smuts and *Erysiphe* spores than those utilizing culturing, partly because samples for culturing were taken only during the day and partly because some of these spore types will either not grow in culture or produce only sterile mycelia or alternative spore forms. By contrast, culture methods allowed specific identification of yeasts, which probably account for many of the ballistospores found on microscope slides, and of *Aspergillus, Penicillium* and other genera with small spherical spores. Overall, more categories of spores have been recognized on inertial spore trap slides than on sedimentation slides.

Different sampling methods are often complementary and there is much value in combining at least two in an investigation, one allowing microscopic assessment of the total bioaerosol and one isolation in culture to identify the predominant viable types.

Siting

Sampler location is at least as important as sampler efficiency[77] and several factors should be considered when selecting a site for air sampling.[99,263]

1. The site should be representative of the area and not be dominated by large spore sources atypical of the area.
2. Contamination by non-biological particles should be avoided, where possible.
3. The site needs to be accessible, especially where samples need to be processed rapidly.
4. It should be free from public liability through injury to careless or curious people, especially children.
5. The site should be free from vandalism and theft.
6. Sampling equipment should be acceptable in the area and not trouble neighbors by its noise and appearance.
7. A source of electric power is often necessary.

Not all these criteria are easily satisfied, especially near towns where pollution from factories and vehicles is heavy. Once selected, the site (location, accessibility, fetch, etc.), sampler numbers and positions, times and duration of sampling and experimental design should all be well described.[99] Meteorological measurements should be collected close to the samplers and at relevant heights.

Height of sampling is important but is largely determined by the information required, e.g., whether bioaerosol generation is from a particular source, general ambient exposure or personal exposure. The concentrations of allergenic particles at breathing height, about 1.75 m,[233] may interest allergists but often samples have been taken from the tops of buildings 10–30 m high so as to be above the level of nearby tree-tops. Sampling 25 m above ground level should minimize the problems but a rooftop 10–15 m high may be more practical. Usually concentrations decrease with altitude but in cities, buildings may interfere with air circulation leading to anomalies. For instance, spore concentrations have been found to be positively correlated with height on the windward sides of buildings but negatively on the leeward sides, due to the downward movement of high altitude air on the windward side and the upward movement of spore rich air from the ground on the leeward.[178] For algae, Tiberg et al.,[364] sampled 4 m above ground level, 100 m from a river and, subsequently at another site at 14 m above ground level. Close to the ground, catches may be dominated by nearby sources while those from a greater height are representative of a greater area. Particle concentration usually decreases with height and as wind speed increases but comparison of catches at different heights have sometimes given conflicting results. Concentrations of ragweed pollen in central Long Island, New York, averaged over a season, did not change appreciably between 1.5 and 108 m above ground level although over shorter periods counts close to the ground were 50–150% of those at 108 m.[304] However, in Turku, Finland, counts were mostly greater at ground level than at 15 m above although counts at the two heights were significantly correlated.[293] The ratios between counts at the two levels ranged from 1.0 to 11.5, with the biggest differences with herbaceous and grass pollens and *Botrytis* and Ustilaginales spores and the smallest with tree pollens and basidiomycete spores. *Artemisia* and grass were detected 1–2 weeks earlier at ground level than on the rooftop. Concentrations of a range of fungal spores also decreased with height above the crop canopy (1 m) to 14 m.[96] However, in other studies, differences between 9 and 30 m were never significant and when no spores were dispersed from local sources, concentrations at all levels were similar.[215]

Windbreaks and other obstructions decrease measurable wind speeds up to 50 times the height (h) of the windbreak downwind; by about 20% to 25 h, and create "quiet" zones in which particles may be concentrated to about 8 h.[150,300] Samplers in open fields should be positioned above the plant canopy at a height not exceeding 1% of the uniform upwind fetch.[361] In cities, buildings, trees and other obstructions affect wind patterns but no parameters have been identified to allow adequate prediction of windspeed. Tops of buildings should be free of obstructions that interfere with airflow past samplers which should be sited away from the edge of the roof to avoid spore rich air carried up from ground level. Typically, the orifice of the Burkard automatic volumetric spore trap is 0.5 m above the ground unless it is mounted on a stand or building. The location of arrays of samplers downwind from a source, and vertically, will enable horizontal and vertical gradients to be determined.

Collection Media

The agar media used in sampler or isolation plates and their incubation temperature determines the species able to grow and, with fungi, which will produce spores. Objectives often differ between different studies depending on whether the total microflora or specific microorganisms are chiefly of interest. Non-selective media, such as blood, plate count or nutrient agars, sometimes at half-strength, have been used widely for total bacteria but selective media, developed for many different groups of bacteria of both clinical and phytopathological importance,[191,260,312,314] have been used in air samplers. By contrast, few selective media have been developed for specific fungi and comparisons of different media for isolating airborne fungi have sometimes reached different conclusions.[46,254,309] Burge et al.[46] found Sabouraud's dextrose, 3.2% malt extract-0.1% yeast extract, V-8 and potato dextrose agars better than modified Mehrlich's, casein hydrolysate, potato rose bengal and rose bengal streptomycin agars for isolating airborne fungi. Morring et al.[254] found no significant differences between inhibitory mould, rose bengal streptomycin and Sabouraud dextrose agar and, like Rogerson,[309] selected rose bengal streptomycin for use. However, such rich media as Sabouraud's, meat peptone and blood agars[168] may give poorer results than weaker media, such as potato-dextrose agar, potato carrot agar, V-8 agar or 2% malt extract agar for fungi and R2A, WSA-NS or one tenth strength casein peptone starch agar for bacteria.[169,379] Other work suggests that some bacteria may be recovered more efficiently using rich media.[260] Fast growing fungi may be suppressed by adding Triton N-101 to the medium[230] or by using dichloran-rose bengal-chloramphenicol agar[189] while xerophilic fungi are favored by DG18 (dichloran-glycerol agar).[159] Coconut agar has been used to isolate *Aspergillus flavus*.[161]

Recoveries of airborne bacteria (*Pseudomonas syringae*) have been enhanced by >63%, except at high (80–90%) humidities, by spreading 1000 units of filter sterilized catalase on the surface of agar plates (Luria-Bertani (LB) agar) for slit or Andersen samplers. Even larger increases (>95%) were found where the same amount of catalase was added to the plates after sampling but before incubation.[236] The osmoprotectant, betaine, has also been used, at 2–5 mM in 10 mM phosphate buffer (pH 7.0), in collection fluids for all glass impingers used to isolate airborne bacteria, in MacConkey and tryptic soy agars for plating impinger samples and in agar for slit samplers.[237] At these concentrations, recovery of bacteria was increased by 21.6–61.3% and the effects of betaine in both collection fluids and enumeration media were additive.

Cells may be protected after capture in impingers and large-volume air samplers, by using collection fluids such as phosphate-buffered saline or 1 M L^{-1} sucrose, glycerol or inositol to minimize osmotic shock and loss of viability.[69,73] Antifoams, e.g., Dow Corning Antifoam B,[332] antifreeze, e.g., 2% methyl cellulose[3] or evaporation inhibitors also may be required. Viruses may require richer collecting fluids including 10% skim milk and olive oil, than bacteria and then are assayed by inoculating the fluid into cell cultures or animals or by hemagglutination or immunoassay.[260]

To isolate algae, steamed soil-water extract,[315,316] Bold's basal medium[42] and the medium described by Staub,[348] modified by adding 0.41 mM TRIS-buffer and 1.5% agar, have been used with 35 mL per plate for the Andersen sampler. After exposure, plates were incubated in continuous light (190 μE m^{-2} s^{-1}; Philips TL 55 and 32; 55 W) at 25°C for 3–4 weeks.[364]

Unwanted organisms can be suppressed by adding antibiotics to media: penicillin (20 units mL^{-1}) and streptomycin (40 units/mL) or chloramphenicol (100 μg mL^{-1}) to suppress bacteria and cycloheximide (50–100 μg mL^{-1}) and/or nystatin (50 μg mL^{-1}) against fungi. Novobiocin (25 μg mL^{-1}) has been used to isolate selectively *Thermoactinomyces* and rifampicin (5 μg mL^{-1}) for *Thermomonospora* spp.[16,74,197]

Drying of the agar surface during long samples can increase concentrations of inhibitors in the medium[260] and decrease sampling efficiency. A monolayer of 0.2% oxyethylene docosanol (OED; $C_{18-22}H_{37-45}OC_2H_4OH$), formed by spreading 0.1 mL over the agar surface[14,243] can limit drying and greatly improve recovery of bacteria from ambient air on the first stage of Andersen samplers.

Incubation temperature will depend on the organisms of interest but 25 or 30°C often is employed for organisms from ambient aerosols. Higher temperatures may be utilized for particular groups of organisms, such as 37°C for potential pathogenics, coliform bacteria and some fungi, e.g., *Aspergillus fumigatus* and even 55°C for thermophilic bacteria and actinomycetes.

Different trapping surfaces have been applied for rotorod, rotoslide and similar samplers and for volumetric spore traps. Gregory[128] found glycerine jelly (gelatine 1 g, glycerine 7 g, water 6 mL, phenol 1%) slightly superior to petroleum jelly and much superior to other trapping surfaces tested in wind tunnel tests. However, petroleum jelly containing 10% paraffin wax, applied molten or dissolved in hexane has been widely used for volumetric spore traps although Käpylä[179] employed a mixture of paraffin wax, petroleum jelly, silicone oil and petroleum ether or toluene (1:2:2:10) for sampling the Finnish tundra. Two silicone greases, white petrolatum and Lubriseal grease differed little in their retention capabilities but silicone grease was most satisfactory overall and Lubriseal grease least.[330] However, performance of silicone grease was decreased, or possibly that of petroleum jelly was enhanced, when rain fell during sampling. Rubber cement and silicone oil provide unsatisfactory trapping surfaces.

Handling the Catch

Usually, spores and pollen grains are counted using lenses giving 40 to 100 times magnification and an eyepiece graticule marked with 1 mm squares to define the area for counting. The width of the area counted, defined by the number of graticule squares included, depends on the magnification and the size and abundance of the pollens or spores of interest and is usually in the range 40–200 μm. Use of a graticule is preferable to counting whole eyepiece fields because the area counted and, the spores to be included, can be determined more precisely and the process is less tiring. Although the decision to include or exclude spores in the count can be made on the basis of whether or not they are more than half within the defined area, it is easier to include all spores that touch one margin (top or right for long and short traverses, respectively) and to ignore any that touch the other (bottom or left), regardless of whether they are mostly in or out of the counting area (Figure 16.5).

For large particles, e.g., pollen grains, counts can be made of the whole trapping surface of rotorod or rotoslide preparations but for fungal spores it is desirable to count traverses across the width of the trapping surface, again defined by a squared eyepiece graticule. With traces from both arms of the sampler cut into four parts and mounted side by side, it is convenient to count eight replicate traverses across the eight sections. The total deposit can then be calculated as:

$$N = n \cdot \frac{2l}{8t}$$

where N = total number of spores deposited, n = number of spores counted, l = length of arm (cm), t = width of trace counted (cm)

Collections on slides from suction traps, e.g., the Hirst and Burkard traps, may be counted in two ways. Counting one or more traverses along the length of the slide allows daily mean spore concentrations to be calculated while counting across the width of the trace gives mean hourly concentrations. Because the trapping surface moves at 2 mm h^{-1} past an orifice 2 mm wide, each point on the tape is exposed for 1 h and the deposit represents a running hourly mean spore concentration. Time discrimination below 1 h thus is not possible. Usually, hourly averages are determined from counts at 2 or 4 mm intervals along the trace, representing 1 or 2 h periods. For instance Larsson et al.[202] counted pollen grains along 12 transverse 200 μm traverses at 450 times magnification, representing 5% of the slide area. However, it is usually impractical to count fungal spores on such a large area, especially when concentrations are high, and traverses of 60 μm or less may be necessary. Exceptionally, Pennycook[278] determined hourly average concentrations by

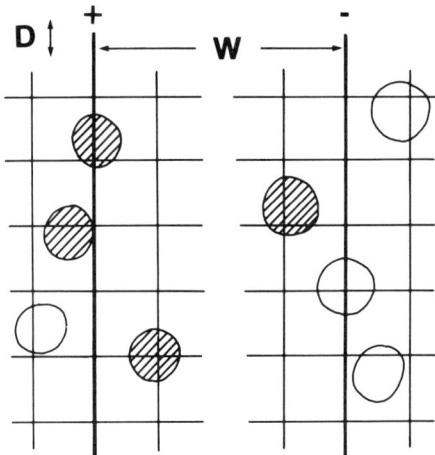

Figure 16.5. Counting criteria for spore trap slides; all particles touching + edge of traverse counted but no particles touching – edge counted (reproduced by permission of Mrs M.E. Lacey). Key: D, direction of counting; W, width of traverse; shaded particles counted, unshaded particles not counted.

counting spores in three fields only along a transverse traverse every hour of sampling while Rubulius[311] counted spores in four squares along a transverse traverse for every 2 h (at 500–800 times magnification). Spore deposition is not uniform along the width or length of the trace but often there are two zones of dense deposit along the long axis, the positions of which change with suction rate and external wind speed.[62,153] Consequently, the mean of 12 or 24 hourly average counts is often greater than the daily average determined directly. Particle deposits, sometimes amounting to 21% of the total[180] may extend beyond the 14 mm width of the trace so that the full width of the tape must be counted. The daily average spore concentration can be calculated using the following equation:

$$N = N_c \times \frac{L \times 10^3}{f} \times \frac{10^3}{24 \times 60 \times F} = \frac{N_c L}{f F} \times 694 = \frac{N_c}{f} \times 972$$

for a standard trap, where N = spore concentration (no. m^{-3}), N_c = mean number of spores counted in one long traverse, L = length of trap orifice (14 mm), f = width of traverse (μm), F = flow rate of sampler (normally 10 l min^{-1}).

To determine the hourly average concentration from counts across short traverses the equation becomes:

$$N = N_c \times \frac{W \times 10^3}{f} \times \frac{10^3}{60 \times F} = \frac{N_c W}{f F} \times 33300 = \frac{N_c}{f} \times 3330$$

for a standard trap, where W = width of trap orifice (2 mm).

A random sampling method was recommended by Mäkinen[232 in 144] to estimate mean diurnal concentrations of pollens while others have counted from the whole trapping area. The three sampling methods (random fields, transverse traverses, longitudinal traverses) have been compared statistically to determine the variance of counts and the sample size necessary to give valid estimates of concentration.[180] Systematic counts from short traverses account for variability in spore deposition across the slide but the greater variability along the length of the trace decreases the efficiency of this method although it is still much superior to random sampling. Sample size depends linearly on the relative variance of the population σ^2/μ^2, where μ is the mean number of particles per sampling unit and σ^2 is the variance of μ, while error depends on the size of the sampling unit (width of trace counted) and on variation of particle numbers along the length of the slide. The coefficient of variation for *Artemisia* pollen (with a narrow peak) fell from about 0.3 to nearly 0.05 when the number of traverses counted was increased from six to twelve but there was little further decrease when 24 traces were counted. *Betula* pollen (with a broader peak) showed less variation, but the coefficient of variation continued to decline to 24 traces counted. Nevertheless, it was concluded that counts from twelve two-hourly traces gave a reliable estimate of daily mean concentration. Counts along longitudinal traverses aggregate large variations with time but may give systematic errors because of irregularities in the distribution of the particles across the slide. Random sampling from the slide measures variability in both directions and therefore requires as many as 100 samples to achieve a reasonable sampling efficiency.

Random sampling requires counting of only 3.6% of the total trace area with a 0.65 mm diameter microscope field, while 16.2% of the area is counted in 12 transverse traverses and 3.4% of the area in one longitudinal traverse of the same width. However, much less time is taken to locate transverse or longitudinal traverses than to select random fields. Often when counting fungal spores, traverses narrower than 0.65 mm must be counted and the sample size may need re-evaluation to give valid results.

Daily maxima and periods of high spore concentration may be as important as daily means in interpreting the significance of spore trap data but their determination presents further problems. If particles enter the orifice with an intensity $\lambda(t)$, which follows a non-stationary Poisson distribution, they are randomly distributed in short air samples. However, when a traverse passes through position t, the most precise information that can be obtained from the collection is the mean concentration for 1 h because t remains under the orifice for 1 h. The observed process has an intensity function:

$$\Lambda(t) = \int_{t-\frac{1}{2}}^{t+\frac{1}{2}} \lambda(s) ds$$

where $\lambda(s)$ is the intensity function for a sample, d is the sample size (trace or field width) and s is the standard deviation of the sample. If λ is locally linear near t, $\Lambda(t) \approx \lambda(t)$. The number of particles in a traverse at time t (d_t mm wide) is distributed according to a Poisson distribution with an intensity $\int \Lambda(s)ds$, where the integral is over the time period of the sample. This integral can be estimated by the number of particles counted (y), its standard deviation by $\sqrt{y}$ and its coefficient of variation by $1/\sqrt{y}$. To limit the coefficient of variation to about 10% for a reliable estimate of the diurnal variation, the number of particles counted during the period of maximum concentration should be at least 50 and preferably nearer 100.

A chemiluminescent technique utilizing 5-amino-1,2,3,4-tetrahydrophthalazine-1,4-dione (luminol) has been used to detect bacteria collected in a Porton type cyclone operated at 1050 L min^{-1} with a collection fluid added at 3.4 mL min^{-1} by peristaltic pump.[138] Samples from the cyclone were continuously extracted (3.4 mL min^{-1}), using the same peristaltic pump, and mixed with NaOH, luminol reagent and perborate at 3.4, 1.2 and 1.2 mL min^{-1}, respectively, before passing through a flow cell in front of a photomultiplier tube to measure light emission which was recorded electronically.[139] Light emission was proportional to bacterial concentration in the cyclone fluid between 10^3 and 10^9 cells mL^{-1} and differed little between different species. The detection limit was 10^3 cells mL^{-1} of cyclone fluid, equivalent to 3×10^3 bacteria m^{-3} air.

Immunochemical detection has been used to detect and quantify airborne allergens of fungi collected on filters[4,196] and of pollen allergens collected in a two stage virtual impactor.[292] Both radioallergosorbent tests and enzyme-linked immunosorbent assays have been used.

Tracking "invisible" particulates as they are dispersed through turbulent air can be difficult and tracers can be useful to describe the movements of an aerosol. Different tracers have been employed, including uranine dye[332] and coliphage f2 alone or with sodium fluorescein (uranine) dye[26] applied through a sprinkler irrigation system carrying waste-water; *Escherichia coli* tagged with tritium;[357] spores of *Bacillus subtilis* var. *niger* (*B. globigii*) with the fluorescent particle tracer zinc-cadmium sulphide[3] or with a green and a yellow fluorescent tracer;[335] and cells of *Serratia marcescens*.[257]

UNDER PLANT CANOPIES AND IN CROPS

Within plant canopies and crops, the main aerobiological concerns are the spread of pollens within and between crops, especially in seed growing areas, and the spread of phytopathogenic fungi and bacteria. Dispersal of spores in and from a crop depends on the conditions prevailing during spore formation and release, transport and deposition. These may be affected by changes in crop structure during growth. Deposition processes are perhaps most important in determining how far spores travel. Air samplers of many different types have been used within crops. Usually samplers have been prepared and samples have been handled as in ambient air sampling but their application has often differed.

The Hirst automatic volumetric spore trap[153] and its variants (Figure 16.6) have been deployed widely in studies within both field and tree crops. They have been utilized to study the occurrence of different fungal spore types and pollens, often in relation to weather parameters and to determine their seasonal and diurnal periodicities, and also in experimental studies of plant disease. Recent studies in England have revealed the presence of airborne ascospores of *Pyrenopeziza brassicae* in oilseed rape crops before the teleomorph was discovered[199] and allowed concentrations of *Sclerotinia sclerotiorum* ascospores to be related to apothecial numbers and infection of sunflowers.[226] *P. brassicae* ascospores were recognized from their characteristic diurnal periodicity and by their frequent occurrence in groups of four on spore trap slides even though morphologically indistinguishable from conidia. The relationship between numbers of *Didymella exitialis* ascospores in the air and the incidence of late summer asthma also has been studied with suction traps.[117,145] In field crops, suction traps usually are placed in an area cleared of plants, with the surrounding plants supported so that free rotation of the windvane is not impeded. The sampler orifice usually is positioned 0.4–1 m above the ground in field crops and 2 m high in tree crops (Table 16.2).[53–54,156]

Different samplers often provide complementary information. For instance, a Burkard automatic volumetric spore trap and arrays of rotorod samplers were used to study the temporal and spatial dispersal of ascospores of *Pyrenopeziza brassicae* and pollen in and over an oilseed rape crop 1 m tall.[224,225] An automatic volumetric spore trap gave measurements of spore concentrations at 0.5 m, averaged over 1 or 24 h, while five rotorod samplers, mounted at heights of 0.5, 1, 1.5, 2, 3 and 4 m above ground level, were operated for about 6 h in the middle of the day to measure vertical spore profiles (Figure 16.6).

Figure 16.6.   (a) An automatic volumetric spore trap (Burkard) and rotorod and meteorological instrument arrays for studying vertical profiles in a field crop. The rotorods are mounted under rainshields that can be omitted in dry weather; (b) rotorods mounted under a rainshield, operated by a rain-sensitive switch so that one samples in dry conditions only and the other during rain (reproduced by permission of Mrs M.E. Lacey and Dr H.A. McCartney).

Wind speeds and temperatures also were measured alongside the samplers while other meteorological parameters were obtained at a nearby meteorological enclosure. Also, horizontal gradients were measured with rotorods placed 1 m above the ground at intervals up to 19 m downwind of the crop. Pollen concentrations followed similar patterns in each of five seasons and a trend was calculated by averaging concentrations relative to the middle of the pollen season in each year and then plotting 11 day running means to smooth the values. The seasonal trend then was removed from each data set by deducting the running mean for each day from the corresponding spore trap count. The interaction between pollen concentration and weather variables could be expressed by the combined linear model:

$$P = A + B.R + C.W$$

Figure 16.6. (continued).

where P = pollen concentration, A, B and C are constants, R = solar radiation (MJ m^{-2}) and W = wind run (km). Hourly concentrations were normalized by dividing each by the average for the whole day. Vertical concentrations were well described by the negative exponential equation:

$$C = A.\exp(-\alpha.z)$$

where C = concentration, z = height and A and α are constants. Horizontal fluxes of pollen grains (F; numbers of pollen grains per unit time per unit width) passing the rotorod masts were calculated from C(z) and wind speed, u(z), as follows:

$$F = \int_{\text{crop height}}^{\infty} C(z).u(z)dz$$

Concentration gradients downwind were fitted by the negative exponential equation:

$$C = B.\exp(-\beta.x)$$

with exponents of 0.17 and 0.12 giving half distances of 4.2 and 5.8 m, respectively.

The dispersal of pollens from line sources of plants of *Ricinus communis*, *Ambrosia trifida* and *Kochia scoparia* have been studied with arrays of rotoslide samplers.[304] These

Table 16.2. Use of Volumetric Spore Traps in the Study of Plant Diseases

Species	Host	Orifice Height (m)	Reference
Venturia inaequalis	Apple	2.0	156
Nigrospora spp.	Banana	3.0	248
Botrytis cinerea	Raspberry	1.0	174
Botrytis cinerea	Grapes	0.8	67
Eutypa armeniacae	Apricot	2.0	53
Tranzschelia discolor	Peach	2.0	54
Ustilago nuda	Barley	0.9	338
Erysiphe graminis	Barley	1.0	339
		0.4	143
Ustilaginoidea virens	Rice	1.0	340
Alternaria padwickii	Rice	1.0	341
Pyricularia oryzae	Rice	1.0	289
Lacellinopsis sacchari	Sugarcane	1.0, 3.0	342
Puccinia spp.	Sugarcane	1.22, 5.79	343
Ustilago scitaminea	Sugarcane	1.22, 5.79	344
Cercospora koepkii	Sugarcane	1.22, 5.79	345
Sphacelotheca sorghii	Sorghum	1.22	321
Puccinia striiformis, Erysiphe graminis, Didymella exitialis	Wheat	0.8	66
Pseudocercosporella herpotrichioides	Wheat	0.5	108
Botrytis fabae	Field beans	0.4	108
Pyrenopeziza brassicae	Oilseed rape	0.4	199, 228
Sclerotinia sclerotiorum	Sunflower	0.5	225
Puccinia antirrhini	Antirrhinum	0.21	55

were mounted on masts 10 m apart in an 8 × 8 square grid pattern at heights of 0.5, 1.5, 3.0 and 4.6 m. The grid was 10 m shorter than the length of the source line. Two horizontal greased slides were placed on the ground at the foot of each mast to measure deposition. Rotoslide samplers and horizontal slides also were used to study dispersal of pollens and fern spores into a forest.[300] The samplers were mounted at heights of 0.5 to 21 m above the ground on towers 20 m apart in three rows and interspersed with shorter supports placed along seven rows 10 m apart, which extended 100 m into the forest from the edge. At all sites, samplers were operated at 1.75 m above the ground and on the towers at other heights also. An alternative sampling pattern was employed for slide-edge cylinder samplers in studies of maize pollen.[303] Samplers were placed in concentric rings 1–60 m from the edges of the source plots at 20° intervals. In these studies, the distribution of particles was plotted as a series of isopleths on plans of the sampling area.

Kramer-Collins volumetric spore samplers have also been deployed to measure vertical spore gradients over crops.[96,215] In one study, the traps were located in a wheat field, 1.5, 9 and 30 m above the ground, and in the other at 1, 3, 6 and 14 m, with other traps at 0.3, 0.45 and 0.65 m above the ground within the canopy. Other studies have employed rotorod samplers at five points on a 2 m × 2 m grid and 1, 2, and 3 m above ground level to study the uniformity of spore distribution.[97] Samplers at the same level did not differ significantly but concentrations of spores from a nearby wheat crop decreased with height in contrast to ascospores and basidiospores from more distant sources.

The dispersal of *Botrytis fabae* conidia in bean crops was studied with vertical cylinders and horizontal slides while deposition of *Erysiphe graminis* conidia in barley crops additionally used simple, single-stage suction traps.[19,22,107,217,221] The slides and rods were used to estimate deposition on leaves and stem and the suction traps to measure spore concentration. The slides, coated with glycerine jelly, were supported horizontally within the crop on wire supports mounted on the frames that held the sticky cylinders, so that all sampled at the same height. The cylinders were of four different diameters, 0.11, 0.33, 0.45 and 0.97 cm, with three of each size mounted on horizontal wooden racks with a wire surround to prevent interference from leaves. The three largest cylinders were metal and coated with a petroleum jelly-paraffin wax mixture on a cellophane strip while the smallest was glass and coated directly with glycerine jelly. The cellophane was removed from the larger rods and mounted in gelvatol on microscope slides to allow spore counting while spores on the smallest were counted directly. The suction samplers consisted of a sampling unit 17 × 21 × 15 mm, with an orifice 16 × 4 mm tapering internally to 14 × 2 mm. The side walls of the unit were grooved to within 0.7 mm of the orifice to take a rectangular acrylic plate, with a semicircle removed from the rear to allow unimpeded air flow (Figure 16.7). The front was either prepared with a strip from a glass coverslip, 15 × 3.5 mm, fastened and made sticky with molten petroleum jelly or in the same way as described for rotorods above. The back plate unit which closed the trap was rectangular, 21 × 15 × 3 mm with a tube, 25 mm long and 9 mm internal diameter mounted centrally. The traps were connected to a vacuum line connected to a petrol driven vacuum pump capable of 500 L min^{-1}. These traps originally were controlled at 10 L min^{-1} by a constricting orifice plate[135] but were operated by McCartney and Bainbridge[221] at 8 L min^{-1}, giving an impaction velocity of 5 m s^{-1}, much greater than the wind speed in the crop but enhancing trapping efficiency to nearly 100%. However, the sampling efficiency of the suction traps changed with wind speed from 54% at 0.82 m s^{-1} to 85% at 3.6 m s^{-1} and 119% at 6.1 m s^{-1}. All the traps were positioned 30 cm above the ground, either between rows or at right angles to them and perpendicular to the mean wind direction and exposed for up to 6 h at a time. The numbers and sizes of clumps and numbers of individual spores deposited were counted microscopically, except that clumps could not be distinguished in samples from the suction traps. Wind speed, solar radiation, sunshine duration and temperatures also were measured at a nearby synoptic weather station during each exposure.

The efficiency of impaction on cylinders (E) depends on the Stokes Number (St) of the impacting particles and the ratio of catches on cylinders of different diameters in non-turbulent air can be used to calculate terminal velocity of particles.[221,222] St is a non-dimensional parameter defined as:

$$St = \frac{(v_s \, u)}{(g \, d)}$$

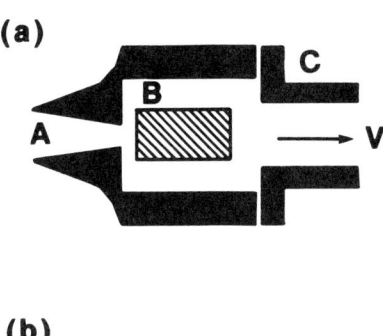

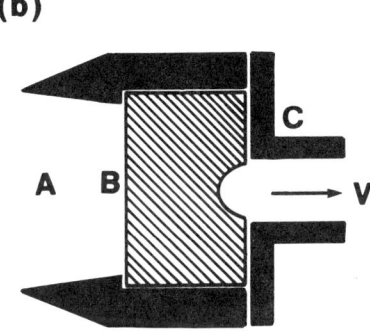

Figure 16.7. Suction sampler unit. (a) median vertical section. (b) median horizontal section though trap (135). Key: A, orifice; B, perspex block bearing trapping surface; C, back-plate unit connecting to vacuum pump, V.

where v_s = terminal velocity, u = wind speed, g = acceleration due to gravity and d = the diameter of cylinder. With the depositing wind speed the same for all cylinders, the ratio depends only on the terminal velocities, v_s, of the particle. McCartney and Bainbridge calculated the impaction:sedimentation ratio for each cylinder diameter using the mean windspeed, u, and the relationship between St and E given by May and Clifford,[244] assuming that the deposition velocity, $v_d = v_s = 1.2$ cm s^{-1} for *Erysiphe* spores. Sedimentation rate (spores per unit area per unit time), S, can be calculated from spore concentration, C as S = C.v_d, Impaction rate, I, is determined by I = C.u.E and the ratio of impaction:sedimentation (R) by:

$$R = \frac{u \, E}{v_d}$$

This ratio should equal the ratio of deposition rates measured on the vertical cylinders and horizontal slides for single spores, under the conventional assumptions about deposition on such traps. However, this method revealed that the measured values of R considerably exceeded those calculated, by 4–5 times for 0.11 cm diameter cylinders and by up to ten times for the 0.97 cm cylinders. Protection of the horizontal slides in large beakers decreased turbulent deposition but v_d was still almost double v_s.[223] The terminal velocity of aggregates of *Erysiphe graminis* spores also could be determined from deposition on the cylinders. In wind tunnel experiments, the relationship between v_s for a clump of spores and that of a single spore in aggregates of up to five spores was calculated to be:

$$\frac{V_{s\ (cluster)}}{V_{s\ (single)}} = 0.73 N^{0.5}$$

where N = the number of spores in the aggregate.

A modification of these methods was used to determine the relative contributions of sedimentation and impaction in the deposition of *Lycopodium* spores dispersed from a line source in a wheat crop.[220] Instead of horizontal slides and vertical cylinders, 17 cm lengths of a 1 cm wide, 0.05 cm thick acetate sheet were bent at right angles at three positions along their length and permanently creased by touching with a heated rod (Figure 16.8). Pairs of strips were placed in paired holders which held the sections at 0, 30, 60 and 90° to the horizontal. One of each pair presented a convex surface and the other a concave. The exposed surfaces were coated with vacuum grease. A 0.6 cm diameter glass rod, also coated with vacuum grease was mounted between the two strips and two suction traps at each of two positions, and heights, determined spore concentration. Deposition on vertical and horizontal surfaces was determined as before, and that on 30 and 60° surfaces from the equation:

$$D(\theta) = D(0).\cos(\theta) + D(90).\sin(\theta)$$

where D(0) is the deposit on the horizontal surface, D(90) the deposit on the vertical and Θ is the angle of the surface to the horizontal. Below the middle of the canopy, where wind velocities were less than a quarter of those at crop height, inertial deposition on vertical surfaces was negligible (D(90) = 0) and D(0) was almost equal to $v_s.C$.

Vertical cylinders, within and at different distances from fields of rye and level with the ears or slightly above, also were used to measure dispersion of pollen from diploid and tetraploid ryes in relation to seed setting and yield.[276] At each trapping position, pairs of cylinders, 0.56 cm in diameter and 10 cm long, were mounted 10 cm apart and 20 cm in front of the axis of rotation, below a rainguard attached to a windvane. After exposure, the numbers of pollen grains occurring in 15 fields, 0.14 mm^2 in area, distributed over the middle 5 cm of the trace, were counted microscopically. A holder was designed to secure the rods to the microscope stage. An automatic volumetric spore trap also was placed in each field to allow calculation of the area dose, using the hourly counts and the corresponding wind velocities

Andersen samplers have been used rarely in studies of plant pathogenic fungi. However, they were used with ethanol-streptomycin agar to detect dispersal of *Verticillium alboatrum* from lucerne fields.[175] During the growing season, 30 min samples were taken either within or 0.5 m above the crop canopy while at harvest, the sampler was placed at the same height above the canopy as before but 10 m downwind of the crop.

A combination of rotorod, Burkard and Andersen samplers has been used to detect and identify aflatoxin-producing *Aspergillus flavus* spores in the air over maize fields.[161] Catches on rotorods and Burkard tapes were stimulated to germinate by placing decapped maize grains on the trapping surfaces in Petri dish humid chambers incubated at 27°C for one week. Infected grains then were analyzed chemically. Isolates from *Aspergillus* differential medium exposed in Andersen samplers were grown on coconut agar at 28°C for 7 days, examined for blue fluorescence and analyzed chemically to confirm aflatoxin production.

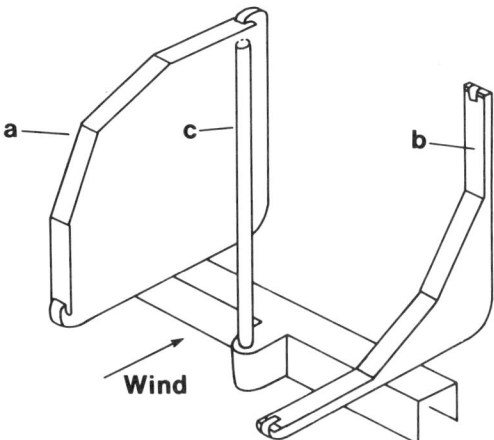

Figure 16.8. Simplified diagram of convex (a) and concave (b) shaped sticky plastic strips and vertical cylinder trap (c) used to determine the relative contributions of sedimentation and impaction of spores (after 220).

Filtration, Andersen and slit samplers have all been used to determine the fungal spore load released during combine harvesting.[76,205] Personal filter samplers, mounted on the harvester, were operated at 2 L min^{-1} using battery operated pumps and Andersen samplers were operated alongside the harvesters with suction, at 25 L min^{-1}, provided by a venturi jet operated from a compressed air bottle.[76] The filter samplers were operated for one revolution of the field and were assessed microscopically while the Andersen sampler could be operated only for a few seconds before it was overloaded. The slit sampler was powered by a petrol-driven generator downwind of the dust plume from combine harvesters and operated for 0.25 h at 17.4 L min^{-1}. Cyclone and filtration samplers occasionally have been used to collect large quantities of dry spores of *Puccinia* spp. and *Pithomyces chartarum* from herbage.[325,360]

A virtual impactor, sampling at 400 L min^{-1}, has been employed to monitor numbers of airborne *Erysiphe* spores by allowing the collected spores to sediment onto leaf segments from susceptible hosts in a chamber below the impactor.[320]

Crop plants are a sensitive means of detecting airborne spores.[155] A field of potatoes, for instance, will detect airborne sporangia of *Phytophthora infestans* (late blight) long before they can be detected with a spore trap and subsequent spread of the disease then can be mapped. Bait plants, grown in disease free conditions in a glasshouse, also can be used to detect airborne spread of disease. Thus, Suzui et al.[351] detected ascospore dispersal of *Sclerotinia sclerotiorum* with kidney bean flowers while Ferrandino and Aylor[106] utilized plants of snap beans, placed in a square grid pattern downwind from the source, to compare rust infection gradients with those obtained from urediniospore deposition on sticky slides placed on the ground, or at a height of 0.25 m. The plants were incubated in a dew chamber overnight and then in a growth chamber for 8–10 days before lesions were counted. By comparison of plant infections and deposits of spore aggregates, it was concluded that more infections occurred than would be expected from dispersal units composed of single urediniospores, and that more dispersal units composed of aggregates of spores germinated than those made up of single spores. Deposition on crop plants also has been measured with stained ragweed (*Ambrosia elatior*) pollen released above the crop, by microscopic counts of spores on acrylic plastic replicas of the leaf surface.[18] Infection of bait plants downwind of a plot of mildewed winter barley permitted determination of when infection occurred and its relationship to weather conditions.[255] Plants were exposed for 7 h, placed in a plastic bag and incubated in a glasshouse before determining the number of lesions on 200 plants. Little dispersal was detected in winter but increased at least tenfold from early April. Spore catches were correlated with maximum temperature, maximum and total solar radiation and maximum potential evapotranspiration and negatively correlated with minimum relative humidity and leaf wetness. Wood sections were used by Rishbeth[307] to detect dispersal of *Fomes annosus* and *Peniophora gigantea*.

In field studies, anisokinetic sampling and yaw are significant sources of sampling error for particles larger than 20 μm although negligible for those smaller than 5 μm. To avoid having to match a variable wind speed for isokinetic sampling, a portable wind tunnel can house isokinetic suction samplers.[57,238] The entrance of the tunnel is bell-shaped so that streamlines entering it curve gently enough for spores and pollens not to be thrown out of the flow.

Although some small droplets carrying spores of splash-dispersed fungi may evaporate so that they can be dispersed through the air like other airborne spores, most splash droplets are too large to form true aerosols. They follow a ballistic trajectory for which impaction samplers are unsuitable. They therefore require different methods of sampling. Horizontal slides, photographic film, funnels draining into beakers and impingers have been found useful.[82,108,110,111] Slides may be placed under rainshields to prevent wash off and can be coated with gelatine containing naphthol green to reveal the positions of splash droplets. Small airborne droplets in low concentrations require more efficient sampling methods than horizontal surfaces exposed near the ground and liquid impingers with a preimpinger[104] and high volume cyclone samplers[37] have been used.

Sampling airborne bacteria presents many of the same problems as sampling splash borne fungi but sometimes aerosols may be formed other than by rainsplash. For instance, the abrasion of leaf surfaces due to plant-to-plant contact may release bacteria into the air[125] while pulverization of potato haulms contaminated with bacteria before harvest may result in aerosols of *Erwinia carotovora*.[280] Air samplers have frequently been employed in epidemiological studies of such airborne bacteria. The presence of *E. carotovora* in aerosols from pulverized potato haulm was demonstrated by exposing Petri dishes containing Stewart's medium (MacConkey agar with a pectate overlay[349]) 1.5 m above the soil and 1.5 m downwind from a pulverizer for 2 min after the machine had passed, then confirming the identity of isolates with physiological tests.[280] Sets of four large plates (total area 4 m^2) filled with Stewart's medium placed 50 and 100 m downwind were used in other tests with an artificial aerosol source. Settle plates have been used also to detect dispersal of bacterial plant pathogens and to monitor their deposition rates[206,209,372] as well as to estimate numbers of airborne bacteria lost to leaf deposition.[64,208] To detect spread of antagonistic bacteria, droplets collected on glass slides, leaves, paper cards, films or long strips of paper have been tested by bioassay against target organisms in the field or laboratory. Dyes or other markers in bacterial suspensions have been utilized to aid tracking.

A simple liquid impinger, with a funnel-shaped intake, and settle plates were able to detect *Erwinia amylovora,* the cause of fireblight disease of pome fruits, in air samples taken in an infected orchard in Germany.[107] The settle plates were filled with a semiselective medium and the impinger suspensions were assayed by immunological techniques including immunofluorescence and ELISA.

Andersen samplers have been chosen more often for studies of airborne bacteria than of fungi. They have been used to study aerosols from crops of snap beans,[208] potato,[280] soybean;[370] tomato and pepper,[229] and cabbage;[193] to detect pathogens and ice nucleation bacteria (*Pseudomonas syringae*, *Ps. fluorescens*, *Erwinia herbicola*) from maize, wheat, pea, and lucerne.[116,207] They have been employed to also monitor dispersal of *Ps. syringae* in a maple tree nursery;[234] *Bacillus thuringiensis* sprayed in a fruit orchard for insect control;[36] to determine fluxes of airborne bacteria above crop plants and bare ground;[64] and dispersal of *Streptomyces* from the surface of a fallow field.[212] A range of media have been applied, including a selective medium for *Ps. syringae* pv. *glycinea*;[370] King's B medium[190] + 100 mg cycloheximide L^{-1} for *Ps. syringae* pathovars,[208,229] DL-lactate based medium for *Ps. syringae* pvs. *syringae* and *tomato* nutrient agar + 100 mg cycloheximide L^{-1} and for *X. campestris* pv. *vesicatoria*;[229] nutrient-starch-cycloheximide agar for *X. campestris* pv. *campestris*;[193] nutrient glycerol agar + 200 mg cycloheximide L^{-1} for ice nucleation bacteria;[64,207] media with 100 mg chlorothanil, 150 mg nalidixic acid and 100 mg rifampicin for *Ps. syringae*;[234] nutrient agar for *Bacillus thuringiensis*;[36] and chitin agar for *Streptomyces*.[212]

The height and orientation of Andersen samplers is not always well documented. Sometimes, they have been operated with or without a rain shield and oriented vertically[207,208,370] or horizontally.[207] Orifices have been positioned from ca 50 cm above fallow ground to detect *Streptomyces*,[212] to canopy height in tomato and pepper fields[229] and at different heights above to 10.2 m to determine vertical fluxes.[64,207,208] Also, sampling strategies have differed between studies from 10 min samples three times a day[229] to 15 min with both 6- and 2-stage Andersen samplers.[193] Usually, the counts for all plates have been totalled to give total concentrations, but Venette and Kennedy[370] disregarded the first stage where tiny splash droplets were collected and used the distribution of colonies on the other stages to determine the aerodynamic sizes of bacterial particles. By contrast, Lindemann et al.[207] omitted the sixth stage from the sampler.

In glasshouse tests, Andersen samplers retrieved more *Ps. syringae* than did liquid impingers, even though colonies on Andersen sampler plates may be derived from aggregates while those from liquid impinger samples are from single cells.[48] However, recovery declined with increasing sampling time and Andersen sampler plates were overloaded when there were more than 10^7 dispersal units m^{-3} air. Most bacteria in tomato and pepper fields were in particles >7 μm[229] while >80% of *E. carotovora* aerosol particles from pulverized potato haulm were <7 μm.[280] Most *X. campestris* pv *campestris* were collected on stages 4–5 of the sampler[193] and 75% of *Bacillus thuringiensis* bacterial particles >20 m from the source were <3.2 μm.[36] More than 80% of *Streptomyces* propagules were associated with soil particles.[212]

The MRE cyclone sampler was tested for collection of airborne *Erwinia carotovora* ssp. *atroseptica* and *carotovora* from open air but a Casella High Volume sampler (700 L min^{-1}) was selected for surveys.[287] The sampler was at a height of 1.3 m at the rear of a Landrover truck, which may have provided some protection from air movement and rain, and collected samples near (10–20 m) and distant from potato fields in 3 years. Soft-rot bacteria readily were detected during rainfall from mid-summer to early winter but not from late-winter to early summer. Graham et al.[126] also could only isolate *E. carotovora* ssp. *carotovora* using the same type of sampler 10 m downwind from an infected potato field during a heavy shower. Aerosols produced by sprinkler irrigation of potatoes were detected only between late-evening and early-morning, never during the day and usually were not found until late in the growing season.[147] Soft-rot erwinias may be dispersed widely in the atmosphere since they have been found in snow collections in the Rocky Mountains, remote from crop producing areas.[114] Melted snow was filtered through diatomaceous earth (Celite) which was resuspended and enriched by anaerobic incubation in pectate-broth before plating on Stewart's medium. Likewise, rain samples nearly all yielded soft-rot bacteria but simultaneous Casella High Volume air samples detected the bacteria only when the relative humidity was high.[115]

Plant viruses potentially can be dispersed as aerosols but demonstrations are few. High-volume Sierra cascade impactors detected tobacco mosaic virus in a field of tobacco most frequently when wind speeds exceeded 25 km h^{-1}.[23] Also, air drawn through moist cotton-wool in a greenhouse near virus-infected plants allowed bioassay of the virus from the cotton.[326] The transport of virus particles by airborne pollen clearly is established for more than 15 plant viruses although few can cause infections in the mother plant.[88] Exceptions occur with raspberries and stone fruits[362] and some cereal crops[52] in which pollen dispersed viruses cause important diseases. Virus-infected pollen may cause infertility in *Paeonia*[162] while *Phaseolus* bean and tobacco pollen sometimes can carry virus passively without becoming infected.[142]

Irrigation with sewage effluent can result in the loss of 0.1 to 1% of the liquid as aerosols,[294] releasing large numbers of microorganisms, including viruses, into the air. These emissions have been studied using a range of sampling methods.[151] Coliform bacteria have been isolated from AGI-30 liquid impinger samples in nutrient broth, by plating on M-Endo, MacConkey or SS agar, and with the same agar media[181] using Andersen samplers. Andersen samplers enabled detection to a greater distance (350 m) than liquid impingers (70 m). Later, a high volume scrubber, supplied with distilled water or "minimal essential medium" at 3 mL min^{-1} for 15–20 min with plating onto violet red bile agar and Buffalo Green Monkey cell lines, was used to detect bacteria and virus, particularly echovirus 7, particles downwind.[356] A nalidixic acid-resistant strain of *E. coli* was used as a tracer in the irrigation water and nalidixic acid was added to the medium to confirm the source of the aerosols. Ten times more bacteria were present at night than by day and nearly half the bacterial particles were respirable. With 1% beef extract as the scrubber collection fluid, airborne *Salmonella* and enteroviruses (poliovirus, echovirus, and coxsackie virus type B) were compared to numbers of airborne coliforms and also to the numbers and types of organisms in the irrigation water.[357] There were few correlations and the value of using coliform bacteria as indicators of airborne contamination by spray irrigation was questioned. Moore *et al.*[253] used a bank of eight high volume electrostatic samplers (Litton-M, 1000 L min^{-1}), to assay air 50 m from a sprinkler wetted area and operating simultaneously for periods of 30 min. The collection fluid was brain heart infusion broth containing 1% Tween 80. Fluid from the eight samplers was pooled, concentrated and then assayed in the laboratory. Coliphage were most numerous but polioviruses 1 and 2 and coxsackie virus B-3 also were detected.

Sorber et al.,[332,333] and Zimmerman et al.[381] deployed AGI-30 liquid impingers, multi-stage liquid impingers, Andersen samplers, high-volume cyclones (600 L min^{-1}) and electrostatic precipitator (LEAP) (900–1000 L min^{-1}) samplers in spray irrigation studies. LEAP sampler lines were disinfected with 0.1% sodium hypochlorite (pH 6.5) and the aluminum collection plate was cleaned with detergent solution and alcohol.[332] Andersen samplers were disinfected with 70% alcohol. Distilled water was used as a collecting fluid for the liquid impingers and plate count broth + 0.1% Tween 80 for the LEAP sampler, which was recirculated to allow concentration of aerosol particles and topped up with distilled water + 0.1% Tween 80 to restore evaporative losses. Endo-agar or peptone-yeast extract-glucose[10] agar were used as isolation media in Andersen samplers and for direct plating of liquids from the LEAP samplers. Alternatively, liquids were filtered and the membrane filters were incubated on M-Endo or M-Enterococcus broth. Bacterial populations were above ambient levels to 198 m from the source, the greatest distance tested. Compared to similar collections made near sites with different disposal techniques, Sorber *et al.*[333] judged sludge dispersal by truck to present a minimal health risk through airborne particles. They used 1% brain-heart infusion broth + 0.1% Tween 80 as the collection fluid in samplers and assayed the suspension by culture, bioassay, enrichment and separation techniques in the laboratory. The efficiency of large volume Litton-type electrostatic air samplers was 45% of that of standard all-glass impingers in a wind-tunnel tests.[333] The samplers were decontaminated using ethanol, buffered sodium hypochlorite and sodium thiosulphate between collections. Multistage liquid impingers, sampling into pH 7.0 phosphate buffer solution, collected only 82% of the bacteria isolated with two-stage Andersen samplers loaded with trypticase soy plus 5% sheep blood, although there was good correlation between the two samplers.[381]

Dewatered municipal sludge may be discarded onto waste ground and with drying and high winds disperse bacteria into the air. Liquid impingers, with their orifices oriented into the wind, sampled air 8 cm above and near the center of a clearcut within a forest that was covered with a 15 cm thick layer of sludge at monthly intervals for six months.[92] Samples were collected for 1 h at only 8 L min^{-1} to minimize foaming of the 1% peptone collection fluid which then was assayed by the most-probable-number technique. Most coliform bacteria were collected when the weather was hot and windy during a dry period. Coliform bacteria were not found in the absence of sludge. Impingers, modified by fitting a 15 cm funnel to the intake, contained phosphate buffer to collect airborne particles generated by a pipe discharging sewage into a cypress swamp.[329] The liquid was concentrated on membrane filters which then were incubated on M-FC broth pads or on nutrient agar.

Groups of two to three Andersen samplers, loaded with Endo or casitone agar and situated at different distances from an irrigation line discharging effluent from a potato processing plant, have been utilized similarly to assess dispersal of bacteria.[272] However, results are difficult to interpret because background populations of bacteria were not determined and isolates were only partially characterized.

IN AND OVER CITIES

Sampling in and over cities, mostly concerned with the incidence of pollen and fungal allergens, has utilized sedimentation slides or plates,[90] automatic volumetric spore traps[79,141,166] and, occasionally, static rods[6,288] and Andersen samplers.[27] Samples mostly have been taken from the roofs of buildings to avoid undue effects from nearby sources and to allow results to be applied over a larger area. Volumetric spore traps used in a series of studies in cities of the European Community were all placed on rooftops, at heights ranging from 17.5 to 54 m.[20,21,58,75,336] Concentrations were smaller than at breathing height, or in rural areas. The methods for preparing, exposing and evaluating the catch are like those for sampling ambient air. Andersen samplers, exposed on rooftops, have been inserted horizontally into a holder and rain shield mounted on a windvane so as to always face into the wind.[27] Three 10-min samples were collected every 3 days for 2 weeks at 0900, 1300, and 1800 h onto a single set of plates which then were incubated at 28°C and colonies counted and identified. Positive hole conversion tables[12,282] were used to give numbers of colony forming units deposited in cubic meters of air.

Studies of airborne bacteria in cities date back to the earliest days of aerobiology.[251,274] Although particles that probably could be classified as bacteria were present in catches from Pasteur's gun cotton filter, these were not identified. Subsequently, Miquel[252] estimated bacterial concentrations by drawing measured volumes of air through liquid medium, which then was partitioned into 50–100 vessels, adjusting the sample size to leave 25–50% of the vessels sterile. More recently, slit and Andersen samplers have been utilized. A Casella slit sampler was used to enumerate bacteria in the outside air of Louisville, KY.[352]

Populations averaged 155 m^{-3} air in daytime samples. An Electrostatic Bacterial Air Sampler (General Electric) and a Bourdillon Slit sampler were both employed to collect bacteria on the roof of the Sun Life Building, Montreal, Canada, onto plates containing either Czapek's agar + 0.1% yeast extract + 0.5% glucose or Lochhead's yeast extract agar (Medium Y).[187] Concentration changed seasonally usually with fewest bacteria in January-March and most in July and November, especially when the sampled air had passed over agricultural areas.

Two-stage Andersen samplers loaded with plastic Petri dishes containing 27 mL trypticase soy agar + 0.5 mg cycloheximide mL^{-1} or Littman's oxgall medium[210] + antibiotics, were used to make weekly collections of airborne bacteria and fungi at 17 stations in suburban Washington, D.C.[176] After collection, colonies were counted directly on the collection medium or transferred to other media by replica plating to aid identification. Sampling location, location characteristics (local traffic, ground cover, etc.), time of day, day of the week, rainfall and relative humidity affected catches significantly, while wind speeds >3.1 m s^{-1} gave larger than average particle concentrations. Again, bacteria were most numerous in summer and autumn, and fewest in winter. Bacteria at parkland and street sites in Stockholm were collected using six stage Andersen samplers, loaded with tryptone glucose agar + 50 μg cycloheximide mL^{-1} medium covered with 0.2% oxyethylene docosanol (OED; 0.1 mL per plate) to prevent drying and operated for 30 or 60 min 2 or 10 m above ground level.[34] A diverse range of bacteria was isolated at 2 m height averaging 763 and 850 cfu m^{-3}, respectively. Half were in particles >8 μm. Concentrations at 10 m were about 90% of those at 2 m. The bacteria were classified by numerical taxonomic analysis of 50 U characters. Few isolates clustered, suggesting that no group predominated but rather that there were small numbers of many types.[11]

Waste disposal, both liquid and solid, provide important sources of airborne microorganisms in urban areas. The processes that disrupt and digest the waste frequently produce aerosols containing bacteria, some pathogenic including fecal coliform bacteria and fecal streptococci, and bacteriophages. To evaluate adequately the health risks from viable microbial aerosols from sewage treatment plants requires: (1.) a volumetric sampler that efficiently traps particles 1–10 μm diameter; (2.) preservation of collected organisms to allow specific identification; (3.) minimum logistical problems, i.e., the sampler is mechanically sound, reliable and operates with allowable tolerances; and (4.) particle size discrimination, enabling the site of deposition in the respiratory tract to be determined.[151] The Andersen sampler was considered to fulfill all these requirements and has been deployed near sewage treatment plants with a range of different media and sampling times. Sampling times downwind of sewage plants have ranged from 5 min to 1 h.[2,256] Cleaning Andersen samplers by swabbing with 70% alcohol between each test has been recommended to minimize cross contamination.[102] Airflow calibration and negative and positive controls are needed to maintain the quality of samples, and results should be analyzed statistically using techniques such as nonparametric (Mann-Whitney U) tests with weighting of geometric means.

Media for isolation and enumeration of bacteria have ranged from non-selective nutrient, casitone and trypticase soy agars, often with the addition of 0.1–0.2% cycloheximide to suppress fungi, to media selective for coliforms, such as Endo and eosin methylene blue agars.[2,70,102,124,160,256] However, selective media in the Andersen sampler have sometimes recovered coliforms less well than nutrient agar, even when colonies on nutrient agar were replica plated onto clean plates of the same selective media.[70] OED monolayers sometimes have been added to prevent desiccation of the medium[70,102] but this can result in wet plates with the agar surface broken by the force of the sampling jets, and greater contamination than with untreated plates.[70] Media have been supplemented in winter with

2.5% carboxyl methyl cellulose or 3% corn starch to prevent freezing.[124] Bacteria have been detected 1.2 km downwind of the source, especially at night, with nearly half of the particles respirable (<5 μm).[2,256]

Slit samplers, aspirated at 10–50 L min^{-1} and loaded with Petri dishes containing plate count agar, have been operated near an activated-sludge sewage plant.[105] The clear plastic top of the sampler was protected with an aluminum foil shield, punctured to allow the orifice to extend into the airstream. Samples on a single, cloudy day with a temperature of 19°C showed most bacteria in walkways between processing areas. Slit samplers collected 22% fewer bacteria than did Andersen samplers but the difference was not significant unlike collections on gelatine filters which were only 3% of the Andersen sampler count.[213]

Impingers, collecting into 1.6 mL phosphate buffer, were used to sample air flowing over an activated sludge tank.[296] The liquid was concentrated by membrane filtration. The filters then were incubated, first on double strength trypticase soy broth, then on MF-Endo broth. Colonies were counted to determine concentrations of airborne bacteria and counts were applied to a Gaussian plume model to estimate dispersion.

Large volume air scrubbers, aspirated at 0.6–0.9 m^3 min^{-1}, have also been employed to collect particles carrying bacteria and viruses. The collection liquid was changed to suit the target organism: trypticase soy broth (25%) for most bacteria; dulcitol selenite broth (50%) for *Salmonella*; phage assay broth (25%) for coliphages; and Hank's balanced salt solution with 25% nutrient broth for enteric viruses. Broths were filtered through membrane filters which then were cultured by standard methods,[10] or bioassays were carried out to detect viruses. Autoclaving of the sampler was recommended to minimize cross contamination.[102]

Large volume Litton-type electrostatic air samplers, with a recirculating phosphate buffer collecting fluid, did not allow animal viruses to be detected in air from waste-water treatment facilities[100] although coliphages had been detected in this way at other plants with recirculating phosphate buffered saline solution, phenol red, antifoam and foetal calf serum collection fluids or with multi-slit impingers with Hank's balanced salt solution, 10% fetal calf serum and tris buffer.[101] Samples were kept on ice until assayed for virus using most-probable-number techniques with tubes of host bacteria.

An army prototype XM2 large volume (1050 L min^{-1}) biological sampler/collector, comprising a particle separator that concentrated particulates 2–12 μm diameter into a small volume of air before they were collected by scrubbing and impaction, collected air samples near an aeration lagoon processing effluent from 13 municipalities and 5 major industries.[38] Bacteria and coliphage were collected in solutions of Earle lactalbumin hydrolysate or in distilled water to detect animal viruses in the hydrolysate solution with four different antibiotics. Samples were kept on ice until they could be assayed. Viruses were not detected but there were appreciable numbers of bacteria in air near the lagoon. The sampler was cleaned and sterilized for 4 h with ethylene oxide after every set of samples and the tubing was immersed in sodium hypochlorite solution overnight then rinsed with sterile water. Alternatively, samplers can be washed with 70% alcohol + 0.01% methylene blue, then with boiling water for 30 min until the effluent temperature reached 56°C,[51] or sterilized with live steam for 15 min before exposing the collection disc to ultraviolet light for 15 min.

Effluent from sewage plants sometimes is utilized in the cooling towers of power plants where insufficient treatment with chlorine gas or chlorine dioxide can allow microbial populations to increase, and form bioaerosols.[281] Such aerosols could be demonstrated with large (15 × 1.5 cm), inverted petri dishes filled with casitone agar exposed for 15 s periods over cooling-tower throats.[1]

Samples in solid waste transfer, recovery and disposal facilities have been taken with AGI-30 liquid impingers, sampling into 20 mL lactose broth for periods of 20 min[203] and with multistage liquid impingers,[241] operated at 55 L min^{-1} for 20 min with quarter-strength Ringer's solution + 2% inositol as a collection fluid, or with polycarbonate membrane filters operated at 4 L min^{-1} for 1 h.[72] Samples from the liquid impingers were kept on ice (ca 3°C) for less than 6 h then plated on violet blue agar to select coliforms. Particles deposited on filters were suspended in an aqueous wash fluid (0.05% Tween 80, 2% inositol, 0.1% peptone) and again diluted in quarter-strength Ringer's solution, plated onto violet red bile glucose agar and incubated at 37°C for 48 h before counting.

Viability of airborne bacteria in urban areas may be affected by the open air factor, (OAF)[87,245] which has been linked to reactions between ozone and olefins.[69] Bacteria differ in their sensitivities to OAF, the potency of which changes with meteorological conditions.[69]

Spore concentrations, determined at three altitudes simultaneously using three aircraft indicated dispersal of microorganisms downwind of San Antonio, Texas, but only when there was a marine-influenced air mass because bacteria and fungi were fewer in the air (<50–700 m^{-3}, except 2 samples) than in land air masses (150–1450 m^{-3}). However, the distribution of types was similar, suggesting that the marine mass had picked up its bacteria and fungi during passage over land.[121]

OVER SEAS

Air over oceans has been sampled for bioaerosols from both ships and aircraft while ZoBell and Mathews[383] have compared the microfloras of land and sea breezes from a building close to the sea shore. Microorganisms from marine sources mostly are bacteria, dispersed into the air from spray droplets released from breaking waves, from foam blown from white-caps and from the bursting of bubbles, which may have been enriched by scavenging during their passage through sea water.[31] Many studies have utilized sedimentation plates but precautions have to be taken to avoid contamination from on-board sources. Bacteria, yeasts and molds have been isolated wherever tests have been made but bacteria may have been underestimated as more were isolated on media made with seawater in place of distilled water.[383] Marine bacteria were detected on sedimentation plates up to 48 km inland, and terrestrial forms up to 208 km out to sea. Gram-negative bacteria predominated and fewer than half were spore-formers. Cocci were few and there were no vibrios or spirillas.[382] Groups of three large sedimentation plates, one filled with each of lactose agar + triphenyltetrazolium + turgitol for coliforms, a rich nutrient agar for terrestrial organisms and Oppenheima Zobel agar for marine organisms, were spaced 6–8 m apart in a matrix covering an area 25 × 250 m to detect bioaerosols emanating from a bay polluted with sewage from Nice, France.[17] Many coliform bacteria and terrestrial bacteria were grown but few marine bacteria, contrasting with a less-polluted bay nearby where marine organisms predominated.

There have been few volumetric studies of marine air. Miquel[251,252] aspirated 1000 L air day^{-1} through glass wool plugs in tubes. Particles were resuspended and inoculated into beef extract broth. Mean recoveries were only 1 bacterium m^{-3} overall and 0.6 m^{-3} in samples >100 km from land. Cryptogam spores and pollens were identified microscopically, even 30 km from land. A vacuum cleaner operated filter trap on the masthead of a ship revealed only 0.18 pollens m^{-3} in the North Sea and 0.007 m^{-3} in mid Atlantic, apart from a few occasions when strong winds carried larger concentrations off shore.[94] Later, Sreeramulu[337] operated an automatic volumetric spore trap 20 m above sea level in the Mediterranean, 8–80 km from the coast. Fungal spores averaged 56.4 m^{-3} and pollens 1.6 m^{-3}. The possibility of long distance transport of particles from Australia (Tasmania) to Commonwealth Bay in the Antarctic was studied using a rotorod sampler, exposed on a ship for 30 s to 2 min daily.[81] Few pollen grain were caught more than two days after leaving Hobart, Tasmania, while one collected at 64°08'75"S was associated with a north easterly windflow and a blocking anticyclone in the Tasman Sea.

Airborne trapping has revealed concentrations of fungal spores and pollens initially increasing with height due to erosion of the base of the spore cloud.[157] At altitudes of 2.4–2.7 km over the Atlantic, using techniques described below, Pady and Kelly[270] found most bacteria (1–32 m^{-3}), including *Micrococcus, Sarcina,* Gram-negative rods, Gram-positive pleomorphic rods, and aerobic spore formers, in tropical air masses in June. Fulton[119] tracked microbial populations in an air mass as it passed from Texas over the Gulf of Mexico in March. Three planes made simultaneous collections at altitudes of 0.15, 1.1, and 2.0 km to a distance of 640 km into the Gulf from Houston. Microbial populations at the lowest altitude decreased with distance from land but, at higher altitudes, populations were extremely variable. At the lowest altitude near the coast, bacterial and fungal populations were initially nearly equal, but by 640 km fungi predominated (82%). At the middle and high altitudes, fungi predominated (76–82%) in all samples. The predominant bacteria were *Micrococcus, Bacillus,* and *Corynebacterium.*

UPPER ATMOSPHERE

Different attempts to measure microbial populations in the upper atmosphere have been reviewed by Bruch,[44] Gregory,[130] Proctor and Parker[285] and Bovallius et al.[35] but there is no agreed method of sampling at high altitudes. Airborne microorganisms first were collected from high altitudes using samplers attached to kites and balloons. Membrane filtration has been used during balloon flights to collect microbes at altitudes from 3 to 27 km.[44] Between 18 and 27 km, 265 bacteria were collected, mostly micrococci but also Gram-positive, Gram-negative, and diphtheroid bacteria.

Aircraft were used first in 1921 by Stakman et al.[346] to study movement of rust uredospores across the U.S.A. The forward speed of the slow moving aircraft allowed spores to impact onto sticky surfaces supported by hand. Surfaces used since have included microscope slides,[149,346] small cylinders,[306] Petri dishes[84,271,284] and small square section rods.[15] Neither sample volume nor deposition efficiency were considered and wind speeds were far greater than those tested by Gregory and Stedman,[128,136] when studying deposition on trap surfaces, so that spores may have bounced or blown off the trapping surface. A slide 25 mm wide probably will trap only particles larger than 15 μm diameter at 320 km h^{-1} (90 m s^{-1}) and a 9 cm Petri dish only particles larger than 20–25 μm diameter. However, a rod 1 mm wide will collect 2 μm particles.[242] Even when suction traps were introduced, sampling was usually at velocities very different from the speed of the aircraft.[269] Results therefore rarely are comparable.

The specification for an airborne sampler should include:

1. An aerodynamically designed sampling probe, with minimal frontal area and mounted so that it collects from air still undisturbed by the aircraft or its engines.
2. The sampling probe and sampler mounted so that the aerodynamics and structure of the aircraft are not affected deleteriously.
3. Isokinetic sampling of measurable volumes. If sampling is not isokinetic, sampling efficiency at different air speeds must be known.

4. Efficient collection and trapping of particles 1–15 μm aerodynamic size.
5. Sampling surfaces changeable from within the aircraft.
6. Samples for microscopic and viable determinations preferably, using the same probe.
7. Multiple samples on the same trapping surface between each change.
8. The minimum of handling when changing trapping surfaces, especially if Petri dishes are used.
9. The path for exhaust air after impact allows smooth unobstructed flow.
10. Viable sampling methods must not subject trapped organisms to additional stress that cause viability loss after sampling.

Different methods of sampling meet many but rarely all of these criteria and some do not achieve isokinetic sampling. Some methods that have been used, or suggested, are described below.

Isokinetic Air Sampler[366]

Paired aerodynamic stainless steel probes, with matched orifices (25 mm outside diameter, 16 mm internal diameter) were mounted through the forward window of the bombardiers compartment at the front of B-17 and B-25 aircraft, extending forwards 1.5 m into air undisturbed by the aircraft (Figure 16.9 a). The sample probe passed directly into the aircraft while the pilot probe was only 38 cm long and open ended so that air flow was unrestricted. Air in each probe passed over the holes of the differential pressure measuring system to give a negative pressure in a manometer indicating the resistance of the soluble gelatine foam collection filter. Vacuum was applied to equalize the pressures in both tubes. Airflow depended on the resistance of the filter, the vacuum applied, the velocity of the aircraft and the altitude, with the available vacuum airflows up to 60 L air min^{-1} possible. Cruising at 290 km h^{-1}, with an orifice 0.37 cm diameter this vacuum allowed an air flow of 52 L min^{-1} at standard temperature and pressure (STP). As the mass of air sampled changes with altitude, at 5500 m (0.5 atm) the actual air mass sampled would be only about half that at STP. Results therefore were based on air space volume (ASV), which could be easily determined. The trap was efficient at all altitudes and ambient temperatures. After exposure, the filters were dissolved in water and measured samples were pipetted into Petri dishes. Cooled melted agar (45°C), either X medium (peptone-yeast autolysate) for total microorganisms; X medium + 10 mL L^{-1} 1% cycloheximide, pH 4, for bacteria and yeasts; or M-A agar (dextrose-peptone-ammonium nitrate-potassium phosphate for fungi was added and mixed well with the suspension. Plates were incubated at 30°C for 3–5 days.[121] After sampling the viability of some cells could be affected by dehydration stress and exposure to hot agar.[112]

Isokinetic Sampler for Light Aircraft[298]

This is a retractable sampler that was mounted on a wing strut of a Cessna 182 aircraft. It is simple to operate and allows examination of the filters by light or scanning electron microscopy or by culturing, although again, viability may be lost through dehydration. The sampling array consisted of five parts: (1) the sampler head assembly; (2) the track assembly; (3) a replacement window; (4) the air moving and monitoring unit; (5) a power source (Figure 16.9 b). The base of the filter unit carried a hose and was mounted onto a

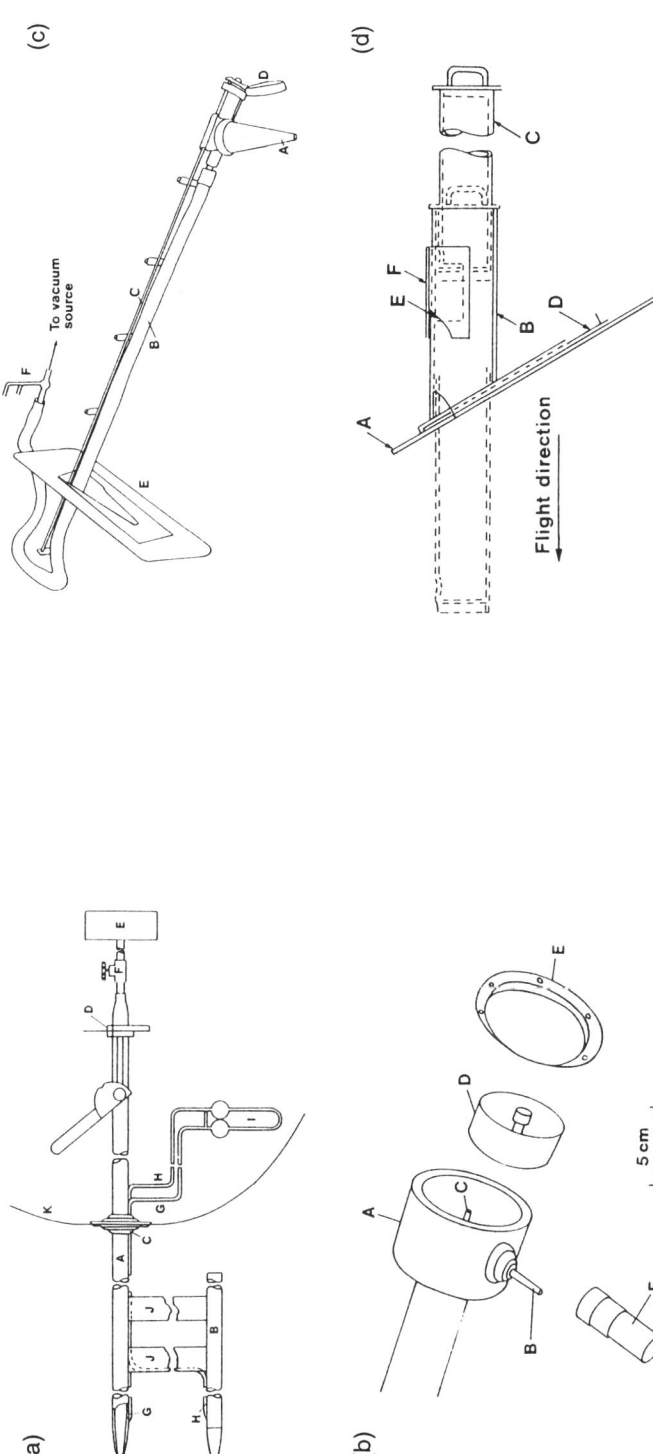

Figure 16.9. Examples of methods used to trap airborne microorganisms from aircraft. **(a)** Isokinetic air sampler.[366] Key: A, sampling probe; B, pitot probe; C, pressure plate ball and socket device; D, filter holder; E, vacuum source; F, vacuum control valve; G, H, 8 mm internal diameter tube welded over two 1.6 mm holes 1.6 mm apart; I, manometer; J, aerodynamic strut; K, bombardier's window. **(b)** Isokinetic sampler.[298] Key: A, head incorporating filter holder; B, flexible hose; C, track assembly; D, clip to attach to wing strut; E, replacement aircraft window; F, pitot tube connected to differential pressure gauge. **(c)** Modified isokinetic impactor.[158] Key: A, sampling head on probe extending 0.6 m from side of aircraft; B, orifice, 3 mm diameter tapering to 10 × 0.33 mm slit; C, spindle for rotating drum; D, drum; E, cover, retained by screws; F, protective cover for orifice when not in use. **(d)** Petri dish sampler showing sampling and loading positions.[188] Key: A, steel plate replacing bombardier's window; B, outer sleeve; C, inner tube with clips to hold Petri dish on end; D, sliding shutter, open for sampling and closed when tube withdrawn to change dish; E, breech for access to Petri dish covered with F, breech cover, during sampling.

stainless steel track for sliding it in and out of aircraft to change filters. A cone, 9 cm internal diameter and 12.6 cm external diameter at the base, 29 cm long with a 7.5° taper, prevented internal turbulence, and had a 2 cm diameter orifice. The orifice which was placed over the filter holder and took 102 mm diameter filters was mounted well forward of the strut and 30 cm below the wing. Suction was provided by the pump of a 24 v DC high-volume air sampler and a miniature pitot tube mounted in the pipe and connected to a differential pressure gauge monitored air flow. The aircraft modification required FAA inspection and approval, and the aircraft was placed in the restricted category. The sampler was designed to match aircraft speeds of 27–38 m s^{-1} but could be used at higher speeds with suitable filters and fully charged batteries. It usually was operated at 35.7 m s^{-1}, giving a flow rate, at STP, of 0.67 m^3 min^{-1}. At high altitudes or non-standard temperatures, the pilot could fly either at the true air speed with the flow rate through the sampler adjusted to match (requiring a knowledge of temperature and pressure but allowing the same volume of air space to be sampled each time) or at a constant indicated air speed and flow rate. True air speed, true flow rate and true entrance speed remain proportional to each other and sampling remains isokinetic but the volume of air space sampled changes with altitude. Samples were collected onto vinyl metricel or polycarbonate (Nuclepore) membrane filters with 8 μm pores. For pollen and spore analysis, the filters were dissolved and particles concentrated by filtering through a smaller filter. The small filter then was mounted in pre-stained glycerine jelly for microscopic examination. Alternatively, particles on polycarbonate filters could be resuspended by reverse flushing and treated as for pollens and spore or could be examined directly by scanning electron microscopy. Culture also would be possible, by using a suitable resuspension fluid, e.g., quarter-strength Ringer's solution + 2% inositol[93] to spores and plating on suitable media.

Modified Isokinetic Impactor[157,158]

A retractable impactor, modified from a design by Pasquill,[273] was mounted on the fuselage of a Hastings or Varsity aircraft and spores were collected on the surface of a drum that could be rotated through 120 positions for successive samples. The orifice was 3 mm diameter but converted into a 10 × 0.33 mm slit situated 0.5 mm from the trapping surface. The sampling head was mounted coaxially on a long tube extending 0.6 m from the side of aircraft and carried on a transverse carriage for it to be retracted into the aircraft for changing (Figure 16.9 c). The inboard end was fitted with a suction outlet, vacuum gauge, and knob for selecting different drum positions, each separated by 3° or 1.62 mm on the circumference of a drum 6.15 cm diameter and 2.4 cm wide. The pressure drop across the orifice was calculated to give theoretically isokinetic sampling at all altitudes up to 3000 m. At 160 and 180 knots (1 knot = 1.853 km h^{-1}, the pressure drops were 152 and 178 mm Hg, respectively, and gave airflows of 35 and 39 L min^{-1}. Spores were trapped on Melinex tape (Burkard Manufacturing Co. Ltd, Rickmansworth, UK) made sticky with petroleum jelly containing 10% paraffin wax (m.p. 54°C) wrapped around the circumference of the drum. Drums were prepared in a room with a filtered air supply and then sealed in boxes. Contamination was measured in areas of the tape where there was no trap deposit but only a few *Cladosporium* spores were found. The airflows through the sampler allowed efficient trapping, even of small particles, and much more dust was collected than in ground level samplers so that it was necessary to limit samples below 1800 m to 1 min (equivalent to 5.56 km at 180 knots), and at greater altitudes to 2 min. Three replicate samples were collected at the same altitude and position to give a detection threshold of 10 spores m^{-3}. A dummy sample was collected between each set of three replicates to remove spores collected in the orifice. The tapes were cut into suitable lengths and mounted on microscope slides in gelvatol as for the Hirst and Burkard traps.[153] A square graticule in the eyepiece aided spore counting along traverses across the width of the deposit. Results were plotted as log (spore concentration + 1) against linear height.

Andersen Drum Sampler[250]

A commercially available Andersen drum sampler was attached to a wing strut of a Cessna 172 aircraft, to collect microorganisms directly onto agar coated drums. Thus it required minimal post collection handling, but was not aerodynamically designed and so sampling was not isokinetic. Also it probably failed to sample undisturbed air. Suction was provided at 1 L min^{-1} using the aircraft vacuum system and power provided through a 24 v DC converter. The inner drum coated with a tryptone glucose yeast extract agar (tryptone 15 g, glucose 1 g, yeast extract 1 g, agar 15 g, water 1 L) moved at 0.25 rpm in a descending spiral past the orifice in the outer drum. Exposed drums were incubated in containers at 35°C for 24 h when bacterial and fungal colonies were counted. Position on the spiral impaction path permitted calculation of time of collection.

Qualitative Collection of Airborne Microorganisms[188]

Petri dishes clipped to the end of a steel shaft were exposed 30 cm in front of the aircraft by pushing them through a steel tube, 10 cm internal diameter, fitted to the inside of a steel plate which replaced a window at the front of a B-29 aircraft (Figure 16.9 d). An airtight cover allowed access to the Petri dish for changing. The Petri dishes contained a modified Czapek's agar with 0.1% yeast extract and 0.5% glucose, and exposed dishes were incubated for 5 d at 25°C. The sampler was neither volumetric nor isokinetic.

McGill G.E. Electrostatic Bacterial Air Sampler[188]

The McGill G.E. sampler is an electrostatic sampler designed for sampling in only a slow moving air flow. It was mounted therefore in a large box, 45 × 45 × 45 cm, into which air entered at the upper left corner through a hose, connected to a right angle pipe, 2.54 cm diameter, projecting through a side window of North Star (DC-6) or C-54 aircraft. Air was exhausted from the lower right corner, through a similar pipe but with the elbow facing backwards. Air movement was provided by the forward movement of the aircraft. The sampler operated from the 24 v DC supply of the aircraft, through a converter to give 110 v AC. The rubber pipes and sample chamber were cleaned on take off by allowing an unrestricted flow of air for 15–20 min. For sampling, the air flow was restricted to 141 L min^{-1} by clamping the hoses. Sampling thus was not isokinetic. Two Petri dishes, containing a modified Czapek's agar with 0.1% yeast extract and 0.5% glucose, could be exposed simultaneously, or one could be replaced by a Petri dish or slide coated with silicone grease allowing simultaneous collection of viable and total particles to determine cell or spore viability. Positive pressure was maintained in the chamber when changing plates or slides to minimize air exchange with the atmosphere of the aircraft. Sample times in summer in continental tropic air masses were 2–5 min and in polar air 4–10 min, while in winter samples were up to 1 h. Flights were mostly at 1.8–3.0 km altitude at ≈ 320 km h^{-1}.[186]

Exposed Petri dishes were incubated for 5 d at 25°C before colonies were identified and counted. Silicone-coated plates and slides were counted microscopically.

Filter Sampler[188]

A filter holder and flow meter were connected in line by hose to aluminum elbows projecting through the fuselage of DC-6 (North Star) or C-54 aircraft as was described above. Filters consisting of 16 discs of glass fiber and five discs of lens paper were cut with a 2.86 cm diameter cork borer and mounted in a brass hose coupling before steam sterilization. Filters were exposed anisokinetically at a flow rate of 42–85 l min^{-1}. Air was sampled for 10–20 min in continental tropic air masses and 20–40 min in polar air in summer, and for up to 1 h in winter. Flights mostly were at 1.8–3.0 km at ≈ 320 km h^{-1}.[186] After exposure, the filter discs were shaken in 50 mL sterile water for 10 min and 2 or 5 mL samples were placed in Petri dishes and mixed with cooled agar (a modified Czapek's agar with 0.1% yeast extract and 0.5% glucose). Plates were incubated at 25°C for 5 d before colonies were identified and counted. Recovery of bacteria was poor,[186] through poor trapping efficiency or through loss of viability or binding to the filter material.

Straight Tube Ram Jet Impactor[242]

The straight tube ram jet impactor was proposed as a high-volume sampling which was automatically isokinetic and which would collect small particles efficiently. Air would be sampled through an aerodynamically designed, forward directed probe which carried air to the trapping surface from where it exhausted smoothly without obstruction. An orifice 1.4 cm diameter would sample at 800 L min^{-1}, at an air speed of 320 km h^{-1}, and sample particles down to 2 μm while a jet 3 mm diameter would collect 60% of particles 1 μm diameter. However, a circular deposit is difficult to scan and a 2.5 × 0.5 cm jet may be more practical, sampling at 700 L min^{-1} at 320 km h^{-1}. The jet should be separated from the trapping surface by 1/2 to 1/3 of the jet diameter. The trapping surface either could be a loop of suitably coated 35 mm cine film moved from inside the aircraft, or agar in small dishes pushed through the probe shaft to the impact point. Deposits on film could be divided and mounted on glass slides for microscopic examination and those on agar plates cultured suitably. However, there have been no reports of the testing and use of this sampler.

Membrane Filtration

Membrane filters in an isokinetic sampler mounted on a Cessna 180 aircraft have been used to trap marine algae, bacteria, pollen, and spores in air masses.[235,247]

Cyclone Sampler[367]

Cyclone samplers (350 L min^{-1}) carried on a twin-engine Beachcraft have been employed with a scrubbing liquid of glycerol (50 mL), Tris (hydroxy methyl) aminomethane (0.364 g) Polyglycol P2000 (2 drops) in 1 L of distilled water supplied to the air inlet at ca. 1 mL min^{-1}, which was subsequently cultured on tryptone glucose extract agar + cycloheximide.

Sampling from Rockets[171]

An ingenious method of trapping onto films that had been glued together, sterilized and then loaded aboard rockets was used to sample airborne particulates at altitudes up to 85 km. A special device separated the films, exposed the inner surface of one to the atmosphere as it moved through the exposure area at ca. 4.5 m sec^{-1} and then rerolled and resealed it. After recovery, the films were cultured, revealing airborne bacteria, mainly *Mycobacterium luteum* and *Micrococcus albus*, at altitudes of 48–77 km.

Culture Media and Incubation Temperatures for Airborne Sampling

A wide range of different culture media have been employed for airborne sampling including:

X medium (peptone-yeast autolysate) for total microorganisms, X medium + 10 mL L^{-1} 1% cycloheximide, pH 4, for bacteria and yeasts and M-A agar (dextrose-peptone-ammonium nitrate-potassium phosphate for fungi);[121] tryptone glucose yeast extract agar (tryptone 15 g, glucose 1 g, yeast extract 1 g, agar 15 g, water 1 L);[250] and modified Czapek's agar with 0.1% yeast extract and 0.5% glucose.[188]

Other media worth testing for their ability to isolate organisms, either directly during high altitude sampling or indirectly from catches on other substrates might include, for fungi:

Malt extract agar (2% malt extract, 1.5–2% agar + 20 IU penicillin and 40 units streptomycin mL^{-1} or alternative antibiotics to suppress bacteria;[197] dichloran rose bengal chloramphenicol agar (Oxoid);[189] and DG18 agar (Oxoid)[159] especially suitable for *Aspergillus* and *Penicillium* spp. but will grow other species while the glycerol content may counteract freezing.

and for bacteria:

Half-strength nutrient agar (Nutrient agar, Oxoid, at half recommended strength + 1% agar + 100 µg cycloheximide (Sigma) mL^{-1});[197] tryptone soya agar; tryptone yeast extract agar (Oxoid + 50 µg cycloheximide mL^{-1});[34] and casein peptone starch agar (at 1/10th strength + 50 µg cycloheximide mL^{-1}).[379]

Meticulous attention to contamination control is necessary when preparing plates.[127]
Surprisingly few problems with freezing of exposed medium have been recorded although icing caused frequent changes of plates when sampling over the Arctic.[284] The altitude at which freezing temperatures occur depends on altitude, season and latitude. In Britain, this is above 1–5 km in summer and 0–3 km in winter.[157] Silicone coated plates have been used for airborne sampling in winter.[188,267] Cooled, molten agar then was added on returning to the laboratory although this could kill some sensitive organisms.[112] Addition of 2.5% carboxyl methyl cellulose or 3% corn starch to media also could prevent freezing.[124] Low humidities, < 25% in anticyclonic weather could cause desiccation of

media and microorganisms but could be limited by the addition of 0.1 mL 0.2% OED to the agar surface.[34]

A range of incubation temperatures and periods have been applied from 25°C for 5 d,[188] through 30°C for 3–5 d[121] to 35°C for 24 h.[250] However, lower temperatures, perhaps even 5–15°C for up to 6–8 weeks, would minimize the shock of transfer from the external environment.

Sampling Plans

Most airborne sampling has been along transects at uniform altitudes. Kelly and his colleagues[186,188,267–269] flew mostly at 1.8–3.0 km at ≈ 320 km h^{-1} while Fulton[118–120] used three aircraft sampling simultaneously at three different altitudes, either 152, 1066 and 1981 m (over sea), 365, 1280 and 2500 m or 690, 1600 and 3127 m (over land). For the over-sea samples, Fulton[119] sampled initially at right angles to the wind direction, at increasing distances from the coast, and only on the return flights were successive samples taken parallel to the wind direction. Hirst et al.[157,158] used three different types of sampling patterns to determine concentrations of fungal spores at different altitudes, to intercept immigrant spore clouds from continental Europe to the British Isles, and to obtain vertical sections through spore clouds:

1. Spiral ascents and descents with straight level sampling positions at different altitudes;
2. Level flights at 600 m for 185–555 km perpendicular to the wind direction over the sea to detect immigrant spore clouds with, near the center of the transect, a spiral ascent from 300–3000 m;
3. Repeated ascents and descents along a transect parallel to the wind direction with a maximum altitude of 1800–3000 m, taking samples at different altitudes on both ascents and descents.

Sampling should be related to meteorological conditions and to atmospheric layers, i.e., troposphere, stratosphere, since material crossing the tropopause into the stratosphere (~10 km) is likely to remain airborne for considerable periods. The altitude of the tropopause changes with meteorological conditions and latitude from 6–8 km at the poles to 17 km at the equator. Thus, no one height is representative of any atmospheric layer and in meteorology to define layers it is usual to employ constant pressure levels where the physical properties of the air mass are comparable. For comparison with other studies, a sampling altitude of about 3000 m, or a constant pressure level of either 700 mb (about 3000 m) or 500 mb (5–6000 m) should be used. Spiral ascents and descents should be inserted to relate to meteorological phenomena. Full records must be kept of meteorological parameters, including the positions of fronts relative to samples, convective cloud systems, origins of air masses, temperature, pressure, relative humidity, wind speed and direction, etc. If possible, samples should allow assessment of both total and viable numbers.

LARGE VS. SMALL BIOAEROSOL PARTICLES: CHOICE OF SAMPLERS

The choice and positioning of samplers and the timing of samples must depend on species studied, spore size, mode of dispersal, abundance in the air and time of spore release as well as the objective of the study. Large particles especially >100 μm diameter, are difficult to trap, although the vertical axis rotating arm sampler of Hameed et al.[140] was designed to eliminate the effects of both wind speed and particle size, collecting particles 35–250 μm diameter. Large pollens and fungal spores may be underestimated by suction samplers because of the effects of anisokinetic sampling and yaw and because the rotation of the samplers rarely keeps pace with changes in wind direction.[246] For these large spores, impaction on cylinders is quite efficient but there has to be air movement for spores to be caught. Rotorod samplers trap efficiently down to about 10 μm diameter, are independent of wind direction and speed and have high sampling rates. However, they cannot be operated continuously for long periods without overloading. Different species release their spores at different times of day depending on the method of spore liberation. Species requiring drying conditions for liberation, e.g., *Phytophthora infestans*, *Botrytis fabae,* are most abundant during early morning, those requiring water, e.g., many ascomycetes, mostly at night and those released mechanically, e.g., *Erysiphe graminis*, around midday when wind speeds and turbulence are greatest. Species such as *E. graminis* and *Ustilago* spp. have well-defined seasons for spore release, corresponding respectively, to the availability of green leaf tissue and inflorescences for infection. Consequently, for a particular species it may not be necessary to trap or to record spore concentrations over the whole 24 h period, or the whole year. Measurement of dispersal gradients, both vertical and horizontal, requires arrays of traps that need to be easily constructed and inexpensive.

By contrast, bacterial and viral particles may be smaller than 10 μm and require more efficient methods of sampling with sufficiently small cut-off diameters that allow recovered microbes to be grown, e.g., slit samplers, Andersen samplers, impingers and large volume air samplers. However, often these cannot be used away from a power source and filtration may be necessary, provided viability is not affected.

Often in aerobiological studies, the physical characteristics of samplers and their requirements for optimum sampling have been neglected. Often, samplers have been inappropriate for the particle or environment being studied, have had a poor collection efficiency or have not been operated isokinetically, so that results from different studies are not comparable.

SUMMARY

Many methods are available for outdoor sampling but often these have been applied inappropriately so that they yield incomplete or biased results. Few methods either allow isokinetic sampling or are independent of air movement. No one method is suitable for all purposes but different methods used with care can yield complementary results. Sedimentation slides and settle plates have been used widely and are simple to operate but they are affected by particle size and air movement, and their results are difficult to interpret and compare, except when conditions are still and the particles of interest are uniformly sized. Vertical cylinders trap particles efficiently, provided there is some air movement, and efficiency is less affected by particle size. Rotorods are inexpensive, portable, easily constructed samplers, independent of wind speed with high trapping efficiency down to about 10 μm that can be set out in large arrays to determine dispersal gradients. Also they can be operated sequentially, to provide time discrimination, or intermittently, to prevent overloading and provide daily mean concentrations. Suction traps also can trap efficiently down to about 10 μm, and with a moving trapping surface allow time discrimination and unattended sampling for up to 7 days. Changes in wind direction cause the body of the trap to rotate so that the orifice faces into the wind but the suction speed is a compromise between isokinetic sampling in a changing wind speed and the requirements for efficient impaction. However, microscopic assessment of fungi on spore trap slides is laborious and requires skill, while many spore types fail to grow or sporulate in culture or produce alternative spore forms. Andersen samplers provide efficient sampling down to <1 μm and growth in culture, with size discrimination, but plates are easily overloaded in heavily contaminated atmospheres. Liquid impingers and cyclones are most suitable for sampling airborne bacteria. They are not limited by overloading since catches can be diluted before plating and liquid impingers retain particles down to ca. 1 μm efficiently, although some types may subject cells to strong shear forces that may affect viability. Filtration provides an inexpensive and flexible sampling method that can be assessed in many different ways: by light and scanning microscopy, culture, immunochemically and chemically. However, sampling heads often are not designed well for isokinetic sampling and some microorganisms, especially Gram-negative bacteria, may be subject to desiccation damage. Litton-type electrostatic samplers have been operated successively for the collection of airborne antigens for immunochemical assay, with better recoveries than by high volume filtration, but are expensive and possibly not available commercially. Other types of sampler have been little utilized but may be useful for some purposes. There is no agreed method for airborne sampling and a range of methods have been applied. However, when sampling from aircraft few allow isokinetic sampling and sometimes they even fail to sample air undisturbed by the aircraft movement. Sampling of airborne microorganisms out of doors, whether for allergens or for human, animal or plant pathogens, and whether at ground level or at high altitudes, requires a broad knowledge of sampling methods and their limitations.

ACKNOWLEDGMENTS

We thank Dr. H.A. McCartney, Mrs. M.E. Lacey, Dr. Harriet Burge and Dr. Mary Kay O'Rourke for their help with discussions, illustrations and access to unpublished material used in the preparation of this chapter.

REFERENCES

1. Adams, A.P., H.B. Garbett, H.B. Rees, and B.G. Lewis. "Bacterial aerosols from cooling towers," *J. Water Pollut. Contr. Fed.* 50:2362–2369 (1978).
2. Adams A.P. and J.C. Spendlove. "Coliform aerosols emitted by sewage treatment plants," *Science* 169:1218–1220 (1970).
3. Adams, A.P., J.C. Spendlove, E.C. Rengers, and M. Garbett. "Emission of microbial aerosols from vents of cooling towers. I. Particle size, source strength and downwind travel," *Dev. Ind. Microbiol.* 20:769–779 (1979).
4. Agarwal, M.K., J.W. Yunginger, M.C. Swanson, and C.E. Reed. "An immunochemical method to measure atmospheric allergens," *J. Allergy Clin. Immunol.* 68:194–200 (1981).
5. Agashe, S.N. "Airborne entomophilous pollen—potential source of allergy," *Rec. Res. Ecol. Environ. Pollut.* 3:153–157 (1985).
6. Agashe, S.N. and J.N. Abraham. "Pollen calendar of Bangalore City: Part I," *Ind. J. Allergy Appl. Immunol.* 1:35-38 (1988).
7. Akesson, N.B. and W.E. Yates. "Problems relating to application of agricultural chemicals and resulting drift residues," *Annu. Rev. Entomol.* 9:285–318 (1964).
8. Al-Doory, Y. and J.F. Domson. *Mould Allergy.* (Philadelphia, PA: Lea and Febiger, 1984).

9. Allen, M.F., L.E. Hipps, and G.L. Wooldridge. "Wind dispersal and subsequent establishment of VA mycorrhizal fungi across a successional arid landscape," *Landscape Ecol.* 2:165–171 (1989).
10. American Public Health Association. "Standard methods for the examination of water and wastewater," 14th Edition (Washington, DC: American Public Health Association, 1976).
11. Ånäs, P. and Å. Bovallius. "Investigation of natural airborne bacteria by numerical taxonomy," in *Proceedings of the First International Conference on Aerobiology,* Munich, 13–15 August 1978, (Berlin: Eric Schmidt Verlag, 1980), pp. 452–455.
12. Andersen, A.A. "New sampler for the collection, sizing and enumeration of viable airborne particles," *J. Bacteriol.* 76:471–484 (1958).
13. Andersson, R., B. Bergström, and B. Bucht. "Air dispersal of bacteria from a sewage treatment plant," *Vatten* 2:117–123 (1973).
14. Andersson, R., B. Bergström, and B. Bucht. "Outdoor sampling of airborne bacteria; results and experiences," in *Airborne Transmission and Airborne Infection,* J.F.P. Hers and K.C. Winkler, Eds. (Utrecht: Oosthoek Publishing Company, 1973), pp. 58–62.
15. Asai, G.N. "Intra- and inter-regional movement of black stem rust in the upper Mississippi Valley," *Phytopathology* 50:535–541 (1960).
16. Athalye, M., J. Lacey, and M. Goodfellow. "Selective isolation and enumeration of actinomycetes using rifampicin," *J. Appl. Bact.* 51:289–297 (1981).
17. Aubert, M. and J. Aubert. "Transfer of microorganisms by marine aerosols," in *Airborne Transmission and Airborne Infection,* J.F.P. Hers and K.C. Winkler, Eds. (Utrecht: Oosthoek Publishing Company, 1973) pp. 66–71.
18. Aylor, D.E. "Deposition of particles in a plant canopy" *J. Appl. Meteorol.* 14:52–57 (1975).
19. Aylor, D.E., H.A. McCartney, and A. Bainbridge. "Deposition of particles liberated in gusts of wind," *J. Appl. Meteorol.* 20:1212–1221 (1981).
20. Bagni, N., H. Charpin, R.R. Davies, N. Nolard, and E. Stix. "City spore concentrations in the European Economic Community (EEC)," *Clin. Allergy* 6:61–68 (1976).
21. Bagni, N., R.R. Davies, M. Mallea, N. Nolard, F.T. Spieksma, and E. Stix. "Sporenkonzentrationen in Städten der Europäischen Gemeinschaft (EG)," *Acta Allergol.* 32:118–138 (1977).
22. Bainbridge, A. and O.J. Stedman. "Dispersal of *Erysiphe graminis* and *Lycopodium clavatum* spores near to the source in a barley crop," *Ann. Appl. Biol.* 91:187–198 (1979).
23. Banttari, E.E. and J.R. Venette. "Aerosol spread of plant viruses: Potential role in disease outbreaks" *Ann. N.Y. Acad. Sci.* 353:167–173 (1980).
24. Bassett, I.J., C.W. Crompton, and J.A. Parmelee. *An Atlas of Airborne Pollen Grains and Common Fungus Spores of Canada.* Research Branch Canada Department of Agriculture Monograph No. 18 (1978).
25. Batchelder, G.L. "Sampling characteristics of the Rotorod, Rotoslide and Andersen machines for atmospheric pollen and spores," *Ann. Allergy* 39:18–27 (1977).
26. Bausum, H.T., S.A. Schaub, K.F. Kenyon, and M.J. Small. "Comparison of coliphage and bacterial aerosols at a wastewater spray irrigation site," *Appl. Environ. Microbiol.* 43:28–38 (1982).
27. Beaumont, F., H.F., Kauffman, T.H. van der Mark, H.J. Sluitter, and K. de Vries. "Volumetric aerobiological survey of conidial fungi in the north-east Netherlands," *Allergy* 40:173–180 (1985).
28. Benaim, C. "Microculture method for some viable airborne particles," *J. Allergy Clin. Immunol.* 55:203–206 (1975).

29. Billing, E. "The taxonomy of bacteria on the aerial parts of plants," in *Microbiology of Aerial Plant Surfaces,* C.H. Dickinson and T.F. Preece, Eds. (London: Academic Press, 1976), pp. 223–273.
30. Blackley, C.H. *Experimental Researches on the Causes and Nature of Catarrhus Aestivus (Hay Fever or Hay Asthma)* (London: Baillière, Tindall and Cox, 1873).
31. Blanchard, D.C. "Bubble scavenging and the water-to-air transfer of organic material in the sea," in *Advances in Chemistry Series,* No. 145 (Albany, NY: Atmospheric Sciences Research Center, State University of New York, 1975), pp. 360–387.
32. Blomquist, G., U. Palmgren, and G. Ström. "Improved techniques for sampling airborne fungal particles in highly contaminated environments," *Scand. J. Work Environ. Health* 10:253–258 (1984).
33. Bourdillon, R.B., O.M. Lidwell, and J.C. Thomas. "A slit sampler for collecting and counting airborne bacteria," *J. Hyg., Camb.* 41:197–224 (1941).
34. Bovallius, Å., B. Bucht, R. Roffey, and P. Ånäs. "Three year investigation of the natural airborne bacterial flora at four localities in Sweden," *Appl. Environ. Microbiol.* 35:847–852 (1978).
35. Bovallius, Å., R. Roffey, and E. Henningson, "Long-range transmission of bacteria," *Ann. N.Y. Acad. Sci.* 353:186–200 (1980).
36. Bovallius, Å. and P. Ånäs. "Bacterial spores in the air and on the ground in connection with biological insect control," in *Proceedings of the First International Conference on Aerobiology, Munich, 13–15 August 1978* (Berlin: Eric Schmidt Verlag, 1980), pp. 227–231.
37. Brennan, R.M. "Observations on the splash dispersal of *Leptosphaeria nodorum* spores in the laboratory and field," Ph.D. Thesis, University of Manchester (1985).
38. Brenner, K.P., P.V. Scarpino, and C.S. Clark. "Animal viruses, coliphages, and bacteria in aerosols and wastewater at a spray irrigation site," *App. Environ. Microbiol.* 54:409–415 (1988).
39. Broady, P.A. "Example of a research proposal for an antarctic field project: Algal colonization of the McMurdo Dry Valleys region," in *BIOTAS Manual of Methods,* D.D. Wynn-Williams, Ed. (Cambridge: Scientific Committee on Antarctic Research, 1992), 16 pp.
40. Brown, H.M. "An automatic volumetric culture plate slit sampler," in *Airborne Transmission and Airborne Infection,* J.F.P. Hers and K.C. Winkler, Eds. (Utrecht: Oosthoek Publishing Company, 1973), pp. 57–58.
41. Brown H.M. and F.A. Jackson. "Aerobiological studies based in Derby," *Clin. Allergy* 8:589–597 (1978).
42. Brown, R.M. "Studies of Hawaiian fresh-water and soil algae. I. The atmospheric dispersal and fern spores across the island of Oahu, Hawaii," in *Contributions in Phycology.* B.C. Parker and R.M. Brown, Eds. (Lawrence, KS: Allen Press Inc., 1971), pp. 175–188.
43. Brown, R.M., D.A. Larson, and H.C. Bold. "Airborne algae: Their abundance and heterogeneity," *Science* 143:583–585 (1964).
44. Bruch, C.W. "Microbes in the upper atmosphere and beyond," in *Airborne Microbes.* P.H. Gregory and J.L. Monteith, Eds. (*Symp. Soc. Gen. Microbiol.* 17) (Cambridge: The University Press, 1967), pp. 345–374.
45. Burge, H.P., J.R. Boise, J.A. Rutherford, and W.R. Solomon. "Comparative recoveries of airborne fungus spores by viable and non-viable modes of volumetric collection, *Mycopathologia* 61:27–33 (1977a).
46. Burge, H.P., W.R. Solomon, and J.R. Boise. "Comparative merits of eight popular media in aerometric studies of fungi," *J. Allergy Clin. Immunol.* 60:199–203 (1977b).
47. Burrows, R., "Excretion of foot and mouth disease virus prior to the development of lesions," *Vet. Rec.* 82:387–388 (1968).

48. Buttner, M.P. and L. Stetzenbach. "Evaluation of four aerobiological sampling methods for the retrieval of aerosolized *Pseudomonas syringae*," *Appl. Environ. Microbiol.* 57:1268–1270 (1991).
49. Cadham, F.T. "Asthma due to grain rusts," *J. Am. Med. Assn.* 83:27 (1924).
50. Campbell, C.C. "(Philosophical) review of air currents as a continuing vector, *Ann. N.Y. Acad. Sci.* 353:123–139 (1980).
51. Carnow, B., R. Northrop, R. Wadden, S. Rosenberg, J. Holden, A. Neal, L. Sheaff, P. Scheff, and S. Meyer, "Health effects of aerosols emitted from an activated sludge plant," *U.S. Environmental Protection Agency Report*, 600/1-79-019, 215 pp. (1979).
52. Carroll, T.W., and D.E. Mayhew. "Anther and pollen infection in relation to the pollen and seed transmissibility of two strains of barley stripe mosaic virus in barley," *Can. J. Bot.* 54:1604-1621 (1976).
53. Carter, M.V. "Ascospore deposition in *Eutypa armeniacae*," *Aust. J. Agric. Res.* 16: 825-836 (1965).
54. Carter, M.V., W.J. Moller, and S.M. Pady. "Factors affecting uredospore production and dispersal in *Tranzschelia discolor*," *Aust. J. Agric. Res.* 21:905-914 (1970).
55. Carter, M.V., A.S.J. Yap, and S.M. Pady. "Factors affecting uredospore liberation in *Puccinia antirrhini*," *Aust. J. Agric. Res.* 21:921-925 (1970).
56. Chamberlain, A.C. "Transport of *Lycopodium* spores and other small particles to rough surfaces," *Proc. Roy. Soc., A* 296:45-70 (1966).
57. Chamberlain, A.C. and R.C. Chadwick. "Deposition of spores and other particles on vegetation and soil," *Ann. Appl. Biol.* 71:141-158 (1972).
58. Charpin, H., R.R. Davies, N. Nolard, F.T.M. Spieksma, and E. Stix. "Concentration urbaine des spores dans les pays de la communauté économique européenne. III. Les Urticacées," *Rev. Fr. Allergol.* 17:181-187 (1977).
59. Chatigny, M.A., J.M. Macher, H.A. Burge, and W.R. Solomon. "Sampling airborne microorganisms and aeroallergens," in *Air Sampling Instruments for Evaluation of Atmospheric Contaminants*, 7th ed., S.V. Hering, Ed. (Cincinnati, OH: American Conference of Governmental Industrial Hygienists, 1989), pp. 199-220.
60. Chatterjee, J. and F.E. Hargreave. "Atmospheric pollen and fungal spores in Hamilton in 1972 estimated by the Hirst automatic volumetric spore trap" *CMA J.* 110:659-663 (1974).
61. Cherry, E. and C.E. Peet. "An efficient device for the rapid collection of fungal spores from infected plants," *Phytopathology* 56:1102-1103 (1966).
62. Christensen, L.B., O. Egelund, C. Goldberg, A.H. Johansen, P. Nielsen, and G. Poulsen. "Registration of airborne pollen with the Burkard trap," in *Pollen and Mould Spore Counts in Denmark: Description of Methods*, B.N. Petersen, L. Larsen, O. Egelund, H. Buch, F. Jurgensen, and S. Gravesen, Eds. (Hillerød: Pharmacia, 1986), pp. 4-13.
63. Cocke, E.C. "Calculating pollen concentration of the air," *J. Allergy* 8:601-606 (1937).
64. Constantinidou, H.A., S.S. Hirano, L.S. Baker, and C.D. Upper. "Atmospheric dispersal of ice nucleation-active bacteria: The role of rain," *Phytopathology* 80:934-937 (1990).
65. Converse, J.L. and R.E. Reed. "Experimental epidemiology of coccidioidomycosis," *Bact. Rev.* 30:678-694 (1966).
66. Corbaz, R. "Études des spores fongiques captées dans l'air. I. Dans un champ de blé," *Phytopath. Z.* 66:69-70 (1969).
67. Corbaz, R. "Etudes des spores fongiques captées dans l'air. II. Dans un vignoble," *Phytopath. Z.* 74:318-328 (1972).
68. Cour, P. "Nouvelles techniques de detection des flux et des retombées polliniques: Étude de la sédimentation des pollens et des spores a la surface du sol," *Pollen et Spores* 15 (1974).

69. Cox, C.S. *The Aerobiological Pathway of Microorganisms.* (Chichester: John Wiley, 1987).
70. Crawford, G.V. and P.H. Jones. "Sampling and differentiation techniques for airborne organisms emitted from wastewater," *Water Res.* 13:393-399 (1979).
71. Crook, B. Unpublished results.
72. Crook, B., S. Higgins, and J. Lacey, "Airborne gram negative bacteria associated with the handling of domestic waste," in *Advances in Aerobiology,* G. Boehm and R.M. Leuschner, Eds. *(Experientia Suppl.* 51), (Basel: Birkhauser, 1987), pp. 371-375.
73. Crook, B., S. Higgins, and J. Lacey. "Methods for sampling airborne microorganisms at solid waste disposal sites," in *Biodeterioration 7,* D.R. Houghton, R.N. Smith, and H.O.W. Eggins, Eds. (London: Elsevier Applied Science, 1988), pp. 791-797.
74. Cross, T. "Thermophilic actinomycetes," *J. Appl. Bact.* 31:36-53 (1968).
75. D'Amato, G., J. Mullins, N. Nolard, F.T.M. Spieksma, and R. Wachter. "City spore concentrations in the European Economic Community (EEC). VII. Oleaceae (Fraxinus, Ligustrum, Olea)," *Clin. Allergy* 18:541-547 (1988).
76. Darke, C.S., J. Knowelden, J. Lacey, and A.M. Ward. "Respiratory disease of workers harvesting grain," *Thorax* 31:294–302 (1976).
77. Davids, D.E., and D.C. O'Connell. "A man-operated particulate aerosol sampler," in *Airborne Transmission and Airborne Infection,* J.F.P. Hers and K.C. Winkler, Eds. (Utrecht: Oosthoek Publishing Company, 1973), pp. 42-47.
78. Davies, R.R. "A study of air-borne *Cladosporium,*" *Trans. Br. Mycol. Soc.* 40:409-414 (1957).
79. Davies, R.R., M.J. Denny, and L.M. Newton. "A comparison between the summer and autumn air-spores at London and Liverpool," *Acta Allergol.* 18:131–147 (1963).
80. Davies, R.R. "Air sampling for fungi, pollens, and bacteria," in *Methods in Microbiology,* Vol. 4, C. Booth, Ed. (London: Academic Press, 1971), pp. 367–404.
81. De Pagny, J.M. and J.A. Peterson. "Long distance transport of pollen and spores in the Antarctic," *Abstracts, Antarctic Science Advisory Committee Conference and Workplace on Antarctic Weather and Climate, Flinders, South Australia,* 5–7 July, 1989, pp. 14–15 (1989).
82. Deadman, M.L. and B.M. Cooke. "An analysis of rain-mediated dispersal of *Drechslera teres* conidia in field plots of spring barley," *Ann. Appl. Biol.* 115:209–214 (1989).
83. Delany, A.C. and S. Zenchelsky. "The organic component of wind-erosion-generated soil-derived aerosol," *Soil Sci.* 121:146–155 (1976).
84. Dillon Weston, W.A.R. "Observations on the bacteria and fungal flora of the upper air," *Trans. Br. Mycol. Soc.* 14:111–117 (1929).
85. Dimmick, R.L., H. Wolochow, and M.A. Chatigny. "Evidence that bacteria can form new cells in airborne particles," *Appl. Environ. Microbiol.* 37:924–927 (1979).
86. Donaldson, A.I. "Factors influencing the dispersal, survival and deposition of airborne pathogens of farm animals," *Vet. Bull.* 48:83–94 (1978).
87. Druett, H.A., and K.R. May. "Unstable germicidal pollutant in rural air," *Nature, Lond.* 220:395–396 (1968).
88. Dunez, J. "Le pollen, un vecteur de virus," *Rev. Cytol. Biol. Végétales, Le Botaniste* 5:21–29 (1982).
89. Durham, O.C. "The volumetric incidence of atmospheric allergen. II. Simultaneous measurements by volumetric and gravity slide methods. Results with ragweed pollen and *Alternaria* spores," *J. Allergy* 15:226–235 (1944).
90. Durham, O.C. "The volumetric incidence of atmospheric allergens. IV. A proposed standard method of gravity sampling, counting and volumetric interpolation of results," *J. Allergy* 17:79–86 (1946).
91. Edmonds, R.L. "Collection efficiency of rotorod samplers for sampling fungus spores in the atmosphere," *Pl. Dis. Reptr.* 56:704–708 (1972).

92. Edmonds, R.L. and W. Littke. "Coliform aerosols generated from the surface of dewatered sewage applied to a forest clearcut," *Appl. Environ. Microbiol.* 36:972-974 (1978).
93. Eduard, W., J. Lacey, K. Karlsson, U. Palmgren, G. Ström, and G. Blomquist. "Evaluation of methods for enumerating microorganisms in filter samples from highly contaminated occupational environments," *Am. Ind. Hyg. Assn. J.* 51:427–436 (1990).
94. Erdtman, G. "Pollen grains recovered from the atmosphere over the Atlantic," *Acta Hort. Gothoburg.* 12:185–196 (1937).
95. Errington, F.P. and E.O. Powell. "A cyclone separator for aerosol sampling in the field," *J. Hyg. Camb.* 67:387-399 (1969).
96. Eversmeyer, M.G. and C.L. Kramer. "Vertical concentrations of fungal spores above wheat fields," *Grana* 26:97-102 (1987a).
97. Eversmeyer, M.G. and C.L. Kramer. "Single versus multiple sampler comparisons," *Grana* 26:109-112 (1987b).
98. Ewetz, L. and J. Lundin. "Luminol chemiluminescence technique for detection of microorganisms," in *Airborne Transmission and Airborne Infection,* J.F.P. Hers and K.C. Winkler, Eds. (Utrecht: Oosthoek Publishing Company, 1973), pp. 23–26.
99. Fannin, K.F. "Methods for detecting viable microbial aerosols," in *Wastewater Aerosols and Disease,* H. Pahren and W. Jakubowski, Eds., EPA-600/9-80-028 (Cincinnati, OH: U.S. Environmental Protection Agency, 1980), pp. 1-22.
100. Fannin, K.F., J.J. Gannon, K.W. Cochran, and J.C. Spendlove. "Field studies on coliphages and coliforms as indicators of animal virus contamination from wastewater treatment facilities," *Water Res.* 11:181-188 (1977).
101. Fannin, K.F., J.C. Spendlove, K.W. Cochran, and J.J. Gannon. "Airborne coliphages from wastewater treatment facilities," *Appl. Environ. Microbiol.* 31:705-710 (1976).
102. Fannin, K.F., S.C. Vana, and W. Jakubowski. "Effect of an activated sludge wastewater treatment plant on ambient air densities of aerosols containing bacteria and viruses," *Appl. Environ. Microbiol.* 49:1191-1196 (1985).
103. Fattal, B. and B. Telsch. "Viruses in wastewater aerosols," *Environ. Int.* 7:35 (1982).
104. Faulkner, M.J. and J. Colhoun. "An automatic spore trap for collecting pycnidiospores of *Leptosphaeria nodorum* and other fungi from the air during rain and maintaining them in a viable condition," *Phytopath. Z.* 89:50-59 (1977).
105. Fedorak, P.M. and D.W.S. Westlake. "Airborne bacterial densities at an activated sludge treatment plant," *J. Water Pollut. Contr.* 52:2185–2192 (1980).
106. Ferrandino, F.J. and D.E. Aylor. "Relative abundance and deposition gradients of clusters of urediniospores of *Uromyces phaseoli,*" *Phytopathology* 77:107-111 (1987).
107. Ficke, W., F. Ehrig, M. Nachtigall, K. Naumann, K. Richter, H.-J. Schaefer, and R. Zielke. "Möglichkeiten der Erfassung in der Luft befindlicher Zellen des Feuerbranderregers *(Erwinia amylovora* (Burrill) Winslow et al.). (Possibilities for detecting of the causal agent of fireblight *(Erwinia amylovora* (Burrill) Winslow et al.) in the air space of orchards," *Zentlbl. Mikrobiol.* 145:121-133 (1990).
108. Fitt, B.D.L. and A. Bainbridge. "Dispersal of *Pseudocercosporella herpotricoides* spores from infected wheat straw," *Phytopath. Z.* 106:214-225 (1983).
109. Fitt, B.D.L., N.F. Creighton, and A. Bainbridge. "The role of wind and rain in the dispersal of *Botrytis fabae* conidia," *Trans. Br. Mycol. Soc.* 85:307-312 (1985).
110. Fitt, B.D.L., M. Lysandrou, and R.H. Turner. "Measurement of spore-carrying splash droplets using photographic film and an image-analyzing computer," *Plant Pathol.* 31:19-24 (1982).
111. Fitt, B.D.L., H.A. McCartney, and P.J. Walklate. "The role of rain in dispersal of pathogen inoculum," *Annu. Rev. Phytopath.* 27:241-270 (1989).
112. Flannigan, B. "The use of pour plates for yeast counts," *J. Stored Prod. Res.* 10:61-64 (1974).

113. Fontanges, R. and P. Isoard. "A new apparatus with large air intake for the sampling of atmospheric microconstituents," in *Airborne Transmission and Airborne Infection* J.F.P. Hers and K.C. Winkler, Eds. (Utrecht: Oosthoek Publishing Company, 1973), pp. 62-66.
114. Franc, G.D., M.D. Harrison, and D. Maddox. "The presence of *Erwinia carotovora* in snow and surface water in the United States," in *Report of the International Conference on Potato Blackleg Disease 26-29 Jun 1984,* D.C. Graham and M.D. Harrison, Eds. (Edinburgh: Royal Society of Edinburgh, 1985a), pp. 46-47.
115. Franc, G.D., M.D. Harrison, and M.L. Powelson. "The presence of *Erwinia carotovora* in ocean water, rain water and aerosols," in *Report of the International Conference on Potato Blackleg Disease 26-29 Jun 1984* D.C. Graham and M.D. Harrison, Eds. (Edinburgh: Royal Society of Edinburgh, 1985b), pp. 48-49.
116. Franc, G.D., M.D. Harrison, and M.L. Powelson. "The dispersal of phytopathogenic bacteria," in *The Movement and Dispersal of Agriculturally Important Biotic Agents,* D.R. MacKenzie, C.S. Barfield, G.G. Kennedy, and R.D. Berger, Eds. (Baton Rouge, LA: Claitor's Publishing Div., 1985c), pp. 37-49.
117. Frankland, A.W. and P.H. Gregory. "Allergenic and agricultural implications of airborne ascospore concentrations from a fungus *Didymella exitialis,"* *Nature Lond.* 245:336-337 (1973).
118. Fulton, J.D. "Micro-organisms of the upper atmosphere. III. Relationship between altitude and micropopulation," *Appl. Microbiol.* 14:237-240 (1966a).
119. Fulton, J.D. "Micro-organisms of the upper atmosphere. IV. Micro-organisms of a land air mass as it traverses an ocean," *Appl. Microbiol.* 14:241-244 (1966b).
120. Fulton, J.D. "Micro-organisms of the upper atmosphere. V. Relationship between frontal activity and micropopulation at altitude," *Appl. Microbiol.* 14:245-250 (1966c).
121. Fulton, J.D. and R.B. Mitchell. "Micro-organisms of the upper atmosphere. II. Micro-organisms in two types of air masses at 690 meters over a city," *Appl. Microbiol.* 14: 232-236 (1966).
122. Ganderton, M.A. *"Phoma* in the treatment of seasonal allergy due to *Leptosphaeria,"* *Acta Allergol.* 23:173 (1968).
123. Gillespie, V.L., C.S. Clark, H.S. Bjornson, S.J. Samuels, and J.W. Holland. "A comparison of two-stage and six-stage Andersen impactors for viable aerosols," *Am. Ind. Hyg. Assn. J.* 42:858-864 (1981).
124. Goff, G.D., J.C. Spendlove, A.P. Adams, and P.S. Nicholes. "Emission of microbial aerosols from sewage treatment plants that use trickling filters," *Health Serv. Rep.* 88: 640-652 (1973).
125. Grace, J. "Plant response to wind," in *Windbreak Technology,* J.R. Brandle, D.L. Hintz, and J.W. Sturrock, Eds. (Amsterdam: Elsevier Science Publishers, 1988), pp. 71-89.
126. Graham, D.C., C.E. Quinn, and L.F. Bradley. "Quantitative studies on the generation of aerosols of *Erwinia carotovora* var. *atroseptica* by simulated raindrop impaction on blackleg-infected potato stems," *J. Appl. Bact.* 43:413–424 (1977).
127. Greene, V.W., P.D. Pederson, D.A. Lundgren, and C.A. Hagberg. "Microbiological exploration of stratosphere: Results of six experimental flights," in *Proc. Atmospheric Biology Conference,* H.M. Tsuchiya and A.H. Brown, Eds. (Minneapolis: University of Minnesota, 1964), pp. 199–211.
128. Gregory, P.H. "Deposition of air-borne *Lycopodium* spores on cylinders," *Ann. Appl. Biol.* 38:357–376 (1951).
129. Gregory, P.H. "The construction and use of a portable volumetric spore trap," *Trans. Br. Mycol. Soc.* 37:390–404 (1954).
130. Gregory, P.H. *The Microbiology of the Atmosphere,* 2nd ed. (Aylesbury: Leonard Hill Books, 1973).

131. Gregory, P.H. "Aerobiology: past, present and future," *Int. Aerobiol. Newsl.* 17:9–16 (1983).
132. Gregory, P.H., E.D. Hamilton, and T. Sreeramulu. "Occurrence of the alga *Gloeocapsa* in the air," *Nature Lond.* 176:1270 (1955).
133. Gregory, P.H. and J.M. Hirst. "The summer air-spora at Rothamsted in 1952," *J. Gen. Microbiol.* 17:135–152 (1957).
134. Gregory, P.H. and M.E. Lacey. "The discovery of *Pithomyces chartarum* in Britain," *Trans. Brit. Mycol. Soc.* 47:25–30 (1964).
135. Gregory, P.H., T.J. Longhurst, and T. Sreeramulu. "Dispersion and deposition of airborne *Lydopodium* and *Ganoderma* spores," *Ann. Appl. Biol.* 49:645–658 (1961).
136. Gregory, P.H. and O.J. Stedman. "Deposition of airborne *Lycopodium* spores on plane surfaces," *Ann. Appl. Biol.* 40:651–674 (1953).
137. Hall, D.J. and S.L. Upton. "A wind tunnel study of the particle collection efficiency of an inverted frisbee used as a dust deposition gauge," *Atmospheric Environment* 22:1383–1394 (1988).
138. Hallin, P., G. Linfors, and G. Sandström. "Effectiveness of device for detection of biological aerosols in the laboratory," FOA Report C40175-B2 (1983).
139. Hallin, P., G. Linfors, and G. Sandström. "A device for the rapid detection of bacteria in air by a chemiluminescent technique," in *Nordic Aerobiology,* S. Nilsson, Ed. (Stockholm: Almquist and Wiksell, 1984), pp. 47–50.
140. Hameed, R., P.H. McMurry, and K.T. Whitby. "A new rotating coarse particle sampler," *Aerosol Sci. Technol.* 2:69–78 (1983).
141. Hamilton, E.D. "Studies in the air spora," *Acta Allerg.* 13:143–175 (1959).
142. Hamilton, R.I., E. Leung, and C. Nichols. "Surface contamination of pollen by viruses," *Phytopathology* 67:395–399 (1977).
143. Hammett, K.R.W. and J.G. Manners. "Conidium liberation in *Erisiphe graminis.* I. Visual and statistical analysis of spore trap records," *Trans. Br. Mycol. Soc.* 56:387–401 (1971).
144. Hanninen, H. "A method for identifying the flowering season from aerobiological data," in *Nordic Aerobiology,* S. Nilsson, Ed. (Stockholm: Almquist and Wiksell, 1984), pp. 51–57 (1984).
145. Harries, M.G., J. Lacey, R.D. Tee, G.R. Cayley, and A.J. Newman Taylor. "*Didymella exitialis* and late summer asthma," *Lancet* 1:1063–1066 (1985).
146. Harrington, M.A., G.C. Gill, and B.R. Warr. "High efficiency pollen samplers for use in clinical allergy," *J. Allergy* 48:357–375 (1959).
147. Harrison, M.D. "Aerosol dissemination of bacterial plant pathogens," *Ann. N.Y. Acad. Sci.* 353:94–104 (1980).
148. Hasenclever, H.F. and W.R. Piggott. "Air sampling for *Histoplasma capsulatum*," in *Ecological Systems Approaches to Aerobiology. II. Development Demonstration and Evaluation of Models* (US/IBP Program Handbook No. 3; NTIS No. AP-USIBP-H-73-3), R.L. Edmonds, and W.S. Benninghoff, Eds., (Ann Arbor, MI: University of Michigan, 1973), pp. 159–162.
149. Heise, H.A. and E.R. Heise. "The distribution of ragweed pollen and *Alternaria* spores in the upper atmosphere," *J. Allergy* 19:403–407 (1948).
150. Heisler, G.M., and Dewalle, D.R. "Effects of windbreak structure on wind flow," in *Windbreak Technology,* J.R. Brandle, D.L. Hintz, and J.W. Sturrock, Eds., (Amsterdam: Elsevier Science Publishers, 1988), pp. 41–69.
151. Hickey, J.C.S. and D.C. Reist. "Health significance of airborne microorganisms from wastewater treatment process," *J. Water Poll. Control Fed.* 47:2758–2773 (1975).
152. Hidy, G.M. *Aerosols an Industrial and Environmental Science* (New York, NY: Academic Press, 1984).
153. Hirst, J.M. "An automatic volumetric spore trap," *Ann. Appl. Biol.* 39:257–265 (1952).

154. Hirst, J.M. "Changes in atmospheric spore content: diurnal periodicity and the effects of weather," *Trans. Br. Mycol. Soc.* 36:375–393 (1953).
155. Hirst, J.M. "Spore liberation and dispersal," in *Plant Pathology—Problems and Progress 1908–1958*, C.S. Holton, Ed. (Madison, WI: University of Wisconsin Press, 1959) pp. 529–538.
156. Hirst, J.M. and O.J. Stedman. "The epidemiology of apple scab *(Venturia inaequalis* (Cke) Wint.)," *Ann. Appl. Biol.* 49:290–305 (1961).
157. Hirst, J.M., O.J. Stedman, and W.H. Hogg. "Long-distance spore transport: methods of measurement, vertical spore profiles and the detection of immigrant spores," *J. Gen. Microbiol.* 48:329–355 (1967).
158. Hirst, J.M. and G.W. Hurst, "Long-distance spore transport," *Symp. Gen. Microbiol.* 17:307–344 (1967).
159. Hocking, A.D. and J.I. Pitt. "Dichloran-glycerol medium for enumeration of xerophilic fungi from low-moisture foods," *Appl. Environ. Microbiol.* 39:488–492 (1980).
160. Holden, J.A. and L.R. Babcock. "The use of plume dispersion modelling for viable aerosols from an activated sewage sludge treatment plant," *Environ. Pollut.* (Ser. B) 9:215–235 (1985).
161. Holtmeyer, M.G. and J.R. Wallin. "Identification of aflatoxin-producing atmospheric isolates of *Aspergillus flavus*," *Phytopathology* 70:325–327 (1980).
162. Huang, B., G.J. Hills, and N. Sunderland. "Virus-infected pollen grains on *Paeonia emodi*," *J. Exp. Bot.* 34:1392–1398 (1983).
163. Hudson, H.J. "Aspergilli in the air-spora at Cambridge," *Trans. Br. Mycol. Soc.* 52: 153–159 (1969).
164. Hudson, H.J. "Thermophilous and thermotolerant fungi in the air-spora at Cambridge," *Trans. Br. Mycol. Soc.* 60:596–598 (1973).
165. Hugh-Jones, M.E., W.H. Allen, F.A. Dark, and J.G. Harper. "The evidence for the airborne spread of Newcastle disease," *J. Hyg. Camb.* 71:325 (1973).
166. Hyde, H.A. "Volumetric counts of pollen grains at Cardiff, 1954–57," *J. Allergy* 30: 219–234 (1959).
167. Hyde, H.A. and D.A. Williams. "Studies in atmospheric pollen. II. Diurnal variation in the incidence of grass pollen," *New Phytol.* 44:83–94 (1945).
168. Hýsek, J., Z. Fišar, and B. Binek. "Long-run monitoring of bacteria, yeasts and other micromycetes in the air of an industrial conurbation," *Grana* 30:450–453 (1991).
169. Hyvärinen, A.M., Martikainen, and A.I. Nevalainen, "Suitability of poor medium in counting total viable airborne bacteria," *Grana* 30:414–417 (1991).
170. Ibach, M.J., H.W. Larsh, and M.L. Furcolow. "Isolation of *Histoplasm capsulatum* from the air," *Science, N.Y.* 119:71 (1954).
171. Imshenetsky, A.A., S.V. Lysenko, and G.A. Kazakov. "Upper boundary of the biosphere," *Appl. Environ. Microbiol.* 35:1–5 (1978).
172. Isoard, P., G. Valla, P. Didillon, J. Michel-Brun, I. Achard, J. Coudert, and R. Fontanges. "Influence des conditions météorologiques sur les micromycètes atmosphériques, au cours d'une étude cinétique de cent heures," *Mykosen* 14:213–224 (1971).
173. Jaenicke, R. "Physical aspects of the atmospheric aerosol," in *Aerosols and Their Climatic Effects*, H.E. Gerber and A. Deepak, Eds., (Hampton, VA: A. Deepak Pub., 1984), pp. 7–34.
174. Jarvis, W.R. "The dispersal of spores of *Botrytis cinerea* Fr. in a raspberry plantation," *Trans. Br. Mycol. Soc.* 45:549–559 (1962).
175. Jimenez-Diaz, R.M. and R.L. Millar. "Sporulation on infected tissues, and presence of airborne *Verticillium albo-atrum* in alfalfa fields in New York, *Plant Pathol.* 37:64–70 (1988).
176. Jones, B.L. and J.T. Cookson. "Natural atmospheric microbial conditions in a typical suburban area," *Appl. Environ. Microbiol.* 45:919–934 (1983).

177. Jones, W., K. Morring, P. Morey, and W. Sorenson. "Evaluation of the Andersen viable impactor for single stage sampling," *Am. Ind. Hyg. Assoc. J.* 46:294–298 (1985).
178. Käpylä, M. "The variation of airborne pollen concentrations around a big building in a town," in *Nordic Aerobiology,* S. Nilsson, Ed. (Stockholm: Almquist and Wiksell, 1984), pp. 39–42.
179. Käpylä, M. "Adhesives and mounting media in aerobiological sampling," *Grana* 28: 215–218 (1989).
180. Käpylä, M. and A. Penttinen. "An evaluation of the microscopical counting methods of the tape in Hirst-Burkard pollen and spore trap," *Grana* 20:131–141 (1981).
181. Katzenelson, E. and B. Teltch. "Dispersion of enteric bacteria by spray irrigation," *J. Water Poll. Control Fed.* 48:710–716 (1976).
182. Kauppinen, E.I., A.V.K. Jäppinen, R.E. Hillamo, A.H. Rantio-Lehtimäki, and A.S. Koivikko. "A static particle size selective bioaerosol sampler for the ambient atmosphere," *J. Aerosol. Sci.* 20:829–838 (1989).
183. Keith, L.H., Ed. *Principles of Environmental Sampling.* (Washington, D.C.: American Chemical Society, 1988).
184. Keith, L.H. *Practical Guide for Environmental Sampling and Analysis.* (Chelsea, MI: Lewis Publishers, 1990).
185. Keith, L.H. "Environmental sampling—a summary," *Environ. Sci. Technol.* 24:610–617 (1990).
186. Kelly, C.D. and S.M. Pady. "Microbiological studies of air over some nonarctic regions of Canada," *Can. J. Bot.* 31:90–106 (1953).
187. Kelly, C.D. and S.M. Pady, "Microbiological studies of air masses over Montreal during 1950 and 1951," *Can. J. Bot.* 32:591–600 (1954).
188. Kelly, C.D., S.M. Pady, and N. Polunin. "Aerobiological sampling methods from aircraft," *Can. J. Bot.* 29:206–214 (1951).
189. King, D.A., A.D. Hocking, and J.I. Pitt. "Dichloran rose bengal medium for enumeration and isolation of moulds from foods," *Appl. Environ. Microbiol.* 37:959–964 (1979).
190. King, E.O., M.K. Ward, and D.E. Raney. "Two simple media for the demonstration of pyocyanin and fluoresin," *J. Lab. Clin. Med.* 44:301–307 (1954).
191. Kingston, D. "Selective media in air sampling: A review," *J. Appl. Bact.* 34:221–232 (1971).
192. Kramer, C.L. and S.M. Pady. "A new 24-hour spore sampler," *Phytopathology* 56:517–520 (1966).
193. Kuan, T.L., G.V. Minsavage, and N.W. Schaad. "Aerial dispersal of *Xanthomonas campestris* pv. *campestris* from naturally infected *Brassica campestris,"* *Plant Dis.* 70: 409–413 (1986).
194. Lacey, J. "Airborne spores in pastures," *Trans. Br. Mycol. Soc.* 64:265–281 (1975).
195. Lacey, J. "The aerobiology of conidial fungi," in *The Biology of Conidial Fungi,* G.T. Cole and W.B. Kendrick, Eds. (New York, NY: Academic Press, 1981), pp. 373–416.
196. Lacey, J. Unpublished results.
197. Lacey, J. and J. Dutkiewicz. "Methods for examining the microflora of mouldy hay," *J. Appl. Bact.* 41:13–27 (1976).
198. Lacey, M.E. "The summer air-spora of two contrasting adjacent rural sites," *J. Gen. Microbiol.* 29:485–501 (1962).
199. Lacey, M.E., C.J. Rawlinson, and H.A. McCartney. "First record of the natural occurrence in England of the teleomorph of *Pyrenopeziza brassicae* on oilseed rape," *Trans. Br. Mycol. Soc.* 89:135–140 (1987).
200. Langmuir, I. "The production of rain by a chain reaction in cumulus clouds at temperatures above freezing," *J. Meteorol.* 5:175–192 (1948).

201. Larsen, L.S. "A three-year survey of microfungi in the air of Copenhagen 1977–1979," *Allergy* 36:15–22 (1981).
202. Larsson, K.-A., G. El-Ghazaly, P. El-Ghazaly, S. Nilsson, and T. Wictorin. "Pollen incidence in Eskiltuna, Sweden, 1976-82. A comparison between traps and different stations," in *Nordic Aerobiology,* S. Nilsson, Ed. (Stockholm: Almquist and Wiksell, 1984), pp. 74–84.
203. Lembke, L.L. and R.N. Kniseley. "Coliforms in aerosols generated by a municipal solid waste recovery system," *Appl. Environ. Microbiol.* 40:888–891 (1980).
204. Leuschner, R.M. and G. Boehm. "Individual pollen collector for use of hayfever patients in comparison with the Burkard trap," *Grana* 16:183–186 (1977).
205. Lighthart, B. "Microbial aerosols: estimated contribution of combine harvesting to an airshed," *Appl. Environ. Microbiol.* 47:430–432 (1984).
206. Lindemann, J., D.C. Arny, S.S. Hirano, and C.D. Upper. "Dissemination of bacteria, including *Pseudomonas syringae,* in a bean plot," *Phytopathology* 71:890 (1981) (abstr.).
207. Lindemann, J., H.A. Constantinidou, W.R. Barchet, and C.D. Upper. "Plants as sources of airborne bacteria, including ice nucleation-active bacteria," *Appl. Environ. Microbiol.* 44:1059–1063 (1982).
208. Lindemann, J. and C.D. Upper. "Aerial dispersal of epiphytic bacteria over bean plants," *Appl. Environ. Microbiol.* 50:1229–1232 (1985).
209. Lindow, S.E., G.R. Knudsen, R.J. Seidler, M.V. Walter, V.W. Lambou, P.S. Amy, D. Schmedding, V. Prince, and S. Hern. "Aerial dispersal and epiphytic survival of *Pseudomonas syringae* during a pretest for the release of genetically engineered strains into the environment," *Appl. Environ. Microbiol.* 54:1557–1563 (1988).
210. Littman, M.T. "A culture medium for the primary isolation of fungi," *Science* 106:109–111 (1947).
211. Liu, B.Y.H., D.Y.H. Pui, X.Q. Wang, and C.W. Lewis. "Sampling of carbon fiber aerosols," *Aerosol Sci. Technol.* 2:499–511 (1983).
212. Lloyd, A.B. "Dispersal of streptomycetes in air," *J. Gen. Microbiol.* 57:35–40 (1969).
213. Lundholm, I.M. "Comparison of methods for quantitative determinations of airborne bacteria and evaluation of total viable counts," *Appl. Environ. Microbiol.* 44:179–183 (1982).
214. Lurie, H.I. and M. Way. "The isolation of dermatophytes from the atmosphere of caves," *Mycologia* 49:178–180 (1957).
215. Lyon, F.L., C.L. Kramer, and M.G. Eversmeyer. "Vertical variation of airspora concentrations in the atmosphere," *Grana* 23:123–125 (1984).
216. Lyon, F.L., C.L. Kramer, and M.G. Eversmeyer. "Variation of airspora in the atmosphere due to weather conditions," *Grana* 23:177–181 (1984).
217. McCartney, H.A. "Deposition of *Erysiphe graminis* conidia on a barley crop. II. Consequences for spore dispersal," *J. Phytopathol.* 118:258–264 (1987).
218. McCartney, H.A. "The dispersal of plant pathogen spores and pollen from oilseed rape crops," *Aerobiologia* 6:147–152 (1990).
219. McCartney, H.A. "Dispersal mechanisms through the air," in *Species Dispersal in Agricultural Habitats,* R.G.H. Bunce and D.C. Howard, Ed. (London: Pinter Publishers Ltd., 1990), pp. 133–158.
220. McCartney, H.A. and D.E. Aylor. "Relative contributions of sedimentation and impaction to deposition of particles in a crop canopy," *Agri. Forest Meteorol.* 40:343–358 (1987).
221. McCartney, H.A. and A. Bainbridge. "Deposition of *Erysiphe graminis* conidia on a barley crop. I. Sedimentation and impaction," *J. Phytopathol.* 118:243–257 (1987).
222. McCartney, H.A., A. Bainbridge, and D.E. Aylor. "The importance of wind gusts in distributing fungal spores among crop foliage," *EPPO Bull.* 13:133–137 (1983).

223. McCartney, H.A., A. Bainbridge, and O.J. Stedman. "Spore deposition velocities measured over a barley crop," *Phytopath. Z.* 114:224–233 (1985).
224. McCartney, H.A. and M.E. Lacey. "The production and release of ascospores of *Pyrenopeziza brassicae* Sutton et Rawlinson on oilseed rape," *Plant Pathol.* 39:17–32 (1990).
225. McCartney, H.A. and M.E. Lacey, "Wind dispersal of pollen from crops of oilseed rape *(Brassica napus* L.)," *J. Aerosol Sci.* 22:467–477 (1991a).
226. McCartney, H.A. and M.E. Lacey. "The relationship between the release of ascospores of *Sclerotinia sclerotiorum,* infection and disease in sunflower plots in the United Kingdom," *Grana* 30:486–492 (1991b).
227. McCartney, H.A. and M.E. Lacey. Personal communication.
228. McCartney, H.A., M.E. Lacey, and C.J. Rawlinson. "Dispersal of *Pyrenopeziza brassicae* spores from an oil-seed rape crop," *J. Agric. Sci., Camb.* 107:299–305 (1986).
229. McInnes, T.B., R.D. Gitaitis, S.M. McCarter, C.A. Jaworski, and S.C. Phatak. "Airborne dispersal of bacteria in tomato and pepper transplant fields," *Plant Dis.* 72:575–579 (1988).
230. Madelin, T.M. "The effect of a surfactant in media for the enumeration, growth and identification of airborne fungi." *J. Appl. Bact.* 63:47–52 (1987).
231. Magill, P.L., E.A. Lumpkins, and J.S. Arveson. "A system for appraising airborne populations of pollens and spores," *Am. Ind. Hyg. Assn. J.* 29:293–298 (1968).
232. Makinen, Y. "Random sampling in the study of microscopic slides," *Rep. Aerobiol. Lab. Univ. Turku* 5:27–43 (1981).
233. Malik, P., A.B. Singh, C.R. Babu, and S.V. Gangal. "Atmospheric concentration of pollen grains at human height." *Grana* 30:129–135 (1991).
234. Malvick, D.K. and L.W. Moore. "Survival and dispersal of a marked strain of *Pseudomonas syringae* in a maple nursery," *Plant Pathol.* 37:573–580 (1988).
235. Mandrioli, P., M.G. Negrini, G. Cesari, and G. Morgan. "Evidence for long range transport of biological and anthropogenic aerosol particles in the atmosphere," *Grana* 23:43–53 (1984).
236. Marthi, B., B.T. Shaffer, B. Lighthart, and L. Ganio. "Resuscitation effects of catalase on airborne bacteria," *J. Appl. Environ. Microbiol.* 57:2775–2776 (1991).
237. Marthi, B. and B. Lighthart. "Effects of betaine on enumeration of airborne bacteria," *Appl. Environ. Microbiol.* 56:1286–1289 (1990).
238. May, K.R. "The cascade impactor: an instrument for sampling coarse aerosols," *J. Sci. Instrum.* 22:187–195 (1945).
239. May, K.R. "Fog-droplet sampling using a modified impactor technique," *Q. J. R. Met. Soc.* 87:535–548 (1961).
240. May, K.R. "Calibration of a modified Andersen bacterial aerosol sampler," *Appl. Microbiol.* 12:37–43 (1964).
241. May, K.R. "A multi-stage liquid impinger," *Bact. Rev.* 30:559–570 (1966).
242. May, K.R. "Physical aspects of sampling microbes," In *Airborne Microbes,* P.H. Gregory and J.L. Monteith, Eds., *(Symp. Soc. Gen. Microbiol.* 17), (Cambridge: Cambridge University Press, 1967), pp. 60–80.
243. May, K.R. "Prolongation of microbiological air sampling by a monolayer on agar gel," *Appl. Microbiol.* 18:513–514 (1969).
244. May, K.R. and R. Clifford. "The impaction of aerosol particles on cylinders, spheres, ribbons and discs," *Ann. Occup. Hyg.* 10:349–361 (1967).
245. May, K.R. and H.A. Druett. "A microthread technique for studying the viability of microbes in a simulated airborne state," *J. Gen. Microbiol.* 51:353–366 (1968).
246. May, K.R., N.P. Pomeroy, and S. Hibbs. "Sampling techniques for large windborne particles," *J. Aerosol Sci.* 7:53–62 (1976).

247. Maynard, N.G. "Significance of airborne algae," *Zeitschr. Allg. Mikrobiol.* 8:225–226 (1968).
248. Meredith, D.S. "Atmospheric content of *Nigrospora* spores in Jamaican banana plantations," *J. Gen. Microbiol.* 26:343–349 (1961).
249. Michel, F.B., P. Cour, L. Quet, and J.P. Marty. "Qualitative and quantitative comparison of pollen calendars for plain and mountain areas," *Clin. Allergy* 6:383-393 (1976).
250. Mill, R.A., J.M. Robertson, and B. Walker. "A technique for airborne aerobiological sampling," *J. Environ. Health* 35:51–53 (1973).
251. Miquel, P. *Les organismes vivantes dans l'atmosphere* (Paris: Gauthier-Villars, 1883).
252. Miquel, P. Annual reports, in *Annu. Obs. Montsouris.* (1885, 1886).
253. Moore, B.E., B.P. Sagik, and C.A. Sorber. "Procedure for the recovery of human enteric viruses during spray irrigation of treated wastewater," *Appl. Environ. Microbiol.* 38:688–693 (1979).
254. Morring, K.L., W.G. Sorenson, and M.D. Attfield, "Sampling for airborne fungi: A statistical comparison of media," *Am. Ind. Hyg. Assoc. J.* 44:662–664 (1983).
255. Munk, L. "Dispersal of *Erisiphe graminis* conidia from winter barley," *Grana* 20:215–217 (1981).
256. Napolitano, P.J. and D.R. Rowe. "Microbial content of air near sewage treatment plants," *Water Sewage Works.* 113:480–483 (1966).
257. Nikodemusz, I., I. Vedres, and M. Balatoni. "Studien uber mikrobielle Kontamination durch Eisenbahnzüge," *Zentralbl. Bakteriol. Mikrobiol. Hyg., Abt. B.* 187:70–74 (1988).
258. Nilsson, S., Ed. *Atlas of airborne fungal spores in Europe* (Berlin: Springer-Verlag, 1983).
259. Nilsson, S., J. Praglowski, and L. Nilsson. *Atlas of airborne pollen grains and spores in Northern Europe* (Stockholm: Bokförlaget Natur och Kultur, 1977).
260. Noble, W.C. "Sampling airborne microbes—handling the catch," in *Airborne Microbes*, P.H. Gregory and J.L. Monteith, Eds. *(Symp. Soc. Gen. Microbiol.* 17) (Cambridge, The University Press, 1967), pp. 81–101.
261. O'Rourke, M.K. Personal communication.
262. Ogden, E.C. and G.S. Raynor. "A new sampler for airborne pollen: The rotoslide." *J. Allergy* 40:1–11 (1967).
263. Ogden, E.C., G.S. Raynor, J.V. Hayes, D.M. Lewis, and J.H. Haines. *Manual for Sampling Airborne Pollen* (New York, NY: Hafner Press, 1974).
264. Ogunlana, E.O. "Fungal air spora at Ibadan, Nigeria," *Appl. Microbiol.* 29:458–463 (1975).
265. Olenchock, S.A. "Quantitation of airborne endotoxin levels in various occupational environments," in *Proceedings of the Fourth Finnish-U.S. Joint Symposium on Occupational Safety and Health with Swedish Participation. Scand. J. Work Environ. Health* 14, Suppl.:72–73 (1988).
266. Orr, C.C., and O.H. Newton. "Distribution of nematodes by wind," *Pl. Dis. Reptr.* 55: 61–63 (1971).
267. Pady, S.M. and L. Kapica. "Air-borne fungi in the arctic and other parts of Canada," *Can. J. Bot.* 31:309–323 (1953).
268. Pady, S.M. and L. Kapica. "Fungi in air over the Atlantic Ocean," *Mycologia* 47:34–50 (1955).
269. Pady, S.M. and C.D. Kelly. "Studies of microorganisms in arctic air during 1949 and 1950," *Can. J. Bot.* 31:107–122 (1953).
270. Pady, S.M., and C.D. Kelly, "Aerobiological studies of fungi and bacteria over the Atlantic Ocean," *Can. J. Bot.* 32:202–212 (1954).
271. Pady, S.M., C.D. Kelly, and N. Polunin. "Arctic aerobiology. II. Preliminary report on fungi and bacteria isolated from the air in 1947," *Nature Lond.* 162:379-381 (1948).

272. Parker, D.T., J.C. Spendlove, J.A. Bondurant, and J.H. Smith. "Microbial aerosols from food processing waste spray fields," *J. Water Pollut. Contr. Fed.* 49:2359–2365 (1977).
273. Pasquill, F. "Preliminary studies of the distribution of particles at medium range from a ground level point source," *Porton Tech. Paper, No.* 498 (1955).
274. Pasteur, L. "Mémoire sur les corpuscles organisés qui existent dans l'atmosphère. Examen de la doctrine des générations spontanées," *Ann. Sci. Nat. (Zool.)*, 4e sér. 16:5–98 (1861).
275. Peck, R.M. "Efficiency tests on the Tauber trap used as a pollen sampler in turbulent water flow." *New Phytol.* 71:187–198 (1972).
276. Pedersen, P.N., H.B. Johansen, and J. Jorgensen. "Pollen spreading in diploid and tetraploid rye. II. Distance of pollen spreading and risk of intercrossing," *Yearbook, Roy. Vet. Agric. Coll., Copenhagen* 1961:68–86 (1961).
277. Pedgley, D.E. *Windborne Pests and Diseases: Meteorology of Airborne Microorganisms,* (Chichester: Halstead Press, 1982).
278. Pennycook, S.R. "The air spora of Auckland city, New Zealand. 1. Seasonal and diel periodicities," *N. Z. J. Sci.* 23:27–37 (1980).
279. Perkins, W.A. "The rotorod sampler," *2nd Semiannual Rept. Aerosol Lab.,* Dept. Chem. and Chem. Engng., Stanford University. CML., 186 (1957).
280. Perombelon, M.C.M., R.A., Fox, and R. Lowe. "Dispersion of *Erwinia carotovora* in aerosols produced by pulverization of potato haulm prior to harvest." *Phytopath. Z.* 94: 249–260 (1979).
281. Peterson, E.W. and B. Lighthart. "Estimation of downwind viable airborne microbes from a wet cooling tower-including settling," *Microb. Ecol.* 4:67–79 (1977).
282. Peto, S. and Powell, E.O. "The assessment of aerosol concentration by means of the Andersen sampler," *J. Appl. Bact.* 33:582–598 (1970).
283. Platts-Mills, T.A.E., P.W. Heymann, J.L. Longbottom, and S.R. Wilkins. "Airborne allergens associated with asthma: Particle sizes carrying dust mite and rat allergen measured with a cascade impactor," *J. Allergy Clin. Immunol.* 77:850–857 (1986).
284. Polunin, N. and C.D. Kelly. "Arctic aerobiology: fungi and bacteria, etc., caught in the air during flights over the geographical North Pole," *Nature, Lond.* 170:314-316 (1952).
285. Proctor, B.E. and B.W. Parker. "Microorganisms in the upper air," in *Aerobiology,* Moulton, F.R. Ed., *Am. Assoc. Advanc. Sci. Pub.* 17:48–54 (1942).
286. Puschkarew, B.M. "Über die Verbreitung der Süsswasserprotozöen durch die Luft," *Arch. Protistenk.* 28:323–362 (1913).
287. Quinn, C.E., I.A. Sells, and D.C. Graham. "Soft rot *Erwinia* bacteria in the atmospheric aerosol," *J. Appl. Bact.* 49:175–181 (1980).
288. Ramalingam, A. "The construction and use of a simple air sampler for routine aerobiological surveys," *Environ. Health* 10:61–67 (1968).
289. Ramalingam, A. "Further studies on the effective dispersal of *Pyricularia oryzae* Cav.," *J. Mysore Univ. (N.S.) Sect. B* 22:39–46 (1968-1969).
290. Ramalingam, A. "A rotary drum miniature spore trap," *Sci Culture* 44:366–367 (1978).
291. Rantio-Lehtimäki, A. "Research on airborne spores in Finland," *Grana* 16:163–165. (1977).
292. Rantio-Lehtimäki, A., E. Kauppinen, and A. Koivikko. "Efficiency of a new bioaerosol sampler in sampling *Betula* pollen for antigen analyses," in *Advances in Aerobiology,* G. Boehm and R.M. Leuschner, Eds. *(Experientia Suppl.* 51) (Basel: Birkhauser, 1987), pp. 383–390.
293. Rantio-Lehtimäki, A., A. Koivikko, R. Kupias, Y. Mäkinen, and A. Pohjola, "Significance of sampling height of airborne particles for aerobiological information," *Allergy* 46:68–76 (1991).

294. Rao, V.C. "Introduction to environmental virology," in *Methods in Environmental Virology,* C.P. Gerba and S.M. Goyal, Eds., (New York, NY: Marcel Dekker, Inc., 1982), pp. 1–13.
295. Rati, E. and Ramalingam, A. "Air-borne aspergilli at Mysore," *Aspects Allergy Appl. Immunol.* 9:139–149 (1976).
296. Raygor, S.C., and K.P. Mackay. "Bacterial air pollution from an activated sludge tank," *Water Air Soil Pollution* 5:47–52 (1975).
297. Raynor, G.S. and E.C. Ogden. "The swing-shield: an improved shielding device for the intermittent rotoslide sampler," *J. Allergy* 45:329–332 (1970).
298. Raynor, G.S. "An isokinetic sampler for use on light aircraft," *Atmos. Environ.* 6:191–196 (1972).
299. Raynor, G.S. "Sampling techniques in aerobiology," in *Aerobiology: The Ecological Systems Approach,* R.L. Edmonds, Ed. (Stroudsburg, PA: Dowden, Hutchison and Ross, 1979), pp. 151–172.
300. Raynor, G.S., J.V. Hayes, and E.C. Ogden. "Particulate dispersion into and within a forest," *Boundary-Layer Meteorol.* 7:429–456 (1974).
301. Raynor, G.S., J.V. Hayes, and E.C. Ogden. "Temporal variability in airborne pollen concentrations," *Ann. Allergy* 36:386–396 (1976).
302. Raynor, G.S., E.C. Ogden, and J.V. Hayes. "Variation in ragweed pollen concentration to a height of 108 metres," *J. Allergy Clin. Immunol.* 51:199–207 (1970).
303. Raynor, G.S., E.C. Ogden, and J.V. Hayes, "Dispersion and deposition of corn pollen from experimental sources," *Agron. J.* 64:420–427 (1972).
304. Raynor, G.S., E.C. Ogden, and J.V. Hayes. "Dispersion of pollens from low-level, crosswind line sources," *Agric. Meteorol.* 177–195 (1973).
305. Reddi, C.S. "A comparative survey of atmospheric pollen and fungus spores at two places twenty miles apart," *Acta Allergol.* 25:189–215 (1970).
306. Rempe, H. "Untersuchungen über die Verbreitung des Blutenstaubes durch der Luftstromungen," *Planta* 27:93–147 (1937).
307. Rishbeth, J. "Dispersal of *Fomes annosus* Fr. and *Peniophora gigantea* (Fr.) Massee," *Trans. Br. Mycol. Soc.* 42:243–269 (1959).
308. Roffey, R., Å. Bovallius, P. Ånäs, and E. Könberg. "Semicontinuous registration of airborne bacteria at an inland and a coastal station in Sweden," *Grana* 16:171-177 (1977).
309. Rogerson, C.T. "Kansas aeromycology. I. Comparison of media," *Trans. Kansas Acad. Sci.* 61:155–162 (1958).
310. Rosas, I., C. Calderon, S. Gutierrez, and P. Mosino. "Airborne fungi isolated from rain water collected in Mexico City," *Contam. Ambient* 2:13–23 (1986).
311. Rubulius, J. "Airborne fungal spores in Stockholm and Eskiltuna, central Sweden," in *Nordic Aerobiology,* S. Nilsson, Ed. (Stockholm: Almquist and Wiksell, 1984), pp. 85–93.
312. Rudolph, K., M.A. Roy, M. Sasser, D.E. Stead, M. Davis, J. Swings, and F. Gossele. "Isolation of bacteria," in *Methods in Phytobacteriology,* Z. Nement, K. Rudolph, and D.C. Sands, Eds., (Budapest: Akademiai Kiado, 1990), pp. 43–94.
313. Sattar, S.A. and M.K. Ijaz. "Spread of viral infections by aerosols," *CRC Crit. Rev. Environ. Control* 17:89–130 (1987).
314. Schaad, N.W., Ed., *Laboratory Guide for the Identification of Plant Pathogenic Bacteria,* 2nd ed. (St. Paul, MN: American Phytopathological Society Press, 1988).
315. Schlichting, H.E. "Viable species of algae and protozoa in the atmosphere," *Lloydia* 24: 81–88 (1961).
316. Schlichting, H.E. "Meterological conditions affecting the dispersal of airborne algae and protozoa," *Lloydia* 27:64–78 (1964).
317. Schlichting, H.E. "Airborne algae and protozoa," *Carolina Tips* 33:33–34 (1970).

318. Schlichting, H.E. "Algae and protozoa," in *Aerobiology: The Ecological Systems Approach,* R.L. Edmonds, Ed., *(US/IBP Synthesis Series* 10), (Stroudsburg, PA: Dowden, Hutchison and Ross, Inc., 1979), pp. 22–24.

319. Schlichting, H.E., and J.E. Hudson. "Radio-controlled model boat samples air and plankton," *Science* 156:238–239 (1967).

320. Schwarzbach, E. "Monitoring airborne populations of cereal mildew," in *Plant Disease Epidemiology,* P.R. Scott and A. Bainbridge, Eds., (Oxford: Blackwell Scientific Publications, 1978) pp. 55–62.

321. Shenoi, M.M. and A. Ramalingam. "Aerial dissemination of grain smut of sorghum *(Sphacelotheca sorghi* (Link) Clint)," *Proc. Ind. Nat. Sci. Acad.* 42, sect. R-194-204 (1976).

322. Shenoi, M.M. and A. Ramalingam. "A two year study of the precipitated microbes at Mysore," *IV Int. Palynol. Conf. Lucknow* 3:453–457 (1981).

323. Smith, C.V. "Some evidence for the windborne spread of fowl pest," *Meteorol. Mag.* 93:257–263 (1964).

324. Smith, E.G. *Sampling and Identifying Allergenic Pollens and Molds* (San Antonio, TX: Blewstone Press, 1984).

325. Smith, J.D., W.E. Crawley, and F.T. Lees. "Collection and concentration of spores of *Pithomyces chartarum* from herbage and harvesting from ryecorn cultures," *N. Z. Agric. Res.* 4:725–733 (1961).

326. Smith, K.M. "An air-borne plant virus," *Nature Lond.* 139:370 (1937).

327. Smith, L.P. and M.E. Hugh-Jones. "The weather factor in foot and mouth disease epidemics," *Nature Lond.* 223:712–715 (1969).

328. Smith, W.H. *Air Pollution and Forests* 2nd ed. (New York, NY: Springer-Verlag, 1990).

329. Smith-Holmes, W. and G. Britton. "Evaluation of bacterial aerosols generated in a cypress dome following wastewater application," in *Cypress Swamps.* K.C. Ewel and H.T. Odum, Eds. (Gainesville: University Presses of Florida, 1986), 472 p.

330. Solomon, W.R., H.A. Burge, and J.R. Boise, "Performance of adhesives for rotating-arm impactors," *J. Allergy Clin. Immunol.* 65:467–470 (1980).

331. Solomon, W.R. and J.A. Gilliam. "A simplified application of the Andersen sampler to the study of airborne fungus particles," *J. Allergy* 45:1–13 (1970).

332. Sorber, C.A., H.T. Bausum, S.A. Schaub, and M.J. Small. "A study of bacterial aerosols at a wastewater irrigation site," *J. Water Poll. Control Fed.* 48:2367–2379 (1976).

333. Sorber, C.A., B.E. Moore, D.E. Johnson, H.J. Harding, and R.E. Thomas. "Microbiological aerosols from the application of liquid sludge to land," *J. Water Pollut. Contr. Fed.* 56:830–836 (1984).

334. Spendlove, J.C. "Industrial, agricultural and municipal microbial aerosol problems," *Dev. Ind. Microbiol.* 15:20–27 (1974).

335. Spendlove, J.C., A.P. Adams, P.E. Carlson, and E.C. Rengers. "Emission of microbial aerosols from vents of cooling towers. II. Diffusion modeling," *Dev. Ind. Microbiol.* 20:781–789 (1979).

336. Spieksma, F.T.M., H. Charpin, N. Nolard, and E. Stix. "City spore concentrations in the European Community (EEC)," *Clin. Allergy* 10:319–329 (1980).

337. Sreeramulu, T. "Spore content of air over the Mediterranean sea," *J. Ind. Bot. Soc.* 37:220–228 (1958).

338. Sreeramulu, T. "Aerial dissemination of barley loose smut *(Ustilago nuda),*" *Trans. Br. Mycol. Soc.* 45:373–384 (1962).

339. Sreeramulu, T. "Incidence of conidia of *Erysiphe graminis* in the air over a mildew-infected barley field," *Trans. Br. Mycol. Soc.* 47:31–38 (1964).

340. Sreeramulu, T. and B.P.R. Vittal. "Periodicity in the air-borne spores of the rice false smut fungus, *Ustilaginoidea virens*," *Trans. Br. Mycol. Soc.* 49:443–449 (1966a).
341. Sreeramulu, T. and B.P.R. Vittal. "Some aerobiological observations on the rice stackburn fungus, *Trichoconis padwickii*," *Ind. Phytopathol.* 19:215–221 (1966b).
342. Sreeramulu, T. and B.P.R. Vittal. "Concentrations of *Lacellinopsis sacchari* spores in the air at two heights," *Ind. Phytopathol.* 22:301–305 (1969).
343. Sreeramulu, T. and B.P.R. Vittal. "Periodicity in the uredospore content of air within and above a sugarcane field," *J. Ind. Bot. Soc.* 50:39–44 (1971).
344. Sreeramulu, T. and B.P.R. Vittal. "Spore dispersal of the sugarcane smut *(Ustilasgo scitaminea),*" *Trans. Br. Mycol. Soc.* 58:301–312 (1972).
345. Sreeramulu, T., B.P.R. Vittal, and V. Ramakrishna. "Aerobiology of *Cercospora koepkei* causing the yellow-spot disease of sugarcane," *Ind. J. Agric. Sci.* 41:655–662 (1979).
346. Stakman, E.C., A.W. Henry, G.C. Curran, and W.N. Christopher. "Spores in the upper air," *J. Agric. Res.* 24:599–606 (1923).
347. Starr, J.R., and B.J. Mason. "The capture of airborne particles by water drops and simulated snow crystals," *Q. J. R. Met. Soc.* 92:490–499 (1966).
348. Staub, R. "Ernährungphysiologisch-autoökologische Untersuchungen an der planktischen Blaualge *Oscillatoria rubescens* D.C.," *Schweiz Z. Hydrol.* 23:82–198 (1961).
349. Stewart, D.J. "A selective-diagnostic medium for the isolation of pectinolytic organisms in the Enterobacteriaceae," *Nature Lond.* 195:1023 (1962).
350. Strange, R.E. "Rapid detection of airborne microbes," in *Airborne Transmission and Airborne Infection,* J.F.P. Hers and K.C. Winkler, Eds. (Utrecht, Oosthoek Publishing Company, 1973), pp. 15–23.
351. Suzui, T. and T. Kobayashi. "Dispersal of ascospores of *Sclerotinia sclerotiorum* (Lib.) de Bary on kidney bean plants. Part 2. Dispersal of ascospores in the Tokachi District, Hokkaido," *Hokkaido Nat. Agric. Exp. Stn. Res. Bull.* 102:61–68 (1972).
352. Taft, R.A. *The Louisville air pollution study—A technical report on the joint study of air pollution in Louisville and Jefferson County, Kentucky, 1956-1957* (Washington D.C.: U.S. Dept. Health, Education and Welfare, 1961).
353. Tate, K.G., J.M. Ogawa, W.E. Yates, and G. Sturgeon. "Performance of a cyclone spore trap," *Phytopathology* 70:285–290 (1980).
354. Tauber, H. "Differential pollen dispersion and the interpretation of pollen diagrams," *Danmarks Geologiske Undersøgelse,* II Raekke, Nr. 89, pp. 1–69 (1965).
355. Tauber, H. "A static non-overload pollen collector," *New Phytol.* 73:359–369 (1974).
356. Teltsch, B. and E. Katzenelson. "Airborne enteric bacteria and viruses from spray irrigation with wastewater," *Appl. Environ. Microbiol.* 35:290–296 (1978).
357. Teltsch, B., S. Kedmi, L. Bonnet, Y. Borenzstajn-Rotem, and E. Katzenelson. "Isolation and identification of pathogenic micro-organisms at wastewater-irrigated fields: Ratios in the air and wastewater," *Appl. Environ. Microbiol.* 39:1183–1190 (1980).
358. Tervet, I.W., and R.C. Cassell. "The use of cyclone separators in race identification of cereal rusts," *Phytopathology* 41:286–290 (1951).
359. Tervet, I.W. and E. Cherry. "A simple device for collection of fungus spores," *Pl. Dis. Reptr.* 34:238 (1950).
360. Tervet, I.W., A.J. Rawson, E. Cherry, and R.B. Saxon. "A method for the collection of microscopic particles," *Phytopathology* 41:282–285 (1951).
361. Thom, A.S. "Momentum, mass and heat exchange of plant communities," in *Vegetation and the Atmosphere,* J.L. Monteith, Ed. (London: Academic Press, 1975), pp. 57–109.
362. Thresh, J.M. "The epidemiology of plant virus diseases," in *Plant Disease Epidemiology,*" P.R. Scott and A. Bainbridge, Eds. (Oxford: Blackwell Scientific Publications, 1978) pp. 79–91.

363. Tiberg, E. "Microalgae as aeroplankton and allergens," In *Advances in Aerobiology*, G. Boehm and R.M. Leuschner, Eds. *(Experientia Suppl.* 51) (Basel: Birkhauser, 1987), pp. 171–173.
364. Tiberg, E., B. Bergman, B. Wictorin, and T. Willen. "Occurrence of microalgae in indoor and outdoor environments in Sweden," in *Nordic Aerobiology*, S. Nilsson and B. Raj, Eds. (Stockholm: Almquist and Wiksell International, 1984) pp. 24–29.
365. Tilak, S.T. and R.L. Kulkarni. "A new air sampler," *Experientia* 26:443–444 (1970).
366. Timmons, D.E., J.D. Fulton, and R.B. Mitchell. "Microorganisms of the upper atmosphere. I. Instrumentation for isokinetic air sampling at altitude," *Appl. Microbiol.* 14: 229–231 (1966).
367. Täägårdh, C. "Sampling of aerobiological material from a small aircraft," *Grana* 16: 139–143 (1977).
368. Tyldesley, J.B. "Long-range transmission of tree pollen to Shetland." *New Phytol.* 72: 175–181 (1973).
369. van Overeem, M.A. "On green organisms occurring in the lower troposphere," *Rec. Trav. Botan. Neerl.* 34:389–439 (1937).
370. Venette, J.R. and B.W. Kennedy, "Naturally produced aerosols of *Pseudomonas glycinea,*" *Phytopathology* 65:737–738 (1975).
371. Voisey, P.W. and I.J. Bassett. "A new continuous pollen sampler," *Can. J. Plant Sci.* 41:849–853 (1961).
372. Wadje, S.S., and K.S. Deshpande. "Studies on air-borne bacteria in cotton field." *Ind. Phytopathol.* 30:506–508 (1977).
373. Ware, G.W. *The Pesticide Book.* 3rd ed. (Fresno, CA: Thompson Publications, 1989).
374. Warner, N.J., M.F. Allen, and J.A. MacMahon. "Dispersal agents of vesicular-arbuscular mycorrhizal fungi in a disturbed arid ecosystem," *Mycologia* 79:721–730 (1987).
375. Willeke, K., and P.A. Baron. "Sampling and interpretation errors in aerosol monitoring," *Am. Ind. Hyg. Assn. J.* 51:160–168 (1990).
376. Winkler, K.C. "The scope of aerobiology," in *Airborne Transmission and Airborne Infection*, J.F.P. Hers and K.C. Winkler, Eds. (Utrecht: Oosthoek Publishing Company, 1973) pp. 1–11.
377. Winkler, W.G. "Airborne rabies virus isolation," *Bull. Wildlife Dis. Assoc.* 4:36–40 (1968).
378. Wright, P.B. "Effects of wind and precipitation on the spread of foot-and-mouth disease." *Weather* 24:204–213 (1969).
379. Wynn-Williams, D.D. "Techniques used for studying terrestrial microbial ecology in the maritime Antarctic," in *Cold Tolerant Microbes in Spoilage and the Environment*, A.D. Russell and R. Fuller, Eds. (London: Academic Press, 1979) pp. 67–82.
380. Wynn-Williams, D.D. "Aerobiological samplers currently in use," in *BIOTAS Manual of Methods,* D.D. Wynn-Williams, Ed. (Cambridge: Scientific Committee on Antarctic Research, 1992), 12 pp.
381. Zimmerman, N.J., P.C. Reist, and A.G. Turner. "Comparison of two biological aerosol sampling methods," *Appl. Environ. Microbiol.* 53:99–104 (1987).
382. ZoBell, C.E. "Microorganisms in marine air," in *Aerobiology*, F.R. Moulton, Ed., *Am. Assn. Adv. Sci. Pub.* 17, 55–68 (1942).
383. ZoBell, C.E. and H.M. Mathews. "A qualitative study of the bacterial flora of sea and land breezes," *Proc. Nat. Acad. Sci.* 22:567–572 (1936).

CHAPTER 17

Safety Cabinets, Fume Cupboards and Other Containment Systems

R.P. Clark

INTRODUCTION

Nowadays there is world-wide awareness of the need to control the generation and spread of potentially hazardous biological aerosols in laboratories and manufacturing processes, and to provide systems with quantifiable containment to help achieve this control. This awareness, in many countries, has been highlighted by the introduction of legislation which requires both assessment of risks and the provision of suitable control measures to prevent exposure to dangerous organisms or toxic materials. For example in the UK there are the COSHH (Control of Substances Hazardous to Health) Regulations[1] introduced in 1988 which provide a framework for the assessment of hazard in relation to appropriate control measures.

In many biological research laboratories, besides the paramount need to protect the workers from infection, there is also an important requirement to maintain clean or sterile conditions for the work being handled. In many areas of pharmaceutical manufacture the protection of the product against bioaerosol contamination is also the subject of legislation (viz. UK Medicines Inspectorate).

Control indoors of the spread of biological aerosols in such situations generally is achieved by the use of microbiological safety cabinets. An important development over the last 10–15 years has been the production of various National Standards that attempt to ensure that such safety cabinets have a measurable and consistent degree of operator/product protection in regard to bioaerosols.

With advances in molecular and genetic manipulation techniques, distinctions between bioaerosols and chemical particulates become blurred. For example, bacterial toxins and inactivated cells (human/animal) contain elements of both bioaerosol and chemical particles. Substances like bacterial toxins, even though they come from microbiological origin, are not rendered safe by methods of decontamination appropriate to bacteria or viruses (disinfection and fumigation).

Because of this, variations of the microbiological safety cabinet have appeared that enable these materials to be handled in a way that reduces the possibility of airborne and surface contamination. Such facilities generally have been based on modifications to Class II microbiological safety cabinets with HEPA filters placed to ensure that aerosols are removed from the air streams as soon as possible after generation. Special "no-touch" arrangements then ensure the safe removal of the filters for destruction.

In situations where biological and toxic aerosols may be associated with harmful gases or vapors, hybrid containment facilities have been developed which have been based on both safety cabinet and fume cupboard (hood) technology. In such situations, the particulate phase of the contamination will be removed by appropriate filtration, and the gaseous phase by high level discharge to atmosphere with suitable dilution or removal by methods such as absorption by activated charcoal materials.

With all these engineering control measures, it has been necessary to develop tests to ensure their effectiveness at preventing the spread of bioaerosols to either the environment or the area surrounding a clean product.

SOURCES OF BIOAEROSOLS

There are several sources of indoor bioaerosols that the control measures seek to limit. Procedures such as opening bottles and vials within a safety cabinet can produce airborne droplets that may contain pathogenic microorganisms as the central nucleus. In situations where such containers may have been subjected to variation in temperature, or when biological activity producing gas may pressurize the container, the potential for aerosolization, when they are opened, is enhanced. The actions of using pipettes to remove or introduce biological solutions to vessels also can produce aerosols, particularly when there is a pneumatic element to the equipment used for these manipulations.

Besides potential dangers associated with these activities, there is the obvious hazard when material spills or vessels break releasing large quantities of potentially hazardous material. Such spillages will produce surface contamination but also there will be an element of aerosolization—methods for cleaning such spills also may aerosolize material.

In some instances, more disruptive techniques are used with biological materials. Examples of these are milling and macerating and also centrifugation that has the potential for producing aerosols from non-airtight containers.

In the disciplines of microbiology and virology the technical accomplishments of operators should be such that they are able to perform techniques and manipulations with minimum aerosol generation. However, any failure of technique or equipment may produce large concentrations of bioaerosol.

It will be appreciated that assessing and controlling the dangers of pathogenic organisms must take account of both any aerosol and surface contamination hazard, and their interaction.

ASSESSMENT OF HAZARD

As well as a clear requirement to assess the effectiveness of the various control measures, it is important also to categorize the pathogenicity of the organisms being used. For example, a hazard criteria framework may take account of factors such as

1. pathogenicity of the organism to man,
2. the hazard to laboratory workers and to the community,
3. the likelihood of transmission,
4. availability of effective prophylaxis or treatment.

The UK Advisory Committee on Dangerous Pathogens[2] in considering such a framework produced the following definition of hazard groups for microorganisms:

Hazard Group 1 -   An organism that is most unlikely to cause human disease.

Hazard Group 2 - An organism that may cause human disease and which might be a hazard to laboratory workers but is unlikely to spread to the community. Laboratory exposure rarely produces infection and effective prophylaxis or treatment usually is available.

Hazard Group 3 - An organism that may cause severe human disease and presents a serious hazard to laboratory workers. It may present a risk of spread to the community but usually there is available effective prophylaxis or treatment.

Hazard Group 4 - An organism that causes severe human disease and is a severe hazard to laboratory workers. It may present a high risk of spread to the community and there is usually no effective prophylaxis or treatment.

Having defined methods for assessing the hazard groups of various organisms, then it is important to match the hazard group with the containment level. Where pathogens that cause severe human disease are known to infect by the airborne route, primary containment using microbiological safety cabinets, and the provision of secondary containment using appropriate ventilation, has been recommended by the ACDP in defining specific containment levels for these organisms.

CONTAINMENT SYSTEMS

It is convenient to list the following systems as typical of those used for controlling aerosol exposure to hazardous material:

1. Totally enclosed processes and handling systems,
2. Plant or processes or systems of work which minimize the generation of, or suppress or contain, the hazardous material and that limit areas of contamination in the event of spills/leaks,
3. Partial enclosure with exhaust ventilation,
4. Local exhaust ventilation.

Microbiological Safety Cabinets

Microbiological safety cabinets of Classes I, II and III are defined in a number of National Standards[3-8] and Figure 17.1 illustrates diagrammatically these three classes. A microbiological safety cabinet is intended to offer quantifiable protection to the user and the environment from potential aerosol hazards that may be generated when handling infective or other biological material. Some types of cabinets also protect materials being handled from environmental contamination.

Class I Microbiological Safety Cabinet

A safety cabinet with an opening through which the operator can carry out manipulations inside the cabinet. Air is exhausted continuously from the cabinet at a rate sufficient to minimize the escape of airborne particulate contamination generated within the cabinet.

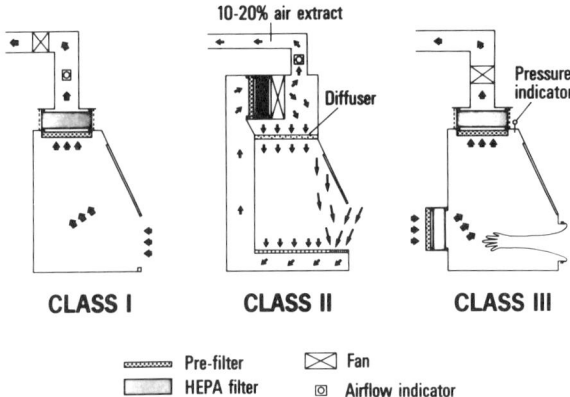

Figure 17.1. Diagram showing the basic air flow configurations in Class I, II & III safety cabinets. The drawings are not definitive in any constructional aspects but illustrate the principles of the three types.

Class II Microbiological Safety Cabinet

A partially enclosed cabinet in which the work space is flushed with a uni-directional, so called laminar, sterile downward airflow and which is so constructed as to minimize, by means of an inward air flow through the working aperture, escape of airborne particulate contamination generated within the cabinet.

In some countries, notably the USA, there are various types of Class II cabinets characterized by the number and position of filters in relation to the working area and exhaust, and whether various compartments within the air circuit are under positive or negative pressure.

Class III Microbiological Safety Cabinet

A safety cabinet by which the operator is separated from the work by a barrier, e.g., by gloves mechanically attached to the cabinet, and from which the escape of any airborne particulate contamination is prevented by an exhaust filtration system.

Convertible Cabinets

Some Class III cabinets can be converted to a Class I cabinet by removal of the glove port front. In such situations the cabinet should be regarded primarily as a Class III type and built to the requirements of Class III cabinets that are specified in National Standards.

Class I safety cabinets are designed to give operator protection only, whereas, Class II types give both operator and product protection. Class III safety cabinets, or glove boxes, differ from the other types by having a physical barrier between the worker and the work. They provide both operator and product protection. Flexible thin film isolator systems also are examples of containment facilities where there is a physical barrier between the work and the worker (Figure 17.2).

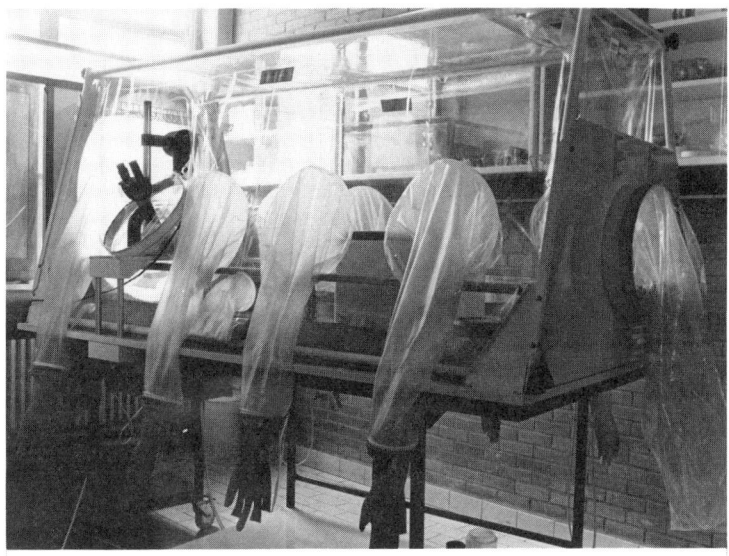

Figure 17.2. An example of a flexible film isolator.

Class I and Class II cabinets rely on moving air to entrain any aerosol contaminants and to carry them to a filter system. The ability of these air streams to maintain "safe" working conditions in the vicinity of the operator is characterized by the *operator protection factor* described later in this chapter.

It should be noted that potentially hazardous biological material used in microbiological safety cabinets generally can be rendered safe by decontamination methods such as fumigation with formaldehyde gas.

Safety Cabinets for Non-Microbiological Use

When the hazardous material is chemical in nature, even though it may have biological origins, it cannot be made safe by the decontamination methods that are suitable for microorganisms. Where carcinogenic, toxic or radioactive particulate and gaseous contamination is present, different methods have to be employed. In situations where hazardous particulate material needs to be handled in clean conditions, modifications to Class II safety cabinets can be used: these are often called cytotoxic cabinets. In these cabinets the main exhaust air filter is situated as near to the work area as possible and is of a type that either can be sealed with an appropriate compound or bagged without the operator coming into contact with the dirty side of the filter. Bagged or sealed filters, once removed from a safety cabinet, generally are disposed of by incineration. At present there is no British Standard for such cabinets although appropriate parts of BS 5726 (1992)[3] may be used.

In cases where contaminants are present in both particulate and gaseous phases, the gas will not be removed by filtration. In these circumstances, dilution of effluent to produce concentrations well below occupational exposure limits[9] is necessary before discharge to atmosphere. In these cases the principles appropriate to fume cupboard extract systems can be used. Figure 17.3 shows diagrammatically a cabinet system that can provide operator and product protection for both gaseous and particulate toxic/infective materials.

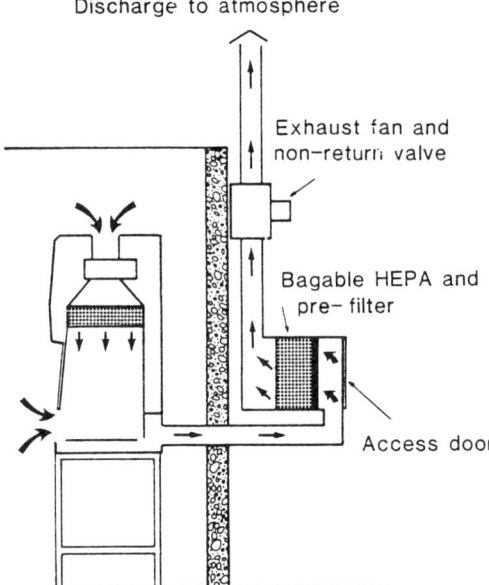

Figure 17.3 An illustration of the principles involved in modifying a Class II safety cabinet for "total dump." The cabinet produces clean working conditions as well as operator protection. There is no re-circulation and all of the air is exhausted through a bagable or sealable HEPA filter before discharge to atmosphere where toxic vapors can be diluted. Such a system is suitable for handling carcinogenic vapors and particulates or radiopharmaceuticals where both operator and product protection are required for particulate and gaseous phase contaminants.

Fume Cupboards (Hoods)

Laboratory fume cupboards (hoods) are found in practically every science laboratory. The recently published standard BS 7258[11] defines a fume cupboard as "a partially enclosed work space that limits the spread of fume to operators and other personnel. It is ventilated by an induced flow of air through an adjustable working aperture that dilutes the fume, and by means of an extract system provides for the release of fume remotely and safely." Unfortunately this British Standard fails to specify important features such as inflow air velocity or containment performance and testing.

Management strategies must make a clear distinction in the use of safety cabinets and fume cupboards. In general, safety cabinets are used for microbiological and not chemical work and ordinary laboratory fume cupboards must be used for chemical and not microbiological work.

Many types of fume cupboards are available. General purpose laboratory fume cupboards do not have filtration systems but an analogy with Class I microbiological safety cabinets, in terms of air flow and operator protection, is useful. The inflowing air streams at the front aperture have to provide a specified level of operator protection as for a safety cabinet. The main difference in a fume cupboard generally is the size of the front opening that is frequently wider than for a safety cabinet and has a greater and variable height, usually under operator control. Variations in fume cupboard design include "walk in" types with a front opening large enough for a person to pass through, systems with constant or variable air volume, systems with vertical or horizontal sliding sashes (or a combination of both), and hybrid systems where filtration may be used to remove particulate contamination.

Absorption Fume Cupboards

Some fume cupboards work on the principle of using activated carbon/charcoal filters for absorbing gases and vapors in the exhaust. Effluent air cleaned in this way sometimes is recirculated to the laboratory (after giving due regard to the potential hazards of the material involved). These devices are not used widely for potentially hazardous material due to the difficulty of knowing when the charcoal filters are loaded to capacity and when subsequent use could lead to the danger of breakthrough of contamination. New instrumentation systems are under investigation to overcome this problem.

Besides being useful for "smelly" rather than harmful chemicals, absorption fume cupboards do have a place for removal of fumes and vapors under controlled conditions (e.g., formaldehyde or glutaraldehyde during equipment sterilization) and for "cleaning" effluent air that ultimately may be discharged to atmosphere.

Local Exhaust Systems (LEV)

There are many other containment systems that come under the heading of LEV. For example, hoods and bench extraction systems to remove formaldehyde fumes are common in pathology laboratories. Figures 17.4, 17.5 and 17.6 illustrate several such systems. Ventilated sinks provide another example of local exhaust ventilation (Figure 17.7).

Rooms as Containment Facilities

In some circumstances, rooms that house containment facilities are themselves at negative pressure with respect to surrounding areas. In this way, if there is an accident within the negative pressure area but outside the containment facility, or if there is an escape of hazardous material from the facility itself, then the negative pressure within the room contributes to containing that hazard and preventing it from reaching corridors and other laboratory areas. One recent example of such a containment laboratory for research has been described with aerosols of hazard group II pathogens, which implies containment at group III level during aerosol production, storage, sampling and animal infection.[10] Figure 17.8 illustrates three ways of achieving negative pressure in a room, either by the use of the containment facility itself to produce the required negative pressure, or by the combined use of input/extract air systems together with the containment facility or, by the use of air lock systems (depending on the level of "cleanliness" required in the laboratory).

In clean room technology applied to a wide range of industries, from electronic and pharmaceutical manufacture to food preparation, the room is at positive pressure with regard to the surroundings. In this way, a very close control can be exercised over the particulate "cleanliness" of the working environment. There is an increasingly recognized need for containment facilities to be used within clean rooms when an operator protection requirement exists. Such containment facilities must be compatible with clean room conditions. Examples of this are to be found in the electronics industry, for both production and research, where clean conditions for assembly are required but where there are some chemicals that may produce hazardous aerosols or toxic gases.

Table 17.1 summarizes the most common types of containment systems and their use.

Figure 17.4 Example of local exhaust ventilation system.

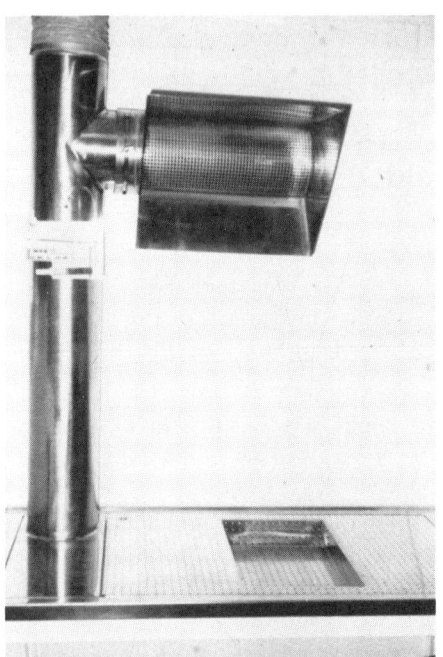

Figure 17.5 Example of local exhaust ventilation system.

CONTAINMENT PRINCIPLES

The majority of containment facilities in use are open fronted (viz. microbiological safety cabinets in classes I and II, cabinets used with non-microbiological material and facilities based on fume cupboards). In these devices the protection—operator and/or product—is achieved by moving air streams or air curtains. For class III safety cabinets, or glove boxes and related devices, containment is provided by a physical barrier between the work and the worker. Methods for testing these physical barriers often are based on assessing the ability of the device to maintain a given pressure for a specified time.

Containment assessment for the open fronted facilities is less simple and test procedures can be divided into three areas: i) operator protection; ii) product protection; and iii) filter/seal integrity.

SAFETY CABINETS, FUME CUPBOARDS AND OTHER CONTAINMENT SYSTEMS 481

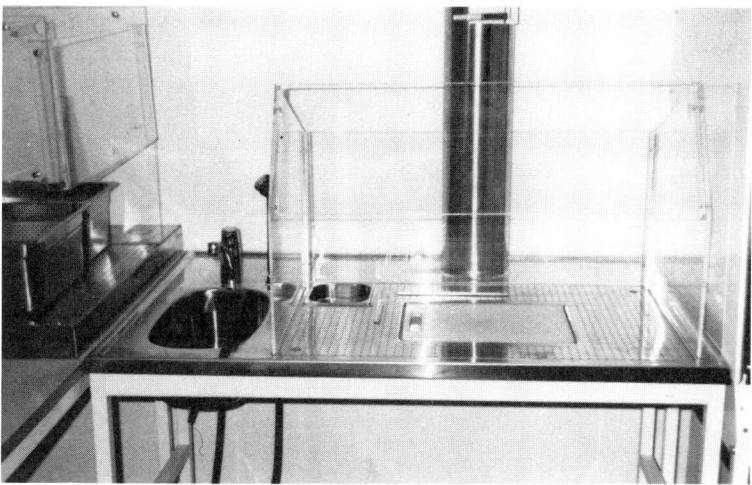

Figure 17.6 Example of local exhaust ventilation system.

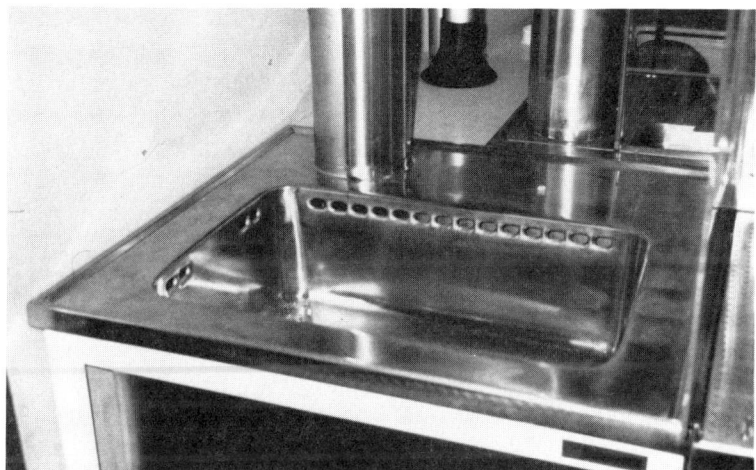

Figure 17.7 An example of a ventilated sink for use with pathology specimens which may contain formaldehyde, etc.

The overall "safety" of open fronted containment systems depends on the following performance parameters:

1. The integrity of filters and associated seals when these are fitted (e.g., safety cabinets);
2. The effective dilution, dispersion or absorption of non-filterable gases and vapors (e.g., for fume cupboards);
3. The ability of the air curtain at the working aperture to prevent escape of aerosols/vapors from the working area into the laboratory (all open fronted containment systems).

For items 1 and 2 above, actual in-service performance depends on factors such as HEPA filter testing in-situ and/or adequate dilution and exhaust velocity for gaseous effluent. All the above performance criteria rely on satisfactory methods for assessing each

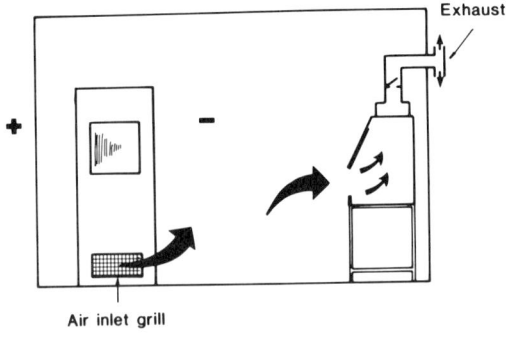
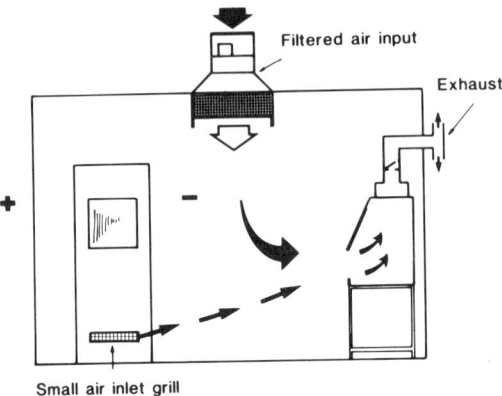
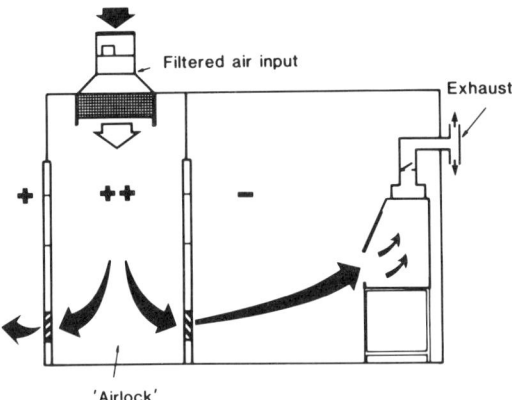

Figure 17.8 Diagram illustrating three ways in which a room housing containment systems (safety cabinets or fume cupboards) may be maintained at negative pressure with regard to the surroundings. The upper figure shows negative pressure maintained in a laboratory by a safety cabinet ducted to the outside. The central illustration shows negative pressure and "semi" clean room conditions achieved in relatively simple way with make-up air partly supplied from the surroundings and from a filtered air input unit. The lower illustration shows an air lock system which supplies positive pressure clean air to the laboratory which is still maintained at negative pressure with regard to the surrounding areas.

Table 17.1 Types of Containment Facility

Type of Facility	Operator Protection	Product Protection	Use
Microbiological safety cabinets			Microbiological work with bacteria and virus's only.
Class I	Yes	No	NOT for chemical work or harmful gases
Class II	Yes	Yes	
Class III	Yes	Yes	
Flexible film isolators	Yes	Yes	Alternative to rigid Class III cabinets for some work
Cabinets for carcinogenic/ radiopharmaceutical material	Yes	Generally Yes	For material not made safe by fumigation; incorporate "safe change" filters; dilution of exhaust if toxic vapors present
Fume cupboards	Yes	No	For gases and in some cases particulates
			NOT for microbiological work
Absorption type fume cupboards	Yes	No	Careful attention needed to filter loading
Local exhaust ventilation	Yes	No	Generally for gases and vapors
Laboratories at negative pressure	Protects surrounding areas	No	Provides additional isolation for work handled in containment systems

parameter in both "type test" situations and also "on-site" at subsequent commissioning and routine maintenance periods.

Filter testing methods, although playing a vital part in the safety performance of many containment systems are outside the scope of this chapter. However, type testing, batch testing and on-site installation and monitoring techniques for HEPA[12] and other filters are well established and in routine use.

In open fronted containment facilities, measurement of containment at the working aperture is of paramount importance and techniques for measuring this are relatively new. Inflowing airstreams, or a combination of inflow and downflow, are the means by which hazardous material is prevented from escaping back into the laboratory. In many instances air velocity alone has been used as a criterion of containment but, as has been widely publicized in recent years, this cannot be taken as an accurate indicator of containment.

Air velocity measurements can all be within specification but still can be inadequate to detect subtle flow anomalies that can dominate containment performance. However, airflow velocity measurement and visualization are techniques that have their place in overall performance testing and complement direct measurements of containment.

Airflow visualization techniques are extremely useful in determining the "base line" performance of containment facilities. Oil smoke (such as DOP or Ondina), fume produced by substances such as titanium tetrachloride, incense sticks, cigarette smoke and more recently dense water fog from ultrasonic nebulizers have been used to visualize and trace the movement of air in containment systems.[13]

Even with adequate airflows in microbiological safety cabinets, BS 5726 still requires that airflow visualization tests demonstrate inward airflow over the whole of the front aperture.

One of the most elegant ways of demonstrating airflow in a safety cabinet is by the use of a Schlieren optical system.[14] Figure 17.9 shows diagrammatically the way in which a safety cabinet would be arranged in the parallel beam of the Schlieren system. This method is sensitive to changes of refractive index of air within the working space that occurs with increased temperature in re-circulating Class II cabinets and by "seeding" the air with carbon dioxide in Class I cabinets or other facilities where temperature alone does not produce a sufficient change in refractive index.

The Schlieren system has been able to demonstrate the interaction between the human micro-environment (the air streams generated around the human body by virtue of its higher than ambient temperature) and safety cabinet airflows with additional disturbances such as those caused by operator movement and by bunsen or other burners.

OPERATOR PROTECTION FACTORS

One of the most important developments in recent years has been the introduction of the *operator protection factor* that quantitatively characterizes the performance of open fronted containment systems. This was first specified in BS 5726 (1979) where the operator protection factor was defined as "...the ratio of the exposure to airborne contamination generated on the open bench to the exposure resulting from the same dispersal within the cabinet."

This definition comes from the concept of a *transfer index* first proposed by Lidwell in 1960.[15] This transfer index mathematically describes the exposure experienced by a worker when a known amount of tracer substance is liberated. One transfer index may be evaluated when this tracer is liberated on a bench and a second index can be found when the aerosol is generated within the containment system under test. The ratio of these indices defines the protection factor. In order that the index may be standardized it is necessary to define reference conditions for the open bench situation, that is, the exposure experienced by an operator working at the bench in a ventilated room but without using a containment system.

The simplest equation describing the ventilation of an enclosed space such as a laboratory, occurs when the room is mechanically ventilated so that the whole volume is continuously and completely purged by the incoming air. In this case, if a tracer substance is liberated in the room, the concentration (C_o) will be uniform throughout the space and, when the tracer liberation stops, it would decline according to the expression:

$$C_t = C_o e^{-Rt}$$

Where C_t is the concentration at time t and R is a constant equivalent to the ventilation rate expressed as the number of air changes supplied to the room in unit time.

If there is only partial mixing in the room the requirement for uniform concentration of the tracer substance throughout the whole space is not met. The concentration measured at any site will depend on the position(s) where the tracer is liberated and sampled.

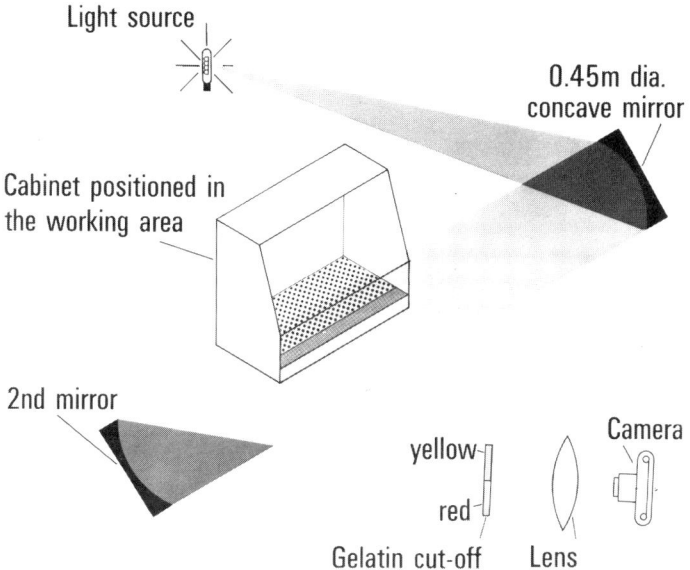

Figure 17.9 Diagram of the schlieren optical system used to visualize air patterns in and around safety cabinets.

Because of eddying and non-uniform turbulence within a room, the sample concentration also will depend on the time at which the tracer is liberated. Because of these complex interactions, the simplest parameter that could describe ventilation between the two places in a room is the integrated concentration of tracer sampled at one place after liberation of unit quantity at the other. This is the transfer index with dimensions that are the reciprocal of volume per unit time (time/length^3). The index is therefore the amount of tracer sampled (n) per unit sample volume flow rate (s) per unit amount of tracer liberated (N) and is expressed by n/Ns.

In defining the operator protection factor for a microbiological safety cabinet the reference open bench conditions are specified as a room with a ventilation air supply of, say, 10 m^3/min and complete mixing. In this specific case the ventilation rate (changes of air/min) $R = v/V$ and

$$T = 1/n \int_0^\infty C_t dt \tag{17.1}$$

and

$$C_t = C_o e^{-Rt} = (N/V)e^{-Rt} \tag{17.2}$$

From equations 1 and 2 we get

$$T = 1/V \, e^{-Rt} dt = 1/VR = 1/v \tag{17.3}$$

where V = volume of the room, m^3
        v = rate of supply ventilating air, m^3/min
        N = the quantity of tracer liberated, number of particles
        C_t = the concentration of tracer at the sampling point at time t, particles/min.
        T = the transfer index, min/m^3

The transfer index of the reference room is therefore $1/v = 1/10$ min/m^3 which, when divided by the transfer index with the tracer released in the cabinet in a room with partial mixing (n/Ns), gives the operator protection factor as Ns/10n; this becomes Ns/10^4n when the tracer sampling rate is expressed in L/min. The protection factor, being the ratio of the two transfer indices, is non-dimensional.

This is the theoretical basis for the definition of operator protection factor and as such makes no reference to any particular particle, method of generating it or sampling. It is now possible to discuss practical methods that are available to test containment facilities.

THE MICROBIOLOGICAL METHOD

BS 5726 requires that 5 replicate tests for operator protection factor be carried out and that each must measure greater than $10.^5$ This requirement means that for every 100,000 challenge particles in the working area of the cabinet not more than 1 should escape. This requirement is therefore 3 times more stringent than that for HEPA filters which have a penetration efficiency of 99.997%.

Various methods of testing have been devised. They all rely on generating a challenge aerosol within the working area at a sufficient concentration that the permitted quantity escaping can be sampled and assessed reliably. One method, specified in NSF 49 and BS 5726, is to generate a cloud of microorganisms by an air driven nebulizer placed strategically within the working space. The nebulizer produces a cloud or mist of droplets of fluid which contains, as a nucleus, one or more organisms of a specified type.

The challenge produced must approximate to a mono-dispersed aerosol consisting mainly of single organisms. The nebulizer is required to be of a reflux blast type operating from an air supply of 20kPa and spraying approximately 0.2 mL/min with no more than 10 L/min of free air being discharged. A Collison type is most suitable.[16,17]

In order to make the test as realistic as possible, and at the same time reproducible, the introduction of an artificial "arm" is specified in the British Standard. This consists of a cylindrical metal tube placed within the cabinet in such a way as to represent a worker's arm. Air flows within the cabinet then will be disturbed in a way representative of disturbances that may be caused by the operator.

The aerosol forming the challenge is liberated at specific sites within the cabinet in relation to this artificial arm and to the front opening; the generation point for a Class II cabinet is level with and 100 mm behind, the top of the front aperture and, in Class I cabinets it is again 100 mm behind the plane of the aperture and mid-way between the work surface and the underside of the artificial arm.

Suitable bacterial samplers are placed in the laboratory, in a position representative of that occupied by a worker at a cabinet. These would be either all glass impingers[18] or microbiological slit samplers[19] with their inlets (or extensions) placed at positions described in the various standards. Such assessments of containment are meant to be representative of the performance that would occur in practice. From a knowledge of the number of particles produced by the nebulizer and the number recovered by the samplers at a known sampling rate, it is possible to evaluate the operator protection factor. In order for the test to have adequate selectivity as well as sensitivity, it is necessary to ensure that the organisms at the test aerosol are characteristically different from any that may be found in the general ambient air of the laboratory in which the test is being carried out. As a consequence, typical aerosols used in these tests have been *Serratia marcessens* (a pigmented organism easily identified and having a relatively short viable life (see also Chapter 5) or *Bacillus subtillis* var niger spores, a relatively harmless, easily identifiable organism but one which has a long viable life that can cause potential contamination problems in both the laboratory and the facility being tested.

Considerable expertise is required to carry out correctly these microbiological tests. The production of a "clean" spore suspension together with the filtration procedures and dilutions necessary for determining the challenge dose, require scrupulous attention to detail and cleanliness to avoid contamination. Background counts to determine levels of any test organism in the laboratory air are necessary since any escape increases the background airborne concentration. The protection factor tests may be rendered invalid should these levels be too high.

The effective challenge to the air curtain at the front of the safety cabinet is therefore a droplet of a size consistent with the spectrum of sizes produced by the air driven nebulizer. It is becoming increasingly difficult to justify the use of such microbiological challenges for use other than in special test facilities. For example, *Serratia marcessens* has been implicated in various septicaemic and/or bacteraemic infections and is regarded as a low grade pathogen. *B. subtilis* spores, while not having a history of pathogenicity, can, in sensitive individuals, be allergenic.

The use of organisms is possible when the tests are confined to "type tests" of prototype or production examples of safety cabinets in well controlled test conditions. However, there can be no guarantee that any particular safety cabinet will, once installed, necessarily produce the same operator protection factor as the one subjected to the "type test." There are many reasons for this. Although nominally similar, the component parts of a safety cabinet can come from different suppliers—notably the fans and motors—and can handle air in subtly different ways and produce containment values different to those measured on a production prototype. Perhaps more importantly, the installation can dominate performance. The effects of draughts caused by door movements and personnel traffic and the influence of ventilation input and extract ducts, the presence of windows, etc., can all provide environmental conditions that adversely affect the performance of a containment system. Therefore, it has become of paramount importance to not only "type test" facilities but also to test them at commissioning and during routine maintenance. Such tests for the real level of containment achieved, therefore can take into account environmental and managerial/working practices. However, these on-site tests cannot practically be carried out using microbiological challenges because of the above mentioned hazards.

THE POTASSIUM IODIDE KI-DISCUS METHOD

The need for a non-microbiological "on-site" test became apparent very soon after the first publication of BS 5726 in 1979 when many "type" tested cabinets following installation were shown to perform poorly. Many safety cabinets are used for tissue culture work and situated in areas where contamination by test organisms is unacceptable. In such circumstances, commissioning and routine containment tests must be performed by methods that do not contaminate either the cabinet or the laboratory and which can be carried out quickly and without the aid of a microbiologist.

A suitable test[20] was developed by the author and his colleagues, and is based on a method originally used for tracing particle movements throughout an air-conditioned hospital. It relies on an aerosol challenge of potassium iodide particles produced by a small spinning disc generator (see also Chapter 7). One advantage of this method is that the sizes and quantity of the generated particles are more uniform and easily controlled than from an air driven nebulizer. Another advantage is that the complete test equipment is available commercially.

In the latest equipment, potassium iodide solution is fed from a peristaltic pump and reservoir onto a thin plastic disc rotating at approximately 28,500 rpm. A primary droplet is formed at the circumference of the spinning disc and is projected away by centrifugal force; the sprayed solution consists of 1.5% potassium iodide in alcohol and droplets quickly evaporate to leave a solid particle of potassium iodide having a volume equal to 1.5% that of the primary drop size. By accurate control of the quantity of solution dispersed, the speed of the spinning disc and the concentration of the solution, an aerosol is produced with known characteristics. When a safety cabinet is challenged by such an aerosol any that escapes is collected by special centripetal air samplers and deposited onto fine filter membranes. At the end of a test, these membranes are removed from the samplers and placed in a solution of palladium chloride. Within a few seconds, the chemical reaction between palladium chloride and the potassium iodide produces well defined grey-brown dots of palladium iodide that are easily identifiable and counted using a magnifier or low power microscope. From a knowledge of the number of particles liberated in the challenge, the numbers recovered in the air samplers and the sampling rate, the protection factor of the cabinet can be evaluated quickly. The operator protection factor is evaluated from the equation $PF = Nv/10^4{}_n$ where N is the number of potassium iodide particles generated [$3.1 \times 10^7 \times M$ where M is the quantity of potassium iodide in mL dispersed by the aerosol generator (20 mL)]. This expression incorporates a constant derived from the droplet size, flow rate and speed of rotation of the disc. The sampling rate v is 100 L/min when the suction pressure on the samplers is 200 mm water gauge. For a test that is to pass the requirement for an operator protection of 10^5 there must therefore be no more than 62 spots on the developed membranes.

The KI-Discus test system (Figures 17.10 and 17.11) has now become the Industry Standard throughout the UK and also is being used in many mainland European countries. It is now fully described in BS 5726 (1992) for testing microbiological safety cabinets; in addition, it is widely used to measure the operator protection of a whole range of other containment systems including laboratory fume cupboards. The reasons for its acceptance include ease of use, ability to determine quickly if containment levels are satisfactory, and if not, to provide re-test data immediately after adjustments have been made to the containment facility or environmental conditions.

In 1981 it was established that a close relationship existed between the results of containment tests carried out by the microbiological method and the KI-Discus system and that both of these methods were practical systems for achieving the implementation of the "operator protection factor" concept.[21,22]

TESTS BASED ON GAS TRACERS

There are other containment measuring systems that rely on sampling, identifying and quantifying a mass/time relationship of tracer escaping from a containment facility. One

SAFETY CABINETS, FUME CUPBOARDS AND OTHER CONTAINMENT SYSTEMS

Figure 17.10 The KI-Discus containment system.

such system was incorporated in BS DD80 (1982)[23] for evaluating fume cupboard containment and introduced a containment index defined by the equation:

$$L = \log (r/C_o) - \log (hwv) \qquad (17.4)$$

where L = containment index (dimensionless)
       r = release rate of test gas inside the containment facility
       C_o = concentration of test gas detected in the plane of the aperture, averaged over space and time, mL/m^3
       h = height of the sash opening, m
       w = aperture width, m
       v = average face velocity, m/s

Equation 17.4 approximates to:

$$L = \log C_i/C_o \qquad (17.5)$$

where C_i is the concentration of the test gas inside the facility.

In such tests, there often are conflicting requirements of high sensitivity of the detector, low release rate of the test gas and high capital and operating costs.

The gas used generally for these tests is sulphur hexafluoride (SF_6) and is specified in a subsequent BS Draft for Development (DD191[24]—associated with BS 7258 for fume cupboards) and DIN 12 924[25]—also for fume cupboards. The test protocols described in these documents are currently under evaluation.

In Australia, a test using DOP (dioctylphthlate) as a challenge aerosol and a light scattering aerosol photometer for sampling, was proposed for assessing Class II safety cabinets (AS 2252.2 - 1985).[7] However, the sensitivity was such that operator protection factors only up to 10^4 could be assessed reliably.

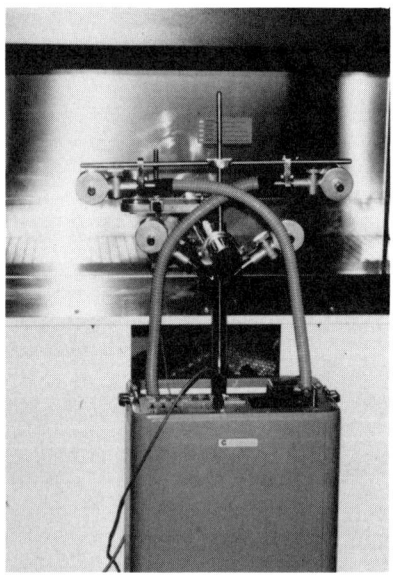

Figure 17.11 The KI-Discus containment system shown set up in front of a Class II safety cabinet.

EXAMPLES OF THE USE OF CONTAINMENT MEASUREMENTS

1. Characterization of containment performance of Class I and II cabinets

The measurement of operator protection factor for safety cabinets at the "type test" stage, at installation and during routine maintenance is now well established. However, a manufacturer/designer can acquire more comprehensive containment performance information by carrying out tests at various settings of inflow (for Class I & II cabinets) and downflow (Class II cabinets). The relationship between inflow velocity and operator protection factor in a Class I cabinet gives an indication of likely performance as the exhaust filter becomes loaded. Design changes, both large or small, can be evaluated in terms of any change in containment performance that may result (Figure 17.12). Containment can be reduced if inflow air velocity is too fast. In this case eddies around the aperture may be produced and the "bounce" effect at the rear wall may further disrupt the airflow and cause a reduction of containment (Figure 17.13).

The same general remarks apply to Class II cabinets although characterization is more complex because of the interaction between inflow and downflow. One approach is to carry out a series or operator protection factor measurements at different inflows while keeping the downflow fixed. By repeating these tests for a range of downflows, a family of diagrams can be constructed providing comprehensive performance information (Figure 17.14).

2. Measurement of protection factors with vertical aperture height for a fume *cupboard*

Airflows have many similar features at the front apertures of Class I safety cabinets and fume cupboards (hoods). The main difference is in aperture height which may be up

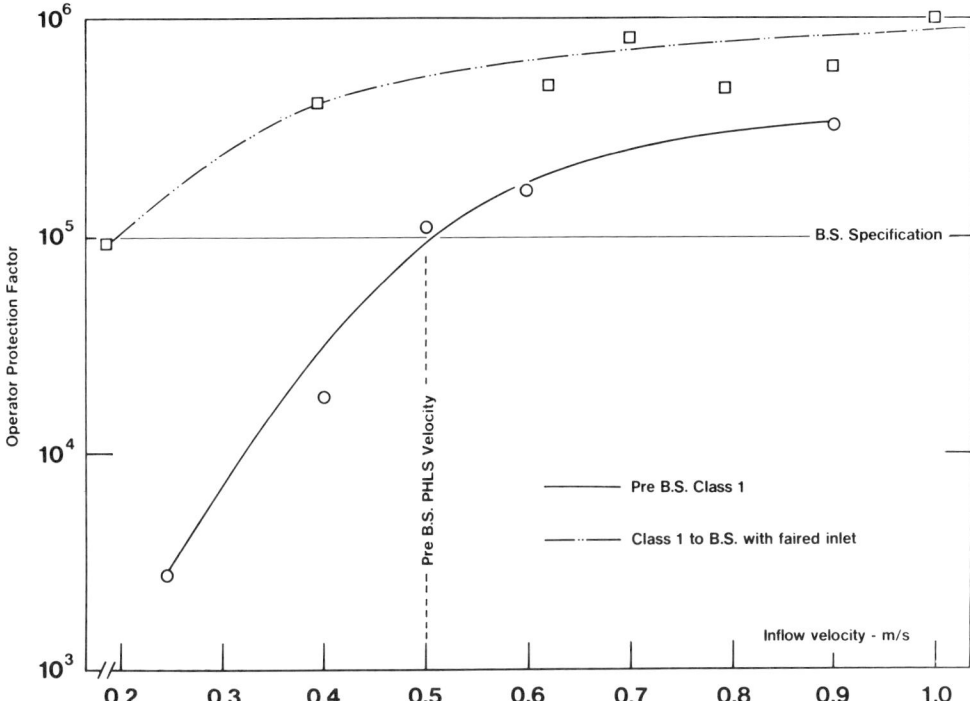

Figure 17.12 Graph showing the variation of operator protection factor with inflow air velocity for a pre-BS 5727 Class I safety cabinet (lower curve) and a BS 5726 Class I cabinet with an aerodynamic faired inlet to enhance performance (upper curve). Such a diagram illustrates the way in which a designer can determine the effectiveness or otherwise of cabinet constructional features and how these may influence performance as for instance, air velocity changes with filter blockage.

to 0.5 m with a fume cupboard. It is important to ensure that operator protection is maintained for the whole aperture height. The KI Discus measuring technique for a Class I safety cabinet can be applied several times at increasing height through the aperture to measure operator protection factors and verify performance (Figure 17.15 a,b).

3. Measurement of room containment in relation to room negative pressure

Containment systems such as safety cabinets and fume cupboards may be installed in laboratories that are at negative pressure with regard to their surroundings. One reason for this is to reduce the escape of any potentially dangerous material made airborne within the laboratory. By releasing challenge aerosol within the laboratory and detecting any escape in surrounding areas, the laboratory can be characterized as a "containment system." The results can be expressed in terms of a protection factor as for a safety cabinet or fume cupboard and may be repeated for different ventilation air flow rates or pressures (Figure 17.16 a,b).

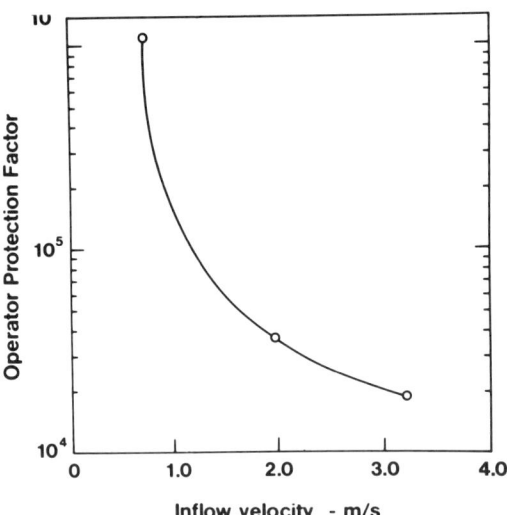

Figure 17.13 The variation of operator protection factor with increasing inflow air velocity above 1 m/sec for a Class I cabinet. Containment is reduced if air velocity is too high due to factors such as "bounce back" of air from the rear wall and the generation of turbulent eddies which can cause leakage from the working area of the cabinet.

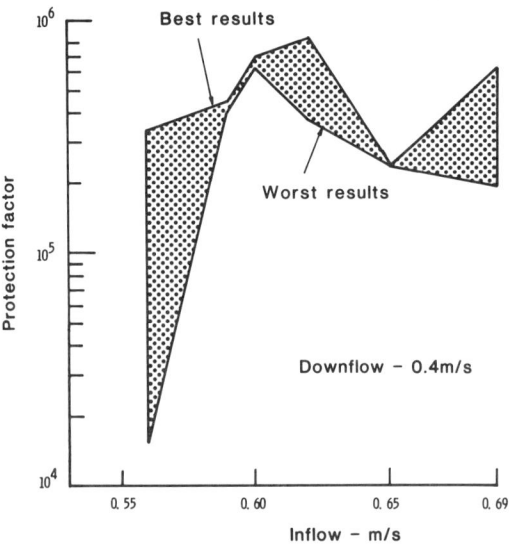

Figure 17.14 Diagram showing the containment "performance envelope" for a Class II cabinet plotted for a series of inflow air velocities and constant downflow. Comprehensive information can be obtained if such envelopes are produced for a range of downflows. Diagrams such as these can be used to define optimum air velocities and safe operating ranges. They can also indicate combinations of inflow and downflows that produce stable containment conditions in the working area.

Figure 17.15 a) A typical high performance aerodynamic fume cupboard, and b) the way in which operator protection factors can be measured at varying heights at the front aperture. Good fume cupboards can achieve operator protection factors in excess of 10^5 over the whole aperture height up to at least 0.5 m.

EFFECT OF LABORATORY VENTILATION SYSTEMS

Difficulties with safety cabinet performance can arise when cabinets are installed in laboratories with very high ventilation rates. In the last 2–3 years a number of new facilities have been built in the UK for handling the human immunodeficiency virus (HIV) where the throughput of ventilation air has been substantially greater than in previous designs. The usual design criteria for the number of air changes in a laboratory is dependent on several factors. The ventilation air has to dissipate heat arising from solar load (factors such as height and location of the building are accounted for) together with heat sources from equipment within the laboratory such as incubators, refrigeration compressors etc. There is also the requirement to maintain the laboratory at negative pressure with respect to the surroundings.

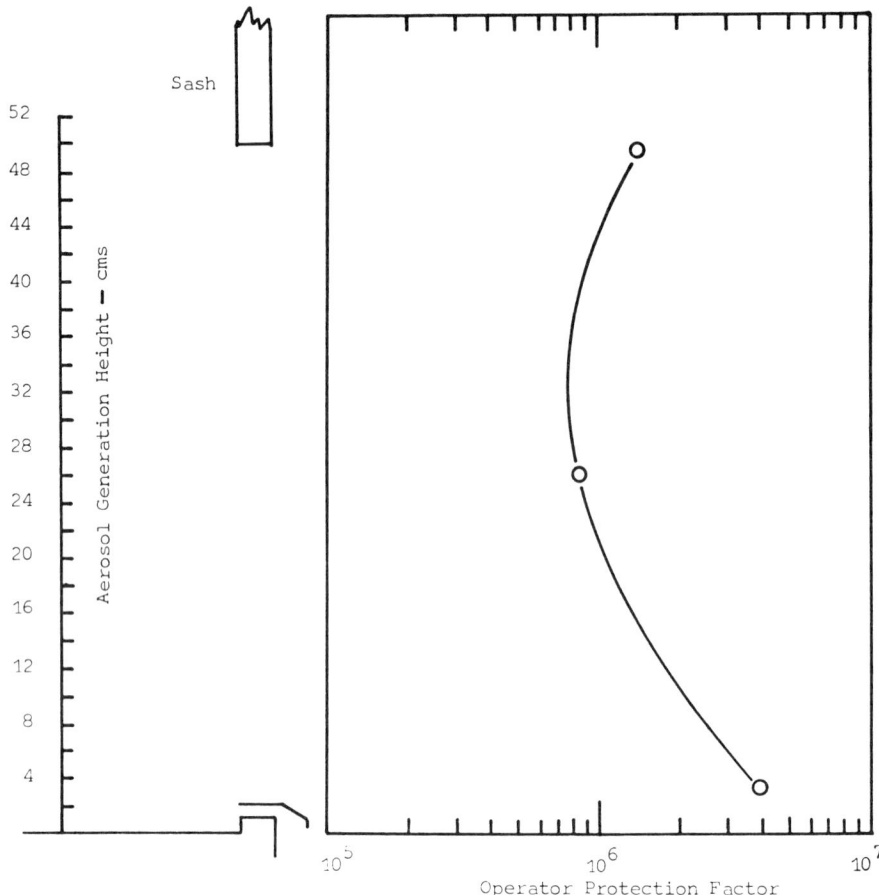

Figure 17.15 (continued).

However, in mechanical plant engineering and architectural terms, design specifications frequently are incomplete as they take no account of interaction of the ventilation air flows with those at the front apertures of containment systems that may be housed in laboratories. The concept of a "safety function" for the combined air systems is vital at the design stage if incompatibilities between safety cabinet/fume cupboard performance and ventilation air flow rates are not to arise at commissioning. When problems do arise, they can result in expensive retro-fit procedures and delays in commissioning.

In one such laboratory specially built for HIV work the room ventilation rate was set originally at some 60 air changes an hour with a negative pressure between the laboratory and the adjacent corridor of 8 mm water gauge.[26] Further demands were put on the system when the UK Health and Safety Executive (HSE) decided that the Class I and II safety cabinets within this laboratory should pass more stringent containment tests than those specified for "type testing" in BS 5726 (1979). The additional tests specified were as follows:

1. Five replicate operator protection factor tests were to be carried out with the test equipment positioned as for BS 5726 (1979) but with two extra air samplers on either side of the aperture center line and level with the bottom edge of the windscreen—these tests initiated the use of 4-headed sampling for the new BS 5726 (1992).[3] (See Figure 17.17).

SAFETY CABINETS, FUME CUPBOARDS AND OTHER CONTAINMENT SYSTEMS   495

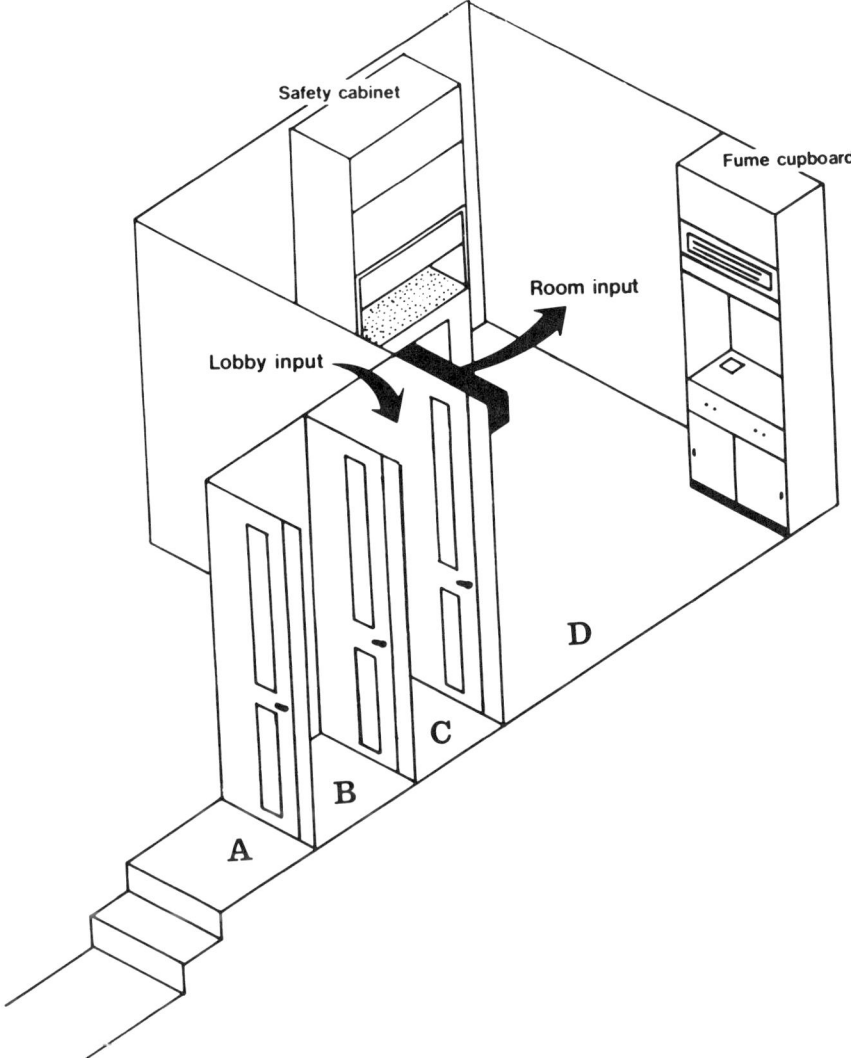

Figure 17.16 a) Diagram of a room constructed for handling carcinogenic material and containing a Class II safety cabinet and a fume cupboard. An external door and an air lock separated the room from the rest of the building. The room D was maintained at negative pressure with regard to the outer area A. b) Diagram showing the protection factor for the room D with respect to the corridor A for different levels of room negative pressure. In this case the room D was considered as the "containment system" and potassium iodide liberated. Samplers in the corridor area A sampled any challenge escaping from the room. From such tests a room containment protection factor could be determined. For example, if the room was required to have the same containment as the safety cabinet or fume cupboard (viz. 10^5) then this would be achieved with a room negative pressure of just over 2 mm water gauge.

2. The same tests as above (but without the artificial arm) with an operator in position with both hands and arms remaining within the cabinet throughout the tests. The operator was to perform repetitive and continuous pipetting procedures

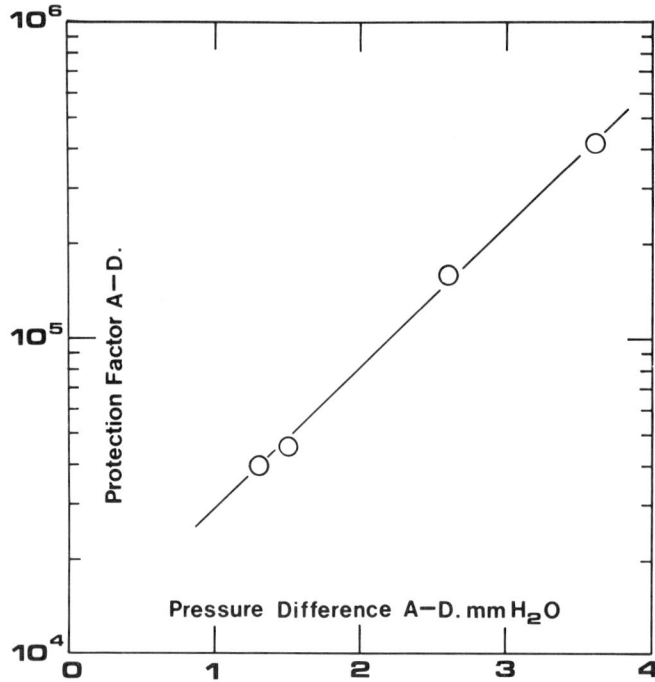

Figure 17.16 (continued).

  (e.g., pipetting to and from a multi-well plate and a bottle with a suitable automatic pipette).
3. Same as 2 above but with a second person wearing a laboratory coat walking backwards and forwards between the operator and the bench at 100 paces a minute.
4. Same as 2 above but with the operator moving away from the cabinet on a swivel chair, taking a bottle from the bench opposite and returning to the cabinet and continuing to pipette. This procedure was to be repeated 30 seconds after each return to the cabinet throughout the duration of the test.

The initial high ventilation rates caused poor containment in the Class II cabinets by producing cross-drafts across the front aperture that acted adversely with cabinet airflows. At the end of a long series of adjustments and modifications to the ventilating air flows (which finished much lower at around 20 air changes an hour), the results from both the Class I and II cabinets were considered satisfactory. Operator protection factors above 10^5 were achieved in all the extended tests with the exception of the last (where the operator moved away from the cabinet) where results were between 2.5 and 5.1×10^4. Such an operator procedure would be regarded as bad laboratory practice and would be described as such in the management guidelines for working in this laboratory. Nevertheless, 10^4 would be a reasonable protection factor in these circumstances.

In some installations it has been possible to have safety cabinets with operator protection factors above 10^5 for this most stringent test as can be seen in Figure 17.18 which summarizes the performance of two Class II and one Class I cabinets installed in another laboratory having 23 air changes an hour.

Besides demonstrating that open fronted safety cabinets can perform satisfactorily under difficult operating conditions and in well ventilated environments, the results of such

Figure 17.17 Sampling heads for the KI Discus containment test system set up in a "4-headed" configuration in front of a Class I safety cabinet.

tests have demonstrated that, properly set up and tested, there was effectively no difference in containment performance of good quality Class I and II cabinets. Subsequent tests within these laboratories also demonstrated that this good cabinet performance, once achieved (by matching the cabinets to the environment), was sustained as long as the air flow control was maintained for both laboratory and cabinets.

AUTOMATED ANALYSIS EQUIPMENT AND POSSIBLE BIOAEROSOL HAZARDS

Tests sometimes are required to determine whether automated clinical chemistry equipment produces potential aerosol hazards, and if so, what containment measures would be appropriate.

A number of micro-processor controlled automatic analyzer machines are available in which patient samples are presented sequentially to devices such as automated syringes that may, according to pre-set programs, dispense aliquots of samples into a number of containers ready for chemical analysis. Samples may proceed via various conveyor systems to dispensers for addition of chemical reagents. After some time, the conveyor moves the samples containers to positions where calorimetric analysis of the chemical reaction can take place. Following this procedure, the sample containers together with the mixture of reagent and patient samples may be washed automatically, cleaned and dried. Depending on the nature of the patient samples being handled, it may be necessary to assess the potential for such machines to produce airborne droplets that may emanate from the infected patient material. In the event that a potential aerosol hazard is possible, the machine either can be completely enclosed by a containment system or the potential risk from any airborne droplets can be reduced by use of local exhaust ventilation.

Figure 17.19 shows an example of an automated analyzer with air samplers (in this case those of the potassium iodide system and a bacterial slit sampler) positioned to detect any

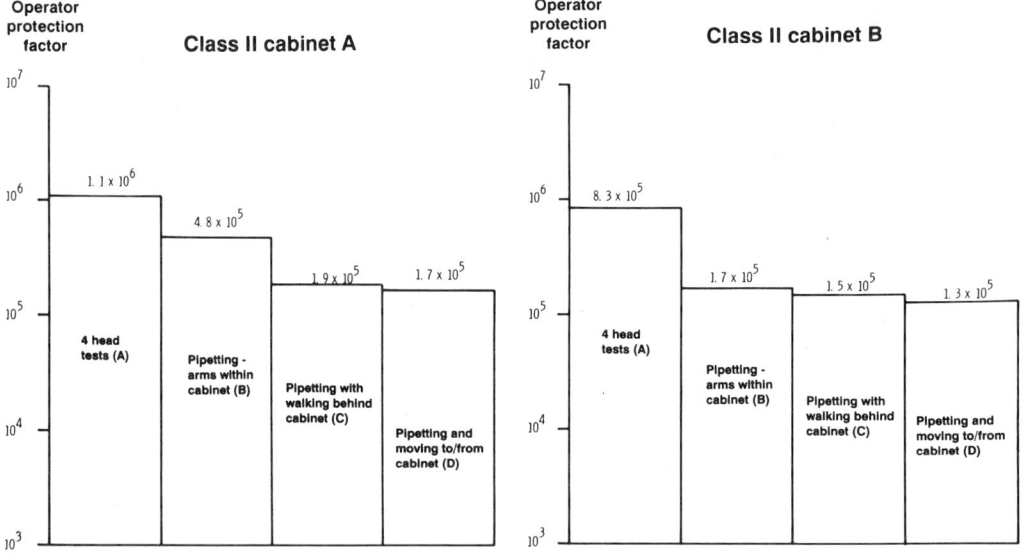
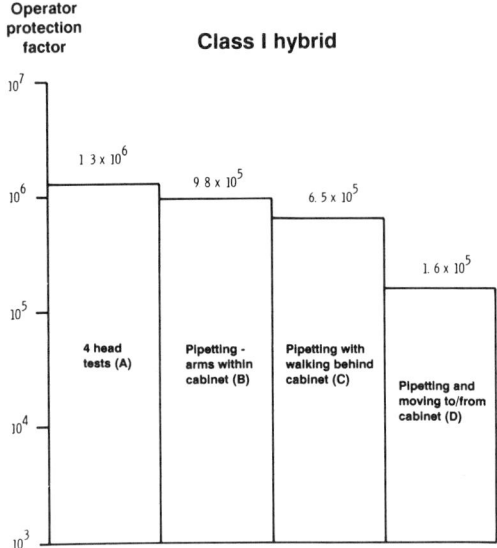

Figure 17.18 A summary of the containment performance of 2 Class II safety cabinets and a Class I hybrid cabinet installed in an HIV facility. The following tests were performed on each cabinet: 4 headed tests; pipetting with the operator's arms remaining within the cabinet; pipetting with a second subject walking to and fro behind the operator; and operator pipetting and moving to and from the cabinet.

airborne material near to the automated syringe injection area. In order to detect any aerosols, tests were made with both chemical and microbiological techniques. With the microbiological method, a bacterial suspension of *B. subtilis* var globigii was made up

SAFETY CABINETS, FUME CUPBOARDS AND OTHER CONTAINMENT SYSTEMS   499

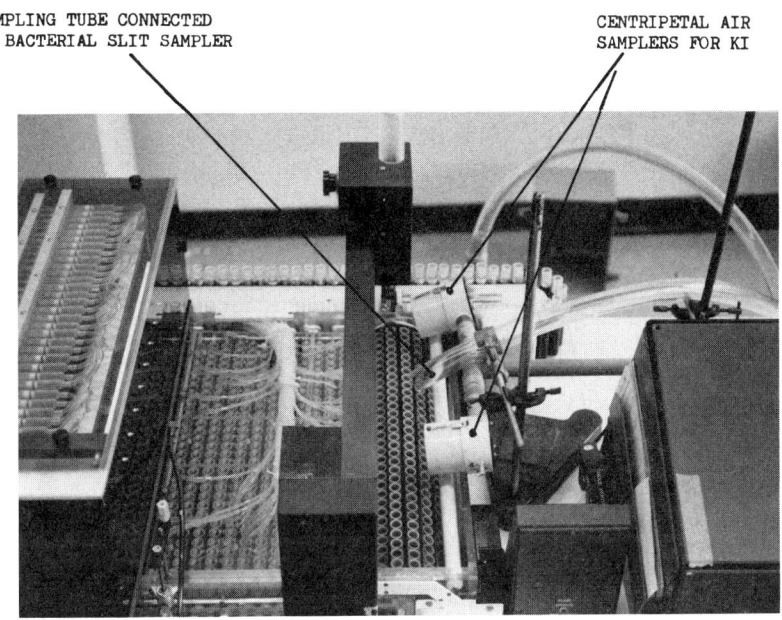

KI AND BACTERIAL SAMPLING POSITIONS AT THE
SAMPLING AND INJECTION AREA.

VIEW SHOWING THE PROXIMITY OF
THE AIR SAMPLERS TO THE SAMPLE
AND REAGENT INJECTION TRACK.

Figure 17.19 Method of sampling for tracer aerosols from an automated clinical chemistry analyzer. Potassium iodide and bacterial sampling positions are shown near to the analyzer sampling an injection area where aerosols could potentially be produced.

together with a chemical solution of 1.5% potassium iodide solution in either water or alcohol. These test solutions represented contaminated patient samples. In this experiment 100 patient sample tubes were filled with 1 mL of either the bacterial suspension or the potassium iodide solution and the machine set to dispense an aliquot of marker into all of the reagent containers. During the 50 min taken to process this material, air near to the specimen injection position was collected by either a bacterial air sampler or one specifically for collecting potassium iodide particles onto filter membranes (similar to the KI-Discus safety cabinet containment testing system). Similarly air samplers were placed at the far end of the machine where the wash process took place. The results of this experiment showed that both bacterial and chemical tracer techniques could demonstrate that marker contaminants could become airborne at various stages in the automated process. With the bacterial marker, the airborne recovery rate was higher as the concentration of the organisms was increased, as might have been expected. In view of the basic operating principle of such machines, it is difficult to see how aerosols could be eliminated totally using a syringe-based mechanism for sampling and dispensing. It would be a relatively easy matter to conduct specific tests on particular machines and to determine the reduction in potential airborne contamination that would be achieved through local containment or exhaust ventilation measures.

SURFACE CONTAMINATION

Surface contamination can result from aerosols that escape from a containment system. One way to assess such contamination is to attach filter membranes to surfaces under investigation (including the protective clothing of workers) and to use a biological and/or chemical marker such as potassium iodide to detect any airborne contamination landing on the membranes. Figure 17.20 shows such an experimental set-up with an operator wearing an apron with filter membranes attached, to enable the identification of any airborne challenge landing on them.

A novel way of investigating surface contamination resulting from aerosol escape from a containment system (or that generated in an "open bench" situation) has been developed.[27] If a microbiological marker such as *B. subtilis* or potassium iodide is generated at a source point (within a containment system or at a non-contained point in the laboratory) and if a number of appropriate samplers are placed around the laboratory (on the floor, on benches etc.) then a spatial distribution of the contamination can be determined. Clearly this can be carried out only at a finite number of positions in the room, but iterative interpolation can determine likely contamination at sites in-between those sampled. Figure 17.21 shows such an example which can identify the likely density of surface contamination from an aerosol generation incident. Factors such as containment and effects of laboratory ventilation on the containment pattern also can be assessed.

SUMMARY

The potential exposure of laboratory personnel to hazardous aerosols such as microorganisms (bacteria and viruses) or toxic particulates or fumes can be minimized by the use of safety cabinets, fume cupboards, and local exhaust ventilation systems. These systems have become an intrinsic and important part of most laboratories and there is now a generally accepted view that the effectiveness of these containment systems needs to be demonstrated at regular intervals.

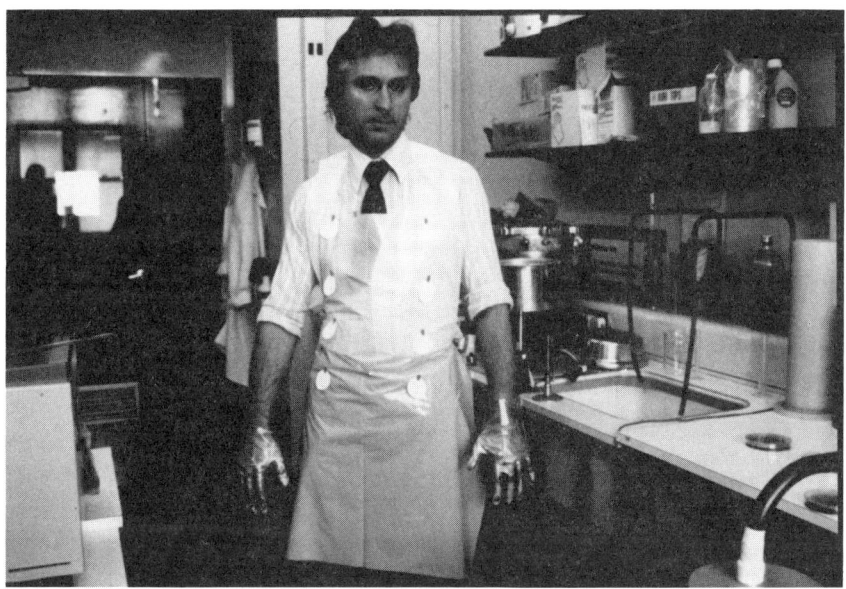

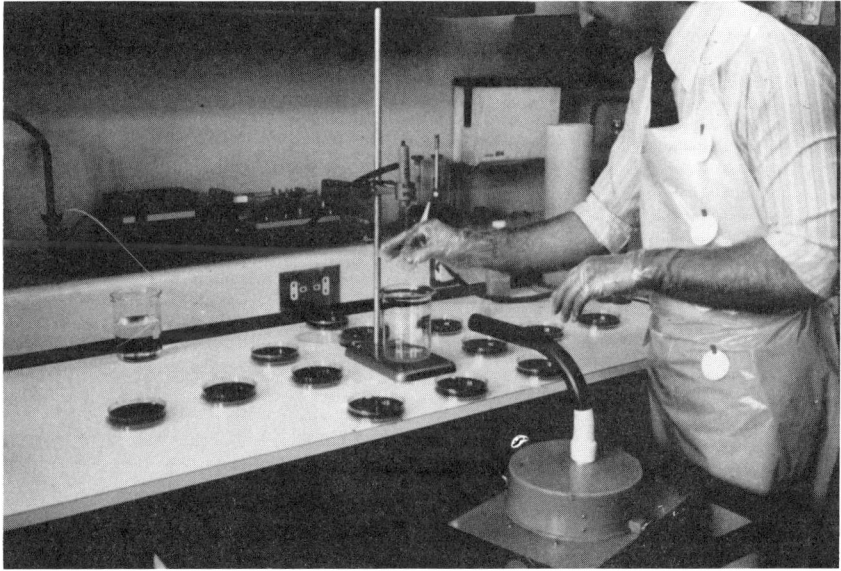

Figure 17.20 a) An example of the way in which surface contamination resulting from bioaerosols can be assessed during laboratory manipulations. A subject is wearing an apron with filter membranes attached to detect airborne contamination. b) Subject carrying out a manipulation using a solution containing potassium iodide and tracer bacterial spores. Settle plates containing blood agar are positioned around the experimental area on the bench. The membranes on the operator's apron, when treated with palladium chloride, identify any potassium iodide component of the contamination. Such experiments can give indications of the potential spread of airborne contamination from a procedure that may subsequently contaminate work surfaces and produce a hazard if touched.

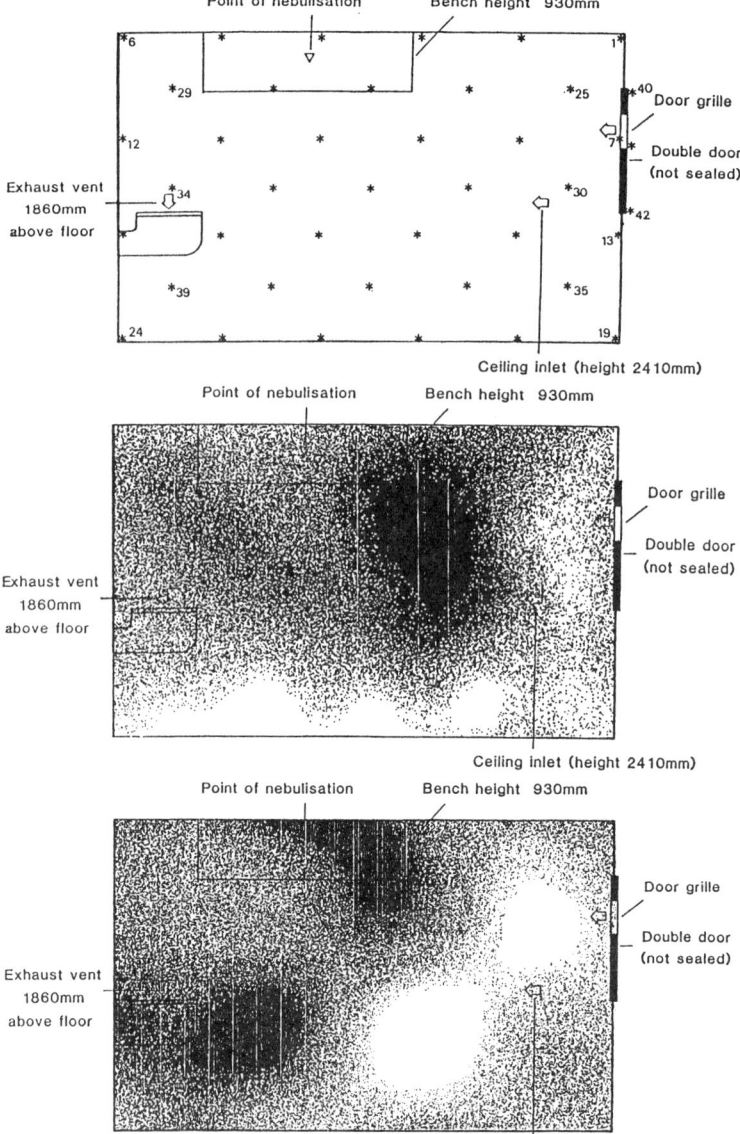

Figure 17.21 Examples of the way that individual measurements of surface contamination produced from a challenge of aerosol particles can be computer analyzed to give a density distribution map of the likely surface contamination. The upper figure shows the layout of a test laboratory with the positions of agar settle plates indicated by *. In this case 42 settle plates were distributed throughout the laboratory with 3 just outside the door. After a bacterial challenge had been nebulized on the bench a period of time was allowed for the contamination to settle on to the plates. Once these were incubated and the colonies counted, these data were fed into a computer program which produced a two-dimensional dot density map of the contamination. The center picture shows such a map for the laboratory without the ventilation system operating. The lower picture is the resultant contamination when the inlet and exhaust air flows to the laboratory are working.

National Standards for safety cabinets and microbiological safety cabinets specify methods for assessing operator protection and these techniques can be used for risk assessment as is required by many regulatory authorities in many countries. The potassium iodide KI-Discus system has gained general acceptance in western Europe for testing safety cabinets, fume cupboards and exhaust ventilation systems in terms of operator protection factors that characterize the ability of a containment system to provide operator protection. The effectiveness of a containment system is a function of good design and manufacture and also of good installation practice in order that the working environment does not adversely affect containment performance and therefore safety of protection systems.

REFERENCES

1. The Control of Substances Hazardous to Health Regulations 1988 (ISBN 0 11 086757 1) and Approved Codes of Practice. (ISBN 0 11 885468 2) HMSO London.
2. Advisory Committee on Dangerous Pathogens. Categorization of pathogens according to hazard and categories of containment 2nd Edition 1990. HMSO London. ISBN 0-11 885564 6.
3. BS 5726 Microbiological safety cabinets. 1992. British Standards Institution London.
4. DIN 12 950 Teil 10 Laboratory furniture; safety cabinets for microbiological and biotechnological work; requirements and testing. Deutsche Norm. 1991
5. NF X 44 201 Microbiological safety cabinets, definitions, classification, characteristics, safety requirements, tests. Association Francais de Normalisation (AFNOR) 1984.
6. NSF 49 Class II (Laminar Flow) Biohazard Cabinetry. USA National Sanitation Foundation. 1976.
7. AS 2252.2 Biological Safety Cabinets; Part 2-Laminar Flow Biological Safety Cabinets (Class II) for personnel and product protection. The Standards Association of Australia 1985.
8. AS 2252.1 Biological Safety Cabinets; Part 1-Biological Safety Cabinets (Class I) for Personnel Protection. The Standards Association of Australia 1981.
9. Occupational Exposure Limits 1992. Guidance Note EH 40/92. Health and Safety Executive. HMSO London.
10. Wathes, C.M. and Johnson, H.E, Physical protection against airborne pathogens and pollutants by a novel animal isolator in a level 3 containment laboratory. 1991. Epidemiology and Infection, 107, 157-170.
11. BS 7258 Laboratory fume cupboards 1990. British Standards Institution London.
12. BS 3928 Method of sodium flame test for air filters for air filters (other than for supply to I.C. engines and compressors). British Standards Institution London.
13. Kennedy, D.A., Water fog as a medium for visualisation of airflows. *Annals of Occupational Hygiene* (1987). 31, 255–259.
14. Clark, R.P. and Mullan, B.J., Airflows in and around linear downflow "safety" cabinets. 1978. *Journal of Applied Bacteriology,* 45, 131–136.
15. Lidwell, O.M., The evaluation of ventilation. *Journal of Hygiene, Cambridge,* (1960). 58, 297–305.
16. Collison, W.E., Inhalation Therapy Technique. 1938. Heinemann, London.
17. May, K.R., The Collison nebulizer: description, performance and application. *Journal of Aerosol Science* 1973. 4, 235–243.
18. May, K.R. and Harper. The efficiency of various liquid impinger samplers for bacterial aerosols. *British Journal of industrial Medicine* (1957). 14, 287.
19. Bourdillon, R.B., Lidwell, O.M. and Thomas, J.C. A slit sampler for collecting and counting airborne bacteria. *Journal of Hygiene Cambridge* (1941). 34, 172.

20. Clark, R.P. and Goff, M.R. The potassium iodide test system for open fronted microbiological safety cabinets. *Journal of Applied Bacteriology* (1981). 51, 439–460.
21. Clark, R.P., Elliott, C.J. and Lister, P.A. A comparison of methods to measure operator protection factors in open fronted microbiological safety cabinets. *Journal of Applied Bacteriology* (1981). 51, 461–473.
22. Clark, R.P. The performance, installation, testing and limitations of microbiological safety cabinets. Occupational Hygiene Monograph No. 9, 1983. Reprinted 1989. Science Reviews Ltd., Leeds. ISBN 0-905927-16-8.
23. DD80 Draft for Development. Laboratory fume cupboards. 1982. British Standards Institution, London.
24. DD191. Draft for Development. Method for determination of the containment value of a laboratory fume cupboard. 1990. British Standards Institution, London.
25. DIN 12 924 Part 1. Laboratory furniture. Requirements for Fume Cupboards. General Purpose Fume Cupboards. 1988. German Standards Institute.
26. Clark, R.P., Osborne, R.W., Pressey, D.C., Grover, F., Keddie, J.R. and Thomas, C. Open fronted safety cabinets in ventilated laboratories. *Journal of Applied Bacteriology,* (1990). 69, 338–358.
27. Reed, P.J. Measurement and models of microbial contamination. 1986. Thesis for Fellowship of the Institute of Laboratory Sciences.

The KI-Discus containment testing equipment is manufactured by Mediscus Products Ltd, Wareham, Dorset, BH20 4SP, United Kingdom.

CHAPTER **18**

Problem Buildings, Laboratories and Hospitals

J.M. Macher, A.J. Streifel and D. Vesley

INTRODUCTION

In this chapter we will discuss bioaerosol sampling in three settings: 1) problem buildings (in which people complain of ill health or discomfort), 2) laboratories, and 3) hospitals. Because microbiological contamination can occur anywhere, we discuss problem buildings first, as general examples, followed by remarks specific to laboratories and hospitals. Chapters 19, 20 and 21, respectively, consider sampling in industrial workplaces, animal houses and residences.

In *problem buildings*, investigators may use air and material sampling: 1) to identify sources of bioaerosols (e.g., air handling systems contaminated with bacteria or fungi), and 2) to check hypotheses generated by epidemiological and medical evaluations (e.g., that disease-causing organisms are, or are not, present in sufficient numbers to account for the symptoms that exposed people report). In biomedical and research *laboratories*, safety specialists may monitor the environment: 1) to measure bioaerosol concentrations in clean rooms (e.g., as part of quality control programs), and 2) to evaluate the potential for laboratory-acquired infections (e.g., by monitoring the release of infectious agents during high-risk procedures). In *hospitals*, infection control personnel may collect samples from people and the environment: 1) to identify sources of contamination during investigations of hospital-acquired infections, 2) to verify air quality in critical care areas prior to patient occupancy, and 3) to monitor air quality during demolition, construction and remodelling.

In this chapter we primarily address air sampling for microorganisms that grow on synthetic media under laboratory conditions, i.e., viable (or perhaps more correctly termed "culturable") bacteria and fungi (see also Chapters 6, 13 and 14). Investigators frequently study viable building-associated bioaerosols because culture techniques are well established and allow identification of a range of environmental isolates. Similar collection techniques can be used to sample airborne pollen, protozoa, microbiological toxins, and allergens other than viable microorganisms, but detection methods other than culture are used to analyze these bioaerosols (see also Chapters 11, 12, 13, 14 and 15).

In addition to describing air sampling methods in this chapter, we also briefly discuss procedures for collecting surface and bulk material specimens, and offer some guidance on interpreting the information obtained from air and material samples. The reader should not be surprised to find that besides describing where, when, why and how to monitor the environment we also discuss reasons *not* to monitor. This is because: 1) sampling can be expensive and time consuming, 2) unless investigators comprehensively consider all types of biological hazards, they may not detect the agents responsible for people's complaints, 3) the findings from improperly designed or executed monitoring programs are frequently difficult to interpret, and 4) the information needed to solve a problem often can be obtained by other means.

It is not uncommon for a study of building-associated illness to become controversial. Therefore, investigators (both in-house inspectors and outside consultants) must guard against developing biases that may interfere with their work.[2] The interests of building occupants, laboratory staff, health-care workers and hospital patients understandably may differ from those of building managers, laboratory directors and hospital administrators. If the results of a study refute beliefs that one of the above groups holds strongly, the group may challenge an investigator's methods or findings. Therefore, we suggest that investigators follow established practices and consult available guidelines when assessing bioaerosol hazards,[12–13,62,64,66] conducting broader indoor air quality studies[10,43,49,77] or designing investigations of laboratory- and hospital-acquired infections.[34,58–59]

SOURCES AND CONSEQUENCES OF BIOAEROSOL EXPOSURES

Edmonds[27] described the main components in the "aerobiological pathway" as microorganism *source, release, dispersion, deposition* and *impact*. This section focusses on the first and last of these. Some potential sources of indoor bioaerosols are outlined in Table 18.1. In addition to these, the reader may refer to Chapter 2 of this book for a general discussion of sources, and to Chapter 16 for a description of outdoor bioaerosols that may enter buildings through ventilation systems and open windows and doors.

Problem Buildings

The term *"sick"* or *"tight" building syndrome* is used to describe non-specific symptoms (most commonly headache, fatigue, eye, nose and throat irritation, and odor annoyance) when they occur in a number of a building's occupants, e.g., ≥20%. These symptoms sometimes are associated with new or recently renovated buildings. Typically people's symptoms increase the longer they remain in a problem building and resolve when they leave, which may make diagnosis difficult unless it is done on site.[63] Although bioaerosols have not been associated conclusively with sick building syndrome, microbiological contamination and volatile microbial by-products may provoke or contribute to such symptoms.[14–15]

The term *"building-related illness,"* on the other hand, refers to a clinically diagnosed disease in one or more of the occupants of a building.[43,63] In general, building-related bioaerosol illnesses can be divided into *respiratory infections* and *hypersensitivity diseases*, as discussed below. Building occupants also may be exposed to endotoxins or mycotoxins as a result of indoor contamination, but such exposures appear to be rare (or rarely are considered) in the non-industrial settings discussed in this chapter.

Respiratory complaints are the leading reason employees give for absence from work.[46,90] Not surprisingly, the incidence of respiratory infections rises in winter when people spend more time indoors and the supply of outdoor air is reduced to save energy. Poor ventilation system design or operation can contribute to the spread of respiratory infections for which *people* are the sources, e.g., measles and tuberculosis.[81] The bacteria responsible for Legionnaires' disease and Pontiac fever [*Legionella* species (spp.)], on the other hand, are not transmitted from person-to-person but originate from *environmental sources*. Outbreaks of legionellosis are caused by bacteria aerosolized from cooling towers, evaporative condensers, atomizing humidifiers, spas, whirlpool baths, shower heads and other warm (20°C to 45°C) water systems.[23,24,26,28,38,75]

Table 18.1 Potential Sources of Indoor Bioaerosols

Outdoor Air
- Soil, decaying vegetable material, stagnant water, bird droppings and wind-pollinated plants
- Construction, excavation and agricultural operations
- Cooling towers and evaporative condensers
- Building exhausts and sanitary vents

Heating, Ventilating and Air-Conditioning (HVAC) Systems
- Air washers and humidification systems
- Condensed water
- Accumulated dirt and debris
- Contaminated insulation
- Microbial growth on filters

Occupied Spaces
- People
- Housekeeping, remodelling and demolition
- Water-damaged and damp materials, e.g., ceiling tiles, wall coverings, insulation material
- Cool mist and ultrasonic humidifiers
- Potted plants and plant containers

The second type of response to indoor bioaerosols, hypersensitivity disease, includes allergic rhinitis, asthma, hypersensitivity pneumonitis (also called extrinsic allergic alveolitis) and humidifier fever. Microorganisms and arthropods are major indoor sources of airborne antigens in non-industrial workplaces. In contrast to airborne dust which may irritate the mucous membranes of all exposed people, an airborne allergen affects only people previously exposed and sensitized to the material. After a person develops a sensitivity to an allergenic material, exposure to even a small amount of the substance may elicit a hypersensitivity response.

Laboratories

Bioaerosols in laboratories have potential *quality control* and *safety* consequences. Quality control in a laboratory is compromised when cultures, cell lines, media, reagents or instruments become contaminated. From the safety perspective, it is clear that laboratory workers have been exposed accidentally to bioaerosols even when they could not remember an obvious breach of safe protocol.[74] Bioaerosols may be released during many routine laboratory practices, e.g., the use and flame-sterilization of transfer loops, the handling of pipettes and syringes, the opening of tubes and bottles, the operation of centrifuges and blenders, the separation of needles from syringes, and accidental breaks and spills (see also Chapter 17).[20,85] In microbiology laboratories, inhalation can be a route of infection for agents not normally transmitted in this manner, e.g., rickettsial diseases. In fact, laboratory workers may face a higher risk of exposure to some infectious agents than the public at large.

Rapid advances in genetic engineering, cell fusion, enzyme production and the handling of nucleic acid fragments have added new dimensions to quality control and safety questions in biomedical and research laboratories. Increasingly, technicians from disciplines other than microbiology handle microorganisms, blood products and tissues. Unfortunately, some of these people may not appreciate the potential hazards of their work and may not receive adequate safety training or supervision, which increases the likelihood of accidents.

Beyond the laboratory itself, industrial-scale production of cultures and biological products creates situations in which an accident could result in very high exposures and difficult problems of decontamination and clean-up. Improper disposal of biological waste also presents opportunities for the generation of bioaerosols, potentially exposing glassware handlers, housekeeping staff and maintenance personnel. In addition to the cultures and other laboratory materials workers handle, research animals and their bedding and feed are potential sources of pathogenic and allergenic bioaerosols[45,53] (see also Chapter 20). Animal handlers need to take precautions with both intentionally infected animals and those carrying natural pathogens.

Hospitals

The control of bioaerosols in hospitals is important to protect patients and health-care workers. People are the most important reservoirs of hospital-associated pathogens. Current diagnostic methods facilitate the rapid identification of infected *patients*, allowing their prompt isolation in properly ventilated, negative-pressure rooms. It may be more difficult to identify *health-care workers* who carry infectious agents or whose improper work practices endanger patients and other staff members.

A *nosocomial infection* in a hospital patient is one that was not present or incubating when the person was admitted.[44] These cases are identified by routine infection control surveillance, and by observant patient-care providers and laboratory staff.[32] The microorganisms responsible for nosocomial infections may be endogenous (i.e., part of a patient's normal flora), or exogenous (i.e., part of a patient's environment, including hospital personnel and other patients, medical devices and the ambient air). Exogenous infections can occur: 1) by contact transmission (e.g., skin and wound infections), 2) via a common vehicle (e.g., through food, blood or blood products, medications, diagnostic reagents or respiratory therapy equipment), or 3) by airborne transmission. Although direct transmission of a nosocomial disease agent requires close contact, airborne transmission may occur across considerable distances,[44] which can make identification of the source difficult.

The number of patients who are susceptible to opportunistic pathogens has risen due to the advent of organ and tissue transplant procedures, the increasing number of people infected with the human immunodeficiency virus, and the expanding treatment of malignant diseases with agents that suppress the immune system.[48] The microorganisms of greatest threat to these patients are environmental mycobacteria and thermotolerant filamentous fungi, specifically those in the genera *Aspergillus, Fusarium, Mucor* and *Rhizopus*.[48,65,72,79,91] For immunocompromised patients, exposure to even very low numbers of opportunistic pathogens may present unacceptable risks. Consequently, procedures usually considered innocuous (such as the sweeping of floors and changing of bed linens) may generate excessive concentrations of bioaerosols.[80] To protect these patients, appropriate precautions must be in place before starting demolition or construction projects, and before inspecting or cleaning heating, ventilating and air-conditioning (HVAC) systems.[87]

Laser surgical tools, introduced in the last decade, have raised concerns among operating teams about possible exposure to viable cells, virus particles and fragments of intact DNA, as well as irritating chemicals in laser plumes. Many dentists and orthopedic surgeons now take precautions to avoid exposure to aerosolized blood and body fluids.

Environmental monitoring has been used to estimate job-related hazards for medical personnel and to evaluate the effectiveness of measures taken to protect health-care workers. Although many hospital laboratories employ automated instruments to test blood, urine and other body fluids, considerable opportunities still exist for accidental exposures to bioaerosols. Under "universal body substance precautions," health-care workers are advised to consider all patients and patient specimens potentially infectious. Adherence to these precautions has helped reduce hospital- and laboratory-associated infections.[40,71,78]

The above are but a few examples of bioaerosol sources in buildings, laboratories and hospitals. The conditions and events necessary to result in human exposure to bioaerosols may be summarized as: 1) the presence of a reservoir that can support the growth of microorganisms or other biological material, 2) multiplication of the contaminating organisms in the reservoir, 3) generation of aerosols containing biological material, and 4) dissemination of bioaerosols into the occupied space.[12] It is important for investigators to realize that although sources are abundant, microorganisms may present little hazard if they do not multiply, or if there is no means for material to become airborne (in particles capable of reaching the target organs in the body, e.g., the upper or lower respiratory tract), or if aerosolized material is not transported to places where susceptible people could be exposed. Our next topic is deciding when it is reasonable to use environmental monitoring to study exposure to biological materials.

DECIDING WHEN ENVIRONMENTAL MONITORING IS APPROPRIATE

Strauss[85] suggested two ways to evaluate bioaerosols: 1) monitor their *effects*, or 2) monitor their *presence*. The first is used in nosocomial disease surveillance and to study outbreaks of respiratory infections (e.g., legionellosis) in office and institutional buildings. When people manifest symptoms shortly after transmission of an infectious agent, it may be possible to identify where and when cases were exposed by comparing risk factors for people who were and were not ill. For many pathogens, infectious disease specialists consider a skin test conversion, serological antibody rise or a single high titer with history of compatible illness proof of recent infection. Although source sampling may be useful in these investigations, air sampling seldom is needed because infectious diseases generally have well-characterized symptoms and the cases serve as evidence that the causative agents are or were present. Investigators often use the second approach, environmental monitoring, to determine the presence and concentration of saprophytic microorganisms (i.e., those that live on dead organic matter) and other biological materials that can cause opportunistic infections and hypersensitivity diseases.

In some cases, mold growth on ceilings or walls merely indicates poor housekeeping and does not affect human health. In other instances, the release of spores or the production of odorous compounds can lead to problems. Without bothering to identify the type of mold present, an investigator can recommend that contaminated materials be cleaned and that damaged furnishings be replaced. In high-risk areas in laboratories and hospitals, certain types of decorations and furnishings (e.g., potted plants and difficult-to-clean drapes, carpets and upholstered chairs) are inappropriate and should be prohibited whether or not environmental testing demonstrates that they release bioaerosols. Similarly, investigators can recommend removal of standing water regardless of the concentrations or types of microorganisms present.

Although contaminated water that has accumulated in a drain pan within a ventilation system is not hazardous unless it is aerosolized, odors may be a problem if microorganisms or algae begin to grow, or the moisture supports fungal colonization of surrounding materials. After the water evaporates, the dried residue may become suspended and be disseminated when air moves through the ventilation system. Some drain pans have overflow ports that extend several centimeters above the bottom of the basin to allow scale and sediments to settle and thus prevent blockage of the drain pipes. This design may not be suitable where microbiological contamination must be minimized. At least annual inspection of HVA equipment and frequent cleaning of drain pans and pipes is recommended to prevent problems.

When a physician sees a confirmed case of Legionnaires' disease or Pontiac fever, the patient's doctor, employer or local health department should attempt to identify other cases (e.g., among family members and co-workers, or in other hospitalized patients) and to locate the source of exposure.[28] However, it can be counter-productive (and even misleading) to monitor a case's environment for *Legionella* spp. if the laboratory did not isolate the bacterium from the patient or did not identify any isolates to the species and serogroup levels. These bacteria grow well in surface waters and soil, and frequently find their way into drinking water supplies and non-potable waters used in recirculating heat-transfer systems. Therefore, if investigators scrutinized an environment closely enough, they very likely would find some species of *Legionella*. Without an isolate from a patient for the laboratory to compare with any recovered from the suspect environment, the investigators could not identify the site of exposure conclusively. To control legionellae, good system design and construction, careful placement of equipment that may release aerosols relative to air intakes and high-risk populations, proper operation and maintenance of the equipment, and routine inspection and cleaning would provide greater assurance of safe operation than would periodic sampling without these precautions.[22–24,28,38,75]

The appropriate degree of caution *laboratory workers* should use depends on: 1) the nature of the operation (i.e., how likely it is a procedure will produce aerosols), 2) the pathogenicity of the microorganisms the laboratory handles, and 3) the availability of effective immunization against these microorganisms and of therapeutic treatments in case of infection.[9,17,82] Despite the potential hazards involved in handling concentrated cultures of known pathogens in laboratories, standard practices, protective clothing and the proper use of containment equipment can minimize the risks.[9,20] Therefore, experts do not recommend environmental monitoring on a routine basis in laboratories. Two exceptions to this general rule are quality control programs in laboratories where extremely low bioaerosol concentrations must be maintained, and safety evaluations in high containment laboratories to ensure that integrity has not been breached.[84,88] Most biological safety cabinets contain aerosols very effectively. Although many manufacturers check the operation of prototype cabinets with test aerosols of bacteria, users typically monitor the inflow of air at the open cabinet face with smoke tubes, flutter strips or pressure sensors. Annual recertification of cabinets can provide even greater assurance of proper functioning (see also Chapter 17).

Likewise in *hospitals*, the most important element for contamination control is not environmental monitoring but proper staff training and supervision, and adherence to tested procedures for equipment disinfection and maintenance.[29,30,34] In the U.S., inpatient hospital facilities must have infection control programs in order to be accredited.[47] When (as part of such a program) an investigation of a cluster of infections implicates certain equipment, procedures or staff members, infection control personnel may initiate systematic testing to determine the mode of disease transmission. Again, however, unless the laboratory has isolates from nosocomial cases, it generally is not informative to collect environmental samples.[73] To chart air quality on a routine basis in patient isolation areas, direct-reading particle monitors may provide more useful information than would bioaerosol sampling (see also Chapter 8). Particle monitors can sample the air continuously and report the data immediately.[86] Therefore, when the concentration of dust in the respirable size range (1 to 10 μm) exceeds a desired limit, health-care workers more readily can identify the cause of the excess aerosols and take prompt corrective action.

In both laboratories and hospitals, prevention and early identification of problems through medical surveillance is preferable to extensive environmental testing after a work-related problem develops.[8,59] In some situations, employees should have medical examinations before assignment to a new job and annually thereafter. It also might be reasonable to collect and store serum specimens which can be tested in the event of a possible exposure or work-related illness.[8] Immunization against the agents laboratory workers handle also may be advisable.

As described above, safety specialists monitor certain laboratory and patient-care settings on a routine basis for quality assurance, but more commonly investigators consider testing when workers complain about a building's air quality, after cross-contamination occurs in a laboratory or when a possible building- or work-related infection is recognized. In the authors' opinion, investigators too often turn to bioaerosol sampling when they cannot find a ready explanation for employees' complaints, or simply to appear to be doing something about a problem. Not surprisingly, this type of testing seldom provides information helpful in solving possibly building-related bioaerosol problems.

If asked how to prevent or control infectious, allergenic and toxic bioaerosols, an investigator can recommend that: 1) architects designers buildings and contractors construct them so that problems of excess moisture do not occur, 2) engineers design HVAC systems so that biological contaminants do not multiply and are not disseminated, 3) facility managers maintain buildings to prevent and control contamination, 4) building owners clean up contamination promptly when they learn of it, 5) laboratory and hospital workers adhere to safety precautions to protect themselves and others, and 6) administrators take action (e.g., relocate sensitive individuals) when other control measures are not effective. Should an investigator decide there is sufficient reason to examine a building for possible biological contamination, we next offer some suggestions on how to proceed.

INVESTIGATING BUILDING-RELATED BIOAEROSOL PROBLEMS AND EVALUATING EXPOSURES

Most investigators adopt a staged approach when inspecting buildings. To begin an investigation, they gather *qualitative* data, e.g., by interviewing building managers and occupants, visually inspecting the site for potential sources and using the gathered information to develop hypotheses that could explain how the problem came about. When appropriate, investigators follow this evaluation with a *quantitative* study in which they test their hypotheses, and either accept or reject them or formulate new hypotheses.[94] At this stage, investigators might collect surface and material samples from potential reservoirs, and sample the air to document that the isolated agents could be disseminated from the identified sources. A third step (too often overlooked) is a *follow-up* study to determine the success of control measures. After interventions are implemented, investigators should re-examine and perhaps again monitor the study sites. Similarly, they should repeat their epidemiological and medical evaluations to ensure that the remedial measures actually reduced occupant illnesses and complaints.

Before beginning monitoring, we advise investigators to prepare a standard data sheet on which to record field observations and measurements (see Table 18.2).[10] Often it is useful to draw a map of the test areas and to record where one made observations and collected samples. The proper use of such a form can reduce the chance that inspectors will

Table 18.2 Field Data Sheet Information

- Collector's name
- Date
- Study site, e.g., street address or building name
- Sample identification number
- Sample type, e.g., air, surface, liquid or bulk material
- Sample collection site (mark location on a map or drawing of the test area, or photograph site with equipment in place)
- Air sampler identification number
- Sample pump identification number
- Sampling air flow rate
- Sample start and finish times
- Volume of air collected
- Indoor and outdoor air temperature
- " " " relative humidity
- " " " CO_2 concentrations
- Number of people and activity at sample site
- Weather conditions
- Barometric pressure
- Wind direction and velocity
- Sample storage conditions
- Sample transportation method/conditions
- Type of analysis requested
- Date/time samples delivered to laboratory

overlook important details and will help ensure that everyone on an investigation team collects comparable information.[10] All team members also should understand the basic operating principles of the sampling equipment and should follow written protocols for collecting air and material samples. With each batch of specimens, inspectors should include *blank samples* (which they handle in the same manner as other samples except that they do not collect any air or material) to detect improper sample handling in the field or laboratory.

Occasionally the managers of a building wish to draw attention to the fact that an investigation is underway, either to reassure worried building occupants or to remind laboratory and hospital staff to adhere to safe practices. However, attracting too much attention may invite interference and other problems. During a study, investigators should make a reasonable effort to prevent disgruntled employees and mischievous on-lookers from intentionally contaminating material samples or generating artificial aerosols, and to uncover conditions facility managers may have tried to hide, e.g., by replacing dirty air filters or water-damaged ceiling tiles, or by removing standing water or treating it with a biocide just before an inspector arrives.

Visual Inspection

Before an initial site visit, investigators should request copies of the building's blueprints or floor plans and make arrangements to examine the HVAC equipment. It may not be possible to gain entry to some sections of an HVAC system if the access panels cannot be removed easily or if the system was designed to be "maintenance-free," i.e., never to need cleaning or repair. Nevertheless, these often are the areas the examiners most need to inspect. Using a fiber-optic scope (which requires only a small access port), investigators can view the interiors of closed HVAC systems and ductwork, and they can collect samples of suspicious-looking materials and dust accumulations through these same openings.

Examples of clues that might lead one to suspect biological contamination in a building are reports of hypersensitivity symptoms similar to those described earlier, as well as musty or moldy odors, visible fungal growth, poor general housekeeping, evidence of past water accumulation or flooding and excessive water in drain pans. If investigators do not find any obvious bioaerosol sources, they may be justified in doubting the likelihood that a biological contaminant is responsible for the occupants' complaints. On the other hand if potential sources are found, further testing may be warranted, as described in the following section.

Surface and Material Sampling

A laboratory can examine surface, water and bulk material samples to confirm suspected building contamination, and to provide isolates for comparison with microorganisms recovered from air samples and infected people. To check building occupants for evidence of immunological sensitization, a medical laboratory may be able to make extracts from isolated bacteria and fungi or other materials for skin and serological testing.[49] Source samples are especially important for fragile and fastidious organisms (such as *Legionella* spp.) for which air sampling seldom is practical or effective.[12]

The turbidity of a *liquid* sample is not always a good indicator of the number of microorganisms it contains. Apparently clear water may contain elevated levels of bacteria or algae whereas cloudy water may contain suspended material other than microorganisms.[24] An investigator can test a liquid anticipated to yield $\geq 10^3$ colony-forming units (cfu)/mL with a dip slide (i.e., a commercially available paddle-shaped slide coated with agar-based medium) by immersing the slide briefly in the test liquid and then covering it for incubation. Investigators also can collect water samples with sterile pipettes or beakers and transfer the samples to sterile leak-proof containers for transport to a testing laboratory. With sterile spoons or spatulas, investigators can collect *sediments* found in drain pans and in humidifiers. Testers can remove specimens of water-damaged *materials* (such as duct lining, carpets, ceiling tiles and wall coverings) with sterile scalpels or scissors, and transport the specimens to the laboratory in new plastic or paper bags for culturing or examination with a dissecting microscope. Standard references outline detailed methods for processing liquid and bulk material samples[21,84] (see also Chapters 11, 12, 13, 14 and 15).

To obtain representative specimens, investigators should sample several areas at each suspected site. Compositing sub-samples reduces the number of items a laboratory must analyze. However, the laboratory should test specimens from apparently clean and contaminated areas separately for comparison. Investigators should submit a sufficient volume of water (preferably ≥ 100 mL), or amount of sediment or bulk material, i.e., several grams. If samples can be delivered to the laboratory within 24 hours, they can be shipped at ambient indoor temperatures. If transport will take longer than one day or if the investigators or laboratory personnel fear the organisms either may multiply or lose viability, the samples should be refrigerated (but not frozen) during storage and transport.

The basic methods for enumerating microorganisms on *surfaces*, in the settings we are discussing, are the swab, rinse, agar contact, vacuum and settle methods.[8,21,58] In the first method, a tester rubs a moistened *swab* over a known surface area (e.g., 50 cm^2), then streaks the swab across a plate of agar-based medium. In the *rinse* method, swabs and other specimens are washed in sterile liquid to release collected organisms, and the liquid is plated on agar-based culture medium.

In the *agar contact* surface sampling method, a tester presses a special plate of raised agar-based medium or a dip slide with a hinged paddle against an impervious surface. This method is limited to test areas that are only moderately contaminated, i.e., those that yield ≤ 10 cfu/cm^2 of sampled surface. In another contact procedure, one can press clear adhesive tape against a surface showing visible mold growth. To identify mold structures, the tape is transferred to a glass slide, stained and examined with a light microscope[14] (see also Chapter 11).

Membrane filters are the collecting substrate when a tester uses a *vacuum probe* to sample fabric surfaces (such as carpets, drapes and upholstered furniture) or to determine how much material a moving airstream could dislodge from a surface. After sampling, the filters are treated as rinse samples or are placed face-up on growth agar. This method was developed in conjunction with the U.S. National Aeronautics and Space Administration's (NASA) planetary quarantine program.[68]

The NASA program also developed a technique to study the microorganisms that accumulate on exposed surfaces. In this procedure, a tester exposes 13 cm^2 sterile stainless-steel strips for varying periods of time, then washes the strips, as described above. Similarly, *settle plates* are used to assess particle fallout.[21] It is assumed that the particles in approximately 30 L of still air will settle on the surface of a standard 9-cm wide culture plate during an exposure time of 15 min.[36] However, because particles settle as a function of their aerodynamic diameter, large and dense particles are over-represented in these samples (see also Chapter 4). Therefore, although settle methods may reflect surface contamination, investigators should not use them to measure bioaerosol concentrations.

Air Sampling

When setting up an air sampling program, investigators must decide: 1) what sampling equipment and procedures they will use, 2) what locations they will monitor, 3) how frequently and at what times they will collect samples, and 4) what detection and analytical methods they will apply.[5,18,85]

Sampling Procedures

No single sampling method allows recovery of all bioaerosols or is ideal in every test situation.[19] Rather, the chosen device must be the most efficient one available for the bioaerosol of interest, considering: 1) the anticipated particle concentration and aerodynamic diameter, 2) the circumstances at the test site, and 3) the available detection methods.[18] For infectious agents, the collection method must protect viability to allow subsequent growth for identification and perhaps for assessment of infectivity (see also Chapters 13, 14 and 15). The condition of the collected material may matter less for detection and identification of non-viable bioaerosols (see also Chapters 11 and 12).

When sampling indoors, investigators may ignore some of the factors that determine the survival of microorganisms and that affect aerosol movement outdoors because: 1) temperature and humidity ranges are limited, 2) ultraviolet radiation levels are lower, and 3) the air often is nearly still (see also Chapters 4 and 6). For accurate particle collection from moving air streams (e.g., inside ventilation ducts and at the faces of supply air diffusers and return or exhaust grilles) investigators should choose sampling probes that match the velocity of the test air stream. During sample collection, an instrument should be left in one place, not waved around in the air, because such movement could interfere with efficient particle collection. Whether a tester should place a sampler with the inlet horizontal or vertical to the floor will depend on the design of the orifice and on air movement at the sampling site.

Samplers investigators commonly use indoors to collect viable bioaerosols include slit, multiple-hole (sieve) and centrifugal impactors, impingers, cyclones, and cassette and high-volume filters[18,19] (see also Chapters 9 and 10). Impaction directly onto agar-based culture medium is the most commonly chosen method to sample viable bacteria and fungi. In occupied areas, hand-held battery-operated samplers have the advantages of portability and independence from a power supply. Although they sample at fairly high flow rates, these devices are quiet and therefore inconspicuous. Depending on the health effects under study, it may be desirable to examine respirable and non-respirable bioaerosol fractions separately. Given typical sampling rates (e.g., 25 to 100 L/min), and available collecting surface areas (e.g., 20 to 180 cm^2), impactor use is limited to environments where bioaerosol concentrations are relatively low (<10^3 cfu/m^3). Closely growing and overlapping colonies and multiple colonies at impaction sites cannot be counted or identified accurately.[69,70] To overcome this shortcoming in some cases, the collected particles can be recovered by washing an impaction surface or dissolving gelatin-based collecting medium.

To avoid overloading impactor samples when investigators expect bioaerosol concentrations $\geq 10^3$ cfu/m^3, liquid impingers or wetted cyclones are recommended. Laboratory analysts can dilute collected particle suspensions from these samplers before culturing, inoculate several types of agar-based media from one liquid sample, and prepare replicate plates for incubation at different temperatures. On the other hand, analysts use filtration to recover microorganisms in liquids containing <10 cfu/mL.[84] However, sampling of *air* onto filters to recover viable microorganisms is only reliable for spore-forming bacteria and fungi.[12]

Sampling Sites

At a minimum, investigators should collect samples: 1) outdoors near air intakes, 2) indoors in the environments of people experiencing symptoms or where contamination is suspected (as determined by interviews and visual inspections), and 3) indoors in comparison or control locations, i.e., in non-complaint areas or sites where no contamination was observed.[5,94] Sampling with and without people present and before and after HVAC equipment is turned on will help assess the bioaerosol contributions from these sources. In buildings served by more than one air-handling unit, questionnaire data may identify the areas warranting inspection and sampling. To locate sources of contamination, it may be necessary to collect samples up- and downstream of heat exchangers, humidifiers, mixing boxes and inaccessible stretches of ductwork, or to sample the supply and return air separately. Investigators can agitate suspected bioaerosol sources (e.g., carpets or banks of filters) during sampling to simulate the disturbance and bioaerosol release that would occur during typical use or operation.[12] However, this type of aggressive sampling should not be done in hospitals and laboratory clean rooms without appropriate precautions to protect patients and sensitive work.

Investigators can collect *area samples* near representative subjects (e.g., several people who have complaints and several who do not) to estimate occupant exposures in general indoor locations. Alternatively, investigators can take *personal samples* (collected with monitors subjects wear on their lapels or otherwise in their breathing zones) to examine the exposure of workers whose specific job locations or activities contribute to their individual exposures and may explain why they have symptoms when other workers do not. Personal samplers are available for collecting particles for microscopic examination, immunoassay, endotoxin measurement and total cell counts and, to a limited extent, for collecting viable microorganisms.[19] Shrouding the inlet of a lapel sampler limits entry of microorganisms a wearer may shed from the skin, scalp, facial hair or clothing.

Sampling Frequency, Timing and Volume

Sampling *frequency* depends on the purpose of the study and on how rapidly the parameters that influence bioaerosol release change. Therefore, investigators should collect multiple samples at close intervals when they expect that conditions vary quickly and the goal is to identify peak exposures rather than long-term averages.[5] After one has established the concentration range and variability, less frequent sampling would suffice for routine surveillance.

As much as possible, the *time* at which samples are collected should reflect the conditions the investigators wish to study. Frequently, it is useful to sample at several different times of day and on multiple days. If occupant symptoms occur only during particular periods, sampling at other times may miss the cause. HVAC system operation, humidity conditions, outdoor sources and occupant activities may vary hourly, daily, weekly and seasonally.

When investigators have no reason to expect that the bioaerosol concentration will be unusually high, they should adjust the sampling time to collect a total *volume* of at least 0.1 m^3.[12] With this sample volume one can detect concentrations on the order of 100 cfu/m^3. Except in ultra-clean areas, one seldom encounters concentrations <10 cfu/m^3 in occupied buildings. Knowing that low bioaerosol concentrations are expected in high-containment laboratories and hospitals, investigators may need to use air samplers with flowrates $\geq$25 L/min to monitor these areas. Samplers with flowrates <5 L/min are not practical when the bioaerosol concentration is <10^2 cfu/m^3 unless they can be operated for several hours. In high concentration areas, on the other hand, samplers with flowrates of 1 to 5 L/min are more reliable than are higher flowrate, direct impactors with which one must use unreasonably short sampling times to avoid overloading. This is especially true when the total sample volume is close to the volume of dead-space within an instrument.

Collecting several samples, each for a different length of time, increases the chances that at least one will yield a countable number of colonies by the culture method or particles for viewing under a microscope, i.e., neither too few nor too many to enumerate reliably. A microbiologist accurately can count between 25 and 250 medium-size colonies on a standard 9-cm wide culture plate.[21] However, laboratory personnel may prefer to have $\leq$100 cfu on a plate when they must examine spreading colonies or pick colonies for subculture and identification. Corresponding guidelines apply to the counting of fungal spores, pollen grains, etc., by microscopy (see also Chapter 11).

The number of replicate samples one should collect depends on the sensitivity and reliability of the analytical method and on the required certainty of detection.[5] At times, it may be more informative to collect one or two samples at each of many sites or a few sites on several occasions than it would be to determine very precisely the bioaerosol concentration at fewer locations or times. At each sampling site, an investigator should collect at least two samples for each culture medium and incubation temperature (or other assay) to be used if the samples cannot be subdivided in the laboratory.

Before beginning, investigators should decide how they will summarize the resulting data. Bioaerosol concentrations rarely follow normal distributions, which precludes the use of standard parametric statistical tests unless the data are transformed. Summarizing lognormally distributed samples by their geometric means and standard deviations is more appropriate than is averaging replicate measurements (which can give undue weight to extreme high values).[5] When measurements include samples from which nothing was detected, reporting the median concentration and concentration range would be appropriate. Environmental measurements often show great variability even for repeated samples from the same sites, e.g., coefficients of variation frequently exceed 100% when the number of samples is small. As a result, only large differences between test and control sites can be detected from the small number of samples that investigators typically collect.

Detection Methods

A variety of methods are available to detect, identify and quantify biological materials (e.g., culture, direct examination by light or electron microscopy, and immunochemical or other assays), depending on the material under study and the type of information needed (see also Chapters 11, 12, 13, 14 and 15). Investigators may select culture media for air and source sample analysis either to support as many culturable microorganisms as possible or to grow selectively specific types if present. Table 18.3 lists media frequently used in indoor investigations.[12,21,67,84] When they have no particular reason to suspect that certain microorganisms will be present, investigators might consider testing air, surface, bulk material and liquid samples for environmental and thermophilic bacteria, and for saprophytic fungi. Sampling for human-commensal bacteria (i.e., those people shed normally from their skin and scalp) may be useful in hospitals, but in other occupied areas the enumeration of these bacteria may be less informative.

A low-nutrient medium (such as R2A agar) and a moderate incubation temperature (e.g., 20 to 30°C) generally yield higher plate counts of environmental bacteria than do rich media and incubation at body temperature.[84] Improved recovery of environmental bacteria has been reported by addition of betaine to reduce osmotic stress and catalase to enhance cell resuscitation.[55,56] The addition of antibiotics (e.g., 0.05 mg/mL cycloheximide) to media for the isolation of bacteria can reduce the growth of fungi.[67] When attempting to isolate *Legionella* spp. from waters that probably contain other microorganisms, the samples should be acid- or heat-treated and plated on media with and without antibiotics.[76,84]

Saprophytic fungi grow better on minimal media kept at room temperature, and adjusting the pH of fungal media to 4.5 to 5.0 inhibits bacterial growth. To identify opportunistic fungal pathogens, mycologists should incubate samples at 35 to 37°C, the temperature at which the fungi must grow to infect a person. Xerophilic fungi (i.e., those that grow better on materials with low available water content) may only be recovered from environmental samples if a laboratory includes culture media with additional NaCl, sucrose or dichloran-glycerol.

Laboratories routinely should check culture media for sterility and ability to support the growth of targeted microorganisms. When a laboratory changes its supplier or the type of media it uses, the replacement media should be tested to ensure that they give comparable results. Disposable plastic culture plates are lighter and safer for field use and postal shipment. Glass or non-stacking disposable plates work best in the six-stage multiple-hole impactor. If an impactor's stage height cannot be adjusted, the laboratory must pour culture plates to the specifications of the sampler's manufacturer because the amount of medium determines the hole- or slit-to-agar distance and, consequently, particle collection efficiency. For other samplers and for general laboratory use, commercially prepared plates generally are adequate.

A single laboratory should process all samples for a study, and if both patient and environmental specimens will be tested, it would be appropriate to choose a licensed clinical laboratory. When a project generates only environmental samples, investigators

Table 18.3 Typical Growth Media

Medium	Incubation Temperature	Target Microorganisms
FUNGI		
• Malt extract agar	20–25°C/35–37°C	Saprophytic/potentially pathogenic fungi (colony counts and identification)
• Inhibitory mold agar	" / "	as above (bacterial suppression, colony counts and identification)
• Rose bengal agar	" / "	as above (bacterial suppression, total counts)
• Malt extract agar with NaCl, sucrose or dichloran-glycerol	20–25°C	Xerophilic fungi (colony counts and identification)
BACTERIA		
• R2A with cycloheximide	20–30°C	Environmental bacteria (fungal suppression, colony counts and identification)
• Soybean-casein digest agar	20–30°C	as above (colony counts and identification)
• Soybean-casein digest agar	50–55°C	Thermophilic bacteria (as above)
• Heart infusion blood agar	35–37°C	Human-commensal bacteria (as above)
• Buffered charcoal yeast extract agar	"	*Legionella* spp. (identification)

may find that laboratories that routinely culture food, soil or water samples have more experience with saprophytic microorganisms than do laboratories specializing in the identification of human pathogens. Many food and water laboratories can modify their standard dilution, plating, counting, identification and reporting methods to handle air samples and environmental water and material specimens.[21,84] Whenever possible, investigators should provide laboratory personnel with references discussing the types of microorganisms possibly present at the study site.[12,62,69,70]

The standard information a laboratory should report for environmental cultures is listed in Table 18.4. The corrected counts (i.e., impactor plate counts adjusted for multiple-particle colonies) and concentrations can be calculated from the total and differential colony counts and the sample volumes. Investigators must take care not to report results with more significant digits than are correct when converting counts to concentrations or calculating median values, e.g., 25 cfu/0.03 m^3 of air should be reported as approximately 830 cfu/m^3, not as 833.3 cfu/m^3. Air sample concentrations typically are reported in cfu/m^3, liquid concentrations in cfu/mL and material concentrations in cfu/g. In some cases, a microbiologist identifies only the predominating microorganisms on a plate. Other times, the investigators might decide that the laboratory should identify all bacteria or fungi present at concentrations greater than a certain level, e.g., 75 cfu/m^3 in air samples, 10^6 cfu/mL in water samples and 10^6 cfu/g in material specimens.[12] Investigators certainly would want to examine microorganisms isolated from both source and air samples, and those present in complaint or suspect sampling areas but absent from control locations.

Table 18.4 Laboratory Data Sheet Information—Microbiological Cultures

- Submitter's name
- Sample identification number
- Study site
- Sample source
- Type of analysis requested
- Date and time sample received
- Description of sample container, amount of material submitted and sample appearance
- Sample storage conditions
- Date and time sample processed
- Microbiologist's name
- Diluent and culture media
- Incubation temperatures
- Times plates examined and observations
- Total plate count
- Corrected total plate count
- For each isolate type:
  — Number of colony-forming units
  — Corrected count (proportional from total)
  — Colony appearance
  — Gram stain
  — Other test results
  — Identification (dated and initialed)

Based on colony morphology, Gram stain, and oxidase and catalase tests, an experienced microbiologist can identify frequently isolated bacteria to the *genus* level, e.g., the genera *Bacillus, Micrococcus, Streptococcus* and *Corynebacterium* (see also Chapter 13). When identification to the *species* level is needed, a microbiologist can consult a reference manual to learn the distinguishing tests.[6] To determine how a patient was infected in a nosocomial disease investigation, the laboratory can compare isolates of the same bacterial species from the patient, suspected staff members and environmental sources by using antibiotic resistance patterns (antibiograms), biochemical profiles (biotypes), serological type, bacteriocin type, plasmid profile, bacteriophage susceptibility (phage types), chromosomal DNA and many newer methods.[58,73] No microbiologist can establish, beyond a doubt, that separate isolates originated from a common source or carrier. Rather, an analyst uses the tests listed above either to decide that isolates clearly differ, or to conclude that they cannot be distinguished and therefore could have had the same origin. The same analyst should test all of the isolates on the same day, and should include known positive and negative controls, as well (if available) as isolates of the same bacterial species from epidemiologically unrelated patients and environments.[73]

While trained microbiologists can recognize frequently seen genera of fungi, only experienced mycologists can correctly identify uncommon fungi and can identify isolates to the species level (see also Chapter 14). Before sampling for viruses, we recommend reading Chapter 13 and Hierholzer's directions,[42] and consulting closely with an experienced virologist.

Sampling only for *viable microorganisms* often underestimates the total concentration of potential *aeroallergens*[11] and does not assess possible exposure to microbiological toxins (see also Chapters 13, 14 and 15). Pollen grains, mold spores, and mite and insect fragments collected from air onto glass slides, or other substrates, can be examined directly by microscope using 100-X magnification. Epifluorescent stains for total cell counts[21,84] are rapid and sensitive, but do not indicate viability or allow one to differentiate bacterial cell types, and it may be difficult to distinguish cells from debris. Antigens and endotoxins are analyzed using *in vitro* immunoassays and biochemical tests, as are mycotoxins, for which *in vivo* assays also may apply.[11,61] Labeled antibody tests are available for the rapid identification of *Legionella pneumophila* in water samples.[26] Other sensitive and specific detection methods (e.g., nucleic acid probes and polymerase chain reaction techniques) are available for some agents (see also Chapters 13, 14 and 15). Wider use of rapid detection methods (with which potential sources can be identified in a few hours rather than a few days) will greatly enhance the ability of laboratories to provide essential information during investigations of outbreaks of infectious diseases.

An appropriate sampling strategy helps ensure that investigators obtain useful, believable and statistically valid results.[85] Unfortunately, investigators generally must work with limited resources. Whenever possible (to avoid costly mistakes), investigators should conduct pilot surveys to estimate air concentrations, to field test sampling equipment and procedures, and to evaluate the analytical methods they plan to use. A study of the preliminary results also will give investigators a chance to consider how they will use their findings to draw conclusions about the test site.

INTERPRETING FINDINGS

Ideally, investigators use the results from environmental sampling either to confirm or refute hypotheses formulated from the information gathered during earlier visual inspections and epidemiological studies. Previously published reports can provide information on contaminant concentrations associated with disease outbreaks.[12,13,62,64] Nevertheless, for the majority of organisms likely to be recovered during routine air and source sampling, very little is known about the minimum dose needed to pose a hazard and there are no guidelines specifying what concentrations of biological material may be considered acceptable in air or material samples.[3] Because people differ in immune status and sensitivity to infectious, allergenic and toxic agents, two people may respond very differently to the same exposure. Consequently, investigators must decide *before sampling* what results would lead them to conclude, for example, that expensive clean-up and control measures are warranted (such as replacing contaminated HVAC equipment or moving patients during hospital construction projects), or that a laboratory procedure or hospital activity potentially may generate excessive bioaerosols.

Explaining sampling results to lay people may present one of the greatest challenges an investigator ever faces. It can be difficult, for instance, to convince the occupants of a building that working there presents little danger when inspectors recovered numerous bacteria and fungi from air samples, some of them potential pathogens. Investigators without extensive training in microbiology should engage an experienced aerobiologist or a medical consultant to help interpret the significance of the microorganisms found. These specialists can determine which of the isolated bioaerosols are known allergens or pathogens, and can describe typical exposure routes and numbers of microorganisms needed to cause disease, if known. Experts also can provide information on common sources and necessary environmental conditions for the growth or occurrence of the identified materials, and recommend control measures.

While positive findings may suggest that microorganisms multiplied in a building and could be disseminated, investigators should repeat overloaded and otherwise questionable samples, duplicating at the same time the corresponding outdoor, test site and control samples. Investigators should consider negative results inconclusive, and should report such samples as "no microorganisms *isolated*" or "no antigen *detected*," naming the target biological material if there was one and reporting the method's limit of detection. Investigators must consider possible reasons for not recovering microorganisms or other biological materials that actually were present, such as unsuitable collection or assay methods. Investigators also should consider the possibility that a source may not have been active when they collected samples or that the concentration was below the sample or assay method's detection limit. Problem buildings may not be investigated for months, sometimes even years, after symptoms first were noticed. Conditions at the time samples are collected may not represent those existing when people initially were exposed and sensitized, or became ill. In a report of their findings, investigators also may wish to list other possible chemical or biological agents that could be responsible for the types of symptoms seen but for which the inspectors did not test.

Whether or not the bioaerosol and environmental monitoring results are considered significant, the investigating team might find it revealing to compare test and control areas for other building-related conditions that affect occupant comfort and well-being. Examples of such factors are air temperature and relative humidity, rate at which outdoor air is supplied, occupancy levels, types of occupant activities, amount of time spent in the building and job satisfaction.[10,77] The concentrations of other air pollutants should not exceed recommended indoor exposure limits.[1,7,89]

Liquid and Material Samples

While it is true that a few microorganisms can be isolated from almost any indoor surface or accumulated water, the presence of a single species of a microorganism in a specimen, rather than a mixture, or the finding of vegetative and spore-bearing fungal structures indicates active, or once-active, growth. The presence of more than 10^5 fungi or 10^7 bacteria per milliliter in a water sample or per gram in a material sample suggests that growth took place but does not necessarily identify the source as a health hazard. Dry soil itself contains 10^5 to 10^8 viable cfu/g.[57] Cell or spore counts from direct stains are almost always higher than plate counts because the first method detects both culturable and non-culturable organisms. The recovery of a potentially pathogenic environmental microorganism from a swab sample of a supply air diffuser or from the interior of a ventilation duct should not cause undue alarm. There is no reliable way to determine when the microorganism was deposited or how frequently it was present in the air supplied to the building. Investigators should consider the possibility of outdoor sources or of upstream contamination within an HVAC system and should examine this equipment more closely if they have reason to think it may be contaminated.

Legionella spp. are found routinely in low concentrations in municipal water supplies and often are present in cooling tower water, apparently without causing disease. Gram-negative, oxidase-positive bacteria (e.g., *Pseudomonas* spp.) often multiply in standing water and may release endotoxins when the cells die and lyse. Although these and other bacteria are found in many waters, to pose a hazard for humans there must be a mechanism for bacterial multiplication and for aerosolization of cells or cell by-products. Several documents discuss methods to monitor cooling waters and to maintain systems so that microorganisms do not multiply.[22–24,28,38,75]

Air Samples

Problem Buildings

Except in hospitals and some laboratories, concentrations of human-commensal bacteria and saprophytic microorganisms (from building occupants, and the outdoor air and building-associated sources) must be fairly high to be of health significance. However, the presence, in a complaint area, of an organism the laboratory did not find in samples of the outdoor air suggests there probably is a building-associated source. Likewise, an explanation should be sought when bioaerosols are found at the test, but not the control sites, or when the total bioaerosol concentration at a test site is many times higher than elsewhere.

The concentration of *fungal spores* usually exceeds that of other bioaerosols in outdoor samples, although pollen, bacteria, algae and insect fragments also are present. Outdoor concentrations of saprophytic fungi in the range of 10^3 cfu/m^3 are not uncommon, especially during summer and in rural areas (see also Chapter 16). In naturally ventilated buildings (in the absence of building-associated sources), indoor fungal air concentrations may nearly equal outdoor concentrations. In mechanically ventilated buildings, the filters of the HVAC system trap some airborne spores and others are removed by sedimentation and impaction within the air distribution system, resulting in lower indoor concentrations of organisms originating from the outdoor air.

Investigators routinely recover molds of the genera *Alternaria, Aspergillus, Aureobasidium, Cladosporium, Epicoccum, Fusarium, Mucor, Penicillium* and *Rhizopus*, and *Rhodotorula* and other yeasts from indoor and outdoor air samples. Because mycotoxin-producing fungi (e.g., *Aspergillus flavus*) only form their toxins under certain growth conditions, investigators must consider the hazard associated with their presence with some care.[12] Unless people are exposed to very high numbers of airborne spores or amounts of contaminated debris (as has occurred in certain agricultural settings) the risk of toxin exposure is probably low. Nevertheless the use of personal protection (e.g., respirators and disposable garments) and of air monitoring to measure exposure levels may be advisable during the removal or cleaning of extensively contaminated materials.

The concentration of *bacteria* in occupied buildings routinely exceeds the outdoor concentration.[12,62] In the absence of building-associated sources, human-commensal bacteria predominate indoors, e.g., *Micrococcus* and *Staphylococcus* spp., which people shed from their skin and hair, and *Streptococcus* spp., which people expel in respiratory secretions during sneezing, coughing and talking. While these Gram-positive cocci generally are not harmful, the recovery of very high concentrations may indicate over-crowding, high activity levels or an insufficient supply of outdoor air. A simpler measure of crowding and of inadequate outdoor air supply is carbon dioxide (CO_2) concentration. People exhale CO_2 and, while it is not itself harmful at these levels, in crowded and poorly ventilated buildings the concentration typically exceeds 800 ppm.

A high concentration of Gram- and oxidase-negative, glucose-fermenting bacteria in the indoor environment may suggest there is a sewer leak or that air exhausted from a sanitary vent is being re-entrained.[12] Higher indoor than outdoor air concentrations of *Bacillus* spp. should prompt a search for an indoor source, e.g., damp wall boards or stored papers. Outdoor endotoxin concentrations may range between 0.5 and 5 endotoxin units (EU)/m^3, as compared to ≤ 0.1 EU/m^3 inside uncontaminated, mechanically ventilated buildings.[60] Actinomycetes are unusual in non-agricultural, indoor environments and their presence indicates that there probably is a contaminated building source when these bacteria are not found in outdoor air samples at similar concentrations.

Laboratories

To assess contamination in laboratories, some quality control programs have adopted the NASA standards as references.[21] These recommendations specify that the viable particle count in a laminar flow (i.e., ultra-clean) room should be <4 cfu/m^3 in air and <2 cfu/cm^2 on test strips exposed for one week.[68] The concentrations for air-conditioned areas are <18 cfu/m^3 in air and <7 cfu/cm^2 on surfaces; and those for industrial areas are <88 cfu/m^3 in air and <32 cfu/cm^2 on surfaces. To be considered acceptable, 9-cm-wide settle plates exposed for 15 min (to monitor surface contamination by particle fallout in laboratories) should recover ≤ 15 cfu.[21]

Hospitals

After establishing baseline measurements in hospitals, periodic environmental sampling can identify changes in air quality. Ranking isolates by concentration for each sampling date and location makes it easier to identify areas deserving further attention. In the cleanest areas of a hospital (e.g., operating rooms and patient isolation areas) the concentrations of saprophytic fungi in samples incubated at room temperature routinely should remain below 15 cfu/m^3.[87] Maintaining the concentrations of opportunistic fungal pathogens below 0.1 cfu/m^3 would limit the number of spores patients could inhale to one or two per day (assuming a person daily breathes 10 to 20 m^3 of air).[91]

Recovery of excessive concentrations of airborne microorganisms in a protected area should prompt investigators to look for inappropriate behavior such as uncontrolled traffic by staff or visitors or for inadequate precautions during housekeeping. The hospital investigator who occasionally recovers small numbers of opportunistic pathogens in air or material samples is faced with a difficult problem of interpretation and may appear (in the eyes of other members of an infection control committee) to be over- or under-reacting. As mentioned earlier for problem-building studies, such a finding may indicate the presence of an indoor or outdoor source of contamination, or might reflect a sporadic or isolated event.

MODELING

In earlier sections of this chapter we explained the need to locate bioaerosol *sources* and mentioned the *impact* of some bioaerosols, but we have not dealt in detail with the *release, dispersion* and *deposition* portions of Edmond's aerobiological model (see also Chapter 4).[27] The movement of air and of suspended bioaerosols in a room depends on the space's size and shape, the types and placement of furnishings, the locations of air supply diffusers and return air grilles, HVAC system operation and the activities of the occupants. Conditions outside a building, e.g., wind speed and direction, also affect building pressurization and particle infiltration and exfiltration.[25] Researchers have attempted to model these effects in order: 1) to identify the major factors that influence bioaerosol behavior, 2) to evaluate the risk of adverse health effects, and 3) to determine the efficacy of intervention measures (see also Chapters 3 and 4).[16,19,25,85,94]

Tracer materials (e.g., gases and surrogate bioaerosols) are useful to track the potential transport of agents that cannot themselves be monitored efficiently because their concentrations are too low, their release sporadic, or detection methods unavailable, expensive or insufficiently sensitive. To determine how much outside air an HVAC system supplies and how the air is distributed throughout a building, ventilation engineers can release and monitor tracer gases (such as halocarbons and sulfur hexafluoride) at strategic locations.[33,35] Although particles move differently in air than vapors, tracer gases are more convenient to use (and in some cases safer) than tracer bioaerosols to demonstrate the potential indoor spread of airborne infectious agents.[52]

Heat-resistant bacterial spores (e.g., *Bacillus subtilis* and *B. stearothermophilus*) have been used to test the efficiency of infectious waste incinerators.[83] Testers have used air samples (collected from exhaust stacks with impingers and filters) to determine the incinerator temperature and residence time needed to ensure that the units do not release infectious agents into the ambient air. In the laboratory, safety specialists have substituted tracer bacteria, bacteriophages and radioactive or fluorescent materials to demonstrate the generation of aerosols and leakage from equipment during routine procedures.[20] Models for studying indoor air quality and HVAC system performance are coming into wider use.[92,93] Epidemic models that summarize the progression of disease through a community also can be applied to the occupants of a building.[81]

VENTILATION FOR BIOAEROSOL CONTROL

Although poorly maintained HVAC systems have been associated with many types of building-related illnesses and complaints, more typically, ventilation improves indoor air quality.[94] *General* or *dilution ventilation* is used to achieve recommended air supply rates, to control temperature, humidity and odors, and to maintain pressure differences that reduce the movement of contaminants into or out of rooms. In office and commercial buildings, dilution ventilation can reduce the indoor concentrations of bioaerosols that the occupants generate. For offices, experts recommend a supply of outdoor air of 10 L/s per person, assuming that the maximum occupancy is 7 people per 100 m^2 of floor area.[89] Local codes for public buildings may specify other outdoor air requirements, and laboratory and hospital codes may mandate positive or negative room pressurization and special handling of exhaust air. *Local exhaust ventilation* is designed to trap aerosols close to a point of generation, e.g., inside a biological safety cabinet[31] or a sputum induction booth (see also Chapter 17).

Because people's activities and movements can be restricted in laboratories and hospitals and because aerosol-generating procedures can be contained, one can achieve better control of bioaerosols in these institutions than in other types of buildings.[9,17,25,41] To protect workers and to restrict the release of organisms from high containment laboratories, regulatory agencies have established a hierarchy of biological safety levels. In these schemes, laboratory workers may only handle microorganisms spread via the aerosol route in laboratories equipped with biological safety cabinets and under negative pressure relative to adjacent areas.[9,17,82]

The amount of supply air needed in a laboratory is determined by room size, physical layout, number of occupants and type of exhaust system. The necessary amount of exhaust ventilation is determined by the type, number, size and operating frequency of safety cabinets, fume hoods and autoclave canopies.[51] The locations of doors, windows and supply air diffusers must be considered when deciding where to place a biological safety cabinet to ensure proper containment at the access port.[4] Thermostat-controlled, variable-air-volume ventilation systems (in which air is supplied only when the space needs heating or cooling) may not be appropriate in laboratories because resulting air pressure changes could affect safety cabinet performance.[51]

Precautions for isolating patients with active tuberculosis are typical of the control measures used in hospitals for cases requiring *respiratory isolation*. Current guidelines recommend ≥6 total air changes per hour (ACH—including at least 2 ACH of outdoor air), negative room pressure relative to adjacent areas, and direct exhausting of room air to the outside and away from outdoor air intakes.[37,39,41] Even higher air change rates are recommended in emergency and autopsy rooms. To protect patients with surgical wounds, burn victims and immunocompromised people, on the other hand, operating rooms and other *reverse isolation* areas should be under positive pressure. For these areas, filtered laminar flow air and high air change rates (e.g., 15 ACH) may be recommended.[41,87] Ventilation is more effective for reducing bioaerosol contamination when the system introduces clean air at the ceiling and removes dirty air near the floor.[37] In hospitals and laboratories, clean and contaminated areas should be ventilated separately, i.e., patient care areas and laboratories should be ventilated independently of laundries, kitchens, and waste collection and disposal areas.[41]

As described above, engineers use ventilation to control air quality because it can be adapted easily to a variety of situations and because (unlike the wearing of respirators and disposable clothing) it does not interfere with occupant activities. Other means of bioaerosol control (e.g., particle removal with in-room air cleaners, and bacterial or viral inactivation with germicidal ultraviolet radiation) have been attempted with some success.[81] Because modification of an existing ventilation system can be more expensive than these alternative control measures, increasingly they are being considered as ways to reduce the spread of airborne infections in high-risk settings such as health-care facilities, shelters for the homeless, correctional facilities and detention centers.[39,81]

SUMMARY

Environmental monitoring in problem buildings, laboratories and hospitals is a valuable investigative tool: 1) to identify sources of contamination, 2) to assess the potential for adverse health effects from aerosolized infectious agents and allergens, and 3) to evaluate the effectiveness of control measures. Ideally, investigators should employ bioaerosol monitoring only after: 1) a medical expert considers the types of infections or hypersensitivity diseases the affected people may suffer, 2) the investigators have no more direct means to obtain the information they need to resolve a problem, 3) a visual inspection has been conducted to identify potential sources, and 4) the investigators formulate testable hypotheses and develop criteria for how they will interpret sampling results.

When investigators determine that environmental monitoring is justified, we agree with Kundsin[50] that they have the greatest chances for success if they can state their purpose clearly, plan the study well, use data collection methods that are scientifically sound, analyze and interpret their results correctly and communicate their findings effectively to all involved parties. Although there are no enforceable upper limits on bioaerosol concentrations in public buildings, laboratories or hospitals, a number of documents offer recommendations and expert opinions.[12,13,62,64] A wide variety of bioaerosol sampling instruments are available, but no single collection or detection method is suitable for all situations. Rapid detection methods (which are becoming more widely available) eventually may overcome some of the limitations inherent in procedures that rely on viable culturing and microscopic examination of bioaerosols. Hopefully in the near future, professional associations will standardized criteria for bioaerosol sampler performance and develop protocols outlining minimum evaluation strategies.[54] Even as bioaerosol sampling becomes simpler and more routine, investigators should expect that they still often will have to rely on their own judgement and experience.

ACKNOWLEDGMENTS

The authors gratefully acknowledge the contributions of J. Cairns, M. First, C. Hines, D. Milton and P. Morey. R. Hashimoto, S. Hayward, J. Wesolowski and K. Willeke also reviewed this manuscript.

REFERENCES

1. *Air Quality Guidelines for Europe.* WHO Regional Publications, European Series No. 23 (Copenhagen, Denmark: World Health Organization, 1987).

2. Allard, P.F. "Project and Problem Definition in Building Air Quality Investigations," in *Biological Contaminants in Indoor Environments,* P.R. Morey, J.C. Feeley and J.A. Otten, Eds. (Philadelphia, PA: American Society for Testing and Materials, 1990), pp. 73–79.
3. American Conference of Governmental Industrial Hygienists. Biologically-Derived Airborne Contaminants, in: *1994–1995 Threshold Limit Values for Chemical Substances and Physical Agents and Biological Exposure Indices.* ACGIH, Technical Information Office, Cincinnati, Ohio, pp. 9–11.
4. *American National Standard for Laboratory Ventilation.* (1993) ANSI/AIHA Z9.5-1992 Fairfax, VA 22031: American Industrial Hygiene Association [2700 Prosperity Avenue, Suite 250].
5. Ayer, H.E. "Occupational Air Sampling Strategies," in *Air Sampling Instruments for Evaluation of Atmospheric Contaminants.* S.V. Hering, Ed. (Cincinnati, OH: American Conference of Governmental Industrial Hygienists, 1989), pp. 21–31.
6. *Bergey's Manual of Systematic Bacteriology,* Volumes 1 to 4. (Baltimore, MD: Williams & Wilkins, 1984 to 1989).
7. Besch, E.L. "Regulation and Its Role in the Prevention of Building-Associated Illness," *Occupational Medicine: State of the Art Reviews,* 4(4):741–752 (1989).
8. *Biohazards Reference Manual.* (Akron, OH: American Industrial Hygiene Association, 1985).
9. *Biosafety in Microbiological and Biomedical Laboratories,* 3rd ed. HHS Pub. No. (CDC) 93-8395. (Washington, DC: U.S. Department of Health and Human Services, Government Printing Office, 1993).
10. *Building Air Quality: A Guide for Building Owners and Facility Managers.* S/N 055-000-00390-4. (Washington, DC: U.S. Environmental Protection Agency/U.S. Department of Health and Human Services, 1991).
11. Burge, H.A. and W.R. Solomon. "Sampling and Analysis of Biological Aerosols," *Atmos. Env.* 21(2):451–456 (1987).
12. Burge, H.A., J.C. Feeley, Sr., K. Kreiss, D. Milton, P.R. Morey, J.A. Otten, K. Peterson, J.J. Tulis and R. Tyndall. *Guidelines for the Assessment of Bioaerosols in the Indoor Environment.* (Cincinnati, OH: American Conference of Governmental Industrial Hygienists, 1989).
13. Burge, H.A., M. Chatigny, J. Feeley, K. Kreiss, P. Morey, J. Otten and K. Peterson. "Guidelines for Assessment and Sampling of Saprophytic Bioaerosols in the Indoor Environment," *Appl. Ind. Hyg.* 2(5):R10–R16 (1987).
14. Burge, H.A. "The Fungi," in *Biological Contaminants in Indoor Environments,* P.R. Morey, J.C. Feeley and J.A. Otten, Eds. (Philadelphia, PA: American Society for Testing and Materials, 1990), pp. 136–162.
15. Burge, S., A. Hedge, S. Wilson, J.H. Bass and A. Robertson. "Sick Building Syndrome: A Study of 4373 Office Workers," *Ann. Occup. Hyg.* 31(4A):493–500 (1987).
16. Burleigh, J.R., R.L. Edmonds, J.B. Harrington, Jr., B. Lighthart, R.E. McCoy, C.J. Mason, G.H. Quentin, A.M. Solomon and J.R. Wallin. "Modelling of Aerobiological Systems," in *Aerobiology: The Ecological Systems Approach.* R.L. Edmonds, Ed. (Stroudsburg, PA: Dowden, Hutchinson & Ross, 1979), pp. 279–371.
17. *Categorization of Pathogens According to Hazard and Categories of Containment.* Advisory Committee on Dangerous Pathogens (London: HMSO, 1984).
18. Chatigny, M.A. and R.L. Dimmick. "Transport of Aerosols in the Intramural Environment," in *Aerobiology: The Ecological Systems Approach.* R.L. Edmonds, Ed. (Stroudsburg, PA: Dowden, Hutchinson & Ross, 1979), pp. 95–109.
19. Chatigny, M.A., J.M. Macher and H.A. Burge. "Sampling Airborne Microorganisms and Aeroallergens," in *Air Sampling Instruments for Evaluation of Atmospheric Contaminants.* B.S. Cohen, Ed. (Cincinnati, OH: American Conference of Governmental Industrial Hygienists, 1994, in press).

20. Collins, C.H. *Laboratory-Acquired Infections: History, Incidence, Causes and Prevention.* 1st ed. (London: Butterworths, 1983), pp. 1–60; 2nd ed. (1988), pp. 1–103.
21. *Compendium of Methods for the Microbiological Examination of Foods.* 3rd ed. (Washington, DC: American Public Health Association, 1992), pp. 51–134.
22. *Control of Legionella in Cooling Towers.* Summary Guidelines. (Madison, WI: Wisconsin Division of Health, 1987).
23. *Control of Legionella in Health Care Premises: A Code of Practice.* HMSO. ISBN 0 11 321208 9 (UK: Department of Health and Social Security, 1989).
24. *Control of Legionellosis Including Legionnaires' Disease.* HS(G)70 HMSO ISBN 0 11 885660 X (Sheffield/London, UK: Health and Safety Executive, 1991).
25. Cox, C.S. *The Aerobiological Pathway of Microorganisms.* (New York: John Wiley & Sons, 1987), pp. 108–171.
26. Dennis, P.J.L. "An Unnecessary Risk: Legionnaires' Disease," in *Biological Contaminants in Indoor Environments.* P.R. Morey, J.C. Feeley and J.A. Otten, Eds. (Philadelphia, PA: American Society for Testing and Materials, 1990), pp. 84–98.
27. Edmonds, R.L. "Introduction," in *Aerobiology: The Ecological Systems Approach.* R.L. Edmonds, Ed. (Stroudsburg, PA: Dowden, Hutchinson & Ross, 1979), pp. 1–10.
28. "Epidemiology, Prevention and Control of Legionellosis: Memorandum from a WHO Meeting," *WHO Bull. OMS.* 68:155–164 (1990).
29. Favero, M.S. "Chemical Disinfection of Medical and Surgical Materials," in *Disinfection, Sterilization and Preservation.* S.S. Block, Ed. (Philadelphia, PA: Lea & Febiger, 1983), pp. 469–492.
30. Favero, M.S. and W.W. Bond. "Sterilization, Disinfection, and Antisepsis in the Hospital," in *Manual of Clinical Microbiology.* 5th ed., A. Balows, Ed. (Washington, DC: American Society for Microbiology, 1991), pp. 183–200.
31. First, M.W. "Ventilation for Biomedical Research, Biotechnology and Diagnostic Facilities," in *Biohazards Management Handbook.* D.F. Liberman and J.G. Gordon, Eds. (New York, NY: Marcel Dekker, 1989), pp. 45–72.
32. Fleming, D.O. "Hospital Epidemiology and Infection Control: The Management of Biohazards in Health Care Facilities," in *Biohazards Management Handbook.* D.F. Liberman and J.G. Gordon, Eds. (New York, NY: Marcel Dekker, 1989), pp. 171–182.
33. Fortman, R.C., H.E. Rector and N.L. Nagda. "A Multiple Tracer System for Real-time Measurement of Interzonal Airflows in Residences," in *Design and Protocol for Monitoring Indoor Air Quality.* (Philadelphia, PA: American Society for Testing and Materials, 1989), pp. 287–297.
34. Garner, J.S. and T.G. Emori. "Nosocomial Infection Surveillance and Control Programs," in *Manual of Clinical Microbiology.* 4th ed. E.H. Lennette, Ed. (Washington, DC: American Society for Microbiology, 1985), pp. 105–109.
35. Grimsrud, D.T. "Tracer Gas Measurements of Ventilation in Occupied Office Spaces," *Ann. Am. Conf. Gov. Ind. Hyg.* 10:71–75 (1984).
36. Gröschel, D.H.M. "Air Sampling in Hospitals," in *Airborne Contagion.* R.B. Kundsin, Ed. (New York, NY: The New York Academy of Sciences, 1980), pp. 230–240.
37. *Guidelines for Construction and Equipment of Hospitals and Medical Facilities.* (Waldorf, MD: American Institute of Architects, Committee on Architecture for Health, 1987).
38. *Guidelines for the Control of Legionnaires' Disease.* (Melbourne, Australia: Health Department Victoria, 1989).
39. "Guidelines for Preventing the Transmission of Tuberculosis in Health-care Settings, with Special Focus on HIV-Related Issues," *MMWR* 39(RR-17):1–29 (1990).
40. "Guidelines for Prevention of Transmission of Human Immunodeficiency Virus and Hepatitis B Virus to Health-Care and Public-Safety Workers," Centers for Disease Control, *MMWR* 38(Suppl 6):1–37 (1989).

41. "Health Facilities," in *Heating, Ventilating, and Air-Conditioning Applications*. (Atlanta, GA: American Society of Heating, Refrigerating and Air-Conditioning Engineers, Inc., 1991), pp. 7.1–7.12.
42. Hierholzer, J.C. "Viruses, Mycoplasmas as Pathogenic Contaminants in Indoor Environments," in *Biological Contaminants in Indoor Environments*. (Philadelphia, PA: American Society for Testing and Materials, 1990), pp. 21–49.
43. Hodgson, M.J. "Clinical Diagnosis and Management of Building-Related Illness and the Sick Building Syndrome," *Occupational Medicine: State of the Art Reviews* 4(4):593–606 (1989).
44. Hughes, J.M. and W.R. Jarvis. "Epidemiology of Nosocomial Infections," in *Manual of Clinical Microbiology.* 4th ed. E.H. Lennette, Ed. (Washington, DC: American Society for Microbiology, 1985), pp. 99–104.
45. Hunskaar, S. and R.T. Fosse. "Allergy to Laboratory Mice and Rats: A Review of the Pathophysiology, Epidemiology and Clinical Aspects," *Lab. Animals* 24:358–374 (1990).
46. *Indoor Air Quality: Biological Contaminants*. WHO Regional Publications, European Ser. No. 31, Report on WHO Meeting, 1990.
47. *Joint Commission Accreditation of Hospitals. Manual for Hospitals.* AMH/86 (Chicago, IL: Joint Commission on Accreditation of Hospitals, 1986).
48. Klimowski, L.L., C. Rotstein, and K.M. Cummings. "Incidence of Nosocomial Aspergillosis in Patients with Leukemia Over a Twenty-year Period," *Infection Control Hosp. Epidemiol.* 10:299–305 (1989).
49. Kreiss, K. "The Epidemiology of Building-Related Complaints and Illness," *Occupational Medicine: State of the Art Reviews.* 4(4):575–592 (1989).
50. Kundsin, R.B. "Microbiological Monitoring of the Hospital Environment," in *Infection Control in Health Care Facilities: Microbiological Surveillance*. K.R. Cundy and W. Ball, Eds. (Baltimore, MD: University Park Press, 1977).
51. "Laboratories," in *Heating, Ventilating, and Air-Conditioning Applications*. (Atlanta, GA: American Society of Heating, Refrigerating and Air-Conditioning Engineers, Inc., 1991), pp. 14.1–14.17.
52. LeClaire, J.M., J.A. Zaia, and M.J. Levin, et al. "Airborne Transmission of Chicken Pox in a Hospital," *New Engl. J. Med.* 302:450–453 (1980).
53. Lipman, N.S., and C.E. Newcomer. "Hazard Control in the Animal Research Facility," in *Biohazards Management Handbook*. D.F. Liberman and J.G. Gordon, Eds. (New York, NY: Marcel Dekker, 1989), pp. 107–149.
54. Macher, J.M., and K. Willeke. Performance criteria for bioaerosol samplers. *J. Aerosol Sci.* 23:S647–650 (1992).
55. Marthi, B., and Lighthart, B. Effects of betaine on enumeration of airborne bacteria. *Appl. Environ. Microbiol.* 56(5):1286–1289 (1990).
56. Marthi, B., B.T. Shaffer, B. Lighthart, and L. Ganio. Resuscitation effects of catalase on airborne bacteria. *Appl. Environ. Microbiol.* 57(9):2775–2776 (1991).
57. McCoy, J.W. *Microbiology of Cooling Water*. (New York, NY: Chemical Publishing Co., 1980).
58. McGowan, J.E. "Role of the Microbiology Laboratory in Prevention and Control of Nosocomial Infections," in *Manual of Clinical Microbiology.* 4th ed. E.H. Lennette, Ed. (Washington, DC: American Society for Microbiology, 1985), pp. 110–122.
59. Miller, B.M. *Laboratory Safety: Principles and Practices*. (Washington, DC: American Society for Microbiology, 1986).
60. Milton, D. Personal communication. Department of Environmental Science and Physiology, Harvard University, Boston, MA, USA (1992).
61. Milton, D.K., H.A. Feldman, D.S. Neuberg, R.J. Bruckner and I.A. Greaves. Environmental endotoxin measurement: the kinetic Limulus assay with resistant-parallel-line estimation. *Environ. Res.* 57:212–230 (1992).

62. Morey, P.R., J.C. Feeley and J.A. Otten, Eds. *Biological Contaminants in Indoor Environments.* (Philadelphia, PA: American Society for Testing and Materials, 1990).
63. Morey, P.R. and J. Singh. "Indoor Air Quality in Nonindustrial Occupational Environments," in *Patty's Industrial Hygiene and Toxicology.* 4th Ed., Vol. 1, Part A. G.D. Clayton and F.E. Clayton, Eds. (New York, NY: John Wiley & Sons, Inc., 1991). pp. 531–594.
64. Morey, P., J. Otten, H. Burge, and M. Chatigny, et al. "Airborne Viable Microorganisms in Office Environments: Sampling Protocol and Analytical Procedures," *Appl. Ind. Hyg.* 1(1):R19–R23 (1986).
65. Murray, J.F. and J. Mills. "Pulmonary Infectious Complications of Human Immunodeficiency Virus Infection, Parts I and II," *Am. Rev. Respir. Dis.* 141:1356–1372, 1582–1598 (1990).
66. Nagda, N.L. and J.P. Harper, Eds. *Design and Protocol for Monitoring Indoor Air Quality.* (Philadelphia, PA: American Society for Testing and Materials, 1989).
67. Nash, P. and M.M. Krenz. "Culture Media," in *Manual of Clinical Microbiology.* 5th ed., A. Balows, Ed. (Washington, DC: American Society for Microbiology, 1991), pp. 1226–1288.
68. National Aeronautics and Space Administration. *NASA Standards for Clean Rooms and Work Stations for the Microbially Controlled Environment.* NHB 5340.2 (Washington, DC: U.S. Government Printing Office, 1967).
69. Nevalainen, A., J. Pastuszka, F. Liebhaber and K. Willeke. "Performance of Bioaerosol Samplers: Collection Characteristics and Sampler Design Considerations," *Atmos. Environ.* (1991).
70. Nevalainen, A., K. Willeke, F. Liebhaber, J. Pastuszka, H. Burge and E. Henningson. "Bioaerosol Sampling," Chapter 21 in *Aerosol Measurement.* K.W. Willeke and P. Baron, Eds. (New York, NY: Van Nostrand Reinhold, 1992).
71. *Occupational Exposure to Bloodborne Pathogens.* Department of Labor, Occupational Safety and Health Administration. 29 CFR 1910, *Fed. Regist.* 54:23042–23139 (1989).
72. Peterson, P.K., P.B. McGlave and N.K. Ramsey. "A Prospective Study of Infectious Disease Following Bone Marrow Transplantation: Emergence of Aspergillus and Cytomegalovirus as the Major Causes of Mortality," *Infect. Control.* 4:81–89 (1983).
73. Pfaller, M.A. "Typing Methods for Epidemiologic Investigations," in *Manual of Clinical Microbiology.* 5th ed., A. Balows, Ed. (Washington, DC: American Society for Microbiology, 1991), pp. 171–182.
74. Pike, R.M. "Laboratory-associated Infections: Incidence, Fatalities, Causes, and Prevention," *Ann. Rev. Microbiol.* 33:41–66 (1979).
75. *The Prevention or Control of Legionellosis (Including Legionnaires' Disease). Approved Code of Practice.* HMSO ISBN 011 885659 6 (Sheffield/London, UK: Health and Safety Executive, 1991).
76. *Procedures for the Recovery of Legionella from the Environment.* National Center for Infectious Diseases (Atlanta, GA: U.S. Department of Health and Human Services, 1992).
77. Quinlan, P., J.M. Macher, L.E. Alevantis and J.E. Cone. "Protocol for the Comprehensive Evaluation of Building-Associated Illness," *Occupational Medicine: State of the Art Reviews,* 4(4):771–797 (1989).
78. "Recommendations for Prevention of HIV Transmission in Health-care Settings," Centers for Disease Control, *MMWR* 36(Suppl):1–18 (1987).
79. Rhame, F.S. "Nosocomial Aspergillosis: How Much Protection for Which Patients?" *Infect. Control Hosp. Epidemiol.* 10:296–298 (1989).
80. Rhame, F.S., A.J. Streifel, J.H. Kersey and P.B. McGlave. "Extrinsic Risk Factors for Pneumonia in the Patient at High Risk of Infection," *Am. J. Med.* 76:42–51 (1984).
81. Riley, R.L. and E.A. Nardell. "Clearing the Air. The Theory and Application of Ultraviolet Air Disinfection," *Am. Rev. Respir. Dis.* 139:1286–1294 (1989).

82. *Safety Measures in Microbiology. Minimum Standards of Laboratory Safety.* World Health Organization Weekly Epidemiological Record. No. 44:340–342 (1979).
83. Segall, R.R., G.C. Blanschan, W.G. DeWees, K.M. Hendry, K.E. Leese, L.G. Williams, F. Curtis, R.T. Shigara and L.J. Romesberg. "Development and Evaluation of a Method to Determine Indicator Microorganisms in Air Emissions and Residue from Medical Waste Incinerators," *J. Air Waste Manage. Assoc.* 41:1454–1460 (1991).
84. *Standard Methods for the Examination of Water and Wastewater.* 17th ed. (Washington, DC: American Public Health Association, American Water Works Association and Water Pollution Control Federation, 1989), pp. 9.4–9.8, 9.52–9.66.
85. Strauss, H.S. "Controlling Laboratory and Environmental Releases of Microorganisms," in *Biohazards Management Handbook.* D.F. Liberman and J.G. Gordon, Eds. (New York, NY: Marcel Dekker, 1989), pp. 345–375.
86. Streifel, A. and F.S. Rhame. "Hospital air filamentous fungal spore and particle counts in a specifically designed hospital." *Proceedings of Indoor Air '93,* July 2 to 6, 1993, Helsinki, Finland, Vol. 4, pp. 161–165.
87. Streifel, A.J., D. Vesley, F.S. Rhame and B. Murray. "Control of Airborne Fungal Spores in a University Hospital," *Environ. Internat.* 15:221–117 (1989).
88. Van Houten, J. "New Frontiers in Biosafety: The Industrial Perspective," in *Biohazards Management Handbook.* D.F. Liberman and J.G. Gordon, Eds. (New York, NY: Marcel Dekker, 1989), pp. 183–208.
89. *Ventilation for Acceptable Indoor Air Quality.* ASHRAE Standard 62-1989 (Atlanta, GA: American Society for Heating, Refrigerating and Air-Conditioning Engineers, 1989).
90. *Vital and Health Statistics. Current Estimates From the National Health Interview Survey: United States, 1988.* U.S. Department of Health and Human Services. Public Health Service. Centers for Disease Control. National Center for Health Statistics. (Hyattsville, MD: DHHS Publication No. (PHS) 89-1501, 1989).
91. Walsh, T.J. and D.M. Dixon. "Nosocomial Aspergillosis: Environmental Microbiology, Hospital Epidemiology, Diagnosis and Treatment," *Eur. J. Epidemiol.* 5(2):131–142 (1989).
92. Woods, J.E. "Measurement of HVAC System Performance," *Ann. Am. Conf. Ind. Hyg.* 10: 77–92 (1984).
93. Woods, J.E. and D.R. Rask. "Heating, Ventilation, Air-conditioning Systems: The Engineering Approach to Methods of Control," in *Architectural Design and Indoor Microbial Pollution.* R.B. Kundsin, Ed. (New York, NY: Oxford University Press, 1988), pp. 123–153.
94. Woods, J.E., P.R. Morey and D.R. Rask. "Indoor Air Quality Diagnostics: Qualitative and Quantitative Procedures to Improve Environmental Conditions," in *Design and Protocol for Monitoring Indoor Air Quality.* N.L. Nagda and J.P. Harper, Eds. (Philadelphia, PA: American Society for Testing and Materials, 1989), pp. 80–98.

CHAPTER 19

Industrial Workplaces

B. Crook and S.A. Olenchock

INTRODUCTION

The presence of bioaerosols in industrial workplaces may have two consequences. The first is their effect on material quality. Airborne dissemination of microorganisms can lead to colonization of materials and a resulting deterioration in structure or quality, with economic losses.[1] The second consequence is the potential health hazards for exposed workers. The aim of this chapter is to give an overview of bioaerosols, created largely by manual or mechanical activity, in indoor work environments and to describe strategies for their measurement from the viewpoint of investigating health effects. When contamination problems are to be investigated, measurement strategies will be similar. Bioaerosols in industrial workplaces are different in terms of the nature and the potential concentrations from those in hospitals and laboratories, or those in domestic and office environments, all of which have been described in Chapter 18. Industrial workplaces described in this chapter encompass agriculture and food production (although bioaerosols associated with intensive livestock rearing will be covered in Chapter 20), factory-based work and biotechnology. The outdoor environment (Chapter 16) also may affect sampling strategies, especially in agricultural environments.

HEALTH EFFECTS OF BIOAEROSOLS IN THE WORKPLACE

Bioaerosols may present the following occupational hazards in the industrial workplace:

1. Microbial infection.
2. Allergic reaction or respiratory sensitization to microorganisms.
3. Allergic reaction or sensitization to non-microbial proteinaceous materials.
4. Toxicological reaction to microbial products or cellular components.

Microbial Infection

Infectious material generally may be regarded as being handled in hospital or laboratory environments, but there is the potential for workers outside these areas, for example, waste treatment personnel (sewage and refuse handling), to be in contact with infectious material.[2]

Opportunist pathogens include those present naturally or as contaminants in the work environment but under normal circumstances their presence poses little risk to an exposed but healthy worker. However, if a susceptible individual comes into contact with the pathogen, infection may arise. An example of this is Legionnaires' disease caused by aerosols of *Legionella pneumophila*. This bacterium may be present in small numbers in many hot water systems, when many people are exposed to low levels of aerosolized droplets supporting bacterial cells. Although the infective dose is not known, the evidence from recent outbreaks has suggested that few viable cells are needed to cause infection when deposited in the lung of a susceptible individual. Risk of infection is significantly increased for older males, smokers, those with high alcohol consumption, diabetics or those undergoing immunosuppressant treatment, i.e., where the body's natural immune system is compromised.[3]

Zoonoses. Infections that spread from animals to man may occur chiefly in farming environments, including microbial aerosols in animal houses (Chapter 20), but in other industries where animal products are being handled there may be the potential for workers to be exposed to airborne microorganisms capable of causing zoonoses. Historically, this includes *Bacillus anthracis,* the bacterium causing anthrax. Although now extremely rare in developed countries, infections have been recorded not only among farm workers and veterinarians, but also among those in contact with blood, bones, hides or wool (anthrax was the cause of an infection referred to as wool sorter's disease).[4] In poultry processing, workers may be at risk from ornithosis caused by aerosols from birds infected by the bacterium *Chlamydia psittaci*.[4]

Allergy to Airborne Microorganisms

Occupational respiratory allergy to inhaled microorganisms includes allergic rhinitis, bronchitis, asthma and extrinsic allergic alveolitis (hypersensitivity pneumonitis) (see also Chapter 15). In addition, organic dust toxic syndrome (ODTS) may result from intense exposure to spores in airborne dust (see below). Mostly, these may result from inhalation of fungal and actinomycete spores, and farming or farm-related activities are the predominant source. A well-known example is farmer's lung disease, a form of allergic alveolitis caused chiefly by inhalation of spores of the actinomycete *Faenia rectivirgula*.[5] However, airborne microorganisms present a risk of respiratory allergy for other work activities in food production, or when handling stored products in factories (e.g., when exposed to contaminated metalworking fluids or air conditioners) or in biotechnology (e.g., when handling microbial cells during downstream processing).[6]

Allergy to Non-Microbial Airborne Materials

Respiratory sensitization can result from inhalation exposure to non-microbial, especially proteinaceous materials present as bioaerosols in the workplace. Mostly these are caused by mechanical handling that creates an aerosol of work material, and typical examples include airborne flour allergens in bakeries,[7] shellfish allergens aerosolized during seafood processing[8] and storage mites in stored grains.[9]

Toxicosis

Endotoxins are heat stable lipopolysaccharide-protein complexes that are integral parts of the outer membrane of Gram-negative bacteria.[10] They are released into the environment after lysis of the bacterial cell or during active cell growth[11] and intact bacterial cells can be phagocytized by macrophages, processed and the endotoxins released with increased toxicity.[12] Both humoral and cellular host mediation systems are profoundly affected by endotoxins.[11,13,14] Complement and coagulation systems are affected, as are a myriad of human cell types. The primary target cell for endotoxin-induced damage after inhalation is the pulmonary macrophage,[15] and human alveolar macrophages are extremely sensitive to the effects of endotoxins *in vitro*.[16] Systemic signs and symptoms that are suggestive of exposure to airborne endotoxins have been reported. Chest tightness, cough, shortness of breath, fever and wheezing have been found in workers in sewage treatment facilities,[17] swine confinement buildings,[18] and poultry units.[19] Controlled exposures of human volunteers to cotton dusts laden with endotoxins resulted in the observation of an association between decreased acute pulmonary function and the airborne level of endotoxins in the dusts.[20] Associations between chronic lung disease and endotoxin concentrations in the air of cotton textile mills also were recorded.[21] Gram-negative bacteria and their endotoxins are ubiquitous, being found in soil, water and other living organisms throughout the world. They are found commonly in various materials in agricultural work sites where large amounts of organic dusts are generated.[22] However, they can be found also in many other occupational environments, including office buildings and libraries especially where humidification systems are in operation.[23]

Mycotoxins are secondary metabolites that may be carcinogenic to humans and are produced by fungi in particular environmental conditions.[24] It is recognized that their ingestion can cause illness and death.[25] Potentially they could be present wherever fungal growth has occurred and may contribute to chronic health problems,[26] but little is known of their toxicity via the inhalation route.[27] Liver cancer has been attributed to occupational exposure to the mycotoxin aflatoxin, produced by the fungus *Aspergillus flavus* in contaminated peanuts, but it is difficult to distinguish between inoculation via inhaled dusts and ingestion.[28]

SOURCES OF BIOAEROSOLS

There are three possible origins of workplace bioaerosols:

1. Contaminants of the material that personnel are handling. These are chiefly infectious, allergenic or toxic microorganisms. However, non-microbial contaminating allergens also may be present, for example, storage mites in grains, or allergenic contaminating plant material, e.g., castor bean. Dust residues in recycled sacks previously used to transport castor beans have contributed to allergy of workers handling sacks of coffee beans,[29] and of workers using old sacks in the manufacture of furniture.[30]
2. Contaminants of the workplace other than work material. This includes processes used to control the work environment, such as air conditioning systems or humidifiers. Bioaerosols from these can present a wide range of occupational hazards from microbial infection, such as *Legionella* contamination, to microbial allergens or toxins in humidifier sludges.[31]
3. Biologically active material as an integral part of the workplace material. This includes non-microbial allergenic bioaerosols created during food processing. Some microorganisms are a necessary part of an industrial process, such as those in composts produced for mushroom cultivation. However, aerosols of actinomycetes in such composts can be generated during handling and spores from the cultivated mushrooms can become airborne during picking, and both may be allergenic.[32] These processes can be considered biotechnology, i.e., the use of

microorganisms to prepare a product. This "traditional" biotechnology is distinguishable from newer and more technologically advanced forms of biotechnology, such as industrial scale fermentations,[33] although the latter also can create hazardous bioaerosols. These may be in the form of microbial allergens or toxins from the process microorganisms, or allergens which were derived from the process microorganisms but are in themselves non-microbial, such as enzyme or antibiotic products. In general, the modern biotechnology industry has a well-controlled approach and few cases of work-related ill-health have been reported. Risk from microbial infection is likely to be small because process organisms are disabled strains and therefore non-infectious.[34]

There follows a summary of some workplaces and work materials where occupational exposure to bioaerosols is known to affect workers' health. These have been reviewed also in detail by Dutkiewicz et al[23] and Lacey and Crook.[6]

Agriculture/Food Production

Bacterial infections: Zoonoses affecting farmers (e.g., leptospirosis from cows), also slaughterhouses and meat processors (e.g., *Chlamydia* infection from ducks and other poultry, Newcastle disease virus from poultry, *Coxiella* infection from sheep or cattle) (see also Chapter 20).[4]

Microbial allergy: Actinomycetes and fungi in stored hay, grains, malted barley, sugar cane bagasse, wood bark chips, mushroom composts (also the mushrooms themselves; *Agaricus*,[35] *Pleurotus*,[36] shiitake[37]); mold contaminants in mushroom growing houses;[32] mushrooms used in soup manufacture; fungal contaminants in sausage production; cheese production; storage of tobacco, spices, herbs, nuts, coffee, tea; contaminants in greenhouses (see also Chapters 14 and 15).

Non-microbial allergy: Wheat, barley, rye, soya, etc., flour, also egg powder and amylase, as well as storage mites, in bakeries; storage mites in stored hay or grains on farms; proteins in sugar beet[38] and potato[39] aerosolized during processing; proteins from seafood processing[8]); coffee (coffee dust and castor bean dust from re-cycled sacks[29]) and tea warehouses; laboratory animal house dusts (dander and urinary proteins).

Toxicosis: Endotoxins in stored grains, feedstuffs such as silage, hays, straw, animal bedding material,[40] composted wood chips,[41] stored timber,[42] tobacco, bulk cottons,[43] *Agaricus bisporus* mushrooms including processing materials such as manure, compost, and spawn,[44] bulk and airborne dusts in swine confinement units[45] and poultry confinement and processing facilities,[46] and in horse and dairy cow barns;[47,48] mycotoxins in stored fodder, grains, nuts, etc.; role of microbial toxins in ODTS after respiratory exposure to high concentrations of organic dusts,[49,50] among farmers,[51] or other end-users of agricultural materials.[52]

Factory/Industrial

Infection: Zoonoses such as *Coxiella,* anthrax from hide/leather processing; legionellosis or Pontiac fever from humidifiers, air conditioners, cooling towers, other process waters.

Microbial allergy: Bacterially contaminated metalworking fluids;[53] bacteria, actinomycetes, fungi, protozoa in humidifiers, air conditioners, cooling towers, papermills and printing works slime, other process waters, waste disposal, sewage and sewage composting; fungal and actinomycete contaminants of wood, cork, bagasse, cotton, wool, flax, hemp, archive material.

Non-microbial allergy: Dusts from wood, cotton, wool, flax, hemp, silk; pharmaceutical or cosmetic preparations ("natural" products, i.e, not biotechnology-produced, e.g., ipecacuanha, senna, henna,[54] castor beans); colophony in solders; furniture making (wood dust, dust from recycled castor bean sacks[30]).

Toxicosis: Endotoxins from humidifiers, air conditioners, cooling towers and other process waters;[55,56] dusts generated during processing of cotton, wool and flax dusts;[20,21,57,58] machining fluids;[59] waste disposal, sewage and sewage composting.[60,61]

Biotechnology

Infection: Pharmaceuticals/therapeutics (*E. coli* often used as a process organism, but a disabled strain; therefore, infectivity is likely to be low).

Microbial allergy: (see also food production) Fungi and bacteria used in antibiotic, enzyme production, single cell proteins.

Non-microbial allergy: Products such as antibiotics, enzymes, detergents, other pharmaceuticals/therapeutics.

Toxicosis: Endotoxicosis from Gram-negative bacterial process organisms, such as *E. coli, Pseudomonas syringae*.[62,63]

BIOAEROSOL SAMPLERS USED IN INDUSTRIAL WORKPLACES

Airborne Microorganisms

A wide range of samplers have been employed to collect microorganisms from the air in workplaces. Table 19.1 gives examples of studies made in a range of industrial workplaces, the sampling methods used and typical numbers of microorganisms found in the air. In most studies, viable microorganisms have been enumerated using one-, two- or six-stage versions of the Andersen microbial impactor. Many other sampling methods (as described in Chapters 9 and 10) also have been used, although few instances have been reported where the RCS centrifugal sampler—often applied to monitoring the air in hospital and laboratories[82]—has been deployed in industrial workplaces. To count total numbers of airborne spores, samples have been collected by filtration or onto microscopic slides for examination by direct observation by light or scanning, epifluorescent or scanning electron microscopy.[65,73,77,79]

Non-Microbial Bioaerosols

To collect airborne allergenic bioaerosols in the industrial workplace, the most frequent method is high-volume filtration followed by extraction into liquid.[83] Alternative methods of high-volume collection directly into liquid, using electrostatic precipitators and glass cyclone samplers with liquid feed, have also been employed.[8,84] High-volume sampling mostly is required because the quantity of airborne allergen is small compared to the

Table 19.1 Examples of Airborne Microbial Contamination Levels in Some Industrial Work Environments

Work Environment	Sampling Method	Bacteria and Actinomycetes	Fungi	Reference
Grain stores	Slit	10^5	10^4	64
Hay, grain on farms	CAMNEA*	10^8	10^8	65
Mushroom farm:				
Compost	Andersen	10^7	10^5	32
Picking	Andersen	10^3	10^5	32
Citrus warehouse	Nuclepore, slit	—	10^5–10^7	64
Sugar beet	Andersen	10^5	10^3	37
Tea factory	Andersen	10^2	10^3	67
Tobacco factory	Andersen	10^3	10^2	67
Domestic waste:	Andersen	—	10^4	68
Tipping hall	Andersen, MSLI Nuclepore	10^4	10^5	69
Recovery plant	Andersen, AGI-30	10^4–10^5	10^3–10^4	70
Composting	Andersen	10^3	10^6	71
Landfill	Andersen	10^5	10^4	72
Cotton mills	Andersen, cascade*	10^4–10^5	10^3–10^4	67,73
		10^6	10^5	73
Woollen mills	Andersen	10^4	10^3	67
Humidifiers	Andersen, N-6	10^2	10^1	74
	RCS biotest	10^2–10^3	10^1–10^2	75
Sawmills	Filter; viable	—	10^4	76
	Filter; SEM*	—	10^6	77
Paper mills	Andersen	10^6	10^2	78
Cork factory	Andersen, cascade*	—	10^6	79
Metalworking: fluids	Impingers	10^6	—	80
Fermenters	Andersen 2-stage	10^2–10^4	—	81
Outdoor, upwind of source	MSLI, Nuclepore	10^2	10^3	69

Numbers = colony-forming units m^{-3} air sampled for viable sampling, except for * non-viable sampling counts by microscopy = cells m^{-3} air.

threshold of sensitivity of available assays, despite being sufficiently large to elicit immunological response in exposed workers. For example, measured airborne antigen concentrations associated with an outbreak of allergic alveolitis (hypersensitivity pneumonitis) in a textile mill were 500 to 15,000 ng m^{-3} air,[85] while in seafood processing factories airborne antigen concentrations were as small as 6 to 115 ng m^{-3}, but still a quarter of workers reported work-related respiratory symptoms.[8] In some instances, however, it may be feasible to develop more sensitive immunoassays involving enzyme-linked immunosorbent assays employing monoclonal antibodies, and this may permit lower volume samplers, including personal filtration samplers.

Endotoxins

Bulk materials, water and settled dusts can be analyzed for endotoxin content. Total dust, respirable and inspirable dusts, and multi-staged size-fractionated dusts also can be tested. Sampling strategies, flow rate, or sampling times would depend on the environment under study. For collecting airborne samples generally polyvinyl chloride (PVC) filters, PVC-co-polymer filters, and glass fiber filters are common while filters of different composition are being examined for their suitability for collecting airborne dusts for endotoxin analyses. Although a concern has been raised relating to the interaction of filter media with lipopolysaccharide,[86] the type of filter employed to collect airborne dusts traditionally was felt to be of little consequence.[87] In general, some modification of the *Limulus* amoebocyte lysate test is used to quantify endotoxins,[86] and the chromogenic modification has been endorsed by two groups of scientists in the study of agriculturally related lung diseases.[87–89]

PRACTICAL CONSIDERATIONS

Sampling Strategies

Bioaerosol sampling in the workplace for occupational health investigation aims at determining the location and concentration of bioaerosol to which the workforce is exposed. This information can form part of an environmental hygiene study, a longitudinal monitoring program or an investigation of an outbreak of ill health. The strategy for sampling therefore is dictated by these needs. The ways in which samples are taken in the workplace have been placed into three categories: random, systematic and judgmental.[90] Random and systematic sampling both involve placing samplers in the whole area to be studied, at points chosen randomly or at specific locations after the area to be monitored has been subdivided by a grid pattern. With judgmental sampling, experience is used to predict where a sampler should be sited to pinpoint and measure a bioaerosol emission.

Random sampling is the most statistically valid method of measurement, although in practice strict adherence to this method involves a much larger amount of sampling, that can be unworkable both from cost and logistics of sample collection and analyses.[90] In addition, such a method of sampling may fail to measure the bioaerosol at its point of emission, which, because peak exposures could be an important factor, may be a primary objective.

The sampling strategies described above apply mostly to area samplers and a different approach may be needed if personal sampling is to be done. Personal sampling may give a more accurate picture of a worker's exposure as samples are taken from the breathing zone during his normal work activity. In practice, however, few samplers currently available for collecting bioaerosols other than filtration devices are suitable for personal sampling (Chapters 9 and 10). However, if personal monitoring devices, such as filtration, can be used then the personnel to wear samplers should be chosen at random, or judgement used to choose those most likely to be exposed.

In practice, a combination of each method described above is the approach most widely adopted to sample bioaerosols in the work environment. Area samplers are located near to a perceived bioaerosol emission, with other samplers positioned randomly in the work area for general ambient measurements, or systematically downwind to measure dispersal of the emission. For comparison, samples also should be taken upwind if out of doors, or in an area separate from the bioaerosol, to provide a background count.

For personal sampling, those whose work activities make them most likely to be exposed to a bioaerosol are monitored and, for comparison, other workers are chosen at random. This sampling regime may be supplemented by general area sampling, including upwind or background samples as described above.

Sampler Positioning

The height at which samples are taken also may be important. Mostly this will be dictated by the bioaerosol source, but for general area sampling, the most appropriate height is one corresponding to the worker breathing height, i.e., 1.5 to 1.7 m above the floor level. In outdoor environments, where winds and convection currents may have an effect, sampling height has been found to influence concentrations of bioaerosols.[91] In the indoor work environment, the geography of the workplace (e.g., siting of large pieces of equipment, presence and position of open windows, exhaust vents or air conditioners, movement of people or vehicles, height of ceilings) will affect both movements and concentrations of bioaerosols.

An important consideration when deciding on a sampling strategy is a walk-through examination of the workplace.[92,93] This may assist in locating potential sources, and possibly the scale, of the bioaerosol, as well as influential factors described above. To assist in source location and the siting of samplers, a powerful light source can be useful for illuminating points of dust or aerosol generation, as can air velocity meters and smoke tubes for determining air movements.

Sampler position is further influenced by practical considerations. If the sampler (personal or general area) affects normal working routines, then the result may be atypical. If sampling is conducted for a long time without supervision, there is the danger that equipment may be switched off (accidentally or otherwise), the air flow through the sampler to the vacuum source become restricted (e.g., by a blocked or kinked tube), the sampler become overloaded, or samples to be interfered with. Consequently, vacuum pumps with elapse timers and memories are an asset, but interference with equipment may be more difficult to control.

When to Sample

Timing of sampling is important. If work activity is sporadic or rapidly changing then the concentration, source and/or nature of the bioaerosol also may change. Samples can be taken continuously to give an overall picture irrespective of these changes, or at specific times to measure bioaerosol release corresponding to a particular activity. Sampling also may be appropriate before work activity starts for determining background values, and after it has finished for measuring the time taken for bioaerosols to disperse.

Handling the Catch

Biological material undergoes various stresses when aerosolized (see also Chapter 6) including dehydration, exposure to UV light and airborne pollutants, and these affect its integrity.[92] For instance, they may kill microbial cells or cause a "switched-off" metabolic state in which they are viable but cannot be cultivated, referred to as non-culturable but viable (NCBV).[95] Post-sample handling must aim to overcome, or allow for, these stresses.

For microorganisms, the main requirement usually is to recover viable cells for cultivation. If an aqueous suspension is prepared from the collected cells, the choice of medium is important if damaged cells are to repair. An isotonic aqueous medium, such as quarter-strength Ringers solution or peptone water, is common while the addition of simple sugars, such as inositol, has been found to enhance recovery of cells stressed by aerosolization.[96,97] After sampling, the aim must be to transfer the catch from the sampler to a suitable transportation receptacle without contamination (sometimes difficult in a "dirty" work environment such as a farm or a domestic waste site) then to store the sample so that viable cells neither multiply nor die before analysis. For suspensions in liquid, this usually involves refrigeration and as short a time delay as possible. More robust cells, collected on filters, may be stored for several days at room temperature.[98,99]

Choice of medium for viable recovery is also important. Some microorganisms, especially when stressed, have exacting growth requirements, so that a range of media and incubation temperatures may be needed.[100] Also, the selective agents of some specific isolation media can have an inhibitory effect, even on the microorganisms targeted, if they are stressed by aerosolization.[101-102] Non-cultivation methods of analysis of microorganisms, e.g., fluorescence microscopy,[103] may be important for deriving the total number of cells present, and sample handling then is less exacting. In contrast, to detect specific microorganisms in the environment, molecular biological techniques, such as polymerase chain reaction (PCR) and gene probe technology may be appropriate.[104] Filtration has been shown to be a successful way of concentrating cells from liquids for PCR analysis,[105] and this may be adapted to concentrate specific microorganisms collected from the airborne environment.

For non-microbial bioaerosols such as airborne allergens, sample handling once again may be less exacting than that required for viable microorganisms. However, aerosolization can induce some structural changes, e.g., to protein configuration,[94] and this may alter the binding potential for an antigen and therefore its characteristics for subsequent immunoassay. If an immunoassay has been prepared to detect a protein specifically, detection may be harder if that protein has been altered structurally following aerosolization.

Recovery also may be enhanced by addition of protective or resuscitative agents. For example, adding oxyethylene docosenol to agar plates protects them from freezing during sampling with Andersen impactors at low ambient temperatures,[106] while adding betaine to collection fluid and agar,[107] or catalase to agar plates[108] enhances repair of aerosol-damaged cells.

SUMMARY

This chapter has described some methods and sampling strategies appropriate for the measurement of bioaerosols of a microbiological, allergenic or toxigenic nature in the industrial workplace. Airborne concentrations typically found in a number of workplaces have been given.

At present no occupational exposure standards or threshold limit values exist for comparison with workplace concentrations of airborne microorganisms or aeroallergens.

Such standards would need to take account of the content as well as the total concentration, for example, the number of airborne microorganisms within that total that represented a health hazard. As few longitudinal studies have been made so far that relate workers' health to workplace exposure to bioaerosols, little information is available on dose-response relationships on which to set any standards. However, as described above, collection of data on background concentrations, and use of a regular monitoring regime to determine typical workplace bioaerosol concentrations under normal, well-controlled circumstances, can provide a useful point of comparison and a basis to devise an action level. For airborne microorganisms, for instance, this action level could be a concentration of total airborne microorganisms, or of a species recognized as being a health hazard, above which there is an indication of poor microbiological control and a need for remedial action. Continued monitoring of workplace exposures, both quantitative and qualitative, eventually will allow practicable and meaningful standards to be set.

REFERENCES

1. Holt, D.M. "Microbiology of paper and board manufacture," in *Biodeterioration 7* (London: Elsevier Publ., 1988), pp. 493–506.
2. McCunney, R.J. "Health effects of work at waste water treatment plants: a review of the literature with guidelines for medical surveillance," *Am. J. Ind. Med.* 9:271–279 (1986).
3. Mayand, C. and E. Dournon. "Clinical features of legionnaires disease," in *A Laboratory Manual for Legionella* (Chichester, UK: Wiley-Interscience, 1988), pp. 5–11.
4. Bell, J.C., S.R. Palmer and J.M. Payne, Eds. *The Zoonoses* (London: Edward Arnold, 1988), p. 241.
5. Lacey, J., J. Pepys and T. Cross. "Actinomycete and fungus spores in air as respiratory allergens," in *Safety in Microbiology*, Soc. Appl. Tech. Ser. No. 6 (London: Academic Press, 1972), pp. 151–184.
6. Lacey, J. and B. Crook. "Fungal and actinomycete spores as pollutants of the workplace and as occupational allergens," *Ann. Occup. Hyg.* 32:515–533 (1988).
7. Musk, A.W., K.M. Venables, B. Crook, A.J. Nunn, R. Hawkins, G.D.W. Crook, B.J. Graneek, R.D. Tee, N. Farrer, D.A. Johnson, D.J. Gordon, J.H. Darbyshire and A.J. Newman Taylor. "Respiratory symptoms, lung function and sensitisation to flour in a British bakery." *Br. J. Ind. Med.* 46:636–642 (1989).
8. Griffin, P., F.P. Roberts and M.D. Topping. "Measurement of airborne antigens in a crab processing factory" in Proceedings 7th International Symposium on Inhaled Particles, Edinburgh (1994), in print.
9. Blainey, A.D., M.D. Topping and R.J. Davies. "Storage mite allergy in UK grain workers," *J. Allergy Clin. Immunol.* 77:173 (1986).
10. Windholz, M., S. Budvari, L.Y. Stroumtsos and M.N. Festig, Eds. *The Merck Index, Ninth Edition* (Rahway, NJ, Merck and Company, 1976), p. 469.
11. Bradley, S.G. "Cellular and molecular mechanisms of action of bacterial endotoxins," *Ann. Rev. Microbiol.* 33:67–94 (1979).
12. Duncan Jr., R.L., J. Hoffman, B.L. Tesh, and D.C. Morrison. "Immunologic activity of lipopolysaccharides released from macrophages after the uptake of intact *E. coli* in vitro," *J. Immunol.* 136:2924–2929 (1986).
13. Galanos, C., M.A. Freudenberg, O. Luderitz, E.T. Rietschel, and O. Westphal. 1979. "Chemical, physicochemical and biological properties of bacterial lipopolysaccharides," in *Biomedical Applications of the Horseshoe Crab (Limulidae)*. Cohen, E., Ed. (New York: Alan R. Liss, 1979), pp. 321–332.

14. Morrison, D.C. and R.J. Ulevitch. "The effects of bacterial endotoxins on host mediation systems," *Am. J. Pathol.* 93:527–617 (1978).
15. Rylander, R. and M.-C. Snella. "Endotoxins and the lung: cellular reactions and risk for disease," *Prog. Allergy* 33:332–344 (1983).
16. Davis, W.B., I.S. Barsoum, P.W. Ramwell and H. Yeager, Jr. "Human alveolar macrophages: effects of endotoxin in vitro," *Infect. Immun.* 30:753–758 (1980).
17. Lundholm, M. and R. Rylander. "Work related symptoms among sewage workers," *Brit. J. Indust. Med.* 40:325–329 (1983).
18. Donham, K.J., D.C. Zavala, and J.A. Merchant. "Respiratory symptoms and lung function among workers in swine confinement buildings: A cross-sectional epidemiological study," *Arch. Environ. Health* 39:96–101 (1984).
19. Thelin, A., O. Tegler and R. Rylander. "Lung reactions during poultry handling related to dust and bacterial endotoxin levels," *Europ. J. Resp. Dis.* 65:266–271 (1984).
20. Castellan, R.M., S.A. Olenchock, K.B. Kinsley and J.L. Hankinson. "Inhaled endotoxin and decreased spirometric values," *New Eng. J. Med.* 317:605–610 (1987).
21. Kennedy, S.M., D.C. Christiani, E.A. Eisen, D.H. Wegman, I.A. Greaves, S.A. Olenchock, T-T. Ye and P-L. Lu. "Cotton dust and endotoxin exposure-response relationships in cotton textile workers," *Am. Rev. Resp. Dis.* 135:194–200 (1987).
22. Olenchock, S.A. "Endotoxins in various work environments in agriculture," *Dev. Indust. Microbiol.* 31:193–197 (1988).
23. Dutkiewicz, J., L. Jablonski and S.A. Olenchock. "Occupational biohazards: a review," *Amer. J. Ind. Med.* 14:605–623 (1988).
24. Scott, P.M., W. van Walbeek, B. Kennedy and D. Anyeti. "Mycotoxins and toxigenic fungi in grains and other agricultural products," *J. Food Chem.* 20:1103–1109 (1972).
25. Miller, J.D. "Fungi as contaminants in indoor air" in *Indoor Air '90*, Proceedings 5th International Conference on Indoor Air Quality and Climate, Ottawa, Vol. 5 (1990), pp. 51–64.
26. Croft, W.A., B.B. Jarvis and C.S. Yatawara. "Airborne outbreak of trichothecene toxicosis," *Atmos. Environ.* 20:549–552 (1986).
27. Flannigan B., E.M. McCabe and F. McGarry. "Allergenic and toxigenic microorganisms in houses," *J. Appl. Bacteriol.* 70:61s–73s (1991).
28. Peers, F.G., G.A. Gilman and C.A. Linsell. "Dietary aflatoxins and liver cancer: a study in Swaziland," *Int. J. Cancer* 17:167–176 (1976).
29. Thomas, K.E., C.J. Trigg, P.J. Baxter, M.D. Topping, J. Lacey, B. Crook, P. Whitehead, J.B. Bennett and R.J. Davies. "Factors relating to the development of respiratory symptoms in coffee process workers," *Brit. J. Ind. Med.* 48:314–322 (1991).
30. Topping, M.D., R.T.S. Henderson, C.M. Luczynska and A. Woodmass. "Castor bean allergy among workers in the felt industry," *Allergy* 37:603–608 (1982).
31. Baur, X., J. Behr, M. Dewair, W. Ehret, G. Fruhmann, C. Vogelmeier, W. Weiss and V. Zinkernagel. "Humidifier lung and humidifier fever," *Lung* 166:113–124 (1988).
32. Crook B. and J Lacey. "Airborne allergenic microorganisms associated with mushroom cultivation," *Grana* 30:446–449 (1991).
33. National Economic Development Council. "New life for industry; Biotechnology, industry and the community in the 1990's and beyond." (London: Nat. Econ. Development Office: 1991).
34. Bennett, A.M., S.E. Hill, J.E. Benbough and P. Hambleton. "Monitoring safety in biotechnology," in *Genetic Manipulation,* Soc. Appl. Bacteriol. Tech. Ser. (London: Edward Arnold: 1991), pp 361–376.
35. Sanderson W., G. Kullman, J. Sastre, J. Olenchock, A. O'Campo, K. Musgrave and F. Green. "Outbreak of hypersensitivity pneumonitis among mushroom farm workers." *Amer. J. Ind. Med.* 22:859–872 (1992)

36. Cox, A., H.T.M. Folgering and L.J.L.D. van Griensven. "Extrinsic allergic alveolitis caused by spores of the oyster mushroom Pleurotus ostreatus," *Eur. Resp. J.* 1:466–468 (1988).
37. Lenhart, S.W. and E.C. Cole. "Respiratory illness in workers of an indoor shiitake mushroom farm," *Appl. Occup. Environ. Hyg.* 8:112–119 (1993).
38. Forster H.W., B. Crook, B.W. Platts, J. Lacey and M.D. Topping. "Investigation of organic aerosols generated during sugar beet slicing," *Am. Ind. Hyg. Assoc. J.* 50:44–50 (1989).
39. Hollander, A., D. Heederik and H. Kauffman. "Acute respiratory effects in the potato processing industry due to a bioaerosol exposure." *Occup. Environ. Med.* 51:73–78 (1994).
40. Olenchock, S.A., J.J. May, D.S. Pratt, L.A. Piacitelli and J.E. Parker. "Presence of endotoxins in different agricultural environments," *Am. J. Indust. Med.* 18:279–284 (1990).
41. Olenchock, S.A., W.G. Sorenson, G.J. Kullman and W.G. Jones. "Biohazards in composted wood chips," in *Biodeterioration and Biodegradation 8.* Rossmore, H.W., Ed. (London: Elsevier Applied Science, 1990), pp. 481–483.
42. Dutkiewicz, J., W.G. Sorenson, D.M. Lewis, and S.A. Olenchock. "Microbial contaminants of stored timber as potential respiratory hazards for sawmill workers," in *Proceedings of the VIIth International Pneumonoconioses Conference.* DHHS (NIOSH) Publication 90-108. pp. 712–716 (1990).
43. Olenchock, S.A., D.C. Christiani, J.C. Mull, T-T. Ye and P-L. Lu. "Endotoxins in baled cottons and airborne dusts in textile mills in The People's Republic of China," *Appl. Environ. Microbiol.* 46:817–820 (1983).
44. Olenchock, S.A., D.M. Lewis, J.J. Marx, Jr., A.G. O'Campo and G.J. Kullman. "Endotoxin contamination and immunological analyses of bulk samples from a mushroom farm," in *Biodeterioration Research 2.* O'Rear, C.E. and G.C. Llewellyn Eds. (New York: Plenum Press, 1989), pp. 139–150.
45. Donham, K., P. Haglind, Y. Peterson, R. Rylander and L. Belin. "Environmental and health studies of farm workers in Swedish swine confinement buildings," *Brit. J. Indust. Med.* 46:31–37 (1989).
46. Lenhart, S.W., P.D. Morris, R.E. Akin, S.A. Olenchock, W.S. Service and W.P. Boone. "Organic dust, endotoxin, and ammonia exposures in the North Carolina poultry processing industry," *Appl. Occup. Environ. Hyg.* 5:611–618 (1990).
47. Olenchock, S.A., S.A. Murphy, J.C. Mull and D.M. Lewis. "Endotoxin and complement activation in an analysis of environmental dusts from a horse barn," *Scand. J. Work Environ. Health* 18:58–59 (1992).
48. Siegel, P.D., S.A. Olenchock, W.G. Sorenson, D.M. Lewis and T.A. Bledsoe. "Histamine and endotoxin contamination of hay and respirable hay dust," *Scand. J. Work Environ. Health* 17:276–280 (1991).
49. doPico, G.A. "Report on diseases," *Am. J. Indust. Med.* 10:261–265 (1986).
50. Parker, J.E., E.L. Petsonk and S.L. Weber. "Hypersensitivity pneumonitis and organic dust toxic syndrome," *Immunol. and Allergy Clinics of North America* 12:279–290 (1992).
51. Husman, K., E.O. Terho, V. Notkola and J. Nuutinen. "Organic dust toxic syndrome among Finnish farmers," *Am. J. Indust. Med.* 17:79–80 (1990).
52. Brinton, W.T., E.E. Vastbinder, J.W. Greene, J.J. Marx, Jr., R.H. Hutcheson and W. Schaffner. "An outbreak of organic dust toxic syndrome in a college fraternity," *J. Amer. Med. Assoc.* 258:1210–1212 (1987).
53. Mattsby Baltzer I., L. Edebo, B. Jarvholm, B. Lavenius and T. Soderstrom. "Subclass distribution of IgG and IgA antibody response to Pseudomonas pseudoalcaligenes in

humans exposed to infected metal-working fluid," *J. Allergy Clin. Immunol.* 86:231–238 (1990).
54. Griffin, P., K. Wiley, H. Forster and R.G. Crane. "An investigation of occupational respiratory allergy in a cosmetics factory." *Clin. Exp. Allergy,* 24:177 (1994).
55. Rylander, R., P. Haglind, M. Lundholm, I. Mattsby and K. Stenqvist. "Humidifier fever and endotoxin exposure," *Clin. Allergy* 8:511–516 (1978).
56. Flaherty, D.K, F.H. Deck, M.A. Hood, C. Liebert, F. Singleton, P. Winzenburger, K. Bishop, L.R. Smith, L.M. Bynum and W. Byron Witmer. "A *Cytophaga* species endotoxin as a putative agent of occupation-related lung disease," *Infect. Immun.* 43: 213–216 (1984).
57. Ozesmi, M., H. Aslan, G. Hillerdal, R. Rylander, C. Ozesmi, and Y.I. Baris. "Byssinosis in carpet weavers exposed to wool contaminated with endotoxin," *Brit. J. Indust. Med.* 44:479–483 (1987).
58. Rylander, R. and P. Morey. "Airborne endotoxin in industries processing vegetable fibers," *Am. Ind. Hyg. Assoc. J.* 43:811–812 (1982).
59. Kennedy, S.M., I.A. Greaves, D. Kriebel, E.A. Eisen, T.J. Smith, and S.R. Woskie. "Acute pulmonary responses among automobile workers exposed to aerosols of machining fluids," *Am. J. Indust. Med.* 15:627–641 (1989).
60. Lundholm, M. and R. Rylander. "Work related symptoms among sewage workers," *Brit. J. Indust. Med.* 40:325–329 (1983).
61. Lundholm, M. and R. Rylander. "Occupational symptoms among compost workers," *J. Occupat. Med.* 22:256–257 (1980).
62. Olenchock, S.A. "Quantitation of airborne endotoxin levels in various occupational environments," *Scand. J. Work Environ. Health* 14:72–73 (1988).
63. Palchak, R.B., R. Cohen, M. Ainslie and C.L. Hoerner. "Airborne endotoxin associated with industrial-scale production of protein products in gram-negative bacteria," *Am. Indust. Hyg. Assoc. J.* 49:420–421 (1988).
64. Dutkiewicz, J. "Exposure to dust borne bacteria in agriculture. i. Environmental studies," *Arch. Environ. Health* 33:250–259 (1978).
65. Malmberg, P., A. Rask-Andersen, U. Palmgren, S. Hoglund, B. Kolmodin-Hedman and G. Stalenheim. "Exposure to microorganisms, febrile and airway-obstructive symptoms, immune status and lung function of Swedish farmers," *Scand. J. Work Environ. Health* 11:287–293 (1985).
66. Strom, G. and G. Blomquist. "Airborne spores from mouldy citrus fruit, a potential occupational health hazard," *Ann. Occup. Hyg.* 30:455–460 (1986).
67. Cinkotai, F.F., M.G. Lockwood and R. Rylander. "Airborne microorganisms and the presence of byssinotic symptoms in cotton mills," *Amer. Ind. Hyg. Assoc. J.* 38:554–559 (1977).
68. Huuskonen, M.S., K. Husman, J. Jarvisalo, O. Korhonen, M. Kotimaa, T. Kuusela, H. Nordman, A. Zitting and R. Mantyjarvi. "Extrinsic allergic alveolitis in the tobacco industry," *Brit. J. Ind. Med* 41:77–83 (1984).
69. Crook B. and J. Lacey. "Enumeration of airborne microorganisms in work environments," *Environ. Tech. Letters* 9:515–520 (1988).
70. Lembke, L.L. and R.N. Kniseley. "Airborne microorganisms in a municipal solid waste recovery system," *Can. J. Microbiol.* 31:198–205 (1985).
71. Clark, C.S., R. Rylander and L. Larsson. "Levels of Gram negative bacteria, *Aspergillus fumigatus,* dust and endotoxin at compost plants," *Appl. Env. Microbiol.* 45:1501–1505 (1983).
72. Rahkonen, P., E. Matti, M. Laukkanen and M. Salkinoja-Salonen. "Airborne microbes and endotoxins in the work environment of two sanitary landfills in Finland," *Aerosol Sci. Technol.* 13:505–513 (1990).

73. Lacey, J. and M.E. Lacey. "Microorganisms in the air of cotton mills," *Ann. Occup. Hyg.* 31:1–19 (1987).
74. Kateman, E., D. Heederik, T.M. Pal, M. Smeets, T. Smid and M. Spitteler. "Relationship of airborne microorganisms with the lung function and leucocyte levels of workers with a history of humidifier fever," *Scand. J. Work Env. Health* 16:428–433 (1990).
75. Grillot, R., S. Parat, A. Perdrix and J. Croize. "Contamination of air systems: 1987–1988 assessment and prospects of the Grenoble intervention group," *Aerobiologia* 6:58–65 (1990).
76. Hedenstierna, G., R. Alexandersson, L. Belin, K. Wimander and G. Rosen. "Lung function and rhizopus antibodies in wood trimmers," *Int. Arch. Occup. Environ. Health* 58:167–177 (1986).
77. Eduard, W., P. Sandven, B.V. Johansen and R. Bruun. "Identification and quantification of mould spores by SEM analysis of filter samples collected in Norwegian saw mills," *Ann. Occup. Hyg.* 32 (Suppl. 1):447–455 (1988).
78. Niemala, S.I., P. Vaatanen, J. Mentu, A. Jokinen, P. Jappinen and P. Sillanpa. "Microbial incidence in upper respiratory tracts of workers in the paper industry," *Appl. Environ. Microbiol.* 50:163–168 (1985).
79. Lacey, J. "The air spora of a Portuguese cork factory," *Ann. Occup. Hyg.* 10:223–230 (1973).
80. Travers Glass, S.A., P. Griffin and B. Crook. "Bacterially contaminated oil mists in engineering works: a possible respiratory hazard," *Grana* 30:404–406 (1991).
81. Martinez, K.F., J.W. Sheehy, J.H. Jones and L.B. Cusik. "Microbial containment in conventional fermentation process," *Appl. Ind. Hyg.* 3:177–181 (1988).
82. Casewell, M.W., N. Desai and E.J. Lease. "The use of the Reuter centrifugal air sampler for the estimation of bacterial air counts in different hospital locations," *J. Hosp. Infec.* 7:250–260 (1986).
83. Agarwal, M.K., J.W. Yunginger, M.C. Swanson and C.E. Reed. "An immunochemical method to measure atmospheric allergens," *J. Allergy Clin. Immunol.* 68:194–200 (1981).
84. Griffin, P., B. Crook, J. Lacey and M.D. Topping. "Airborne scampi allergen and scampi peelers asthma," in *Aerosols, their Generation, Behaviour and Applications* (London: The Aerosol Society, 1988), pp. 347–352.
85. Reed, C.E., M.C. Swanson, M. Lopez, A.M. Ford, J. Major, W.B. Witmer and T.B. Valdes. "Measurement of IgG antibody and airborne antigen to control an industrial outbreak of hypersensitivity pneumonitis," *J. Occup. Med.* 25:207–210 (1983).
86. Milton, D.K., R.J. Gere, H.A. Feldman and I.A. Greaves. "Endotoxin measurement: Aerosol sampling and application of a new Limulus method," *Am. Ind. Hyg. Assoc. J.* 51:331–337 (1990).
87. Popendorf, W. "Report on agents," *Am. J. Indust. Med.* 10:251–259 (1986).
88. Olenchock, S.A. "Endotoxins," in *Biological Contaminants in Indoor Environments,* Morey, P.R., J.C. Feeley, Sr., and J.A. Otten, Eds. (Philadelphia: American Society for Testing and Materials, 1990), pp. 190–200.
89. Rylander, R. "Role of endotoxins in the pathogenesis of respiratory disorders," *Europ. J. Respir. Dis.* 71:136–144 (1987).
90. Keith, L.H. "Environmental sampling: a summary," *Envir. Sci. Tech.* 24:610–617 (1990).
91. Rantio Lehtimaaki, A., A. Koivikko, R. Kupias, Y. Makinen and A. Pohjola. "Significance of sampling height of airborne particles for aerobiological information," *Allergy* 46:68–76 (1991).
92. Burge, H.A., J. Otten, M. Chatigny, J. Feeley, K. Kreiss, P. Morey and K. Peterson. "Guidelines for assessment and sampling of saprophytic bioaerosols in the indoor environment," *Appl. Ind. Hyg.* 2:R-10 (1987).

93. Corn, M. "Assessment and control of environmental exposure," *J. Allergy Clin. Immunol.* 72:231–241 (1983).
94. Cox, C.S. *The Aerobiological Pathway of Microorganisms.* (Chichesrter, UK: Wiley Interscience, 1987), p. 293.
95. Colwell, R.R., P.R. Braton, D.J. Grimes, D.R. Roszak, S.A. Hugo and L.H. Palmer. "Viable but non culturable *Vibrio cholerae* and related pathogens in the environment; implications for the release of genetically engineered microorganisms," *Biotechnology* 3:817–820 (1985).
96. Webb, S.J. "The influence of oxygen and inositol on the survival of semi-dried microorganisms." *Can. J. Microbiol.* 13:733–742 (1967).
97. Crook, B., S. Higgins and J. Lacey. "Methods for sampling airborne microorganisms at solid waste disposal sites," in *Biodeterioration 7.* D.R. Houghton, R.N. Smith and H.O.W. Eggins, Eds. (London: Elsevier, 1987), pp. 791–797.
98. Nordisk Minsterrad. "Harmonisation of sampling and analysis of mould spores," Nordic Council of Ministers (1988).
99. Blomquist, G., U. Palmgren and G. Strom. "Improved techniques for sampling airborne fungal particles in highly contaminated environments," *Scand. J. Work Environ. Health* 10:253–258 (1984).
100. Burge, H.A, W.R. Solomon and J.R. Boise. "Comparative merits of eight popular media in aerometric studies of fungi," *J. Allergy Clin. Immunol.* 60:199–206 (1977).
101. Burge, H.A. "Bioaerosols: prevalence and health effects in the indoor environment," *J. Allergy Clin. Immunol.* 86:687–701 (1990).
102. Bovallius A., B. Bucht, R. Roffey and P. Anas. "Three year investigation of the natural airborne bacterial flora of four localities in Sweden," *Appl. Env. Microbiol.* 35: 847–852 (1978).
103. Palmgren, U., G. Strom, G. Blomquist and P. Malmberg. "Collection of airborne microorganisms on Nuclepore filters—estimation and analysis—CAMNEA method," *J. Appl. Bacteriol.* 61:401–406 (1986).
104. Pickup, R.W. "Development of molecular methods for the detection of specific bacteria in the environment," *J. Gen. Microbiol.* 137:1009–1019 (1991).
105. Bej, A.K., M.H. Mahbubani, J.L. Dicesare and R.M. Atlas. "PCR-gene probe detection of microorganisms by using filter-concentrated samples." *Appl. Env. Microbiol.* 57: 3529–3534 (1991).
106. May, R.R. "Prolongation of microbiological air sampling by a monolayer on agar gel," *Appl. Microbiol.* 18:513–514 (1969).
107. Marthi, B. and B. Lighthart. "Effects of betaine on enumeration of airborne bacteria," *Appl. Env. Microbiol.* 56:1286–1289 (1990).
108. Marthi, B., B.T. Shaffer, B. Lighthart and L. Ganio. "Resuscitation effects of catalase on airborne bacteria," *Appl. Env. Microbiol.* 57:2775–2776 (1991).

CHAPTER 20

Bioaerosols in Animal Houses

C.M. Wathes

> *"I keep six honest serving-men
> (they taught me all I knew);
> their names are what and why and when
> and how and where and who."*
> Rudyard Kipling, 1902: Just So Stories—The Elephant Child

INTRODUCTION

Kipling's servants provide a sound methodological framework for practitioners of the art of bioaerosol sampling in any environment. Animal houses *per se* do not present any novel challenges additional to those found elsewhere: indeed, some aspects are simpler because the system under study is well defined, which can make this environment appealing as a model system. If the six honest serving men can be satisfied, then sampling of bioaerosols in animal houses is straightforward.

The aerobiology of animal houses is a science and technology that is relevant practically to veterinarians and farmers. The catalogue of microbial pathogens that are known or thought to be transmitted through the air is lengthy and covers many of the diseases of farm livestock of economic importance, e.g., infectious diseases of pigs and poultry (Table 20.3).[78] Dusts arising from several other sources are the second type of bioaerosol in animal houses and pose less problems in sampling because there is little need to preserve their biological integrity, i.e., viability or infectivity *in vivo*. With this one important exception, the technical and logistical requirements for sampling both types of bioaerosol are similar and no special distinction between them will be made in this chapter.

WHY?—THE REASONS FOR BIOAEROSOL SAMPLING IN ANIMAL HOUSES

The objective of sampling bioaerosols in an animal house is clear, viz. to collect a representative sample and to characterize its biological action. Bioaerosols include dusts and the air spora, defined by the pioneering aerobiologist P.H. Gregory as analogous to flora and fauna and taken to include airborne viruses, bacteria, protozoan cysts, spores of fungi and lichens, small algae, plant spores and pollen grains.[38] However, this apparently simple purpose of sampling is often confused and many studies appear to have been undertaken because of the availability of the sampler and its operator. This confusion of purpose arises partly because it is only comparatively recently that the significance (or its lack) of air pollution in animal houses and of aerially transmitted diseases of farm livestock has been appreciated. There is growing evidence from laboratory and epidemiological research that poor air hygiene in livestock buildings affects adversely animal and human health by a variety of mechanisms.[28,71,88,93] Similarly, the aerial route is the natural pathway for transmission of the hardy species of fungi and some bacteria, as well as those more fragile vegetative microbes that happen to be launched into the air and dispersed among a large flock or herd of susceptible farm livestock, perhaps totalling 30,000 broilers (domestic fowl destined for poultry meat) or 1000 fattening pigs.

Bioaerosols can play an essential part in the etiology of animal disease. Some infectious diseases of farm livestock have a simple etiology and obey Koch's postulates of causation. An important example is foot and mouth disease. Others have a far more complex etiology in which a host of environmental and biological factors combine to tip the balance in favor of disease. Examples of these are shown in Table 20.1.[85] A third important class comprises non-infectious diseases, e.g., allergies such as obstructive pulmonary disease of horses. This simplified analysis of disease categories helps to determine the purpose of bioaerosol sampling in animal houses (Table 20.2).

If demonstration of the presence in the air of potential microbial pathogens in an infective state is required, as for category I diseases, then every effort must be made to preserve the integrity of the catch throughout the entire process of sampling, collection and assay. Equally difficult may be separation of the pathogen from the rest of the air spora: the task is the aerobiological equivalent of searching for needles in haystacks. Category II diseases require a different approach because their multifactorial etiology demands a full assessment of the air spora, i.e., the concentrations and types of different viable and non-viable microorganisms, inert organic and mineral dusts, and often gaseous pollutants that also may be implicated. A range of samplers and assay techniques is needed. Ostensibly the simplest objective is quantification of one or more potential pathogens as in category III conditions. In many cases, however, painstaking analytical work is required to distinguish and identify unambiguously various spores and other allergens.

In summary, the budding observer of bioaerosols in animal houses must set himself very clear objectives for his bioaerosol sampling if his labors are to be justified and frustration avoided.

WHAT?—THE NATURE OF BIOAEROSOLS IN ANIMAL HOUSES

Bioaerosols in animal house are often rich in both variety and number, especially in those systems of intensive livestock production with poor standards of husbandry. All of the different constituents of the air spora will be found to a greater or lesser extent in one or other type of livestock building. For example, the spora in equine stables mainly comprises fungal spores or hyphal fragments: hay and bedding materials are the dominant sources from which over 50 species of fungal spores have been recovered.[15] Conversely, bacteria will dominate by numbers the air of a poultry house.[54] In general, bioaerosol concentrations will be higher in animal houses than in other industrial, residential or ambient settings.

The natural history of the air spora of animal houses has been reviewed regularly over the past decade or more.[9,23,25,78,88] The major sources of bioaerosols are the animals themselves by way of their various secretions and excretions, their feed and bedding, and material brought indoors by the incoming airstream. The nature and concentration of the air spora represent a dynamic equilibrium between these sources on the one hand and sinks,

Table 20.1 Common Infectious Diseases of Farm Animals Attributed in Major Part to the Housing Environment

Host	Disease	Pathogens	Environment
Pigs	Atrophic rhinitis	*Bordetella bronchiseptica* *Pasteurella multocida*	) crowding,) poor ventilation,
	Enzootic pneumonia	*Mycoplasma suipneumoniae*)	poor drainage, high relative humidity
	Diarrhea	Rotavirus, *E.Coli*, etc.	weaning, hygiene, cold
Cattle	Pneumonia	(*Mycoplasma bovis, dispar* (RSV, PI3	) crowding, poor) feeding, changes,
	"Shipping Fever"	*P. haemolytica*, etc.	) high relative humidity, "stress"
	Environmental mastitis	*E. coli, Strep. uberis*	contaminated bedding, stage of lactation
Horses	Obstructive pulmonary disease	*Mycropolyspora faeni* *Aspergillus fumigatus*	dusty feed and bedding, poor ventilation

Source: Webster (1982).[85]

such as dilution by ventilation, sedimentation and impaction, and death *in situ* for viable microorganisms on the other. Simple steady state models of their equilibrium have been derived[78] but assume perfect mixing, constant emission rates and a single compartment. It is unlikely that these simplifying assumptions will hold in the complex physical environment of a livestock building.

Recently, more complex theoretical models of the temporal and spatial distribution of aerial pollutants in livestock buildings have been formulated in which the equations governing the transport of momentum and energy of the ventilation air flow and the mass transport of the pollutant have been solved numerically.[52,56] These models have advanced our understanding of bioaerosol dynamics in ventilated air spaces because they can account for turbulent transport and examine scales ranging from the microscopic to the macroscopic. However, there are few data collected at full scale that can be used for validification and verification, though current research is addressing this deficiency.

The major benefit of dynamic models of bioaerosols is that different control strategies can be examined and problems identified.[79] For example, what is the optimum location and design of air inlets? Are stagnant zones of poor air quality present and how can they be eliminated? Will feeding troughs form significant barriers to airflow? What are the direct and indirect consequences of raised stocking densities for thermal and aerial environments? These questions are practical in nature but may have profound consequences for animal health and enterprise profitability.

Airborne Pathogens in Animal Houses

Many pathogens of farm livestock are known to be transmitted through the air. Table 20.3 lists examples for pigs and poultry. For some hardy spores, e.g., *Aspergillus fumigatus*, the aerial route is the natural pathway for transmission. Other microorganisms are more

Table 20.2 A simplified Categorization of Diseases of Farm Animals in Which Bioaerosols Play an Important Part

Category	Example	Objective of Bioaerosol Sampling
I. Simple infections	Foot and Mouth disease of cattle, sheep and pigs Newcastle disease of poultry Infectious bronchitis of poultry	Recovery and identification of causative microorganisms are necessary. Quantitative estimates of specific infective pathogens are useful but not essential.
II. Complex infections	Atrophic rhinitis of pigs Enzootic pneumonia of pigs and calves	Complete quantitative and qualitative analysis of air spora and other bioaerosols are essential. Estimates of numbers of viable and non-viable microbes are necessary.
III. Non-infectious conditions	Obstructive pulmonary disease of horses Aspergillosis in poultry	Quantification of putative pathogen(s), e.g., allergens or endotoxin, is required.

Source: Originally conceived at an EC workshop on aerosol sampling in animal houses.[83]

Table 20.3 Common Pathogens of Pigs and Poultry Known to Be Transmitted Aerially

Bacteria

Bordetella bronchiseptica	Mycobacterium tuberculosis
Brucella suis	Mycoplasma gallisepticum
Corynebacterium equi	Mycoplasma hyorhinus
Erysipelothrix rhusiopathiae	Mycoplasma suipneumoniae
Escherichia coli	Pasteurella multocida
Haemophilus gallinarus	Pasteurella pseudotuberculosis
Haemophilus parasuis	Salmonella pullorum
Haemophilus pleuropneumoniae	Salmonella typhimurium
Listeria moncytogenes	Straphylococcus aureus
Leptospira pomona	Streptococcus suis type II
Mycobacterium arium	

Fungi

Aspergillus flavus	Coccidioides immitis
Aspergillus fumigatus	Cryptococcus neoformans
Aspergillus nidulans	Histoplasma farcinorum
Aspergillus niger	Rhinosporidium seeberi

Rickettsia

Coxiella burnetii

Protozoa

Toxoplasma gondii

Viruses

African swine fever	Infectious nephrosis of fowls
Avian encephalomyelites	Infectious porcine encephalomyelitis
Avian leukosis	Marek's disease
Foot-and-mouth disease	Newcastle disease
Fowl plague	Ornithosis
Hog cholera	Porcine enterovirus
Inclusion body rhinitis	Swine influenza
Infectious bronchitis of fowls	Transmissible gastroenteritis of swine
Infectious laryngotracheitis of fowls	

Source: After Wathes (1987)[78] with additions.

fragile yet sufficient numbers survive the physiological and physical trauma of aerosolization and airborne travel to initiate disease. Airborne transmission is not confined to respiratory disease nor does it imply that inhalation is the sole route of inoculation since pathogenic microbes may be ingested or enter the host via other portals such as the conjunctiva or wounds (e.g., *Salmonella typhimurium* in guinea pigs and calves;[63,84] *Escherichia coli* in piglets,[82] *Salmonella enteritidis* in hens).[5]

Several primary pathogens of farm animals have been recovered from the air of animal houses, e.g., foot and mouth disease virus,[27] Newcastle disease virus,[48] mycoplasmas of pigs,[74] Marek's disease virus,[10] *Pasteurella multocida*[2] and Aujeszky's disease virus of pigs.[6] This list appears short but merely reflects the extent to which attempts have been made to isolate specific pathogens. Once a pathogen has been shed in an animal's excretions, secretions or exhaled in breath, then the likelihood of recovering it from the air will be reasonable given a gentle and appropriate sampler and suitable assay. The important point is that few direct studies have been made: instead, airborne transmission has been implicated by indirect, circumstantial evidence. For example, only about 40% of secondary infections of *S. typhimurium* among calves that are penned individually can be attributed to contagion: transmission among the rest is due to fomites, vermin, stockmen and airborne transmission.[42]

There is only one disease of farm animals for which the aerobiological pathway has been determined completely.[26] Speculation on airborne transmission of foot and mouth disease virus among cattle, sheep and pigs began in the early 1900s but it was not until the late 1960s that experimental and epidemiological studies provided substantive evidence. The virus is shed in exhaled breath, feces, urine and secretions at rates dependent on the stage of disease.[27] Virus survival in aerosol is humidity dependent (see also Chapter 6) but sufficient numbers survive to transmit over long distances, ≈ 10 km or more. A computational model has been devised[37] to forecast the likely spread of the virus and utilizes experimental data on shedding rates, aerosol survival and infective doses for the major host species; meteorological and topographical information also are included. The model has been validated in field outbreaks and is in current operational use in the UK.[37] The model is eminently suitable for other diseases, e.g., Newcastle disease of poultry, but lack of biological data on shedding rates, aerosol survival and infective doses makes such applications impossible.[36]

Typical Constituents of Bioaerosols in Animal Houses

The main constituents of bioaerosols in animal houses are whole cells and fragments of the air spora (viruses, bacteria, fungal spores, etc.) and non-specific, organic and inorganic dusts. A complete quantitative and qualitative analysis is only necessary for category II diseases (Table 20.2). For these environmental diseases of farm livestock that have a multifactorial etiology, analysis of the particulate and gaseous composition of the animal house air is also necessary because certain irritant or noxious pollutants, such as dusts and ammonia, may also act synergistically with bioaerosols in exacerbating or provoking respiratory diseases.[41] For example Robertson et al,[71] have demonstrated a strong association between poor air hygiene in the farrowing and first-stage weaner houses and the severity of atrophic rhinitis in young pigs and suggest that the mass or number of inspirable dust particles and airborne bacteria may compromise the local defenses of the upper respiratory tract, thereby facilitating colonization by *Bordetella bronchiseptica* and *Pasteurella multocida*.

The bioaerosol hazard in animal houses can be evaluated and expressed according to the usual quantities, viz. toxicity, allergenicity and infectivity, though in the last case it is rare to determine host infectivity directly but instead to infer its level from the viability *in vitro* of the microorganism in question. Quantification covers not only the presence or absence of a particular bioaerosol but also (i) the mean and maximum number, surface area and mass concentration; (ii) the size distribution, though this may be simplified to respirable and inspirable size fractions; (iii) the temporal nature of exposure, either acute, chronic or episodic; and (iv) a chemical analysis of the various compounds, e.g., endotoxin assay or elemental analysis. Clearly, coverage of all of these quantities requires much analytical effort and this is rarely undertaken in studies of bioaerosols in animal houses.

As an example of the nature of bioaerosols in animal houses, the results of two studies of them in a poultry house are described in detail.[17,54] In the first study,[54] day old cockerels in duplicate groups of 250 were kept either on litter wood shavings or plastic-netting floors raised 25 cm above a concrete floor. Quantitative and qualitative measurements of bioaerosol concentration and constituents were recorded over the normal growing period of eight weeks. All bioaerosol samples were taken in the breathing zone of the birds, i.e., ≈10–15 cm above floor level. Number concentration of dust particles was measured with an optical particle counter (Rion RCO1A). Airborne fungal and actinomycete propagules were collected with a six-stage Andersen sampler and airborne bacteria with a May three-stage liquid impinger. Dust particles were measured microscopically from a visual examination of the composition of samples collected with a May cascade impactor (*v. infra* for a description of samplers). Details of the experimental protocol and methods are given in the original paper.[54]

Table 20.4 and Figures 20.1 to 20.3 inclusive present the results of this experiment. Briefly, the concentration of respirable particles (0.5–5.0 μm diameter) was always high, typically 10^7 to $10^8/m^3$, while large particles (> 5 μm) were fewer in number (Figure 20.1). Microscopic examination showed that skin squames accounted for most particles, with a small quantity of down or feather fragments and some food and fecal debris. Fungal spores never formed more than 5% of the total dust burden (by particle number). Peak concentration of viable fungal propagules exceeded $10^6/m^3$ (Figure 20.2) and a wide range of species were present (Table 20.4). However, the concentrations of airborne bacteria were higher (Figure 20.3), though fewer species were involved. Overall, the bioaerosol hazard was reduced greatly for cockerels kept on a raised netting floor, which demonstrates the role of the litter as both a source and reservoir of bioaerosols.

The second study[17] was undertaken in a large, commercial poultry house holding initially 10,000 pullets on deep litter. A similar approach to the first study was employed for bioaerosol measurements on eight occasions at 12 sampling sites (Figure 20.4). In addition, the mass concentration of inspirable and respirable dust was measured gravimetrically with filter samplers. Microbial samples were restricted to airborne bacteria and species identification was not undertaken.

Bioaerosol concentrations were high and typical of the poor standard of air quality in animal houses. For example, the mean number concentrations of fine (respirable) and coarse (> 5 μm) particles were in the range $2–10 \times 10^7/m^3$ and $6–30 \times 10^6/m^3$, respectively (Figures 20.5 and 20.7). On a mass basis, concentrations were approximately 1.0 and 10.0 mg/m^3, respectively (Figures 20.6 and 20.8), levels of which approach current tolerable exposure limits for farm animals and man.[79] Typically, $10^7/m^3$ airborne bacteria were recovered from the air (Figure 20.9). A full analysis of these data showed no effect of sampling position (or other experimental factors such as operation of the punka fans or number of birds) on any bioaerosol variable; i.e., the spatial distribution was uniform with the corollary that all sites could be considered representative of the overall concentration of any one bioaerosol.

There are numerous other studies of bioaerosols in animal houses (Table 20.5). Invariably they demonstrate high concentrations of bioaerosols by comparison with levels encountered in other environments, a multitude of fungal and bacterial species, and

Table 20.4   Microorganisms Identified in the Air of Poultry Houses, in Newly Opened Food, and in Clean Wood-Shavings (Madelin and Wathes, 1989)[54]

	Weeks Present, Out of Total of 8			
	L	N	Food	Wood Shavings
Fungi				
Acremonium sp.	1	1	+	-
Acremonium strictum	0	0	+	-
Aspergillus spp. (total)	7	6	+	-
A. amstelodami	1	0	-	-
A. candidus	4	4	-	-
A. echinulatus	1	1	-	-
A. flavipes	1	1	-	-
A. flavus	7	6	+	-
A. fumigatus	3	1	-	-
A. glaucus	6	6	+	-
A. nidulans	1	1	-	-
A. repens	7	6	+	-
A. sydowii	1	0	-	-
A. terreus	1	1	-	-
Aureobasidium pullulans	1	3	-	-
Basidiomycetes with clamps	6	6	-	-
Botrytis cinerea	1	1	-	-
Candida fennica	5	4	-	-
Cladosporium spp.	5	7	-	-
Debaryomyces hansenii	7	7	-	-
Fusarium sp.	2	3	-	-
Malbranchea sulfurea	0	1	-	-
Mucor racemosus	3	2	+	-
Paecilomyces variotii	5	2	-	+
Penicillium spp. (total)	8	8	+	+
P. brevicompactum	8	8		
P. chrysogenum	8	8		
P. corylophilum	8	8	not identified to species	
P. granulatum	0	1		
P. purpurogenum	3	3		
Rhizopus nigricans	3	1	-	+
Rhizopus rhizopodiformis	6	4	-	+
Rhodotorula sp.	2	2	+	-
Scopulariopsis brevicaulis	8	8	+	-
Syncephalastrum racemosum	5	5	-	-

Table 20.4 Microorganisms Identified in the Air of Poultry Houses, in Newly Opened Food, and in Clean Wood-Shavings (Madelin and Wathes, 1989)[54]

	Weeks Present, Out of Total of 8			
	L	N	Food	Wood Shavings
Stemphylium sp.	1	0	-	-
Verticillium sp.	0	0	+	-
Actinomycetes				
Streptomycete (white)	1	1	+	-
Thermoactinomyces thalpophilus	4	1	not tested	
Thermoactinomyces vulgaris	4	4	not tested	
Bacteria				
Staphylococcus spp. (total)	7	7	+	-
S. hominis	7	7	not identified to species	
S. saprophyticus	7	7		
S. xylosus	7	7		
Micrococcus sp.	2	2	-	-
Bacillus sp. (gram+ve)	0	0	-	-

L = litter bedding; N = net floors; + = present; - = not present.

complex relationships between physical factors, such as air temperature and humidity, and biological factors, e.g., method of feeding and type of bedding. Nevertheless the complex web can be untangled given proper experimental designs.

HOW?—GENERAL PRINCIPLES OF BIOAEROSOL SAMPLING IN ANIMAL HOUSES

Few, if any of the bioaerosol samplers that are employed routinely in animal houses were designed specifically to this end. Most instruments were devised originally for studies in occupational medicine, plant pathology, clean room technology and warfare. In many cases, a bioaerosol sampler may serve several purposes but the nature of bioaerosols in animal houses can pose some logistic and technical difficulties in bioaerosol sampling and assay. These include (i) a high background concentration of bioaerosols comprising many microbial species and other constituents, which may overload some samplers unless a diluter is utilized or the catch is handled appropriately; (ii) a general requirement for three or more sampling methods if a comprehensive quantitative and qualitative analysis is needed, as for category II diseases (Table 20.2); (iii) practical constraints due to the overall uncleanliness and spatial restrictions of animal houses, which demand high standards of equipment cleaning and disinfection as well as ingenuity in protecting fragile equipment and sampling lines from the aggressive attentions of the animals under study, particularly pigs; and (iv) a requirement for small, robust, portable samplers that are silent in operation and preferably battery powered to minimize disturbance to—and hence unnecessary activity of—the animals, which will alter artificially bioaerosol emission rates.

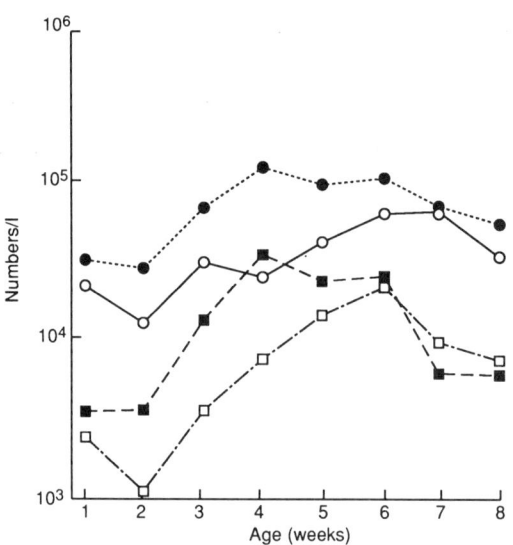

Figure 20.1 Numbers of airborne particles in rooms of broilers kept on either wood shavings litter (closed symbols) or raised netting floors (open symbols): particles 0.5 to 5 µm diameter ● ○, particles > 5 µm diameter ■ □. Madelin and Wathes (1989).[54]

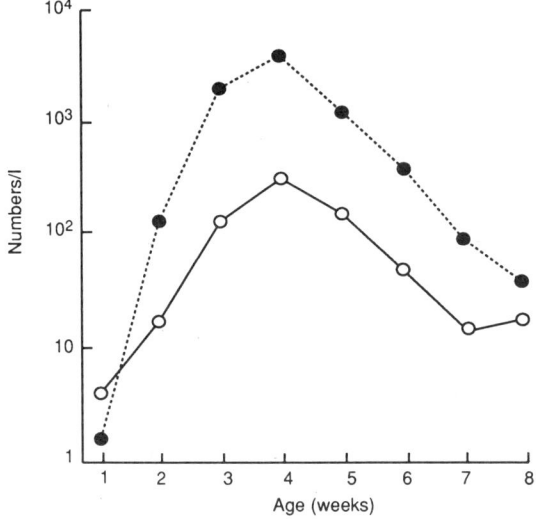

Figure 20.2 Numbers of fungal cfu (colony-forming units) grown at 25°C, trapped by means of an Andersen sampler from the air in rooms of broilers kept on wood shavings litter ●, or on raised netting floors ○. Madelin and Wathes (1989).[54]

Other requirements for instrument calibration (for flow rate, collection efficiency and size separation) follow traditional practices (see Chapters 7, 8, 9 and 10). Similarly, many of the rules for good sampling practice are general.

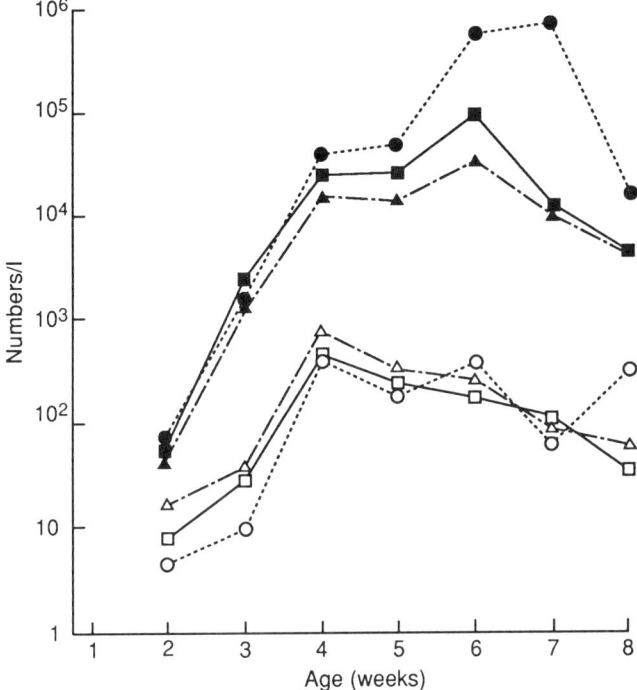

Figure 20.3 Numbers of bacterial cfu (colony-forming units) grown at 37°C, collected by means of May liquid impingers from the air in rooms of broilers kept on wood shavings litter (closed symbols) or raised netting floors (open symbols): cfu from particles > 10 μm diameter ● ○, particles > 10 μm diameter ■ □, < 4 μm ▲ △. Madelin and Wathes (1989).[54]

Physical Processes Involved in Sampling Bioaerosols

First Step—Take a Representative Sample

All bioaerosol sampling can be separated into four main physical processes. First, a representative sample of the bioaerosol must be taken from the air of the animal house. Almost inevitably this sample will be only a small proportion, typically 10^{-5}–10^{-7} vol/vol, of the building's atmosphere. The exact location and number of sampling sites will be determined by considerations of the building layout, ventilation system, method of feeding and manure disposal, animal species, etc. (see below), but the overall aim is to take a sample that is both biologically and physically representative of the system under study. Diurnal and annual rhythms also may be present in the bioaerosol type and concentration. Statistical considerations of acceptable counting accuracy for particle numbers, instrumentation errors and the normal variation in bioaerosol distribution in an animal house also must be applied. Finally, any interference by the probe, and its associated shields, must be minimized through good design of the sampling train.

In other environments there is usually a requirement for bioaerosol sampling to be both isoaxial and isokinetic to ensure that the sample is physically representative (see Chapter 4). Within an animal house, the ventilation air flows are turbulent and slow

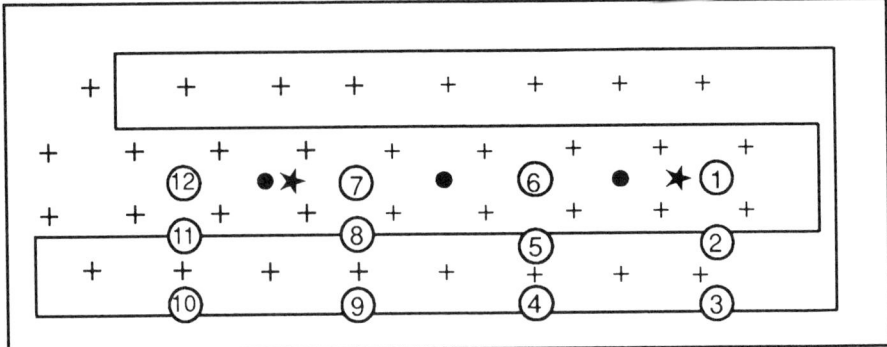

Figure 20.4   Layout of the poultry house and location of 12 sampling sites. ○ sampling location; ★ punka fan; + water drinker; — chain feeder; ● brooder. The building dimensions (l × w × cross h × ridge h) were 33 × 13 × 1.8 × 3.0 m. Conceição et al. (1989).[17]

moving with typical air speeds of 0.2–1.0 m/s. Although the overall air flow can be controlled and predicted, air movement within the animal's breathing zone and elsewhere is unpredictable (Figure 20.10). In consequence isoaxial sampling with an orientated probe is not feasible. Similarly, the errors of anisokinetic sampling increase with particle size[59] and for bioaerosols that are typical of animal houses, then the reduction in collection efficiency at the inlets of common samplers is of the order of 20%.

While there may be some uncertainty involved with the errors of anisoaxial sampling, it is clear that representative samples must be taken rigorously and isokinetic sampling is unnecessary.

Second Step—Transport the Sample to the Collector

Once the bioaerosol sample has been taken, then it must be transported, sometimes with conditioning, to the collector. In many instruments this process may be omitted, but sampling trunks have been used in animal houses where it has proved impracticable to position the samplers within the animal pens. Instead, a large bore duct (≈0.1–0.5 m diameter, 2–5 m in length) equipped with an air mover can be suspended over the pen and samples transported out with the pen or building. Sampling probes are inserted then within the duct or trunk and samples collected isoaxially and isokinetically (since flow is laminar and at a different velocity to the air in the building).

Losses during transport occur by physical processes, such as diffusion, electrostatic precipitation, sedimentation, inertial impaction and turbulent deposition, and are dependent on particle size, trunk length and diameter, and flow rate. They are least for particles with aerodynamic diameters of about 1 μm (see Chapter 4). The main precaution to be taken in minimizing losses during transport is to use short, vertical, metallic ducts with few bends (and avoid aerosol concentrations of 10^{13} particles/m^3, though these are unlikely to occur within animal houses).

Third Step—Condition the Sample

Conditioning a bioaerosol sample before collection includes dilution to an acceptable concentration, modification of air temperature and relative humidity, and electrical neutral-

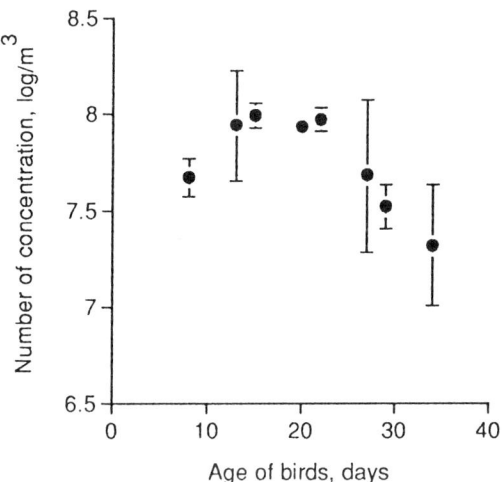

Figure 20.5 Mean (± s.d., n = 12 sites) number concentration of respirable dust particles in a pullet house. Samples collected with a RION KCO1A optical particle sizer. Conceição et al. (1989).[17]

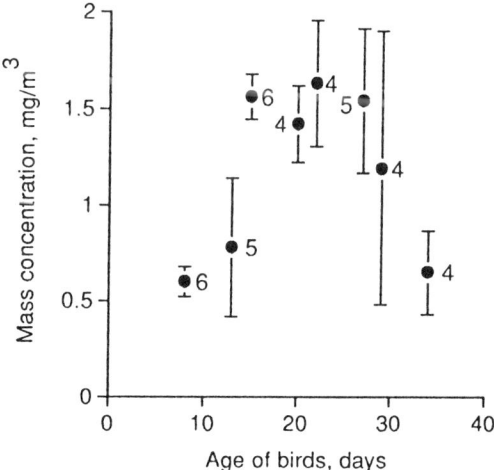

Figure 20.6 Mean (± s.d., n = 4 to 6 sites) mass concentration of respirable dust particles in a pullet house. Samples collected with Cassella gravimetric sampler. Conceição et al. (1989).[17]

ization. Dilution is essential at high number concentrations, $\geq 10^8/m^3$, especially microbial aerosols collected with an impactor, such as the Andersen sampler. Some optical particle counters can be equipped with a diluter at 100:1 or 10:1 (see Chapter 8), but there is a

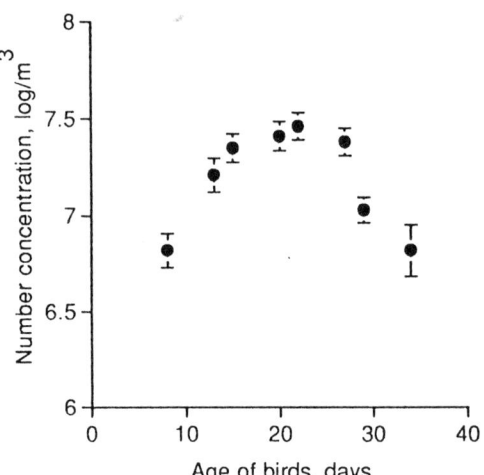

Figure 20.7 Mean (± s.d., n = 12 sites) number concentration of large dust particles in a pullet house. Samples collected with a RION KCO1A optical particle sizer. Conceição et al. (1989).[17]

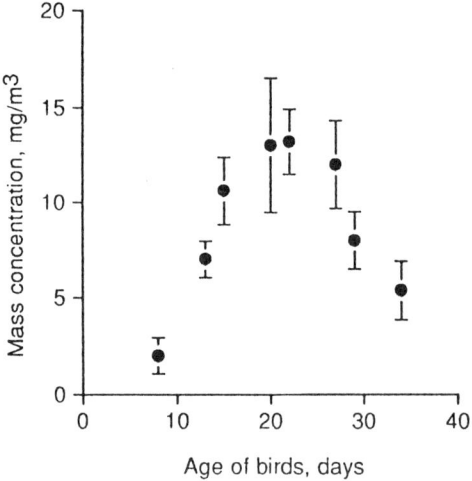

Figure 20.8 Mean (± s.d., n = 12 sites) mass concentration of inspirable dust particles in a pullet house. Samples collected with glass fiber filter samplers. Conceição et al. (1989).[17]

general need for a diluter for other samplers. An unsatisfactory alternative is to shorten the sampling duration, even to the extent that the instrument's dead space is not cleared!

One of the major limitations of all bioaerosol samplers is that the total process does not mimic the gentle vapor phase rehumidification and warming that occurs in the mammalian or avian respiratory system. Conjointly, these changes may (i) alter the size of

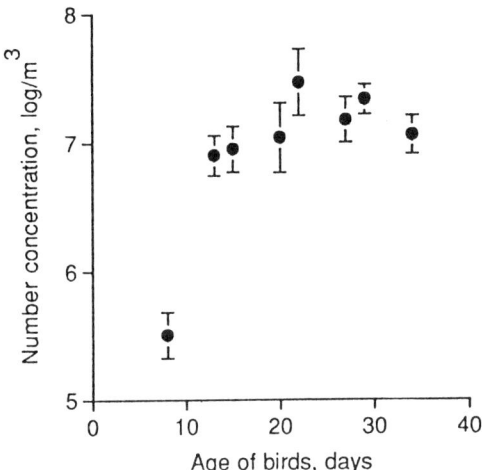

Figure 20.9 Mean (± s.d., n = 12 sites) number concentration of airborne bacteria in a pullet house. Samples collected with AGI-30 liquid impingers. Conceição et al. (1989).[17]

hygroscopic particles, thereby altering the likely landing site within the respiratory tract, and (ii) affect the biological properties of the bioaerosol, particularly its infectivity (see Chapter 6). At the present time, there are no bioaerosol samplers available that incorporate temperature and humidity control.

Finally, the electrical charge distribution of many artificial and some natural aerosols may not follow the Boltzman distribution. This deficiency may affect the coagulation rate of bipolar aerosols and particle deposition. The solution is to attempt to take samples in humid atmospheres, > 70% relative humidity, and/or to employ a charge neutralization chamber. However, this latter course has rarely been followed in animal house studies and more research is needed on the importance of electrical charges of bioaerosols.

Fourth Step—Collect the Bioaerosol with a Sampler

Once the bioaerosol sample has been taken, transported and conditioned, then the final step is to collect it with an appropriate sampler. The choice of samplers is vast (see Chapters 7, 8, 9 and 10 especially). Each sampler has its own advocates but none is perfectly suited for the complex nature of bioaerosols found in animal houses. Instead, several different types usually are employed according to the study's aims (Table 20.2).

In summary, the four basic processes of bioaerosol sampling in animal houses are common to those of other environments but iso-axial and isokinetic sampling is unnecessary. A useful addition to the sampler's armory would be a simple conditioner that dilutes, neutralizes electrically and modifies air temperature and humidity to levels found in the respiratory system before sample collection. However, it is not necessary to develop a new sampler specifically for animal houses: the existing range of instruments more than encompasses current study aims and objectives.

Table 20.5 Studies of Bioaerosols in Animal Houses (1970 Onward)

Species	Study	
Poultry	Clark et al, 1983; [14]	Cravens et al, 1981; [18]
	Conceição et al, 1989; [17]	Dennis and Gee, 1973; [24]
	Feddes et al, 1992; [31]	Feddes et al, 1992; [32]
	Feddes et al, 1982; [33]	Glennon et al, 1984; [35]
	Feddes and Licsko, 1993; [34]	Madelin and Wathes 1989 [54];
	Maghirang et al, 1989; [55]	Maghirang et al, 1990; [56]
	Maghirang et al, 1991; [57]	Qi et al, 1992; [68]
	Pinello et al, 1977; [66]	Wathes et al, 1991 [81]
Pig	Baekbo and Wolstrup, 1989; [2]	Bourgueil et al, 1992; [6]
	Bundy and Veenhuizen, 1987; [7]	Butera et al, 1991; [8]
	Chiba et al, 1987; [11]	Chiba et al, 1985; [12]
	Clark et al, 1983; [14]	Clark and McQuitty, 1987; [13]
	Crook et al (in press); [20]	Curtis et al, 1975; [21]
	Curtis et al, 1975; [22]	Elliott et al, 1976; [30]
	Grunloh et al, 1971; [39]	Heber et al, 1988; [43]
	Heber et al, 1988; [44]	Hill and Kenworthy, 1970; [45]
	Honey and McQuitty, 1979; [47]	Meyer and Bundy, 1991; [62]
	Pickrell et al, 1993; [65]	Robertson, 1992; [70]
	Robertson et al, 1990; [71]	Underdahl et al, 1982; [76]
	Van't Klooster, 1993; [51]	Welford et al, 1992; [87]
Cattle	Jones and Webster, 1981; [50]	Pritchard et al, 1981; [67]
	Wathes et al, 1984; [80]	
Horse	Clark et al, 1987; [13]	Crichlow et al, 1980; [19]
	Webster et al, 1987; [86]	Woods et al, 1993; [89]
	Zeitler, 1985; [90]	
General including reviews	Carpenter, 1986; [9]	De Boer and Morrison, 1988; [23]
	Donaldson, 1978; [25]	Donham, 1987; [28]
	Gustafsson and Martensson, 1990; [40]	Honey and McQuitty, 1976; [46]
	Wathes, 1987; [78]	Wathes, 1994; [79]
	Whyte, 1993; [88]	

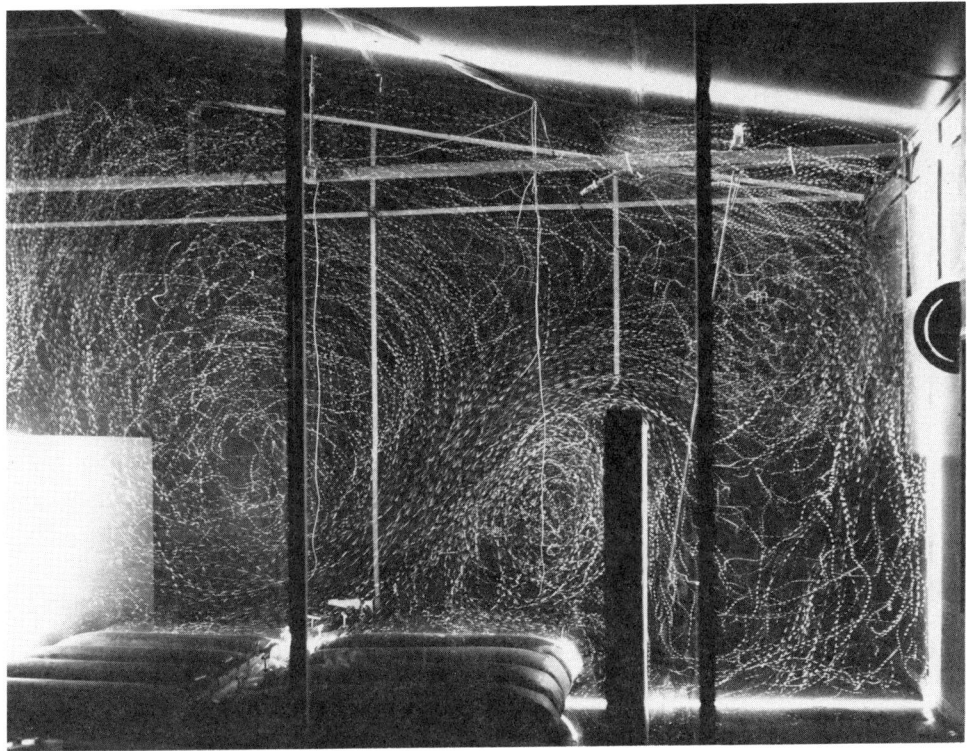

Figure 20.10 Airflows in an animal house are nearly always turbulent and slow moving with typical air speeds from 0.2-1.0 m/s. The overall flow pattern can be controlled (and predicted) from a specification of the ventilation system and ambient conditions, but the local patterns are less predictable in each axis. A cross-section of a piggery has been illuminated and neutral density tracer bubbles reveal the chaotic air flows. (Photo courtesy of J.M. Randall, Silsoe Research Institute, UK.)

Choice of Bioaerosol Samplers for Use in Animal Houses

The first principle is that the choice of sampler must be determined from a critical analysis of the sampling objective, particularly the disease category (Table 20.2). While many instruments have been utilized in animal houses, the most common choices are the Andersen,[1] May[60] and Burkard impactors, the AGI-30[61] and May multi-stage liquid impingers,[58] gravimetric filtration samplers (see Chapter 10) and various optical particle counters. Comparisons of samplers for microbial bioaerosols in the environment of animal houses are rare. Recently, Thorne et al[75] assessed the suitability of the Andersen (in the one- or two-stage configuration), AGI-30 and a filtration-elution method using Nucleopore filters in sampling bioaerosols in a piggery. Using the criteria of samplers yielding data within the limits of detection, intraclass reliability and correlation between methods, they concluded that the AGI-30 was best for total counts of bacteria and fungi while the Andersen was preferred for enteric bacteria. A similar comparison was undertaken with eight bioaerosol samplers.[49] Studies of this type are most valuable and should be extended to other types of animal house.

Optical Particle Counters

Optical particle counters, such as the Rion KCO1A, may be employed in animal houses to determine the size distribution of airborne dusts. Their main advantage is rapid analysis of bioaerosol characteristics with good statistical accuracy. Their chief limitations are size [since they are often sited among the animals (Figure 20.11) to avoid the use of long sampling trains and to minimize animal disturbance] and expense, which restricts the number employed in any one study. They suffer too from a general weakness that calibration nearly always utilizes experimental aerosols that may have different optical properties from those under study (see Chapters 7 and 8). Diluters are available and often may be needed when particle concentrations reach $10^8/m^3$.

Impactors

The Burkard single-stage impactor has been employed in a number of studies in stables,[15,16] principally for qualitative microscopic analysis of the air spora. This slit sampler is portable, inexpensive and small; practical use is therefore simple and rapid. However, its collection efficiency is dependent on particle size,[15] though this limitation may not be serious if the bioaerosol size distribution is narrow. A modified version has been developed that collects particles onto a single miniature Petri dish rather than a microscope slide (Burkard Manufacturing Co., Rickmansworth, UK).

The Andersen six-stage impactor is probably the most common bioaerosol sampler used in animal houses. Its principal advantages are a well-characterized collection efficiency, size separation into six fractions in the respirable range, utilization of standard-size plastic Petri dishes with direct transfer of the catch to an incubator for culture and assay, and wide-spread adoption, which facilitates comparisons between studies. However, it is cumbersome to use, costly and its high flow rate usually means that sampling times must be short if the sampler is not to be overloaded in the heavily contaminated air of most animal houses. The minimum sampling time should be at least 10 s and preferably 30 s. Finally, while glass Petri dishes are recommended in preference to plastic[1] to avoid problems with electrostatic charge and biased collection, few adhere to this advice, mainly because of the inconvenience and cost of glass dishes. It is important to state the type of plate utilized and to ensure that the depth of agar is sufficient to maintain the critical clearance between the agar surface and the underside of the impactor jets. For plastic dishes, approximately 35 mL of agar is needed, which is almost twice the volume usually dispensed.[54]

The May 'ultimate' seven-stage impacter[60] collects particles onto conventional microscope slides for visual examination and particle sizing. Alternatively, particles can be washed off the slides and assayed. The overriding disadvantage of its use in animal houses is the effort and expertise needed to examine each slide. Automated image analysis systems rarely can be used due to the heterogeneous composition of the bioaerosols, and manual methods must be employed.[53] However, particles are collected according to aerodynamic size and the sampler is probably best used to determine the relative proportions of different types of particles.

Impingers

Both the AGI (All Glass Impinger) and the May three-stage liquid impinger have been employed to sample the air spora in animal houses. The collection characteristics of each

Figure 20.11 Bioaerosol sampling in the crowded, cramped conditions cf an aviary for laying hens requires ingenuity in sampler positioning and protection. The cage contains an Andersen six-stage impactor, a May ultimate cascade impactor and an AGI-30 liquid impinger. The hens are perching on a Rion KCO1A optical particle counter.

are known.[58,61] In practice, both samplers are cheap to purchase—though the three-stage impinger requires skill in glass construction to meet the exacting specifications for various critical dimensions—and overloading is not a problem since the collection fluid can be diluted and its contents assayed on various microbiological media. Furthermore, long sampling durations are feasible (≈ 30 min) and flow rate is controlled via a critical orifice (see below), though this feature could be incorporated usefully in other types. The principal disadvantages are fragility because of the glass construction and limited knowledge of size distribution. [Naive users are reminded that impingers provide details of the number of viable microorganisms (which are dispersed within the collection liquid) while impactors, such as the Andersen sampler, record the number of viable particles, each one of which may contain one or more organisms.] Collection stresses on certain microbial species also may be higher in liquid impingers (see Chapter 6).

Filtration Samplers

Filtration samplers are simple to deploy, cheap and well characterized, e.g., the IOM (Institute of Occupational Medicine, Edinburgh, UK) inspirable dust sampler. Size separation into respirable and inspirable fractions is routine and instruments with high- and low-volume flow rates can be obtained. The latter are usually designed for personal sampling and are therefore lightweight, portable and battery powered.

Filtration samplers are not recommended for microbial bioaerosols because of desiccation stresses that occur as air flows through the filters. However, they are widely utilized for sampling dusts, whence filters are assayed gravimetrically or microscopically. Other principal limitations are variable air flows, which decrease as the filter becomes laden, and difficulties in measuring low mass concentrations of airborne dust.

Gravimetric analyses of dust can be based on filter samples that either are desiccated at 105°C for 14 h or are exposed to a constant humidity and temperature pre- and post-sampling.[83] The latter method is less rigorous and uniform due to the varying moisture contents of aerosol particles of different types. In either case, details of the method must be reported.

Calibration for Particle Size and Flow Rate and Other Good Practices

Good laboratory practice demands that bioaerosol samplers be calibrated regularly for particle size and flow rate. Sampling in animal houses is no different in this respect.

Details of size calibration methods are given in Chapter 8 and these techniques are normally suitable. In some cases, e.g., optical particle counters, the detection of an aerosol particle is determined by light scattering, which implies that the particle has known optical properties. The heterogeneous nature of bioaerosols in animal houses makes it unlikely that these properties will be known (or even be different). There are, however, no special caveats with regard to calibration of flow rate though it is surprising how few samplers employ critical orifices to control flow rate despite their advantages, which were presented some 40 years ago.[29] Construction techniques for critical orifices based on micropipettes are given.[77]

All samplers should be cleaned regularly between sampling sessions (i) to avoid contamination, especially microbial; and (ii) to prevent alteration of the sampler's collection efficiency, which may occur if substantial deposits of dust accumulate.[69]

Supplementary Information to Be Reported

In addition to details of the techniques for sampling, assay, analysis and calibration, it is helpful to record the following supplementary information when bioaerosols are sampled in animal house environments:[83]

(1) Animals—species, type, number, age, stocking density (weight, m^2/animal, m^3/animal), history of clinical or sub-clinical disease, behavior (especially abnormal) and activity.
(2) Building—orientation, proximity to other buildings, overall dimensions, volume, layout, position, type of floor and pen walls, ventilation system (vent and fan position and sizes).
(3) Feeding—method, equipment, feeding times and duration, type of feed (meal, pellet, liquid), fat and water content.
(4) Manure—bedding (type, amount, frequency of replenishment), manure removal system, quantity and quality.
(5) House and ambient environment—temperature, relative humidity, wind speed and direction, house ventilation rate.

WHERE?—LOCATION OF BIOAEROSOL SAMPLERS IN ANIMAL HOUSES

In most studies of bioaerosols in animal houses, static bioaerosol samplers should be located within or close to the animals' breathing zone, i.e., usually within representative pens at approximate head height. Occasionally other locations will be chosen where the primary aim is to understand the dynamics of the bioaerosols, e.g., the relative strengths of the various sources and sinks. Recently, several authors have reported the use of personal samplers for horses,[4,89] similar to those employed in human occupational medicine. Burdens of bioaerosols were many times greater than those recorded with static samplers. This innovation is timely but is unlikely to be feasible for active aggressive animals, such as pigs, or small poultry. Furthermore, information provided by personal samplers is clearly of direct relevance to (occupational) daily exposure during normal behaviors, e.g., feeding, lying, grooming and mating, while static samplers give details of the burden of bioaerosols at discrete locations and operations.

Many of the practical difficulties associated with bioaerosol sampling in animal houses arise from the need to place the sampler in or close to the animal's breathing zone (Figure 20.11). Samplers and their associated power cables and air lines need to be protected from the vigorous attentions of the livestock, which usually are curious about these novel objects introduced into their environment. Provision of a mobile laboratory is often useful, especially where remote gas analyzers and vacuum pumps are employed (Figure 20.12). Within the animal house in this example, the umbilical hose is split into seven separate sections, each equipped with filtration samplers for airborne dust, a sampling port for various gases and a combined sensor for temperature and humidity with signals transmitted by radiotelemetry (Figures 20.13, 20.14 and 20.15). The sampling heads may be suspended from the roof just above head height (and at other positions of interest) as in a cattle shed (Figure 20.14). Alternatively, the heads may be attached to a convenient support, such as the stanchion shown in an aviary for laying hens (Figure 20.15). Often an adaptable fixing method will be needed for the protective cage. Umbilical hoses can be cumbersome to handle, but are convenient to clean and disinfect and require access via the buildings walls or openings.

Location and Number of Sampling Sites

The location and number of sampling sites within an animal house are governed by the need for the samples to be representative of the building and the system of animal husbandry. (An outside location upwind of the animal house can act as a control if needed.) Ideally, static samplers should be located within the animal's breathing zone, but animals move freely within their pen, adopt several postures and engage in many activities. In practice a compromise therefore must be reached in determining the location of the samplers, which must be stated in all reports.

Table 20.6 gives recommendations for the height of bioaerosol samplers in animal houses. These heights are within the breathing zone for the majority of animals with typical patterns of activity and posture. The horizontal location of the sampling sites will be governed chiefly by the building and pen layouts and ventilation system. If possible, sites adjacent to walls or barriers should be avoided, but their use is sometimes necessary to provide a secure anchorage for protective shields. The layout of the 12 sampling sites shown in Figure 20.4 assumes lateral symmetry, but this assumption may not always be justified, e.g., in a fattening piggery with a part-slatted floor. Furthermore, the distribution

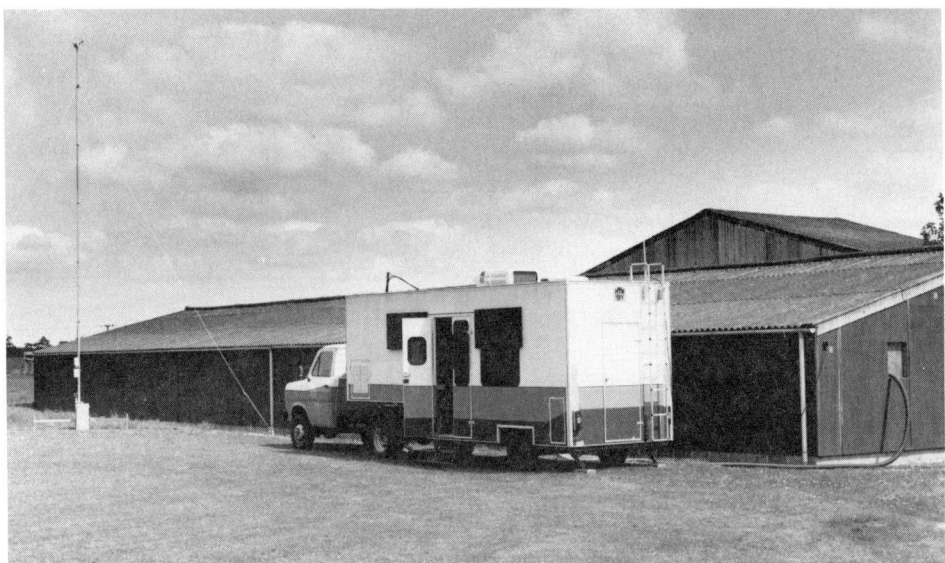

Figure 20.12 This mobile laboratory houses gas analyzers, radiotelemetry recorders (for measurement of temperature and relative humidity), vacuum pumps and a data logger. The umbilical hose (or elephant's trunk) encases the gas sampling lines and vacuum lines in a heated sheath to eliminate condensation in cold weather: it is also convenient for handling and cleaning. (Photo courtesy of M.R. Holden, Silsoe Research Institute, UK.)

of birds and the principal equipment (feeders, drinkers and ventilation openings) was also uniform, but this may not always be the case. Vigorous mixing of air within animal houses by turbulent diffusion ensures that vertical and horizontal gradients of bioaerosols are likely to be small with the corollary that pathogenic and other bioaerosols will disperse rapidly (within hours) throughout a building after emission.

Given a uniform source strength, the number of sampling locations should be determined from statistical requirements of reproducibility (see also Chapter 5). Occasionally, only one location is employed, but this practice usually results from limited availability of equipment rather than ignorance of good sampling procedures. In the poultry study described earlier,[17] statistical analysis of the spatial distribution of bioaerosol concentration showed few significant differences between the 12 sampling sites because of the uniform distribution of sources and sinks: in other words, each site could be considered representative of the concentration of common bioaerosols in a deep-litter poultry house of the type under study.

A similar study of the spatial variability of airborne and settled dust in a fattening piggery was undertaken by Barber et al (1991).[3] Measurements over 16 days at 16 locations in the horizontal and vertical planes showed a distribution consistent with the likely locations of dust sources and airflow patterns. The distribution was heterogeneous because of the building's particular layout, e.g., pen barriers. The spatial distribution *per se* did not change appreciably day-to-day, but the overall dust burdens varied substantially between days.

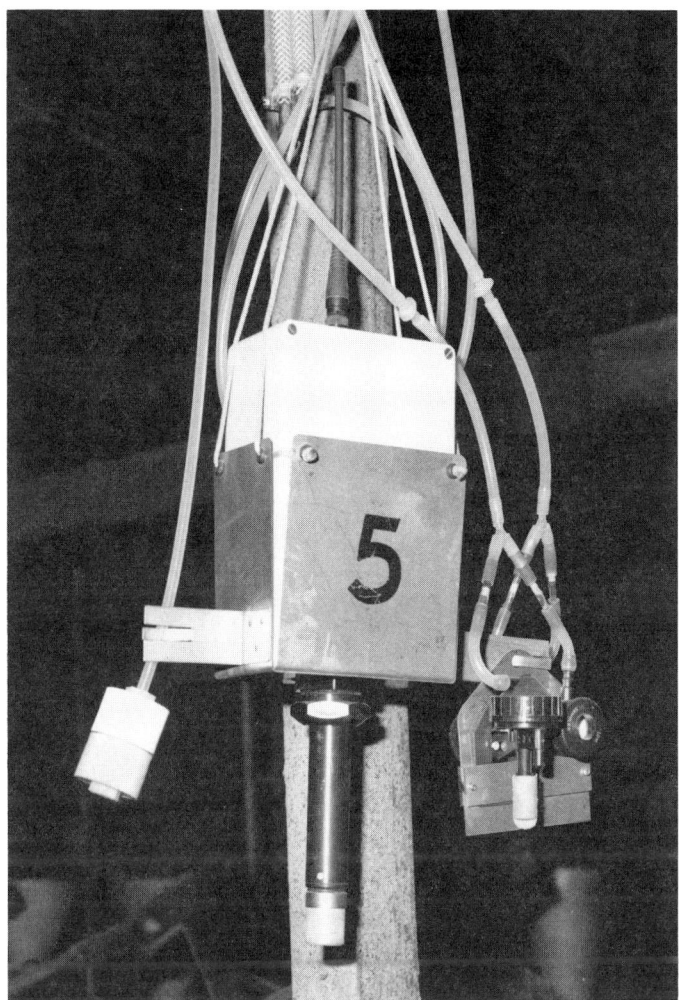

Figure 20.13 Within the animal house, the umbilical hose is split into seven separate sections, each equipped with filtration samplers for airborne dust, a sampling port for various gases (NH_3, N_2O, CO_2 and CH_4 in this study) and a combined temperature and humidity radiotelemetric sensor. (Photo courtesy of M.R. Holden, Silsoe Research Institute, UK.)

If the study's objective focuses on the burden of bioaerosol in the breathing zone of animals within a building, the airspace is relatively well mixed and the source strength is uniform, then the general statistical procedure described by Sokal and Rohlf[73] can be employed to determine the minimum number of sampling sites necessary for a predetermined level of statistical accuracy. For each bioaerosol, information is needed about the true standard deviation (σ) of the population, the smallest difference (δ) between two means, which it is desired to detect, the level of significance (α) and the power of the test (ρ). Using this method, Conceição et al.,[17] calculated that a minimum of 8 separate locations were needed for sampling bioaerosols typical of a deep-litter poultry house ($\sigma/\delta = 0.5$, $\alpha = 0.001$, $\rho = 0.8$). This minimum is increased if higher levels of significance or

Figure 20.14 A typical UK cattle house for dairy cows kept in cubicles. Bioaerosol samplers can be seen at six locations. (Photo courtesy of M.R. Holden, Silsoe Research Institute, UK.)

test powers are required. Calculations of this sort are essential in any experimental study if the limitations and failures of inadequate sample sizes are to be avoided.

Conversely, if the study's objective is an understanding of bioaerosol dynamics, e.g., identification of source and sink strengths, then a statistical criterion cannot be used exclusively. Additionally, the sites for sampling must be selected after careful study of source locations and airflow patterns.[3] In this latter study, placing the dust sampler near the front of the pig pen in the building's center gave the best indication of the hazard to the stockman and the worst conditions for the pigs. It must be noted that not all animal houses exhibit the symmetry shown in the studies described here.

WHEN?—FREQUENCY OF BIOAEROSOL SAMPLING IN ANIMAL HOUSES

Dynamic Behavior of Bioaerosols

The dynamic behavior of bioaerosols in animal houses is due to temporal fluctuations in the strengths of individual sources and sinks and the turbulent nature of the airflow. The two most important cycles are circadian and annual. The circadian cycle arises from the regular, daily tasks of animal husbandry in livestock buildings, including distribution of feed, removal and replenishment of soiled bedding and inspection of stock. Animal movement and feeding contribute much to the varying flux of bioaerosols, principally by

Figure 20.15 Robust shields are needed when bioaerosol samplers are sited within the animal's breathing zone, as in this aviary for laying hens. The equipment is the same as that in Figure 20.12. (Photo courtesy of M.R. Holden, Silsoe Research Institute, UK.)

disturbance of bedding and feed, and these activities occur mostly in daylight hours. Some species, e.g., broilers, are kept on a 23:1-hour light:dark cycle and temporal fluctuations will be correspondingly smaller than those reared on natural light cycles, e.g., calves. For example, Pedersen[64] showed a positive linear correlation between pig activity as measured with a passive, infra-red detector (as used in security systems) and dust number concentration, which varied by a factor of three over a 20-hour period.

It is often convenient to divide a day into sampling periods according to animal activity. For example, for fattening pigs fed a restricted ration and allowing for changeover periods these will typically comprise (i) feeding, cleaning and inspection in the morning of ≈2.5 hours; (ii) a daytime quiescent period of ≈3 hours; (iii) feeding and inspection in the afternoon of ≈2.5 hours; and (iv) a nighttime rest period of ≈11 hours.[72] Using this protocol, aerial concentrations of inspirable and respirable dust were found to be affected as much by pig activity at certain times as by ventilation rate.[72]

Annual cycles in bioaerosol burdens in animal houses arise primarily through changes in climate, ventilation rate and airflow patterns. In many livestock buildings air temperature, but not humidity, is maintained within set limits, according to the animal species, by adjusting the building ventilation rate. Physical dilution will obviously lower bioaerosol concentration, while changes in air humidity may affect the survival of some airborne

Table 20.6 Recommended Heights Above the Floor for Sampling Bioaerosols in Animal Houses Using Static Samplers.

Species	Recommended Height, m
Humans	1.5
Laying hens in cages and aviaries	0.5 - mid tier
Broiler chickens on litter	0.2
Beef cattle and dairy cows	1.5
Calves	0.5
Adult pigs	0.5
Piglets	0.2
Horses	1.5

Source: Wathes and Randall (1989)[83] with amendment.

microbes. These mechanisms explain the common observation that dust concentrations are higher in winter than in summer.

Housed animals often are disturbed by the entry of unfamiliar humans, especially if accompanied by experimental baggage. It is good practice to delay the start of any bioaerosol sampling until normal animal activity has resumed; the delay should be at least 15 minutes.

Duration and Frequency of Sampling

The minimum requirement for the duration of sampling bioaerosols is for measurements to be made over 24 hours, sub-divided into periods of defined animal or husbandry activities. This will integrate short-term fluctuations in bioaerosol concentrations and types and will facilitate comparisons with other studies for each individual sampler. There will be minimum durations of sampling according to (i) the accuracy and precision of the assay and (ii) the transit time of air through the instrument to 'wash out' the dead space. Maximum sampling times may also be needed to avoid overloading, e.g., in impactor samplers such as the Andersen.

Replication of sampling over several days may be needed and a similar statistical procedure should be adopted to that described above for determining the minimum number of sampling sites within an animal house. Other sampling frequencies will depend on the study's objectives.

WHO?—STUDENTS OF BIOAEROSOLS IN ANIMAL HOUSES

Students of bioaerosols in animal houses require diverse skills in, and knowledge of, aerosol physics, microbiology, veterinary science and animal husbandry. Animal houses can be dirty, cramped and noisy, and ingenuity is needed in siting and operating samplers within the animal's breathing zone. Remote control and operation of samplers is preferable but this may not be possible with microbial samplers such as impingers and impactors. However, the tenacious student will be rewarded by the knowledge that his studies should promote our understanding of aerobiology, increase the efficiency of animal production and raise standards of animal health.

CONCLUSIONS

Solid progress has been made over the past decade in studies of bioaerosols in animal houses. First, the relative importance of the roles of bioaerosols of all types in animal disease has been recognized. Second, steady improvements have been made in the methodology of bioaerosol sampling in animal houses such that it is now on a firm foundation of aerosol physics and aerobiology. Third, research now is being directed toward ways of controlling bioaerosol burdens to agreed targets with several promising techniques proposed. This chapter has attempted to guide the novice in the art and science of bioaerosols in animal houses; hopefully, it may also aid the experienced observer.

REFERENCES

1. Andersen, A.A. "New sampler for the collection, sizing and enumeration of viable airborne particles." *J. Bact.* 76:471–484 (1958).
2. Baekbo, P. and J. Wolstrup "Aerosol sampling in pig fattening units" in *Aerosol Sampling in Animal Houses*, C.M. Wathes and J.M. Randall, Eds. (Commission of the European Communities, Luxembourg EUR 11877, 1989) pp. 27–32.
3. Barber, E.M., J.R. Dawson, V.A. Battams, and R.A.C. Nicol "Spatial variability of airborne and settled dust in a piggery." *J. Agric. Eng. Res.*, 50:107–127 (1991).
4. Bartz, J., and J. Hartung "Dust measurements on a horse using an 'equine personal sampler'" in *Proceedings Fourth International Livestock Environment Symposium* (Warwick, UK, American Society of Agricultural Engineers, 1993), pp. 742–746.
5. Baskerville, A., T.J. Humphrey, R.B. Fitzgeorge, R.W. Cook, H. Chart, B. Rowe, and A. Whitehead "Airborne infection of laying hens with *Salmonella enteritidis* phage type 4". *Vet. Rec.* 130:395–398 (1992).
6. Bourgueil, E., E. Hutet, R. Cariolet and P. Vannier "Air sampling procedure for evaluation of vival excretion level by vaccinated pigs infected with Aujeszky's disease (*pseudorabies*) virus". *Res. Vet. Sci.* 52:182–186 (1992).
7. Bundy, D.S. and M.A. Veenhuizen "Dust and bacteria removal equipment for controlling particulates in swine buildings" in *Proceedings Latest Developments in Livestock Housing* (Commission Internationale du Génie Rural, Illinois, 1987) pp. 137–145.
8. Butera, M., J.H. Smith, W.D. Morrison, R.R. Hacker, F.A. Kains, and J.R. Ogilvie "Concentration of respirable dust and bioaerosols and identification of certain microbial types in a hog-growing facility". *Can. J. Anim. Sci.* 71:271–277 (1991).
9. Carpenter, G.A. "Dust in livestock buildings—review of some aspects", *J. Agric. Eng. Res.*, 33:227–241 (1986).
10. Carrozza, J.H., T.N. Fredrickson, R.P. Prince, and R.E. Luginbuhl "Role of desquamated epithelial cells in transmission of Marek's disease". *Avian Diseases*, 17:767–781 (1973).
11. Chiba, L.I., E.R. Peo and A.J. Lewis "Use of dietary fat to reduce dust, aerial ammonia and bacterial colony forming particle concentrations in swine confinement buildings". *Trans. Am. Soc. Agric. Engrs* 30:464–468 (1987).
12. Chiba, L.I., E.R. Peo, A.J. Lewis, M.C. Brumm, R.D. Fritschen and J.D. Crenshaw "Effect of dietary fat on pig performance and dust levels in modified-open-front and environmentally regulated confinement buildings." *J. Anim. Sci.* 61:763–781 (1985).
13. Clark, P.C. and J.B. McQuitty "Air quality in six Alberta commercial free-stall dairy barns." *Can. Agric. Eng.*, 29:77–80 (1987).
14. Clark, S., R. Rylander and L. Larsson "Airborne bacteria endotoxin and fungi in dust in poultry and swine confinement buildings." *Am. Ind. Hyg. Assn J.* 44:537–541 (1983).

15. Clarke, A.F. and T. Madelin "Technique for assessing respiratory health hazards from hay and other source materials". *Eq. Vet. J.* 19:442–447 (1987).
16. Clarke, A.F., T.M. Madelin, and R.G. Allpress "The relationship of air hygiene in stables to lower airway disease and pharyngeal lymphoid hyperplasia in two groups of thoroughbred horses". *Eq. Vet. J.* 19:524–530 (1987).
17. Conceição, M.A.P., H.E. Johnson and C.M. Wathes "Air hygiene in a pullet house: spatial homogeneity of aerial pollutants." *Brit. Poult. Sci.,* 30:765–776 (1989).
18. Cravens, R.L., H.J. Beaulieu and R.M. Buchan "Characterization of the aerosol in turkey rearing confinements." *Am. Ind. Hyg. Assoc. J.* 42:315–318 (1981).
19. Crichlow, E.C., K. Yoshida and K. Wallace "Dust levels in a riding stable". *Eq. Vet. J.* 12:185–188 (1980).
20. Crook, B., J.F. Robertson, S.A. Travers, E.M. Botheroyd, J. Lacey and M.D. Topping "Airborne dust, ammonia, microorganisms and antigens in pig confinement houses and the respiratory health of exposed farm workers". *Am. Ind. Hyg. Assoc. J.* (in press).
21. Curtis, S.E., J.G. Drummond, D.J. Grunloh, P.B. Lynch and A.H. Jensen "Relative and qualitative aspects of aerial bacteria and dust in swine houses". *J. Anim. Sci.* 41:1512–1520 (1975).
22. Curtis, S.E., J.G. Drummond, K.W. Kelley, D.J. Grunloh, V.J. Meares, H.W. Norton and A.H. Jensen "Diurnal and annual fluctuations of aerial bacterial and dust levels in enclosed swine houses". *J. Anim. Sci.* 41:1502–1511 (1975).
23. De Boer, S. and W.D. Morrison "The effects of the quality of the environment in livestock buildings on the productivity of swine and safety of humans" (University of Guelph, Ontario, 1988) pp. 121.
24. Dennis, C. and J.M. Gee "The microbial flora of broiler-house litter and dust." *J. Gen. Microbiol.* 78:101–107 (1973).
25. Donaldson, A.I. "Factors influencing the dispersal, survival and deposition of airborne pathogens of farm animals." *Vet. Bull.* 48:83–94 (1978).
26. Donaldson, A.I. "Airborne foot-and-mouth disease". *Vet. Bull.* 49:653–659 (1979).
27. Donaldson, A.I., K.A.J. Herniman, J. Parker and R.F. Sellers "Further investigations on the airborne excretion of foot-and-mouth disease virus". *J. Hyg.* 68:557–564 (1970).
28. Donham, K.J. "Human health and safety for workers in livestock housing" in *Proceedings Latest Developments in Livestock Housing* (Commission Internationale du Génie Rural, Illinois, 1987) pp. 86–95.
29. Druett, H.A. "The construction of critical orifices working with small pressure differences and their use in controlling airflow". *Brit. J. Ind. Med.* 12:65–70 (1955)
30. Elliott, L.F., T.M. McCalla and J.A. Deshazer "Bacteria in the air of housed swine units". *Appl. Environ. Microbiol.* 32:270–273 (1976).
31. Feddes, J.J.R., H. Cook and M.J. Zuidhof "Characterization of airborne dust particles in turkey housing." *Can. Agric. Eng.* 34:273–280 (1992).
32. Feddes, J.J.R., B.S. Koberstein, F.E. Robinson and C. Riddell "Misting and ventilation rate effects on air quality, and heavy farm turkey performance and health". *Can. Agric. Eng.,* 34:177–181 (1992).
33. Feddes, J.J.R., J.J. Leonard and J.B. McQuitty "Heat and moisture loads and air quality in commercial broiler barns in Alberta" (University of Alberta Research Bulletin 82-2, 1982) 83 pp.
34. Feddes, J.J.R. and Z.J. Licsko "Air quality in commercial turkey housing." *Can. Agric. Eng.* 35:147–150 (1993).
35. Glennon, C.R., J.B. McQuitty and P.C. Clark "Heat and moisture loads and air quality in commercial poultry rearing facilities" (Canadian Society of Agricultural Engineering, 1984) Paper No. 84-405.
36. Gloster, J. "Factors influencing the airborne spread of Newcastle disease." *Brit. Vet. J.* 139:445–451 (1983).

37. Gloster, J., R.M. Blackall, R.F. Sellers and A.I. Donaldson "Forecasting the airborne spread of foot-and-mouth disease." *Vet. Rec.* 108:370–374 (1981).
38. Gregory, P.H. "Airborne microbes: their significance and distribution." *Roy. Soc. Series B.* 177:469–483 (1971).
39. Grunloh, D.J., S.E. Curtis, A.H. Jensen, J. Simon and B.G. Harmon "Airborne bacterial particles in farrowing rooms" *J. Anim. Sci.* 33:1139–1140 (1971).
40. Gustafsson, G. and L. Mårtensson "Gases and dust in poultry houses" (Severiges Lantbruksuniversitit, Inst. für Lantbrukets Byggnadsteknik, Lund, Sweden, 1990 Report No. 68) 88 pp.
41. Hamilton, T.D.C., J.M. Roe, F.G.R. Taylor, G. Pearson and A.J.F. Webster "Aerial pollution: an exacerbating factor in atrophic rhinitis of pigs" in *Proceedings Fourth International Livestock Environment Symposium* (American Society of Agricultural Engineers, Warwick, 1993) pp. 895–903.
42. Hardman, P.M., C.M. Wathes and C. Wray "Salmonella transmission among calves penned individually." *Vet. Rec.* 129:327–329 (1991).
43. Heber, A.J., M. Stroik, J.M. Faubion and L.H. Willard "Size distribution and identification of aerial dust particles in swine finishing buildings." *Trans Am. Soc. Agric. Engrs*, 31:882–887 (1988).
44. Heber, A.J., M. Stroik, J.L. Nelssen and D.A. Nichols "Influence of environmental factors on concentrations and inorganic content of aerial dust in swine finishing buildings." *Trans. Am. Soc. Agric. Engrs*, 31:875–881 (1988).
45. Hill, I.R. and R. Kenworthy "Microbiology of pigs and their environment in relation to weaning". *J. Appl. Bact.* 33:299–316 (1970).
46. Honey, L.F. and J.B. McQuitty "Dust in the animal environment" (University of Alberta Research Bulletin 76-2, 1976) 66 pp.
47. Honey, L.F. and J.B. McQuitty "Some physical factors affecting dust concentrations in a pig facility." *Can. Agric. Engng*, 21:9–14 (1979).
48. Hugh-Jones, M., W.H. Allan, F.A. Dark and G.J. Harper "The evidence for the airborne spread of Newcastle disease". *J. Hyg.* 71:325–339 (1973).
49. Jensen, P.A., Todd, W.F., Davis, G.N. and Scarpino P.V. "Evaluation of eight bioaerosol samplers challenged with aerosols of free bacteria." *Am. Ind. Hyg. Assn. J.* 53:660–667 (1992).
50. Jones, C.R. and A.J.F. Webster "Weather induced changes in airborne bacteria within a calf house." *Vet. Rec.* 109:493–494 (1981).
51. van't Klooster, C.E., P.F.M.M. Roelofs and L.A. Den Hartog "Effects of filtration, vacuum cleaning and washing in pig houses on aerosol levels and pig performance." *Live. Prod. Sci.* 33:171–182 (1993).
52. Liao, C.M. and J.J.R. Feddes "Modeling and analysis of the dynamic behavior of airborne dust in a ventilated airspace" (American Society of Agricultural Engineers Paper PNR 89-402, Penticton, British Columbia, 1989).
53. Madelin, T.M. "Aerosol sampling methods appropriate for stables and poultry houses" in *Aerosol Sampling in Animal Houses*, C.M. Wathes and J.M. Randall, Eds. (Commission of the European Communities, Luxembourg EUR 11877, 1989) pp. 69–74.
54. Madelin, T.M. and C.M. Wathes "Air hygiene in a broiler house: comparison of deep litter with raised netting floors." *Brit. Poult. Sci.* 30:23–37 (1989).
55. Maghirang, R.G., H.B. Manbeck, J.M. Carson, D.J. Meyer, F.V. Muir and W.B. Roush "Spatial distribution of particulates in layer houses" (American Society of Agricultural Engineers Paper No. 89-4555, New Orleans, Louisiana, 1989).
56. Maghirang, R.G., H.B. Manbeck, D.J. Meyer, W.B. Roush and F.V. Muir "Air contaminant distribution in a commercial layer house" (American Society of Agricultural Engineers Paper No. 90-4058, Columbus, Ohio, 1990).

57. Maghirang, R.G., H.B. Manbeck, W.B. Roush and F.V. Muir "Air contaminant distributions in a commercial laying house." *Trans. Am. Soc. Agric. Engrs,* 34:2171–2180 (1991).
58. May, K.R. "Multistage liquid impinger." *Bac. Rev.* 30:559–570 (1966).
59. May, K.R. "Physical aspects of sampling airborne microbes" in *Airborne Microbes,* P.H. Gregory and J.L. Monteith, Eds. (17th Symposium of the Society for General Microbiology, Cambridge University Press, London, 1967) pp. 60–80.
60. May, K.R. "An "ultimate" cascade impactor for aerosol assessment". *J. Aero. Sci.* 6:413–419 (1975).
61. May, K.R. and G.J. Harper "The efficiency of various liquid impinger samplers in bacterial aerosols". *Brit. J. Ind. Med.* 14:287–297 (1957).
62. Meyer, V.M. and D.S. Bundy "Farrowing building air quality survey". (American Society of Agricultural Engineers, Paper 91-4012, 1991).
63. Moore, B. "Observations pointing to the conjunctiva as the portal of entry in salmonella infection of guinea-pigs". *J. Hyg.* 55:414–433 (1957).
64. Pedersen, S. "Time based variation in airborne dust in respect to animal activity" in *Proceedings Fourth International Livestock Environment Symposium* (Warwick, UK, American Society of Agricultural Engineers, 1993) pp. 718–725.
65. Pickrell, J.A., A.J. Heber, J.P. Murphy, S.C. Henry, M.M. May, D. Nolan, F.W. Oehme, J.R. Gillespie and D. Schoneweis "Characterization of particles, ammonia and endotoxin in swine confinement operations." *Vet. Hum. Toxic.* 35:421–428 (1993).
66. Pinello, C.B., J.L. Richard and L.H. Tiffany "Mycoflora of a turkey confinement brooder house." *Poult. Sci.* 56:1920–1926 (1977).
67. Pritchard, D.G., G.A. Carpenter, S.P. Morzaria, J.W. Harkness, M.S. Richards and J.I. Brewer "Effect of air filtration on respiratory disease in intensively housed veal calves." *Vet. Rec.* 109:5–9 (1981).
68. Qi, R., H.B. Manbeck and R.G. Maghirang "Dust net generation rate in a poultry layer house." *Trans. Am. Soc. Agric. Engrs,* 35:1639–1645 (1992).
69. Robertson, J.F. "Aerosol sampling in pig houses" in *Aerosol Sampling in Animal Houses,* C.M. Wathes and J.M. Randall, Eds. (Commission of the European Communities, Luxembourg EUR 11877, 1989) pp. 95–101.
70. Robertson, J.F. "Dust and ammonia in pig buildings". *Farm Build. Progr.* 110:19–24 (1992).
71. Robertson, J.F., D. Wilson and W.J. Smith "Atrophic rhinitis: the influence of the aerial environment". *Anim. Prod.* 50:173–182 (1990).
72. Smith J.H., C.R. Boon, and C.M. Wathes "Dust distribution and airflow in a swine house". in *Proceedings Fourth International Livestock Environment Symposium* (Warwick, UK, American Society of Agricultural Engineers, 1993) pp. 657–662.
73. Sokal, R.R. and F.J. Rohlf, *Introduction to Biostatistics,* Second Edition (W.H. Freeman, New York, 1987).
74. Tamási, G. "Mycoplasma isolation from the air" in *Airborne Transmission and Airborne Infection,* J.F.Ph. Hers and K.C. Winkler, Eds. (Oosthoek Publishing Company, Utrecht, 1973) pp. 68–71.
75. Thorne, P.S., M.S. Kiekhaefer, P. Whitten and K.J. Donham "Comparison of bioaerosol sampling methods in barns housing swine." *App. Env. Microb.* 58:2543–2551 (1992).
76. Underdahl, N.R., M.B. Rhodes, T.E. Socha and D.D. Sholte "A study of air quality and respiratory infections in pigs raised in confinement." *Live. Prod. Sci.* 9:521–529 (1982).
77. Vaughan, N.P. "Construction of critical orifices for sampling applications". in *Third Annual Conference of the Aerosol Society,* (West Bromwich, UK, 1989), 77–82.
78. Wathes, C.M. "Airborne micro-organisms in pig and poultry houses" in *The Environmental Aspects of Respiratory Disease in Intensive Pig and Poultry Houses Including the Implications for Human Health,* J.M. Bruce and M. Sommer, Eds. (Commission of the European Communities, Luxembourg, 1987) pp. 53–71.

79. Wathes, C.M. "Air and surface hygiene" in *Livestock Housing,* C.M. Wathes and D.R. Charles, Eds. (CAB International, Wallingford, 1994) in press.
80. Wathes, C.M., K. Howard, C.D.R. Jones and A.J.F. Webster "The balance of airborne bacteria in calf houses." *J. Agric. Eng. Res.* 30:81–90 (1984).
81. Wathes, C.M., H.E. Johnson and G.A. Carpenter "Air hygiene in a pullet house: effects of air filtration on aerial pollutants measured *in vivo* and *in vitro*". *Brit. Poult. Sci.* 32:31–46 (1991).
82. Wathes, C.M., B.G. Miller and F.J. Bourne "Cold stress and post-weaning diarrhoea in piglets inoculated orally or by aerosol". *Anim. Prod.* 49: 483–496 (1989).
83. Wathes, C.M. and J.M. Randall, Eds. *Aerosol Sampling in Animal Houses.* (Commission of the European Communities, Luxembourg, 1989) EUR 11877, 141 pp.
84. Wathes, C.M., W.A.R. Zaidan, G.R. Pearson, M. Hinton and J.N. Todd "Aerosol infection of calves and mice with *Salmonella typhimurium.*" *Vet. Rec.* 123:590–594 (1988).
85. Webster, A.J.F. "Improvements of environment, husbandry and feeding" in *The Control of Infectious Diseases in Farm Animals*, H. Smith and J.M. Payne, Eds. (BVA Trust, London, 1982) pp. 28–35.
86. Webster, A.J.F., A.F. Clarke, T.M. Madelin and C.M. Wathes "Air hygiene in stables. 1. Effects of stable design, ventilation and management on the concentration of respirable dust." *Eq. Vet. J.* 19:448–453 (1987).
87. Welford, R.A., J.J.R. Feddes and E.M. Barber "Pig building dustiness as affected by canola oil in the feed" *Can. Agric. Engng,* 34:365–373 (1992).
88. Whyte, R.T. "Aerial pollutants and the health of poultry farmers". *World's Poult. Sci. J.* 49:139–156 (1993).
89. Woods, P.S.A., N.E. Robinson, M.C. Swanson, C.E. Reed, R.V. Broadstone and F.J. Derksen "Airborne dust and aeroallergen concentration in a horse stable under two different management systems." *Eq. Vet. J.* 25:208–213 (1993).
90. Zeitler, M.H. "Konzentration und Korngrößenverteilung von luftgetragenen Staubpartikeln in Pferdestllen". (Berliner und Münchiner Tierärztliche Wochenschrift, 1985) pp. 241–246.

CHAPTER 21

Bioaerosols in the Residential Environment

Harriet A. Burge

Houses usually are considered to be protective environments. The outside air is often well-supplied with fungal spores and pollens that can cause allergic disease in some people.[1] A clean, enclosed home can protect against these bioaerosols.

The indoor environment, however, can become contaminated with biological particles that present different and sometimes more serious risks than those related to outdoor exposures. For example, infectious diseases, such as influenza, are not readily transmitted outdoors because of the large air mass available to dilute the aerosols, and because the ultraviolet light of the sun kills many microorganisms. However, many of these same microorganisms can survive and accumulate to dangerous levels in indoor air, and many bioaerosol-related diseases are associated primarily with indoor environments.

The kinds of organisms and their effluents that accumulate in indoor air, the diseases they cause, and factors that control their prevalence in residential environments are the subjects of this chapter.

RELATIONSHIPS BETWEEN HUMAN DISEASES AND BIOAEROSOLS INDOORS

The bioaerosol-related diseases that are associated with residential environments are summarized in Table 21.1. Of these, the airborne viral and bacterial infectious diseases, asthma related to arthropods and mammals, allergic alveolitis, and the diseases produced by exposure to microbial (bacterial and fungal) toxins occur primarily in indoor environments.

Infections

The vectors of airborne contagious disease, by definition, always are transmitted from one human to another or, rarely, from an animal to a human.[2] Infectious diseases that are commonly associated with residential aerosols are influenza, the childhood viruses (measles, mumps, chickenpox) and some common cold viruses. These diseases are transmitted readily in residences because of the close contact that occurs between occupants and because residences tend to be poorly ventilated.[3] Currently, tuberculosis, a contagious bacterial disease, is becoming of increasing concern in clinics and in residences for the homeless.[4]

Table 21.1 Examples of Disease-Bioaerosol Relationships for Indoor Air

Source Organism	Kinds of Disease	Examples of Disease	Examples of Agent	Examples of Reservoirs
Viruses	Infectious	Measles	Rubella	Human
		Influenza	Influenza	Human
Bacteria	Infectious	Tuberculosis	*Mycobacterium tuberculosis*	Human
		Legionellosis	*Legionella pneumophila*	Hot water heaters
	Hypersensitive	Allergic alveolitis*	*Thermoactinomyces* species	Humidifiers Clothes dryers
	Toxic	Humidifier fever	Gram-negative bacteria	Humidifiers
Fungi	Infectious	Aspergillosis	*Aspergillus fumigatus*	Wet organic material
	Hypersensitive	Allergic alveolitis	*Penicillium Aspergillus Sporobolomyces Merulius*	
		Asthma	*Alternaria* Basidiospores	
	Toxic	Stachybotryo toxicosis	*Stachybotrys atra*	
		Aflatoxicosis	*Aspergillus flavus*	
Dust mites	Hypersensitive	Asthma	*Dermatophagoides pteronyssinus*	Bedding Carpeting

* Hypersensitivity pneumonitis.

Some agents of infectious disease inhabit environmental reservoirs. The environmental source pathogens that occupy residential niches are almost all opportunistic and only cause disease in people with some underlying immune system defect (for example, AIDS and transplant patients, and heavy smokers). *Legionella pneumophila* is one of these opportunists. It can grow in domestic hot water systems and some kinds of humidifiers. At least 4% of the American population has anti-Legionella antibodies.[5] More than 20,000 community-acquired cases probably occur each year, many as a result of exposure to residential sources.[6]

Aspergillus fumigatus is one of the most common of the fungal opportunists. It is one of the few fungi that can grow at body temperature, and it secretes toxins that allow tissue invasion. Like most fungi, *A. fumigatus* will grow on almost any carbon-containing substrate as long as adequate water is present, and is common in both indoor and outdoor reservoirs.[7]

Hypersensitivity

Many agents that cause hypersensitivity disease also can contaminate residential environments. Some kinds of bacteria (i.e., thermophilic actinomycetes), fungi, arthropods (mites and cockroaches), and mammals (cats, dogs, mice, gerbils, etc.), produce airborne materials (antigens) that cause allergies (see also Chapters 9, 10, 14 and 16). Diseases that result from such airborne exposure include hay fever (allergic rhinitis), asthma and allergic alveolitis (hypersensitivity pneumonitis). The development of the symptoms of hypersensitivity diseases is always a two-step procedure: sensitization occurs as the result of one or more exposure events, and is followed by symptom development at subsequent exposure events. The presence of allergen sources in homes apparently plays a role in both of these steps.[8]

Recent epidemiological studies have found correlations between childhood respiratory disease and dampness indicators in homes as reported on questionnaires by occupants.[9-12] Other studies have demonstrated a connection between dampness indicators and concentrations of airborne fungi.[13] Symptoms of hay fever and asthma as well as immediate skin reactivity have also been related to fungus spore levels in homes.[13-17]

Similar epidemiological studies comparing actual exposure and symptoms have not been reported for other residential allergens. However, case reports have tied exposure to thermophilic actinomycetes in homes to the development of hypersensitivity pneumonitis.[18,19] Also, asthma has been related to the presence of arthropod pests such as house dust mites (*Dermatophagoides pteronyssinus, D. farinae*) and cockroaches (*Blattella germanica, Periplanata americana*) in homes by relating presence of the organisms or their effluents in dust samples to skin reactivity and symptoms in occupants,[20-26] and assays have been developed that allow measurement of antigen in air samples.[27-29]

Birds and mammals that often share the home environment as pets also shed allergens that lead to hypersensitivity disease. As many as 30% of allergic people may be sensitive to domestic animals,[30-33] and 57% of asthmatic children are sensitive to at least one animal species.[34] Cats shed salivary proteins that are allergenic, remain airborne for long periods of time, and are difficult to remove from an environment where a cat has been.[35] Dog allergens are also common in homes.[36,37] As many as 14% of unselected adolescents have been reported as reactive to skin prick tests with dog allergens.[38] Birds kept indoors as pets shed antigens in their droppings that can cause hypersensitivity pneumonitis, although this is probably rare except in people who breed birds in their home environment as a hobby.

Toxicoses

Finally, both bacteria and fungi can produce chemical toxins that may affect human health, although the incidence of such effects are probably quite rare, especially in residences. Bacteria produce both exotoxins and endotoxin. Exotoxins have not been studied with respect to indoor aerosols. Endotoxin, a lipopolysaccharide that forms part of the cell wall in Gram-negative bacteria, can cause fever, chest tightness and general malaise with high-level respiratory exposure.[39,40] The fungi produce toxins that can be carcinogenic, mutagenic, immunosuppressive, acute nervous system toxins or merely irritants. One group of fungi (the *Aspergillus flavus* group) produces aflatoxins among which are the most potent known carcinogens.[41] *A. flavus* has been reported from homes, and speculatively linked with house-associated leukemia. Other fungi (e.g., *Stachybotrys, Fusarium*) produce trichothecene toxins, which have acute, immunosuppressive effects.[42] Severe contamination with *Stachybotrys atra* has been associated with acute toxic symptoms in one home.[43] It should be noted that very severe contamination of a home environment is probably necessary before these toxic effects are likely to occur.

The sick building syndrome is a set of non-specific symptoms that may be related to microbial toxins[44-47] or to macromolecular components of dust that are presumed to be of biological derivation.[48] Although given little publicity, the characteristic symptoms of sick building syndrome can occur in residences as well as in office buildings and schools.

RESERVOIRS, AMPLIFIERS AND DISSEMINATORS OF BIOAEROSOLS

The distribution of organisms and their effluents that cause disease can be studied with respect to reservoirs, amplifiers and disseminators. A reservoir is the immediate source of organisms that are available for colonization of amplifiers. An amplifier is a site where one or more organisms can multiply. Disseminators are mechanisms for aerosolizing organisms or their effluents. It is important to note that reservoirs for most of the common bioaerosol components are ubiquitous, and investigations are usually directed toward interactions between amplifiers and disseminators. Table 21.2 lists reservoirs, amplifiers and disseminators for some common residential bioaerosols.

Outdoor Air

The outdoor air is the ultimate reservoir for many bioaerosols and for organisms that eventually contaminate indoor reservoirs. Components of the outdoor aerosol include pollens and other green plant spores, fungus spores, bacteria and bacterial spores, algae, and probably other organisms and their effluents (see also Chapters 14 and 16).

Factors that control the nature and concentration of outdoor aerosols include the presence or absence of reservoirs, conditions that allow amplification within the reservoirs and factors that produce dissemination. The reservoir for most outdoor fungal aerosols is dead plant material. Major factors that control the abundance of dead plant material include meteorological variables (temperature, rainfall, humidity), natural ecology (presence or absence of lakes, oceans, forests, mountains, etc.), and land use (field crops, forests, suburban, urban, etc.).[49,50] Similar variables control outdoor bacterial aerosols, although most are probably shed from the surfaces of living plants.[47]

Factors that control the amplification of fungi and bacteria in these reservoirs can be similar to those that control the distribution of the reservoirs (e.g., temperature, humidity, rainfall), or can be imposed by man (e.g., presence or absence of fungicides). Dissemination from these amplified reservoirs can result from air movement (wind), or animal or human disturbance (e.g., harvesting field crops).[51,52] These interactions are reflected in the differing fungal aerosols that are found both indoors and out in farming and urban environments.[53,54]

Because homes are usually ventilated with untreated outdoor air, the indoor bioaerosol is often dominated by that outdoors. In fact, in environments without indoor reservoirs, amplifiers or disseminators, the indoor aerosol should reflect the outdoor aerosol, differing only in concentration.[55,56] If the indoor aerosol concentration is significantly higher than that outdoors, or if different species are present indoors and out, then indoor reservoirs and/or amplifiers and disseminators are present.[13]

The major factor that controls the indoor/outdoor ratio of bioaerosol concentration for environments with no indoor amplifiers is the nature of ventilation for the indoor space. In naturally ventilated homes, summer aerosols freely enter from outdoors, and the ratio moves toward unity. In air conditioned homes, particles cannot penetrate so easily and the ratio is usually <1. This ratio is lowest for homes with central air conditioning rather than room units.[57] In winter, when outdoor aerosols are negligible, the indoor/outdoor ratio may be >1, although actual levels are usually low. This is because dust in the indoor environment acts as a reservoir for fungus spores that have entered in the summer and are disseminated with activity during the winter.

Table 21.2 Examples of Reservoirs, Amplifiers and Disseminators

Aerosol	Organism	Reservoir	Amplifier	Disseminator
Influenza virus	Virus	Humans	Humans	Cough, sneeze
Legionella	Bacterium	Outdoor substrates	Cooling towers	Action of the cooling tower
Outdoor pollen	Plant	Plant	Plant	Air movement
Outdoor fungus spores	Fungi	Dead organic material	Dead organic material	Air movement; human activity
Indoor fungus spores	Fungi	Outdoor air	Damp carpet	Human activity
Mite antigen	Mites	House dust	House dust/ dampness	Human activity
Cat antigen	Cats	Cats	Cats	Cat/human activity

Human Sources

While it is possible and desirable to have indoor spaces free of fungus reservoirs, bacterial and probably viral aerosols are an intrinsic part of occupied indoor environments, and the bacterial (and viral) aerosol in an occupied interior is usually qualitatively different than the outdoor aerosol. This is because human beings act as the dominant reservoir, amplifier and disseminator of bacteria and viruses in indoor environments.[58]

Human occupants are the primary source for most of the infectious agents as well as skin surface bacteria in residences. Factors controlling the kinds of organisms that an individual harbors include the status of the individual's immunity, exposure history and internal physiological factors. Amplification occurs when resident organisms are able to multiply and cause disease. When disease symptoms include coughing or sneezing, the individual becomes a disseminator.

Bioaerosol components that have human reservoirs, amplifiers and disseminators include the influenza viruses, many types of common cold viruses, the "childhood" viruses (measles, chicken pox, mumps) and the bacterium that causes tuberculosis. In addition to these disease-causing microorganisms, human beings also shed skin scales at the rate of more than a million per minute. These skin scales are the primary component of house dust and provide food for both mites and fungi. In addition, each skin scale carries one or more bacteria such as *Staphylococcus epidermidis*, and *Corynebacterium* and *Micrococcus* species, among others. In some people, fungi colonize the secretions in the nasal passages.[59] Dissemination from this source has not been studied. Fungi from outdoor air or, for example, agricultural environments, also may be present as passengers on clothing, skin and hair.[54] This is probably the reason why some farm homes have a more concentrated fungus aerosol than urban homes. In this case, the human is acting as a reservoir and disseminator, but not an amplifier.

Factors that control shedding of microorganisms and skin scales from people are primarily related to the site from which dissemination occurs (i.e., the respiratory tract, skin or clothing) and to human activity patterns. Respiratory organisms are usually spread by coughing, sneezing, singing or even talking. Skin scales and skin surface organisms are shed as clothes are removed or donned, and during other abrasive activities, as are the organisms being carried on clothing. Concentrations of human-source aerosols in a defined space depend on the number of people present, the kinds and numbers of organisms they bear, the activities of the people and the ventilation type and rate.

House Dust

House dust is a complex mixture that contains (among other things) human skin scales; fibers; silicates; particulate combustion products; living arthropods and/or their parts and droppings; sometimes urinary, skin and salivary proteins from mice, rats, cats, dogs and birds. In addition, microorganisms are present that have accumulated from outdoor air intrusion or from other reservoirs in the space, and microorganisms that are actually growing on the organic material in the dust. The skin scales that dominate house dust are hygroscopic and can absorb water from the air at relatively low water vapor levels, allowing arthropods (mites) and fungi to use the skin scales both as a nutrient and as a water source.

House dust is well known as a source for allergens, although only recently has the nature of some of the allergens been revealed. Probably the most wide-spread, and possibly the most allergenic of house dust components are those related to house dust mite populations.[60] In humid areas as many as 90% of homes have mite populations high enough to cause sensitization in a susceptible person (i.e., greater than 100 mites/gram of dust).[61] In drier climates, e.g., Denver, Colorado, the central part of northern California, etc., mite levels usually remain below 100/gram of dust. Much of the eastern and central United States have humid summers and dry winters. Mite populations tend to be cyclic in these areas, increasing during the summer, and decreasing in the winter. Within each of these geographic areas, mite prevalence varies between homes for reasons that are as yet unclear.[62]

Population densities of mites in house dust depend on water that is present in the microenvironment of house dust (i.e., deep in carpeting or bedding), the availability of skin scales and fungi for food, and the presence or absence of the predator mite Chyletus.[63,64] Humidity in the house dust microenvironment may be related to ambient humidity, or to other factors. For example, carpeting that remains damp supports large mite populations, and carpeting installed on poorly ventilated grade-level concrete is especially conducive to mite growth. Factors that control ambient humidity also affect the levels of mites and mite allergens in homes. Use of humidifiers, and lowering ventilation rates by keeping windows closed and sealing cracks and crevices to prevent heat loss are associated with increased rates of respiratory illness, some of which may be due to increased levels of mite antigen in the environment.[65]

Cockroaches are another important source of house dust allergen. Cockroaches live on inadequately stored human or animal food, or food remnants that are stored indoors in open garbage containers. Given an adequate water supply, cockroaches can be abundant even when no apparent food supply is present, and residents in the south, where humidity is consistently high, wage a continual battle against the pests. While it is possible to keep a single-family residence free of cockroaches, apartment dwellers are at the mercy of their neighbors, and, especially in inner city areas, a significant proportion of patients with asthma are sensitive to cockroach derived proteins.[66-69] The German cockroach (*Blatella germanica*) is probably the most common sensitizer in the U.S. *Periplaneta americana* (the American cockroach), *Blatta orientalis*, *Periplaneta australisiae* and *Supella supelledium* can also become locally abundant and are probably sensitizing.

Fungal spores that are carried into the house with ventilation air or on the occupants' clothing accumulate over time and are an inevitable part of house dust. On the other hand, a specialized population of fungi can develop in house dust providing adequate water is present. When humidity in the house dust microenvironment is high (>60%) skin scales absorb enough water to support the growth of the xerophilic fungi including members of the *Aspergillus glaucus* group (*A. repens, A. amstelodami*), *A. versicolor, Penicillium chrysogenum, P. crustosum, P. thomii, Wallemia sebi* and members of the Mucorales.[70–72]

Several studies have linked respiratory disease to numbers and kinds of fungi isolated from dust samples.[73,74] Sick building syndrome also has been linked to a macromolecular component of house dust, which is assumed, at least in part, to be derived from fungi.[48]

Although bacteria are common components of house dust, especially the skin surface cocci and *Bacillus* species, they have not been well-studied. Members of the genus *Bacillus* produce endospores that are extremely resistant, and probably can survive for months or years, allowing them to gradually accumulate over time in dust. The skin surface organisms that are recovered from dust samples were probably recently shed.

In order to have an effect on air quality, components of house dust must become aerosolized. This can happen when floors and carpets are swept or vacuumed, when children crawl on floors, or when any other activity occurs that disturbs a dust reservoir.[56] The type of flooring has a significant effect on respiratory exposure to mite antigen. According to Price et al.,[8] wool carpets allow more antigen to enter the air than synthetic carpets or bare floors, even when mite antigen per gram of dust remains constant. House dust can be spread through ventilation systems, although data from one study indicate that primarily small particles circulate in ventilation ductwork.[75]

Other Substrates for Fungi

Fungi can inhabit any reservoir where a carbon source and water are present and can degrade a wide variety of carbonaceous materials including cellulose (cloth, paper), lignin (wood and wood products), urea formaldehyde (insulation, some filter media), carpet backing, particle board, cork board, and even soap films and other organic material on ceramic tile and in grout.[76–79] Virtually all homes and apartment dwellings have most of these potential substrates either as structural material (wood, wood products), as interior finishing, or as contents. In other words, fungi can utilize most of the building and decorating materials in homes. The factor that prevents our homes from being jungles of fungi is water. As long as organic substrates remain dry, the fungi will not grow. Dry, however, is a relative term. Some fungi can use water that has been absorbed from the air by hygroscopic materials (e.g., skin scales, cotton materials). Others require free water such as is provided by leaks and floods, or by the condensation that occurs when relative humidity is consistently high, such as occurs in basements and on poorly insulated external wall and ceiling surfaces. Only a few fungi (yeasts, slimy-spored fungi) can grow in standing water; most are inhibited by the lack of oxygen in this environment.

Fungi on walls, woodwork or in insulation material may release volatile organic compounds that cause irritation or illness with no agitation. To produce a spore aerosol, however, some kind of disturbance is usually required. With sporulating fungi such as *Penicillium*, the disturbance can be as slight as the air movement that occurs when a person enters the room. With other fungi (e.g., *Aureobasidium*) more vigorous disturbance is necessary, such as cleaning, construction, demolition activities, etc.[56,80] If the fungus is inhabiting the ventilation system, air movements within the system can produce fungus spore aerosols.[75] Fungi on upholstered furniture release spores into the air when one sits down.

Many fungi have active spore discharge mechanisms. The wood-rotting basidiomycetes can grow in the structural wood of houses, causing the wood to lose its structural integrity.[81] These fungi can produce fruiting bodies in the indoor space that forcibly discharge billions of spores. These spore clouds have been related to home-associated asthma and hypersensitivity pneumonitis.[82,83]

Water Reservoirs and Other Sites for Bacteria

Bacteria, including those that cause opportunistic infections, can inhabit residential reservoirs that contain standing water, including hot water tanks, humidifiers, refrigerator drip pans, sumps, etc. For example, *Legionella pneumophila* has been isolated from residential water systems.[84,85] The controlling factor for this type of contamination is water temperature. According to Lee et al.,[85] water temperatures below 48.8°C were significantly associated with colonization by *Legionella pneumophila*. Other water reservoirs in homes that can support the growth of bacteria include humidifiers, dehumidifiers, refrigerator drip pans and cooling-coil drip pans associated with climate control equipment. The Gram-negative bacilli (e.g., *Pseudomonas*, *Flavobacterium*) are most common in these environments. These organisms contain endotoxin and can cause opportunistic infections.[39,86,87]

There is a large literature on other kinds of bacteria that can inhabit drinking water systems, both hot and cold. Their contribution to the residential bioaerosol and to human respiratory disease has not been investigated. The thermophilic actinomycetes are bacteria that require temperatures in excess of 50°C to survive, and occupy sites such as humidifiers that are attached to furnaces, refrigerator drip pans and clothes-dryer exhaust ducts.[19]

As with the fungi, the organisms in bacterial reservoirs need some kind of activity to become aerosolized. Water systems, showers and water faucets produce aerosols with use.[88] Likewise, some humidifiers (e.g., ultrasonic and cool mist humidifiers) are designed to produce a water aerosol. The water droplets carry entrained cells and other materials derived from growth in the water reservoirs.[39] Droplets falling into any water-filled reservoir can produce an aerosol that contains bacteria (e.g., in dehumidifiers, drip pans). The warm air rushing through a dryer vent pipe, or through contaminated heating systems spreads the dry thermophilic actinomycete spores.

Pets and Pests

It is estimated that approximately 100 million domestic animals reside in the U.S., the most common being cats and dogs.[89] All of these animals produce bioaerosols that can be allergenic. Cats produce allergenic proteins in their saliva and in skin secretions. The saliva is spread onto the fur by the cat, and flakes off with activity, becoming aerosolized in very small particles that remain airborne for long periods of time.[35] Cat allergens also can be carried into homes on the clothing of visiting pet owners. Dog allergens also become airborne, although less is known about the dynamics of bioaerosol production.

Caged pets (hamsters, gerbils, mice, birds) shed urine and fecal material indoors as well as skin scales and saliva. Those people handling the pets and especially those that clean the cages are exposed to intense animal bioaerosols from these sources, as well as bacteria and fungus spore aerosols.[90] Individual cases of extreme sensitization to pet rodents (such as rabbits, mice or guinea pigs) are well recognized as causes of rhinitis or contact urticaria.[91]

Some mammals can be household pests (rats, mice), and rodents are often concentrated in laboratory animal facilities. All of these animals shed proteins and occasionally bacteria or viruses into the environment. In areas of the U.S. where mice and rats are major pests, a significant proportion of allergic patients have positive skin test to rodent urinary proteins, although there have been no epidemiological surveys to confirm the importance of allergic reactions to rodent urine as a risk factor for asthma

The allergens shed by birds and mammals can become airborne directly from the animals or by disturbance of settled dust as described above for house dust. Factors affecting the abundance of animal allergens in the air of homes include numbers of animals, time spent indoors, ventilation rates, furnishings that can act as reservoirs, and activity of both people and the animals.

SAMPLING

The mode of sampling chosen to characterize bioaerosols in a residence depends on the reason for sampling, and can be as simple as visual observation or as complex as air sampling (see also Chapters 9 and 10). Visual observation is an important part of any investigation designed to detect sources of exposure related to existing disease. Although visual recognition of a bioaerosol amplifier/disseminator is not proof of causation of disease, it is a likely cause, and remediation can be attempted before the environment is studied using more complex methods.

On the other hand, research that seeks to characterize completely an environment with respect to reservoirs, amplifiers, and disseminators of bioaerosols, or attempts to establish correlations between specific bioaerosols and disease states, requires the use of a constellation of sampling techniques. There are three parts to bioaerosol sampling in residences and elsewhere: collecting a valid sample, analyzing the sample and interpreting the results.

Collecting a Valid Sample

Collecting a valid visual sample involves the development and use of a list of factors (reservoirs, amplifiers, disseminators) that are known to be indicative of biological contamination and bioaerosol production. For example, a series of questions related to animal allergens might be: Are there pets? What kind(s)? How many? Are droppings and/or urine present (e.g., in cage or litter box)? Do the animals spend time outdoors?

Odors can indicate the presence of active microbial growth. Because many volatile, odoriferous compounds tend to deaden the sense of smell, it is important to record observations of ambient odors immediately upon entering a home. Disturbing suspected reservoirs temporarily may increase the levels of volatile compounds and provide information as to sources of reported transient odors.

While sensory investigations allow the entire residence to be evaluated, all other approaches require that a small environmental sample be collected and analyzed. Deciding how to collect a relevant sample is a significant challenge, and depends on the reason for sampling, the kind of sample to be collected and the nature of the contamination of interest. For any type of sample, it is necessary to know the extent of variability in space and time in order to decide how many samples must be collected and how large each sample must be. Unfortunately, for most biological agents this kind of information is not yet available, and most sampling protocols will have to include multiple samples distributed over both time and space.

Sampling for biological agents can be accomplished by surface sampling, bulk sampling, and/or air sampling. Surface sampling (swabs, contact plates) can provide information on the kinds of organisms present and, in some cases, some indication of relative numbers per unit of surface area sampled. Collecting a reliable sample involves choosing appropriate surfaces for sampling and collecting enough replicates of each surface so that some indication of variability can be determined. This method is often used to evaluate conditions in hospital operating rooms (contact plates),[92] and to allow identification of organisms that have been visually detected on surfaces (plates, swabs or adhesive tape collections for microscopic examination).

Bulk collections from reservoirs can be quantitatively analyzed and provide data on the numbers of each kind of organism present in the reservoir. For liquid bulk samples from stagnant water reservoirs, a mixed sample can be obtained by carefully mixing the reservoir water before sample collection. Another method involves taking a number of small samples from different parts of a reservoir and mixing them before analysis. This is often done to save money on analysis. Finally, individual samples can be collected from a reservoir and analyzed separately.[72] This allows determination of spatial and temporal variability within the reservoir.

Finally, a volumetric air sample can provide data on numbers and kinds of biological particles in the air. Note that gravimetric air samples only provide presence/absence data and some information on the kinds of particles present. It can be argued that the important information is available only with air sampling, and, in fact, properly collected air samples probably most closely represent respiratory exposure. They are essential if statistical correlations with symptoms or other disease markers are to be made. However, if residential sampling is being done to identify reservoirs that are contributing to the aerosol so that effective remediation can be undertaken, then bulk samples from potential reservoirs may be most representative.

Collecting a valid air sample represents the greatest challenge, because it is almost certain that the bioaerosol will change during sampling and because each of the available air samplers collects a different part of the aerosol. Most residential air samples are collected using impaction onto culture media[93,94] or onto an adhesive surface.[95] Filtration is also occasionally used, especially for the study of airborne allergens.[27] The samplers employing culture plates collect short-term grab samples (usually less than 5 minutes), and multiple samples need to be taken over time at each site to obtain a representative picture of the bioaerosol. Samplers impacting particles onto adhesives, or removing particles from an air mass by filtration, can be operated over longer periods and provide an integrated sample (filtration) or time-discriminated (Burkard) sample at each sampling site.[96] On the other hand, the sieve plate impactors tend to be more efficient at trapping particles than the adhesion impactors, while filtration captures all the particles that enter the device that are larger than the pore size.[97] If the efficiency of a particular sampler is known, one can sometimes adjust counts from a less efficient device for comparison purposes. Factors that control the choice of a bioaerosol sampler include the type of analysis to be used (e.g., cultural, microscopic, immunoassay, etc.), the expected size of the aerosol particles and the expected concentration of the aerosol.[98,99]

Sample Analysis

Samples can be analyzed by culture (fungi, bacteria, protozoa), by microscopy (fungi, bacteria, mites), by bioassay (endotoxin), by immunoassay (any allergen for which specific antibodies are available) or by chemical assay (e.g., mycotoxins). There are essentially no upper limits to the levels that can be measured by any of these methods, since most samples can be diluted. Lower limits are different depending on the assay of choice, the sample size, and the agent to be studied.

Culture of bulk or air samples is the most commonly used method for evaluating samples for bacterial and fungal content.[58,72] It should be emphasized, however, that no bioaerosol contains only viable and culturable organisms, no set of culture conditions is ideal for all organisms,[100,101] and recoveries on cultured samples depend on the viability and culturability of the organisms, the kind of culture medium used, incubation conditions, and, often, on the presence or absence of organisms that produce biocides.

Microscopy is the only currently available method for evaluating levels of fungus spores that do not germinate or that fail to produce distinctive spore-bearing structures in culture.[93,95] Bacterial counts can be made microscopically with fluorescent stains.[102] However, this method works best for samples that contain high levels of bacteria, or that can be concentrated. Use of fluorescence staining for fungal spores is more difficult because the dark pigment in some spores masks the stain.

Immunoassay of bulk samples has been the method of choice for evaluating arthropod and mammalian allergens in homes, although the same analytical methods can be used for air samples. Monoclonal antibodies are available for cat, mite, cockroach, etc.[25,26,103] Monoclonals have been developed for several fungal allergens. However, the many different kinds of fungi that can be present in dust make it necessary to use culture and microscopic examination for most investigations.

Toxins must usually be analyzed by bioassay or by chromatographic methods.[104,105] Although monoclonal antibodies have been produced that are specific for some mycotoxins, they have yet to be used in the analysis of residential samples.

DATA ANALYSIS AND INTERPRETATION

Data interpretation can be directed at an evaluation of health risk or characterization of the environment. For most bioaerosols, there are inadequate dose/response data to allow accurate risk analysis.

With respect to allergens, one health-based standard has been proposed. For dust mite antigen, 2 μg Der p I (one of the major dust mite allergens) per gram of dust has been proposed to represent a safe level, and >20 μg/g is a level where sensitization will occur in susceptible occupants.[61] However, there is evidence that measures of allergen per gram of dust are less predictive of sensitization than measures of airborne antigen, or numbers of mites per square meter of carpet.[8]

Su et al.[106] have found a consistent correlation between hay fever symptoms and log colony-forming unit concentrations of several dark-spored fungi at geometric mean levels well below 500 cfu/m^3. They also report a relationship between asthma and log concentrations of *Aspergillus* at similar concentrations. The alternative approach is to characterize a kind of environment (e.g., residences in a community), define a range of organisms or their effluents that normally occurs, and define a normal level based on the mean value. Ideally, the information in the data base should be collected from randomly selected homes. Reponen et al.[107] have used this method to define upper normal limits for fungi and bacteria in non-complaint homes in Finland. For airborne bacteria, this group has recommended 4500 cfu/m^3 as the highest normal level (using the N6 Andersen with tryptone glucose yeast agar). For airborne fungi, they have recommended a highest normal level of 500 cfu/m^3 (using the N6 Andersen and modified Hagem samplers). These numbers are based on the mean value of recoveries from 71 homes plus two standard deviations. They stress that these levels should not be used for health risk determinations, but as an indication that an unusual exposure situation exists.

It is entirely possible that "normal" levels that are defined on the basis of existing conditions will be well above what is actually safe from a health perspective. The Su[106] data are a case in point. These were randomly selected homes in which the mean value for, e.g., *Epicoccum* is well within the range where correlations were found with symptoms. In addition, risks associated with bioaerosol exposure differ for different aerosols and for different people. The fallacy in depending on a total number for bacteria or fungi should be obvious. Levels of non-pathogenic bacteria in excess of 10^6 may not constitute a significant risk whereas a single cell of *Mycobacterium tuberculosis* is sufficient to initiate an infection. On the other hand, 10^5 *Aspergillus fumigatus* spores per cubic meter may not constitute a significant health risk for a normal person, but a single spore of this fungus could initiate an infection in, for example, a transplant patient on immunosuppressive medication. Until dose/response data are available, a combination approach to data interpretation must be used that takes into consideration the kinds of organisms present, the kinds of people exposed, the numbers of specific organisms present and their source.

PREVENTION AND REMEDIATION

Prevention of Intrusion

Several studies have demonstrated that keeping homes closed (and using mechanical cooling) will lower the levels of airborne fungi in homes[57] and automobiles.[108] Sources for bioaerosols that can be kept out of homes and should always be excluded from public places include pets, spray humidifiers and dead organic material such as firewood. Unfortunately, until the disease becomes insupportable, many people will not give up either their pet or their humidifier, and instead seek medical resolutions to their problems.

Control of Potential Reservoirs

House dust control has been recommended and successfully practiced for many years for prevention of asthma.[109] Recommended methods essentially control house dust mite populations, as well as reducing the amount of other dust components that patients inhale. Climate control is the best overall method to limit amplification in indoor reservoirs of fungi and dust mites. If humidity is maintained below 50% throughout a home throughout the year, contamination with these organisms is unlikely.[110] Note that humidity inside plush carpeting may be much higher than that in ambient air, particularly if water-based methods have been used to clean carpeting in-situ.[63]

Careful use of appliances that are susceptible to contamination with microorganisms also can prevent the amplification and production of bioaerosols. Humidifiers, for example, almost always are contaminated with bacteria. In order to prevent (or at least significantly reduce) such contamination, humidifiers that spray water into the air (cool mist vaporizers, ultrasonic humidifiers) must be emptied and dried every day, and cleaned with either dilute bleach or dilute hydrogen peroxide every third day. Clothes dryers, a potential reservoir for thermophilic actinomycetes, should always be vented to the outdoors. Refrigerator drip pans should dry completely each day; the presence of standing water in these pans is an indication that the unit is not operating efficiently, and this warm water can easily support the growth of a number of organisms, probably including *Legionella pneumophila*.

Furnace filters should be changed at the end of the heating system. Dirty filters absorb water from the air and may support fungal growth.[78]

Remediation

Some indoor bioaerosols are inevitable or, in some cases, unpredictable and difficult to control. These include human source bacterial and viral aerosols. There is some evidence that increasing outdoor air ventilation rates will lessen the risk of disease related to rhinoviruses[111] and dust mites.[65] However, very few North American homes are mechanically ventilated. Most rely on infiltration through leaks in the building envelop to provide ventilation air. Often, especially during cooling and heating seasons, homes are sealed for energy conservation so that infiltration supplies as little as 0.1 air changes per hour.

Air cleaners have been recommended for removal of particulate bioaerosols from residential environments. However, it appears that some air cleaners have very little effect on allergen aerosols[112] and even the best are not adequate to clean air where actively emitting sources are present.[113]

Cleaning contaminated reservoirs can include removal of microbial growth alone or removal of the growth along with substrates. Such cleaning is often followed by the application of biocides. A large industry has grown up that focuses on cleaning accumulated dust from ventilation system duct work in homes. Although duct work can accumulate dust that includes biological agents, and duct cleaning does remove these accumulations, there has been no organized effort to date to evaluate the impact of this process on the health of occupants.

There are biocides available that will kill, or render inactive, most biological agents. Oxidizing agents (chlorine, ozone, peroxide) have been used to prevent and control microbial contamination of water (e.g., in cooling towers) and to clean contaminated reservoirs. Unfortunately, use of biocides in reservoirs that disseminate bioaerosols is likely to cause dissemination of the biocide unless care is taken to deactivate the dissemination mechanism until all of the biocide has been removed.

SUMMARY

It is clear that some kinds of human respiratory disease are associated with bioaerosol-related conditions in the home environment. However, we have only begun to document the extent of this problem and the efficacy of means used to control the problem. Research is needed that evaluates the role of specific bioaerosols in producing respiratory disease using personal sampling techniques that accurately measure exposure. Dose/response relationships need to be evaluated for most bioaerosols, and the efficacy of existing methods as well as the development of new methods of preventing exposure to bioaerosols needs investigation. These efforts will require multi-national and multi-disciplinary cooperation.

REFERENCES

1. Chapman JA, Williams S. 1984. Aeroallergens of the southeast Missouri area: a report of skin test frequencies and air sampling data. *Ann Allergy* 52(6):411–418.

2. Dick EC, Jennings LC, Mink KA, Wartgow CD, Inborn SL. 1987. Aerosol transmission of rhinovirus colds. *J Inf Dis* 156:442 448.
3. Spengler JDS, Sexton K. 1983. Indoor air pollution: a public health perspective. *Science* 221–229.
4. Nolan CM, Elarth AM, Barr H, Saeed AM, Risser DR. 1991. An outbreak of tuberculosis in a shelter for homeless men. A description of its evolution and control. *Am Rev Respir Dis* 143(2):257–261.
5. Winn WC Jr. 1985. Legionella and Legionnaires' disease: a review with emphasis on environmental studies and laboratory diagnosis. *CRC Crit Rev Clin Lab Sci* 21(40):323–381.
6. Meyer RD. 1983. Legionella infections: a review of 5 years of research. *Rev Inf Dis* 5:258–278.
7. Solomon WR, Burge HP, Boise JR. 1978. Airborne *Aspergillus fumigatus* levels outside and within a large clinical center. *J Allergy Clin Immunol* 62(1):56–60.
8. Price JA, Pollock I, Little SA, Longbottom JL, Warner JO. 1990. Measurement of airborne mite antigen in homes of asthmatic children [see comments]. *Lancet* 336(8720):895–897.
9. Dekker C, Dales R, Bartlett S, Brunekreef B, Zwanenburg H. 1991. Childhood asthma and the indoor environment. *Chest* 100(4):922–926.
10. Brunekreef B, Dockery DW, Speizer FE, Ware JH, Spengler JD, Ferris BJ. 1989. Home dampness and respiratory morbidity in children. *Am Rev Respir Dis* 140:1363–1367.
11. Waegemaekers M, Van Wageningen N, Brunekreef B, Boleij JS. 1989. Respiratory symptoms in damp homes. A pilot study. *Allergy* 44(3):192–198.
12. Dales R, Zwanenburg H, Burnett R. 1990. The Canadian air quality health survey: influence of home dampness and molds on respiratory health. In: Proceedings of the Fifth International Conference on Indoor Air Quality and Climate, Toronto, Canada, Vol 1:145–147.
13. Brunekreef B, de Rijk L, Verhoeff AP, Samson R. 1990. Classification of dampness in homes. In: Proceedings of the Fifth International Conference on Indoor Air Quality and Climate, Toronto, Canada, Vol 2:15–20.
14. Su HJ, Burge HA, Spengler JD. 1992. Association of airborne fungi and wheeze/asthma symptoms in school-age children. *J Allergy Clin Immunol* 89(1)pt 2.
15. Roby RR, Sneller MR. 1979. Incidence of fungal spores at the homes of allertgic patients in an agricultural community. II. Correlations of skin tests with mold frequency. *Ann Allergy* 43(5):286–288.
16. Strachan DP, Flannigan B, McCabe EM, McGarry F. 1990. Quantification of airborne molds in the homes of children with and without wheeze. *Thorax* 45(5):382–387.
17. Flannigan B, McCabe EM, McGarry F. 1990. Wheeze in children: an investigation of the air spora in the home. In: Proceedings of the Fifth International Conference on Indoor Air Quality and Climate, Toronto, Canada, Vol 2:27–32.
18. Acierno LJ et al 1985. Acute hypersensitivity pneumonitis related to forced air systems—a review of selected literature and a commentary on recognition and prevention. *J Environ Health* 48(3):138–141.
19. Fink JN, Banaszak EF, Barboriak JJ, Hensley GT, Kurup VP, Scanlon GT, Schlueter DP, Sosman AJ, Thiede WH, Unger GF. 1976. Lung disease from forced-air systems. *Clin Notes Respir Dis* 15(3):10–12.
20. Menon V, Menon P, Hillman B, Stankus R, Lehrer S. 1989. American (*Periplaneta americana*) and German (*Blatella germanica*) cockroach skin test reactivity in atopic asthmatics. *J Allergy Clin Immunol* [abstract] 83:265

21. Voorhorst R, Spieksma FTHM, Varenkamp H, Leupen MJ, Lyklema AW. 1967. The house dust mite (*Dermatophagoides pteronyssinus*) and the allergens it produces: Identity with the house dust allergen. *J Allergy* 39:325.
22. Voorhorst R. Spieksma FThM, Varekamp N. 1969. House dust atopy and the house dust mite *Dermatophagoides pteronyssinus* (Troussart, 1897). Leiden: Stafleu's Scientific Publishing.
23. Feinberg AR, Feinberg SM, Benaim-Pinto C. 1956. Asthma and rhinitis from insect allergens. I. Clinical importance. *J Allergy Clin Immunol.* 27:437–444.
24. Tovey ER, Chapman MD, Platts-Mills TAE. 1981. Mite faeces are a major source of house dust allergens. *Nature* 389:592–593.
25. Pollart SM, Smith TF, Morris EC, Gelber LE, Platts-Mills TAE, Chapman MD. 1991. Environmental exposure to cockroach allergens: analysis with monoclonal antibody-based enzyme immunoassays. *J Allergy Clin Immunol* 87:505–510.
26. Chapman MD, Aalberse RC, Brown MJ, Platts-Mills TAE. 1988. Monoclonal antibodies to the major feline allergen Fel d I. II. Single-step affinity purification of Fel d I, N-terminal sequence analysis, and development of a sensitive two-site immunoassay to assess Fel d I exposure. *J Immunol* 140:812–818.
27. Swanson MC, Agarwal MK, Reed CE. 1985. An immunochemical approach to indoor aeroallergen quantitation with a new volumetric air sampler: studies with mite, roach, cat, mouse, and guinea pig antigens. *J Allergy Clin Immunol* 76:724–729.
28. deBlay F, Heymann PW, Chapman MD, Platts-Mills AE. 1991. Airborne dust mite allergens: comparison of group II allergens with group I mite allergen and cat-allergen Fel d I. *J Allergy Clin Immunol* 88(6):919–925.
29. Sakaguchi M, Inouye S, Yasueda H, Irie T, Yoshizawa S, Shida T. 1989. Measurement of allergens associated with dust mite allergy. II. Concentrations of airborne mite allergens (Der I and Der II) in the house. *Int Arch Allergy Appl Immunol* 90:190–193.
30. Barbee RA, Brown GB, Kaltenborn WS, Halonen M. 1981. Allergen skin test reactivity in a community population sample: correlation with age, histamine skin reaction and total serum immunoglobulin E. *J Allergy Clin Immunol* 68:15.
31. Fontana VJ, Wittig H, Holt LE. 1963. Observations on the specificity of the skin test. The incidence of positive skin tests in allergic and nonallergic children. *J Allergy* 34:348.
32. Ohman JL, Kendall S, Lowell FC. 1977. IgE antibody to cat allergens in an allergic population. *J Allergy Clin Immunol* 60:317.
33. Ohman JL. 1978. Allergy in man caused by exposure to mammals. *J Am Vet Med Assoc* 172:1403.
34. Kjellman B, Pettersson R. 1983. The problem of furred pets in childhood atopic disease. Failure of an information program. *Allergy* 38:65.
35. Luczynska CM, Arruda LK, Platts-Mills TAE. 1990. Airborne concentrations and particle size distribution of allergen derived from domestic cats (*Felis domesticus*): measurements using cascade impactor, liquid impinger, and a two-site monoclonal antibody assay for Fel d I. *Am Rev Respir Dis* 141:361–367.
36. Lind P, Norman PS, Newton M, Lowenstein H, Schwartz B. 1987. The prevalence of indoor allergens in the Baltimore area: house dust mite and animal dander antigens measured by immunochemical techniques. *J Allergy Clin Immunol* 80:541–547.
37. Schou C, Hansen GN, Lintner T, Lowenstein H. 1991. Assay for the major dog allergen Can f I: investigation of house dust samples and commercial dog extracts. *J Allergy Clin Immunol* 88(6):847–853.
38. Haahtela T, Jaakonmaki I. 1981. Relationship of allergen-specific IgE antibodies, skin prick tests, and allergic disorders in unselected adolescents. *Allergy* 36:215–216.
39. Burke GW, Carrington CB, Strauss R, Fink JN, Gaensler EA. 1977. Allergic alveolitis caused by home humidifiers. *JAMA* 238:2705.

40. Heederik D, Brouwer R, Biersteker K, Boleij JS. 1991. Relationship of airborne endotoxin and bacteria levels in pig farms with lung function and respiratory symptoms of farmers. *Int Arch Occup Environ Health* 62(8):595–601.
41. Baxter CS, Wey HE, Burg WR. 1981. A prospective analysis of the potential risk associated with inhalation of aflatoxin-contaminated grain dusts. *Food Cosmet Toxicol* 19:763–769.
42. Shank RC. 1981. *Mycotoxins and N-Nitroso Compounds: Environmental Risks.* CRC Press, Boca Raton, FL.
43. Croft WA, Jarvis BB, Yatawara CS. 1986. Airborne outbreak of trichothecene toxicosis. *Atmosph Environ* 20:549–552.
44. Auger PL, St Onge M, Aberman A, Irwin J, Lamanque M, Miller D. 1990. Pseudo chronic fatigue syndrome in members of one family which was cured by eliminating the *Penicillium brevicompactum* molds found in their home. In: Proceedings of the Fifth International Conference on Indoor Air Quality and Climate, Toronto, Canada, July 1990, Vol 1:181–185.
45. Harrison J, Pickering CAC, Faragher EB, Austwick PKC. 1990. An investigation of the relationship between microbial and particulate indoor air pollution and the sick building syndrome. In: Proceedings of the Fifth International Conference on Indoor Air Quality and Climate, Toronto, Canada, Vol 1:149–154.
46. Lundholm M, Lavrell G, Mathiasson L. 1990. Self-leveling mortar as a possible cause of symptoms associated with "sick building syndrome." *Arch Environ Health* 45(3):135–140.
47. Nevalainen A, Heinonen-Tanski H, Savolainen R. 1990. Indoor and outdoor air occurrence of *Pseudomonas* bacteria. In: Proceedings of the Fifth International Conference on Indoor Air Quality and Climate, Toronto, Canada, Vol 2:51–53.
48. Gravesen S, Skov P, Valbjorn O, Lowenstein H. 1990. The role of potential immunogenic components of dust (MOD) in the sick-building-syndrome. In: Proceedings of the Fifth International Conference on Indoor Air Quality and Climate, Toronto, Canada, Vol 1:9–13.
49. Burge HA, Solomon WR. 1991. Outdoor Allergens. In: Allergen Immunotherapy, eds. Lockey RF, Bukantz SC, Marcel Dekker, New York.
50. Al-Doory Y. 1985. The indoor airborne fungi. *N Engl Reg Allergy Proc* 6(2):140–149.
51. Sneller MR, Roby RR. 1979. Incidence of fungal spores at the homes of allergic patients in an agricultural community. I. A 12 month study in and out of doors. *Ann Allergy* 43(4):225–228.
52. Sneller MR, Roby RR, Thurmond LM. 1979. Incidence of fungal spores at the homes of allergic patients in an agricultural community. III. Associations with local crops. *Ann Allergy* 43(6):352–355.
53. Burge HA, Muilenberg ML, Chapman J. 1991. Crop plants as a source for medically important fungi. In: *Microbiol Ecology of Leaves.* eds. Andrews J, Hirano S, Springer Verlag, New York, pp. 222–236.
54. Pasanen AL, Kalliokoski P, Pasanen P, Salmi T, Tossavainen A. 1989. Fungi carried from farmers' work into farm homes. *Am Ind Hyg Assoc J* 50(12):631–633.
55. Bunnag C, Dhorranintra B, Plangpatanapanichya A. 1982. A comparative study of the incidence of indoor and outdoor mold spores in Bangkok, Thailand. *Ann Allergy* 48(6):333–339.
56. O'Rourke MK, Quackenboss JJ, Lebowitz MD. 1990. Indoor pollen and mold characterization from homes in Tucson Arizona, USA. In: Proceedings of the Fifth International Conference on Indoor Air Quality and Climate, Toronto, Canada, Vol 2:9–14.
57. Pan PM, Burge HA, Su HJ, Spengler JD. 1992. Central vs room air conditioning for reducing exposure to airborne fungus spores. *J Allergy Clin Immunol* 89(1)pt 2.

58. Nevalainen A. 1989. Bacterial aerosols in indoor air. National Public Health Institute, Helsinki, Finland.
59. Kauffman C, Burge H, Solomon W. 1988. Air and human-borne *Aspergillus fumigatus* in patient care areas. *J Allergy Clin Immunol* 81(1):273 [Abstract].
60. Platts-Mills TAE, de Weck AL. 1989. Dust mite allergens and asthma—a worldwide problem [international workshop]. *J Allergy Clin Immunol* 83:416–427.
61. Platts-Mills TAE, Chapman MD. 1987. Dust mites: immunology, allergic disease, and environmental control. *J Allergy Clin Immunol* 80:755–775.
62. Arlian LG. 1989. Biology and ecology of house dust mite, *Dermatophagoides* spp and *Euroglyphus spp. Immunol Allergy Clin N Am* 9:339–356.
63. Wassenaar DP. 1988. Effectiveness of vacuum cleaning and wet cleaning in reducing house dust mites, fungi and mite allergen in a cotton carpet: a case study. *Exp Appl Acarol* 4(1):53–62.
64. Irie T. Hara M, Hasegawa Y, Yoshikawa M. 1990. Measurement of living mites in private wooden dwelling, Japan. In: Proceedings of the Fifth International Conference on Indoor Air Quality and Climate, Toronto, Canada, Vol 2:61–66.
65. Sundell J, Wickman M, Nordvall L, Pershagen G. 1990. Building hygiene and house dust mite infestation. In: Proceedings of the Fifth International Conference on Indoor Air Quality and Climate, Toronto, Canada, Vol 1:27–29.
66. Bernton HS, Brown H. 1970. Insect allergy: the allergenicity of the excrement of the roach. *Ann Allergy* 28:543–547.
67. Twarog FJ, Picone FJ, Strunk RS, So J, Colten HR. 1977. Immediate hypersensitivity to cockroach: isolation and purification of the major antigens. *J Allergy Clin Immunol* 59:154–160.
68. Kang BC, Chang JL, Johnson J. 1989. Characterization and partial purification of the cockroach antigen in relation to house dust and house dust mite (D.f.) antigens. *Ann Allergy* 63:207–212.
69. Hulett AC, Dockhorn RJ. 1979. House dust, mite (*D. farinae*), and cockroach allergy in a midwestern population. *Ann Allergy* 42:160–165.
70. Saad R, el-Gindy AA. 1990. Fungi of the house dust in Riyadh, Saudi Arabia. *Zentralbl Mikrobiol* 145(1):65–68.
71. Samson RA. 1985. Occurrence of molds in modern living and working environments. *Eur J Epidemiol* 1(1):54–61.
72. Gravesen S. 1978. Identification and prevalence of culturable mesophilic microfungi in house dust from 100 Danish homes. Comparison between airborne and dust-bound fungi. *Allergy* 33(5):268–272.
73. Booij-Noord H, de Vries K, Sluiter HJ, Orie NGM. 1972. Late bronchial obstructive reaction to experimental inhalation of house dust extract. *Clin. Allergy* 2:43.
74. Tarlo SM, Radkin A, Tobin RS. 1988. Skin testing with extracts of fungal species derived from the homes of allergy clinic patients in Toronto Canada, *Clin Allergy* 18:45–52.
75. Jantunen MJ, Bunn E, Pasanen P, Pasanen AL. 1990. Does moisture condensation in air ducts promote fungal growth? In: Proceedings of the Fifth International Conference on Indoor Air Quality and Climate, Toronto, Canada, Vol 2:73–78.
76. Strom G, Palmgren U, Wessen B, Hellstrom B, Kumlin A. 1990. The sick building syndrome: an effect of microbial growth in building constructions? In: Proceedings of the Fifth International Conference on Indoor Air Quality and Climate, Toronto, Canada, Vol 1:173–178.
77. Bisset J. 1987. Fungi associated with urea-formaldehyde foam insulation in Canada. *Mycopathologia* 99(1):47–56.
78. Bernstein RS, Sorenson WG, Garabrant D, Reaux C, Treitman RD. 1983. Exposures to respirable airborne *Penicillium* from a contaminated ventilation system: clinical

environmental and epidemiological aspects. *American Industrial Hygiene Assoc J* 44(3):161–169.

79. Gallup J, Kozak P, Cummins L, Gillman S. 1987. Indoor mold spore exposure: characteristics of 127 homes in southern California with endogenous mold problems. *Experientia Suppl* 51:139–142.

80. Streifel AJ, Vesley D, Rhame FS. 1990. Occurrence of transient high levels of airborne fungal spores. In: Proceedings of the Fifth International Conference on Indoor Air Quality and Climate, Toronto, Canada, Vol 1:207–212.

81. Duncan CG, Eslyn WE. 1966. Wood-decaying Ascomycetes and Fungi Imperfecti. *Mycologia* 58:642–645.

82. Stone CA, Johnson GC, Thornton JD, Macauley BJ, Holmes PW, Tai EH. 1989. *Leucogyrophana pinastri*, a wood decay fungus as a probable cause of an extrinsic allergic alveolitis syndrome. *Aust NZ J Med* 19(6):727–729.

83. O'Brien IM, Bull J, Creamer B, Sepulveda R, Harries M, Burge PS, Pepys J. 1978. Asthma and extrinsic allergic alveolitis due to *Merulius lacrymans*. *Clin Allergy* 8:535.

84. Stout JE, Yu VL, Muraca P. 1987. Legionnaires' disease acquired within the homes of two patients. Link to the home water supply. [published erratum appears in JAMA 1987 May 15:257(19):2595] *JAMA* 257(9):1215–1217.

85. Lee TC, Stout JE, Yu VL. 1988. Factors predisposing to *Legionella pneumophila* colonization in residential water systems. *Arch Environ Health* 43(1):59–62.

86. Burge HP, Solomon WR, Boise JR. 1980. Microbial prevalence in domestic humidifiers. *Appl Environ Microbiol* 39(4):840–844.

87. Patterson R, Fink JN, Miles WB, Basich JE, Schleuter DB, Tinkelman DG, Roberts M. 1981. Hypersensitivity lung disease presumptively due to Cephalosporium in homes contaminated by sewage flooding or by humidifier water. *J Allergy Clin Immunol* 68:128–132.

88. Fraser DW. 1984. Sources of legionellosis. In: Legionella, Proceedings of the 2nd International Symposium, eds. Thornsberry C, Balows A, Feeley JC, Jakubowski W. American Society of Microbiology, Washington DC, pp. 277–280.

89. Knysak D. 1989. Animal allergens. *Immunology and Allergy Clinics of North America* 9(2):357–364.

90. Burge HP, Solomon WR, Williams P. 1979. Aerometric study of viable fungus spores in an animal care facility. *Laboratory Animals* 13:333–338.

91. Cockroft A, McCarthy P, Edwards J, et al. 1981. Allergy in laboratory animal workers. *Lancet* 1:827.

92. Fauero MS. 1968. Microbiological sampling of surfaces. *J Appl Bact* 31:336–343.

93. Burge HP, Boise JR, Rutherford JA, Solomon WR. 1977. Comparative recoveries of airborne fungus spores by viable and nonviable modes of volumetric collection. *Mycopathologia* 61(1):27–33.

94. Pan P, Hoyer M, Muilenberg M, Burge H, Solomon W. 1987. Characteristics of the SAS culture plate sampler for assessing indoor fungi. *J Allergy Clin Immunol* 79(1):210 [Abstract].

95. Muilenberg M, Pan P, Hoyer M, Burge H, Solomon W. 1987. Incident particulate sampling for indoor bioaerosols: the Burkard Personal spore trap. *J Allergy Clin Immunol* 79(1):209 [Abstract].

96. Burge HA. 1986. Some comments on the aerobiology of fungus spores. *GRANA* 25:143–146.

97. Eduard W, Lacey J, Karlsson K, Palmgren U, Strom G, Blomquist G. 1990. Evaluation of methods for enumerating microorganisms in filter samples from highly contaminated occupational environments. *Am Ind Hyg Assoc J* 51(8):427–436.

98. Stedman OJ. 1978. A seven-day volumetric spore trap for use within buildings. *Mycopathologia* 66(1–2):37–40.

99. Burge H, Solomon W. 1987. Sampling and analysis of biological aerosols. Atmosph. Science 21(2):451.
100. Verhoeff AP, van Wijnen JH, Boleij JS, Brunekreef B, van Reenen-Hoekstra ES, Samson RA. 1990. Enumeration and identification of airborne viable mould propagules in houses. A field comparison of selected techniques. *Allergy* 45(4):275–284.
101. Burge HP, Solomon WR, Boise JR. 1977. Comparative merits of eight popular media in aerometric studies of fungi. *J Allergy Clin Immunol* 60(3):199–203.
102. Palmgren LL, Strom G, Blomquist B, Malmberg P. 1986. Collection of airborne microorganisms on Nuclepore filters, estimation and analysis—CAMNEA method. *J Appl Bacteriol* 61(5):401–406.
103. Chapman MD, Heymann PW, Wilkins SR, Brown MJ, Platts Mills TAE. 1987. Monoclonal immunoassays for the major dust mite (Dermatophagoides) allergens, Der p I and Der f I, and quantitative analysis of the allergen content of mite and house dust extracts. *J Allergy Clin Immun* 80:184–194.
104. Milton DK, Feldman HA, Neuberg DS, Bruckner RJ, Greaves IA (1992): Environmental endotoxin measurement: the kinetic limulus assay with resistant-parallel-line estimation. *Envron Res* 57:212–230.
105. Rodricks JV, Hesseltine CW, Mehlman MA. 1977. Mycotoxins in human and animal health. Pathotox Publishers Inc, Park Forest South IL.
106. Su HJ, Burge HA, Spengler JD. 1990. Indoor saprophytic aerosols and respiratory health. *J Allergy Clin Immunol* 85(1) pt 2:248 [abstract]
107. Reponen T, Nevalainen A, Jantunen M, Pellikka M, Kalliokoski P. 1990. Proposal for an upper limit of the normal range of indoor air bacteria and fungal spores in subarctic climate. In Proceedings of the Fifth International Conference on Indoor Air Quality and Climate, Toronto, Canada, Vol 2:47–50.
108. Muilenberg ML, Skellenger WS, Burge HA, Solomon WR. 1991. Particle penetration into the automotive interior. I. Influence of vehicle speed and ventilatory mode. *J Allergy Clin Immunol* 87(2):581–585.
109. Murray AB, Ferguson AC. 1983. Dust-free bedrooms in the treatment of asthmatic children with housedust or housedust mite allergy: a controlled trial. *Pediatrics* 71:418
110. Korsgaard J. 1983. House dust mites and absolute indoor humidity. *Allergy* 38:85.
111. Brundage JF, McN Scott R, Ledenar WM, Smith DW, Miller RN. 1988. Building-associated risk of febrile acute respiratory diseases in army trainees. *JAMA* 259(14): 2108–2112.
112. Nelson HS, Hirsch SR, Ohman JL, Platts-Mills TAE, Reed CE, Solomon WR. 1988. Recommendations for the use of residential air-cleaning devices in the treatment of allergic respiratory diseases. *J Allergy Clin Immunol* 82::661–669.
113. Nelson HS, Skufca RM. 1991. Double-blind study of suppression of indoor fungi and bacteria by the PuriDyne biogenic air purifier. *Ann Allergy* 66(3):263–266.

Index

Acridine dyes 340, 342
Acridine orange 274, 298–300, 318, 347, 350, 371, 394
Acridine orange stain 347
Actinoycetes 83, 84, 264, 362, 370, 377–380, 383, 409, 421, 426, 455, 458, 522, 533, 534, 536, 555, 580, 581, 586, 590
Additives 338
Adenovirus 345
Adhesives for sticky traps 364
AEA Lisatek 216, 224
Aerial route 373, 548, 549
Aeroallergens 363
Aerobiological model/pathway 506, 523
Aerobiology 354, 387
Aerodynamic equivalent diameter 366
Aerodynamic particle sizer 57, 110, 144, 147, 153, 155, 156, 243, 366, 381
Aerometrics P/DPA 216, 224
Aeromycological assays 363
Aerosol beam 319
Aerosol sampling 178
Aerosols (see also Bioaerosols) 5–9, 11–14, 15, 17–19, 21, 25, 32, 48–53, 57, 60–62, 65, 66, 74, 75, 78–80, 83, 84, 88, 89, 101–107, 109–117, 119–127, 129–137, 139–143, 145–148, 150–155, 157, 158, 161, 162, 165–175, 177–179, 183, 184, 188, 191, 192, 195–198, 201, 202, 204, 205, 208, 209, 211–213, 215, 218, 219, 221, 222, 228, 230–233, 235, 237, 238, 240–247, 255, 265–267, 271,–273, 275, 276, 279, 281–283, 285, 287, 292, 296, 303, 317–319, 323–325, 328, 330, 332–340, 342–344, 346, 354–356, 365, 366, 376, 381–383, 403, 404, 406, 407, 414, 418, 430, 438–440, 455, 458, 459, 461–467, 469, 471, 474, 475, 477, 479, 484, 486–489, 491, 497, 500, 501, 503, 514, 524, 528, 529, 532, 538, 539, 543, 544, 550, 552, 558, 566, 572–577, 582, 583, 585, 586, 588, 592
Aflatoxicosis 376
Aflatoxin 375, 376
Agar plates 339, 340
Agarwal and Liu criteria 180
Air
 flow 8, 19–22, 27, 109, 110, 124, 129, 152, 184, 237, 248, 261, 262, 270, 277, 288, 340, 434, 446, 448, 449, 476, 486, 491, 494, 497, 538, 549, 558
 laminar 8, 18–22, 124, 184, 476, 524, 558
 visualization 483, 484
 hygiene 548
 movements
  indoors 85, 288, 367, 388, 515, 548
  modeling 523
  outdoors 85, 288, 371, 514, 515, 582
 pollution 548
 spora 286, 362–367, 372, 409, 410, 412, 415, 418–421, 423, 547, 548, 552, 564
 velocity 12, 20, 22, 27, 30, 126, 179, 181, 250, 255, 257, 365, 446, 478, 483, 490, 514, 538, 558
Airborne
 bacteria 553
 mycoses 373
 transmission 552
Aircraft-mounted samplers 366
Alcohols 338
Algae 6, 83, 285, 299, 300, 407–409, 411, 414, 418–422, 424–426, 450, 456, 466, 468, 469, 509, 513, 521, 547, 582
Aliquots 341
Alkaline phosphatase 351
All Glass Impingers (AGI) 78, 79, 82, 255, 256, 258–260, 264, 279, 369, 421, 440, 444, 536, 558, 563–565
Allergens 6, 7, 10–12, 14, 77, 78, 84, 91, 97, 98, 262, 271, 276, 281, 282, 335, 375, 377–379, 385, 391, 395–406, 410, 412, 420, 421, 430, 441, 454, 458, 467, 471, 505, 520, 525, 532–534, 539, 540, 544, 550, 581, 584, 586–589, 592–597
 air conditioning 11, 260, 533, 582, 594
 allergenicity 5, 7, 10, 15, 17, 78, 84, 364, 377, 380, 390, 394, 403, 552, 595
 animals 10, 13, 338, 377, 395, 400, 410, 532, 548, 550, 581, 586, 587
 bacterial 84, 97, 430, 441, 454, 519, 532, 534, 586, 589
 coffee 533, 535, 583, 591
 cosmetics 543
 detergents 11, 341, 535
 egg 5, 292, 342–344, 534
 factory 12, 15, 237, 261, 267, 283, 531, 534, 536, 540, 543, 544
 feedstuffs 534

599

flour 532, 534, 540
fungal 10, 269, 276, 281, 363, 375, 377–379, 401, 410, 420, 430, 441, 532–534, 548, 584, 589
human 6, 363, 375, 379, 395, 397, 410, 454, 533, 548, 581, 584, 586
humidifiers 377, 506, 507, 513, 515, 533–536, 580, 584, 586, 590, 593, 596
microbial 97, 98, 251, 270, 280, 335, 454, 532–534, 539, 548, 581, 587
mites 13, 532–534, 580, 581, 583, 584, 588–591, 595, 597
pharmaceutical 535
plant 6, 11, 12, 395, 397, 398, 401, 410, 420, 430, 441, 454, 505
pollen 10, 11, 84, 391, 395–401, 410, 412, 420, 430, 441, 454, 505
poultry 3, 354, 410, 532–534, 547–554, 558, 562, 567–569, 573–577
protein 78, 84, 97, 98, 396, 398, 532, 539
sewage 7, 249, 285, 286, 408–410, 440–444, 454, 455, 459, 460, 462, 466, 531, 533, 535, 541, 543, 596
shellfish 532
sub-micron 6, 102, 104, 126, 127, 137, 155, 161, 164, 166, 179, 186, 201, 209, 215, 218, 231, 235, 238, 285, 296, 391, 392, 401
waste disposal 256, 258, 267, 282, 383, 407, 409, 442, 458, 535, 545
Allergic aspergillosis 375
Allergic bronchopulmonary aspergillosis 377
Allergic rhinitis 377
Allergies 387
Alternaria species 362, 377, 380
Alveoli 2, 391
Amadori rearrangement 90, 91
Ambrosia 396, 397, 401, 408, 411, 432, 437
Amherst Process Instruments Aerosizer 208
Amino acids 338
Amplification 351
Analytical Electron Microscopies (ASEM and ATEM) 320
Andersen Mark-II cascade impactor 201
Andersen Mark-III 'in-stack' impactor 201
Andersen sampler 253, 254, 256, 258, 263, 264, 266, 276, 369, 376, 383, 419, 422, 426, 437, 439, 442, 443, 467, 469, 553, 556, 559, 565
Angiosperms 387, 403
Anhydrobiotic 82, 97–99, 338
Animal (see also aviary, poulty house, stables) 6, 7, 11, 12, 15, 16, 25, 97, 99, 102, 263, 264, 282, 292, 293, 296, 297, 322, 343, 346, 358, 363, 366, 369, 373–375, 381, 401, 409, 410, 443, 454, 456, 459, 473, 479, 503, 505, 508, 528, 532, 534, 547–550, 552, 553, 555, 557, 558, 561, 562–564, 566–568, 570–573, 575–577, 579, 581, 582, 584, 586, 587, 593, 596, 597
bedding 7, 13, 376, 508, 534, 548, 549, 555, 566, 570, 571, 580, 584
feeding 13, 106, 549, 555, 557, 566, 567, 570, 571, 577
houses 11, 12, 363, 409, 505, 532, 547–550, 552, 553, 555, 558, 561–564, 566–568, 570–573, 575–577
inoculation 343
movement 558, 570, 582, 583, 585
pathogens 373
Animals' breathing zone 567
Anisokinetic sampling 182
Annual cycles 571
Antibiograms 519
Antibiotic effects of airborne fungi 376
Antibiotics 340, 342
Antibodies 79, 274, 292, 298, 300, 304, 312–314, 322, 335, 344, 346, 348, 350, 351, 373, 374, 378, 379, 385, 395, 397, 399, 402, 405, 537, 544, 580, 588, 589, 593
monoclonal 292, 298, 310, 350, 351, 402, 405, 537, 589, 593, 597
polyclonal 350, 405
Antibody titres 346
Antigen extracts 378
Antigens 267, 281, 283, 337, 348, 350, 351, 359, 377–379, 385–387, 391, 392, 396–398, 401, 404, 405, 454, 507, 519, 540, 574, 580, 581, 593, 595
 in the atmosphere 378
Antihistamines 400
Aqueous phases 335
Arbovirus 346
Area samples 515
Argon 81, 117, 135, 207, 216, 301, 323, 353
Arithmetic/midpoint diameter 58, 59
Arithmetic progression 66
Aspergillomas 375
Aspergillosis 373
Aspergillus 375, 380
Aspergillus flavus 371, 375
Aspergillus fumigatus 370, 375, 378
Aspergillus glaucus 370, 376
Assay of fungal aerosols 364
Asthma 7, 10, 282, 286, 377, 395, 398, 399, 402, 430, 456, 457, 461, 467, 507, 532, 544, 579–581, 584, 586, 587, 589, 590, 592, 593, 595, 596

Atmosphere–surface exchange 33
Atmospheric diffusion 50
   layers 36, 94, 327, 358, 376, 387, 422, 452
   stability 11, 50, 77, 80, 82, 92, 97, 98, 104, 109, 119, 166, 233, 298, 356
Atomic spectroscopy 328
Aurasperone 375
Aureobasidium species 380
Aureobasidium pullulans 361, 362, 377
Autoradiography 292
Average sizes 59
Aviary 565, 567, 568

Bacillus anthracis 346
Bacillus piliformis 343, 344, 346
Bacillus species 422, 585
Background 219, 237, 244, 274, 289, 296, 297, 299, 307, 329, 340, 353, 405, 441, 487, 538, 540, 555
Bacteria in chains or clusters 347
Bacterial colony-forming particles 57
Bacterial membranes 339
Bacterial spores 361
Bacteriophage 99, 265, 267, 283, 357, 410, 519
   coliphage 78, 79, 92, 99, 430, 443, 455
   head-tail complex 92
   typing by 344
Basidiomycetes 370, 554, 586
Below-cloud scavenging 42
Bergeron process 43
Bessel functions 218
Betaine 342, 357, 426, 465, 517, 528, 539, 545
Betula 388, 391, 396, 397, 403, 408, 420, 429, 467
Bi-polar ions 124
Bioaerosol 5, 7–19, 21, 23, 24, 27, 39, 55, 57, 60, 62, 64, 66, 67, 69, 70, 72, 73, 77–79, 81–85, 88, 89, 93, 177, 178, 205, 227, 240, 252, 256, 260, 262, 263, 266, 270, 280, 285290, 294, 295, 298–300, 309–311, 313, 317, 319, 322, 325, 328–330, 347, 350, 351, 353, 354, 357, 391, 392, 403, 407–409, 419, 421, 423, 424, 452, 463, 467, 473, 474, 497, 505, 506, 509–511, 513–516, 520, 521, 523–525, 528, 529, 535, 537, 538, 540, 542, 547–550, 552, 553, 555, 557, 558, 560, 561, 563–573, 575, 576, 579, 580, 582, 583, 586–591
   agricultural 9, 12, 47, 48, 54, 75, 249, 276, 336, 355, 364, 407, 408, 442, 454, 460, 464, 469, 507, 522, 531, 533, 534, 541, 542, 573–576, 583, 592, 594
   benefits 6
   collection 82–84, 279–280
   colonization 6, 7, 407, 456, 509, 531, 552, 582, 586, 596
   cyclone 206–208, 259–264, 271, 322, 325, 421, 535
   definition 5, 9, 13
   deposition of 27, 33, 34, 36, 38, 419, 422, 423
   dispersity 60, 62, 65, 66, 70, 102, 113, 124, 139
   dynamic 208, 227, 298, 548, 549, 570
   essential organisms of decay 2
   generation 15, 18, 77, 78, 84, 97, 101, 262, 423, 424, 473–475, 508, 509, 523, 524, 538
   inactivation 98, 335, 338–340, 524
   inertial forces 10, 19, 21, 24, 277
   settling 6, 8, 15–18, 152, 211, 285, 286
   sizing 55, 73
   sources 4, 11, 12, 14, 15–18, 29, 69, 177, 325, 401, 407–410, 424, 444, 474, 505–509, 511, 513, 515, 516, 520, 521, 523, 525, 533, 538, 547, 548, 567, 568, 570, 583, 586, 587, 590, 591
   storage 16, 78, 85, 94, 263, 287, 301, 303, 339, 357, 367, 370, 371, 376, 395, 479, 512, 513, 519, 532–534, 540
   terminal velocity 16, 17, 44, 195, 210, 231, 368, 410, 411, 434, 435
   tracer 51, 53, 121, 142, 148, 430, 440, 484–486, 488, 498, 500, 501, 523, 527, 563, 564
   viability 7, 8, 10, 12, 15, 17, 24, 78, 81–84, 87–89, 92, 95–98, 102, 205, 240, 255, 269, 278–280, 288–300, 311, 317–319, 337, 339–344, 350, 354, 356, 364, 368, 369, 371, 372, 380, 384, 389, 390, 393–395, 401, 403, 405, 407, 408, 411, 418, 420, 424, 426, 444, 446, 449, 450, 453, 454, 465, 513, 514, 519, 547, 552, 589
Bioaerosol distribution 557
Bioaerosol monitoring methods
   atomic absorption 19, 24, 79, 297, 318, 321, 345, 474, 479, 481
   chemiluminescence 322, 329, 332, 349, 353, 355, 459,
   conductivity 109, 271
   electron microscopy 13, 99, 143, 235, 246, 273, 274, 282, 290, 292, 315, 316, 320, 347, 348, 350, 358, 400, 446, 448, 517, 535
   emission spectroscopy 328, 334
   half-life 17, 18

hanging drop 394
immunoelectrophoresis 397
immunofluorescence 300, 314, 315, 322, 325, 329, 345, 348, 350, 373, 379, 402, 438
infra-red spectroscopy 323
laser absorption spectroscopy 19, 24, 79, 297, 318, 321, 345, 474, 479, 481
microbiological 6, 73, 204, 211, 257, 260, 261, 264, 265, 267, 270, 272, 275, 277, 282, 288, 311, 313, 314, 317, 327, 331–334, 336, 338–341, 343, 344, 349, 352–355, 357–359, 380, 383, 460, 463, 465, 469, 473, 475–478, 480, 481, 484–488, 498, 500, 503–506, 510, 519, 526, 528, 539, 540, 545, 565, 596
pyrolysis 115, 312, 319–321, 323, 324, 328, 331, 334
Raman 19, 317, 319, 321, 323, 329, 330, 332
SDS-PAGE 348, 351, 397
volumetric 9, 12, 118, 126, 134, 148, 179, 180, 185, 187, 189, 190, 198, 201, 202, 206, 209, 247, 264, 281, 314, 364–370, 381–383, 390, 391, 393, 402, 408, 410–412, 415, 417, 425–427, 430, 431, 434, 436, 441, 442, 444, 449, 455–458, 460–462, 588, 593, 596
X-ray fluorescence 10, 14, 19, 273, 274, 276, 281, 282, 289, 290, 292, 297–301, 310, 312, 315, 317–322, 325, 326, 328, 329, 331–333, 350, 358, 371, 372, 394, 436, 539, 589
Bioaerosol particle sizer 57, 110, 144, 147, 153, 155, 156, 162, 231, 234, 243, 246, 366, 381, 558
Bioaerosol samplers 8, 15, 16, 19, 79, 252, 256, 266, 286, 288, 528, 529, 535, 555, 560, 561, 563, 566–568, 567, 575
Andersen drum 13, 449
AGI-30 78, 79, 82, 255, 256, 258–260, 264, 279, 421, 440, 444, 536, 558, 563, 565
Biotest 260, 261, 536
Bourdillon 265, 442, 456, 503
bubble 341, 456
Burkard 12, 252, 253, 365, 367, 369, 370, 376, 381, 385, 390, 391, 393, 403, 404, 415–417, 419, 420, 425, 427, 430, 431, 436, 448, 449, 457, 463, 464, 563, 564, 588, 596
calibration 8, 564, 566
capillary 105, 133, 134, 296, 320, 321, 332
cascade cyclone 32, 49, 143, 193, 207, 206
cascade impactor 55, 252, 391

Casella 200, 250, 288, 367, 369, 416, 418, 439–441
centrifuge 104, 151–153, 170, 193, 209, 211, 212, 244
cold trap 391
conifuge 210, 243
continuous flow 166, 233, 246
cyclone
   liquid feed inlet 260
Durham microscope slide 390
efficiency 8, 252, 256, 263, 564, 566, 588
electrostatic 535, 564
filters 256, 286, 563, 566
filtration 256, 287, 535, 537–539, 563, 566–568, 588
gravity slide 393, 458
heaters 421, 580
high volume 185, 262, 421, 438–440, 454
Hirst 5, 12, 14, 264, 286, 314, 365, 368, 381, 390, 403, 416, 419, 422, 427, 430, 449, 452, 457, 461–463
impaction 252, 256, 263, 285–288, 588
   cylinder 132, 271, 310, 390, 411–413, 417, 433, 435, 437
   plate 22, 23, 31, 57, 144
   rod 32, 248, 287, 364, 390, 436, 445
   slide 286–288
impactor 252, 286, 288, 535
   Burkand 564
   SSBAS 391, 392
   virtual 147–150, 201–204, 242
impinger 79, 256
   AGI-30 78, 79, 82, 255, 256, 258–260, 264, 279, 421, 440, 444, 536, 558, 563, 565
   micro 201, 202, 205, 237, 242, 256, 257, 281, 282, 302, 314–316, 319, 323, 325, 329, 332, 357, 358, 360, 460, 470, 484, 497, 500, 576
   midget 279
   MSLI 267–259, 264, 536
   size fractionating 9, 257, 258
inertial 8, 19, 21
inlet 8, 256, 257
inspirable dust 552, 558, 565
LEAP 277, 440
Litton 277, 421, 440, 441, 443, 454
May 79, 112, 115, 200, 365, 366, 369, 376, 553, 563, 564
midget 279
miniature continuous monitor 378
moving slide 415
personal 252, 256, 391, 567
pre-impinger 48

pressure drop and   277
ram jet   450
Reuter centrifugal   260, 267, 544
rotoslide   382, 413, 426, 427, 432, 433, 455, 466, 468
rotorod   13, 32, 248, 249, 263, 287, 366, 367, 382, 390, 413, 414, 426, 426, 430, 431, 432, 434, 436, 445, 453, 455, 458, 467
   net   15, 390, 393, 420, 422, 555, 576
sequential   280, 364, 413
settle plates   13, 247, 378, 411, 438, 453, 501, 514, 522
Shipe   255, 266, 356
sieve   13, 21, 58, 248, 251–253, 514, 588
   stacked   13, 21, 248, 251
slit   248, 251, 288, 340, 367, 369, 418, 426, 437, 443, 453
spore trap   248, 250, 263, 264, 286, 287, 314, 368, 369, 381, 403, 415, 417, 419, 424, 425, 428–431, 436, 437, 444, 454, 457, 459–461, 463, 467, 470, 596
static rods   412
surface   79, 287, 288, 564, 588
tape   248, 249, 287, 327, 364, 365, 381, 403, 412, 413, 415–418, 420, 427, 428, 448, 463, 513, 588
Tauber trap   411, 467
thermal precipitators   18, 280
ultimate cascade   251, 265, 565
whirling arm   13, 248, 249, 390, 412, 414, 423
wind tunnel   30, 32, 365
XM2   443
Bioaerosol sampling (see also Sampling)
   anisokinetic   27, 29, 33, 182–184, 189m 438, 453, 558
   efficiency of   9, 572
   in the field   256
      home   8, 592
   isokinetic   9, 13, 419, 557, 561
   nozzles   27, 28, 48, 366
   overload   13, 206, 406, 470, 555
   particle bounce   23, 30, 31, 39, 41, 49, 112, 143, 145–147, 193, 204, 242, 254
   positioning   15, 260, 452, 538, 565
   stagnation point   365, 420
   still air   16, 30, 31, 49, 285, 361, 410, 419, 514
Bioaerosol sources   506–509
Biological activity   56
Biological decay   337, 338
Biological stains   300
Bioluminescence   322

Bioluminescence or Chemiluminescence   349
Bioluminescence photometer   353
Biomass   300, 310, 315, 316, 331, 353
Biotin-streptavidin   351
Blastomyces dermatitidis   373, 374
Blastomycosis   373
Blood agar   342
Boltzmann distribution   124, 233
Boron   394
Botryodiplodia ricinicola   371
Botrytis   380
Botulism toxin   344
Boundary layer   36, 187, 361
   laminar   8, 18, 19, 36, 39–41, 58, 124, 184–187, 195, 210, 231, 241, 476, 503, 522, 524, 558
   turbulent   8, 10, 19–21, 36–39, 41, 46, 52, 53, 132, 184–187, 211, 247, 285, 361, 430, 434, 435, 467, 491, 549, 557, 558, 563, 564, 568, 570
Boundary layer of still air   361
Breathing zone   251, 271, 378, 412, 537, 553, 558, 567–569, 572
British Standard Graticule   295
Bronchi   377
Bronchitis   377
Brownian diffusion   39
Brownian motion   15, 27, 39, 43, 194, 297
Brownian movement   11
Brucella   346
Building-related illness   506, 507
Buildings   6, 7, 13, 15, 362, 367, 380, 424, 425, 441, 505, 506, 509, 511, 515, 516, 520–522, 524, 525, 533, 541, 542, 548, 549, 566, 567, 570, 571, 573–576, 581, 596
   acquired infections   505, 506, 526
   animal houses   6, 7, 11, 12, 15, 16, 25, 363, 409, 505, 532, 547–550, 552, 553, 555, 558, 561–564, 566–568, 570–573, 575–577
   hospitals   7, 8, 11, 15, 261, 286, 505, 508–511, 515–517, 520–522, 524, 525, 527, 528, 531
   illness associated   506, 507
   investigations of   333, 505, 506, 519
   laboratories   7, 8, 11, 15, 53, 140, 144, 173, 241, 246, 296, 345, 346, 351, 376, 399, 473, 479, 481, 491, 493, 494, 497, 500, 504, 505, 507, 509–511, 516, 517, 519, 521, 522, 524–526, 528, 531
   monitoring   367, 505, 508–511, 520, 522, 525
   problem   15, 505, 506, 520, 521, 525

sick 377, 506
tight 506
Bureau of Community Reference (BCR) 137
Burkard impactor 564
Burkard portable hand-held sampler 365
BVD virus 345

Cahn microbalance 167
Calcofluor White 371
Calibration
  aerosizer 154
  aerosol mass/monitors concentration 167
  APS33B 153
  condensation nuclei counters (CNCs) 166
  electrical aerosol analyzers 162
  gas cyclones 142
  impactors 142
  inertial analyzers 142
  inertial spectrometers 150
  Inspec aerosol spectrometer 152
  optical particle counters 157
  purposes 142
  real-time aerodynamic particle sizers 153
  Stöber spiral duct centrifuge 152
  Timbrell spectrometer 150
  time-of-flight aerodynamic particle sizers 153
  virtual impactors 147
Calibration aerosols
  gas-phase reactions 136
Calibration methods 4, 566
Calibrations
  mass based 142
  number based 142
California Instruments QCM impactor 201
Calm-air sampling 179
Candida 380
Capsule 297
Capture assay 351
Carbohydrates 338
Carbol fuschin stain 292
Cargille 299
Cascade
  cyclone 32, 143, 191, 193, 206–208
Cascade impactors 30, 197, 340
  losses due to electrostatic charge 205
  particle bounce 204
  particle fragmentation 205
  precautions 204
  use 204
  viability 205
  wall losses 205
Casella Airborne Bacteria Sampler 369
Casella bacterial slit samplers 367

Catastrophe theory 87, 89
  denaturation kinetics 87
  termperature effects 88
Catalase
  low pressure 126, 133, 200–202, 242
Centrifugation 342
Centripeter 31
Characteristic curve
  cyclones 143
  impactors 143
Chemical methods 317
'Chen' virtual impactor 202
Chernobyl 42, 51
Chi-square 66
  goodness-of-fit test 66
Chickens 346
Chlamydia species 344, 346
Chlorella 408, 409
Chlorococcum 408, 409
Chromatographic methods 320
Chromatography 266, 316–318, 320, 332, 334, 347, 349, 352, 354, 360, 376, 379, 402
  gas 349, 352
  liquid 317–319, 347, 349, 352
Chromogenic substrate tests 352
Chromogenic/fluorogenic enzyme substrate tests 349
Cladosporium 362, 370, 377, 380
Cladosporium restinae 371
Class mark 57
Clearance mechanisms 37
Climet CI-800 215
Climet CL 3060 217
Climet CL 6300 217
Closed-ended 60
Clostridium species 346, 353
Cloud condensation nuclei 43
Coat of viruses 339
Coccidioides immitis 374
Coccidioidomycosis 373, 374
Coefficient of variation 60
Coincidence 155, 156, 161, 209, 219, 220, 222, 224, 295, 334, 366
Collecting
  fluids 81, 95, 341–343, 345, 347, 350–353, 369, 426
    antibiotic 288, 300, 374, 376, 519, 534, 535
    hypertonic 81, 95, 342
  media
    differential 352
    selective 342
Collection efficiency
  cyclones 145

impactors   145
Collection errors   9
Colonization of barren of environmentally altered substrates   6
Colony-forming unit (CFU)   338
Concentration   7, 9, 12, 16, 17, 30, 33, 36–38, 43–47, 49, 52, 79, 81, 85, 87, 89–91, 101, 102, 104, 106, 113, 116–120, 123, 126, 127, 129, 132, 135, 137, 141, 142, 152, 155, 156, 160, 161; 163, 166–168, 178, 188, 189, 191, 195, 209, 211, 215, 219–221, 226, 233, 235, 237, 238, 240, 242, 243, 245, 246, 249, 262, 263, 266, 270, 294, 299, 309, 326, 332, 338, 343, 350, 351, 358, 362, 364, 365, 367, 368, 372, 383, 391, 394, 401, 412, 413, 423, 425, 427–432, 434–436, 440, 442, 449, 457, 465, 467–469, 484–489, 500, 509, 510, 514, 516, 519–522, 537, 538, 540, 548, 553, 555, 557, 558, 568, 571, 573, 577, 582, 588
   particle number   57, 126, 141, 143, 145, 178, 226, 233, 235, 553
Condensation aerosol generator   117, 122
   types   116
Connector protein   78
Containment   490, 491
   gas tracers   488
   KI-Discus method   487
   operator protection factors   484, 489, 491, 492, 496, 503, 504
   principles   480
   surface contamination   53, 461, 473, 474, 500, 501, 514, 522
   systems   8, 473, 475–481, 483, 484, 488, 491, 494, 497, 500, 503
   tracers   17, 54, 267, 430, 488
Contamination colonies   340
Contamination   6, 7, 12, 29, 45–47, 52, 53, 218, 231, 264, 267, 273, 274, 282, 292, 335, 340, 342, 357, 367, 368, 371, 378, 402, 404, 408, 416, 418, 424, 440, 442–444, 448, 451, 459, 461, 473–477, 479, 484, 487, 500, 501, 504–506, 510–515, 521–525, 531, 533, 536, 539, 542, 544, 566, 581, 586, 587, 590, 591
   ground   45
   surface   51–53, 473, 474, 500, 501, 514, 522
Continuously recording samplers   365
Corkworker's lung   377
Corona discharge   277, 278
Corynebacterium   359, 445, 518, 551, 583
Coulter counter   55, 58, 66, 67, 312, 347, 358
Coxiella   534, 551

Critical orifice   189, 201, 209, 255, 258, 565, 566
Cryptococcosis   374
Cryptococcus neoformans   362, 374
Cryptostrome   380
Crystal growth   140
Cultural assays   344
Culture media
   for indoor samples   517, 518
Culture media for molds   370
Cumulative distributions   66
Cumulative percentage   63
Cunningham correction factor   29, 39
Cunningham slip correction factor   16, 124, 183, 231
Curvularia   379
Cyanobacteria   299, 300
Cyclone calibration
   non-ideal behavior   146
Cyclone sampler   207, 259, 261, 367, 369, 439
   liquid feed inlet   260
Cyclone separator   267, 271, 272, 383, 459
Cyclones   32
Cytochromes   353
Cytopathic effects (CPE)   345
Cytoplasmic
   inclusions   292
   membrane   86, 93–95

Dactylis glomerata   397, 400, 404, 405
Dairy plant air   69
Damaged membranes   339
Damp-air spora   362
Dantec DPA   224
Dantec PDA   216
Davies criteria   179
Deagglomeration   132
Dehydration   11, 24, 78, 79, 82, 84, 86, 87, 91, 92, 94–98, 262, 269, 276, 280, 292, 338–340, 354, 446, 539
Denaturation   24, 87–89, 94, 98, 339
Deposition   7, 9–11, 18, 24, 27, 30, 33, 34, 36–43, 45–54, 75, 109, 132, 141, 147, 150–152, 167, 186, 187, 195, 197, 204, 205, 209, 231, 241, 247, 250, 251, 254, 256, 258, 263, 269, 272, 273, 335, 341, 343, 346, 364, 369, 372, 381, 384, 390, 404, 408, 410–412, 414, 419, 422, 423, 428–430, 433–438, 442, 445, 455, 457–461, 464, 465, 468, 506, 523, 558, 561, 574
   dust   7, 33, 46, 75, 132, 167, 247, 251, 273, 422, 423, 437, 558
   gauges   33
   flux   33, 36–38, 50, 53, 214, 457, 570

in lungs 369
on surfaces 86, 285, 340, 364, 513, 522, 588
rate 33, 36, 46, 341, 435, 438
wet 33, 41
Deposition rates of fungal particles 364
Deposition velocity 36
Desiccation 6, 12, 77–84, 88, 89, 91, 92, 95–98, 287–289, 318, 361, 442, 451, 454, 566
Detection methods
for microorganisms in indoor samples 517–519
Tween-80 341
Deviations from lognormality 66
Diagnosis of mycotoxicoses 376
Diameter of average quantity 60
Differential mobility particle sizer (DMPS) 162, 231, 234
Diffusion 36
Diluters 209, 264, 55, 564
Dilution ventilation 524
Dimers 86, 97
induced by desiccation 81
Diploid cell lines 344
Direct examination 347, 348
Disease 11, 12, 45, 77, 79, 86, 89, 93, 98, 263, 266, 278, 282, 286, 335, 345, 346, 354, 355, 360, 373–377, 381, 384, 402, 408, 410, 420, 430, 437, 438, 455, 456, 458–460, 462, 465, 469–471, 474, 475, 505–510, 519–521, 523, 527, 529, 530, 532–534, 540, 541, 543, 548–552, 563, 566, 573–576, 579–583, 585–588, 590–593, 595, 596
categories 548
environmental 552
infectious 335, 374, 509, 520, 547, 548, 579, 580
Dispersal of fungal spores 361
Dispersal unit 7
Disposable water 338
Distributions 11, 54, 55, 57–60, 62–67, 69, 70, 72–74, 140, 141, 167, 193, 196, 201, 204, 208, 211, 215, 228, 244, 245, 247, 295, 303, 309, 310, 313, 516, 576
lognormal 59, 64, 66–71
normal 62–65, 67, 69
Poisson 429
Diurnal periodicity of spore release 364
DNA 24, 79, 86, 92, 97, 99, 274, 298–300, 318, 326, 348, 351, 352, 359, 508, 519
damage 79, 92, 97
dimers 86, 97
gyrase 300

repair 86, 97
strand breaks 86, 92, 97, 99
DNA/RNA hybridization 348
Downwind travel 16, 454
Drag force 16, 39, 231
Drag forces 55
Drechslera 380
Droplets 33, 43–45, 104–108, 110, 112, 113, 115, 117–123, 126, 130, 145–149, 154, 161, 162, 166, 170, 193, 208, 231, 243, 285, 312, 319, 361, 367, 391, 420, 422, 438, 439, 444, 459, 474, 486, 488, 497, 532, 586
Dry
bulb 24
deposition 33, 41
dissemination 18, 82, 83, 373–375, 461, 464, 469, 509, 531, 582, 583, 591
Dry powder disperser (DPD) 133
Dry-air spora 362
Drying stress 340
Duration of sampling 572
Durham microscope slide 390
Dust
Arizona road 140, 167
generators 131
limitations 132
MIRA 140
organic 376
Dynamic shape factor 109, 155, 195, 199, 208, 231
Dynamics 549

Eagle's minimum essential medium 342
EAST 313, 379, 455
ED_{50} 346
Eddy correlation 38
Eddy diffusivities 37
Effective cut-off diameter (ECD) 145
Egg yolk-enriched solutions 342
EID_{50} 346
Einstein equation 15
Electrical
gradient 18, 36–38, 50, 123, 138, 279, 367, 413
mobility 102, 103, 124, 142, 162, 165, 166, 192–194, 230–235
Electrochemical 319, 326, 330
impedance 326
LAPS 327
methods 326
sensors 168, 201, 326, 327, 510
Electromagnetic radiation 6, 19, 24, 174, 178, 244

scattering 19, 23, 25, 58, 142, 158, 174, 193, 214–216, 218, 219, 221, 222, 224, 226, 244, 304, 332, 366, 489, 566
Electrometer 125, 126, 166, 231–233
  Faraday-cup 166, 231, 233
Electrophoresis 351
Electrophoretic mobility 400
Electroprecipitation samplers 341
Electrostatic
  classifier 124
  precipitators 18, 280, 319, 535
  samplers 277, 440, 449, 454
ELISA 310, 312, 316, 351, 379, 391, 392, 438
Elutriators 12, 132, 193, 195
Embryonated chicken eggs 343, 344, 346
EMC virus 79
Encephalitis virus 86, 346
Endotoxin 266, 335, 337, 353, 355, 360, 409, 466, 515, 522, 528, 533, 537, 541–544, 550, 553, 573, 576, 581, 586, 588, 593, 597
Enterococcus 342
Enterovirus 345, 346, 551
Environmental diseases 552
Environmental monitoring 14, 508–510, 520, 525
Enzyme allergosorbent tests (EAST) 379
Enzyme immunoassay/Radioimmunoassay 348
Enzyme immunoassays 345, 350
Enzyme-linked immunosorbent assay (ELISA) 379
Epicoccum 380
Epidemiological investigations/studies of building-related illness 505, 511, 520
Epidiascope 290, 303, 306, 307, 309, 311
Epithelial cells 67
Equilibration 340
Equivalent diameters 55, 57
Ergot alkaloids 376
Ergotism 376
Errors of sampling 8
Erysiphe graminis 372
Escherichia coli 81, 83, 88–91, 99, 357, 359, 430, 551, 552
Estrogenic syndrome 376
Ethidium bromide 352
Europium (III) thenoyltrifluoroacetonate 372
Exine 387, 400
Exothermic reactions 322
Experimentally generated bioaerosols 336
Exploding wire aerosol generators 12, 135
Exposure limits 7, 503, 521, 553
Extreme habitats 338
Extrinsic allergic alveolitis 377

Eyepiece graticules 294, 295

Facial eczema 376
Faenia rectivirgula 378, 380
Fairs graticule 295
Faraday-cup electrometer 166, 231, 233
Farmer's lung 377, 385, 386, 532
Fern 292, 407, 408, 411, 414, 420, 433, 456
Ferrous metal beads 351
Field data collection form 512
Filar 295
Filar micrometer 295
Filobasidiella neoformans 362, 374
Filter 13, 32, 124, 133, 143, 150, 161, 166–168, 177, 188, 189, 191, 198, 204, 211, 232, 237–240, 260, 269–273, 275–277, 279–282, 287, 289, 298–300, 303, 320, 331, 332, 341, 347, 357, 367, 369, 373, 378, 390, 391, 420, 421, 426, 437, 441, 444, 446–448, 450, 459, 477, 478, 480, 481, 483, 488, 490, 491, 500, 501, 536, 537, 544, 545, 553, 558, 566, 585, 596
  cellulose 204, 237, 270, 273, 275, 276, 287, 412–414, 442, 451, 585
  fibrous 13, 144, 237, 269, 270, 308
  flat 13, 143, 168, 251, 270, 271, 274, 275, 277, 295, 411, 412, 416
  gelatin 270, 275, 276, 281, 355, 368, 369, 412, 413, 426, 438, 443, 446
  glass 30, 146, 204, 237, 240, 255–260, 270, 271, 276–279, 339, 376, 450, 537
  holder 167, 237, 271–273, 275, 280, 287, 298, 447, 448, 450
  paper 84, 146, 240, 242, 270, 277, 279, 421
  PTFE 270, 276
  PVC 270, 537
  samplers 341, 565
  testing, 8, 12, 140, 261, 332, 352–354, 379, 394, 397, 450, 451, 478, 480, 481, 483, 486, 488, 494, 500, 503, 504, 509–511, 513, 517, 525, 526–530, 544, 595
Filter sampling methods 341
Filters for air and surface sampling 514, 515, 523
Filtration samplers 565
Fingerprint techniques 351
Flame photometry 328
Flavobacterium 86, 88, 89, 586
Flow cytometry 314, 316, 319, 325, 329, 333
Fluidized bed aerosol generators 130
Fluorescein diacetate 371
Fluoresceine isothiocyanate 350
Fluorescence-activated cell sorter (FACS) 350
Fluorochromes 298–301, 312, 329, 383

Fluorogenic enzyme substrate tests 352
Foaming 341
Foot-and-mouth disease virus 86, 345, 548, 574
Fourier analyses 55
Fourier transform 221, 323, 332
Fractal dimension 56, 74
Fragility 78, 79, 84, 193, 565
Francisella pestis 78
    phage 78, 85, 97, 443, 519, 573
Francisella tularensis 346
Free radicals 89, 93, 97, 98, 339
    and Open Air Factor 11, 85, 89
Frequency of sampling 572
Friction velocity 37
Fritsch Analysette 22, 223
Fulvia 380
Funigaclavine 375
Fungal spores 7, 9, 126, 247, 250, 256, 260, 265, 273, 274, 276, 278, 281, 282, 286, 287, 289, 294, 361–364, 366–369, 371, 375, 380, 383, 384, 390, 407, 409, 410, 416, 420, 422, 425, 427, 429, 445, 452, 453, 457, 459, 464, 466, 468, 516, 521, 530, 548, 552, 553, 579, 584, 589, 592, 594, 596, 597
Fungi 6, 7, 10–12, 16, 83, 97, 126, 252, 258, 263–266, 276, 282, 285, 286, 296, 300, 344, 361–364, 366–377, 379–384, 409, 412, 418, 421–426, 430, 436, 438, 442, 444–446, 451, 454–456, 458, 459, 462–468, 471, 505, 508, 513, 515, 517–523, 526, 533–536, 541, 545, 547, 548, 551, 554, 563, 573, 580–586, 588–590, 592, 594–597
Fusarium 376, 380

Galai CIS-I 216
Galai time-of-interaction analyzer 214
Gas-liquid chromatography 349, 352
Gaseous phases 335
Gaudin-Schuhman power-law distribution 65
Gelatin filters 340
Gelatin-phosphate 342
General limitations 336
Genetic 6, 7, 10, 359, 405, 473, 507, 541
    engineering 6, 9, 14, 243, 266, 474, 494, 507, 530, 544, 574
Genetic diversity 6
Geometric mean diameter 59
Geometric midpoint diameter 58
Geometric progressions 66
Geometric standard deviation 60
Geotrichum 380

Germ 368, 390, 394, 397
    pores 96, 237, 270, 341, 345, 387, 390, 448
    tube 22, 105, 115, 117, 122, 124, 126, 131, 133, 138, 147, 157, 167, 187, 189, 190, 241, 255, 257, 289, 301, 302, 304, 306, 320, 322, 323, 368, 373, 394, 395, 397, 401–404, 412, 429, 434, 447–450, 486, 538
Germination 367, 368, 376, 383, 394–396, 401, 402, 404, 422
    percentage 63, 65, 67–69, 234, 294, 309, 394, 395, 400
    tests of 66
Giemsa stain 347
Glass slides 339
Glycerol 340
Glycoproteins 379, 396
Gold particles 350
Gram stain 347
Gram-negative 80, 81, 92, 93, 96–98, 292, 353, 355, 360, 409, 444, 445, 454, 521, 532, 533, 535, 543, 580, 581, 586
    bacteria 80, 92, 93, 96–98, 292, 353, 409, 444–445, 454, 521, 522, 532, 533, 535, 581, 586
    stain 274, 292, 296, 298–300, 304, 347, 372, 518, 519, 589
Graphium 380
Gravitational field 10, 16
Gravitational sedimentation 364
Gravity-settling culture plates 368
Guinea pigs 346
Gymnosperms 387
Gypsy moth virus 410

Haemocytometer 347
Haemocytometer chamber 347
Haemophilus species 341, 344
Hamsters 346
Haptens 397
Harmonic mean diameter 59
Harmonic midpoint diameter 58
Harvesting conditions 339
Hatch-Choate equations 55, 71
Hay fever 399, 464
Hazard 6, 7, 45, 49, 241, 375, 473–475, 479, 497, 501, 503, 509, 520–522, 526, 528, 540, 543, 544, 552, 553, 570
    assessment 48, 473–475
        GROUP 1 474
        GROUP 2 475
        GROUP 3 475
        GROUP 4 475

Health effects   15, 457, 515, 523, 525, 531, 540, 545
Health hazards   55
Health-care workers   506, 508, 509, 511
Helium   81, 301, 353
Hemagglutinating antibodies   346
HEPA   473, 478, 481, 486
HEPA filters   473, 486
"Hering' low pressure impactor   202
Herpes simplex virus   345
Herpes virus   341
Herpesviridae   346
Heteroploid continuous cell lines   345
HIAC-Royco   4, 102, 215
High
    altitude transport   422
    volume samplers   271, 276, 367, 420, 421, 537
High-performance liquid chromatography   352
Hirst continuous volumetric spore trap   368
Hirst-Burkard sampler   365
Histamine   377, 399, 542, 593
Histograms   58, 66, 74, 325
    size   66
Histoplasma capsulatum   373, 374
Histoplasmosis   373, 374
HIV   346, 493, 494, 497, 527, 529
Horseradish peroxidase   351
Hospital staff/hospital workers   510–512, 519, 523
Hospitals   7, 9, 11, 15, 261, 286, 505, 506, 508–511, 515–517, 520–522, 524, 525, 527, 528, 531
    bioaerosols in   508, 509
    monitoring in   509–511
Human mycoses   373
Human respiratory system   369
Humidity control   138, 561
    measurement   24, 25
    relative   10, 11, 19, 23, 24, 78, 98, 287, 338, 361, 438, 440, 442, 452, 512, 521, 549, 558, 561, 566, 568, 585
HVAC equipment   512, 515, 520
HVAC systems   508, 510–512, 515, 516, 520, 521, 523, 524
3-Hydrate   372
Hydrates   87, 89, 94
Hydrogen bonds   93
Hydrolytic enzymes   396
Hydrophobic bonds   93
Hypersensitivity   16, 377, 378, 380, 385, 386, 506, 507, 509, 512, 525, 532, 537, 541, 542, 544, 580, 581, 586, 592, 595, 596
Hypersensitivity pneumonitis   377, 378, 380

Identification of air spora constituents   367
Image analyzer
    arc measurements   309
    aspect ratio   309
    binary   308–310
        editor   308
        image   308–310
    circularity   309
    colony counting   310, 311
    detected area   308, 309
    equivalent circle diameter   309
    edge detection   307
    height   309
    mouse, and   593
    optical density   309
    orientation   309
    particle   309, 310, 311
        shape analysis   308–310
        size analysis   312
    perimeter   308, 310
    object area   308–310
    pixel transformations   307
    total biomass   310
    width   309
    video camera, and   305, 306
Immersion oil   273, 274, 299
    Cargille   299
Immune-deficiency diseases   375
Immune electron microscopy   348, 350
Immunoassay   286, 337, 345, 348, 350, 351, 359, 426, 515, 539, 588, 589, 593
    enzyme   88, 90, 97, 292, 312, 337, 345, 348–353, 379, 385, 391, 392, 394, 398, 430, 507, 534, 535, 537, 593
    radio   359, 366, 392, 420, 421, 469
Immunobiological techniques   327
Immunocompromised patients   508
Immunoelectroblot   348, 351, 359
Immunoelectroblot techniques   351
Immunofluorescence (Epifluorescence microscopy)   348
Immunofluorescence tests   350
Immunoglobulin E (IgE)   395
Immunologic tests   373
Immunological methods   379
Immunosuppressive drugs   375
Immunotherapy   395, 399, 400, 594
Impaction   6, 0, 14, 21–23, 30–32, 39, 40, 42, 43, 48, 49, 57, 126, 144, 167, 198, 241, 243, 248–253, 256, 261, 263, 265, 276, 285–288, 320, 324, 334, 340, 341, 364–366, 369, 381, 390, 412, 415, 418, 419, 422, 434–438, 443, 449, 453, 454, 460, 464, 465, 515, 549, 558, 588

efficiency 365
Impactor
    low pressure 201
    overloading 201, 204, 390, 413, 419, 453, 454, 515, 516
Impactor calibration
    non-ideal behavior 146
Impactor samplers 514–518
Impactors
    electrostatic charges 147
    particle bounce 146
    wall losses 146
Impingement 242, 255–258, 264, 285, 369
Impingers 514, 515, 523
    continuous 319, 341, 368, 426
In-cloud scavenging 42
In situ methods 318
Incubation temperature 339, 340
Individual sampling 391
Indoor air 363
Inert atmospheres 81, 82
Inertial
    forces 6, 7, 10, 16–19, 21, 23, 24, 47, 55, 78, 93, 105, 112, 132, 178, 180, 185, 277, 341, 454
Inertial impaction 364
Inertial spectrometer 32
    non-ideal behavior 152
Infectivity 5, 7, 10, 11, 15, 17, 77–80, 83, 88, 98, 318, 336, 337, 339–345, 354, 356, 364, 372, 373, 380, 420, 514, 535, 547, 552, 561
    and particle size 157, 182, 218, 245, 453, 593
    assay 10, 337, 341, 345, 364, 372, 420, 552
    of microbes 77, 82, 84, 98, 285, 336, 338–342, 351, 354, 465
    of plant pathogenic fungi 372
Infectivity loss 339
Influenza virus 345, 346
Infrared spectroscopy 323
Inhalation 8, 10, 30, 45–47, 256, 337, 338, 354, 355, 373–377, 379, 384, 391, 409, 503, 507, 532, 533, 551, 594, 595
    of infective material 373
    of mycotoxins 375
    of organic dust 376
Injured microbes 99, 356
Inlet
    nozzle 27, 28, 30, 120, 153, 182, 189, 206, 208, 323, 365
    probe 179, 180, 182–186, 189, 190, 301, 302, 316, 357, 358, 414, 415, 445–447, 450, 514, 539, 545, 557, 558

Inocula 275, 368
Inorganic acids 394
Insitec PCSV 216, 225
Inspirable dust sampler 565
Instrument calibration 141
Intensity deconvolution analyzer 214
Interception 40
Interface 335
Interpreting sample results 520–522
    air samples/sampling 521, 522
    in hospitals 522
    in laboratories 522
    in problem buildings 521
    liquid samples 521
    material samples 521
Intine 387, 396
Ionic strength 339, 340
Ions 93, 124, 169, 231, 277, 324, 325, 327, 400, 404
    bi-polar 124
Irregular particles 55
Isoaxial sampling 182, 558
Isokinetic sampling 9, 27, 28, 181, 182, 185, 189, 190, 365, 390, 419, 438, 445, 446, 448, 453, 454, 558, 561

KI-Discus method 487
Klebsiella 341
Klebsiella pneumoniae 81–83, 85
Kotrappa and Light centrifuge 211
Kramer-Collins sampler 365
Kratel Parto-scope 217

Laboratories 7, 8, 15, 53, 140, 144, 173, 241, 246, 296, 345, 346, 351, 376, 399, 473, 479, 481, 491, 493, 494, 497, 500, 504, 505, 507, 509–511, 516, 517, 519, 521, 522, 524–526, 528, 531
    bioaerosols in 507, 508
    monitoring in 509–517
Laboratory data/report form 519
Laboratory staff/laboratory workers 506, 507, 510–512, 524
Lag-phases 339
Laminar flow 18, 19, 58, 124, 185, 186, 195, 210, 231, 503, 522, 524
    horizontal 9, 18, 32, 33, 37, 69, 95, 105, 106, 112, 185, 186, 193, 195–197, 241, 295, 296, 309, 367, 410, 411, 413, 414, 416, 418, 425, 431–436, 438, 453, 479, 514, 567, 568
    vertical 9, 17, 22, 32, 33, 36–38, 47, 95, 104, 106, 109, 112, 116, 122, 185, 186, 193, 206, 248, 287, 295, 301, 364, 365,

402, 411–414, 422, 425, 430–432, 434–437, 439, 452, 453, 459, 462, 464, 479, 490, 514, 558, 568
Laser   13, 126, 138, 153, 154, 157, 172, 177, 208, 209, 214–219, 221–225, 227, 228, 244, 245, 290, 301, 312, 317–321, 323–325, 329–333, 366, 508
   active-cavity   214, 218, 219
   anemometry   216, 222, 245
   diffractometer   221, 222
   Doppler   13, 157, 214, 216, 222, 224, 225, 227, 245, 366
   surgical tools   508
Laser diffractometers   214
Laser Doppler velocimetry   366
Laser spectrometry   321
LD_{50} value   346
Leeds and Northrup Microtrac   223
Legionella   344, 346, 510, 513, 517, 519, 521
Legionella species   346
Legionellosis   506, 509
Legionnaires' disease   506, 510
Light
   inactivation   80, 98, 335, 338–340, 524
   intensity deconvolution   13, 214, 216, 225, 226
   microscopy   237, 273, 274, 289, 320, 323, 364
   monochromatic   157, 218, 221, 323
   polychromatic   218, 348
Light microscopic recognition   10
Light scattering
   principles   218
Limulus polyphemus   353
Limulus test   349, 353
Linear regression   63
Links with manufacturers   7
Lipid   23, 78, 80, 92, 93, 96, 98, 292, 297, 339, 341, 353, 394
   and relative humidity   24, 442, 521, 558, 568
   and temperature   11, 24, 78, 79, 86, 189, 211, 216, 338, 566
   containing viruses   339
   oxidation   138, 353, 383
   protein complexes   80, 532
Lipid membranes   339
Lipopolysaccharides (LPS)   338
Liquid samples/sampling   513, 515, 517
Liquid solid chromatography   352
Liquids   340
Livestock   6, 13, 286, 375, 531, 547–549, 552, 567, 570, 571, 573–577
Local exhaust ventilation   524

Lognormal distribution   59, 64, 66–71
Long-range pollination   389
Losses during transport   558
Low pressure impactors   201
LPS   349, 351
LPS detection   353
Lung   11, 12, 335, 344, 360, 363, 374, 375, 377, 380, 385, 386, 532, 533, 537, 540, 541, 543, 544, 592, 593, 596
Lycoperdon spores   411, 422
Lyophilization   338, 342

Macrophages   335, 533, 540, 541
MAGE condensation aerosol generator   120
MAGE operating conditions   123
Maillard reactions   11, 86, 88–93, 96–99, 339
   and free radicals   339
   and oxygen   86, 97, 299, 322, 339
   and radiation   97, 98
Malvern 2600   215
Malvern Master Sizer   215, 223
Mass moment mean diameter   60
Mass spectrometry   321
Mass spectroscopy   312, 316, 321, 325, 332–334, 376
Material samples/sampling   505, 511–513, 520, 521, 523
May liquid impinger   369
May/RE cascade impactor   365, 376
Media composition   340
Media selective   339, 340
Membrane damage   95, 340, 343
   vesicles   84, 95
   virus   13, 79, 80, 86, 88, 96, 98, 249, 256, 265, 266, 278, 282, 285, 335, 337, 339, 341, 343–346, 350, 354, 356, 358, 410, 421, 440, 443, 456–459, 462, 469–471, 481, 493, 508, 527, 529, 534, 552, 573, 574, 583
Membrane filter spore traps   369
Membrane filters   13, 66, 69, 167, 228, 237, 246, 270, 275, 276, 281, 331, 344, 357–359, 367, 421, 440, 441, 443, 444, 448, 450, 514
Met One   205, 207
Metabolic activity   338
Methylene blue   109, 110, 115, 144, 145, 292, 442, 443
Mice   346
Micro-orifice impactors   201
Micrococci   343, 344
Micrococcus   344
Micrococcus radiodurans   85

Micrometer 115, 295, 309
  stage 5, 13, 30–32, 49, 69, 70, 79, 109, 119, 142, 143, 145–147, 158, 165, 166, 179, 195, 197–201, 203, 204, 206, 233, 240, 242, 243, 248, 250–253, 256–259, 265, 266, 289, 293, 295, 301, 302, 305, 307, 309, 310, 365, 376, 377, 382, 391, 407, 414, 415, 419–421, 426, 430, 434, 436, 439–442, 460, 463, 465, 490, 494, 511, 517, 535, 536, 549, 552, 553, 563–565
Microorganisms 67
Microscope 10, 14, 55, 57, 58, 69, 98, 133, 150, 166, 170, 195, 211, 216, 228, 236, 237, 247–251, 263, 273, 274, 279, 286–210, 313, 314, 316, 320, 323, 348, 350, 365–367, 390, 393, 410, 412–419, 424, 429, 434, 436, 445, 449, 488, 513, 516, 519, 564
  analytical 8, 55, 90, 141, 272, 316–321, 328, 331, 334, 352, 353, 376, 514, 516, 520, 529, 548, 553, 589
  bright field 14, 289–291, 298, 300
  confocal 14, 290, 301, 302, 313, 315, 331
  dark field 14, 296
  depth of field 290, 313
  electron 13, 14, 55, 69, 98, 99, 143, 170, 235–237, 246, 273, 274, 282, 283, 286, 288–290, 292, 297, 301–306, 313, 315, 316, 320, 324, 347, 348, 350, 358, 362, 400, 446, 448, 517, 535
  ASEM 320
  ATEM 320
  EDXA 320
  scanning 10, 14, 55, 69, 216, 234, 237, 246, 273, 274, 282, 286, 287, 290, 301–304, 312, 313, 315, 316, 320, 324, 331, 362, 400, 446, 448, 454, 535
  mode 59, 63, 64, 67, 69. 166. 167, 228, 233, 302, 310, 319, 452, 510, 587, 597
  transmission 6, 14, 45, 80, 166, 188, 237, 286, 289, 290, 297, 298, 300, 304, 305, 309, 311, 313, 316, 320, 355, 356, 455, 456, 458–460, 470, 471, 474, 508–510, 527–529, 548, 549, 551, 552, 573, 575, 576, 592
  epi-illumination 289, 298, 299, 301, 311
  excitation filter 298
  eyepiece 289, 293–295, 298, 306, 427, 449
  fluorescence
    scanning 10, 14, 55, 69, 216, 234, 237, 246, 273, 274, 282, 286, 287, 290, 301–304, 312, 313, 315, 316, 320, 324, 331, 362, 400, 446, 448, 454, 535
  graticule 293–295, 427, 449
  interference 222, 223, 290, 296, 297, 299, 307, 340, 345, 347, 434, 512, 538, 557
  light
    sources 7, 11, 12, 14, 16, 18, 29, 47, 53, 69, 103, 117, 126, 177, 186, 188, 242, 314, 325, 361, 362, 367, 368, 375, 384, 400, 401, 405, 407–410, 424, 425, 432, 438, 441, 442, 444, 464, 468, 474, 493, 505–509, 511, 513, 515, 516, 519–523, 525, 533, 538, 547, 548, 567, 568, 570, 580, 581, 583, 586, 587, 590, 591, 596
  mounting medium 299, 300
  objectives 8–11, 289, 290, 298, 299, 425, 548, 561, 572
  oil immersion 290
  phase contrast 14, 290, 296, 297, 299, 300
  resolving power 154, 211, 290, 296, 297, 304
  scanning
    electron 13, 14, 55, 69, 98, 99, 143, 170, 235–237, 246, 273, 274, 282, 283, 286, 288–290, 292, 297, 301–306, 313, 315, 316, 320, 324, 347, 348, 350, 358, 362, 400, 446, 448, 517, 535
    optical 12, 13, 16, 57, 58, 105, 137, 139–141, 143–145, 153, 155, 157, 158, 162, 177, 178, 188, 192, 193, 213–222, 227, 228, 230, 235, 237, 244–246, 290, 298, 301, 303, 306, 309, 310, 315, 484, 485, 553, 558, 559, 563–566
  UV 77–79, 86, 93, 298, 301, 329, 330, 348, 400, 539
Microscopic examination of spores 368
Microscopical detection 320
Microscopy
  collection filter types 237
  particle counting 237
  recommended magnifications 239
  sample collection 237
Microspheres 103, 110, 115, 132, 137, 151, 161, 195, 211, 230
Microsporum 423
Microthread technique 465
Midpoint diameter 70
Mie scattering 19
Mie theory 157, 161, 218
Miller graticule 293
Minimum number of sampling sites 569
Miniram 167
Mold growth inhibitors 370
Mold incubation temperatures 370
Moment average 60
Moment ratios 61
Monkeys 346

Monoclonal antibodies   292, 298, 351, 402, 405, 537, 589, 593
  stains   14, 286, 290–292, 300, 306, 313, 348, 350, 371, 372, 384, 394, 416, 519, 521, 589
Monodispersity   101,m 112, 114, 115, 118, 123, 166, 170, 230
Morphology
  particle   102, 296, 301, 303, 364, 367, 387, 400, 519
Morphometry   285, 289, 292, 295, 313
Moss   51, 285, 384, 407, 408, 414, 420
Molds   357, 463
MRE gravimetric dust sampler   167
Mucor   508, 522, 554
Multimodal distributions   64
Multiplication   338
Multiplication cycles   338
Multiply   339
Multistage liquid impingers   369
Mushroom worker's lung   377
Mushrooms   533, 534
Mutation   86
Mycetoma   373
Mycetomas   375, 377
Mycobacterium species   341, 344, 346, 347
Mycobacterium tuberculosis   346
Mycological culture techniques   371
Mycoplasma   341, 344
Mycoplasma pneumoniae   83
Mycotoxin   375, 376, 384, 522, 533
Myxomycetes   285, 363, 380, 381
Myxovirus   343

Nannochloris atomus   83
NASA standards   522, 529
National Institute of Standards and Technology (NIST)   137
Natural bioaerosols   423
Naturally occurring bioaerosols   336
Nebulizers
  compressed air   126
  ultrasonic   127
Necropsy   346
Neissers stain   292
Neodiprion sertifer virus   410
Neutralization test   345
Newcastle disease   346
Newcastle disease virus   346, 410, 534, 552
Newton's rings   296
Nitrocellulose   351
Nitrogen   8, 81, 83, 88–90, 116, 121, 137, 278, 353
Non-cultural methods   337

Normal distribution   62–65, 67, 69
Normal probability graph   63
Norsolorinic acid   375
Nosocomial infection   508, 527
'Novick and Alvarez' virtual impactor   204
Nozzles
  sharp-edged   27, 28, 49, 182, 185, 206
Nucleic acid hybridization   351
Nucleic acids   339
Nutrient   339
Nutrient agar plates   342

Obscuration   215, 222
Occupation-related respiratory allergies   363
Ochratoxicosis   376
Ochratoxin   376
Olefin   85
Oleic acid   145–148
On-line methods   318
Open air factor   11, 79, 85, 89, 98, 99, 338, 444
  and free radicals   339
  and olefins   90, 444
  and ozone   278, 400
Open-ended classification   66
Open-endedness   72
Operator protection factor   477, 484–488, 490, 491, 494
  transfer index   484–486
Optical   12, 13, 16, 57, 58, 105, 137, 139–141, 143–145, 153, 155, 157, 158, 162, 177, 178, 188, 192, 193, 213–222, 227, 228, 230, 235, 237, 244–246, 290, 298, 301, 303, 306, 309, 310, 315, 484, 485, 553, 558, 559, 563–566
  sensing volume   219, 366
Optical particle counter calibration
  inertial pre-collector method   158
Optical particle counters   57, 214
  border-zone error   219
  coincidence effects   219
  non-spherical particles   219
  particle refractive index   219
Organic dust toxic syndrome   376
Osmotic shock   78, 399, 426
Ouchterlony   379, 386
Ouchterlony's gel diffusion technique   379
Outdoor air   362
Outer coat   339
Oxygen   11, 77, 79–86, 89, 90, 92, 97–99, 116, 136, 299, 318, 322, 338–340, 373, 545, 585
Oxygen concentration   338
Oxygen tension   339, 340

Ozone 51, 79, 85, 90, 98, 278, 318, 400, 444, 591
Paracoccidioides brasiliensis 373, 374
Paracoccidioidomycosis 374
Parainfluenza virus 345
Particle
  agglomeration 43, 55, 135, 167
  atlas 235, 246, 281, 357, 358, 455, 466, 545
  biogenic 286, 287, 292, 294, 296, 299, 597
  bounce 23, 30, 31, 39, 41, 49, 112, 143, 145–147, 193, 204, 242, 254
  blow-off 23, 39
  charge 18, 25, 104, 124, 125, 130, 132, 147, 148, 167, 205, 230–233, 277, 561, 564
  coagulation 119, 130, 135, 167, 285, 533, 561
  cylinders 9, 40, 48, 93, 365, 381, 390, 412, 434–436, 445, 453, 460, 465
  density 16, 17, 20, 29, 73, 110, 195, 208, 243
    from vibrating orifice generator 109
  diameter 16, 20, 29, 33, 38, 39, 46, 57, 59, 73, 101, 124, 163, 169, 192, 194, 214
    definitions 5, 6, 58, 73, 194, 344, 503
    geometric mean 59, 62, 64, 68–70, 102, 589
    mean 15, 16, 18, 19, 29, 38, 44, 59–64, 67–71, 101, 102, 106, 114, 115, 126, 137, 142, 163, 165, 184, 208, 245, 270, 286, 294, 306, 322, 365, 388, 408, 409, 412, 418, 427–429, 431, 434, 435, 444, 453, 552, 553, 558, 589, 590
    volume mean 60, 64
  distribution 65, 419
  fragmentation 86, 205, 319, 321
  inertia 103, 180, 182, 340
  inspirable 49, 537, 552, 553, 558, 565, 571
  length 73
  losses 178, 180, 185, 206, 219, 221
  mass 57, 73, 117, 204, 287, 309
  monodisperse 12, 44, 62, 101–103, 105–107, 110, 112, 113, 116, 119–122, 124–127, 137, 139, 143, 145, 149, 151, 152, 154, 163, 165–168, 170, 197, 230, 233, 243, 244, 334
  morphology 102, 296, 301, 303, 355, 364, 367, 387, 400, 518
  multiplet 104, 151
  nucleation 43, 45, 49, 409, 439, 457, 464
  polymer latex 12, 102–106, 127, 134, 145, 146, 148, 151, 153, 154, 157, 230
  polystyrene 103, 105, 132, 137, 139, 147, 151, 195, 211, 219, 230, 351
  refractive index 19, 24, 102, 105, 140, 157, 170, 214, 216, 218, 219, 221, 224, 228, 230, 484
  respirable 53, 133, 167, 168, 178, 237, 251, 259, 260, 271, 272, 281, 366, 398–440, 443, 510, 515, 537, 542, 553, 558, 564, 565, 571, 573, 577, 595
  rods 32, 289, 364–367, 390, 412–414, 434, 436, 445
  sedimentation 12, 14, 17, 23, 27, 30, 33, 36, 39–41, 43, 46, 47, 132, 178, 186, 191–193, 195–198, 209, 211, 229, 237, 241, 247, 264, 285, 286, 288, 340, 364, 390, 410, 411, 423, 424, 435, 436, 437, 441, 444, 453, 457, 464, 522, 549, 558
  shape 5, 7, 8, 10, 11, 13, 24, 32, 55–57, 63, 73, 74, 109, 123, 132, 140, 141, 143, 146, 147, 154–156, 158, 170, 180, 184, 195, 199, 208, 216, 219, 228, 231, 235, 238, 245, 272–274, 285, 286, 300, 309, 317, 347, 364, 408, 523
  spheres 23, 48, 55, 57, 66, 105, 108, 111, 139, 195, 199, 219, 231, 241, 244, 465
  standards 8, 12, 101, 104, 123, 134, 137, 139–141, 154, 157, 167, 172, 182, 188–190, 293, 353, 473, 476, 486, 503, 504, 522, 529, 539, 540, 548, 555, 572
  surface area 7, 56–61, 72, 73, 115, 366, 513, 552, 588
  titanium dioxide 126
  trajectory 22, 225, 233, 438
  volume 66, 67, 187
Particle monitoring
  in hospitals 510, 511
Particle quantity 56
Particle size 10, 11, 12, 13, 15–18, 23, 27, 30, 39, 40, 46, 50, 59, 60, 62, 66, 67, 74, 75, 101, 102, 105, 116, 124, 132, 139, 140, 142, 143, 145, 146, 148, 152, 154, 157, 158, 167, 168, 170–173, 177–180, 182, 185, 188, 189, 191, 193, 195, 197, 206, 208, 212, 214, 216, 218, 219, 221, 224, 225, 228, 231, 233, 235, 239–241, 243–247, 250–252, 260, 267, 286, 310, 312, 314, 324, 340, 369, 391, 392, 398, 403, 413, 418, 422, 442, 453, 454, 463, 558, 564, 566, 593
  analyzer 332
  and infectivity 11, 77, 78, 318, 337, 339–342, 354, 356, 552
  and survival 6, 361, 382
  average 11, 33, 47, 57, 59, 60, 62, 71, 73, 74, 113, 185, 186, 199, 219, 344, 362, 375, 390, 408–410, 422, 428, 432, 442, 489

distribution
  bimodal 66, 69
  Chi-square 66
  frequency polygon 63, 67, 69
  goodness of fit 11, 66, 74
  lognormal 59, 64, 66–71
  mass-size 141, 142, 201, 211
  normal 62, 66, 67, 113, 115, 167, 198, 204, 233, 289, 300, 302, 306, 338, 399, 508, 516, 553, 557
  number-size 141, 142, 193, 222, 233
  Rosin-Rammler 11, 62, 65, 69, 70, 73, 221
  standard deviations 516, 589
  variance 11, 55, 60, 67, 73, 428, 429
  weighted 55, 59–62, 67, 69, 72, 211, 222
  fractionation 30, 32, 47, 247, 251, 253, 259, 260, 333
  geometric 56–60, 62, 64, 66, 68–70, 72, 73, 102, 121, 142, 194, 307, 442, 516, 589
  on generation 18
  measurement 8, 12, 14, 25, 30, 37, 44, 74, 177, 188, 193, 208, 219, 230, 240, 241, 244–246, 256, 266, 267, 282, 283, 286, 287, 308, 309, 325, 333, 349, 359, 363, 453, 459, 483, 490, 491, 527, 530, 539, 540, 544, 568, 581, 592, 593, 595
  parameter 7, 44, 57, 59, 145, 214, 218, 219, 224, 228, 308, 335, 366, 434, 483, 485
  standards 188
Particle size analysis 197
  cascade cyclones 206
  centrifugal spectrometers 209
  electrical mobility techniques 230
  elutriators 195
  GALAI CIS analyzer 227
  impactors 197
  intensity deconvolution technique 225
  laser diffractometers 221
  laser-Doppler systems 222
  methods 191
  microscopy 235
  optical particle counters 214
  optical techniques 213
  phase-Doppler systems 222
  real-time analyzers 208
  sedimentation techniques 195
Particle sizers 366
  real-time 12, 153, 170, 201, 208, 209, 242, 243, 323, 334, 337, 527
Pathogen 327, 337, 351, 372, 373, 380, 459, 464, 487, 532, 548, 550, 552
Pathogenesis 346
Pathogenicity 286, 346, 373, 474, 487, 510

Pathogenicity tests 373
Pathogens 549
Pathomorphological examination 346
Patient isolation 524
Pattern recognition 321
Patterson globe graticule 295
PCR analysis 539
Penicillium 375, 380
Penicillium camembertii 362
Penicillium species 375
Peroxides 339
Personal exposure 8
Personal samplers 378
Personal samples 515
Pesticide 471
Petri dish 247, 249, 253, 287, 288, 311, 369, 411, 418, 419, 436, 445, 447, 449, 564
  desiccation of 442
Phages 344
Phase changes 11, 24, 92, 93, 96–98
  and temperature 7, 24, 78, 79, 86, 189, 211, 216, 338, 566
Phase-Doppler analyzers 214
Phloxine 372
Phosphoglycerides 92, 94
  oxidation 12, 138, 353, 383
Phospholipids 79, 92, 94, 97, 318, 325
  aggregation 18, 273, 281, 332, 345, 348, 366
  complexes 78, 80, 350, 532
  dimensional changes 95
  oxidation 16, 138, 353, 383
  phase changes
    separations 95, 237
Photophoresis 19
Physical decay 17
Phytophthora infestans 362
Piezobalance 167, 168
Pitot tubes 188, 189
Plants as samplers of infective fungi 372
Plaque assay 345
Plasmolysis 339
Plastic beads 351
Plate counting 343
Pleurotus 380
PMS LAS-X 217
PMS LPC-555 217
PMS-LAS 250X 215
PMS-LASX 215
Poisson distribution 429
Polio virus 79, 345
  fragments 5, 7, 10, 12, 13, 90, 292, 294, 319, 324, 351, 361, 375, 377, 380, 391, 398, 401, 507, 519, 521, 548, 552, 553

network   393, 403
toxic   81, 82, 93, 318, 355, 364, 372, 375, 376, 400, 473, 477, 478, 480, 481, 500, 511, 520, 532, 533, 542, 576, 580, 581
Pollen information services   388
Pollenkitt   396
Pollination   387, 389, 396, 403, 405, 407
  anemophilous   387, 396, 401, 408
  entomophilous   387, 396, 408, 454
Pollinosis   395, 396, 401, 403, 404
Pollutants   6, 8, 11, 43, 53, 77, 85, 97, 318, 400–402, 408, 423, 503, 521, 539, 540, 548, 548, 549, 552, 574, 577
Polyclonal anitbodies   405
Polydispersity   17
Polymer latex particles   12, 102–105, 127, 134, 145, 146, 148, 151, 153, 154, 157, 230
  aerodynamic properties   105
  aerosol generation   104
  density   105
  dilution of stock   105
  optical properties   105
  suppliers   103
Polymerase chain reaction (PCR)   348, 351
Polysaccharide capsules   341
Polysaccharides   351
Polystyrene microtube wells   351
Polystyrene particles   151
Polytec   157, 161, 162, 216–219, 244
Polytec HC-15   217
Polytec HC-15/70   216
Polyvinylidene fluoride   351
Pontiac fever   506, 510
Pore sizes   341
Porins   96
Porton   6, 255, 295, 429, 467
  graticule   293–295, 427, 449
Poultry house   548
Powdery mildews   373
Pre-collector   158, 160–162, 170
Precipitation   14, 18, 33, 41–45, 50–54, 244, 269, 277–281, 283, 285, 342, 357, 422, 471, 558
Predamaged microorganisms   341
Pressure fluctuations   193, 211
Primary cell cultures   344
Probit   64
Probit weighting coefficients   65
Problem buildings   15, 505, 506, 520, 521, 525
  bioaerosols in   506, 507
  monitoring in   509–511
Prodi inertial spectrometer   211
Profilins   398
Project objectives   8, 10

Propidium iodide   371
Protozoa   363, 407, 410, 420, 421, 468, 469, 505, 534, 551, 588
Pseudogermination   394
Pseudorabies virus   345
Psittacosis   83
Public health safety   336
Puccinia   77, 384, 411, 414, 422, 431, 437, 457
Pulmonary aspergillosis   375
Pyrogram   321

Qualitative methods   336
Quantitative methods   336

Rabbits   346
Rabies virus   278, 282, 410, 421, 471
Radioactive labeling   342
Radiographic films   351
Radioimmunoassays   350
Radioimmunoassays include radioallergosorbent tests (RAST)   379
Raffinose   82, 83, 91
Rain-actuated processes   10
Rain-dispersed air spora   367
Raindrop spectrum   44
Raman microprobe   330, 332
Raman spectroscopy   323
Rapaport-Weinstock condensation aerosol generator   119
RAĬT   378, 379, 406, 421
Rats   346
Rayleigh scattering   218
Re-entrainment   33, 45, 143, 193, 254
Reaction order
  first   87, 119, 160, 162, 324, 351, 399, 418–420, 426, 439, 521
  second   87, 88, 142, 148, 160, 324, 399, 418, 420
Real-time aerodynamic particle sizers
  density effects   155
  shape effects   155
Real-time analyzers
  co-incidence   208
  density effects   208
  refractive index effects   208
  shape effects   208
Real-time control   337
Recovery medium   357
Reference particles   137
Refractive index
  imaginary component   19, 219
  of particles   19, 24, 105, 140, 157, 170, 214, 216, 218, 219, 221, 228, 230
  real component   19

Refrigeration 341
Rehumidification 78, 79, 97, 340, 345, 560
    and repair 97, 337
Relative humidity 10, 11, 19, 23, 24, 78, 98, 287, 338, 361, 438, 440, 442, 452, 512, 521, 549, 558, 561, 566, 568, 585
Relaxation time 28, 39, 179
Repair mechanisms 337
Replication cycles 338
Representative sample 557
Reproducibility 337
Respirable particles 553
Respiratory allergy 369, 377
Respiratory complaints 506
Resuspension 10, 33, 45–48, 50, 52–54, 102, 187, 275, 401, 448
Resuspension factor 46
Resuspension rate 46
Reviews
    calibration 102
Reynold's Numbers 39
    for particle 39
Rhizopus 508, 522, 544, 554
Rhodamine 350
Rhodamine B 298
Richardson Number 37
Richardson plot 56
Rickettsia 83, 341, 358, 551
RNA 79, 274, 298, 299, 348, 351
Rosin-Rammler 11, 62, 65, 69, 70, 73, 221
Rosin-Rammler distribution 65
Rotating drum 24, 200, 415, 447
Rotating rod samplers 32
Rotorod sampler 249, 366, 382, 445, 467
Rotoslide sampler 366
Rouchness length 37
Rous sarcoma virus 345
Routine monitoring 337
Royco 227 217
Royco 236 217
Royco 5100 217

Saccharomonospora viridis 370, 380
Safety 9, 12, 241, 266, 281, 336, 393, 466, 473–478, 480, 481, 483–485, 487–497, 500, 503–505, 507, 508, 510, 511, 523, 524, 527–529, 540, 541, 574
    cabinets 473, 475–478, 480, 481, 484–491, 493, 494, 496, 500, 510, 524
        CLASS II 473, 476–478, 481, 484, 486, 489–492, 496, 497, 503
        CLASS III 46, 480, 481
        convertible 476
    fume cupboards 473, 478–481, 488–492, 500, 503, 504
    hoods 478, 479, 490, 524
    isolation 336, 337, 342, 344, 346, 420, 424, 425, 440, 442, 508, 510, 517, 523, 539
    local exhaust systems 479
    tracer
        gas 488
    transfer index 484–486
    ventilation 7, 11–13, 15. 85, 378, 392, 475, 479, 480, 481, 484, 485, 487, 491, 493, 494, 496, 497, 500, 501, 503, 506, 509, 510, 514, 521, 523, 524, 526, 527, 530, 549, 557, 563, 564, 566–568, 571, 574, 577, 582, 584, 585, 587, 591, 595
Saliva 77, 82, 586
Salix 387, 396
Salmonella species 342
Saltating material 46
Sample volume 513, 516
Sampling 6–10, 12–14, 16, 18, 25, 27–32, 39, 48–50, 57, 74, 75, 78, 79, 83, 84, 92, 97–99, 133, 161, 167, 170, 172, 177–191, 193, 195, 197, 198, 201, 202, 204–206, 211, 218, 228, 237, 238, 240–244, 246–249, 251, 252, 254–256, 259–267, 269–277, 279–283, 285–289, 293–295, 303, 309, 311, 313, 318–320, 323, 330, 331, 336, 337, 339–343, 348, 349, 351, 354–357, 362, 364–370, 376, 378, 381–383, 390, 391, 393, 402, 404, 405, 407, 408, 410, 413–415, 417–424, 426–430, 433, 434, 437–442, 444–461, 463–469, 471, 479, 486–489, 494, 495, 498, 500, 505, 509–520, 522, 525–527, 529, 531, 535–539, 544, 545, 547, 548, 550, 553, 555–558, 560, 561, 563–573, 575–577, 587, 588, 591, 596
    bacteria 6, 79, 92, 97, 98, 252, 256, 260, 261, 264, 276, 285, 286, 288, 303, 318, 336, 337, 339–343, 348, 354, 362, 369, 376, 407, 418, 420–422, 424, 426, 429, 430, 438–442, 444–446, 450, 451, 454, 500, 505, 510, 513, 515, 517, 518, 520, 523, 535, 536, 547, 548, 553, 555, 558, 563, 587, 588
    calm air 9, 49, 181, 240
    circadian cycle 570
    condition 16, 27, 167, 182, 189, 208, 325, 363, 377, 459, 514, 558
    duct 12, 27, 32, 170, 177, 184, 185, 188–191, 206, 209, 211, 513, 558, 591
    duration 572
    efficiency 178
    errors 8

frequency 516, 572
fugitive emissions 185
houses, in 554
inlet 9
isoaxial 182, 558
  large particles 13, 14, 29, 30, 32, 33, 39, 45, 47, 50, 149, 167, 182, 184, 186, 187, 208, 218, 263, 280, 411, 412, 415, 416, 420, 427, 453, 553
  lines 88, 89, 96, 178, 180, 185–188, 190, 237, 343–345, 416, 417, 440, 507, 555, 567, 568
line losses 185
location 16, 24, 45, 90, 152, 167, 188, 189, 205, 206, 210, 237, 304, 309, 337, 362, 407, 424, 425, 442, 493, 512, 522, 537, 538, 549, 557, 558, 567, 568
losses due to diffusion 186
losses due to sedimentation 186
losses due to turbulent deposition 187
methods 390
non-inertial 4, 9, 269, 279, 280
period 93, 163, 247, 260, 263, 285, 297, 340, 345, 364, 365, 372, 401, 412, 414, 417–420, 429, 441, 453, 501, 528, 553, 571
plans 434, 452, 512
position and 566
probe 179, 180, 182–186, 189, 190, 301, 302, 316, 357, 358, 414, 415, 445–447, 450, 514, 539, 545, 557, 558
probe misalignment 183
random 15, 19, 62, 144, 270, 293, 311, 365, 393, 428, 429, 465, 537, 538
representative 8, 15, 57, 167, 178, 224, 237, 238, 240, 263, 365, 366, 368, 423–425, 452, 486, 513, 515, 547, 553, 557, 558, 567, 568, 588
site 24, 29, 51, 254, 256, 263, 287, 288, 361, 374, 393, 416, 424, 425, 442, 455, 456, 469, 483, 484, 487, 506, 510–514, 516, 518–521, 539, 561, 568, 582, 583, 588, 593
siting 424, 538, 572
stacked 21, 248, 251
standards 188
systematic 81, 82, 143, 178, 182, 237–239, 365, 428, 429, 510, 526, 537
times 516
timing of 452, 538
umbilical hose 567, 568
velocity 12, 15–18, 20–24, 27, 30, 36–40, 44, 46, 47, 50, 51, 54, 126, 127, 132, 154, 179, 180–182, 185–187, 189, 190, 194, 195, 198, 206, 209–211, 214, 216, 222, 223, 230, 231, 245, 250, 255, 257, 286, 294, 365, 366, 368, 369, 410, 411, 415, 418–420, 434, 435, 446, 478, 481, 483, 489–491, 512, 514, 538, 558
windbreaks 425
Sampling from flowing gas streams 182
Sampling from pipes 188
Sampling from stacks 188
Sampling sites 515, 567
Sandwich assay 351
Scanning electron microscope 69
Scavenging 52
  ratio 18, 19, 21, 22, 42, 43, 47, 55, 56, 58, 61, 64–66, 79, 104, 105, 123, 143, 182, 189, 195, 293, 309, 317, 326, 342, 345, 370, 411, 434, 435, 484, 486, 582
Scavening coefficient 43
Scavenging ratio 43
Schiff's base 90–92
Sclerotinia sclerotiorum 371
Sclerotium sclerotiorum 372
Scopulariopsis 380
SDS-PAGE 348, 351, 397
SDS-PAGE/Immunoelectroblot 348
Seasonal mold allergies 377
Secalonic acid 375
Secondary standards 140
Secondary-Ion-Mass-Spectroscopy 327
Sedimentation
  particle 17, 23, 27, 30, 38–42, 46, 132–134, 178, 186, 195, 197, 209, 237, 285, 286, 288, 340, 411, 423, 435, 453, 522, 558
  samplers 195, 410–412
Sedimentation battery 196
Sedimentation velocity 30, 39
Selective culture media 342, 371
Selectivity of media 370
Selenite broth 342
Semliki forest virus 79, 86, 88
Sensitivity 342
Sequential
  samplers 264, 280, 369, 413, 453
Serial dilution 342
Serratia marcescens 78, 79, 81, 85, 89, 90, 99, 430
Settling 6, 8, 15–18, 57, 152, 153, 179, 180, 186, 194, 195, 206, 211, 241, 247, 264, 285, 286, 294, 366, 368, 467
  of particles 179, 180
  plates 247
  velocity 179, 180, 194, 206, 286, 294, 366
Shape factors 56
Shape frequency distribution 55

Shape standards   140
Shear forces   78, 341, 454
  in samplers   441
Shigella   353
Shipe impinger   255, 266, 356
Sick building syndrome   377, 506
Sierra-Andersen cyclone trains   207
Signature waveforms   56
Silicon micromachining   140
Silver staining   351
Sinclair-LaMer condensation aerosol
    generator   116
Single particle analysis   319
Six-stage Anderson sampler   69
Size distribution   564
Size fractionation   30
Size range
  calibration aerosols   103
Size-categorization of aerosols   365
Size-dependent   341
Size-limited lognormal distributions   66
Skim milk   342
Skin cells   67
Skin scales   583–586
Skin tests   346, 374, 379, 399, 592, 593
Slaframine   376
Slime molds   363
Slit samplers   248, 251, 288, 340, 367, 369,
    418, 426, 437, 443, 453
Slobber syndrome   376
Small pox virus   346
Small-scale powder disperser (SSPD)   133
Sneezing   7, 396, 522, 583
Snowflakes   33, 43
Sodium chloride   111, 121, 152, 153, 164, 231,
    301
  flame test   503
Sodium dodecyl-polyacrylamide gel
    electrophoresis (SDS-PAGE)   351
Solid media   342
Southern blotting   352
Specific antibodies   344
Specificity   337, 340, 342
Spectrometer
  anti-Stokes Raman   317
  atomic   14, 54, 244, 317, 321, 327, 328
    emission   321, 328
  chemiluminescence   322, 329, 332, 349, 353,
      355, 459
  FABMS   325, 330
  fluorescence   319
  Fourier transform   221, 323, 332
  FTIR   323
  Immunofluorescence   300, 314, 315, 322,
      325, 329, 345, 348, 350, 373, 379, 402,
      438
  inertial   32, 150–153, 168, 205, 211
  infra-red   86, 571
  laser
    pyrolysis   115, 312, 319–321, 323, 324,
        328, 331, 334
    luminescence   297, 319, 321, 322, 328, 329,
        332, 337, 353
    mass   319, 323, 324
    microprobe   302, 313, 323, 330, 332, 333
    molecular   317
    nuclear resonance   317
    particle beam   323, 325, 333
    Prodi   49, 50, 121, 122, 151, 153, 171, 173,
        193, 209, 211, 212, 244
    Raman   19, 317, 319, 321, 323, 329, 330,
        332
    SIMS   245, 327
    Timbrell   150, 152, 153, 173, 195, 196, 241
Spinning disc aerosol generator   112
Spinning top aerosol generator   112, 116
  ceramic microspheres   115
  efficiency   115
  operating conditions   112
  satellite droplets   113
Spirochaetes   341
Splash-disersed air spora   367
Spore traps of the Hirst   12
Spore-bearing bacteria   338
Spores   342
Sporidesmin   376
Sporobolomyces   361, 362, 380, 409, 415, 580
Sporoderm   387
Spray fluids   82
Stables   548, 564, 574, 575, 577
Stachybotryotoxicosis   376
Stachybotryotoxin   376
Stachybotrys atra   376
Stacked sieve samplers   13, 21, 248, 251
Stagnation point sampling   420
Staining   274, 281, 286, 288, 289, 291, 292,
    297–299, 304, 308, 316, 320, 326, 347,
    348, 351, 354, 358, 368, 371, 372, 379,
    383, 391, 394, 397, 589
Staining and light microscopy   348
Standard particles   137
Staphylococci   343
Starch and grain particles   67
Starch granules   394, 399
Statistical accuracy   55
Statistical analysis   7
Statistics   11, 55–57, 70, 74, 178, 530
Stemphylium   379

Sticky slide traps 364
Sticky tape 287, 364, 417
Stigma 396, 397, 402
Stigmatic fluid 394
Stirred settling 17, 18
Stöber spiral-duct centrifuge 211
Stokes Law 16, 195, 285
Stokes Number 28
Strain selection 339
Streamlines 21, 40, 42, 182, 184, 197, 198, 248, 438
Streptococcus species 342
Streptomyces species 370
Stress 14, 77–79, 82, 94, 97, 99, 318, 340, 355, 356, 363, 446, 517, 549, 577, 589
  collection 276, 340, 565
  factors 77, 78
  radiation 77, 338
  shear 37, 78, 92, 132, 133, 144, 341, 454
Structured walk technique 56
Submicron
  allergenic
Surface samplers 7
Surface samples/sampling 505, 511, 513, 517, 521, 522
Surface volume mean diameter 60
Survival 6, 11, 14, 77–80, 79–85, 88–90, 95–99, 263, 275, 292, 318, 331, 335, 338–342, 354–356, 361, 382, 458, 464, 465, 514, 545, 552, 571, 574
  effect of
    growth medium 368, 369, 373
    radiation 10, 19, 24, 77, 78, 86, 93, 97, 98, 174, 178, 244, 298, 318, 323, 338, 361, 432, 434, 438, 514, 524
    rehydration 24, 78, 92, 97, 98, 256, 338, 371
SV 40 345
Swine house dust 69
Swine vesicular disease virus 86
Sympatec Helos 223
Synechococcus 83

T3 coliphage 78, 99
T7 coliphage 78
Target molecules 77, 79
  most probable 77–79
$TCID_{50}$ 345
Temperature 11, 15, 18, 19, 24, 37, 38, 78–80, 86, 88, 89, 94–96, 98, 99, 115, 117, 118, 120–123, 126, 178, 189, 193, 204, 205, 211, 216, 241, 244, 279, 288, 319, 324, 338–340, 342, 344, 353, 370, 395, 403, 411, 425, 426, 438, 443, 446, 448, 452, 474, 484, 512, 514, 516–518, 521, 523, 524, 539, 555, 558, 561, 566–568, 571, 580, 582, 586
  gradient 18, 36–38, 50, 123, 138, 279, 367, 413
  inversion 221
Terminal velocity 16, 17, 44, 195, 210, 231, 368, 410, 411, 434, 435
Tetrazolium stains 372
Texas red 350
Thermal 9, 10, 14, 17, 18, 24, 36, 37, 119, 122, 123, 135, 138, 193, 269, 279, 280, 281, 283, 285, 319, 321, 331, 341, 357, 549
  gradient 18, 36–38, 50, 123, 138, 279, 367, 413
  precipitator 278, 283, 357, 375, 440
Thermal samplers 341
Thermoactinomyces 380
Thermoactinomyces thalpophilus 370
Thermoactinomyces vulgaris 370
Thermophilic actinomycetes 363, 370, 378
Thermophoresis 18, 283
Thermophoretic velocity 18
Thermotolerant molds 370
Thin-layer chromatography 352
Timbrell spectrometer 195
TIS Electrical Aerosol Analyzer (EAA) 231
Tissue culture 343, 344, 487
Tissue tropism 346
Titanium dioxide particles 126
Titration in animals 342
Tobacco mosaic virus 410, 440
Total efficacy 340
Toxicity 10
Toxicological activity 56
Toxicosis 15, 375, 384, 532, 535, 541, 580, 594
Toxins 400
Tracers 17, 54, 267, 430, 488
  use of in building investigations 523
Translocation 346
Transmissibility 346
Traveller's Allergy Service Guide 392, 406
Trehalose 82, 91, 95, 97–99, 342, 357
Trichophyton 380
Trichothecene toxicosis 376
Trichthecenes 376
Trichothecium 380
Tryptose-saline 342
Tryptose-soy-agar 342
TSI 3755 217
TSI APS33B (Aerodynamic Particle Sizer) 208

TSI Differential Mobility Particle Sizer (DMAs) 233
TSI Scanning Mobility Particle Sizer (SMPS) 234
TSI small-scale powder disperser (SSPD) 134
TTC 394
Turbulence 7, 19, 21, 33, 36, 40, 184–186, 347, 361, 411, 412, 422, 448, 453, 485
Turbulent surface layer 36
Tween 80 275, 419, 440, 441, 444

UKAEA virtual impactor 203
Ulocladium 379
University of Washington 'in-stack' impactors 201
Urticaria 396, 586
Ustilago 362
UV radiation 78

Vaccinia 346
Vaccinia virus 346
Vapor phase rehydration 338, 371
Variance of size 60
Venezuelan equine encephalomyelitis virus 79, 265
Ventilation 7, 11–13, 15, 85, 378, 392, 475, 479, 480, 481, 484, 485, 487, 491, 493, 496, 497, 500, 501, 503, 506, 509, 510, 514, 521, 523, 524, 526, 527, 530, 549, 557, 563, 564, 566–568, 571, 574, 577, 582, 584, 585, 587, 591, 595
Ventilation for bioaerosol control 524, 525
Venturi scrubbers 374
Venturia inaequalis 367
Vertical rod traps 365
Verticillium 380
Vesicular stomatitis virus 79
Viability 7, 8, 10, 12, 15, 17, 24, 78, 81–84, 87–89, 92, 95–98, 102, 205, 240, 255, 269, 278–280, 288, 289, 300, 311, 317–319, 337, 339–344, 350, 354, 356, 364, 368, 369, 371, 372, 380, 384, 389, 390, 393–395, 401, 403, 405, 407, 408, 411, 418, 420, 424, 426, 444, 446, 449, 450, 453, 454, 465, 513, 514, 519, 547, 552, 589
Viability of fungal spores 368
Vibrating orifice aerosol generator
  efficiency 109
  fluorescent particles 112
  operating conditions 106
  particle shape 110
  precautions 108

Vibrating orifice aerosol generator (VOAG) 105
Video cameras 14, 290, 304, 306
Virtual impactors 203
Virulence 346
Virus 13, 79, 80, 86, 96, 98, 249, 256, 265, 266, 278, 282, 285, 335, 337, 339, 341, 343–346, 350, 354, 356, 358, 410, 421, 440, 443, 456–459, 462, 469–471, 481, 493, 508, 527, 529, 534, 552, 573, 574, 583
Virus particles 341
Viruses 339, 341
Visual building inspections 512
Vital staining 368, 371, 383, 394
Vital staining techniques 371
Volumetric assays 365
von Karman constant 37
Vortex 189, 259

Wallemia sebi 370, 380
Wastewater treatment 258, 459, 461
Water-saturated atmosphere 338
Wearing some small sampler 12
Weighted averages 55
Wet
  and dry bulb method 24
  deposition 33, 41
Wind 8, 9, 13, 27, 30, 32, 33, 36–38, 40, 41, 46–49, 52–54, 134, 140, 185, 272, 285, 287, 361, 362, 365, 367, 387, 390, 396, 401, 405, 407–415, 418–420, 422, 425, 427, 428, 431, 432, 434–436, 438, 440–442, 445, 452–455, 458–461, 464–466, 471, 507, 512, 523, 566, 582
  bulb 24
  direction 27, 185, 390, 411–413, 418, 434, 452, 453
Wind tunnel 30, 32, 365
Windvane 415, 422, 430, 436, 441
WRAC sampler 32
Wright dust feed 132

X-rays 79, 86, 93, 301
Xerophilic storage fungi 376

Yeasts 361, 362, 373, 380, 415, 424, 444, 446, 451, 462, 522, 585

Zearalenone 376
Zero plane displacement 37
Zerostat 167
Ziehl-Neelsen acid-fast stain 347
Zoonoses 532, 534, 540